FORMULAS

5280 ft = 1 mile

Distance formula	The distance between two points (x_1, y_1) and$$d = \sqrt{(x_2 - x_1)}$$
Midpoint formula	The midpoint of the line segment connecting two points (x_1, y_1) and (x_2, y_2) is the point $(\bar{x}, \bar{y})$ where $$\bar{x} = \frac{x_1 + x_2}{2} \quad \text{and} \quad \bar{y} = \frac{y_1 + y_2}{2}$$
Quadratic formula	If $a \neq 0$, then the solution(s) of $ax^2 + bx + c = 0$ are $x = \dfrac{-b \pm \sqrt{b^2 - 4ac}}{2a}$
Simple interest	If a principal of P dollars is deposited in an account paying simple interest at a rate r, then the balance B after t years is given by $B = P(1 + rt)$.
Compound interest	If a principal of P dollars is deposited in an account paying interest at a rate r compounded n times per year, then the balance B after t years is given by $$B = P\left(1 + \frac{r}{n}\right)^{nt}$$
Arithmetic series	If $a_1, a_2, a_3, \ldots, a_n$ is an arithmetic sequence, then $\displaystyle\sum_{k=1}^{n} a_k = \frac{n}{2}(a_1 + a_n)$.
Geometric series	If $a_1, a_2, a_3, \ldots, a_n$ is a geometric sequence with common ratio $r \neq 1$, then $$\sum_{k=1}^{n} a_k = \frac{a_1(1 - r^n)}{1 - r}$$
Binomial theorem	$(a + b)^n = \binom{n}{n}a^n + \binom{n}{n-1}a^{n-1}b + \binom{n}{n-2}a^{n-2}b^2 + \cdots + \binom{n}{1}ab^{n-1} + \binom{n}{0}b^n$
Height of a falling object	If an object is thrown into the air from an initial height s_0 feet with an initial velocity v_0 feet per second, its height in feet after t seconds is given by $h(t) = -16t^2 + v_0 t + s_0$.
Universal law of gravitation	The force of attraction between two objects with masses m_1 and m_2 kilograms that are a distance r meters apart is given by $F = G\dfrac{m_1 m_2}{r^2}$, where $G = 6.672 \times 10^{-11}$ Nm^2/kg^2.
Time dilation formula	If a person travels at a velocity equal to v times the speed of light for an elapsed time of T_0 seconds (according to her watch), then the elapsed time on Earth is given by $$T = \frac{T_0}{\sqrt{1 - v^2}}$$

ALGEBRAIC FORMULAS

Sum of cubes
$a^3 + b^3 = (a + b)(a^2 - ab + b^2)$

Difference of cubes
$a^3 - b^3 = (a - b)(a^2 + ab + b^2)$

Difference of squares
$a^2 - b^2 = (a + b)(a - b)$

COLLEGE ALGEBRA

A CONTEMPORARY APPROACH

SECOND EDITION

COLLEGE ALGEBRA
A CONTEMPORARY APPROACH

SECOND EDITION

DAVID DWYER
MARK GRUENWALD

University of Evansville

Brooks/Cole
Thomson Learning™

Australia • Canada • Mexico • Singapore • Spain • United Kingdom • United States

Marketing Team: *Karin Sandberg and Beth Kroenke*
Advertising: *Samantha Cabaluna*
Editorial Assistants: *Joanne Von Zastrow and Emily Davidson*
Production Editor: *Kirk Bomont*
Production Service: *New Leaf Publishing Services*
Manuscript Editor: *Carol Dondrea*
Interior Design: *E. Kelley Shoemaker*
Cover Design: *Roger Knox*
Cover Photo/Illustration: *Minoru Toi/Photonica*

Art Editor: *Kathy Joneson*
Interior Illustration: *Rolin Graphics, Tech Graphics*
Photo Researcher: *Christine Nealy*
Print Buyer: *Vena Dyer*
Typesetting: *UG/GGS Information Services, Inc.*
Cover Printing: *Phoenix Color Corporation*
Printing and Binding: *World Color, Versailles*

For more information, contact:
BROOKS/COLE
511 Forest Lodge Road
Pacific Grove, CA 93950 USA
www.brookscole.com

For permission to use material from this work, contact us by

Web: www.thomsonrights.com
fax: 1-800-730-2215
phone: 1-800-730-2214

Printed in the United States of America

10 9 8 7 6 5 4 3 2 1

Library of Congress Cataloging-in-Publication Data
Dwyer, David.
   College algebra : a contemporary approach / David Dwyer, Mark
Gruenwald.—2nd ed.
      p.   cm.
   Includes index.
   ISBN 0-534-35147-6
   1. Algebra.   I. Gruenwald, Mark (Mark Edward)   II. Title.
QA152.2.D89   2000
512.9—dc21
                                                        99-41146
                                                        CIP

*Dedicated to our families, especially our wives,
Carrie Dwyer and Margaret Gruenwald*

CONTENTS

CHAPTER **7** SYSTEMS OF EQUATIONS AND INEQUALITIES **523**

CHAPTER **8** INTEGER FUNCTIONS AND PROBABILITY **619**

PREFACE

Mathematics, the science of pattern and the universal language of quantification, is a living, breathing subject with profound implications for nearly every aspect of life. It is woven into the fabric of art, astronomy, chemistry, cinema, law, literature, music, religion, philosophy, physics, politics, sports, and technology. Moreover, mathematics is a work in progress with its frontiers ever expanding and new areas of application arising daily. Ozone depletion, rain forest destruction, global warming, the spread of contagious disease, nuclear proliferation, poverty, and overpopulation are just a few of the world's problems illuminated by mathematical techniques. Arguably, the destiny of Earth itself hinges upon our ability to make intelligent decisions based on mathematical models of these real-world phenomena.

But far too many students view mathematics in general, and college algebra in particular, in an entirely different light. To them math is a soulless game of symbolic manipulation, a lifeless subject passed down through generations without alteration, its arcane formulas and rules to be memorized and quickly forgotten. They see mathematics as inaccessible, insignificant, and utterly irrelevant to their majors, their careers, their lives. Not surprisingly, mathematics courses are often loathed like few others in the college curriculum, for who among us enjoys struggling to learn mind-numbingly boring and yet oppressively difficult material?

In writing this text, we attempt to overcome such attitudes by sharing our vision of mathematics as a vital discipline that is not only widely applicable, but also interesting, entertaining, and even—at times—breathtakingly beautiful. In addition to developing fundamental concepts and skills, we want our readers to experience the thrill of conducting penetrating mathematical explorations with a graphing calculator and sheer force of reason as the only sources of illumination. And though algebra can be challenging, we believe that most of us come equipped with brains hardwired to think mathematically, though such circuitry may lie dormant from disuse and neglect. It is our goal to tap into the student's natural affinity for mathematics by providing motivation, inspiration, and intellectual stimulation through the explanations, examples, and exercises that constitute this text.

FEATURES OF THE TEXT

Chapter Opening Photographs

Each chapter begins with a photograph and an extended caption that connects the image to the subject matter of the chapter. These striking images are designed to generate interest by underscoring the relevance and applicability of the material in that chapter.

Motivating Questions

Each section begins with stimulating questions designed to pique the reader's curiosity and motivate the material in the section. For example, Section 3 of Chapter 5 begins with the questions: *How powerful would the "bombquake" be that would result from simultaneously detonating all of the world's nuclear warheads? How can multiplication and division be performed simply by sliding pieces of paper? How many jamboxes would it take to cause a fatal musical overdose? If the entire population of China simultaneously jumped a foot off the ground, how strong would the resulting earthquake be?* Some answers can be found simply by reading the text, whereas others require the reader to complete one of the exercises at the end of the section.

Examples

Examples are used to illustrate the techniques and theory discussed in the text or, in some cases, to introduce a topic or motivate a theoretical concept. Every attempt is made to include the reader in the problem-solving process, to explain not just *how* each step is executed but also *why* a particular problem-solving strategy is chosen. Multistep examples often include explanatory notes beside each step or detailed explanations between important steps.

Graphing Calculators

The graphing calculator is an invaluable tool for solving equations, exploring mathematical concepts graphically, and tackling complex applications. To make optimal use of the power of the graphing calculator, one must know two things: *how* to use the calculator and, of equal importance, *when* to use the calculator. Throughout the text, we discuss when to use the graphing calculator, when to use symbolic methods, and when to use a combination of symbolic and graphical techniques. All of the important graphing calculator functions are discussed in boxes titled "Calculator Keys." Even more detailed model-specific keystroke instructions can be found in the *Graphing Calculator Manual.*

Art and Graphics

The text includes hundreds of graphs, all of which are computer generated for accuracy. In addition, dozens of computer-generated graphing calculator screen images have been incorporated wherever students will be performing a task with the graphing calculator. Many applied examples and exercises are accompanied by color illustrations or bar graphs. The text also includes dozens of color photographs, selected for their relevance, novelty, and beauty.

Mathematical Notes

To underscore the vitality of the subject, the text includes brief accounts of recent developments in mathematics, such as Andrew Wiles' proof of Fermat's Last Theorem and recent investigations of dynamical systems. In addition, we explore several intriguing applications of mathematics, such as the use of mathematical morphing in movies and television, radiocarbon dating to test the authenticity of the Shroud of Turin, and mathematical models assessing the risk of a cometary collision with Earth.

Warnings and Rules of Thumb

"Warning" boxes are provided to point out common mathematical pitfalls and how they can be avoided. For example, when solving equations, students are cautioned to avoid dividing by expressions containing a variable. "Rule of Thumb" boxes describe helpful hints and tricks of the trade. For example, when solving work problems, students are encouraged to first express the *rate* at which work is done.

Understanding and Mastery Checklists

Each section concludes with a checklist of newly introduced topics and terminology, along with a separate listing of the skills that the student is expected to have mastered upon completion of the section. The checklists are designed to help students assess whether they are ready to start solving exercises.

Exercise Sets

There are some tasks for which pencil and paper techniques are superior, others where the graphing calculator is much more efficient, and still others that are best solved using some combination of algebraic and graphical tools. To develop the ability to select the mathematical tool most appropriate for a given task, many exercises have been included in which the student must first decide whether or not to use a graphing calculator. Thus, exercises that require the use of a graphing calculator are integrated within many of the following categories of exercises.

- **Standard exercises,** graded in difficulty, are included to reinforce the development of important skills and concepts. Many of the standard exercises are similar to examples and are intended primarily to test the student's manipulative ability. Some are unlike any example in the text to ensure that the student has synthesized the underlying concepts and isn't just blindly using examples as templates.

- **Applications,** or applied problems, vary in scope from explorations of topical issues with important social, environmental, and economic ramifications, to problems with whimsical, thought-provoking, and (we hope) entertaining premises. Many involve modeling with real-world data from such diverse sources as governmental agencies, the *World Almanac,* the *Environmental Almanac,* and the *Guinness Book of World Records.* Some involve important notions from the physical and social sciences, such as gravity, projectile motion, earthquakes, the speed of light, time dilation, body mass index, the TNT equivalent of a nuclear warhead, population dynamics, and the spread of disease. On the lighter side are problems involving whimsical premises such as two cows tormenting a train engineer, a cyclist powering a small city, and a hot-air balloon raising the Statue of Liberty. A fictional detective, Inspector Magill, appears throughout the text solving mysteries such as a kidnapping, a hijacking attempt, a carjacking, murders, and a medical mystery.

- **Questions for Discussion or Essay** are intended for use as thought-provoking discussion questions or as subjects for short essays. Some ask students to analyze the premises of application problems. Others require students to connect the material with their experiences in other math classes, other disciplines, and everyday life.

- **Concepts and Critical Thinking** exercises, including true/false and ''give an example of'' questions, are designed to help students assimilate the most important ideas and terminology of a section. These conceptually oriented exercises complement more computationally oriented standard exercises.

- **Projects for Enrichment** are nontrivial multistep problems designed either for group work or as especially challenging problems, possibly for extra credit. Some explore related mathematical topics that go beyond the classical boundaries of college algebra. Examples of such topics include the cubic formula, fractal geometry, and iterates of a function. Other projects are in-depth application problems, such as an investigation of the relationship between alcohol consumption and blood alcohol level, a development of Kepler's third law for planetary motion, an application of systems of equations to balancing chemical equations, and a technique for computing the day of the week for any date in history.

Chapter Reviews and Tests   Each chapter concludes with a chapter review and a chapter test. The chapter reviews consist of comprehensive selections of exercises representative of those previously encountered in the chapter. The chapter tests include a sampling of representative problems but are not intended to be comprehensive. Each includes a section of true/false and "give an example of" problems that are designed to test the student's mastery of important concepts and key definitions.

NEW IN THE SECOND EDITION

The second edition includes improvements and updates to all aspects of the text. Specific changes and additions include:

- The formal treatment of functions is now addressed earlier in the text. This move was motivated by the utility of functions as mathematical models and by the central role functions play in solving algebraic problems graphically.
- "Understanding and Mastery Checklists" are included at the end of each section.
- "Concepts and Critical Thinking" exercises have been added to each section.
- Hundreds of new exercises, including dozens of applied problems and several projects, have been added. In addition, many applied problems from the first edition have been updated.
- Many additional examples and exercises requiring the use of a graphing calculator have been added.
- Introductory paragraphs have been added to the beginning of many sections to help the student place the material in context.
- Every section has been reviewed, and in some cases extensively rewritten, to enhance clarity, readability, and accuracy.

ANCILLARIES FOR THE TEXT

The following ancillaries are available to accompany this text:

Instructor's Solutions Manual, by Kathryn C. Wetzel, includes worked-out solutions to all of the even-numbered text exercises. (ISBN 0-534-35148-4)

Student Solutions Manual, by Kathryn C. Wetzel, includes worked-out solutions to all of the odd-numbered text exercises. (ISBN 0-534-35149-2)

Test Bank with Chapter Tests, by Cheryl Cantwell, contains printed test forms with answers for the instructor. (ISBN 0-534-35153-0)

Thomson Learning Testing Tools™ provides text-specific algorithmic testing options designed to give the instructor greater flexibility. Versions are available for IBM PC and Macintosh systems. (ISBN 0-534-35151-4)

Graphing Calculator Manual, by Dennis Pence, includes text-specific examples and keystroke instructions for the TI-82, TI-83, TI-83 Plus, TI-85, TI-86, TI-89, TI-92, TI-92 Plus, Casio CFX-9850Ga Plus, and the HP-38G. (ISBN 0-534-37432-8)

Text-Specific Video Series, by Mark Basse, features worked-out examples from each chapter. (ISBN 0-534-35157-3)

ACKNOWLEDGMENTS

The candid comments and insightful suggestions of our colleagues were instrumental in the revision of this text. We'd like to thank the following reviewers: Theresa Adsit, University of Wisconsin—Green Bay; Jeff Berg, Arapahoe Community College; Jean Bevis, Georgia State University; Matt Dempsey, Jackson Community College; Barbara Edwards, Oregon State University; Patrick J. Enright, Arapahoe Community College;

Chris Gardiner, Eastern Michigan University; Sudhir Goel, Valdosta State University; Dan Harned, Michigan State University; Karla Karstens, University of Vermont; Donna McCracken, University of Florida; Devilyna Nichols, Purdue University; Yifei Pan, Indiana University/Purdue University; Ernie Solheid, California State University—Fullerton; and Lyndon Weberg, University of Wisconsin—River Falls. In particular, we would also like to thank Dan Shapiro of Ohio State University for work above and beyond the call of duty.

We are especially grateful for the assistance provided by the editorial, marketing, and production professionals on the Brooks/Cole team who have contributed to our texts. Thanks to Karin Sandberg, marketing manager; Seema Atwal and Stephanie Schmidt, ancillary editors; Kirk Bomont, production editor; Curt Hinrichs, editorial; Emily Davidson, editorial assistant; and Vernon Boes, design director. Our appreciation also goes to Carol Dondrea, copyeditor, Kelly Shoemaker, designer, and Christine Nealy, photo researcher, for their dedication to the project. Thanks to Gene Bennett and Sudhir Goel for their many corrections and helpful suggestions, and to Melia Aldridge for her assistance with the manuscript. Special thanks to Nancy Shammas for her tireless efforts in coordinating the production of this text—a most difficult task at a most difficult time. Margot Hanis was instrumental in developing this edition. We consider ourselves fortunate to have worked with such a stellar talent.

David Dwyer
Mark Gruenwald

1 FOUNDATIONS AND FUNDAMENTALS

The Nashua River, desecrated by human waste and bleeding red from industrial dyes, had become a lifeless cocktail of toxins by the 1960s. Today the river teems with life, its pristine beauty restored, thanks to a concerted community campaign. Mathematical models, taking into account such factors as water temperature, levels of pollutants, and flow rates, help ecologists design such cleanup efforts. In Example 11 of Section 7 we will consider a simplified model for predicting the expense associated with removing pollution.

SECTION **1.1**   **NUMBER SYSTEMS AND ESTIMATION**

☐ Which is larger: 0.999... or 1?

☐ In which state could the entire population of Earth fit if each person were allotted a square yard?

☐ How long a convoy of trucks would be required to haul away Mt. Everest?

☐ How many families could Bill Gates feed for a year if he were willing to liquidate his entire fortune?

☐ How far will you walk in your lifetime?

☐ How much gasoline would be saved if every car in the nation got 50 miles per gallon?

NUMBER SYSTEMS

Number systems have developed as a language for expressing how many, how much, and how far. In much the same way as English, French, Mandarin, and other languages have grown to accommodate the expression of new ideas, so too have new number systems been created whenever an earlier system proved inadequate for the expression of a new type of quantity. As we describe the evolution from the natural numbers to the whole numbers, integers, rationals, reals, and, finally, complex numbers, you will see that at each stage a new number system is formed by extending an existing number system. Table 1 gives an overview of the number systems we will explore in this section.

TABLE 1

Number system	Examples	Inadequacy	Extension
Natural	$1, 2, 3, \ldots$	No placeholder or zero	Add zero to create whole numbers
Whole	$0, 1, 2, 3, \ldots$	Cannot subtract certain pairs	Include negatives to create integers
Integer	$\ldots, -2, -1, 0, 1, 2, \ldots$	Cannot divide certain pairs	Include fractions to create rationals
Rational	$-\frac{1}{7}, 5, \frac{13}{2}, 0.333\ldots$	Cannot express certain distances	Include all decimal expansions to create reals
Real	$\sqrt{2}, -\frac{5}{8}, 0, \pi$	Certain equations don't have solutions	Include all numbers of the form $a + bi$ to create complex numbers
Complex	$2 + \sqrt{3}i, 0, 1.7i$		

The **natural numbers** are the numbers $1, 2, 3, \ldots$. They developed out of our need to count—livestock, furs, beads, years—and are thus sometimes called the counting numbers. Zero was introduced not as a number for counting (''I have zero sheep''), but rather as a placeholder for positional systems for writing numbers. The natural numbers together with zero constitute the **whole numbers.** The whole numbers do not allow for the expression of negative quantities, which is a useful tool for computations in which one must distinguish between assets and debts or surpluses and deficits. From a strictly mathematical point view, the whole numbers are thus incomplete in the sense that there are pairs of whole numbers for which the operation of subtraction is not defined. For example, we cannot subtract 2 from 1 and stay within the whole number system.

When we include the negatives of the whole numbers, we form a new number system—the **integers**: ..., $-3, -2, -1, 0, 1, 2, 3, \ldots$. The integers are complete with respect to subtraction in the sense that the difference between any two integers is itself an integer. Note that the integers include the whole numbers, which, in turn, include the natural numbers.

EXAMPLE 1   ▓   **Recognizing Natural Numbers and Integers**

Determine whether each of the given numbers is an integer. If it is, determine whether it is a natural number.

a. $1,234,000,031 - 23,886,986,102$
b. The number of atoms that constitute the planet Jupiter
c. The number of complete minutes from the birth of Jesus until now minus the number of complete seconds from the birth of Jesus until now
d. Your height in feet

SOLUTION
a. Since $1,234,000,031 - 23,886,986,102$ is formed by subtracting two integers, it too is an integer. However, because the second number is larger than the first, the result is negative and hence not a natural number.
b. Although the number of atoms constituting the planet Jupiter is unknown (and no doubt enormous), this number will be a natural number, and hence an integer as well.
c. Since there are 60 times as many seconds in this time interval as there are minutes, the difference between the number of complete minutes and the number of complete seconds will be negative and hence not a natural number.
d. If you happen to be *exactly* 3, 4, 5, 6, 7, or 8 feet tall, then your height in feet will be a natural number (and, of course, an integer also). Otherwise, your height in feet will not be an integer. For example, if you are 5 feet 5 inches or 65 inches tall, then your height in feet will be $65 \div 12$, which is not an integer.

As part d of Example 1 illustrates, the integers are also incomplete. The expression $65 \div 12$ has no meaning if we are confined to the set of integers. To give such expressions meaning, we must extend our number system to the **rational number system,** the set of fractions. In the rational number system, $65 \div 12$ is simply the fraction $\frac{65}{12}$. In fact, whenever we divide any integer by a nonzero integer, the result is a rational number. A rational number can be expressed as the quotient of two integers, or it can be written as a decimal. It will be shown in Exercise 73 that whenever a rational number is expressed as a decimal, the expansion either terminates ($\frac{1}{4} = 0.25$, for example) or repeats ($\frac{1}{11} = 0.0909\ldots$, for example).

E X A M P L E 2   ■   Recognizing Rational Numbers

Which of the following numbers are rational?

a. 22.47
b. −2
c. 0.24343434...
d. The price in dollars of a compact disc
e. 0.101001000100001000001...

SOLUTION

a. 22.47 is a terminating decimal, so it is a rational number. Written as a fraction, $22.47 = \frac{2247}{100}$.

b. −2 is a rational number since it can be expressed as $-2 = \frac{-2}{1}$. (In fact, all integers are rational numbers.)

c. 0.24343434... is a repeating decimal, so it is a rational number. (Using the technique outlined in Exercise 73, we find that $0.24343434... = \frac{241}{990}$.)

d. In the United States, all retail prices are rounded to the nearest cent, which means that the price of the compact disc in dollars could be expressed as a decimal, with two digits to the right of the decimal point, such as 12.95. Thus, all prices are rational numbers.

e. Since 0.101001000100001000001... neither repeats nor terminates, it is not a rational number.

The rational numbers are complete with respect to the four basic operations of addition, subtraction, multiplication, and division. That is, if we perform any of these operations on a pair of rational numbers (with the exception of division by zero), the result will be a rational number. From the point of view of the ancients, all numbers were rational numbers. Seemingly every physical quantity—distance, force, time—could be measured using rational multiples of an appropriate unit. In fact, even today, in practice, physical quantities are given as rational multiples of a given unit. We speak of a distance of $2\frac{5}{8}$ inches, a weight of $155\frac{1}{4}$ pounds, an age of $2\frac{1}{2}$ years, and so forth. But the Greeks made a discovery that shattered their world view: No matter what unit of distance is chosen, some distances cannot be expressed as rational numbers! In particular, if we form a right triangle with legs of length 1, it is a consequence of the Pythagorean Theorem (see Figure 1) that the hypotenuse c will satisfy $c^2 = 1^2 + 1^2 = 2$, so that $c = \sqrt{2}$. But it can be shown (see Exercise 74) that $\sqrt{2}$ is not a rational number. This means that there is at least one distance that is not a rational number. There are many others; for example, $\pi \approx 3.14159...$ is not a rational number, yet it corresponds to the distance around a circle with diameter 1. In fact, there is a sense in which most distances are not rational.

To remedy the inadequacy of the rational numbers for expressing distance, we extend to the **real number system.** The real number system can be described as the set of all numbers with decimal expansions. Since the rational numbers have terminating or repeating decimal expansions, the nonrational real numbers, or **irrationals** as they are called, consist of those numbers with nonterminating, nonrepeating decimal expansions. Examples of irrational numbers include $\sqrt{2}$, π, and 0.1010010001....

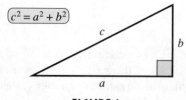

$c^2 = a^2 + b^2$

FIGURE 1

THE REAL NUMBER LINE

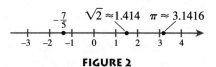

FIGURE 2

The real number system can be represented geometrically by the **real number line,** shown in Figure 2. The point labeled 0 is called the **origin.** The points to the left of the origin correspond to the negative real numbers, and the points to the right of the origin correspond to the positive real numbers. The arrow is used to indicate the direction of increasing real numbers.

The number line establishes a **one-to-one correspondence** with the real numbers in the sense that every point on the number line corresponds to a real number, and every real number corresponds to a point on the number line. As a consequence, we can order the real numbers according to their position on the number line.

Ordering the Real Numbers ☐

> For two real numbers a and b, we say that a **is less than** b (denoted $a < b$) if a lies to the left of b on the number line, and a **is greater than** b (denoted $a > b$) if a lies to the right of b.

The symbols "$<$" and "$>$" are called **inequality symbols.** Two other inequality symbols are "$\leq$", which denotes **less than or equal to,** and "$\geq$", which denotes **greater than or equal to.**

EXAMPLE 3   ■   Ordering Real Numbers

Place the appropriate inequality symbol in the box between the numbers.

a. $2 \; \square \; 5$
b. $-\frac{2}{3} \; \square \; -\frac{5}{6}$
c. $0 \; \square$ The weight of a bacterium in pounds
d. $-2.34 \; \square \; 1.03$
e. $0 \; \square$ The number of dimes a man has in his pocket
f. $100 \; \square$ Your score as a percentage on your next mathematics exam

SOLUTION

a.   Since 2 lies to the left of 5 on the number line, we write $2 < 5$.
b.   Since $-\frac{2}{3} = -\frac{4}{6}$ and $-\frac{4}{6}$ lies to the right of $-\frac{5}{6}$ on the number line, we write $-\frac{2}{3} > -\frac{5}{6}$.
c.   The weight of a bacterium, no matter how small, is positive, so we have

$$0 < \text{The weight of a bacterium in pounds}$$

d.   Every negative number is less than every positive number; thus we have $-2.34 < 1.03$.
e.   It is possible that a man might have no dimes in his pocket, so we write

$$0 \leq \text{The number of dimes a man has in his pocket}$$

f.   Because your score will be a number that is at most 100, we have

$$100 \geq \text{Your score as a percentage on your next mathematics exam}$$

SETS OF REAL NUMBERS

The most direct way to write a set is to simply list all of its elements within braces. For example, the set consisting of the positive even numbers less than 10 could be written as {2, 4, 6, 8}. The set of all prime numbers less than 30 is {2, 3, 5, 7, 11, 13, 17, 19, 23, 29}. The **empty set,** the set with no elements, may be written as { } or ∅.

Although listing is convenient for small sets, it is impractical for large sets and impossible for sets with infinitely many elements. **Set-builder notation** is useful for describing large or infinite sets that are defined by a property. For example, consider the set of real numbers less than 2. In set-builder notation, we write $\{x \mid x < 2\}$. The set of real numbers between and including -3 and 3 can be written $\{x \mid -3 \le x \le 3\}$.

It is often helpful to graph sets of real numbers on a number line. To do this we simply draw a number line and then draw a heavy line through all values that are included in the set. For example, the set $\{x \mid -2 < x \le 4\}$ is graphed as shown in Figure 3. Note that a parenthesis, "(", is used at -2 to indicate that -2 is *not* included, whereas a bracket, "]", is used at 4 to indicate that 4 *is* included.

FIGURE 3

E X A M P L E 4 ■ **Graphing Sets of Real Numbers**

Graph the following sets.

a. $\{x \mid x \ge 3\}$ **b.** $\{x \mid -\frac{3}{2} \le x < 3\}$

SOLUTION

a.

b.

THE DIGITS OF PI

The number π, the ratio of the circumference of a circle to its diameter, is irrational, and hence its decimal expansion neither repeats nor ends. It begins

3.14159265358979323846264338327950288419716939937510582097494459 23…

For most practical purposes, the first few digits of π provide us with a sufficiently accurate approximation. Indeed, the 65 digits just given provide us with sufficient accuracy to estimate the circumference of a circle with a radius equal to the distance of the furthest known object (a quasar some 7.8×10^{22} miles away) to within the length of the smallest hypothesized object in the universe (a "string" approximately 10^{-33} centimeters in length). But mathematicians, who are not constrained by the boundaries of the physical universe, have computed π to what at first glance seems a ridiculous number of decimal places. As of late 1998, the record holders are Yasumasa Kanada and Daisuke Takahashi of the University of Tokyo, who have computed π to over 51 billion decimal places. To put this in perspective: If their decimal expansion were written on a single line with the same size type that is used in this sentence, it would wrap around the Earth's equator more than 8 times!

Mathematicians have been competing to compute the most accurate version of π since ancient times. Although the need for increased accuracy may have been initially dictated by practical concerns, the driving force behind the quests of many modern "digit hunters" is clearly the competition itself. Also, computation of π to billions of decimal places provides an excellent means for testing the speed and accuracy of supercomputers, such as the massively parallel computer shown here.

But there is more here than meets the eye. Although we define π to be the ratio of the circumference of a circle to its diameter, it appears in many settings having nothing to do with circles. It appears in probability theory, calculus, number theory, and physics. To many mathematicians, π is not simply a human convention; π is an integral part of our universe. If we were one day to encounter intelligent beings from another galaxy, then surely they would know about π. To a digit hunter, the digits of π represent a piece of the cosmic code, a divine communiqué containing, perhaps, revelations of nature. And so they compute digits of π, looking for patterns, trying to detect structures. Are the digits truly random? Are there as many strings of consecutive digits as we would expect? These are the sorts of questions that have not yet been answered, but perhaps we will know more once we see the next 51 billion digits.

Sets of real numbers, such as those graphed in Example 4, can be conveniently described using **interval notation.** For example, instead of writing $\{x \mid -\frac{3}{2} \le x < 3\}$, we can simply write $[-\frac{3}{2}, 3)$. Notice that we are following our graphing convention of using a parenthesis to indicate that an endpoint of the interval is not included and a bracket to indicate that an endpoint is included. The set $\{x \mid x \ge 3\}$ can be written as $[3, \infty)$. Note that the symbol "∞" (infinity) does not represent a real number but rather indicates that the interval includes all points to the right of 3. Table 2 summarizes the nine types of intervals.

TABLE 2   Interval notation

Interval	Set-builder notation	Graph
(a, b)	$\{x \mid a < x < b\}$	
$[a, b]$	$\{x \mid a \le x \le b\}$	
$(a, b]$	$\{x \mid a < x \le b\}$	
$[a, b)$	$\{x \mid a \le x < b\}$	
(a, ∞)	$\{x \mid x > a\}$	
$[a, \infty)$	$\{x \mid x \ge a\}$	
$(-\infty, b)$	$\{x \mid x < b\}$	
$(-\infty, b]$	$\{x \mid x \le b\}$	
$(-\infty, \infty)$	$\{x \mid -\infty < x < \infty\}$	

EXAMPLE 5   **Using Interval Notation**

Write the given set using interval notation and sketch its graph.

a. $\{x \mid -7 \le x < 2\}$ **b.** $\{s \mid s < \sqrt{5}\}$

SOLUTION

a. $[-7, 2)$

b. $(-\infty, \sqrt{5})$

EXAMPLE 6   **Converting from Interval Notation to Set-Builder Notation**

Express each of the following sets using set-builder notation.

a. $(2, 5]$ **b.** $(-\infty, 3)$ **c.** $[-4, \infty)$

SOLUTION

a. $\{x \mid 2 < x \le 5\}$ **b.** $\{x \mid x < 3\}$ **c.** $\{x \mid x \ge -4\}$

ESTIMATION

In many real-life situations (and problems that you will encounter in this text), precise information about real number quantities is unavailable. Usually, however, the question at hand can be answered just by knowing a range of values for the quantities.

For example, suppose that you drive by a Maserati (an Italian sports car) with a for-sale sign in the rear window. You know that you have some change in your pocket left over from the grilled chicken sandwich that you just purchased, but you are not sure exactly how much. You wonder, "Do I have enough money to buy this car?" In spite of your lack of precise information, either concerning the cost of the car or the amount of change in your pocket, you deduce that you cannot buy the car. How did you do it? Although you were probably not conscious of this process, your thinking probably went something like this: "I know I have less than $20 in my pocket, and the Maserati must cost way more than $10,000. Guess I'll be riding this moped for a while." Although the approximations used in this example were trivial, the same sort of reasoning can be used to answer much more difficult questions.

EXAMPLE 7   Lifetime Walking—An Estimation Problem

Estimate the distance you will walk in your lifetime.

SOLUTION   We begin by making a rough estimate of the distance you might walk in a single day. It's not unreasonable to assume that if we include walking from room to room, walking to and from the parking lot, and all the other walks that occur in the course of a day, then you walk about 40 or so minutes per day. Now a pace of 20 minutes per mile is typical, so we'll assume that you walk 2 miles in the course of a day.

Next we compute the number of days in your lifetime. If you live 75 years, then (ignoring leap years) you will live $75 \times 365 = 27,375$ days. Multiplying the 2 miles that you walk every day by the number of days in your lifetime, we obtain $2 \times 27,375 = 54,750$ miles—far enough to circle Earth twice!

An important component of numeracy (facility in using and understanding numbers) is intuition regarding the relative sizes of the objects that make up our world. Most of us do respectably well when estimating the sizes of human-sized objects such as cars, refrigerators, desks, and so forth, but tend to ridiculously underestimate the dimensions of large, naturally occurring features such as canyons, volcanos, and mountains. The answer to the following estimation problem may surprise you.

EXAMPLE 8   Leveling Everest—An Estimation Problem

If it were possible to grind up Mt. Everest, fill semitrailers with the contents, and then form the semis into a convoy, how long would the convoy be?

SOLUTION   First we will estimate the volume of Mt. Everest. We look in a reference book (such as an almanac or *The Guinness Book of World Records*) and find that Mt. Everest is approximately 29,000 feet tall. Now, unless we are lucky enough to have

A team of scientists headed by B. Chamoux and A. de Polenza measures Mt. Everest.

a reference that gives the volume of Mt. Everest, we are going to have to make an educated guess as to its shape. Let us assume that the mountain is roughly the shape of a cone with a height of 29,000 feet. We look up the formula for the volume of a cone and find that $V = \frac{1}{3}\pi r^2 h$, where r is the radius of the circle that forms the base of the cone, and h is the height. We are taking h to be 29,000, but we don't know r. Here again, we will make an educated guess that the diameter of the base is roughly twice the height, so that the radius equals the height. Thus, we take r to be 29,000 feet. Upon substituting the values of r and h into the formula for V, we find that

$$V = \frac{1}{3}\,\pi\,(29{,}000\text{ ft})^2\,(29{,}000\text{ ft}) \approx 25{,}540{,}000{,}000{,}000 \text{ cubic feet}$$

Next we estimate the capacity of a semitrailer. They appear to be about 40 feet long, 8 feet wide, and 9 feet tall. This would make the volume of a semitrailer equal to 40 feet $\times$ 8 feet $\times$ 9 feet = 2880 cubic feet. Hence it would take approximately $25{,}540{,}000{,}000{,}000 \div 2880 \approx 8{,}868{,}000{,}000$ semitrailers to haul off Mt. Everest. Now we must figure out how long a convoy this would make. The cab of a semi is about 10 feet in length, so the entire truck would be about 10 feet + 40 feet = 50 feet long. Assuming two truck lengths (or 100 feet) between trucks, the convoy would be about $8{,}868{,}000{,}000 \times 150$ feet = $1{,}330{,}200{,}000{,}000$ feet long. Using the fact that a mile is 5280 feet, we determine that the convoy is $1{,}330{,}200{,}000{,}000 \div 5280 \approx 251{,}900{,}000$ miles long. This, by the way, is long enough to reach to the sun and back, and then make over 137 round-trips to the moon! Although our estimates were crude, even if we are off by a factor of 10 or even 100, it is clear that such a convoy would be impossible to form.

COMPLEX NUMBERS

If a mathematician of only a few hundred years ago were asked to produce the square root of -1 (a number whose square is -1), the response would have been, "There is no such number." And indeed, there is no such *real* number. But strangely, square roots of negative numbers have reared their heads in problems in number theory, geometry, and even physics. Although such solutions were at first discarded, it soon became evident that at the very least, the square roots of negative numbers could be used as a sort of "trick" to solve many problems. Mathematicians found that by temporarily pretending such numbers existed, they were able to produce correct results. Since the square roots of negative numbers were not seen as being full-fledged numbers, they were called **imaginary numbers.** The square root of -1 was called i (so that

$i^2 = -1$), and any other imaginary number could be expressed as a real number times i: $2i$, $5i$, $-1.23i$, and so on.

As mathematics has evolved, the status of the so-called imaginary numbers has been elevated; modern mathematicians view imaginary numbers as no less "real" than the real numbers themselves. The **complex number system,** an extension of the real number system, is formed by including all possible sums of real numbers with imaginary numbers, such as $2 + 3i$ and $3 - 4i$.

EXAMPLE 9   ■   Classification of Complex Numbers

Determine whether the given number is real, imaginary, or neither.

a.   A number that when multiplied by itself gives -25
b.   The distance that light travels in a year expressed in millimeters
c.   A number that when squared gives $2i$

SOLUTION

a.   There is no real number whose square is negative; only imaginary numbers have negative squares. Thus, any number that when multiplied by itself gives -25 is imaginary.
b.   All distances can be expressed as real numbers, so this number is real. (In fact, light travels approximately 9,439,922,663,424,000,000 millimeters in one year.)
c.   When imaginary numbers are squared, the result is a negative real number. When real numbers are squared, the result is a positive real number. Thus, any number whose square is $2i$ is neither real nor imaginary. (In fact, $2i$ has two complex square roots, $1 + i$ and $-1 - i$, neither of which is real or imaginary.)

We now summarize the number systems that have been discussed. The relationships among the systems are diagrammed in Figure 4.

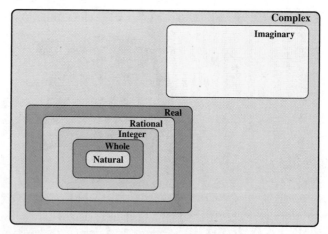

FIGURE 4

Number Systems   □

Natural numbers are the numbers 1, 2, 3, . . .

Whole numbers are the numbers 0, 1, 2, 3, . . .

Integers are the numbers . . . , -2, -1, 0, 1, 2, 3, . . .

> **Rational numbers** are the fractions—that is, numbers of the form *p/q*, where *p* and *q* are integers and $q \neq 0$. Rational numbers have terminating or repeating decimal expansions.
>
> **Real numbers** are the numbers that may be written as decimals. The real numbers that are not rational are called **irrational,** and they have nonterminating, nonrepeating decimal expansions.
>
> **Complex numbers** are numbers of the form *a* + *bi*, where *a* and *b* are real numbers and $i = \sqrt{-1}$. The term **imaginary** is used to describe complex numbers of the form *bi*, where $b \neq 0$.

UNDERSTANDING AND MASTERY CHECKLISTS

CONCEPTS TO UNDERSTAND

- ☐ Natural number system
- ☐ Whole numbers
- ☐ Integers
- ☐ Rational number system
- ☐ Irrational numbers
- ☐ Real number system
- ☐ Imaginary numbers
- ☐ Complex number system
- ☐ The number line
- ☐ Inequality symbols
- ☐ Set-builder notation
- ☐ Interval notation
- ☐ Estimation

SKILLS TO MASTER

- ☐ Determine to which systems a given number belongs.
- ☐ Determine which of two real numbers is greater.
- ☐ Use the symbols < and > appropriately.
- ☐ Express sets of real numbers in set-builder notation.
- ☐ Express sets of real numbers using interval notation.
- ☐ Estimate quantities by making reasonable assumptions.

EXERCISES 1.1

EXERCISES 1–18 *List all the terms that describe the given number. Choose from natural, whole, integer, rational, irrational, real, and complex.*

1. $\frac{6}{3}$

2. -47.134

3. $\sqrt{7}$

4. $\frac{12}{5}$

5. $3 - i$

6. $-\sqrt{3}$

7. The number of goats on Mars

8. 0.8888888…

9. 0.8888888

10. $\sqrt{-4}$

11. 4

12. The ratio of the circumference of a circle to its diameter

13. $-\sqrt{25}$

14. $34{,}343{,}434{,}343 - 35{,}343{,}434{,}343$

15. A number whose square is -7

16. The number of men named Bob living as of 9:43:26.24 A.M. EST on June 17, 2001

17. The height in miles of a 1-foot-tall rooster.

18. The hypotenuse of an isosceles right triangle with legs of length 1.

EXERCISES 19–24 *Place the most appropriate order symbol between each pair of quantities.*

19. 3 ☐ 0

20. -4 ☐ -5

21. $-\frac{6}{5}$ ☐ $-\frac{4}{5}$

22. $\frac{5}{6}$ ☐ $\frac{2}{3}$

23. The number of "Ben Franklins" in your wallet ☐ 0

24. A woman's age in years on or after her 20th birthday ☐ 20

EXERCISES 25–30   *Place the given list of numbers in ascending order (the smallest first, the largest last). Estimate as needed.*

25. $23, -22.9, -23, 22.9$

26. $-\frac{1}{4}, -0.26, -\frac{3}{16}, -0.2$

27. $\frac{22}{7}, \pi, 3.15, 3 + \frac{1}{10} + \frac{4}{100}$

28. $\frac{1}{3}, 0.33333, 0.3334, \frac{1001}{3000}$

29. The number of miles from Earth to the sun, the gross domestic product of the United States in 1998 in dollars, 1,000,000,000, the number of living organisms on Earth (*Hint*: Some of these values can be found in the numerography inside the front cover.)

30. The diameter of the period "." in feet, 0.0001, the weight of this page in tons, Shaquille O'Neal's height in light-years (*Hint*: The numerography inside the front cover may be helpful.)

EXERCISES 31–38   *Write each set using interval notation and sketch the graph.*

31. $\{x \mid -2 < x \le 4\}$

32. $\{x \mid 3 \le x < 8\}$

33. $\{x \mid -7 < x < -\frac{1}{2}\}$

34. $\{x \mid -2 \le x \le \sqrt{3}\}$

35. $\{x \mid x < -3\}$

36. $\{x \mid x \ge -1\}$

37. $\{x \mid x \ge \sqrt{2}\}$

38. $\{x \mid x \le \frac{3}{2}\}$

EXERCISES 39–46   *Write each set using set-builder notation.*

39. $(3, 5)$

40. $(-\infty, 2)$

41. The real numbers 1 through 10^{100}

42. $(0, \infty)$

43. $[1, 2]$

44. The real numbers -100 through 0

45. $(100, \infty)$

46. $(1, 2)$

▪ APPLICATIONS

47. *Cigarette Expense*   Estimate the amount of money that a 3-pack-a-day smoker will spend on cigarettes in 30 years.

48. *Lifting a Van*   Estimate the number of kindergartners required to lift a minivan. What practical difficulties would arise in performing such a stunt?

49. *Earth Population*   Estimate the number of square yards of living space available to each person if the entire population of Earth were crammed into the state of Delaware.

50. *Garbage Weight*   Estimate the weight of all the garbage produced in residences in the United States in a year's time.

51. *Sharing Wealth*   As of July 14, 1999, Bill Gates's personal wealth was estimated to be $107 billion. (*Data source*: Bill Gates Personal Wealth Clock, www.webho.com/WealthClock). Estimate the number of families that Bill Gates could feed for a year if he were willing to spend his entire fortune. If Bill Gates had not existed, would the rest of the world be $107 billion richer?

52. *Gasoline Usage*   Estimate the number of gallons of gasoline that would be saved each year if every passenger car in the United States got 50 miles per gallon under all conditions (highway, city, and so on).

53. *Estimating Words*   Estimate the number of words that the average American speaks in a day. About how many words are spoken every day in the United States?

54. *Papering the Nation*   Total U.S. circulation of daily newspapers in 1996 was 56.983 million and for Sunday newspapers was 60.790 million (*Data source*: Newspaper Association of America). At this rate, how many years would it take to paper over the United States, with an area of 3,618,770 square miles?

55. *Billion Dollar Cube*   Estimate the side length of a cube that could contain $1 billion in $100 bills.

56. *Shrinking Earth*   The Great Wall of China, with a length of approximately 4000 miles, is the only man-made structure visible from space with the naked eye. If you were constructing an extremely detailed globe with a diameter of 10 feet, how long would the Great Wall of China be? How tall would a human be? How much area would Canada cover?

The Great Wall of China appears as an orange line in this satellite image.

57. *Prolific Author*   Dame Barbara Cartland, listed in *The Guinness Book of World Records* as the world's best-selling living author, wrote roughly 600 novels from 1925 to 1996. Estimate the average number of words written per day by Dame Cartland during this period.

58. *The Book of Pi*   Suppose that all the known digits of pi were written on pages the size of this page, and these pages were then bound to form a book. How thick would such a book be? Refer to the math note on the digits of pi for an estimate of the number of known digits.

■ CONCEPTS AND CRITICAL THINKING

EXERCISES 59–64   *Answer true or false.*

59. Every natural number is a whole number.

60. Some integers are irrational numbers.

61. Some complex numbers are real.

62. All rational numbers have a finite number of decimal places.

63. Negative integers are rational numbers.

64. If the square of a number is negative, then the number itself cannot be rational.

EXERCISES 65–69   *Give an example of each.*

65. An integer that is not a natural number

66. A real number that is not a rational number

67. A complex number that is not real

68. A rational number that is not an integer

69. A whole number that is not a natural number

■ QUESTIONS FOR DISCUSSION OR ESSAY

70. It is a fact (see Exercise 73) that $0.999\ldots = 1$. Although mathematicians are in universal agreement, many other people refuse to believe it. What makes this fact so difficult to believe? Does the statement that $0.999\ldots = 1$ seem to defy common sense? Why?

71. People often complain about being treated as if they were "just a number." How many numbers can you think of that are used for identification? To what number system do they belong?

72. When solving estimation problems, it is important to bear in mind the desired degree of accuracy. For example, if someone asks you for the time, he is really asking you for an estimation of the time. If you were to answer either "the twenty-first century" or "at the instant you perceive that I have blinked my left eye, the time will be 3:25.62859 P.M. September the twentieth, A.D. two thousand and one," you would no doubt be institutionalized. How does one determine the degree of precision required of an estimation?

■ PROJECTS FOR ENRICHMENT

73. *Decimal Expansions*   The text states that rational numbers have terminating or repeating decimal expansions. The converse of this statement is also true; that is, if a decimal expansion terminates or repeats, then it represents a rational number. In this project, we learn how to express terminating and repeating decimals as fractions.

    Terminating decimals are converted to fractions as follows. The numerator of the fraction is the integer obtained by removing the decimal point; the denominator is an appropriate power of 10 (10, 100, 1000, and so on). For example, 0.123 may be written as $\frac{123}{1000}$, 1.03 as $\frac{103}{100}$, and so on.

    a. Express each of the terminating decimals as a fraction.

i. 87.096	ii. 232.1
iii. 0.0034	iv. 0.000048

    The following examples illustrate a method for converting *repeating* decimals to fractions.

    Example: Convert $13.454545\ldots$ to a fraction.

    Solution: Let $x = 13.454545\ldots$. Then $100x = 1345.454545\ldots$. We subtract x from $100x$ as follows:

    $$\begin{aligned} 100x &= 1345.454545\ldots \\ -x &= -13.454545\ldots \\ \hline 99x &= 1332.0 \end{aligned}$$

    Dividing both sides by 99, we obtain $x = \frac{1332}{99}$.

    Example: Convert $2.512012012\ldots$ to a fraction.

    Solution: Let $x = 2.512012012\ldots$. Then $10,000x = 25,120.12012012\ldots$ and $10x = 25.12012012\ldots$. Subtracting, we obtain

    $$\begin{aligned} 10,000x &= 25,120.120120120\ldots \\ -10x &= -25.120120120\ldots \\ \hline 9,990x &= 25,095.0 \end{aligned}$$

    Thus, $x = \dfrac{25,095}{9990} = \dfrac{1673}{666}$.

b. Express each of the following repeating decimals as a fraction.

    i. 0.252525...   ii. 0.999...

    iii. 12.0010101...   iv. 344.123123123...

74. *The Irrationality of* $\sqrt{2}$ In this project, we show that $\sqrt{2}$ is irrational. We use the method called *reductio ad absurdum,* or reduction to an absurdity. That is, we assume the opposite of what we think is true and show that this leads to a contradiction.

a. Suppose that $\sqrt{2}$ is rational—that is, that $\sqrt{2} = p/q$ for a pair of integers p and q with no common factors. (If p and q have a common factor, then p/q is not in lowest terms.) Show that $p^2 = 2q^2$.

b. Use part a to show that 2 must be a factor of p^2. If 2 is a factor of p^2, does it follow that 2 is a factor of p?

c. Suppose that 2 divides evenly into p. Then $p = 2 \cdot m$ for some integer m. Show that $4m^2 = 2q^2$.

d. Show that $q^2 = 2m^2$, and hence that 2 divides into q.

e. The result to part d is an absurdity. Why? Since the assumption that $\sqrt{2}$ is rational leads to an absurdity, it follows that $\sqrt{2}$ is irrational.

SECTION 1.2 # INTRODUCTION TO COORDINATE GEOMETRY

> ☐ **What can coordinate geometry reveal about the national debt or the concentration of CO_2 in the atmosphere?**
>
> ☐ **In what way does a city's system of streets and avenues suggest a link between algebra and geometry?**
>
> ☐ **How can the ideas of a Greek mathematician from several hundred years B.C. and a French philosopher from the 17th century A.D. be combined to determine distances without a tape measure?**
>
> ☐ **If the life expectancy for an American is 76 years, then how much longer can someone 75 years and 364 days old expect to live?**

THE CARTESIAN PLANE

In Section 1.1, we saw that real numbers can be represented by points on the number line. René Descartes, a brilliant 17th-century mathematician and philosopher, extended this connection between the real number system and geometry by associating with every point in the plane a pair of real numbers. This simple yet profound step enables us to apply powerful algebraic techniques to solve difficult geometric problems, extending greatly the range of practical problems that can be solved with algebra.

We form the **Cartesian plane** by placing a horizontal number line, called the **x-axis,** and a vertical number line, called the **y-axis,** so that they intersect at the zero point

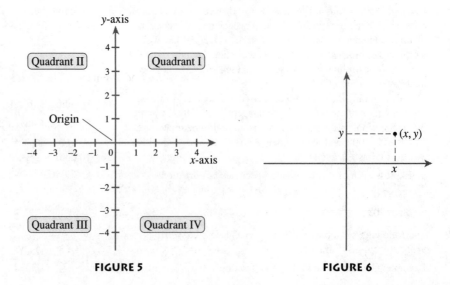

FIGURE 5   **FIGURE 6**

of each line. The point of intersection of the two axes is called the **origin,** and the axes divide the plane into four regions, called **quadrants,** as shown in Figure 5.

Each point in the Cartesian plane corresponds to an ordered pair of numbers (x, y), where x is the number on the x-axis directly above or below the point, and y is the number on the y-axis directly to the right or left of the point, as shown in Figure 6. In other words, the number x indicates how many units to move horizontally from the origin (to the right if x is positive and to left if it is negative); the number y indicates how many units to move vertically (up if y is positive and down if it is negative). The numbers x and y are called the **coordinates** of the point.

EXAMPLE 1   ▨   **Plotting Points**

Plot the points $(2, 3)$, $(-1, -4)$, $(-3, 0)$, and $(\pi, -\sqrt{2})$.

SOLUTION   To plot the point $(2, 3)$, we move 2 units to the *right* of the origin and 3 units *up*. For the point $(-1, -4)$, we move 1 unit to the *left* of the origin and four units *down*. The point $(-3, 0)$ is on the x-axis, 3 units to the *left* of the origin. Finally, the point $(\pi, -\sqrt{2})$ is π units to the *right* of the origin and $\sqrt{2}$ units *down*. See Figure 7.

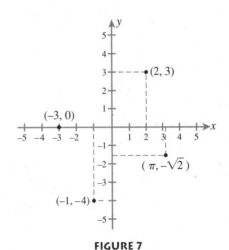

FIGURE 7

The principles of coordinate geometry are involved whenever a grid system is used to identify the locations of objects. Two examples of this are the rectangular grids formed by streets and avenues in cities and the yard lines and hash marks on football fields. Nonrectangular grids are used on radar screens and for marking latitude and longitude on the surface of the Earth. Some of these examples will be considered in the exercises.

Another common application of coordinate geometry is in the area of *data analysis.* Plots of points can be used to organize and display data so that trends can be easily recognized.

EXAMPLE 2   ▨   **Plotting Real-World Data**

The U.S. Department of the Treasury records the following data on the national debt since 1950. Using the year as the x-coordinate and the debt as the y-coordinate, plot the data on a rectangular coordinate system.

Year	1950	1955	1960	1965	1970	1975	1980	1985	1990
Debt (in billions of dollars)	256	273	284	314	370	533	910	1823	3233

SOLUTION   Two possible choices for the scale on the horizontal axis are shown in Figures 8 and 9. The axis in Figure 8 is broken between the origin and the year 1950 to indicate that the years between those dates have been omitted. Figure 9 illustrates the loss of clarity if this is not done. As you will see in Exercise 41, the broken axis technique must be used carefully because it can result in misleading conclusions.

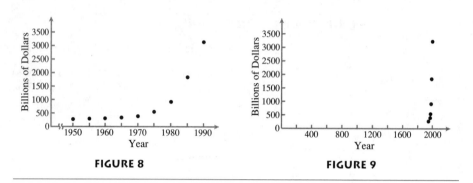

FIGURE 8   **FIGURE 9**

PLOTTING POINTS WITH A GRAPHING CALCULATOR

Most graphing calculators are capable of plotting individual points. However, if more than just one or two points are to be plotted, it is much more convenient to use a scatterplot. This feature is particularly useful when one is looking for trends in data.

CALCULATOR KEYS

Plotting Data Points

A **scatterplot** is simply a graph of a collection of data points (x, y). The creation of such a graph typically involves the following four steps:

1.   Clear any existing graphs.
2.   Clear any existing statistical data.
3.   Enter the points (x, y) as statistical data.
4.   Select the scatterplot option from the list of available statistical plots.

EXAMPLE 3   ■   Using a Scatterplot to Identify Trends in Data

Plot the following points with the scatterplot option on a graphing calculator. Use the graph to estimate the missing y-value.

x	1.4	2.1	2.9	3.5	4.3	5.1
y	1.0	1.4	1.7	2.0	2.4	?

SOLUTION   Before creating a new scatterplot, it is usually necessary to clear any existing graphs and any existing statistical data. Next we enter the new data points as statistical data. Finally, we select the scatterplot option from the list of available statistical plots. The display should look similar to Figure 10. The plot suggests that the points lie on a straight line and that the line is rising from left to right. To estimate the missing y-value, we first move the cursor to the right until the x-coordinate displayed at the bottom of the screen is approximately 5.1, and then we move the cursor up until it looks roughly on a line with the other points. The y-coordinate at this location is approximately 2.6 (Figure 11).

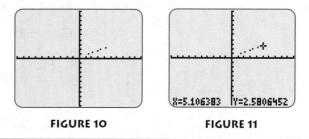

FIGURE 10   **FIGURE 11**

Often it is necessary to adjust the *viewing window* of the graphing calculator so that an appropriate portion of the graph can be seen on the display. This is done by modifying the *window variables*.

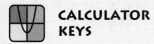 **CALCULATOR KEYS**

Controlling the View

The **viewing window** is the portion of the coordinate system that is displayed by the graphing calculator. The **window variables** define the viewing window. The variables Xmin and Xmax specify the range of x-coordinate values; Ymin and Ymax specify the range of y-coordinate values. The data from Example 3 are shown in Figure 12, with a viewing window defined by Xmin=−10, Xmax=10, Ymin=−10, and Ymax=10. Figure 13 shows the same set of data with a viewing window defined by Xmin=−4.7, Xmax=4.7, Ymin=−3.1, and Ymax=3.1.

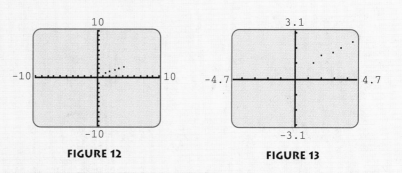

FIGURE 12   **FIGURE 13**

EXAMPLE 4   ▦   Using a Scatterplot to Identify Trends in Real-World Data

The atmospheric concentration of CO_2 over a recent time period is given in the following table. Describe any trends in the data. [*Data source*: Dean Abrahamson, Ed., *The Challenge of Global Warming* (Washington, DC: Island Press, 1989)].

Month	0	3	6	9	12	15	18	21	24
CO_2 concentration (parts per million)	341	343	341	338	342	345	343	340	343

SOLUTION   We enter the data points as statistical data, with x representing the month and y representing the corresponding CO_2 concentration. The scatterplot is shown with two different viewing windows in Figures 14 and 15. The view in Figure 15 suggests that there are regular fluctuations in the CO_2 level and also that there is a slight upward shift in the second year.

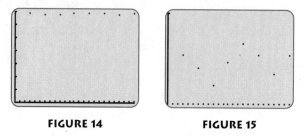

FIGURE 14   **FIGURE 15**

DISTANCES BETWEEN POINTS

The distance between two points on the number line can be found by subtracting the smaller number from the larger. In a similar way, we can find the distance between points that lie on the same horizontal or vertical line. For example, to find the distance between the points $(2, -1)$ and $(2, 5)$ shown in Figure 16, we can compute the difference between the y-coordinates to obtain $5 - (-1) = 6$. Likewise, the distance between $(-6, 1)$ and $(-3, 1)$ is the difference between the x-coordinates—namely, $-3 - (-6) = 3$.

Suppose now that we are interested in the distance between two points that are not on the same horizontal or vertical line, say $(2, 5)$ and $(6, 8)$. From Figure 17, we see that the horizontal distance between the points is $6 - 2 = 4$, and the vertical distance is $8 - 5 = 3$. Thus, the line segment joining the points is the hypotenuse of a right triangle with side lengths of 3 and 4. By the Pythagorean Theorem, the distance between the points is given by $d = \sqrt{3^2 + 4^2} = \sqrt{25} = 5$.

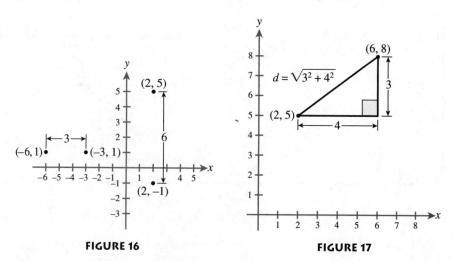

FIGURE 16   **FIGURE 17**

By applying this process more generally to the points (x_1, y_1) and (x_2, y_2), we obtain the following formula for finding the distance between two points.

The Distance Formula □

> The distance d between the points (x_1, y_1) and (x_2, y_2) is given by
>
> $$d = \sqrt{(x_2 - x_1)^2 + (y_2 - y_1)^2}$$

EXAMPLE 5 ■ **Finding the Distance Between Points**

Find the distance between $(3, 5)$ and $(-1, -2)$.

SOLUTION According to the distance formula, we compute the square of the differences between the x- and y-coordinates, add them, and take the square root of the sum.

$$
\begin{aligned}
d &= \sqrt{(-1 - 3)^2 + (-2 - 5)^2} \\
&= \sqrt{(-4)^2 + (-7)^2} \\
&= \sqrt{16 + 49} \\
&= \sqrt{65} \approx 8.06
\end{aligned}
$$

EXAMPLE 6 ■ **Identifying a Polygon**

Show that the points $(-4, 2)$, $(0, 0)$, $(-1, 5)$, and $(3, 3)$ form the vertices of a parallelogram. Is the parallelogram in fact a rhombus? That is, are all four sides of equal length?

SOLUTION The four points are plotted in Figure 18. Since a quadrilateral is a parallelogram if and only if opposite sides have equal length, we must show that $a = c$ and $b = d$.

$$
\begin{aligned}
a &= \sqrt{(-4 - 0)^2 + (2 - 0)^2} = \sqrt{20} \\
c &= \sqrt{(-1 - 3)^2 + (5 - 3)^2} = \sqrt{20} \\
b &= \sqrt{(-1 - (-4))^2 + (5 - 2)^2} = \sqrt{18} \\
d &= \sqrt{(3 - 0)^2 + (3 - 0)^2} = \sqrt{18}
\end{aligned}
$$

Because the opposite sides have the same length, the figure is a parallelogram. However, it is not a rhombus since the four sides are not of equal length.

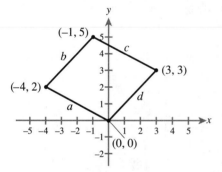

FIGURE 18

THE MIDPOINT FORMULA

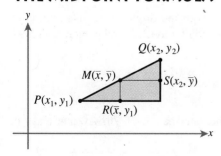

FIGURE 19

Suppose that we are interested in finding the coordinates $(\bar{x}, \bar{y})$ of the midpoint M of the line segment joining two arbitrary points $P(x_1, y_1)$ and $Q(x_2, y_2)$, as in Figure 19. First we note that $\triangle PMR$ and $\triangle MQS$ are congruent. Thus, $PR = MS$ and $MR = QS$. Using the coordinates of the points involved, we can compute $\bar{x}$ and $\bar{y}$.

$$
\begin{array}{cc}
PR = MS & MR = QS \\
\bar{x} - x_1 = x_2 - \bar{x} & \bar{y} - y_1 = y_2 - \bar{y} \\
2\bar{x} = x_1 + x_2 & 2\bar{y} = y_1 + y_2 \\
\bar{x} = \dfrac{x_1 + x_2}{2} & \bar{y} = \dfrac{y_1 + y_2}{2}
\end{array}
$$

We have derived the midpoint formula.

The Midpoint Formula □

The midpoint of the line segment joining the points (x_1, y_1) and (x_2, y_2) is $(\bar{x}, \bar{y})$, where

$$\bar{x} = \frac{x_1 + x_2}{2} \quad \text{and} \quad \bar{y} = \frac{y_1 + y_2}{2}$$

Note that $\bar{x}$ is the average of the *x*-coordinates, and $\bar{y}$ is the average of the *y*-coordinates.

EXAMPLE 7   ▨   **Using the Midpoint Formula**

Find the coordinates of the midpoint of the line segment connecting $(3, 5)$ and $(-1, -2)$.

SOLUTION   According to the midpoint formula, we have

$$\bar{x} = \frac{3 + (-1)}{2} = 1 \quad \text{and} \quad \bar{y} = \frac{5 + (-2)}{2} = \frac{3}{2}$$

Thus, the midpoint has coordinates $(1, \frac{3}{2})$.

EXAMPLE 8   ▨   **An Application Involving the Midpoint Formula**

A subway is to be built connecting points *A* and *B* on the city map in Figure 20. Find the location of a subway stop that is exactly halfway between *A* and *B*.

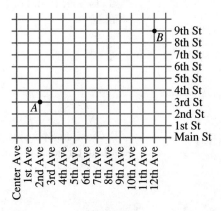

FIGURE 20

SOLUTION   By placing a coordinate system with the origin at the intersection of Center Avenue and Main Street, we can express point *A* as $(2, 3)$ and point *B* as $(12, 9)$. The midpoint of $\overline{AB}$ has coordinates

$$\bar{x} = \frac{2 + 12}{2} = \frac{14}{2} = 7 \qquad \bar{y} = \frac{3 + 9}{2} = \frac{12}{2} = 6$$

So the subway stop should be located at the intersection of 7th Avenue and 6th Street.

UNDERSTANDING AND MASTERY CHECKLISTS

CONCEPTS TO UNDERSTAND

☐ Cartesian plane
☐ Coordinates
☐ Scatterplot
☐ Viewing window
☐ Distance formula
☐ Midpoint formula

SKILLS TO MASTER

☐ Plot points in the coordinate plane.
☐ Determine the quadrant of a given point.
☐ Produce a scatterplot with a graphing calculator.
☐ Find the distance between two points.
☐ Find the midpoint between two points.

EXERCISES 1.2

EXERCISES 1–2 *Plot the given points.*

1. $(-2, 1)$, $(\sqrt{3}, -5)$, $\left(1, \frac{3}{2}\right)$, $(0, -4)$
2. $(1, -1)$, $\left(\frac{2}{3}, 0\right)$, $(-1, 3)$, $(-2, -\sqrt{5})$

EXERCISES 3–6 *Plot the pair of points and find the distance between them.*

3. $(3, 5)$ and $(3, -2)$
4. $(-6, -1)$ and $(-2, -1)$
5. $(-3, -2)$ and $(5, 4)$
6. $(-8, 6)$ and $(4, 1)$

EXERCISES 7–10 *Plot the pair of points, find the distance between them, and find the coordinates of the midpoint of the line segment connecting them.*

7. $(1, 2)$ and $(5, 4)$
8. $(-2, 3)$ and $(4, -1)$
9. $\left(-\frac{4}{3}, 2\right)$ and $\left(\frac{8}{3}, 1\right)$
10. $\left(-2, -\frac{7}{2}\right)$ and $\left(5, \frac{1}{2}\right)$

EXERCISES 11–14 *Show that the given points form the vertices of the indicated polygon.*

11. Parallelogram: $(2, 3)$, $(4, 5)$, $(6, 4)$, $(8, 6)$ (*Hint*: Show opposite sides are of equal length.)
12. Rhombus: $(-1, -3)$, $(1, 5)$, $(8, -1)$, $(-8, 3)$ (*Hint*: Show all four sides are of equal length.)
13. Right triangle: $(1, 0)$, $(2, 3)$, $(4, -1)$ (*Hint*: Use the Pythagorean Theorem.)
14. Square: $(1, 10)$, $(-5, 2)$, $(3, -4)$, $(9, 4)$ (*Hint*: The diagonals must be the same length, as must the adjacent sides.)

EXERCISES 15–18 *Use a scatterplot to graph the given data. Then, use the graph to estimate the missing y-value.*

15.

x	0.7	1.2	1.6	2.0	2.6	3.0
y	2.7	2.5	2.4	2.2	1.9	?

16.

x	1.4	1.8	2.2	2.6	3.0	3.4	3.8
y	1.0	1.7	2.1	2.2	2.0	1.4	?

17.

x	0.7	1.6	2.4	3.0	3.9	4.9	5.3	6.0	6.6	7.1
y	2.5	2.0	1.5	1.9	2.5	1.8	1.5	1.7	2.3	?

18.

x	1.4	2.2	2.6	3.3	3.5	4.1	4.7	5.2	5.6	6.1
y	1.0	3.0	4.1	3.3	3.1	4.3	6.2	5.8	5.2	?

EXERCISES 19–22 *When applying coordinate geometry to real-world problems, the first step is to define the coordinate system by deciding on the location of the origin (and the orientation of the axes). Figure 21 shows a portion of a plane, along with two different possible origins, O and O′. Thus, for example, although the point P has coordinates $(0, 2)$ with respect to O, it has coordinates $(3, 0)$ with respect to O′.*

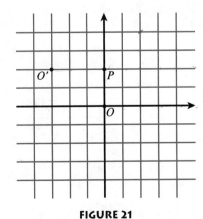

FIGURE 21

19. A point Q with coordinates $(5, 2)$ relative to the origin O has what coordinates relative to the origin O'?
20. A point Q with coordinates $(-1, -2)$ relative to the origin O has what coordinates relative to the origin O'?

21. A point Q with coordinates (5, 2) relative to the origin O' has what coordinates relative to the origin O?

22. A point Q with coordinates $(-1, -2)$ relative to the origin O' has what coordinates relative to the origin O?

APPLICATIONS

23. *Crawling Ant* Suppose that an ant is initially at the point $(-2, 4)$ and that it makes an "antline" for the point $(-7, -4)$. From there it crawls to the point $(3, -2)$, and then returns to $(-2, 4)$. Is the ant traveling clockwise or counterclockwise? What is the total distance traveled by the ant?

24. *Punctuation Mark* Plot the points $(-2, 4)$, $(-1, 5)$, $(0, 5)$, $(1, 4)$, $(1, 3)$, $(0, 2)$, $(0, 1)$, $(0, 0)$, and $(0, -2)$. What punctuation mark does this set of points suggest?

25. *Minimum Wage* The U.S. Department of Labor records the following data on the minimum hourly wage from 1955 to 1995.

Year	1955	1960	1965	1970	1975	1980	1985	1990	1995
Wage	0.75	1.00	1.25	1.60	2.10	3.10	3.35	3.80	4.25

Use the year as the x-coordinate and the wage as the y-coordinate to plot the data on a rectangular coordinate system. On the basis of the data, what would you predict for the minimum wage in 1997? The actual value was $5.15. How far off is your estimate?

26. *Life Expectancy* In the graph in Figure 22, we see that the expected age at death increases with age. For example, a 20-year-old can expect to live to the age of 77.1, whereas a 60-year-old can expect to live to the age of 81.2. Use the data in Figure 22 to find data points (x, y), with x representing current age and y representing life expectancy—the expected number of years until death. Plot these points. Use your plot to estimate the life expectancy and the expected age at death of a 25-year-old and a 90-year-old. (*Data source*: Centers for Disease Control.)

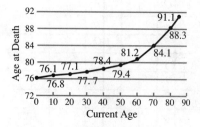

Expected Age at Death

FIGURE 22

27. *City Street Map* A coordinate system is placed on a city street map with the origin located at the intersection of Main Street and Center Avenue, as shown in Figure 23. All streets and avenues, except Lloyd Avenue, run north/south or east/west. What is the shortest route from point $A(-6, -2)$ to point $B(5, 4)$?

Compute the total distance for this shortest route (using a standard city block as 1 unit), and compare it to the distance one would have to travel if Lloyd Avenue were closed for repairs.

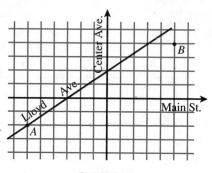

FIGURE 23

28. *City Street Map* Refer to the city street map in Figure 23. A bicycle messenger, currently at point $A(-6, -2)$, receives a message to pick up a package from an office building at the intersection of Main and Center and deliver it to a building at point $B(5, 4)$. Find the shortest route for her to take, and compute the total distance (using a standard city block as 1 unit).

29. *Suspended Ceiling* The suspended ceiling shown in Figure 24 consists of $2' \times 2'$ square tiles. An electrical junction box is to be placed halfway between points A and B. Use the midpoint formula to locate the halfway point.

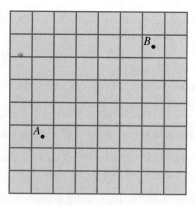

FIGURE 24

30. A drawing utility was used to construct the floor plan shown in Figure 25. The coordinate system for this drawing has the origin placed at the upper left corner of the outer wall, positive x values increasing to the right, positive y values increasing downward,

and units in inches. The coordinates of 8 points along the closet wall are given in Table 3. Find the coordinates of the 8 corresponding points that would result if the width of the closet is decreased by 2 inches and the depth is increased by 3 inches.

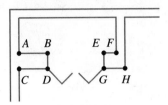

FIGURE 25

TABLE 3   Closet points

Point	Coordinates
A	(4.5, 25)
B	(13, 25)
C	(4.5, 29.5)
D	(13, 29.5)
E	(55, 25)
F	(60.5, 25)
G	(55, 29.5)
H	(65, 29.5)

CONCEPTS AND CRITICAL THINKING

EXERCISES 31–34   *Answer true or false.*

31. The distance between two points is the sum of the differences of the *x*- and *y*-coordinates.

32. A point with coordinates $(x, -x)$ (with $x \neq 0$) must lie in either the second or fourth quadrant.

33. The midpoint of the line segment joining (x_1, y_1) and (x_2, y_2) has coordinates $(\bar{x}, \bar{y})$, where $\bar{x} = \dfrac{x_1 - x_2}{2}$, and $\bar{y} = \dfrac{y_1 - y_2}{2}$.

34. The viewing window is the portion of the coordinate system displayed by a graphing calculator.

EXERCISES 35–38   *Give an example of each.*

35. A pair of points in the first quadrant that are $\sqrt{2}$ units apart

36. A pair of points 4 units apart from one another with midpoint on the *x*-axis

37. A point lying in the third quadrant

38. The coordinates of a point in the fourth quadrant that is 2 units from the *y*-axis

39. A point (x, y) is 4 units from the point $(2, -3)$. Write an equation that expresses this fact.

40. How many coordinates are required to specify a position on Earth? How many coordinates must be specified in order to schedule a meeting?

QUESTIONS FOR DISCUSSION OR ESSAY

41. Figures 26 and 27 illustrate the growth in newspaper circulation for a certain city over a 3-year period. Discuss how the choice of scale and the use of a broken vertical axis could lead to differences in interpretation.

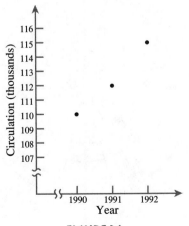

FIGURE 26

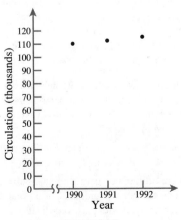

FIGURE 27

42. A punter kicks a football from a point on the 2-yard line approximately 18 yards from the left edge of the field (point A in Figure 28). The ball is picked up on the 47-yard line approximately 5 yards from the right edge of the field (point B). From yard line to yard line, the punt would appear to be $47 - 2 = 45$ yards. Explain why this is not the actual distance traveled by the ball. Given that the width of a football field is $53\frac{1}{3}$ yards, explain how the principles of coordinate geometry can be used to calculate the actual distance of the punt.

43. In Example 4, we observed a regular fluctuation in the atmospheric concentration of CO_2. Study the pattern more closely and determine the length of time between the ''peaks'' and ''valleys.'' What explanation can you give for these fluctuations?

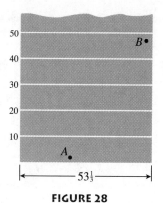

FIGURE 28

PROJECTS FOR ENRICHMENT

44. *Polar Coordinate Systems and Radar Screens*   The rectangular coordinate system formulated by Descartes is not the only way of identifying locations of points. About 40 years after Descartes' discovery, Sir Isaac Newton developed a system of *polar coordinates* that references points by their distance and ''compass direction'' from the origin. To define polar coordinates, we fix the origin at a point O and construct a horizontal ray with O as the initial point, as shown in Figure 29. Each point P is then associated with a pair $(r; \theta)$, where r is the *directed distance* from O to P and θ is the *directed angle measure* from the ray to the segment $\overline{OP}$. Note the use of a semicolon instead of a comma.

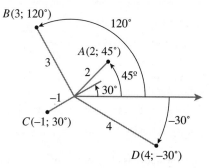

FIGURE 30

b. Find a second representation for each of the points in part a. Use degree measures that are between $-360°$ and $360°$.

By placing a rectangular coordinate system on top of a polar coordinate system, as shown in Figure 32, it is possible to convert coordinates from one system to the other. For example, the point

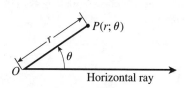

FIGURE 29

In Figure 30, we have plotted the four points $A(2; 45°)$, $B(3; 120°)$, $C(-1; 30°)$, and $D(4; -30°)$. Notice that positive angle measures are measured counterclockwise from the horizontal ray, whereas negative angles are measured clockwise. Notice also that because r refers to a directed distance, the point C with coordinates $(-1; 30°)$ is 1 unit from the origin in the exact opposite direction from that of the point $(1; 30°)$.

Points with polar coordinates are much easier to plot on *polar graph paper*, which consists of concentric circles centered at the origin and rays emanating from the origin at common angle measures, as shown in Figure 31.

a. Photocopy the polar graph paper in Figure 31 and plot the points $(3; 60°)$, $(2; 135°)$, $(4; -90°)$, and $(-5; 30°)$.

One significant difference between rectangular and polar coordinates is that the polar coordinates of a point are not unique. For example, the point with polar coordinates $(2; -30°)$ can also be identified using the coordinates $(2; 330°)$ or even $(-2; 150°)$.

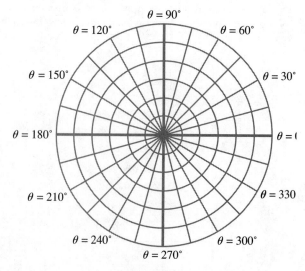

FIGURE 31

with rectangular coordinates (2, 2) has approximate polar coordinates (2.8; 45°). The point with polar coordinates (3; 30°) has rectangular coordinates approximately (2.6, 1.5).

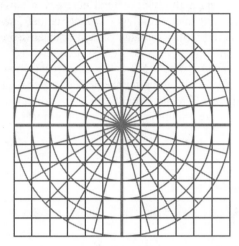

FIGURE 32

c. Photocopy Figure 32 and use it to convert the points given in part a to rectangular coordinates.

d. Convert the points with rectangular coordinates (2, 0), (3, 5), (−4, 4), and (−3, −2) to polar coordinates.

e. A radar screen, such as the one below, reports the *bearing* (compass direction) and *range* (distance) of an object as it relates to the source of the radar signal. Explain why a polar coordinate system would be the system of choice for recording the readings from a radar screen.

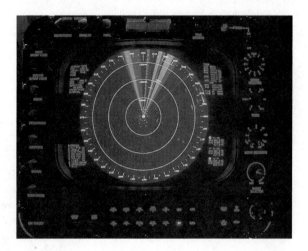

SECTION 1.3   PROPERTIES AND OPERATIONS

> ☐ **What property do addition, multiplication, and marriage have that subtraction, division, and love do not?**
>
> ☐ **Why is division by zero not allowed?**
>
> ☐ **Which discount would you rather have: 30% off, or 10% off of a price already reduced 20%?**
>
> ☐ **How can you rapidly compute the product of two-digit numbers, such as 92 × 97, in your head?**

ADDITION AND MULTIPLICATION OF REAL NUMBERS

The arithmetic operation of addition is a formalization of an idea so natural, so intuitive, that we understood it as children, perhaps before we learned to speak. In fact, there is even evidence to suggest that primates such as apes and chimpanzees employ something like addition: If the animal sees two marbles placed into an empty bowl, and a third marble is added to the bowl, it will appear confused if 3 marbles are not found in the bowl. Addition of real numbers is merely an extension of this primitive notion to the real number system.

Multiplication, although slightly less intuitive than addition, is also firmly grounded in our everyday experience. Multiplication of real numbers is a generalization of multiplication of natural numbers, which can be thought of as repeated addition. Thus, the product 3 · 5 can be thought of as the sum 5 + 5 + 5 consisting of three 5s. Of all the real-world contexts in which multiplication arises, perhaps none captures the essence

FIGURE 33

of multiplication better than the computation of the area of a rectangular region. As Figure 33 suggests, a 3-square by 5-square rectangle will have a total of $5 + 5 + 5 = 3 \cdot 5$ squares in all.

As we gain experience adding and multiplying natural numbers, a few obvious facts emerge. For example, when adding and multiplying natural numbers, neither ordering nor grouping affects the final result (the commutative and associative properties, respectively). These properties of multiplication and addition are preserved when we extend the operations to the set of all real numbers. Moreover, in the real number system (as opposed to the natural number system), every number has a negative and every number except for zero has a reciprocal. The most significant of the addition and multiplication properties for real numbers are summarized in Table 4. Note that some entries have been left for exercises.

TABLE 4   Properties of the real number system

Name of property	Verbal description	Algebraic description
Commutative property of addition	The order doesn't matter when adding.	$a + b = b + a$
Commutative property of multiplication	The order doesn't matter when multiplying.	_____
Associative property of addition	The grouping doesn't matter when adding.	_____
Associative property of multiplication	The grouping doesn't matter when multiplying.	$a(bc) = (ab)c$
Distributive property	_____	$a(b + c) = ab + ac$
Existence of the additive identity	There is a number called 0 (zero) whose sum with any number is that number.	$a + 0 = a$
Existence of the multiplicative identity	There is a number called 1 (one) whose product with any number is that number.	$a \cdot 1 = a$
Existence of the additive inverse	Every real number has an additive inverse; that is, for each number there is a second number that when added to the first gives 0.	$a + (-a) = 0$
Existence of the multiplicative inverse	_____	$a\left(\dfrac{1}{a}\right) = 1$

As you can see, many of these properties are very awkward to state in words. It is much more efficient to use the language of *algebra*. Whenever we use letters to represent numbers, we are using algebra.

E X A M P L E   1   ▨   Translating English into Algebra

Translate the following statement algebraically: Whenever a number is multiplied by the additive identity, the result is the additive identity.

SOLUTION Using x to represent "a number" and writing 0 (zero) for the additive identity, we obtain

$$x \cdot 0 = 0$$

EXAMPLE 2   Computing a Tip with the Distributive Property

Use the distributive property to compute a 15% tip on a restaurant bill of $42.00.

SOLUTION We wish to compute 15% of 42.00. Since 15% = 0.15, we have

$$
\begin{aligned}
0.15 \cdot 42.00 &= (0.1 + 0.05) \cdot 42.00 \\
&= (0.1 \cdot 42.00) + (0.05 \cdot 42.00) \\
&= 4.20 + 2.10 \\
&= 6.30
\end{aligned}
$$

Thus, the tip should be $6.30. Note that this distributive computation can actually be done mentally. Since 15% is 10% plus half of 10%, we compute 10% of $42.00, obtaining $4.20, and add half of that, namely $2.10 to obtain $6.30.

EXAMPLE 3   The Property of Commutativity in Other Contexts

Indicate whether the given operation is commutative.

a. Taking the minimum of two real numbers
b. Forming a hyphenated surname from a couple's surnames prior to their marriage

SOLUTION
a. Taking the minimum of two numbers is commutative: The minimum of x and y is the same as the minimum of y and x.
b. The formation of a hyphenated surname is not commutative; Jackie Kersey-Joiner is not the same as Jackie Joiner-Kersey.

ADDITION AND MULTIPLICATION OF COMPLEX NUMBERS

In Section 1.1, we defined a complex number as one of the form $a + bi$, where a and b are real numbers and $i = \sqrt{-1}$ (that is, $i^2 = -1$). Two complex numbers, $a + bi$ and $c + di$, can be added or multiplied according to the following definitions.

Addition and Multiplication of ☐
Complex Numbers

$$
\begin{aligned}
(a + bi) + (c + di) &= (a + c) + (b + d)i \\
(a + bi)(c + di) &= (ac - bd) + (ad + bc)i
\end{aligned}
$$

Using these definitions, it can be shown (see Exercise 81) that the commutative, associative, and distributive properties hold for the addition and multiplication of complex numbers. Thus, it is not necessary to memorize the definitions. Instead, we can simply apply the properties of complex addition and multiplication to compute the required sum or product.

E X A M P L E   4   ▩   **Adding Nonreal Numbers**

Perform the operation $(2 + i) + (13 + 4i)$.

SOLUTION   By combining the real and imaginary parts, one can probably tell at a glance that the answer is $15 + 5i$. But let us analyze how the properties of complex numbers are used to obtain this answer.

$$(2 + i) + (13 + 4i) = (2 + 13) + (i + 4i) \qquad \text{Commutative and associative properties of addition}$$
$$= 15 + (1 + 4)i \qquad \text{Distributive property}$$
$$= 15 + 5i$$

E X A M P L E   5   ▩   **Multiplying Nonreal Numbers**

Perform the operation $(5 - 2i)(8 + 3i)$.

SOLUTION   We apply the distributive property and the fact that $i^2 = -1$.

$$(5 - 2i)(8 + 3i) = 5(8 + 3i) + (-2i)(8 + 3i) \qquad \text{Distributive property}$$
$$= (40 + 15i) + (-16i - 6i^2) \qquad \text{Distributive property}$$
$$= (40 + 15i) + (-16i - 6(-1)) \qquad \text{Definition of } i \ (i^2 = -1)$$
$$= (40 + 15i) + (6 - 16i)$$
$$= 46 - i \qquad \text{Addition of complex numbers}$$

All of the properties listed in Table 4 also hold for complex numbers. The number $0 + 0i = 0$ is the additive identity, the number $1 + 0i = 1$ is the multiplicative identity, and the additive inverse of a complex number $a + bi$ is the complex number $-a - bi$. The multiplicative inverse $1/(a + bi)$ of a complex number $a + bi$ can be simplified using the **complex conjugate** of $a + bi$—namely; $a - bi$. The process is demonstrated in the following example.

E X A M P L E   6   ▩   **Finding the Multiplicative Inverse of a Complex Number**

Write the multiplicative inverse of $4 - 2i$ in the standard form $c + di$.

SOLUTION   We multiply the numerator and denominator of $1/(4 - 2i)$ by $4 + 2i$, the complex conjugate of the denominator $4 - 2i$.

$$\frac{1}{4 - 2i} = \frac{1}{4 - 2i} \cdot \frac{4 + 2i}{4 + 2i} \qquad \text{Multiplying by } 1 = \frac{4 + 2i}{4 + 2i}$$
$$= \frac{4 + 2i}{16 + 8i - 8i - 4i^2} \qquad \text{Multiplying numerators and denominators}$$
$$= \frac{4 + 2i}{16 - 4(-1)} \qquad \text{Substituting } i^2 = -1$$
$$= \frac{4 + 2i}{20}$$

$$= \frac{4}{20} + \frac{2}{20}\,i$$

$$= \frac{1}{5} + \frac{1}{10}\,i$$

CHECK We can check our answer by multiplying $4 - 2i$ by $\frac{1}{5} + \frac{1}{10}i$ to see if we obtain 1 as the answer.

$$(4 - 2i)\left(\frac{1}{5} + \frac{1}{10}\,i\right) = \frac{4}{5} + \frac{4}{10}\,i - \frac{2}{5}\,i - \frac{2}{10}\,i^2$$

$$= \frac{4}{5} - \frac{1}{5}\,(-1) = 1$$

SUBTRACTION AND DIVISION

Subtraction and division can be defined in terms of addition and multiplication. Subtraction is just addition of the additive inverse, and division is nothing more than multiplication by the multiplicative inverse. More formally, we define

$$x - y = x + (-y)$$

$$x \div y = x \cdot \frac{1}{y} \quad \text{for } y \neq 0$$

Note that division by 0 is not defined. But why is this? Let us consider the expression $\frac{1}{0}$. According to the definition, $\frac{1}{0}$ would be the multiplicative inverse of zero. In other words, if we let $c = \frac{1}{0}$, then $0 \cdot c = 1$. But we know that $0 \cdot c = 0$. This is a contradiction, so there is no such number as $\frac{1}{0}$.

Of course, most of us are well aware that division by zero is undefined, and so we immediately recognize an expression like $\frac{1}{0}$ as being problematic. But divisions by zero can be quite well-camouflaged, and if we don't catch them, we may well find ourselves having "proven" an absurdity.

E X A M P L E 7 ■ **A False Demonstration That 7 = 2**

What is wrong with the following proof that $7 = 2$?

Suppose that $x = 1$. Then $5x - 8 = -3$. We perform the following operations:

$$5x - 8 = -3$$
$$5x - 8 + 2x + 1 = -3 + 2x + 1 \qquad \text{Adding } 2x + 1 \text{ to both sides}$$
$$7x - 7 = 2x - 2 \qquad \text{Simplifying}$$
$$7(x - 1) = 2(x - 1) \qquad \text{Distributive property}$$
$$7 = 2 \qquad \text{Dividing by } (x - 1)$$

SOLUTION The error is in the last step where we divided by $(x - 1)$. We began the proof by supposing that $x = 1$, so that $x - 1 = 0$. Thus, the last step of the proof was a division by 0, which is not allowed.

When dividing complex numbers, we use the complex conjugate of the denominator to write the quotient in the standard form $a + bi$.

EXAMPLE 8   ■   Dividing Complex Numbers

Write the quotient $(3 - 2i)/(1 + 4i)$ in the standard form $a + bi$.

SOLUTION

$$\frac{3 - 2i}{1 + 4i} = \frac{3 - 2i}{1 + 4i} \cdot \frac{1 - 4i}{1 - 4i} \qquad \text{Multiplying by } 1 = \frac{1 - 4i}{1 - 4i}$$

$$= \frac{3 - 12i - 2i + 8i^2}{1 - 4i + 4i - 16i^2} \qquad \text{Multiplying numerators and denominators}$$

$$= \frac{3 - 14i + 8(-1)}{1 - 16(-1)} \qquad \text{Substituting } i^2 = -1$$

$$= \frac{-5 - 14i}{17}$$

$$= -\frac{5}{17} - \frac{14}{17} i$$

ORDER OF OPERATIONS

The associative properties of addition and multiplication tell us that, when either adding or multiplying a string of numbers together, we may group them in any way that we wish. For example, we could evaluate $1 + 2 + 3$ in either of the following ways.

$$(1 + 2) + 3 = 3 + 3 = 6$$
or $$1 + (2 + 3) = 1 + 5 = 6$$

But suppose that we need to evaluate $1 - 2 - 3$. Depending on how the subtractions are grouped, we could obtain either

$$(1 - 2) - 3 = -1 - 3 = -4$$
or $$1 - (2 - 3) = 1 - (-1) = 2$$

Since the grouping affects the result, we conclude that there is no associative property for subtraction. Similarly, there is no associative property for division. In fact, the grouping also matters if we have both addition and multiplication in the same expression. For example, if we wish to evaluate $2 \times 3 + 5$, we might obtain either

$$(2 \times 3) + 5 = 6 + 5 = 11$$
or $$2 \times (3 + 5) = 2 \times 8 = 16$$

The question is, Which result is correct? The answer is simple. We agree that unless grouping symbols (parentheses, brackets, and so forth) indicate otherwise, we will evaluate expressions involving the fundamental operations of negation, addition, subtraction, multiplication, and division according to the following rule.

Order of Arithmetic Operations   ☐

> If no grouping symbols are used, perform operations in the following order:
>
> **1.** Negations (taking the negative)
>
> **2.** Multiplications and divisions from left to right
>
> **3.** Additions or subtractions from left to right
>
> If grouping symbols (such as parentheses) are present, the operations within the grouping symbols are performed first, again following the order just given.

EXAMPLE 9   ▨   Order of Operations

Evaluate the following expressions.

a. $10 - 6 \div 2 \times 5$ b. $(10 - 6) \div (2 \times 5)$

SOLUTION

a.   We first perform divisions and multiplications in left-to-right order.

$$10 - 6 \div 2 \times 5 = 10 - (6 \div 2) \times 5$$
$$= 10 - 3 \times 5$$
$$= 10 - (3 \times 5)$$
$$= 10 - 15$$
$$= -5$$

b.   We perform the operations in parentheses first.

$$(10 - 6) \div (2 \times 5) = 4 \div 10$$
$$= 0.4$$

EXAMPLE 10   ▨   A False Proof That $0 = -4$

What is wrong with the following proof that $0 = -4$?

Let $x = 2 \times 1 + 2 \times (-1)$.

Then we have

$$x = \underbrace{2 \times 1} + 2 \times (-1)$$
$$= \underbrace{2 + 2} \times -1$$
$$= 4 \times -1$$
$$= -4$$

On the other hand, we have

$$x = \underbrace{2 \times 1} + \underbrace{2 \times (-1)}$$
$$= 2 \quad + \quad -2$$
$$= 0$$

Thus, $-4 = 0$.

SOLUTION   The second step of the first evaluation of x is in error since multiplications are to be performed before additions.

CALCULATOR KEYS

Arithmetic Operations

With most scientific and graphing calculators, arithmetic expressions—even those involving complex numbers—are entered intuitively: They are typed the way they are written. The only rule for entering arithmetic expressions that might at first seem unnatural concerns the distinction between the negation and subtraction symbols. When we subtract one expression from another, as in $5 - 3$, for example, we are using the **minus sign.** On the other hand, when we express the negative of a number—such as -2, for example—we are using the **negation symbol.** Distinguishing between the two may seem fussy when writing, but it is a crucial distinction to the calculator. For this reason, we must key in $\boxed{(-)}\,\boxed{2}$ (or $\boxed{+/-}\,\boxed{2}$ on some scientific calculators), *not* $\boxed{-}\,\boxed{2}$ to enter -2.

ABSOLUTE VALUE

The notion of absolute value is used whenever we are interested in the size or magnitude of a real number, without regard to whether it is positive or negative. Thus, 3 and -3 have different values, but have the same *absolute* value—namely, 3. In general, we define the absolute value of a number x to be the distance $|x|$ from zero to x. More precisely, we make the following definition.

Definition of Absolute Value   □

> The **absolute value** of a real number x, denoted $|x|$, is its distance from zero. In symbols,
>
> $$|x| = \begin{cases} x & \text{if } x \geq 0 \\ -x & \text{if } x < 0 \end{cases}$$

Note that since the absolute value of a number represents a distance, it cannot be negative. Thus, the absolute value of a negative number can be obtained by changing the sign of the number, whereas the absolute value of a positive number is the number itself. In other words, we have the following rule.

> **RULE OF THUMB**   When finding the absolute value of a negative number, remove the negative sign. When finding the absolute value of a positive number, leave it alone.

EXAMPLE 11   ■   **Finding the Absolute Value**

Evaluate each of the following.

a. $|14.5|$ **b.** $|-\pi|$ **c.** $|0|$
d. $-|-4|$ **e.** $|x|$, where x is 27 units from the origin

SOLUTION
a.   Since $14.5 \geq 0$, $|14.5| = 14.5$.
b.   Since $-\pi < 0$, $|-\pi| = -(-\pi) = \pi$.
c.   Since $0 \geq 0$, $|0| = 0$.
d. $-|-4|$ means $-(|-4|)$ or "the opposite of $|-4|$." Since $-4 < 0$, $|-4| = 4$. Thus, $-|-4| = -4$. Note that this does not contradict the rule of thumb because it is the negative sign outside of the absolute value that makes the result negative.
e.   Since the absolute value of a number is its distance from the origin, $|x| = 27$. Of course, this means that either $x = 27$ or $x = -27$.

EXAMPLE 12   ■   **Rewriting an Absolute Value Expression**

Write the expression $|x - 2|$ without using absolute value, given that $x < 2$.

SOLUTION   Since $x < 2$, $x - 2 < 0$. Thus, $|x - 2| = -(x - 2) = -x + 2$.

We have seen that the distance between two real numbers can be found by subtracting the smaller from the larger. Using absolute value, the distance between two numbers can be defined without regard to which of the two is larger. For example, the distance between 2 and 6 can be found by computing $|2 - 6| = 4$. More generally, we have the following definition.

Distance ☐

> The **distance** between two real numbers a and b is given by $|a - b|$.

EXAMPLE 13   ■   Finding Distance

Find the distance between $\frac{4}{5}$ and $\frac{9}{11}$.

SOLUTION   The distance is $\left|\frac{4}{5} - \frac{9}{11}\right| = \left|\frac{44}{55} - \frac{45}{55}\right| = \left|-\frac{1}{55}\right| = \frac{1}{55}$.

EXAMPLE 14   ■   Expressing Distance with Absolute Value

Find an expression for the distance between a real number x and -3.

SOLUTION   The distance between x and -3 is given by $|x - (-3)| = |x + 3|$.

UNDERSTANDING AND MASTERY CHECKLISTS

CONCEPTS TO UNDERSTAND

- ☐ Properties of real numbers
- ☐ Operations with complex numbers
- ☐ Complex conjugate
- ☐ Order of operations
- ☐ Absolute value

SKILLS TO MASTER

- ☐ State the commutative, associative, and distributive properties of addition and multiplication.
- ☐ Use the commutative, associative, and distributive properties to simplify expressions.
- ☐ Perform the four arithmetic operations with complex numbers.
- ☐ Evaluate expressions by performing arithmetic operations in the correct order.
- ☐ Compute the absolute value of a real number.

EXERCISES 1.3

EXERCISES 1–4   *See Table 4.*

1. Write an algebraic description of the commutative property of multiplication.

2. Write an algebraic description of the associative property of addition.

3. Write a verbal description of the distributive property.

4. Write a verbal description of the existence of the multiplicative inverse.

EXERCISES 5–14   *Perform the indicated operation.*

5. $(3 + 4i) + (1 - 2i)$

6. $(2 - i) - (5 - 9i)$

7. $7 - (3 + 2i)$

8. $(4 - 8i) + 2i$

9. $2 \cdot (3 + i)$

10. $(4 + i) \cdot 5$

11. $(1 + i) \cdot i$

12. $6i \cdot (2 - 3i)$

13. $(2 + 3i) \cdot (1 - i)$

14. $(3 - 4i) \cdot (-5 + i)$

EXERCISES 15–18 *Write the multiplicative inverse of the given number in the form a + bi. Check your answer by showing that the product of the number with its multiplicative inverse is equal to 1.*

15. $2i$

16. $-3i$

17. $1 + i$

18. $2 - 6i$

EXERCISES 19–24 *Write the quotient in the form a + bi.*

19. $\dfrac{5}{2 + i}$

20. $\dfrac{-13i}{2 - 3i}$

21. $\dfrac{-18 + 13i}{5 - 2i}$

22. $\dfrac{29 - 3i}{3 + 4i}$

23. $\dfrac{8 + 3i}{-4i}$

24. $\dfrac{3 - 2i}{5i}$

EXERCISES 25–30 *Evaluate the given expression.*

25. $5 - 3 \times 4$

26. $6 + 3 \div 2$

27. $1.2 - 2.5 - 3.4 - 1.8$

28. $3 \div 1 + 2 - 4 + 2 \times 3$

29. $4 \times [3 - (4 - 5)]$

30. $12 \div 2 \times 3 \div 2 \times 3$

EXERCISES 31–38 *Evaluate the given expression.*

31. $|-6|$

32. $-|4|$

33. $|0|$

34. $|10.61|$

35. $|3 - 1|$

36. $|-1 - 4|$

37. $\dfrac{|-3.7|}{-3.7}$

38. $\dfrac{|2 - 5|}{2 - 5}$

EXERCISES 39–42 *Write the expression without using absolute values.*

39. $|x - 1|$ if $x > 1$

40. $|t + 4|$ if $t < -4$

41. $|y + 2|$ if $y < -2$

42. $|x - 7|$ if $x > 7$

EXERCISES 43–50 *Find an expression for the distance between the given quantities. Simplify when possible.*

43. -3 and 7

44. -5 and -12

45. $-\frac{1}{3}$ and $-\frac{2}{5}$

46. $\frac{3}{4}$ and $\frac{5}{6}$

47. $3x$ and 2

48. $-x$ and 1

49. $(x + \frac{2}{3})$ and $-\frac{2}{3}$

50. $(2x - 3)$ and $(x + 2)$

APPLICATIONS

51. *Dining Gratuity* Use the distributive property to compute a 15% tip on a restaurant bill of $64.00.

52. *Delivery Tip* Use the distributive property to compute a 15% tip on a pizza delivery of $18.00.

53. *Percentage Discount* One of the authors (really!) went to a fast-food Mexican restaurant with a friend and ordered two meals on the same ticket. The clerk interrupted, informing them that there was a special unadvertised 10% discount on everything in the restaurant and suggesting that in light of this special, the author and his friend should order their meals separately. Seeing the puzzled look on the author's face, the clerk explained that by ordering separately, there would be two 10% discounts—one for each order—for a total of 20%! Suppose that the bill for the author's meal is $3.50 and that of his friend is $2.80.

    a. Compute the bill (not counting tax) if both meals are placed on the same ticket, and then 10% is discounted from the total.

    b. Compute the total bill if the meals are on separate tickets, each of which is discounted by 10%.

    c. Use the distributive law to explain the puzzled look on the author's face.

54. *Employee Discount* An applicant for a position at a clothing store is told by the interviewer that employees receive a 10% discount on all purchases, including sale items. The interviewer gives as an example the following scenario: ''Suppose that we're having a 20% clearance sale. Then you would get an additional 10% off, for a total discount of 30%.'' Explain the flaw in the interviewer's reasoning and compute the actual total discount. Does the answer depend on which discount—the 10% employee discount or the 20% clearance discount—is taken first? What properties of real number arithmetic are you using to answer these questions?

CONCEPTS AND CRITICAL THINKING

EXERCISES 55–60 *Answer true or false.*

55. For all real numbers a, b, and c, $\dfrac{a}{b + c} = \dfrac{a}{b} + \dfrac{a}{c}$.

56. Multiplication is an associative operation.

57. The absolute value of a real number is never negative.

58. The absolute value of a negative number is not defined.

59. 0 is the multiplicative identity.

60. For all real numbers a and b satisfying $a < b$, $|a| < |b|$.

EXERCISES 61–64 *Give an example of each.*

61. A property that addition and multiplication have but that subtraction and division do not

62. An illustration showing that there is no associative law of division

63. Two numbers a and b such that $a - b = b - a$

64. Two numbers a and b such that $\frac{a}{b} = \frac{b}{a}$

65. An operation is said to be commutative if "the order doesn't matter." For example, addition is commutative because $a + b = b + a$. Subtraction, however, can be shown to be non-commutative by considering, for example, $a = 0$ and $b = 1$. Here $a - b = 0 - 1 = -1$, whereas $b - a = 1 - 0 = 1$. The pair of numbers $a = 0$, $b = 1$ are a *counterexample* to the statement that subtraction is commutative. State whether each of the following operations is commutative; if not, give a counterexample.

   a. Division

   b. Multiplication

   c. Marriage (Is the outcome of Bob marrying Sue the same as that of Sue marrying Bob?)

   d. Love (Is the outcome of Dirk loving Camille the same as Camille loving Dirk?)

   e. Taking the maximum of two real numbers

   f. Addressing (Is "Jamal speaks to Ann" the same as "Ann speaks to Jamal"?)

   g. Implication (Is "If A, then B" the same as "If B, then A"?)

   h. Finding the greatest common divisor of two natural numbers

66. Indicate whether the given operation is associative. If not, give a counterexample. [*Hint:* An operation is associative if the grouping doesn't matter. For example, addition is associative because $a + (b + c) = (a + b) + c$.]

   a. Multiplication

   b. Division

   c. Subtraction

   d. Linking chains

   e. Connecting Lego blocks together

67. Place a one dollar bill in front of you right side up. Rotate the front of the bill 90° clockwise so that George Washington's nose points toward you. Now flip the bill over (top to bottom) so that you now see the reverse side of the dollar bill, with the seal of the United States nearest you. Now, return the dollar bill to its original position, and this time, flip the bill first and then rotate it 90° clockwise. Do rotations and flips commute with one another?

68. The distributive property was given in Table 4 as $a(b + c) = ab + ac$. Show that it is also true that $(a + b)c = ab + ac$. (*Hint:* Use the commutative property of multiplication and the distributive property.)

69. Explain how you could compute $999,999 \times 75$ in your head. (*Hint:* Think of 999,999 as $1,000,000 - 1$ and use the distributive property.)

70. Explain how you could compute $1,000,003 \times 90$ in your head.

71. Use the properties of real numbers to show that $a \cdot 0 = 0$. [*Hint:* Use $0 = 1 + (-1)$ and the distributive property.]

72. Suppose that we were unaware of the rules regarding order of operations. List all possible results of the computation $3 \cdot 5 + 6 \div 2$. (*Hint:* There are four in all.)

73. Using Example 7 as a model, construct a false proof that $1 = 1,000,000$.

74. Construct a false proof that you are the same age as your grandfather.

▨ QUESTIONS FOR DISCUSSION OR ESSAY

75. How would you respond to someone who states, "Just because we don't know how to divide by zero now doesn't mean that someday someone might not be able to figure out how to do it."

76. The definition of absolute value as stated in the text is

$$|x| = \begin{cases} x & \text{if } x \geq 0 \\ -x & \text{if } x < 0 \end{cases}$$

Thus, under some circumstances, $|x| = -x$. On the other hand, absolute values are never negative. Explain this apparent contradiction.

77. Are the properties of addition and multiplication discovered or created? Support your claim.

78. Which of the properties discussed in this section do you think are so transparently obvious that virtually everyone who does arithmetic with some regularity—clerks, schoolchildren, bookkeepers, waiters, and so forth—employs them correctly? Which properties are most likely to be misused?

79. To prove that subtraction is *not* commutative, it is sufficient to produce a single pair of numbers (such as 0 and 1) for which we obtain a different answer depending on the order in which we perform the subtraction. Why then *can't* we establish the commutative property of addition by simply producing a single pair of numbers such that the order in which we add these numbers doesn't matter? For example, what is wrong with this "proof" of the commutativity of addition?

"Consider the numbers 2 and 3. Since $2 + 3 = 5$, and also $3 + 2 = 5$, addition is commutative."

80. How is algebra like a language? What mathematical entities correspond to letters, words, and sentences?

■ **PROJECTS FOR ENRICHMENT**

81. *Verifying Properties* Using the properties of addition and multiplication of real numbers, we can prove that similar properties hold for the addition and multiplication of complex numbers. For example, we may establish the commutative property of addition of complex numbers as follows:

$(a + bi) + (c + di)$

$= (a + c) + (b + d)i$	Definition of complex addition
$= (c + a) + (d + b)i$	Commutative property of addition of reals
$= (c + di) + (a + bi)$	Definition of complex addition

Use the properties of real number addition and multiplication to establish each of the following properties for complex number addition and multiplication.

  a. The commutative property of multiplication

  b. The associative property of addition

  c. The associative property of multiplication

  d. The distributive property

82. *The Complex Plane* For every operation with real numbers, there is a corresponding operation with complex numbers. We have seen that complex numbers can be added, subtracted, multiplied, and divided just like real numbers. However, what in the world would the absolute value of a complex number be? In the case of real numbers, the absolute value of a number was defined as being the distance between that number and 0. But how far is $3 + 4i$ from 0? In fact, *where* on a number line is $3 + 4i$ anyway?

Let us answer the last question first: $3 + 4i$ isn't anywhere on a number line; only real numbers are on the number line. But $3 + 4i$ is in the **complex plane.** The complex plane looks a lot like the rectangular coordinate plane, with a few important differences. Instead of the *x*-axis we have the real axis, instead of the *y*-axis we have the imaginary axis, and in place of the origin we have the number 0. The complex number $3 + 4i$, for example, is located in the first quadrant 3 units over and 4 units up. Thus, the complex number $3 + 4i$ corresponds to the point $(3, 4)$ in the rectangular coordinate plane. In general, the complex number $x + yi$ corresponds to the point (x, y) in the complex plane. The complex plane is shown in Figure 34 with several complex numbers plotted.

  a. Indicate the quadrant in which the given complex number lies.

  i. $-2 + 4i$ **ii.** $3 - 7i$

  iii. $1 + i$ **iv.** $-3 - \frac{8}{3}i$

The absolute value of a complex number is defined as being the distance of the number from zero, just as in the case of real numbers. For example, $|3i| = 3$ and $|-4i| = 4$.

  b. Find the absolute value of the given complex number. (*Hint:* It is helpful to first plot the given number in the complex plane. The Pythagorean Theorem may be useful.)

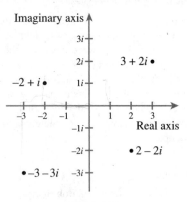

FIGURE 34

  i. $1.5i$ **ii.** $3 + 4i$

  iii. $-4 - 3i$ **iv.** $1 + i$

  c. Describe the set of points in the complex plane with absolute value equal to 1.

83. *Stupid Human Tricks* Although most of us find it difficult to multiply numbers larger than 12 in our heads, there are individuals—some with IQs in the mentally retarded range—capable of extraordinary feats of mental computation. There are documented cases of so-called idiot savants who can mentally multiply 10-digit numbers but are unable to solve a logic problem that would be child's play for a normal 3-year-old. In such cases, the method by which the computation is performed is concealed by the verbal limitations of the performer, although recent research suggests that these individuals rely on a combination of prodigious memories—often *eidetic* or visual memory—and a collection of specialized rules. Fortunately, not all human calculators are mentally challenged. Scott Flansburg (''The Human Calculator'') has written several books in which he describes many of his techniques for rapid mental computation. In this project, we will investigate how and why two of Scott's tricks work.

  a. To multiply two numbers that are almost 100, say, 92 and 97, do the following. First write the two factors in a column, and then form a second column by subtracting each number from 100, as shown in Figure 35. In this case, we obtain 8 and 3. Next compute the ''cross-difference,'' $92 - 3$ or $97 - 8$, and write down the result, 89, at the bottom of the first column. Finally, multiply the entries in the second column, 8 and 3, and write down the product, 24, at the bottom of the second column. The product of the two numbers is then given by the digits in the third column—in this case, 8924.

Figure 36 shows how this technique could be used to multiply 94 and 91.

  i. Perform the following multiplications using the technique illustrated in Figures 35 and 36: $93 \cdot 96$, $89 \cdot 98$, $93 \cdot 95$.

FIGURE 35   **FIGURE 36**

ii. Attempt to do the following multiplications using the same technique, but this time do the computations in your head: $94 \cdot 97$, $92 \cdot 99$, $96 \cdot 92$.

iii. Suppose that we wish to multiply x and y (both of which are close to but less than 100) using Scott's technique. Complete the computations shown in Figure 37.

$$
\begin{array}{cc}
x & 100 - x \\
\times & | \\
\hline
y & \\
\hline
& \\
\end{array}
$$

FIGURE 37

iv. Referring to our first example, the number 8924 is equivalent to $8900 + 24$. Thus, $92 \cdot 97 = 89 \cdot 100 + 24$. Similarly, $94 \cdot 91 = 85 \cdot 100 + 54$. Use this idea and your work from part iii to find an expression for xy.

v. Now use basic properties to confirm the formula found in part iv.

vi. Use algebra to confirm that it does not matter which cross-difference is used to compute the first two digits of a product using this technique.

vii. Explain why this technique is most useful for multiplying numbers that are close to 100.

b. A similar technique allows us to multiply two numbers that are just larger than 100, as shown in Figure 38.

FIGURE 38

i. Using Figure 38 as a guide, write out instructions for multiplying two numbers just larger than 100.

ii. Use algebra to explain why this technique works.

SECTION 1.4 INTEGER EXPONENTS

☐ How thick would a piece of paper be if you folded it in half 52 times?

☐ If 2^3 is 2 multiplied by itself 3 times, then what is 2^{-3}?

☐ Which would you rather have, the money accumulated from Julius Caesar's deposit of $100 at 10% simple interest, or from Ben Franklin's deposit of $1 at 6% interest compounded annually?

☐ An individual doubles in weight between the ages of birth and 6 months. If he continued to double in weight every six months, how much would he weigh on his forty-first birthday?

☐ What system can be used to express numbers as large as the number of atoms in the universe and as small as the weight of an electron?

NATURAL NUMBER EXPONENTS

Suppose that you wish to compute the number of people in a theater with 50 rows and 40 seats in each row. You *could* add the number of seats in each row 50 times:

$$40 + 40 + 40 + 40 + 40 + 40 + 40 + 40 + 40 + 40 +$$
$$40 + 40 + 40 + 40 + 40 + 40 + 40 + 40 + 40 + 40 +$$
$$40 + 40 + 40 + 40 + 40 + 40 + 40 + 40 + 40 + 40 +$$
$$40 + 40 + 40 + 40 + 40 + 40 + 40 + 40 + 40 + 40 +$$
$$40 + 40 + 40 + 40 + 40 + 40 + 40 + 40 + 40 + 40 = 2000$$

... but you wouldn't. Instead, you would simply multiply 40×50 to obtain 2000. As this example illustrates, multiplication is a shorthand notation for repeated addition.

Now let's consider an example in which multiplication is repeated. Suppose that we want to find the thickness of a piece of ordinary notebook paper that has been folded in half 52 times. Since there are 500 sheets of paper in a ream, and a ream is approximately 1.5 inches thick, we'll assume that a single piece of paper is approximately $\frac{1.5}{500} = 0.003$ inch thick. After one folding, it will be 2×0.003 inch, after two foldings it will be $2 \times 2 \times 0.003$ inch, and so on. Thus, after 52 foldings it would be

$$2 \times 2 \times 2 \times 2 \times 2 \times 2 \times 2 \times 2 \times 2 \times 2 \times$$
$$2 \times 2 \times 2 \times 2 \times 2 \times 2 \times 2 \times 2 \times 2 \times 2 \times$$
$$2 \times 2 \times 2 \times 2 \times 2 \times 2 \times 2 \times 2 \times 2 \times 2 \times$$
$$2 \times 2 \times 2 \times 2 \times 2 \times 2 \times 2 \times 2 \times 2 \times 2 \times$$
$$2 \times 2 \times 2 \times 2 \times 2 \times 2 \times 2 \times 2 \times 2 \times 2 \times 2 \times 0.003$$

inches thick. The unreasonable tedium of writing and reading such products motivates us to employ *exponential notation,* a compact way of expressing repeated multiplications. Thus, instead of a notational abomination consisting of fifty-two 2s and fifty-two "times" signs, we write $2^{52} \times 0.003$ inches. (This, by the way, is over 200,000,000 miles, more than twice the distance to the sun!) In much the same way that multiplication expresses repeated additions, exponential notation expresses repeated multiplications.

<table>
<tr><td>***Definition of Natural Number Exponents*** □</td><td>If n is a natural number, then $a^n = a \times a \times \cdots \times a$, where there are n factors of a. The number a is called the **base** and the number n is called the **exponent**.</td></tr>
</table>

E X A M P L E   1   ■   **Evaluating Expressions Involving Natural Number Exponents**

Evaluate each of the following.

a. 2^5 **b.** $(-3)^3$ **c.** $\left(\frac{3}{4}\right)^2$

SOLUTION
a. $2^5 = 2 \cdot 2 \cdot 2 \cdot 2 \cdot 2 = 32$
b. $(-3)^3 = (-3) \cdot (-3) \cdot (-3) = -27$
c. $\left(\frac{3}{4}\right)^2 = \frac{3}{4} \cdot \frac{3}{4} = \frac{9}{16}$

Note that the parentheses in parts b and c of Example 1 indicate that the base is the entire quantity inside the parentheses. What about expressions such as -3^2, where no parentheses are given? Should we assume that -3^2 means $(-3)^2 = 9$ or $-(3)^2 = -9$? In Section 1.3, we discussed the fact that whenever an algebraic expression could have more than one meaning, conventions determine the order in which operations are to be performed. For example, if there are no grouping symbols, then multiplication and division are performed before addition and subtraction. In the case of exponents, the rule is that we perform all exponentiations before we perform negations, multiplications, divisions, additions, and subtractions. Thus, $-3^2 = -(3^2) = -9$.

Order of Operations ☐

If no grouping symbols are used, perform operations in the following order:

1. Exponentiations

2. Negations

3. Multiplications and divisions from left to right

4. Additions or subtractions from left to right

If grouping symbols are present, the operations within the grouping symbols are performed first, again using the order just given.

E X A M P L E 2 ▓ **Evaluating Expressions Involving Natural Exponents**

Evaluate each of the following.

a. $\dfrac{9^2}{3}$ **b.** $[(2^3)^2]^2$ **c.** $-(-4)^2$

SOLUTION

a. We evaluate the exponent first to obtain $\dfrac{9^2}{3} = \dfrac{81}{3} = 27$.

b. We evaluate the innermost exponent first to obtain

$$[(2^3)^2]^2 = [(8)^2]^2 = [64]^2 = 4096$$

c. $-(-4)^2 = -[(-4)^2] = -[16] = -16$

E X A M P L E 3 ▓ **A Doubling Problem**

Suppose that a child doubles in weight in the first six months of his life, and continues to double in weight every six months. How much will he weigh on his forty-first birthday? His fiftieth birthday?

SOLUTION We will assume that he weighs 7 pounds at birth. Then after one 6-month period he will weigh $2 \cdot 7$ pounds, after two 6-month periods he will weigh $2 \cdot 2 \cdot 7 = 2^2 \cdot 7$ pounds, and so forth. Thus, if n represents the number of 6-month periods that have elapsed, he will weigh $2^n \cdot 7$ pounds after n 6-month periods. On his forty-first birthday, $2 \cdot 41 = 82$ 6-month periods will have elapsed. Thus, he will weigh

$$2^{82} \cdot 7 = 33{,}849{,}922{,}949{,}209{,}616{,}891{,}772{,}928 \text{ pounds}$$

more than double the weight of Earth. On his fiftieth birthday he will weigh

$$2^{100} \cdot 7 = 8{,}873{,}554{,}201{,}597{,}605{,}810{,}476{,}922{,}437{,}632 \text{ pounds}$$

more than double the weight of the sun!

The following properties are used for simplifying expressions involving exponents. Each can be proven using the definition of natural number exponents, as suggested by the accompanying illustrations.

Properties of Exponents   ☐

	Property	Example	Illustration
	Assume that m and n are natural numbers.		
1.	$a^m \cdot a^n = a^{m+n}$	$2^3 2^4 = 2^{3+4} = 2^7$	$(2 \cdot 2 \cdot 2) \cdot (2 \cdot 2 \cdot 2 \cdot 2)$ $= 2 \cdot 2 \cdot 2 \cdot 2 \cdot 2 \cdot 2 \cdot 2 = 2^7$
2.	$\dfrac{a^m}{a^n} = a^{m-n}$	$\dfrac{3^5}{3^3} = 3^{5-3} = 3^2$	$\dfrac{3 \cdot 3 \cdot 3 \cdot 3 \cdot 3}{3 \cdot 3 \cdot 3} = \dfrac{\cancel{3} \cdot \cancel{3} \cdot \cancel{3} \cdot 3 \cdot 3}{\cancel{3} \cdot \cancel{3} \cdot \cancel{3}}$ $= 3^2$
3.	$(a^m)^n = a^{mn}$	$(5^2)^3 = 5^{2 \cdot 3} = 5^6$	$(5 \cdot 5) \cdot (5 \cdot 5) \cdot (5 \cdot 5) = 5^6$
4.	$(ab)^n = a^n b^n$	$(3 \cdot 5)^2 = 3^2 5^2$	$(3 \cdot 5) \cdot (3 \cdot 5) = (3 \cdot 3) \cdot (5 \cdot 5)$ $= 3^2 \cdot 5^2$
5.	$\left(\dfrac{a}{b}\right)^n = \dfrac{a^n}{b^n}$	$\left(\dfrac{2}{5}\right)^3 = \dfrac{2^3}{5^3}$	$\dfrac{2}{5} \cdot \dfrac{2}{5} \cdot \dfrac{2}{5} = \dfrac{2 \cdot 2 \cdot 2}{5 \cdot 5 \cdot 5} = \dfrac{2^3}{5^3}$

E X A M P L E 4   ◼   **Simplifying Expressions Involving Natural Number Exponents**

Simplify the following expressions using the rules of exponents.

a. $3^4 \cdot 3^{17}$ **b.** $(x^3)^5$

SOLUTION

a.
$$3^4 \cdot 3^{17} = 3^{(4+17)} \qquad \text{Using Property 1}$$
$$= 3^{21}$$

b.
$$(x^3)^5 = x^{(3 \cdot 5)} \qquad \text{Using Property 3}$$
$$= x^{15}$$

E X A M P L E 5   ◼   **Simplifying an Expression Involving Natural Number Exponents**

Simplify the expression $(2^5 x^3 y^2 z)(2^3 x y^4 z^2)$

SOLUTION

$$(2^5 x^3 y^2 z)(2^3 x y^4 z^2) = 2^{5+3} x^{3+1} y^{2+4} z^{1+2} \qquad \text{Using Property 1}$$
$$= 2^8 x^4 y^6 z^3$$

E X A M P L E 6   ◼   **Simplifying an Expression Involving Natural Number Exponents**

Simplify the expression $\dfrac{7^8 a^5}{7^3 a^3}$.

SOLUTION

$$\frac{7^8 a^5}{7^3 a^3} = \frac{7^8}{7^3} \cdot \frac{a^5}{a^3} \qquad \text{Separating factors}$$
$$= 7^5 a^2 \qquad \text{Using Property 2}$$

E X A M P L E   7 ▧   **Simplifying an Expression Involving Natural Number Exponents**

Simplify the expression $\dfrac{3^4 x^6 y^2}{3^2 x^3 y} \cdot \dfrac{3x^3 y^3}{(3x^2 y)^2}$.

SOLUTION

$$\frac{3^4 x^6 y^2}{3^2 x^3 y} \cdot \frac{3x^3 y^3}{(3x^2 y)^2} = \frac{3^4 x^6 y^2 \cdot 3x^3 y^3}{3^2 x^3 y \cdot 3^2 x^4 y^2}$$ Multiplying fractions and using Property 4 to remove parentheses

$$= \frac{3^5 x^9 y^5}{3^4 x^7 y^3}$$ Using Property 1 in the numerator and denominator

$$= 3x^2 y^2$$ Using Property 2

NEGATIVE AND ZERO EXPONENTS

If exponentiation is repeated multiplication, then what is 2^0? That is, how do you multiply 2 by itself 0 times? The answer is that although "multiplying 2 by itself 0 times" *doesn't* have any meaning, we can still define 2^0 in a sensible way using the properties of exponents just given. For example, according to Property 2 for natural number exponents, 2^0 should have the same value as $2^1/2^1 = 1$; therefore we *define* 2^0 to be 1. Similarly, we can define 2^{-3} to be $2^0/2^3 = 1/2^3$. More generally, we give the following definitions.

Definitions of Negative and Zero Exponents ☐

For $a \neq 0$, and n a positive integer,

$$a^{-n} = \frac{1}{a^n}$$

For $a \neq 0$,

$$a^0 = 1$$

E X A M P L E   8 ▧   **Evaluating Expressions with Nonpositive Exponents**

Evaluate each of the following expressions.

a. 3^{-2} **b.** 10^{-1}

c. $(23457892.09845987234857)^0$

SOLUTION

a. $3^{-2} = \dfrac{1}{3^2} = \dfrac{1}{9}$

b. $10^{-1} = \dfrac{1}{10^1} = \dfrac{1}{10}$

c. $(23457892.09845987234857)^0 = 1$, since anything (except 0) raised to the zero power is 1.

Exponentiation is now defined for integer exponents in such a way that all of the properties of natural number exponents are valid. We use these properties in the following examples.

EXAMPLE 9 ■ **Simplifying an Expression with a Negative Exponent**

Simplify the expression $x^3 \cdot x^{-5}$. Express your answer using only positive exponents.

SOLUTION

$$
\begin{aligned}
x^3 \cdot x^{-5} &= x^{3+(-5)} && \text{Using Property 1} \\
&= x^{-2} && \text{Simplifying} \\
&= \frac{1}{x^2} && \text{Using the definition of negative exponents}
\end{aligned}
$$

EXAMPLE 10 ■ **Simplifying an Expression with Negative Exponents**

Simplify the expression

$$
\frac{2^3 a^{-4} b^2}{2^{-2} a^{-2} b^5}
$$

Express your answer using only positive exponents.

SOLUTION

$$
\begin{aligned}
\frac{2^3 a^{-4} b^2}{2^{-2} a^{-2} b^5} &= \frac{2^3}{2^{-2}} \cdot \frac{a^{-4}}{a^{-2}} \cdot \frac{b^2}{b^5} && \text{Grouping like factors} \\
&= 2^{3-(-2)} \cdot a^{-4-(-2)} \cdot b^{2-5} && \text{Using Property 2} \\
&= 2^5 a^{-2} b^{-3} && \text{Simplifying} \\
&= 2^5 \cdot \frac{1}{a^2} \cdot \frac{1}{b^3} && \text{Using the definition of negative exponents} \\
&= \frac{32}{a^2 b^3} && \text{Simplifying}
\end{aligned}
$$

EXAMPLE 11 ■ **Simplifying an Expression with Negative Exponents**

Simplify the expression

$$
\left(\frac{x^3 y^{-2} z^{-3}}{x^{-5} y z^{-1}} \right)^{-1}
$$

Express your answer using only positive exponents.

SOLUTION This expression can be simplified many different ways. We demonstrate two.

METHOD 1:

$$
\begin{aligned}
\left(\frac{x^3 y^{-2} z^{-3}}{x^{-5} y z^{-1}} \right)^{-1} &= \left(\frac{x^3}{x^{-5}} \cdot \frac{y^{-2}}{y} \cdot \frac{z^{-3}}{z^{-1}} \right)^{-1} && \text{Grouping like factors} \\
&= (x^8 y^{-3} z^{-2})^{-1} && \text{Using Property 2} \\
&= x^{-8} y^3 z^2 && \text{Using Property 4} \\
&= \frac{y^3 z^2}{x^8} && \text{Using the definition of negative exponents}
\end{aligned}
$$

METHOD 2:

$$\left(\frac{x^3y^{-2}z^{-3}}{x^{-5}yz^{-1}}\right)^{-1} = \frac{(x^3y^{-2}z^{-3})^{-1}}{(x^{-5}yz^{-1})^{-1}}$$ Using Property 5

$$= \frac{x^{-3}y^2z^3}{x^5y^{-1}z}$$ Using Property 4

$$= x^{-8}y^3z^2$$ Using Property 2

$$= \frac{y^3z^2}{x^8}$$ Using the definition of negative exponents

SIMPLE AND COMPOUND INTEREST

Financial contracts have existed since the dawn of recorded history. In fact, evidence suggests that writing was invented in Mesopotamia around 3000 B.C. for the purpose of recording financial transactions. For thousands of years, beginning even before the invention of currency, lenders have been charging interest on loaned commodities, and bankers have been paying interest on deposited commodities. Until the early 19th century, the most common form of interest was **simple interest.** With a simple interest account, only the initial amount invested (the **principal**) earns interest. For example, if the simple interest rate is 5% and the principal is $1000, then each year the account will earn interest in the amount of $5\% \cdot \$1000 = \$50$, so that after 1 year, the balance is $1050, after 2 years the balance will be $1100, and so forth. Thus, after t years the balance (in dollars) will be given by

$$B = 1000 + 50t$$

More generally, if we denote the balance after t years by B, the principal by P, and the simple interest rate by r, we have

$$\text{Balance} = \text{Principal} + \text{Interest}$$
$$B = P + (\text{Interest per year}) \cdot (\text{Number of years})$$
$$B = P + (rP) \cdot t$$
$$B = P(1 + rt)$$

With a simple interest account, the principal does all the work of creating new wealth: Interest doesn't earn new interest. It follows that the balance of a simple interest account will grow at a constant rate.

For a **compound interest** account, on the other hand, the entire balance is put to work creating wealth: Both principal and interest alike earn interest. For example, if interest is 5% compounded annually, then after the first year the balance would be $1050, just as with simple interest. But in the second year, the entire balance of $1050 will earn interest, so that interest earned in the second year will be $5\% \cdot \$1050 = \$52.50$, giving us a balance after 2 years of $1102.50. In fact, each year we will earn interest equal to 5% of the balance at the beginning of the year. In other words, for each year we have

$$\text{New balance} = \text{Beginning balance} + \text{Interest}$$
$$= \text{Beginning balance} + 5\% \text{ of Beginning balance}$$
$$= \text{Beginning balance} \cdot (1.05)$$

Since each year the balance is multiplied by a factor of 1.05, the balance after 1 year will be $\$1000 \cdot 1.05$; after 2 years, we will have $\$1000 \cdot 1.05^2$, and so forth. It follows that, because we multiply by a factor of 1.05 each year, the balance after t years will be given by $\$1000 \cdot 1.05^t$. More generally, the balance B after t years for an account with principal P and annual compound interest rate r is given by

$$B = P(1 + r)^t$$

Whereas simple interest accounts grow at a constant rate ($\$50$ per year, for example), compound interest accounts grow at a constant *percentage* rate (such as 5% per year). The difference between these two methods of computing interest can be startling, especially over long time spans, as Table 5 illustrates.

TABLE 5 **Simple versus compound interest**

Year	Balance (5% simple interest)	Balance (5% interest compounded annually)
0	$1,000.00	$1,000.00
1	$1,050.00	$1,050.00
2	$1,100.00	$1,102.50
3	$1,150.00	$1,157.63
4	$1,200.00	$1,215.51
5	$1,250.00	$1,276.28
10	$1,500.00	$1,628.89
15	$1,750.00	$2,078.93
20	$2,000.00	$2,653.30
30	$2,500.00	$4,321.94
50	$3,500.00	$11,467.40
100	$6,000.00	$131,501.26
1000	$51,000.00	$1,546,318,920,731,927,238,984.57

We summarize simple and compound interest as follows.

Simple and Annually Compounded Interest Formulas □

	Simple Interest	Compound Interest
Interest earned by . . .	principal only	entire balance—both principal and interest
Balance grows . . .	at a constant rate (for example, $50 per year)	at a constant percentage rate (for example, 5% per year)
Formula for balance:	$B = P(1 + rt)$	$B = P(1 + r)^t$

EXAMPLE 12 ■ Compound Interest

Find the balance after 20 years if $500 is invested at 6% interest compounded annually.

SOLUTION Using the compound interest formula and the exponent key on a calculator, we obtain

$$B = P(1 + r)^t$$
$$= 500(1.06)^{20}$$
$$\approx 1603.57$$

Thus, the balance after 20 years would be $1603.57

SCIENTIFIC NOTATION

Collectively, the sciences provide us with a window on the universe through which we steal fleeting glimpses of wonders unthinkably small and unimaginably large. From ephemeral subatomic particles dancing into and out of existence, to violently colliding clusters of galaxies; from the minuscule bacteria that teem on this very page, to ecosystems spanning entire continents—science deals with the microscopic, the macroscopic, and everything in between. The decimal system of notation, so useful when dealing with human-scale entities, is stretched past its functional limit when used to express quantities at such vastly different scales, as Table 6 illustrates.

TABLE 6

Quantity	Approximation
Distance light travels in a year in miles	5,878,000,000,000
Weight of a bloodsucking banded louse in pounds	0.00000001101
Mass of an electron in grams	0.000000000000000000000000000910953
Number of atoms in the observable universe	10,000,000,000,000,000,000,000, 000,000,000,000,000,000, 000,000,000,000,000,000, 000,000,000,000,000,000

Scientific notation was developed as a more compact way of expressing such very large and very small numbers. For example, using scientific notation, we would say that there are 5.878×10^{12} miles in a light-year, that a louse weighs 1.1101×10^{-8} pound, an electron weighs 9.10953×10^{-28} gram, and there are about 1×10^{85} atoms in the observable universe. Note that in each case, we have expressed a real number as the product of two numbers: a number between 1 and 10, and an integer power of 10. More precisely, we have the following definition.

Scientific Notation   □

> A number is said to be expressed in **scientific notation** if it is written in the form $a \times 10^n$, where $1 \le a < 10$, and n is an integer.

> **RULE OF THUMB**   When converting from decimal notation to scientific notation, the number between 1 and 10 is obtained by shifting the decimal point to the position immediately after the first nonzero digit. The power of 10 is obtained by counting the number of places that the decimal point has been shifted. If the original number is larger than 1, the exponent is nonnegative. If the original number is smaller than 1, then the exponent is negative.

EXAMPLE 13   ■   Converting to Scientific Notation

Express the following numbers in scientific notation.

a. 234.0023   b. One billion
c. 0.00012   d. _123,456,789.01

SOLUTION

a. $234.0023 = 2.340023 \times 10^2$
b. One billion $= 1,000,000,000 = 1 \times 10^9$
c. $0.00012 = 1.2 \times 10^{-4}$
d. $123,456,789.01 = 1.2345678901 \times 10^8$

EXAMPLE 14   ■   Converting from Scientific Notation to Decimal Notation

Express the following numbers in standard decimal notation.

a. 7.03×10^{-3} b. 9.05×10^5 c. 1.47×10^0

SOLUTION

a. $7.03 \times 10^{-3} = 0.00703$ The decimal point has been moved 3 units to the left.
b. $9.05 \times 10^5 = 905,000$ The decimal point has been moved 5 units to the right.
c. $1.47 \times 10^0 = 1.47$ Since $10^0 = 1$, the decimal point isn't moved at all.

CALCULATOR KEYS   *Scientific Notation*

Numbers expressed in scientific notation can be entered into a calculator using a key that is usually labeled either EE or EXP. For example, to enter the number 2.04×10^{32} we first enter the number 2.04, then press EE (or EXP), and finally enter the number 32.

UNDERSTANDING AND MASTERY CHECKLISTS

CONCEPTS TO UNDERSTAND

☐ Exponential notation
☐ Negative and zero exponents
☐ Simple interest
☐ Compound interest
☐ Scientific notation

SKILLS TO MASTER

☐ Evaluate expressions involving integer exponents.
☐ Simplify expressions involving negative exponents.
☐ Compute the balance in a simple interest account.
☐ Compute the balance in a compound interest account.
☐ Convert from decimal to scientific notation.
☐ Convert from scientific notation to decimal notation.

EXERCISES 1.4

EXERCISES 1–6   *Express each product in exponential form.*

1. $2 \cdot 2 \cdot 2 \cdot 2 \cdot 2 \cdot 2$
2. $(-1) \cdot (-1) \cdot (-1)$
3. $x \cdot x \cdot x \cdot x \cdot x \cdot x \cdot x$
4. $a \cdot b \cdot b \cdot a \cdot a \cdot b \cdot a \cdot b \cdot b$
5. $p \cdot q \cdot p \cdot p \cdot q \cdot p \cdot p$
6. 5

EXERCISES 7–30   *Evaluate the given expression.*

7. $(-1)^5$
8. $(\frac{2}{5})^3$
9. $(-2)^4$
10. $(-\frac{1}{10})^3$
11. 3^{-2}
12. 4^{-3}
13. $(-2)^{-1}$
14. $(-1)^{-4}$
15. $(\frac{1}{3})^{-1}$
16. $(\frac{4}{3})^{-2}$
17. 0.1^5
18. 0.02^3
19. 1.2^{-2}
20. 0.1^{-3}
21. -5^2
22. -3^{-2}
23. $(-\frac{3}{7})^{-2}$
24. $(2^3)^2$
25. $(4^{-2})^3$
26. $2^{(3^2)}$
27. $((2^{-3})^2)^{-2}$
28. $((10^{-1})^2)^{-1}$

29. x^n, where x is the number of ozone molecules in Earth's atmosphere and n is the number of talking rhinos

30. x^n, where x is the number of talking rhinos and n is the number of ozone molecules in Earth's atmosphere

EXERCISES 31–54   *Simplify the given expression. Write your final answer without using negative exponents.*

31. $a^3 a^6$
32. $x^5 x^{-3}$
33. $\dfrac{2x^8}{x^3}$
34. $\dfrac{b^3}{b^1}$
35. $\dfrac{r^5}{r^7}$
36. $\dfrac{x^{-2}}{x^3}$
37. $\dfrac{y^{-1}}{y^{-2}}$
38. $x^2 y^4 x^{-3}$
39. $(a^3 b^{-2})(a^{-2} b^{-4})$
40. $(x^{-1})^2$
41. $(x^2 y^3)^2$
42. $(s^{-4} t^3)^{-2}$

43. $\dfrac{p^{10} q^6}{p^4 q^2}$
44. $\dfrac{2^3 s^5 t^2}{2^2 s^6 t^5}$
45. $(a^3 b^{-1} c^2) \cdot (a^{-2} b^{-3} c^3)$
46. $\dfrac{(a^2 + b^2)^3}{(a^2 + b^2)^2}$
47. $\dfrac{(x - y)(x + y)}{(x - y)^2}$
48. $\dfrac{p^{-2}}{q^{-3}} \cdot \dfrac{q^{-1}}{p^4}$
49. $\dfrac{3^{-1} x^4 y^{-2}}{3^{-2} x^{-2} y}$
50. $\left(\dfrac{p^2}{q^3}\right)^{-1}$
51. $(y^{-1} z^2) \cdot (yz^{-1})^{-1}$
52. $\left(\dfrac{a^2 b^3 c}{a^2 b^{-1} c^2}\right)^2$
53. $\left(\dfrac{xy^{-2} z^{-3}}{2x^{-2} y^1 z^2}\right)^{-2}$
54. $\left(\dfrac{p^2 q^{-1}}{p^4 q^{-2} r^3}\right)^{-3}$

EXERCISES 55–60   *Convert the given number to standard decimal notation.*

55. 3.04×10^{-3}
56. 8.38×10^3
57. 3.001×10^5
58. 2.7×10^{-1}
59. 7.5×10^0
60. 3.2×10^1

EXERCISES 61–74   *Convert the given number to scientific notation.*

61. $4{,}000{,}000$
62. $10{,}000.001$
63. 0.0943
64. 0.0008456
65. Fifty-two trillion
66. One billionth
67. 20.31×10^3
68. 0.00737×10^{-4}
69. $(2.5 \times 10^{-45}) \cdot (4 \times 10^{52})$
70. $(8.1 \times 10^{12}) \div (2.7 \times 10^{-5})$
71. $13{,}176{,}000{,}000{,}000{,}000{,}000{,}000{,}000$ (The weight of Earth in pounds)
72. $92{,}955{,}900$ (The distance from Earth to the sun in miles)
73. 0.000007 (The probability of being struck by lightning in a given year)
74. 0.0003 (Lethal dose of the nerve gas VX in grams)

▪ APPLICATIONS

75. **Simple Interest**   What would the balance be on an account paying 8% simple interest if $100 were deposited for 6 years?

76. **Simple Interest**   Suppose that $750.00 is deposited in an account paying 4% simple interest. How much interest will be earned in the first 10 years?

77. **Compound Interest**   Suppose that in 1776 one dollar was deposited in an account paying 5% interest compounded annually. What would the balance be in the year 2001?

78. **Compound Interest**   What is the balance after 50 years in an account paying 10% interest compounded annually with a principal of 1¢?

79. **Salary Options**   You are hired as a temporary worker to begin work on the first of May; the job is scheduled to be completed on May 31. Your employer gives you the choice of either receiving $300,000 a day for a month, or of receiving 1¢ on the first day of the month, 2¢ on the second day, 4¢ on the third,

doubling the pay on each successive day of the month. Which deal should you take? Justify your answer.

80. *Faded Jeans*   Suppose that a pair of jeans fades in such a way that only $\frac{3}{4}$ of the dye remains after every washing. Assuming that the jeans are washed once a week for a year, how much of the original dye remains after 1 year?

81. *Julius Caesar versus Ben Franklin*   In this problem, we will see the enormous advantage of compound interest over simple interest.

   a. Which would be worth more today, Julius Caesar's deposit of $100 at 10% simple interest or Ben Franklin's deposit of $1 at 6% interest compounded annually?

   b. What approximations did you make in order to solve part a?

   c. What aspects of the premise of part a are unrealistic?

82. *The Bouncing Ball*   Whenever a certain ball is dropped, it bounces to two-thirds of its original height. Suppose the ball is dropped off the Sears Tower in Chicago (height 1454 feet).

   a. How high would the ball rise after one bounce?

   b. How high would the ball rise after two bounces?

   c. How high would the ball rise after n bounces?

   d. Suppose that a film of the bouncing ball is run backwards. The first visible bounce appears to be a height of about 1 inch. How many additional bounces will occur before the ball is seen leaping over the edge of the Sears Tower?

83. *Repeated Multiplication*   Suppose that you enter the number 1.001 in your calculator, and then multiply it by 1.001. You then multiply this number by 1.001 and continue in this fashion at the rate of one multiplication per second. How large a product would be obtained after a 24-hour period of doing this?

84. *Cross-Country Trip*   Suppose that you resolve to drive from Los Angeles to New York (driving distance 2774 miles) according to the following scheme. On the first day you will drive half the distance, and on each successive day you will drive half of the remaining distance. How many miles will remain after 3 weeks of travel?

EXERCISES 85–86   *If inflation were to remain at a steady rate of i%, then, on the average, the price of an item will increase according to the formula $N = P(1 + \frac{i}{100})^n$, where P is the original price of the item and N is the price after n years.*

85. *CD Inflation*   If a CD costs $15.00 in 2000, how much would it cost in 2030 assuming 4% inflation?

86. *Television Inflation*   Given that inflation for the period 1955 to 1998 has averaged about 3.8%, estimate the price in 1955 dollars for a color television that sold for $325 in 1998. In fact, a new color television would have cost over $500 in 1955. Does this fact suggest that our inflationary model is not legitimate?

CONCEPTS AND CRITICAL THINKING

EXERCISES 87–90   *Answer true or false.*

87. For all real numbers x and integers a and b, $x^a x^b = x^{ab}$.

88. $-3^2 = 9$

89. 11.72×10^{13} is not in scientific notation.

90. Over long time periods, accounts paying compound interest will pay significantly more than accounts paying simple interest at the same rate.

EXERCISES 91–94   *Give an example of each.*

91. An illustration showing that $a^b a^c \neq a^{bc}$

92. A number greater than 100 expressed in scientific notation

93. A number less than 1 expressed in scientific notation

94. A reason why scientific notation is used

QUESTIONS FOR DISCUSSION OR ESSAY

95. In this section, we have dealt with many very large numbers. But do such numbers have any significance? Can you think of any context (other than those mentioned in the text) in which numbers larger than a million arise? In particular, what topics discussed on the nightly news might involve very large numbers?

96. In the text, we define $a^{-n} = 1/a^n$ for positive integers n and $a \neq 0$. Why do we exclude $a = 0$?

97. Suppose that the Board of Directors of Simpleton Bank and Trust decides to offer only simple interest in order to cut expenses. How could you foil their efforts by obtaining compound interest from their bank?

98. Example 3 involved the premise that a man continues to double in weight every 6 months for the rest of his life. This, of course, led to ridiculous results. What factors prevent a child from continuing to double in weight every 6 months? The hypothesis that

a child's weight doubles every 6 months is an example of a *mathematical model*. This particular mathematical model was a complete failure. What characteristics should a good mathematical model have? What information would be helpful in constructing a good mathematical model of weight gain?

99. In the text, we discussed folding a piece of paper in half 52 times, creating a piece of paper more than 200,000,000 miles thick. In practice, about how many times can a piece of paper be folded? What factors prevent one from folding a piece of paper 52 times?

■ PROJECTS FOR ENRICHMENT

100. *Exponential Growth*   In this project, we will attempt to fathom the incredible consequences of exponential growth over long periods of time. For dramatic effect, we will take poetic license by pretending that dollars existed 2000 years ago.

  a. Earth is approximately spherical with a radius of about 4000 miles. The volume of a sphere is given by the formula $V = \frac{4}{3}\pi r^3$, where r represents the radius and V is the volume. Estimate the volume of Earth. Express the answer in scientific notation.

  b. Estimate the volume of a dollar bill. (*Hint*: A dollar bill can be viewed as a box with a very small thickness. The formula for the volume of a box is $V = l \cdot w \cdot t$, where l is the length, w is the width, and t is the thickness.)

  c. Using parts a and b, estimate the number of dollar bills that could fit inside Earth, if Earth were hollow. Express the answer in scientific notation.

  d. How much would \$1 deposited in the year A.D. 1 be worth in 2001? Assume interest is 6% compounded annually. Express the answer in scientific notation.

  e. Using the results of parts b and c, compute the number of "Earth banks" that could be filled if the balance from a deposit of \$1 deposited in A.D. 1 were paid in one dollar bills in 2001.

101. *Repeated Exponentiation*   As was mentioned in the text, multiplication represents repeated addition, and exponentiation represents repeated multiplication. Suppose that we take this one step further and define "hyperexponentiation" to be repeated exponentiation. For example, just as $2 \cdot 3$ represents 2 added to itself three times, and 2^3 represents 2 multiplied by itself 3 times, we could define $2\,\boxed{h}\,3$ to be 2 "exponentiated 3 times," or $2^{(2^2)} = 2^4 = 16$. More generally, we could define $m\,\boxed{h}\,n$ to be

$$m^{\left(m^{\left(m^{\left(\cdots^{m}\right)}\right)}\right)}$$

where m appears n times in all.

Compute the following quantities and express the answer in scientific notation.

  a. $5\,\boxed{h}\,2$ b. $2\,\boxed{h}\,4$
  c. $1\,\boxed{h}\,10$ d. $10\,\boxed{h}\,3$

SECTION 1.5   POLYNOMIAL EXPRESSIONS

☐ **How can you estimate the height of a bridge with only a stopwatch and a rock?**

☐ **How can you estimate the number of board feet in a tree that has a circumference of 60 inches?**

☐ **If you are traveling at a rate of 65 mph, and you suddenly notice an obstruction in the road 200 feet ahead of you, will you be able to stop in time?**

☐ **Is there any truth to the oft-repeated saying, It's not the heat; it's the humidity?**

☐ **How can algebra be used to expose a mind reader's "magic"?**

DEFINITION OF POLYNOMIAL

Expressions involving variables raised to natural number powers arise frequently in the study of real-world phenomena. For example, the laws of physics tell us that if an object is dropped near the surface of Earth, it will fall $16t^2$ feet in t seconds. The expression $16t^2$ is an example of a *monomial*.

Definition of Monomial   □

> A **monomial** is the product of a number and one or more variables raised to nonnegative integer powers. The **coefficient** of a monomial is the number preceding the variable(s). The **degree** of a monomial is the sum of the powers.

Table 7 lists some examples of expressions that are monomials and some that are not. For those that are monomials, we identify the coefficient and the degree.

TABLE 7

Expression	Monomial?	Coefficient	Degree	Comment
$2x^3$	Yes	2	3	
$\frac{1}{3}p^2q^5$	Yes	$\frac{1}{3}$	7	The degree is $2 + 5 = 7$, the sum of the powers.
$5y^{-2}$	No			The exponent is negative.
-4	Yes	-4	0	Numbers can be considered to be monomials since, for example, $-4 = -4x^0$.
$1.5t^{1/3}$	No			The exponent is not an integer.

Definition of Polynomial   □

> A **polynomial** is a sum of monomials. The monomials that make up a polynomial are called the **terms** of the polynomial. The **degree** of a polynomial is the highest degree of its terms.

A monomial can be viewed as a polynomial with only one term. Two other categories of polynomials are *binomials* and *trinomials*. As the prefixes *bi* and *tri* suggest, a **binomial** is a polynomial with two terms, and a **trinomial** is a polynomial with three terms.

EXAMPLE 1   ■   Identifying Polynomials

Identify each of the following as a monomial, binomial, trinomial, polynomial, or none of these, and determine the degree of any that are polynomials.

a. $-2a^5 + 4a^3 - 3a + 4$ **b.** $3x - 5$
c. $3x^2y + 4y^{1/3} + 2xy^3$ **d.** $\frac{1}{2}p^2q^3 + \frac{2}{3}p^3q^5 - pq^6$

SOLUTION
a.   Polynomial of degree 5.
b.   Binomial (polynomial also) of degree 1.

   c. This is not a polynomial because the exponent in $y^{1/3}$ is not an integer.

   d. Trinomial (polynomial also) of degree 8.

EVALUATING POLYNOMIAL EXPRESSIONS

The variables in a polynomial expression usually represent numbers. In many instances, we will be interested in evaluating such expressions with specific values for the variables.

EXAMPLE 2   ▇ **Estimating the Height of a Bridge**

An object dropped near the surface of Earth will fall $16t^2$ feet in t seconds. If a rock that is dropped off the edge of a bridge hits the water after 1.2 seconds, how high above the water is the bridge?

SOLUTION   We evaluate the monomial $16t^2$ using $t = 1.2$ to obtain $16(1.2)^2 \approx 23$ feet.

EXAMPLE 3   ▇ **Evaluating a Polynomial in One Variable**

Evaluate the polynomial $3a^2 - 2a + 3$ for $a = 2$.

SOLUTION   Substituting $a = 2$, we have

$$3(2)^2 - 2(2) + 3 = 12 - 4 + 3 = 11$$

EXAMPLE 4   ▇ **Evaluating a Polynomial in Two Variables**

Evaluate the polynomial $p^2q^3 + p^3q^5 - pq^6$ for $p = 3$ and $q = -1$.

SOLUTION   Substituting $p = 3$ and $q = -1$, we obtain

$$(3)^2(-1)^3 + (3)^3(-1)^5 - (3)(-1)^6 = -9 - 27 - 3 = -39$$

Evaluation of polynomials can be concisely expressed using the notation of *functions*—a concept that will be developed more fully in Chapter 3. To illustrate function notation, consider the polynomial $3a^2 - 2a + 3$ from Example 3. If we name the polynomial $f(a)$ (read "f of a"), so that $f(a) = 3a^2 - 2a + 3$, then $f(2)$ represents the value of the polynomial when $a = 2$; that is,

$$f(2) = 3(2)^2 - 2(2) + 3 = 11$$

Similarly, if we write $h(p, q) = p^2q^3 + p^3q^5 - pq^6$ to indicate the polynomial of two variables p and q from Example 4, then the value of this polynomial for $p = 3$ and $q = -1$ is given by

$$h(3, -1) = (3)^2(-1)^3 + (3)^3(-1)^5 - (3)(-1)^6 = -39$$

Notice that because the polynomial was defined as $h(p, q)$, the first number inside the parentheses must be substituted for p, and the second must be substituted for q.

EXAMPLE 5 ■ Evaluating a Polynomial in One Variable Using Function Notation

Given $p(x) = x^2 - 3x + 7$, find $p(-2)$.

SOLUTION $p(-2) = (-2)^2 - 3(-2) + 7 = 4 + 6 + 7 = 17$

EXAMPLE 6 ■ Evaluating a Polynomial in Two Variables Using Function Notation

Given $g(u, v) = u^2v + 5uv - 4v^3$, find $g(-3, 2)$.

SOLUTION For the polynomial $g(u, v)$, the notation $g(-3, 2)$ means we let $u = -3$ and $v = 2$ to obtain

$$g(-3, 2) = (-3)^2(2) + 5(-3)(2) - 4(2)^3 = 18 - 30 - 32 = -44$$

ADDITION AND SUBTRACTION OF POLYNOMIALS

The operations of addition and subtraction of polynomials are performed by *combining like terms*. The mechanics of this process involve the associative, commutative, and distributive properties.

EXAMPLE 7 ■ Adding Polynomials

Find the sum $(2x^3 - 5x^2 + x - 4) + (8x^2 - 3x + 4)$.

SOLUTION

$$(2x^3 - 5x^2 + x - 4) + (8x^2 - 3x + 4)$$
$$= 2x^3 + (-5x^2 + 8x^2) + (x - 3x) + (-4 + 4) \qquad \text{Combining like terms}$$
$$= 2x^3 + 3x^2 - 2x$$

EXAMPLE 8 ■ Subtracting Polynomials

Find the difference $(7m^2 - 4mn + n^2 + 5n) - (3m^2 + 2mn - 2n^2)$.

SOLUTION

$$(7m^2 - 4mn + n^2 + 5n) - (3m^2 + 2mn - 2n^2)$$
$$= 7m^2 - 4mn + n^2 + 5n - 3m^2 - 2mn + 2n^2 \qquad \text{Distributing}$$
$$= (7m^2 - 3m^2) + (-4mn - 2mn) + (n^2 + 2n^2) + 5n \qquad \text{Combining like terms}$$
$$= 4m^2 - 6mn + 3n^2 + 5n$$

MULTIPLICATION OF POLYNOMIALS

Polynomials are multiplied by repeated application of the distributive property, as the following examples demonstrate.

EXAMPLE 9 ■ Multiplying Polynomials

Find the product $3mn(8m^3 - 12m^2n - n^3)$.

SOLUTION

$$3mn(8m^3 - 12m^2n - n^3)$$
$$= 3mn(8m^3) - 3mn(12m^2n) - 3mn(n^3) \qquad \text{Distributing}$$
$$= 24m^4n - 36m^3n^2 - 3mn^4$$

EXAMPLE 10 ▨ **Multiplying Polynomials**

Find the product $(x - 1)(2x + 3)$.

SOLUTION

$$(x - 1)(2x + 3) = x(2x + 3) + (-1)(2x + 3) \qquad \text{Distributing}$$
$$= 2x^2 + 3x - 2x - 3 \qquad \text{Distributing}$$
$$= 2x^2 + x - 3$$

EXAMPLE 11 ▨ **Multiplying Polynomials**

Find the product $(2y - 5)(3y^2 - y + 2)$.

SOLUTION

$$(2y - 5)(3y^2 - y + 2)$$
$$= 2y(3y^2 - y + 2) + (-5)(3y^2 - y + 2) \qquad \text{Distributing}$$
$$= (6y^3 - 2y^2 + 4y) + (-15y^2 + 5y - 10) \qquad \text{Distributing}$$
$$= 6y^3 + (-2y^2 - 15y^2) + (4y + 5y) - 10 \qquad \text{Combining like terms}$$
$$= 6y^3 - 17y^2 + 9y - 10$$

Two shortcuts are often used for multiplying polynomials. The first, the column method, is especially useful when multiplying polynomials with three or more terms. We demonstrate this method for the polynomials from Example 11.

$$
\begin{array}{r}
3y^2 - y + 2 \\
2y - 5 \\
\hline
-15y^2 + 5y - 10 \\
6y^3 - 2y^2 + 4y \\
\hline
6y^3 - 17y^2 + 9y - 10
\end{array}
$$

-5 times $3y^2 - y + 2$
$2y$ times $3y^2 - y + 2$
Adding down the columns

Example 10 suggests a second shortcut for multiplying binomials. From the second step of that example, we construct the diagram shown in Figure 39.

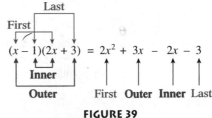

FIGURE 39

Here, $2x^2$ is the product of the *First* terms, $3x$ is the product of the *Outer* terms, $-2x$ is the product of the *Inner* terms, and -3 is the product of the *Last* terms. This is sometimes called the FOIL (**F**irst **O**uter **I**nner **L**ast) method.

E X A M P L E   1 2   Multiplying Binomials Using the FOIL Method

Perform the multiplication $(3t - 5)(t + 3)$ using the FOIL method.

SOLUTION

$$(3t - 5)(t + 3) = 3t^2 + 9t - 5t - 15$$
$$= 3t^2 + 4t - 15$$

Several products of binomials occur so frequently that it is helpful to recognize their forms.

Special Binomial Products □

$$(a + b)(a - b) = a^2 - b^2$$
$$(a + b)^2 = (a + b)(a + b) = a^2 + 2ab + b^2$$
$$(a - b)^2 = (a - b)(a - b) = a^2 - 2ab + b^2$$

E X A M P L E   1 3   Using a Special Binomial Product

Expand $(3x - 5)^2$.

SOLUTION   We use the third special binomial product, replacing a with $3x$ and b with 5 to obtain

$$(3x - 5)^2 = (3x)^2 - 2(3x)(5) + 5^2 = 9x^2 - 30x + 25$$

E X A M P L E   1 4   A Mind-Reading Trick

A magician claims to be able to read minds. She asks you to pick a number, add 2 to it and square the result, subtract the square of your original number, multiply the result by $\frac{1}{4}$, and, finally, subtract your original number. After a dramatic pause, she announces with great pleasure that your result is 1. How did she know?

SOLUTION   The mystery behind such number tricks is easily revealed using polynomial expressions. Letting x represent the number you chose, we can form a polynomial expression using the verbal statements, as shown:

Add 2 and square the result:	$(x + 2)^2$
Subtract the square of the original number:	$(x + 2)^2 - x^2$
Multiply the result by $\frac{1}{4}$:	$\frac{1}{4}[(x + 2)^2 - x^2]$
Subtract the original number:	$\frac{1}{4}[(x + 2)^2 - x^2] - x$

Simplifying this expression, we get

$$\frac{1}{4}[(x + 2)^2 - x^2] - x = \frac{1}{4}[(x^2 + 4x + 4) - x^2] - x$$

$$= \frac{1}{4}(4x + 4) - x$$

$$= (x + 1) - x$$

$$= 1$$

This shows that the polynomial $\frac{1}{4}[(x + 2)^2 - x^2] - x$ is equivalent to 1. In other words, no matter what initial number is chosen for x, the answer will *always* be 1.

COMPLEX ARITHMETIC REVISITED

Recall from Section 1.1 that a complex number has the form $a + bi$, where $i = \sqrt{-1}$. The arithmetic operations of addition, subtraction, and multiplication of complex numbers can be handled very much like the corresponding operations on polynomials. We simply treat the complex numbers as though they were binomials involving the variable i. The only real difference is that $i^2 = -1$. As the following examples illustrate, we can use this property to simplify any natural number power of i.

$$i^3 = i^2 i = (-1)i = -i$$
$$i^4 = i^2 i^2 = (-1)(-1) = 1$$
$$i^6 = (i^2)^3 = (-1)^3 = -1$$
$$i^{13} = (i^2)^6 i = (-1)^6 i = i$$

In fact, all powers of i simplify to ± 1 or $\pm i$. This observation, when combined with the usual rules for adding, subtracting, and multiplying binomials (such as the FOIL method), makes complex arithmetic a simple task.

EXAMPLE 15 ▨ **Subtracting Complex Numbers**

Write $(10 - 3i) - (7 - 6i)$ in the form $a + bi$.

SOLUTION

$$(10 - 3i) - (7 - 6i) = 10 - 3i - 7 + 6i = 3 + 3i$$

EXAMPLE 16 ▨ **Using FOIL to Multiply Complex Numbers**

Write $(2 - i)(3 + 2i)$ in the form $a + bi$.

SOLUTION Applying the FOIL method, we have

$$(2 - i)(3 + 2i) = 6 + 4i - 3i - 2i^2 = 6 + i - 2i^2$$

Since $i^2 = -1$, we can simplify further to obtain

$$6 + i - 2(-1) = 8 + i$$

EXAMPLE 17 ▨ **Simplifying Powers of i**

Write $i^{1007}(4 - 2i)$ in the form $a + bi$.

SOLUTION Since $i^2 = -1$, we wish to determine how many factors of i^2 are in i^{1007}. To do this, we select the largest even number less than 1007—namely, 1006—and write

$$i^{1007} = i^{1006} \cdot i = (i^2)^{503} \cdot i = (-1)^{503} \cdot i = -i$$

Thus,

$$i^{1007}(4 - 2i) = -i(4 - 2i) = -4i + 2i^2 = -4i - 2 = -2 - 4i$$

UNDERSTANDING AND MASTERY CHECKLISTS

CONCEPTS TO UNDERSTAND

☐ Polynomials: coefficients, terms, and degree

☐ Monomials, binomials, and trinomials

☐ Evaluating polynomials

☐ Operations with polynomials

☐ Special binomial products

SKILLS TO MASTER

☐ Recognize a polynomial and determine its degree.

☐ Evaluate polynomials by substituting for variable(s).

☐ Add, subtract, and multiply polynomials.

☐ Compute products of binomials using special formulas.

☐ Perform complex arithmetic using special binomial products.

EXERCISES 1.5

EXERCISES 1–12 *Determine whether or not each expression is a polynomial. For each polynomial, indicate if it is a monomial, binomial, or trinomial, and give its degree.*

1. $2a^2 - 5a + 6$

2. $-z + 2z^3 - 4z^5 + 3$

3. $\frac{7}{2}k^4$

4. $3x^2 - 2x + x^{1/2} - 1$

5. $2u^5 - u^{1/3} + 4u$

6. -10

7. $2x^3y^2 + x^4 - y^4$

8. $x^{-2}y + y^{-2}x$

9. $\sqrt{a^2 + b^2}$ (the length of the hypotenuse of a right triangle with side lengths a and b)

10. $\frac{1}{2}bh$ (the area of a triangle in terms of its base b and height h)

11. $\pi r^2 h$ (the volume of a cylinder in terms of its base radius r and height h)

12. $\frac{1}{16}(2a^2b^2 + 2a^2c^2 + 2b^2c^2 - a^4 - b^4 - c^4)$ (the square of the area of a triangle with side lengths a, b, and c)

EXERCISES 13–26 *Evaluate each polynomial as indicated.*

13. $4x^3 - x^2 + 2x + 1; x = 2$

14. $z^4 + 3z^2 - 6z; z = -1$

15. $t^2 - 2t - 5; t = -\frac{1}{2}$

16. $9y^3 - y + 2; y = \frac{2}{3}$

17. $\frac{1}{6}x^2 - x + 2; x = \sqrt{2}$

18. $s^2 + 4s - 1; s = \sqrt{5}$

19. $x^2 + xy + y^2; x = -2, y = 3$

20. $-\frac{1}{2}gt^2 + v_0t + s_0; t = 2$

21. $z^2 + 4; z = 2i$

22. $z^2 + 10; z = 1 - 3i$

23. $p(x) = x^3 - 5x + 1; p(2)$

24. $q(t) = 2t^3 + t^2 - 4; q(-1)$

25. $f(x, y) = x^3 - 3x^2y + 3xy^2 - y^3; f(-1, 3)$

26. $g(t, s) = t^2 - s^2 - 3t + 4s - 6; g(5, 2)$

EXERCISES 27–56 *Perform the indicated operation. Simplify as much as possible.*

27. $(3x^2 + 5x - 9) + (x^2 - 2x + 1)$

28. $(p^3 + 3p + 2) + (2p^3 - 3p^2 - p + 4)$

29. $(3y^3 - y + 4) - (5y^3 + 2y^2 - 6y)$

30. $(-6t^4 - t^2) - (3t^3 + 7t^2 - 1)$

31. $x(\frac{1}{2}x - 1) + 3x(x + \frac{5}{3})$

32. $k^2(k - \frac{3}{2}) - k(k^2 + 2k)$

33. $(2a + 1)(3a - 2)$

34. $(4x - 2)(x + 6)$

35. $(4 - 3y)(\frac{1}{2} + 2y)$

36. $(\frac{3}{4} + m)(2 - 4m)$

37. $(2t - s)(2t + s)$

38. $(k - \frac{2}{3})(k + \frac{2}{3})$

39. $(x + \sqrt{3})(x - \sqrt{3})$

40. $(t - \sqrt{5})(t + \sqrt{5})$

41. $(2x^2 + 1)(2x^2 - 1)$

42. $(y^3 - 3x)(y^3 + 3x)$

43. $(t + 4)^2$

44. $(p - 3)^2$

45. $(3a + 2b)^2$

46. $(z^2 - 4)^2$

47. $(2x + 1)(x^2 - 6x + 2)$

48. $(t^2 + 2)(2t^3 + t^2 - 5t)$

49. $(x - y + 2)(x + y - 2)$

50. $(x - 2)(x^3 + 2x^2 + 4x + 8)$

51. $(n + 1)(n + 2)(n + 3)$

52. $(2x + 1)(3x - 2)(1 - x)$

53. $(2a + 1)(a - 3) + (4a - 2)(a + 2)$

54. $2u(u - 5) - (u^2 - 3)(u + 1)$

55. $(a - b)^3$

56. $(a + b)^3$

EXERCISES 57–66 *Perform the indicated operation. Write your answer in the form $a + bi$.*

57. $(6 + 2i) - (3 + 4i)$

58. $(9 + 7i) + (5 - 2i)$

59. $(2 + 3i)(3 + 4i)$

60. $(7 - 2i)(5 + 4i)$

61. $(\sqrt{2} - 4i)(\sqrt{2} + 4i)$

62. $(3 - i\sqrt{5})^2$

63. i^{11}

64. $-i^{14}$

65. $-i^{701} \cdot i^2$

66. $i^{2001}(4 - i)$

EXERCISES 67–70 *The following polynomials have complex numbers as coefficients. Perform the indicated operation and simplify as much as possible.*

67. $(z + 2i)(z - 2i)$

68. $(z - 3i)(z + 3i)$

69. $(x + 2 - 3i)(x - 2 + 3i)$

70. $(x - 5 - i)(x + 5 + i)$

EXERCISES 71–74 *Use Example 14 as your guide to reveal the "trick" behind each conclusion.*

71. Choose a number, double it, subtract 4, multiply by 3, divide by 6, and add 2. The result is the number you started with.

72. Choose a number, double it, subtract 1, and square the result. Now add 3 and divide by 4. Finally, subtract the product of the number you started with and one less than that number. The result is 1.

73. Choose two numbers. Subtract the smaller from the larger and multiply the result by the sum of the two numbers. Add the square of the smaller number. The result is the square of the larger number.

74. Choose two numbers. Add them together and square the result. Subtract the square of each number; then divide by two. The result is the product of the two numbers.

■ **APPLICATIONS**

75. *Engine Displacement* The displacement of a gas-powered engine is computed using the formula $\frac{\pi}{4}CB^2S$, where C is the number of cylinders, B is the bore (the diameter of each cylinder in inches), and S is the stroke (the distance in inches that the piston travels one way within the cylinder). Find the displacement of an 8-cylinder Oldsmobile "Big Block" engine with a bore of 4.126 inches and a stroke of 4.254 inches.

76. *Gas Tank Volume* An underground gas tank is formed by attaching a hemisphere to each end of a cylinder with radius r and height h. The volume of each hemisphere is half that of a sphere of radius r, or $\frac{1}{2}(\frac{4}{3}\pi r^3)$. Write out a polynomial that gives the total volume of the tank. What is the volume of a tank formed from a cylinder with height 15 feet and hemispherical ends of radius 5 feet?

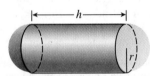

77. *Board Feet* A board foot is a unit of lumber measurement equal to 1 foot square by 1 inch thick. The number of board feet of lumber in a ponderosa pine can be approximated by $0.0015c^3$, where c is the circumference of the tree in inches at waist height. How many board feet of lumber are in a ponderosa pine with a circumference of 60 inches? What about a ponderosa pine with a *diameter* of 60 inches?

78. *Falling Marble* A marble is dropped from the top of the Sears Tower in Chicago. After t seconds, the distance it has traveled is $16t^2$ feet, and its velocity is $32t$ feet per second. If the marble hits the ground after 9.533 seconds, how high is the Sears Tower and what is the velocity of the marble upon impact?

79. *Height of a Ball* A ball thrown in the air with a velocity of 80 feet per second has height $-16t^2 + 80t$ after t seconds. What is the height of the ball after 1 second? After 2 seconds? Continue computing the height for integer values of t until you find the time at which the ball hits the ground.

80. *Compound Interest* If $1500 is deposited in an account that pays interest at a rate i compounded annually, the balance in the account after 3 years is given by the polynomial $1500(1 + i)^3$.

Expand this polynomial to show that it has degree 3. Find the balance for interest rates of 6% and 7%.

81. *Braking Distance* Suppose you are driving on a freeway when the brake lights of the car ahead of you suddenly go on. This begins an important chain of events: Your brain receives the signal to stop the car; it sends a message to your foot, which slams on the brake pedal; and the brakes dissipate energy to bring your car to a stop. From the moment the signal reaches your brain to the moment your car comes to a complete stop, your car will have traveled a certain distance. Studies have shown that this distance (measured in feet) can be approximated with the polynomial $1.1v + 0.05v^2$, where v represents the speed of the car (measured in miles per hour) before the brakes are applied. How far will your car have traveled if your speed was 65 miles per hour?

82. *Triathlon Calorie Expenditure* The total number of calories, C, expended by a triathlete weighing W pounds who swims S hours, cycles B hours, and runs for R hours is approximated by

$$C = (5.0S + 5.4B + 6.4R)W$$

The Ironman triathlon in Kailua-Kona, Hawaii, consists of a 2.4-mile swim, a 112-mile bike ride, and a 26.2-mile run (marathon). A competitive time for the Ironman would combine about 1 hour for the swim, 5 hours for the bike ride, and 3 hours for the marathon. About how many calories would be expended by a 150-pound Ironman competitor?

[*Data sources*: http://www.outsidemag.com/events/ironman96/overall.html, Web page active as of 7/22/99; Henry J. Montoye, Ed., et al. *Measuring Physical Activity and Energy Expenditures* (Champaign, IL: Human Kinetics, 1996).]

83. *Basketball Lineup* Suppose that a basketball coach wishes to try every possible starting lineup of 5 players, without regard to the positions (center, forward, guard) of the players. It can be shown that for a team with n players, the number of possible 5-player starting lineups is given by

$$S(n) = \frac{n(n - 1)(n - 2)(n - 3)(n - 4)}{120}$$

a. Find the number of possible starting lineups if the coach has 6 players.

b. If a team with 12 players played one game per day, how long

would it take for the coach to try every possible starting lineup?

84. *Heat Index*   At high temperatures, our level of discomfort is determined not only by the temperature, but also by the amount of moisture in the air (humidity). The *heat index* is a measure of perceived temperature that accounts for humidity, in much the same way that the windchill factor takes into account the cooling effects of wind. One formula for apparent temperature (in degrees Fahrenheit) is given by

$$T_A = -42.38 + 2.049t + 10.14r - 0.2248tr$$
$$- 0.00684t^2 - 0.0548r^2 + 0.00123t^2r$$
$$+ 0.000853tr^2 - 0.000002t^2r^2$$

where t is the actual temperature (in degrees Fahrenheit) and $r\%$ is the relative humidity. Compute the apparent temperature corresponding to the given temperature–humidity combinations.

a. 90°, 60%

b. 100°, 80%

Sometimes it's not the humidity; it really is the heat. Death Valley has temperatures of up to 130° F and very little moisture.

CONCEPTS AND CRITICAL THINKING

EXERCISES 85–88   *Answer true or false.*

85. The degree of the product of two polynomials is the sum of the degrees of the polynomials.

86. The degree of the sum of two polynomials is the sum of the degrees of the polynomials.

87. For all real numbers a and b, $(a + b)^2 = a^2 + b^2$

88. For all real numbers x and y, $-2(x - y) = -2x - 2y$

EXERCISES 89–92   *Give an example of each.*

89. A fourth-degree binomial

90. A third-degree trinomial

91. A third-degree polynomial in the variables u and v

92. An expression involving x that is not a polynomial, but is such that the product of this expression and x is a third-degree polynomial

93. In Example 1 and Exercises 1–12, you identified a number of expressions as polynomials. Now explain in your own words what is *not* a polynomial. Give some examples of expressions that are not polynomials and explain why they are not.

94. Perhaps you noticed in the examples and exercises that the sum or difference of two polynomials is again a polynomial. How does the degree of the sum or difference of two polynomials compare to the degrees of the two polynomials? Is the product of two polynomials a polynomial? If so, is there a connection between the degrees of the original polynomials and their product?

QUESTIONS FOR DISCUSSION OR ESSAY

95. The column method was discussed in this section as a shortcut for multiplying trinomials. What advantages do you see in using such a method? Does the column method bear any resemblance to a method you have used in the past? If so, comment on the similarities and differences between the methods.

96. The mind-reading trick in Example 14 only works because the verbal description of the operations being performed is awkward to express in English. This awkwardness actually serves as a smoke screen that prevents the listener from "seeing" the trick. By switching to another language, algebra in this case, the smoke screen vanishes and the problem becomes almost trivial. Can you think of other examples where the language of mathematics is superior to English? Explain your examples first using words only and then using mathematics. More generally, to what extent are we limited by the language in which we think?

97. Explain how the distributive property is used "behind the scenes" in the FOIL method. Some mathematicians contend that such shortcuts are the essence of mathematics. Others warn that shortcuts cause important concepts to be overlooked. What do you think about using shortcuts such as the FOIL method? What other mathematical shortcuts do you know?

■ **PROJECTS FOR ENRICHMENT**

98. *A Mind-Reading Trick* Develop your own mind-reading trick that can be explained by using polynomial simplification.

99. *Pythagorean Triples* Three integers, a, b, and c, are called a *Pythagorean triple* if they satisfy the Pythagorean Theorem—that is, if $a^2 + b^2 = c^2$. We denote such a triple by (a, b, c). In this problem we will investigate some ways for finding Pythagorean triples.

  a. Show that the following are Pythagorean triples:

  i. $(3, 4, 5)$ ii. $(6, 8, 10)$

  iii. $(5, 12, 13)$

  b. Find two more Pythagorean triples.

  c. Did you notice that each of the integers in the triple $(6, 8, 10)$ is simply twice the corresponding integer in the triple $(3, 4, 5)$? In general, whenever (a, b, c) is a Pythagorean triple, so is (na, nb, nc) for any integer n. Show this by factoring the expression $(na)^2 + (nb)^2$ and using the fact that $a^2 + b^2 = c^2$.

The difficulty with the method in part c for finding new triples is that you have to know one already. One way to come up with completely new triples is to start with two arbitrary integers m and n, and compute the quantities $m^2 - n^2$ and $2mn$. These quantities will be the first two integers of a Pythagorean triple. For example, if $m = 2$ and $n = 1$, we get $2^2 - 1^2 = 3$ and $2(2)(1) = 4$ as the first two numbers in the triple. To find the third, we can simply compute $\sqrt{3^2 + 4^2}$.

  d. Find the Pythagorean triple corresponding to

  i. $m = 3$ and $n = 1$ ii. $m = 3$ and $n = 2$

  e. Show that $(m^2 - n^2)^2 + (2mn)^2 = (m^2 + n^2)^2$ by simplifying both sides of the equation. Explain why this shows how to find the third number in the Pythagorean triple whose first two integers are $m^2 - n^2$ and $2mn$.

100. *Nested Polynomials* Not all that many years ago, a calculator with a power key such as $\boxed{\wedge}$ or $\boxed{y^x}$ would have been considered a luxury. Even today it is not uncommon to find calculators without such a key. How would one evaluate a polynomial such as $2x^5 - 3x^4 - 7x^3 + x^2 + 5x - 1$ for a given value of x, say $x = 1.2$, without a power key? One way would be to compute all the powers of 1.2 up to 1.2^5 by repeated multiplication, then multiply each of the powers of 1.2 by the appropriate coefficient, and finally, add (subtract) the results to arrive at the answer. But this would either require some pencil and paper to "store" the powers of 1.2 as you compute them—a terrible inconvenience—or you would have to recompute them as you need them. As an alternative, we will develop a method for writing the polynomial in a form that requires no exponentiation, no pencil and paper, and as few multiplications as possible. Let's begin by simplifying the expression $[(4x - 3)x + 6]x - 2$.

$$[(4x - 3)x + 6]x - 2 = [4x^2 - 3x + 6]x - 2$$
$$= 4x^3 - 3x^2 + 6x - 2$$

Note that the expression turns out to be a polynomial. Now compare the number of multiplications in the original expression with the number in the simplified expression. The original has only three: The first is $4x$, the second is $(4x - 3)x$, and the third is where the expression inside the brackets is multiplied by x. In the simplified expression, each power of x requires one less multiplication than the power. Thus, $x^3 = x \cdot x \cdot x$ requires two multiplications, and $x^2 = x \cdot x$ requires one multiplication. We also have three multiplications for the coefficients of each power of x—namely, $4x^3$, $3x^2$, and $6x$. That makes a total of six multiplications. So the simplified polynomial has twice as many multiplications.

  a. Simplify the following polynomials and compare the number of multiplications.

  i. $[(2x - 5)x + 3]x + 7$

  ii. $([(3x + 4)x - 2]x + 9)x - 8$

To avoid exponents and reduce the number of multiplications, we must write polynomials in the *nested* form $(\cdots [(a_n x + a_{n-1})x + a_{n-2}]x + \cdots + a_1)x + a_0$. For example, $8x^3 + 5x^2 - 3x + 2$ can be written as follows:

$$8x^3 + 5x^2 - 3x + 2 = (8x^2 + 5x - 3)x + 2$$
$$= [(8x + 5)x - 3]x + 2$$

  b. Write each of the following polynomials in nested form.

  i. $2x^5 - 3x^4 - 7x^3 + x^2 + 5x - 1$

  ii. $x^6 + 5x^4 - 8x^2 + 4$

  c. Now use the nested forms to evaluate the polynomials in part b for $x = 1.2$.

101. *A Geometric Perfect Square* The expression $(a + b)^2$ can be viewed geometrically as the area of a square with sides of length $a + b$, as shown in Figure 40. Using only the diagram, show that $(a + b)^2 = a^2 + 2ab + b^2$.

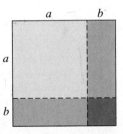

FIGURE 40

102. *Measuring Inflation with the CPI* The Consumer Price Index (CPI) is computed each month by the Bureau of Labor

Statistics. It is a measure of the retail cost of a representative sample of goods and services purchased by American consumers. Table 8 shows the CPI for each month of 1996. We see, for example, that in January the CPI was 154.4. Roughly speaking, this means that goods costing $100 in 1982 would cost $154.4 in January 1996.

TABLE 8

Month (1996)	CPI
January	154.4
February	154.9
March	155.7
April	156.3
May	156.6
June	156.7
July	157.0
August	157.3
September	157.8
October	158.3
November	158.6
December	158.6

Changes in the CPI from one month to the next provide a measure of changes in retail prices paid by consumers. As can be seen in Table 8, the CPI for most months of 1996 increased from the previous month, suggesting that retail prices were rising throughout most of the year. The average of the monthly CPI values for a given year provides an annual measure of retail prices, which can then be used to compare retail prices from one year to the next.

a. Find the average CPI for 1996, and use it to help you complete Table 9. Note that the percent change is computed as follows

$$\frac{|(\text{Current year CPI}) - (\text{Previous year CPI})|}{(\text{Previous year CPI})} \times 100\%$$

Monthly and annual CPI trends are used by government agencies to aid in the formulation of fiscal and monetary policies. They are also used by various agencies, organizations, and companies to keep such things as wages, salaries, pensions, rents, alimony payments, and child support payments in line with changing prices.

Suppose that a company gives salary raises that are based on the CPI. More specifically, suppose that in January of a given year, the percent increase in the average CPI from the previous two calendar years is used to compute new salaries for the

TABLE 9

Year	Average CPI	Percent change
1990	130.7	
1991	136.2	4.21%
1992	140.3	
1993	144.5	
1994	148.2	
1995	152.4	
1996		

coming year. For example, in January 1992 the company would have determined that the average CPI for 1991 had increased 4.21% from that of 1990, and so salaries for 1992 would be increased by 4.21% over the 1991 level.

b. Find the salary in 1992 for an employee who made $18,000 in 1991.

c. Show that if S_0 denotes the salary for a given year and if r denotes the percent increase that is to be applied to that salary, then the salary for the following year, S_1, can be found using the formula $S_1 = S_0(1 + r)$.

d. Compute the salary for each of the years 1993 through 1997 for the employee from part b.

In part d, we were able to find the salary in 1997 by computing the salaries for all of the years between 1991 and 1997. Suppose instead that we wished to find the salary in 1997 without computing any of the intermediate salaries. This can be done with the help of polynomial expressions.

Let $r_1, r_2, \ldots, r_n$ denote n consecutive percent increases in the annual CPI, let S_i denote the annual salary after the percent increase r_i has been applied, and let S_0 denote the salary for the year prior to that in which the percent increase r_1 has been applied. Then, using the formula from part c we have

$$S_1 = S_0(1 + r_1)$$
$$S_2 = S_1(1 + r_2) = S_0(1 + r_1)(1 + r_2)$$
$$S_3 = S_2(1 + r_3) = S_0(1 + r_1)(1 + r_2)(1 + r_3)$$
$$\vdots$$
$$S_n = S_0(1 + r_1)(1 + r_2) \cdots (1 + r_n)$$

So the final salary, S_n, can be found by multiplying the initial salary, S_0, by the polynomial $(1 + r_1)(1 + r_2) \cdots (1 + r_n)$.

e. Use a polynomial expression of the form just given to find the salary in 1997 for employees who had the following salaries in 1991.

i. $20,000   ii. $34,000

Note that you need to compute the product $(1 + r_1)(1 + r_2) \cdots (1 + r_6)$ only once.

f. Use the CPI values in 1990 and 1996 to find the overall percent increase from 1990 to 1996. How does this value compare to the value of the polynomial

$(1 + r_1)(1 + r_2) \cdots (1 + r_6)$ from part e? What does this suggest about finding someone's salary in 1997 if the 1991 salary is known?

g. Do you think the CPI is a reasonable tool for determining raises? Why or why not?

SECTION 1.6   FACTORING POLYNOMIALS

☐ How can $6x^2 + 7x - 20$ be factored correctly with your *first* guess?

☐ Why is it that no prime number could ever be written as the sum of two perfect cubes?

☐ How can polynomial factorizations be used to find a factor of 1,000,001 without a calculator in just seconds?

☐ How can $x^2 - 2$ be factored?

COMPLETE FACTORIZATIONS

The elements in the periodic table and the prime numbers share the property that they are both basic building blocks for much more complex systems. In the case of the elements, the system is the atomic structure of our universe. The prime numbers—that is, numbers such as 2, 7, or 23, which have no factors other than 1 and themselves—are the foundation upon which the natural number system is built. As the Greek mathematician Euclid discovered over 2000 years ago, any natural number except 1 can be factored uniquely as a product of prime numbers. For example, the number 60 can be obtained by multiplying $2 \cdot 2 \cdot 3 \cdot 5$, and this is the only combination of primes that will work. The product $2 \cdot 2 \cdot 3 \cdot 5$ is called the *prime factorization* of 60.

In science as well as mathematics, we can learn a great deal about complex systems by knowing something about their basic building blocks. We have much to gain by decomposing complex structures into simpler elements, and this is certainly the case with polynomials. In fact, the process of factoring a polynomial can be described as decomposing the polynomial into a product of simpler polynomials. Some similarities between factoring polynomials and finding prime factorizations for natural numbers are shown in Table 10. The goal at each stage is to find any other factors; we stop when no more factors can be found.

TABLE 10

Natural number	Polynomial
90	$4x^2y^3 - 6xy^2$
$2 \cdot 45$	$2(2x^2y^3 - 3xy^2)$
$2 \cdot 3 \cdot 15$	$2x(2xy^3 - 3y^2)$
$2 \cdot 3 \cdot 3 \cdot 5$	$2xy^2(2xy - 3)$

We consider a polynomial to be **completely factored over the integers** when it cannot be factored any further using polynomials with integer coefficients. Thus, $2xy^2(2xy - 3)$ is completely factored over the integers since no polynomial with integer coefficients can be factored out of $2xy - 3$. This complete factorization can be obtained in only one step simply by factoring out the greatest common monomial factor $2xy^2$. This is the most direct technique for factoring a polynomial.

E X A M P L E   1   ■   **Factoring Out the Greatest Common Monomial**

Factor out the greatest common monomial in $12a^6 - 3a^5 + 6a^3 + 9a^2$.

SOLUTION   The greatest common factor of each term is $3a^2$. Factoring out $3a^2$ from each term, we get

$$12a^6 - 3a^5 + 6a^3 + 9a^2 = 3a^2(4a^4 - a^3 + 2a + 3)$$

Certain special polynomials with four or more terms can be factored by a technique known as *grouping*. Often these special polynomials involve two variables, as is the case in the following examples.

EXAMPLE 2 ■ **Factoring a Polynomial by Grouping**

Factor the polynomial $xy + 2x + 3y + 6$.

SOLUTION First we group the polynomial into two binomials, each of which can be factored separately. Then we apply the distributive property.

$$
\begin{aligned}
xy + 2x + 3y + 6 &= (xy + 2x) + (3y + 6) \\
&= x(y + 2) + 3(y + 2) && \text{Factoring each binomial} \\
&= (x + 3)(y + 2) && \text{Factoring out the binomial } (y + 2)
\end{aligned}
$$

EXAMPLE 3 ■ **Factoring a Polynomial by Grouping**

Factor the polynomial $2mn^3 + 2m^2n^2 - 4mn^2 - 4m^2n$.

SOLUTION Each term has a common monomial factor of $2mn$, and so we factor that out first.

$$
\begin{aligned}
2mn^3 + 2m^2n^2 - 4mn^2 - 4m^2n &= 2mn[n^2 + mn - 2n - 2m] \\
&= 2mn[(n^2 + mn) - (2n + 2m)] && \text{Grouping binomials} \\
&= 2mn[n(n + m) - 2(n + m)] && \text{Factoring each binomial} \\
&= 2mn[(n - 2)(n + m)] && \text{Factoring out } (n + m) \\
&= 2mn(n - 2)(n + m)
\end{aligned}
$$

RULE OF THUMB As suggested by the previous example, always factor out the greatest common monomial factor before using grouping or other techniques that are discussed in this section.

FACTORING BINOMIALS

Recall that a factorization is considered complete over the integers when no more factoring is possible using integer coefficients. Sometimes it is difficult to tell whether a polynomial is completely factored, especially if the polynomial has many terms. However, for polynomials with three or fewer terms, there are certain clues that make the task much easier. We begin with three special binomial forms.

Special Binomial Forms □

Difference of squares:	$a^2 - b^2 = (a + b)(a - b)$
Difference of cubes:	$a^3 - b^3 = (a - b)(a^2 + ab + b^2)$
Sum of cubes:	$a^3 + b^3 = (a + b)(a^2 - ab + b^2)$

EXAMPLE 4   ■   **Factoring a Difference of Cubes**

Factor $x^3 - 27$.

SOLUTION

$$
\begin{aligned}
x^3 - 27 &= x^3 - 3^3 \\
&= (x - 3)(x^2 + 3x + 3^2) \qquad \text{Applying the difference of cubes form} \\
&= (x - 3)(x^2 + 3x + 9)
\end{aligned}
$$

EXAMPLE 5   ■   **Factoring a Sum of Cubes**

Factor $8z^3 + 125$.

SOLUTION

$$
\begin{aligned}
8z^3 + 125 &= (2z)^3 + 5^3 & \text{Rewriting } 8z^3 \text{ as } (2z)^3 \\
&= (2z + 5)[(2z)^2 - (2z)5 + 5^2] & \text{Applying the sum of cubes form} \\
&= (2z + 5)(4z^2 - 10z + 25) & \text{Simplifying}
\end{aligned}
$$

EXAMPLE 6   ■   **Factoring a Difference of Squares**

Factor $100x^2 - 4y^2$.

SOLUTION

$$
\begin{aligned}
100x^2 - 4y^2 &= 4(25x^2 - y^2) & \text{Factoring out the common monomial} \\
&= 4[(5x)^2 - y^2] & \text{Rewriting } 25x^2 \text{ as } (5x)^2 \\
&= 4[(5x + y)(5x - y)] & \text{Applying the difference of squares form}
\end{aligned}
$$

EXAMPLE 7   ■   **Factoring a Binomial Two Different Ways**

Factor $p^6 - 1$.

SOLUTION   The binomial $p^6 - 1$ can be viewed either as a difference of squares or a difference of cubes. We'll consider both.

Difference of squares:

$$
\begin{aligned}
(p^3)^2 - (1)^2 & \\
= (p^3 + 1)(p^3 - 1) & \qquad \text{Factoring the difference} \\
& \qquad \text{of squares} \\
= [(p + 1)(p^2 - p + 1)][(p - 1)(p^2 + p + 1)] & \qquad \text{Factoring the sum and} \\
& \qquad \text{difference of cubes}
\end{aligned}
$$

Difference of cubes:

$$
\begin{aligned}
(p^2)^3 - (1)^3 &= (p^2 - 1)[(p^2)^2 + p^2 + 1] & \text{Factoring the difference of cubes} \\
&= (p + 1)(p - 1)(p^4 + p^2 + 1) & \text{Factoring the difference of squares}
\end{aligned}
$$

To completely factor the difference of cubes result, you would have to recognize that $p^4 + p^2 + 1 = (p^2 - p + 1)(p^2 + p + 1)$, which is not at all obvious. So in this case, we are better off using the difference of squares first.

FACTORING TRINOMIALS

Some trinomials can be factored by recognizing them as *perfect square trinomials*. A perfect square trinomial is one that can be written as the square of a binomial, and there are two forms you should recognize. They are summarized as follows.

Perfect Square Trinomials   ☐

$$a^2 + 2ab + b^2 = (a + b)^2$$
$$a^2 - 2ab + b^2 = (a - b)^2$$

EXAMPLE 8   ▥   Factoring a Perfect Square Trinomial

Factor $x^2 - 8x + 16$.

SOLUTION   Notice that the outer two terms of $x^2 - 8x + 16$ are perfect squares and the middle term (ignoring the sign) is twice the product of the square roots of the outer terms. So $x^2 - 8x + 16$ is a perfect square trinomial. Thus,

$$x^2 - 8x + 16 = x^2 - 2(4)(x) + 4^2$$
$$= (x - 4)^2$$

EXAMPLE 9   ▥   Factoring a Perfect Square Trinomial

Factor $9y^2 + 30yz + 25z^2$.

SOLUTION   Note that $9y^2 = (3y)^2$ and $25z^2 = (5z)^2$, and the middle term, $30yz$, is twice the product of $3y$ and $5z$. Thus, we have

$$9y^2 + 30yz + 25z^2 = (3y)^2 + 2(3y)(5z) + (5z)^2$$
$$= (3y + 5z)^2$$

Trinomials that are not perfect squares can often be factored by ''educated guessing.'' Here we rely on experience (and sometimes a little trial and error) to guide us to the proper choice of factors. Let's consider the trinomial $x^2 - x - 20$. We are looking for two binomial factors whose first terms multiply to give x^2 and whose last terms multiply to give -20. There are six choices:

$$(x + 1)(x - 20) \quad (x + 2)(x - 10) \quad (x + 4)(x - 5)$$
$$(x - 1)(x + 20) \quad (x - 2)(x + 10) \quad (x - 4)(x + 5)$$

If the constant terms are ''too far apart,'' as in $(x + 1)(x - 20)$, the coefficient of the middle term of their product is much too large, $-19x$ in this case. We can conclude that the constant terms of the two binomial factors should be ''close together.'' The correct choice is

$$(x + 4)(x - 5) = x^2 - x - 20$$

Notice also that the constant terms of the two binomials are the factors of -20 that add up to -1 (the middle term). In general, we have the following rule.

> **RULE OF THUMB**   When factoring trinomials of the form $x^2 + bx + c$, look for the factors of c that add up to b. These will be the constant terms of the binomial factors.

EXAMPLE 10   ▧   **Factoring a Trinomial with Leading Coefficient 1**

Factor $x^2 - 5x - 6$.

SOLUTION   Since the coefficient of x^2 is 1, we look for the factors of -6 that add up to -5. The correct ones are -6 and 1. Thus, $x^2 - 5x - 6 = (x - 6)(x + 1)$.

Unfortunately, the procedure is not as simple for trinomials with a leading coefficient other than 1. However, if all else fails, we can always use trial and error to find the pair of factors that works. We'll investigate a more ''scientific'' method for factoring general trinomials in Exercise 66.

EXAMPLE 11   ▧   **Factoring a Trinomial with a Common Monomial Factor**

Factor $2t^4 + 5t^3 - 3t^2$.

SOLUTION   First we factor out the common factor t^2.

$$2t^4 + 5t^3 - 3t^2 = t^2(2t^2 + 5t - 3)$$

Next we factor $2t^2 + 5t - 3$. The following are the potential factorizations of $2t^2 + 5t - 3$.

$$(2t + 1)(t - 3)$$
$$(2t - 1)(t + 3)$$
$$(2t + 3)(t - 1)$$
$$(2t - 3)(t + 1)$$

By trial and error, we discover that $2t^2 + 5t - 3 = (2t - 1)(t + 3)$. Thus, $2t^4 + 5t^3 - 3t^2 = t^2(2t - 1)(t + 3)$.

EXAMPLE 12   ▧   **Factoring a Fourth-Degree Trinomial with Leading Coefficient 1**

Factor $x^4 - 7x^2 + 12$.

SOLUTION   Since the leading term is x^4 and the middle term involves x^2, we are looking for a factorization of the form $(x^2 + \square)(x^2 + \square)$. The boxed terms must be factors of 12 and they must add up to -7; we select -4 and -3. Thus,

$$x^4 - 7x^2 + 12 = (x^2 - 4)(x^2 - 3)$$
$$= (x + 2)(x - 2)(x^2 - 3)$$

UNDERSTANDING AND MASTERY CHECKLISTS

CONCEPTS TO UNDERSTAND

☐ Complete factorization
☐ Greatest common factors
☐ Special binomial factoring forms
☐ Factoring trinomials

SKILLS TO MASTER

☐ Recognize when an algebraic expression is completely factored.
☐ Factor out the greatest common factor.
☐ Use special binomial forms to factor polynomials.
☐ Factor trinomials by informed trial and error.

EXERCISES 1.6

EXERCISES 1–6 *Factor out the greatest common factor.*

1. $4n^5 - 12n^3 + 24n$
2. $-16t^2 + 40t$
3. $6ab^2 - 8ab + 12a^2b$
4. $3x^2y^2 + 12x^3y - 27x^4y$
5. $w(w + 2) + 5(w + 2)$
6. $(x - 3)^2 + 4(x - 3)$

EXERCISES 7–10 *Factor by grouping. You may need to regroup first.*

7. $9ab + 18b + 5a + 10$
8. $x^2 - 36y + 6xy - 6x$
9. $8p^2 - 9q^2 + 6pq - 12pq$
10. $3t^2 - 2t + 15t - 10$

EXERCISES 11–22 *Factor completely.*

11. $x^2 - 36$
12. $t^2 - 81$
13. $4y^2 - 16z^2$
14. $-12k^2 + 27$
15. $2x^4 - 32$
16. $t^4 - 81$
17. $s^3 - 27$
18. $z^3 + 64$
19. $a^3 + 8b^3$
20. $27n^3 - m^3$
21. $2x^3 + 128$
22. $a^4b - ab^4$

EXERCISES 23–30 *Factor each trinomial.*

23. $x^2 - x - 12$
24. $t^2 + 6t + 8$
25. $y^2 - 9y + 14$
26. $x^2 + 2x - 15$
27. $9m^2 - 12m + 4$
28. $4s^2 + 4s + 1$
29. $2y^2 + 5y - 3$
30. $3a^2 - 10a - 8$

EXERCISES 31–48 *Factor completely.*

31. $2t^2 - 8$
32. $r^2 - rs - 2s^2$
33. $6n^2 - 19n + 10$
34. $8x^3 - 27$
35. $z^4 - 2z^2 + 1$
36. $x^2 + xy + 2x + 2y$
37. $t^6 - 64$
38. $x^2 + x - 42$
39. $p^6 - 125q^3$
40. $16a^2 + 20a + 6$
41. $81x^3 - 3$
42. $t^2 - 7t + 12$
43. $3xz + 6xw - yz - 2yw$
44. $4r^3 + 4s^3$
45. $4x^2 - 12xy + 9y^2$
46. $2ab^2 - 32ac^3$
47. $6x^3y^3 + 48$
48. $x^6 - x^3 - 20$

■ APPLICATIONS

49. *Mystery Box* The volume of a box with a square base and height x inches is given by $x^3 + 4x^2 + 4x$ cubic inches. By how much does the side length of the base exceed the height?

50. *Shattered Window* A square pane of glass shatters into the triangular pieces shown in Figure 41. Add up the given areas of the triangles and use the sum to deduce the side length of the square in terms of x.

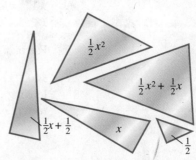

FIGURE 41

▇ CONCEPTS AND CRITICAL THINKING

EXERCISES 51–54   *Answer true or false.*

51. $a^3 - b^3 = (a + b)(a^2 - ab + b^2)$

52. $a^3 + b^3 = (a + b)^3$

53. The complete factorization of $4x^2 - 4y^2$ is $(2x - 2y)(2x + 2y)$.

54. The first step in factoring is to factor out the greatest common monomial.

EXERCISES 55–58   *Give an example of each.*

55. A trinomial for which the greatest common factor is $2x$

56. A binomial that factors as a difference of squares

57. A polynomial of degree 2 for which $x - 3$ is a factor

58. A perfect square trinomial for which $2x + 1$ is a factor

59. Show that one less than the square of a number is divisible by one less than that number.

60. Show that the sum of two perfect cubes is never a prime number.

▇ QUESTIONS FOR DISCUSSION OR ESSAY

61. Explain how the sum of cubes factorization formula can be used to find a factor of 1,000,001 in just seconds without a calculator.

62. Compare and contrast factoring polynomials to "decomposing" chemical substances into elements. Can you think of any processes in other disciplines that involve decomposing complex structures into their basic building blocks?

63. In this section, we've seen a number of special methods for factoring. Each one requires that a polynomial be in exactly the right form before the method can be applied. Discuss the importance of pattern recognition in choosing the right factoring method. Use examples to illustrate your point. Can you think of other situations in which pattern recognition is important, either in mathematics or in life in general?

▇ PROJECTS FOR ENRICHMENT

64. *Factoring Using Irrational and Complex Coefficients*   In all of the examples in this section, we considered a polynomial to be completely factored if it was not possible to factor it again using polynomials with integer coefficients. There are times when we may wish to factor even further, possibly using irrational numbers or even complex numbers in the factors. For example, even though the polynomial $x^2 - 2$ cannot be factored using integer coefficients, we can factor it using irrational coefficients as $(x + \sqrt{2})(x - \sqrt{2})$.

a. Factor the following polynomials into factors of degree 1, using irrational coefficients if necessary.

    i. $y^2 - 20$ **ii.** $2n^2 - 7$

    iii. $4p^3 - 6p$ **iv.** $a^4 - 10a^2 + 25$

Although binomials of the form $a^2 + b^2$ cannot be factored using only real coefficients, they do factor using *complex* coefficients.

b. Use complex multiplication (see Section 1.3) to show that $(a - bi)(a + bi) = a^2 + b^2$.

c. Factor the following polynomials into factors of degree 1, using complex coefficients if necessary.

    i. $x^2 + 25$ **ii.** $2z^4 - 32$

65. *Zeros of Polynomials*   There is a strong connection between factoring a polynomial of one variable and finding its roots or zeros. A number c is called a *root* or *zero* of a polynomial if the value of the polynomial is zero when c is substituted in for the variable. For example, 2 is a zero of the polynomial $x^3 - x^2 + x - 6$ since $2^3 - 2^2 + 2 - 6 = 0$.

a. Show that 1 and -3 are zeros of $x^2 + 2x - 3$. Now factor $x^2 + 2x - 3$. Do you notice any connection between the zeros and the factors? Do you think there are any more zeros?

b. Factor the polynomial $x^2 - 3x - 10$ to help find its zeros.

c. The zeros of the polynomial $x^3 - 13x - 12$ are -3, -1, and 4. Use this information to factor the polynomial.

d. Find a fourth-degree polynomial with zeros -1, 0, 1, and 2.

66. *A Better Educated Guess*   The process of educated guessing can lead to frustration when the coefficients of the first and last terms have many factors. In such situations, the number of possible factors can be overwhelming.

a. Three "possible" factorizations of $6x^2 + 7x - 20$ follow. How many are there all together?

$$(x + 1)(6x - 20), \ (x - 1)(6x + 20), \ (2x + 1)(3x - 20), \ldots$$

Fortunately, there are other ways of obtaining factorizations for trinomials, one of which is an application of the grouping method. We'll illustrate the process with an example. Consider the trinomial $2x^2 - x - 6$. First we multiply the coefficients of the first and last terms, obtaining -12. Next, we find the pairs of factors of -12 that add up to -1, the coefficient of the middle term. The list of all possible pairs is $(-1)12$, $1(-12)$, $(-2)6$, $2(-6)$, $(-3)4$, and $3(-4)$. The two that add up to -1 are -4 and 3. Now we write $2x^2 - x - 6$ as $2x^2 - 4x + 3x - 6$ and apply the grouping technique to obtain the following factorization.

$$2x^2 - x - 6 = (2x^2 - 4x) + (3x - 6)$$
$$= 2x(x - 2) + 3(x - 2)$$
$$= (x - 2)(2x + 3)$$

b. Apply the process just described to factor the following trinomials.

   i. $3x^2 + 2x - 8$ ii. $6x^2 + 7x - 20$

   iii. $4x^2 + 4xy - 15y^2$

This process can even tell us when a trinomial cannot be factored using only integer coefficients. If none of the factors of the prod-

uct of the first and last term add up to the middle term, the trinomial is not factorable.

c. Determine whether any of the following trinomials are not factorable using integer coefficients.

   i. $6x^2 + 14x - 9$ ii. $12x^2 - 7x + 8$

   iii. $20x^2 + 9xy - 4y^2$

You may be wondering how this mysterious factorization method works. Let's consider the product of two arbitrary binomials, $(ax + b)(cx + d) = (ac)x^2 + (ad + bc)x + bd$. If we were to apply the factorization process to the trinomial $(ac)x^2 + (ad + bc)x + bd$, we would multiply the first term ac by the last term bd and then list all the factors, selecting the pair that add up to the middle term $ad + bc$.

d. List all the factors of $(ac)(bd)$ and identify the two that add up to the middle term.

e. Finish the grouping process to show that the polynomial $(ac)x^2 + (ad + bc)x + bd$ does indeed factor as

$$(ax + b)(cx + d).$$

SECTION 1.7   RATIONAL EXPRESSIONS

☐ What possible effect could the planet Mars have on the birth of a child?

☐ How can the manager of a fast-food restaurant determine the average time customers must wait in the drive-through line?

☐ If a ball dropped from 10 feet bounces to a height of 5 feet, and then to a height of $2\frac{1}{2}$ feet, and so forth, how far will the ball have traveled at the end of time?

☐ How can all of the details of an installment loan be computed using only arithmetic?

☐ How is hydrostatic weighing used to calculate body fat percentage?

SIMPLIFYING RATIONAL EXPRESSIONS

A rational expression is simply a quotient of polynomials, such as

$$\frac{x^2 - 4}{x^2 - 3x + 2}$$

In fact, since integers are polynomials, a rational expression can be as simple as the fraction $\frac{2}{3}$. Thus, the rules for operating on rational expressions must also apply to operating on fractions. We review some of these rules in Table 11.

Note that Rule 4 may only be applied when adding fractions that have the same denominator. If the denominators are different, we must first find a *common denomi-*

TABLE 11

Rule	Example	Comment
1. $\dfrac{ac}{bc} = \dfrac{a}{b}$	$\dfrac{16}{24} = \dfrac{2 \cdot 8}{3 \cdot 8} = \dfrac{2}{3}$	Divide out the common factor.
2. $\dfrac{a}{b} \cdot \dfrac{c}{d} = \dfrac{ac}{bd}$	$\dfrac{4}{5} \cdot \dfrac{3}{7} = \dfrac{4 \cdot 3}{5 \cdot 7} = \dfrac{12}{35}$	Multiply numerators and denominators.
3. $\dfrac{a}{b} \div \dfrac{c}{d} = \dfrac{a}{b} \cdot \dfrac{d}{c}$	$\dfrac{2}{3} \div \dfrac{7}{11} = \dfrac{2}{3} \cdot \dfrac{11}{7} = \dfrac{22}{21}$	Multiply by the reciprocal of the divisor. (Invert and multiply.)
4. $\dfrac{a}{b} + \dfrac{c}{b} = \dfrac{a + c}{b}$	$\dfrac{4}{5} + \dfrac{7}{5} = \dfrac{4 + 7}{5} = \dfrac{11}{5}$	Add numerators over a common denominator.
5. $\dfrac{a}{b} - \dfrac{c}{b} = \dfrac{a - c}{b}$	$\dfrac{4}{3} - \dfrac{11}{3} = \dfrac{4 - 11}{3} = \dfrac{-7}{3}$	Subtract numerators over a common denominator.

nator by finding the least common multiple of the two denominators. For example, to add $\frac{5}{12}$ and $\frac{7}{20}$, we note that $12 = 3 \cdot 4$ and $20 = 4 \cdot 5$, so $3 \cdot 4 \cdot 5 = 60$ is the least common denominator. Thus,

$$\frac{5}{12} + \frac{7}{20} = \frac{5 \cdot 5}{12 \cdot 5} + \frac{7 \cdot 3}{20 \cdot 3} = \frac{25 + 21}{60} = \frac{46}{60} = \frac{23}{30}$$

The rules for operations on rational expressions are identical to those in Table 11; we simply allow the variables *a, b, c,* and *d* to represent polynomials. Moreover, just as factoring is the key to working with fractions, so too is factoring the key to working with rational expressions.

E X A M P L E 1 ▦ **Simplifying a Rational Expression by Factoring**

Simplify the rational expression

$$\frac{x^2 - 4}{x^2 - 3x + 2}$$

by dividing out any common factors.

SOLUTION

$$\frac{x^2 - 4}{x^2 - 3x + 2} = \frac{(x + 2)(x - 2)}{(x - 1)(x - 2)} \qquad \text{Factoring both numerator and denominator}$$

$$= \frac{x + 2}{x - 1} \qquad \text{Dividing out the common factor } x - 2$$

WARNING! Although we can divide out common factors of the numerator and denominator of a rational expression, we cannot divide out common *terms*. Thus, for example, we cannot begin the solution of Example 1 by canceling x^2 in both numerator and denominator.

Do cancel common factors . . .

$$\frac{a \cdot b}{a \cdot c} = \frac{\cancel{a} \cdot b}{\cancel{a} \cdot c} = \frac{b}{c}$$

$$\frac{x^2 \cdot y^2}{x \cdot y} = \frac{x^{\cancel{2}} \cdot y^{\cancel{2}}}{\cancel{x} \cdot \cancel{y}} = xy$$

$$\frac{xy}{xyz} = \frac{\cancel{xy}}{\cancel{xy}z} = \frac{1}{z}$$

Don't cancel common terms

$$\frac{a + b}{a + c} \neq \frac{\cancel{a} + b}{\cancel{a} + c} \quad \text{and so} \quad \frac{a + b}{a + c} \neq \frac{b}{c}$$

$$\frac{x^2 + y^2}{x + y} \neq \frac{x^{\cancel{2}} + y^{\cancel{2}}}{\cancel{x} + \cancel{y}} \quad \text{and so} \quad \frac{x^2 + y^2}{x + y} \neq x + y$$

$$\frac{x + y}{x + y + z} \neq \frac{\cancel{x} + \cancel{y}}{\cancel{x} + \cancel{y} + z} \quad \text{and so} \quad \frac{x + y}{x + y + z} \neq \frac{1}{z}$$

EXAMPLE 2 ■ **Simplifying a Rational Expression by Factoring**

Simplify the rational expression

$$\frac{a^2b^3 - a^2}{a^4 - a^4b}$$

by dividing out any common factors.

SOLUTION

$$\frac{a^2b^3 - a^2}{a^4 - a^4b} = \frac{a^2(b^3 - 1)}{a^4(1 - b)} \qquad \text{Factoring both numerator and denominator}$$

$$= \frac{b^3 - 1}{a^2(1 - b)} \qquad \text{Dividing out the common factor } a^2$$

$$= \frac{(b - 1)(b^2 + b + 1)}{a^2(1 - b)} \qquad \text{Factoring the numerator again}$$

$$= \frac{-(1 - b)(b^2 + b + 1)}{a^2(1 - b)} \qquad \text{Recognizing that } (b - 1) = -(1 - b)$$

$$= -\frac{b^2 + b + 1}{a^2} \qquad \text{Dividing out the common factor } (1 - b)$$

The final expressions in Examples 1 and 2 are said to be in **lowest terms** because no other common factors can be divided out. Whenever operations are performed on rational expressions, it is desirable to write the final answer in lowest terms.

EXAMPLE 3 ■ **Multiplying Rational Expressions**

Express the following product in lowest terms.

$$\frac{y - 1}{y + 1} \cdot \frac{y^2 - y - 2}{3y^2 - 3y}$$

SOLUTION

$$\frac{y - 1}{y + 1} \cdot \frac{y^2 - y - 2}{3y^2 - 3y} = \frac{(y - 1)(y^2 - y - 2)}{(y + 1)(3y^2 - 3y)}$$ Multiplying numerators and denominators

$$= \frac{(y - 1)(y + 1)(y - 2)}{(y + 1)3y(y - 1)}$$ Factoring to check for common factors

$$= \frac{y - 2}{3y}$$ Dividing out the common factors $(y + 1)$ and $(y - 1)$

E X A M P L E   4   ■   **Multiplying Rational Expressions**

Express the following product in lowest terms.

$$(n^2 - 16) \cdot \frac{n}{n^3 - 64}$$

SOLUTION

$$(n^2 - 16) \cdot \frac{n}{n^3 - 64} = \frac{n^2 - 16}{1} \cdot \frac{n}{n^3 - 64}$$ Writing $n^2 - 16$ with a denominator of 1

$$= \frac{(n^2 - 16)n}{n^3 - 64}$$ Multiplying numerators and denominators

$$= \frac{(n - 4)(n + 4)n}{(n - 4)(n^2 + 4n + 16)}$$ Factoring to check for any common factors

$$= \frac{(n + 4)n}{n^2 + 4n + 16}$$ Dividing out the common factor $(n - 4)$

$$= \frac{n^2 + 4n}{n^2 + 4n + 16}$$

Note that we **cannot** simplify further by dividing out $n^2 + 4n$ since this expression is not a factor of the denominator.

We divide rational expressions the same way we divide fractions: by multiplying by the reciprocal of the divisor. That is, we "invert and multiply."

E X A M P L E   5   ■   **Dividing Rational Expressions**

Express the following quotient in lowest terms.

$$\frac{r^2 - 36}{r^2 + r} \div \frac{6 - r}{r}$$

SOLUTION

$$\frac{r^2 - 36}{r^2 + r} \div \frac{6 - r}{r} = \frac{r^2 - 36}{r^2 + r} \cdot \frac{r}{6 - r}$$ Multiplying by the reciprocal of the divisor

$$= \frac{(r + 6)(r - 6)}{r(r + 1)} \cdot \frac{r}{-(r - 6)}$$ Factoring and writing $(6 - r)$ as $-(r - 6)$

$$= \frac{r + 6}{-(r + 1)}$$ Dividing out common factors

$$= -\frac{r + 6}{r + 1}$$

EXAMPLE 6   ■   **Dividing Rational Expressions**

Express the following quotient in lowest terms.

$$\frac{\dfrac{k}{k + 4}}{k - 4}$$

SOLUTION

$$\frac{\dfrac{k}{k + 4}}{k - 4} = \frac{\dfrac{k}{k + 4}}{\dfrac{k - 4}{1}}$$ Writing the denominator $k - 4$ as a rational expression with denominator 1

$$= \frac{k}{k + 4} \cdot \frac{1}{k - 4}$$ Multiplying by the reciprocal of the denominator

$$= \frac{k}{(k + 4)(k - 4)}$$

ADDING AND SUBTRACTING RATIONAL EXPRESSIONS

When adding or subtracting rational expressions, such as

$$\frac{1}{t^2 + t} + \frac{2t}{t^2 - 1}$$

we must first find a common denominator. As is the case with adding fractions, we are looking for the least common multiple of the denominators. This can be found by factoring the denominators, choosing the largest power of each factor that appears, and forming the product of all the factors with the appropriate powers.

EXAMPLE 7   ■   **Finding the Least Common Denominator**

Find the least common denominator for

$$\frac{1}{x(x - 2)^3} \quad \text{and} \quad \frac{x + 1}{x^2(x - 2)^2}$$

SOLUTION   Choosing the largest power of each factor, we obtain the least common denominator, $x^2(x - 2)^3$.

EXAMPLE 8   ■   **Finding the Least Common Denominator**

Find the least common denominator for

$$\frac{1}{u^2 + 2u - 3}, \quad \frac{1}{u^2 + 6u + 9}, \quad \text{and} \quad \frac{1}{u^2 - 2u + 1}$$

SOLUTION   We first factor each denominator.

$$u^2 + 2u - 3 = (u + 3)(u - 1)$$
$$u^2 + 6u + 9 = (u + 3)^2$$
$$u^2 - 2u + 1 = (u - 1)^2$$

Now, choosing the largest power of each factor, we obtain the least common denominator, $(u + 3)^2(u - 1)^2$.

EXAMPLE 9 ■ Subtracting Rational Expressions

Write the difference

$$\frac{x}{3-x} - \frac{x+1}{x-3}$$

in lowest terms.

SOLUTION

$$\frac{x}{3-x} - \frac{x+1}{x-3} = \frac{x}{-(x-3)} - \frac{x+1}{x-3} \qquad \text{Recognizing that } 3-x = -(x-3)$$

$$= \frac{-x}{x-3} - \frac{x+1}{x-3} \qquad \text{The common denominator is } x-3$$

$$= \frac{-x-(x+1)}{x-3}$$

$$= \frac{-2x-1}{x-3} \qquad \text{Distributing and simplifying}$$

EXAMPLE 10 ■ Adding Rational Expressions

Write the sum

$$\frac{1}{t^2+t} + \frac{2t}{t^2-1}$$

in lowest terms.

SOLUTION   Since $t^2 + t = t(t+1)$ and $t^2 - 1 = (t+1)(t-1)$, the least common denominator is $t(t+1)(t-1)$. Thus, we proceed as follows:

$$\frac{1}{t^2+t} + \frac{2t}{t^2-1} = \frac{1}{t(t+1)} + \frac{2t}{(t+1)(t-1)} \qquad \text{Factoring the denominators}$$

$$= \frac{1}{t(t+1)} \cdot \frac{t-1}{t-1} + \frac{2t}{(t+1)(t-1)} \cdot \frac{t}{t} \qquad \text{Obtaining the common denominator}$$

$$= \frac{t-1}{t(t+1)(t-1)} + \frac{2t^2}{t(t+1)(t-1)}$$

$$= \frac{2t^2+t-1}{t(t+1)(t-1)} \qquad \text{Adding numerators}$$

To write this expression in lowest terms, we must factor the numerator and divide out any common factors.

$$\frac{2t^2+t-1}{t(t+1)(t-1)} = \frac{(2t-1)(t+1)}{t(t+1)(t-1)}$$

$$= \frac{2t-1}{t(t-1)}$$

EVALUATING RATIONAL EXPRESSIONS

When evaluating a rational expression for particular values of the variables, we must be careful not to allow values that will lead to a zero in the denominator. For example, in the expression

$$\frac{2}{z^2(z - 1)(z + 1)}$$

we would not want to allow z to be 0, 1, or -1 since these values would give 0 in the denominator. In applied contexts, interesting phenomena often occur as we approach values where the denominator is zero.

EXAMPLE 11

 Cleaning Up Pollution

As many cities and companies have discovered in recent years, the cost of cleaning up pollution can be quite high. In fact, the cost typically increases at a greater rate as the percentage of pollutants removed gets closer to 100 percent. One model for the cost of removing pollutants suggests that if it costs \$1,000,000 to remove 50% of the pollutants, then to remove x percent it will cost

$$\frac{1,000,000}{2 - \dfrac{x}{50}} \text{ dollars}$$

Simplify this expression and compute the cost for removing 80%, 90%, and 95% of the pollutants.

Cleanup in the wake of the Chernobyl nuclear disaster of 1986. Total costs may exceed \$60 billion.

SOLUTION

$$\frac{1,000,000}{2 - \dfrac{x}{50}} = \frac{1,000,000}{\dfrac{100 - x}{50}}$$

$$= 1,000,000 \cdot \frac{50}{100 - x}$$

$$= \frac{50,000,000}{100 - x}$$

When $x = 80$, the cost is $\dfrac{50,000,000}{100 - 80} = \dfrac{50,000,000}{20} = 2,500,000$ dollars.

When $x = 90$, the cost is $\dfrac{50,000,000}{100 - 90} = \dfrac{50,000,000}{10} = 5,000,000$ dollars.

When $x = 95$, the cost is $\dfrac{50,000,000}{100 - 95} = \dfrac{50,000,000}{5} = 10,000,000$ dollars

Notice that the cost increases dramatically as we get closer to 100%—it doubles going from 80 to 90%, and then doubles again going from 90 to 95%. Notice also that the cost cannot even be evaluated when $x = 100$.

COMPLEX FRACTIONS

Often we are confronted—as we were in Example 6—with rational expressions whose numerators and/or denominators themselves contain rational expressions. Another example is the expression

$$\frac{\dfrac{1}{x} - \dfrac{1}{y}}{\dfrac{1}{x} + \dfrac{1}{y}}$$

Such expressions are called *complex fractions,* and in the following example we illustrate two methods for simplifying them.

EXAMPLE 12 ■ **Simplifying a Complex Fraction**

Simplify the rational expression

$$\frac{\dfrac{1}{x} - \dfrac{1}{y}}{\dfrac{1}{x} + \dfrac{1}{y}}$$

SOLUTION

METHOD 1: We first simplify the fractions in the main numerator and the main denominator.

$$\text{Main numerator:} \quad \frac{1}{x} - \frac{1}{y} = \frac{y}{xy} - \frac{x}{xy}$$

$$= \frac{y - x}{xy}$$

$$\text{Main denominator:} \quad \frac{1}{x} + \frac{1}{y} = \frac{y}{xy} + \frac{x}{xy}$$

$$= \frac{y + x}{xy}$$

After replacing the main numerator and denominator with their simplified expressions, we invert and multiply.

$$\frac{\dfrac{1}{x} - \dfrac{1}{y}}{\dfrac{1}{x} + \dfrac{1}{y}} = \frac{\dfrac{y - x}{xy}}{\dfrac{y + x}{xy}} \qquad \text{Replacing the main numerator and denominator}$$

$$= \frac{y - x}{xy} \cdot \frac{xy}{y + x} \qquad \text{Inverting and multiplying}$$

$$= \frac{y - x}{y + x} \qquad \text{Simplifying}$$

METHOD 2: We can clear all fractions in one step simply by multiplying the main numerator and main denominator by the least common multiple of all denominators in the expression. In this case, the least common multiple is xy, and so we proceed as follows:

$$\frac{\dfrac{1}{x} - \dfrac{1}{y}}{\dfrac{1}{x} + \dfrac{1}{y}} = \frac{\left(\dfrac{1}{x} - \dfrac{1}{y}\right)xy}{\left(\dfrac{1}{x} + \dfrac{1}{y}\right)xy} \qquad \text{Multiplying numerator and denominator by } xy$$

$$= \frac{\dfrac{1}{x}xy - \dfrac{1}{y}xy}{\dfrac{1}{x}xy + \dfrac{1}{y}xy} \qquad \text{Distributing}$$

$$= \frac{y - x}{y + x} \qquad \text{Simplifying}$$

UNDERSTANDING AND MASTERY CHECKLISTS

CONCEPTS TO UNDERSTAND

- ☐ Rational expression
- ☐ Lowest terms
- ☐ Operations with rational expressions
- ☐ Common denominator
- ☐ Complex fractions

SKILLS TO MASTER

- ☐ Identify a rational expression.
- ☐ Simplify a rational expression by factoring.
- ☐ Add, subtract, multiply, and divide rational expressions.
- ☐ Evaluate a rational expression by substituting.
- ☐ Simplify complex fractions by clearing denominators.

EXERCISES 1.7

EXERCISES 1–12   *Simplify whenever possible.*

1. $\dfrac{9ab}{15bc}$

2. $\dfrac{12xy^2z^3}{16x^2yz^2}$

3. $\dfrac{4t^6 - 10t^4}{2t^2}$

4. $\dfrac{21n^4 + 15n^3}{6n^3}$

5. $\dfrac{x^2 - 4}{x + 2}$

6. $\dfrac{4z^2 - 9}{2z + 3}$

7. $\dfrac{r^2 + 10r + 25}{2r^2 + 4r - 30}$

8. $\dfrac{3a^2 - 6a - 24}{a^2 - 6a + 8}$

9. $\dfrac{y^2 - 3y}{y - 3}$

10. $\dfrac{s^2t^3 - s^5}{st^2 - s^3}$

11. $\dfrac{x^4 - y^4}{x^3 - x^2y + xy^2 - y^3}$

12. $\dfrac{n^3 + 1}{n^2 + 1}$

EXERCISES 13–38   *Perform the indicated operations and write your answer in simplest terms.*

13. $\dfrac{15x^2}{16y^3} \div \dfrac{12y^3}{20x^3}$

14. $\dfrac{8st^2}{25r^2t} \cdot \dfrac{10r^2t}{32s^3t^2}$

15. $\dfrac{1}{a + 1} + \dfrac{1}{a - 1}$

16. $\dfrac{u + v}{u - v} - \dfrac{u - v}{u + v}$

17. $\dfrac{z - 1}{z + 1} - \dfrac{z + 1}{z - 2}$

18. $\dfrac{2q + 1}{q^2 + q} - \dfrac{q + 2}{q + 1}$

19. $\dfrac{x - 1}{x + 2} \cdot \dfrac{x^2 - 4}{x^2 - 1}$

20. $\dfrac{x + 3}{x + 4} \div \dfrac{x^3 + 27}{x^2 - 16}$

21. $\dfrac{1}{x^2 + 2x} - \dfrac{1}{x^2 + 4x + 4}$

22. $\dfrac{a}{a^2 - 4} + \dfrac{1}{2 - a}$

23. $(2a^2 - 8b^2) \cdot \dfrac{b - a}{a^2 - 4ab + 4b^2}$

24. $y + \dfrac{1}{y + 1}$

25. $\dfrac{1 + t}{(1 - t)^2} + \dfrac{1 - t}{t^2 - 1}$

26. $\dfrac{4x + 20}{x^2 + 8x + 16} \cdot \dfrac{x^2 + 4x}{x^2 + 7x + 10}$

27. $\dfrac{\dfrac{r - s}{r + s}}{s^2 - r^2}$

28. $\dfrac{t + 2}{t^3 + t^2} - \dfrac{1}{t + 1}$

29. $\dfrac{a + 1}{a - 2} \cdot \dfrac{a^2 - 3a + 2}{a - 3} \cdot \dfrac{3 - a}{a^2 - 1}$

30. $\dfrac{x^2 + 2xy + y^2}{x^2 - y^2} \div \dfrac{x^2 - xy - 2y^2}{x^2 - 3xy + 2y^2}$

31. $\dfrac{3}{t - 6} + \dfrac{3}{t + 6} - \dfrac{36}{36 - t^2}$

32. $\dfrac{x + 5}{x^3 + 125} - \dfrac{1}{x + 5} + \dfrac{x - 5}{x^2 - 25}$

33. $\dfrac{\dfrac{1}{a} - b}{\dfrac{1}{b} - a}$

34. $\dfrac{\dfrac{1}{x} + \dfrac{1}{y}}{\dfrac{1}{xy}}$

35. $\dfrac{\dfrac{1}{3 + h} - \dfrac{1}{3}}{h}$

36. $\dfrac{\dfrac{1}{x^2} - \dfrac{1}{4}}{x - 2}$

37. $\dfrac{\dfrac{1}{x} - \dfrac{1}{c}}{x - c}$

38. $\dfrac{\dfrac{1}{x + h} - \dfrac{1}{x}}{h}$

EXERCISES 39–40   *Use the operations on rational expressions to reveal the "trick" in the following number games.*

39. Subtract 10 from your age, square the result, subtract 5 times your age, divide by 20 less than your age, and add 5. The result is your age. (There is one age for which the trick fails; what is this age?)

40. Take two numbers, subtract the square of their difference from the square of their sum, and divide by 4 times either number. The result is the other number. (Under what circumstances does this trick fail?)

EXERCISES 41–44   *An expression of the form*

$$a + \cfrac{1}{b + \cfrac{1}{c}}$$

*is an example of a **continued fraction**. For convenience, we define*

$$[a, b, c] = a + \cfrac{1}{b + \cfrac{1}{c}}$$

Thus, for example

$$[2, 5, 3] = 2 + \cfrac{1}{5 + \cfrac{1}{3}}$$

41. Evaluate $[6, 2, 1]$.

42. What is wrong with the expression $[2, 1, -1]$?

43. Write $[a, b, c]$ as a rational expression in lowest terms.

44. What is your guess as to the meaning of $[a, b, c, d]$? Simplify the expression you found and use it to evaluate $[1, 1, 1, 1]$.

APPLICATIONS

45. *Electrical Resistance*   If two resistors with resistance R_1 and R_2, respectively, are connected in parallel, the total resistance in the electrical circuit is given by

$$R = \cfrac{1}{\dfrac{1}{R_1} + \dfrac{1}{R_2}}$$

a. Find the total resistance in a parallel circuit with an 8-ohm and 4-ohm resistor.

b. Show that

$$R = \dfrac{R_1 R_2}{R_1 + R_2}$$

46. *Eternally Bouncing Ball*   A geometric series is a sum of the form $a + ar + ar^2 + ar^3 + \cdots + ar^n$. One application of such a sum is in computing the total distance traveled by a bouncing ball. For example, if a ball is dropped from a height of 10 feet and it rebounds to $\frac{1}{2}$ of its previous height after each bounce, the total distance traveled by the ball while falling, up until the nth bounce, would be

$$10 + 10\left(\dfrac{1}{2}\right)^1 + 10\left(\dfrac{1}{2}\right)^2 + 10\left(\dfrac{1}{2}\right)^3 + \cdots + 10\left(\dfrac{1}{2}\right)^{n-1}$$

Such a sum can be computed using a rational expression, as we'll see.

a. Show that $\dfrac{a(1 - r^2)}{1 - r}$ simplifies to $a + ar$.

b. Show that $\dfrac{a(1 - r^3)}{1 - r}$ simplifies to $a + ar + ar^2$.

c. What is your guess for the sum $a + ar + ar^2 + ar^3$? What about $a + ar + ar^2 + ar^3 + \cdots + ar^{n-1}$?

d. Write a simplified expression for the distance traveled by the ball while falling, up until the nth bounce.

e. Given that the distance traveled by the ball after the 20th bounce is negligible, and that the distance traveled while rising is 10 feet less than that while falling, estimate the total distance traveled by the ball.

47. *Gravitational Pull* Astrologists claim that the alignment of the planets at the time a person is born has an influence on the birth. Let's see what Newton's law of gravitation has to say about this. We will compute the gravitational pull between the planet Mars and a child born on February 12, 1995, which is the date in 1995 on which Mars was closest to Earth. According to Newton's law, the force of attraction in newtons between two bodies of mass m_1 and m_2 kilograms that are a distance r meters apart is

$$\frac{Gm_1m_2}{r^2}$$

where $G = 6.672 \times 10^{-11}$ N·m²/kg² is the universal gravitational constant.

a. Given that the mass of Mars is 6.42×10^{23} kilograms and its distance from Earth on February 12, 1995, is 10^{11} meters, compute the gravitational pull between Mars and a 3.4-kilogram baby born on February 12, 1995.

b. Compute the gravitational pull between an obstetrician with mass 70 kilograms and a 3.4-kilogram baby if the obstetrician is standing 1 meter away.

c. Which of the two bodies (Mars or the obstetrician) exerts more gravitational pull on the baby?

48. *Waiting Time* The manager of a fast-food restaurant is concerned about the amount of time a customer must wait in the drive-through line. One model suggests that, if s denotes the average number of minutes for the drive-through worker to process an order, and if c denotes the average time between customer arrivals, then the average waiting time is given by the rational expression

$$\frac{1}{\dfrac{1}{s} - \dfrac{1}{c}}$$

[*Data source*: Michael Mesterton-Gibbons, *A Concrete Approach to Mathematical Modeling* (Reading, MA: Addison-Wesley, 1989).]

a. Determine the average waiting time for the following values of c and s.

   i. $c = 10, s = 5$ 　　　　ii. $c = 10, s = 6$

   iii. $c = 10, s = 7$ 　　　iv. $c = 10, s = 8$

   v. $c = 10, s = 9$

   What happens to the waiting time as s and c get closer together? Does this make sense?

b. Try some examples to see what happens to the length of time if c is less than s. Does this agree with your intuition about waiting time? Explain. What can you conclude about the waiting-time model?

49. *Percentage Body Fat* Hydrostatic weighing is among the most accurate methods for measuring a person's percentage of body fat. First, the mass of the subject, m, is measured using an ordinary scale. Next, the subject is asked to expel as much air as possible from her lungs and is submerged in water. The submerged weight w is then recorded. The subject's body density is then approximated by

$$d = \frac{m}{m - w - 1}$$

where m is the body mass in kilograms and w is the underwater "weight" in kilograms (technically, a kilogram is a unit of mass). According to one widely used model of body composition, the percentage of body fat is approximated by

$$\text{Percentage body fat} = \frac{495}{d} - 450$$

a. Find a simplified expression for the percentage of body fat in terms of m and w.

b. Estimate the percentage of body fat for a 100-kilogram (220-pound) man with a submerged weight of 7 kilograms.

[*Data source*: William D. McArdle, Frank I. Katch, Victor L. Katch, *Exercise Physiology*, 4th ed. (Baltimore: Williams and Wilkins, 1996), p. 554.]

▮ CONCEPTS AND CRITICAL THINKING

EXERCISES 50–53 *Answer true or false.*

50. $\dfrac{\dfrac{x}{x+1}}{x}$ simplifies to $\dfrac{x^2}{x+1}$

51. $\dfrac{2}{t+1} + \dfrac{3}{t+2}$ simplifies to $\dfrac{5}{2t+3}$

52. $\dfrac{x+1}{x^2+1}$ is in lowest terms

53. $\dfrac{2}{x-1} + \dfrac{2}{1-x}$ simplifies to 0

EXERCISES 54–57 *Give an example of each.*

54. A rational expression consisting of a second-degree polynomial divided by a third-degree polynomial that is equivalent to $\frac{2}{x}$

55. A pair of rational expressions having different denominators and least common denominator $x + 3$

56. A complex fraction that is equivalent to 3

57. A trinomial divided by a binomial that is equivalent to $\dfrac{x+3}{x+2}$

◼ QUESTIONS FOR DISCUSSION OR ESSAY

58. In what instances will the quotient of polynomials be a polynomial?

59. It is not uncommon for modern college algebra students to struggle with the arithmetic of fractions, even though most of us were taught this subject in middle school. Why do you suppose that students of today might have more difficulty with this subject than students of 25 years ago? Now that you've gained some mastery with manipulating rational expressions, do you find fraction arithmetic any easier?

60. In Example 11, we looked at a mathematical model for predicting the cost of cleaning up pollution in terms of the percentage of pollutants to be removed. Do you think this model is valid? What would you say to the oil company executive who states, "Because of the *mathematically proven* unfeasibility of cleaning up 100% of the oil spill, we can only clean 85% of the spill and must leave the rest to mother nature . . ."? More generally, what do you think mathematical models such as this say about the likelihood of removing the existing pollution in our world?

◼ PROJECTS FOR ENRICHMENT

61. *Averages*   The media and advertisers use the term *average* in many different contexts. It's not uncommon to hear or read phrases such as "the average annual income is $40,000" or "the average teenager spends 2 hours each day on the phone." What exactly is meant by the term *average,* and how is it computed in each case? It may surprise you to learn that this term is subject to multiple interpretations. The measure that is most commonly referred to as an average is more properly called the *arithmetic mean.* The arithmetic mean of *n* numbers $a_1, a_2, \ldots, a_n$ is given by

$$A = \frac{a_1 + a_2 + \cdots + a_n}{n}$$

a. During the first three basketball games of the season, Stephanie scored 12, 16, and 13 points. During the next four games, she scored 16, 22, 17, and 19 points. Compute her average for the first three games and then for the next four. How does the average of the two averages compare to the average over the seven games?

b. Assuming that *A* is the arithmetic mean of *m* numbers $a_1, a_2, \ldots, a_m$ and *B* is the arithmetic mean of *n* numbers $b_1, b_2, \ldots, b_n$, find an expression for the arithmetic mean of *A* and *B*. Use the operations on rational expressions to show that the arithmetic mean of *A* and *B* is not necessarily the same as the arithmetic mean of the *m* + *n* numbers $a_1, a_2, \ldots, a_m, b_1, b_2, \ldots, b_n$.

Another type of average is called the *weighted mean.* The weighted mean of *n* numbers $a_1, a_2, \ldots, a_n$ that are weighted by the factors $w_1, w_2, \ldots, w_n$ is given by

$$W = \frac{a_1 w_1 + a_2 w_2 + \cdots + a_n w_n}{w_1 + w_2 + \cdots + w_n}$$

c. David's scores on the three exams in college algebra were 76, 84, and 80, respectively. On the final, he scored 66. If each exam counts 20% and the final counts 40%, David's weighted average is

$$\frac{76 \cdot 20 + 84 \cdot 20 + 80 \cdot 20 + 66 \cdot 40}{20 + 20 + 20 + 40} = \frac{7440}{100} = 74.4$$

Find his weighted average if each exam counts 30% and the final counts 10%.

d. Assuming that each of the weighting factors is the same—*w,* say—find a simplified expression for the weighted mean of *n* numbers $a_1, a_2, \ldots, a_n$. Does it look familiar?

A third type of average is called the *harmonic mean.* The harmonic mean of *n* numbers $a_1, a_2, \ldots, a_n$ is given by

$$H = \frac{n}{\dfrac{1}{a_1} + \dfrac{1}{a_2} + \cdots + \dfrac{1}{a_n}}$$

e. Suppose that you and a friend are given the task of purchasing the beverages for a rather large party. You spend $48 on beverages costing $3 per 12-pack and your friend spends $48 on beverages costing $4 per 12-pack. The average price is *not* ($3 + $4)/2 = $3.50 per 12-pack. Instead, since a total of $96 is spent on $\frac{48}{3} + \frac{48}{4} = 28$ 12-packs, the actual cost is $\frac{96}{28}$ = $3.43 per 12-pack. Show that the harmonic mean of $3 and $4 is equal to $3.43.

f. Show that the harmonic mean of 3 numbers *a, b,* and *c* is

$$H = \frac{3abc}{bc + ac + ab}.$$

62. *Installment Loans*   Have you ever wondered how much the monthly payments would be for a certain car loan? Or have you ever wondered how much you could borrow if you can only afford to pay back a certain amount each month? It may not surprise you to hear that there are standard formulas for determining such things. And it may not surprise you to hear that these formulas are rather complex and would likely intimidate the "algebraically challenged" among us. In a recent book entitled *The Complete How to Figure It* (New York: W. W. Norton, 1996), author Darrell Huff avoids the use of complex algebraic formulas and instead describes verbally how certain loan calculations can be made. In this project, we compare these verbal descriptions with their algebraic counterparts.

An installment loan is a loan that is paid back in regular—usually monthly—equal payments. Each payment includes interest on the remaining unpaid balance. Two examples of installment loans are mortgage loans (for the purchase of a house) and auto loans. The following steps paraphrase those given by Huff to determine the amount of an installment payment if the annual interest rate, the number of payments per year, and the number of years for the loan are known (Huff, p. 123). (Note that we will assume payments are made at the end of each payment period and that interest is compounded when payments are due.)

1. Divide the annual interest rate by the number of payments per year. Save the result for the next step and a future step.
2. Add 1 to the result of step 1.
3. Raise the result of step 2 to the power equal to the total number of payments to be made (the product of the number of years and the number of payments per year.) Save the result for the next step and for a future step.
4. Subtract 1 from the result of step 3.
5. Divide the result of step 3 by the result of step 4.
6. Multiply the result of step 5 by the amount of the original loan.
7. Multiply the result of step 6 by your result from step 1. The answer is the amount of the installment payment.

a. Use these steps to determine the amount of a *monthly* installment payment if you borrow $20,000 at an annual interest rate of 8% for 10 years (a total of 120 payments). Write down the result of each step.

b. Let P denote the original amount of the loan, let r denote the annual interest rate, let n denote the number of payments per year, and let t denote the number of years. Use the seven steps just given to find an algebraic formula for computing the payment amount A.

c. Use the formula you found in part b to find the amount of an installment payment for the following loans:
  i. Monthly payments on $10,000 at 9% for 5 years
  ii. Bimonthly (twice per month) payments on $70,000 at 8.25% for 15 years

The following formula gives the amount of a loan P in terms of the payment amount A, the annual interest rate r, the number of payments per year n, and the number of years t.

$$ P = \frac{An}{r} \left[1 - \frac{1}{\left(1 + \frac{r}{n} \right)^{nt}} \right] $$

d. Find the amount of a loan with the following payments:
  i. A $300 monthly payment for 5 years at 7.75%
  ii. A $250 bimonthly payment for 15 years at 8%

e. Write out a step-by-step verbal procedure for computing the amount of a loan if the payment amount, the annual interest rate, the number of payments per year, and the number of years are known. Avoid the use of any algebraic symbols.

f. Compare and contrast verbal and algebraic methods for describing loan calculations.

SECTION 1.8 **RADICALS AND RATIONAL EXPONENTS**

> ☐ **What does it mean to multiply a number by itself one-third times?**
> ☐ **At what interest rate will $1 grow to $1000 in 15 years?**
> ☐ **How can a woman fall more than 5 miles without a parachute and live?**
> ☐ **The skid marks at the scene of an accident measure 200 feet. How fast was the car traveling?**
> ☐ **A woman on her way to a local restaurant to celebrate her thirtieth birthday picks up her great-great-great-great-grandson at his retirement home, and then stops off at the bank to withdraw $1,000,000 from the account that she established by depositing $1000 on the previous day. How can this be?**

RADICALS

In Section 1.4, we introduced *exponentiation* as a shorthand notation for repeated multiplication of a given base, so that, for example, $2^5 = 2 \cdot 2 \cdot 2 \cdot 2 \cdot 2 = 32$. *Roots* result from reversing this process: An *n*th root of x is a number whose *n*th power is x. Thus, for example, since 32 is the fifth power of 2, it follows that 2 is a fifth root of 32.

Definition of nth Root ☐

> A number y is an **nth root** of x if $y^n = x$. If $n = 2$, the root is called a **square root.** If $n = 3$, the root is called a **cube root.**

Since $2^2 = 4$, and also $(-2)^2 = 4$, both 2 and -2 are square roots of 4. The positive root, 2, is called the **principal square root** of 4. Whenever n is even, there will be two nth roots, one positive and one negative.

Definition of Principal Root ☐

> For n odd and x any real number, the **principal nth root** of x is the real number y such that $y^n = x$. For n *even* and $x > 0$, the **principal nth root** of x is the *positive* real number y such that $y^n = x$. The principal nth root of x is denoted by $\sqrt[n]{x}$.

EXAMPLE 1 ▦ **Finding *n*th Roots**

Find the given roots of 16.

a. 4th   **b.** principal 4th

SOLUTION
a. Since $2^4 = 16$ and $(-2)^4 = 16$, both 2 and -2 are 4th roots of 16.
b. The principal 4th root is the *positive* 4th root—namely, 2.

Radical Notation and ☐
Terminology

> In the expression $\sqrt[n]{x}$, the symbol $\sqrt{}$ is called a **radical,** the number n is called the **index** of the radical, and x is called the **radicand.** If no index is given, it is understood to be 2. Thus $\sqrt{x} = \sqrt[2]{x}$.

EXAMPLE 2 ▦ **Evaluating Radicals**

Evaluate each of the following radical expressions.

a. $\sqrt[3]{27}$ **b.** $\sqrt{121}$

c. $\sqrt[5]{-32}$ **d.** $\sqrt[4]{\dfrac{1}{81}}$

SOLUTION
a. $\sqrt[3]{27} = 3$, since $3^3 = 27$.
b. $\sqrt{121} = 11$, since $11^2 = 121$ and $11 > 0$.
c. $\sqrt[5]{-32} = -2$, since $(-2)^5 = -32$.
d. $\sqrt[4]{\dfrac{1}{81}} = \dfrac{1}{3}$, since $\left(\dfrac{1}{3}\right)^4 = \dfrac{1}{81}$, and $\dfrac{1}{3} > 0$.

> **WARNING!**   The *principal root* always refers to a single number. For example, $\sqrt{25}$ is **not** equal to ± 5, but rather $\sqrt{25}$ is equal to 5.

SIMPLIFYING RADICAL EXPRESSIONS

The rules for simplifying radical expressions can be obtained from the definition of principal root and the properties of exponents. The most commonly used properties of radicals are listed here.

Properties of Radicals ☐

Assume that m and n are positive integers and that x and y are such that all radicals are real.

Property	Examples	Comment
1. $(\sqrt[n]{x})^n = x$	$(\sqrt[3]{2})^3 = 2$	This is just the definition of nth root.
2. For n odd: $\sqrt[n]{x^n} = x$	$\sqrt[3]{7^3} = 7$ $\sqrt[5]{x^{15}} = x^3$	In general, $\sqrt[n]{x^{mn}} = \sqrt[n]{(x^m)^n} = x^m$.
For n even: $\sqrt[n]{x^n} = \lvert x \rvert$	$\sqrt{3^2} = 3$ $\sqrt[4]{(-3)^4} = 3$ $\sqrt{x^2} = \lvert x \rvert$	The absolute value arises because of the fact that the *principal root* of a positive number is positive.
3. $\sqrt[n]{x}\,\sqrt[n]{y} = \sqrt[n]{xy}$	$\sqrt{3}\,\sqrt{5} = \sqrt{15}$	The indices of the radicals must be the same.
4. $\dfrac{\sqrt[n]{x}}{\sqrt[n]{y}} = \sqrt[n]{\dfrac{x}{y}}$	$\dfrac{\sqrt[5]{30}}{\sqrt[5]{5}} = \sqrt[5]{6}$	The indices of the radicals must be the same.
5. $(\sqrt[n]{x})^m = \sqrt[n]{x^m}$	$(\sqrt[3]{5})^2 = \sqrt[3]{5^2}$	This can be derived from Property 3.
6. $\sqrt[m]{\sqrt[n]{x}} = \sqrt[mn]{x}$	$\sqrt[4]{\sqrt[3]{x}} = \sqrt[12]{x}$	This property reminds us of $(x^m)^n = x^{mn}$.

Generally, a radical expression is considered to be completely simplified whenever

- The radicand is as ''small'' as possible; that is, the radicand contains as few factors as possible. For example, $2\sqrt[3]{2}$ is simpler than $\sqrt[3]{16}$. In general, this means that the radicand should contain no factors that are perfect nth powers, where n is the index.
- There are as few **nested radicals** (radicals inside of other radicals) as possible. For example, $\sqrt[15]{x}$ is simpler than $\sqrt[3]{\sqrt[5]{x}}$.
- The index of the radical is as small as possible. For example, $\sqrt{5}$ is simpler than $\sqrt[4]{25}$.
- There are no radicals in denominators. For example, $\sqrt{2}/2$ is simpler than $1/\sqrt{2}$.

E X A M P L E 3 ▨   **Simplifying Radical Expressions**

Simplify the following expressions.

a. $\sqrt{8}$ 　　　　　　　　　　　**b.** $\sqrt[3]{40}$

SOLUTION

a. 　　　　　　$\sqrt{8} = \sqrt{4 \cdot 2}$ 　　　　Factoring out the largest perfect square

　　　　　　　　　$= \sqrt{4}\,\sqrt{2}$ 　　　　Using Property 3

　　　　　　　　　$= 2\sqrt{2}$ 　　　　　Simplifying

b. 　　　　　　$\sqrt[3]{40} = \sqrt[3]{8 \cdot 5}$ 　　　　Factoring out the largest perfect cube

　　　　　　　　　$= \sqrt[3]{8}\,\sqrt[3]{5}$ 　　　　Using Property 3

　　　　　　　　　$= 2\sqrt[3]{5}$ 　　　　Simplifying

E X A M P L E 4 ▨   **Simplifying a Cube Root Involving Variables**

Simplify $\sqrt[3]{16x^7y^9z^5}$.

SOLUTION

$\sqrt[3]{16x^7y^9z^5} = \sqrt[3]{8x^6y^9z^3 \cdot 2xz^2}$ 　　　　Factoring out the largest perfect cubes

　　　　　　　　　$= \sqrt[3]{8}\,\sqrt[3]{x^6}\,\sqrt[3]{y^9}\,\sqrt[3]{z^3}\,\sqrt[3]{2xz^2}$ 　　　　Using Property 3

　　　　　　　　　$= \sqrt[3]{8}\,\sqrt[3]{(x^2)^3}\,\sqrt[3]{(y^3)^3}\,\sqrt[3]{z^3}\,\sqrt[3]{2xz^2}$

　　　　　　　　　$= 2x^2y^3z\,\sqrt[3]{2xz^2}$ 　　　　Using Property 2

Note that this cannot be simplified any further; the radicand contains no perfect cubes.

E X A M P L E 5 ▨   **Simplifying a Square Root Involving Variables**

Simplify $\sqrt{75a^3b^5c^6}$. Assume all variables represent positive quantities.

SOLUTION　　We begin by factoring out the perfect squares.

$\sqrt{75a^3b^5c^6} = \sqrt{25a^2b^4c^6 \cdot 3ab}$ 　　　　Factoring out the perfect squares

　　　　　　　　$= \sqrt{25}\,\sqrt{a^2}\,\sqrt{b^4}\,\sqrt{c^6}\,\sqrt{3ab}$ 　　　　Using Property 3

　　　　　　　　$= 5\,|a|\,|b^2|\,|c^3|\,\sqrt{3ab}$ 　　　　Using Property 2

　　　　　　　　$= 5ab^2c^3\sqrt{3ab}$ 　　　　Assuming all variables represent
　　　　　　　　　　　　　　　　　　　positive quantities

Rational expressions with radicals in the denominator are simplified by **rationalizing the denominator**. Rationalizing the denominator consists of multiplying both numerator and denominator of an expression by a factor *chosen to eliminate any radicals in the denominator*.

E X A M P L E 6 ▨   **Rationalizing the Denominator**

Rationalize $\dfrac{1}{\sqrt{2}}$.

SOLUTION

$$\frac{1}{\sqrt{2}} = \frac{1}{\sqrt{2}} \cdot \frac{\sqrt{2}}{\sqrt{2}} \qquad \text{$\sqrt{2}$ is chosen as the factor since multiplication by $\sqrt{2}$ will eliminate the radical in the denominator}$$

$$= \frac{\sqrt{2}}{2}$$

EXAMPLE 7 ■ **Rationalizing the Denominator**

Rationalize $\dfrac{6}{\sqrt[3]{9}}$.

SOLUTION

$$\frac{6}{\sqrt[3]{9}} = \frac{6}{\sqrt[3]{9}} \cdot \frac{\sqrt[3]{3}}{\sqrt[3]{3}} \qquad \text{$\sqrt[3]{3}$ is chosen to obtain a perfect cube (27) in the radicand of the denominator.}$$

$$= \frac{6\sqrt[3]{3}}{\sqrt[3]{27}} \qquad \text{Using Property 3}$$

$$= \frac{6\sqrt[3]{3}}{3} \qquad \text{Simplifying}$$

$$= 2\sqrt[3]{3} \qquad \text{Simplifying}$$

EXAMPLE 8 ■ **Rationalizing the Denominator by Conjugate Multiplication**

Simplify $\dfrac{7}{4 - \sqrt{2}}$.

SOLUTION We must rationalize the denominator. If we multiply $4 - \sqrt{2}$ by its conjugate $4 + \sqrt{2}$, we obtain $(4 - \sqrt{2}) \cdot (4 + \sqrt{2}) = 4^2 - (\sqrt{2})^2 = 16 - 2 = 14$. Thus, we multiply both numerator and denominator by $4 + \sqrt{2}$ as follows:

$$\frac{7}{4 - \sqrt{2}} = \frac{7}{4 - \sqrt{2}} \cdot \frac{4 + \sqrt{2}}{4 + \sqrt{2}} \qquad \text{Multiplying numerator and denominator by the conjugate $4 + \sqrt{2}$}$$

$$= \frac{7 \cdot (4 + \sqrt{2})}{4^2 - (\sqrt{2})^2} \qquad \text{Using the difference of squares formula to multiply the denominators}$$

$$= \frac{7 \cdot (4 + \sqrt{2})}{16 - 2} \qquad \text{Simplifying}$$

$$= \frac{7 \cdot (4 + \sqrt{2})}{14} \qquad \text{Simplifying}$$

$$= \frac{4 + \sqrt{2}}{2} \qquad \text{Simplifying}$$

When radical expressions are to be added or subtracted, we simplify the individual expressions first and then combine like terms.

EXAMPLE 9 ■ **Addition and Subtraction of Radical Expressions**

Simplify $\dfrac{8}{\sqrt{32}} + \dfrac{10}{\sqrt{5}} - 3\sqrt{20} - 2\sqrt{8}$.

SOLUTION

$$\frac{8}{\sqrt{32}} + \frac{10}{\sqrt{5}} - 3\sqrt{20} - 2\sqrt{8} = \frac{8}{\sqrt{16}\,\sqrt{2}} + \frac{10}{\sqrt{5}} \cdot \frac{\sqrt{5}}{\sqrt{5}} - 3\sqrt{4}\,\sqrt{5} - 2\sqrt{4}\,\sqrt{2}$$

$$= \frac{8}{4\sqrt{2}} + 2\sqrt{5} - 6\sqrt{5} - 4\sqrt{2}$$

$$= \frac{2}{\sqrt{2}} - 4\sqrt{5} - 4\sqrt{2}$$

$$= \frac{2}{\sqrt{2}} \cdot \frac{\sqrt{2}}{\sqrt{2}} - 4\sqrt{5} - 4\sqrt{2}$$

$$= \sqrt{2} - 4\sqrt{5} - 4\sqrt{2}$$

$$= -3\sqrt{2} - 4\sqrt{5}$$

> **WARNING!**   Expressions involving the roots of sums or differences don't generally simplify. For example:
>
> $$\sqrt{x + y} \text{ does not equal } \sqrt{x} + \sqrt{y}$$
> $$\sqrt{x^2 + y^2} \text{ does not equal } x + y$$
> $$\sqrt[3]{x^3 - y^3} \text{ does not equal } x - y$$

SQUARE ROOTS OF NEGATIVE NUMBERS

Recall from Section 1.1 that i was defined to be a number satisfying $i^2 = -1$. Thus, we write $i = \sqrt{-1}$. This allows us to extend the definition of the principal square root to the case where the radicand is negative.

EXAMPLE 10 ■ **Simplification of Square Roots of Negative Numbers**

Simplify each of the following.

a. $\sqrt{-16}$ **b.** $\sqrt{-8}$

SOLUTION

a. $\sqrt{-16} = \sqrt{16}\,\sqrt{-1} = 4i$

b. $\sqrt{-8} = \sqrt{4}\,\sqrt{2}\,\sqrt{-1} = 2\sqrt{2}i$

TIME DILATION

f all the otherworldly phenomena predicted by Einstein's Theory of Relativity, perhaps none is so mind bending as **time dilation**: the tendency for physical processes—clocks—to run more slowly in systems accelerated to high speeds. This experimentally confirmed fact contradicts our intuitive notion of time as an absolute quantity, independent of observer, independent of frame of reference. Of course, our intuition is forged from countless everyday experiences with time, experiences taking place at—to borrow from *Star Trek* terminology—subwarp speeds. It is only at speeds approaching that of light that time dilation emerges as a measurable effect with bizarre consequences.

A full development of time dilation is well beyond the scope of this text, but we can express the relationship between Earth time and traveler time with a simple algebraic formula. In fact, it can be shown that

$$T = \frac{T_0}{\sqrt{1 - v^2}}$$

where T is elapsed time on Earth, T_0 is the time as measured by the traveler, and v is the velocity of the traveler relative to Earth

(expressed as a fraction of the speed of light, 186,000 miles per second). For example, if we were to travel at half the speed of light for 2 years (according to our own clocks), then the time elapsed on Earth would be given by

$$T = \frac{2}{\sqrt{1 - \left(\frac{1}{2}\right)^2}} \approx 2.309 \text{ years}$$

The faster we travel, the greater the discrepancy between Earth time and traveler time. For example, consider the famous **twin paradox.** A 30-year-old astronaut travels to a nearby star and back at great speeds, while her identical twin sister remains on Earth. Because of time dilation, the astronaut's clocks, including the aging processes of her own body, run more slowly than those of her terrestrial twin. Specifically, if our astronaut travels at 99.5% of the speed of light, then after 5 years (astronaut time), approximately $5/\sqrt{1 - 0.995^2} \approx 50$ years will have elapsed on Earth. Thus, when our astronaut triumphantly returns from her celestial sojourn, she will emerge from her starship as a 35-year-old woman with an 80-year-old identical twin sister!

FRACTIONAL EXPONENTS

Many of the properties of radicals bear a striking resemblance to those of exponents. For example, the property $\sqrt[m]{a}\,\sqrt[m]{b} = \sqrt[m]{ab}$ is very similar to the exponential property $a^m b^m = (ab)^m$. There is a good reason for this similarity—radicals can, in fact, be represented by expressions involving *fractional exponents*. To see this, consider the expression $2^{1/3}$. If the properties of integer exponents are to hold for fractional exponents, then we have

$$(2^{1/3})^3 = 2^{(1/3)\cdot 3}$$
$$= 2^1$$
$$= 2$$

Thus, since the cube of $2^{1/3}$ is 2, $2^{1/3}$ is nothing more than the cube root of 2, or $\sqrt[3]{2}$. More generally, we have the following definition.

Definition of $a^{1/n}$ ☐ | For $a > 0$ and n a natural number, $a^{1/n} = \sqrt[n]{a}$.

In the examples and exercises that follow, assume that all variables are positive in expressions involving fractional exponents.

E X A M P L E   11 ■ **Converting from Fractional Exponents to Radicals**

Write each of the following as a radical.

a. $3^{1/10}$ **b.** $2^{1/4}$ **c.** $x^{1/2}$

SOLUTION

a. $3^{1/10} = \sqrt[10]{3}$ **b.** $2^{1/4} = \sqrt[4]{2}$ **c.** $x^{1/2} = \sqrt{x}$

EXAMPLE 12   ▓   **Expressing Radicals in Exponential Form**

Write the following radical expressions in exponential form.

a. $\sqrt{5}$ **b.** $\sqrt[3]{x}$ **c.** $\sqrt[7]{a^2}$

SOLUTION

a. $\sqrt{5} = 5^{1/2}$ **b.** $\sqrt[3]{x} = x^{1/3}$ **c.** $\sqrt[7]{a^2} = (a^2)^{1/7}$

In part c of Example 12, we obtained $\sqrt[7]{a^2} = (a^2)^{1/7}$. But using the properties of exponents, this could be rewritten as $a^{2/7}$. This suggests that we define $a^{2/7}$ to be $\sqrt[7]{a^2}$. More generally, we have the following definition.

Definition of $a^{m/n}$ □ | For $a > 0$ and m and n natural numbers, $a^{m/n} = \sqrt[n]{a^m}$ or $(\sqrt[n]{a})^m$.

EXAMPLE 13   ▓   **Converting from Fractional Exponents to Radicals**

Write each of the following as a radical.

a. $a^{3/7}$ **b.** $2^{4/3}$ **c.** $(x + y)^{16/5}$

SOLUTION

a. $a^{3/7} = \sqrt[7]{a^3}$
b. $2^{4/3} = \sqrt[3]{2^4}$
c. $(x + y)^{16/5} = \sqrt[5]{(x + y)^{16}}$

EXAMPLE 14   ▓   **Expressing Radicals in Exponential Form**

Write each of the following radical expressions in exponential form.

a. $\sqrt[3]{5^2}$ **b.** $\sqrt{q^9}$ **c.** $\sqrt[4]{x^6}$

SOLUTION

a. $\sqrt[3]{5^2} = 5^{2/3}$ **b.** $\sqrt{q^9} = q^{9/2}$ **c.** $\sqrt[4]{x^6} = x^{6/4} = x^{3/2}$

We handle expressions with negative fractional exponents just as we handled expressions with negative integer exponents. Thus, for $a > 0$ and m and n natural numbers,

$$a^{-(m/n)} = \frac{1}{a^{m/n}}$$

EXAMPLE 15 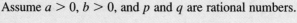 Simplifying an Expression with a Negative Fractional Exponent

Evaluate $4^{-3/2}$.

SOLUTION

$$4^{-3/2} = \frac{1}{4^{3/2}} = \frac{1}{\sqrt{4^3}} = \frac{1}{\sqrt{64}} = \frac{1}{8}$$

 CALCULATOR KEYS

Computing Roots

If your calculator does not have an *n*th root key, roots can be computed by first converting the radical to a fractional exponent and then using the exponent key $\boxed{y^x}$ or $\boxed{\wedge}$. For example, $\sqrt[5]{2}$ can be computed by entering 2∧(1/5).

We now have defined the expression a^q (with $a > 0$) for all rational numbers q. All of the rules for integer exponents given in Section 1.4 hold for rational number exponents as well. We restate them for convenience here.

Properties of Rational Number Exponents □

Assume $a > 0$, $b > 0$, and p and q are rational numbers.

Property	**Example**
1. $a^p \cdot a^q = a^{p+q}$	$2^{1/3} \cdot 2^{4/3} = 2^{5/3}$
2. $\dfrac{a^p}{a^q} = a^{p-q}$	$\dfrac{x^2}{x^{3/4}} = x^{2-(3/4)} = x^{5/4}$
3. $(a^p)^q = a^{pq}$	$(x^{1/2})^{2/3} = x^{(1/2) \cdot (2/3)} = x^{1/3}$
4. $(ab)^q = a^q b^q$	$(2x)^{5/2} = 2^{5/2} x^{5/2}$
5. $\left(\dfrac{a}{b}\right)^q = \dfrac{a^q}{b^q}$	$\left(\dfrac{x}{3}\right)^{1/2} = \dfrac{x^{1/2}}{3^{1/2}}$

EXAMPLE 16 ▦ Simplifying Expressions Involving Fractional Exponents

Simplify the following.

a. $x^{1/2} \cdot x^{3/4}$ **b.** $(a^{2/3})^{3/4}$ **c.** $\dfrac{p^{2/5} q^{-1/3}}{pq^{-2/3}}$

SOLUTION

a. $x^{1/2} \cdot x^{3/4} = x^{(1/2)+(3/4)}$

$\qquad\qquad\quad = x^{5/4}$

b. $(a^{2/3})^{3/4} = a^{(2/3) \cdot (3/4)}$

$\qquad\qquad\quad = a^{1/2}$

c. $\dfrac{p^{2/5}q^{-1/3}}{pq^{-2/3}} = p^{2/5-1}q^{(-1/3)-(-2/3)}$

$\qquad\qquad = p^{-3/5}q^{1/3}$

$\qquad\qquad = \dfrac{q^{1/3}}{p^{3/5}}$

UNDERSTANDING AND MASTERY CHECKLISTS

CONCEPTS TO UNDERSTAND

- ☐ Principal nth root
- ☐ Radical, radicand, and index
- ☐ Properties of radicals
- ☐ Completely simplified radical expressions
- ☐ Complex roots
- ☐ Fractional exponents

SKILLS TO MASTER

- ☐ Evaluate the principal nth root of perfect nth powers.
- ☐ Simplify radical expressions.
- ☐ Rationalize the denominator.
- ☐ Add, subtract, multiply, and divide radical expressions.
- ☐ Evaluate square roots of negative numbers.
- ☐ Convert from radical to exponential notation.
- ☐ Convert from exponential to radical notation.

EXERCISES 1.8

EXERCISES 1–4   *Evaluate the given radical expression.*

1. $\sqrt[3]{27}$

2. $\sqrt{\dfrac{9}{4}}$

3. $\sqrt[4]{\dfrac{1}{16}}$

4. $\sqrt[3]{-8}$

EXERCISES 5–34   *Simplify the given expression. Assume that all variables represent positive real numbers.*

5. $\sqrt{18}$

6. $\sqrt[3]{16}$

7. $\sqrt[4]{100,000}$

8. $\sqrt{5^8}$

9. $\sqrt[3]{3^{17}}$

10. $\sqrt{500}$

11. $\sqrt{a^2b^5}$

12. $\sqrt{x^3y^7}$

13. $\sqrt{28x^4y^3z^2}$

14. $\sqrt{p^9}$

15. $\sqrt{50a^{30}b^{20}c^{11}}$

16. $\sqrt{(a+b)^6}$

17. $\sqrt{x^7(y+z)^4}$

18. $\sqrt[3]{a^7}$

19. $\sqrt[3]{54s^6t^7}$

20. $\sqrt[5]{32x^{11}y^{17}z^2}$

21. $\sqrt[3]{x^3+3x^2y+3xy^2+y^3}$

22. $\sqrt{x^6+y^6}$

23. $\sqrt[3]{x-y}$

24. $\sqrt{\dfrac{1}{5}}$

25. $\sqrt{\dfrac{a^4}{b^2}}$

26. $\dfrac{3}{\sqrt{7}}$

27. $\dfrac{6}{\sqrt[3]{4}}$

28. $\dfrac{a^2}{\sqrt[3]{a}}$

29. $\dfrac{21}{\sqrt{3}}$

30. $\dfrac{3}{\sqrt{5a}}$

31. $\dfrac{2}{3-\sqrt{5}}$

32. $\dfrac{6}{a+\sqrt{3}}$

33. $\dfrac{a-b}{\sqrt{a}-\sqrt{b}}$

34. $\dfrac{\sqrt{x+y}}{1-\sqrt{x+y}}$

EXERCISES 35–38   *Rationalize the numerator. (Eliminate all radicals in the numerator.)*

35. $\dfrac{\sqrt{2}}{6}$

36. $\dfrac{\sqrt[3]{4}}{2}$

37. $\dfrac{\sqrt{x}-\sqrt{a}}{x-a}$

38. $\dfrac{1+\sqrt{8}}{3}$

EXERCISES 39–56   *Perform the indicated operation and simplify.*

39. $\sqrt{3}\cdot\sqrt{15}$

40. $\sqrt{5a^3b}\cdot\sqrt{10ab^3}$

41. $\sqrt[3]{2xyz}\cdot\sqrt[3]{20x^2y^4z}$

42. $(\sqrt{8})^3$

43. $\dfrac{\sqrt{10}}{\sqrt{2}}$

44. $(\sqrt[3]{a})^5$

45. $\sqrt{\dfrac{x^3 y^2}{xy^3}}$ 46. $\dfrac{\dfrac{3}{\sqrt{5}}}{\dfrac{\sqrt{5}}{4}}$

47. $(\sqrt{x} + 3)(\sqrt{x} - 3)$ 48. $(\sqrt{3} + \sqrt{6})^2$
49. $\sqrt{6} + \sqrt{24}$ 50. $\sqrt{5} + 3\sqrt{5}$
51. $\sqrt{8} + \dfrac{1}{\sqrt{2}}$ 52. $\sqrt[3]{9} - \dfrac{2}{\sqrt[3]{3}}$

53. $\sqrt{a} + \dfrac{1}{\sqrt{a}}$

54. $\sqrt{32a^3 b} +$
    $\sqrt{18ab} - \sqrt{2ab}$

55. $\sqrt[3]{\sqrt{x}}$ 56. $\sqrt[5]{\sqrt[3]{x^5}}$

EXERCISES 57–64   *Convert from radical notation to exponential notation. Assume all variables are positive.*

57. $\sqrt[3]{x}$ 58. $\sqrt[5]{y^2}$
59. $\sqrt{17}$ 60. $\sqrt[3]{a^2}$
61. $\sqrt[4]{a^3 b^6}$ 62. $\sqrt[5]{(x + y)^2}$
63. $\dfrac{1}{\sqrt{x}}$ 64. $\dfrac{1}{\sqrt[3]{5}}$

EXERCISES 65–68   *Simplify the given expression.*

65. $\sqrt{-25}$ 66. $\sqrt{-18}$

67. $\dfrac{2}{\sqrt{-8}}$ 68. $\sqrt{-10}\,\sqrt{-5}$

EXERCISES 69–76   *Convert the given expression into radical form.*

69. $2^{3/2}$ 70. $3^{-1/5}$
71. $b^{2/3}$ 72. $x^{4/7}$
73. $(a - b)^{3/2}$ 74. $(x^2 + y^2)^{1/2}$
75. $x^{1.3}$ 76. $y^{2.5}$

EXERCISES 77–80   *Simplify.*

77. $16^{-1/2}$ 78. $27^{1/3}$
79. $32^{2/5}$ 80. $8^{-2/3}$

EXERCISES 81–90   *Simplify the given exponential expression. Express the answer in exponential form without negative exponents.*

81. $x^{1/3} x^{2/3}$ 82. $a^{1/4} a^{1/3}$
83. $(x^{1/2})^{2/3}$ 84. $(x^{-1/2})^{-1/3}$
85. $(81a^4 b^{12})^{1/4}$ 86. $(\tfrac{1}{8} u^6 v^9)^{2/3}$
87. $\dfrac{z^2}{z^{3/2}}$ 88. $\dfrac{x^{3/4}}{x^4}$
89. $\dfrac{x^2 y^{1/3}}{x^{1/2} y}$ 90. $\left(\dfrac{p^{2/3} q^{1/2} r}{p^{1/3} q^{1/4}}\right)^{-2}$

■ **APPLICATIONS**

91. *Terminal Velocity*   According to *The Guinness Book of World Records,* Vesna Vulovic of Yugoslavia survived a fall of 33,330 feet after the DC-9 on which she was a flight attendant blew up on January 26, 1972. If gravity were the only force acting on her, Vulovic's velocity at impact would be given by $v = 8\sqrt{s}$, where s is the distance fallen in feet and v is the velocity in feet per second. Find Vulovic's velocity at impact using this formula. The terminal velocity of a human in free-fall is roughly 180 feet per second. How do you explain the discrepancy between this value and your answer? In fact, Vulovic fell in a portion of the fuselage. How might this have affected her terminal velocity?

92. *Braking Distance*   The velocity at which an automobile is traveling immediately before braking can be estimated from the braking distance using the formula $v = \sqrt{20d}$, where v is the velocity of the car in miles per hour and d is the distance in feet required to brake. If skid marks measure 200 feet, then how fast was the car traveling?

93. *Hypotenuse Length*   An isosceles right triangle with leg length x is shown in Figure 42. Express the length of the hypotenuse in terms of x.

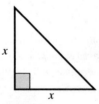

FIGURE 42

94. *Diagonal Length*   A cube with side length x is shown in Figure 43. Express the length of line segment AD in terms of x.

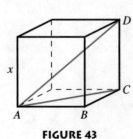

FIGURE 43

95. *Hang Time*   The maximum height h of a jump, the hang time t, and the initial speed v_0 with which a jumper leaves the ground are related by the formulas $v_0 = \sqrt{64h}$ and $h = 4t^2$, where h is in feet, t is in seconds, and v_0 is in feet per second.

a. Use these equations to find a formula for v_0 in terms of t.

b. It is estimated that Michael Wilson, former University of Memphis basketball star and Harlem Globetrotter, had a hang time of 1.036 seconds. Find his initial speed and maximum height.

c. Wilson is reputed to have dunked on a 12-foot goal. Is this consistent with your findings in part b?

96. *Unknown Interest Rate*   If a deposit of P dollars grows to a balance of B dollars after n years, then the annually compounded interest rate r is given by

$$r = \sqrt[n]{\frac{B}{P}} - 1$$

At what interest rate will \$1 grow to \$1000 in 15 years?

EXERCISES 97–98   *Time dilation is the tendency of physical processes to run more slowly in systems accelerated to speeds approaching that of light. Consequently, a person traveling at speeds approaching that of light would age more slowly than a counterpart on Earth. The precise relationship between Earth time and the time of the traveler is given by*

$$T = \frac{T_0}{\sqrt{1 - v^2}}$$

where T is elapsed time on earth, T_0 is the time elapsed by the traveler, and v is the ratio of the traveler's speed to the speed of light. (For further discussion, refer to the math note box on p. 86.)

97. *Fountain of Youth?*   The official air speed record of 2200 miles per hour (3.28554×10^{-6} of the speed of light) is held by the Lockheed SR-71 Blackbird. If a pilot were to stay aloft at this speed for 50 years (according to his watch), then how much younger will he be than if he had stayed on the ground? (Give your answer in seconds.)

98. *Banking on Time*   Suppose that a woman deposits \$1000 into a bank account earning 4% interest compounded annually and she then travels at 99.99999999% of the speed of light for 1 day (according to her watch).

a. Use the time-dilation formula to compute the amount of time that would have elapsed on Earth.

b. Compute the balance that she would find in her account after returning to Earth.

CONCEPTS AND CRITICAL THINKING

EXERCISES 99–104   *Answer true or false.*

99. $\sqrt{(-2)^2} = 2$

100. $\sqrt{x^2 + 25}$ simplifies to $x + 5$

101. $\sqrt[3]{100x^3}$ simplifies to $10x$

102. $(\pm 6)^2 = 36$

103. $\sqrt{36} = \pm 6$

104. $\sqrt[3]{xy}$ is equivalent to $\sqrt[3]{x}\,\sqrt[3]{y}$

EXERCISES 105–108   *Give an example of each.*

105. A pair of numbers x and y such that $\sqrt{x + y} = \sqrt{x} + \sqrt{y}$ (This formula does not hold in general.)

106. A number x such that $\sqrt{x^2} \neq x$

107. A number x such that the principal square root of x is equal to the principal cube root of x

108. A pair of numbers x and y for which $\sqrt{x^2 - y^2} = x - y$ (This formula does not hold in general.)

109. Explain why $\sqrt{25} \neq \pm 5$.

110. Students frequently ask the question, "Why is it considered simpler for the radical to be in the numerator rather than the denominator?" Although we have not explicitly stated the reason for rationalizing the denominator, it can be inferred from one of the examples. Explain.

QUESTIONS FOR DISCUSSION OR ESSAY

111. Time dilation allows us, in a sense, to travel into the future. Are there any logical difficulties with such time travel? What about traveling into the past?

112. It is not uncommon for algebra students to view algebra as a sort of game with arbitrary rules that the "player" must memorize. From their point of view, the statements

$\sqrt{ab} = \sqrt{a}\,\sqrt{b}$ and $\sqrt{a + b} = \sqrt{a} + \sqrt{b}$ seem equally valid. In fact, when told that the second equality is not in general true, they might respond, "Why not? It's just like the first one." How would you answer that question? Why isn't the second equality true? What would it take to prove that it isn't true? Who decides which algebraic statements are true and which aren't?

113. Some of the problems in this section look extremely difficult at first glance. Take, for example, the simplification of

$$\left(\frac{p^{2/3}q^{1/2}r}{p^{1/3}q^{1/4}}\right)^{-2}$$

Some students, however, find such problems easy (although rather long). Typically, successful algebra students remark that they solved the problem by "breaking it up into little pieces." What do they mean by this? Explain how we use this problem-solving strategy to accomplish such everyday tasks as navigating, cooking an omelet, dressing, and looking up a telephone number.

▣ PROJECTS FOR ENRICHMENT

114. *Heron's Formula* Heron's formula gives the area of a triangle in terms of its side lengths. If the triangle has sides of length *a*, *b*, and *c*, and *A* represents the area of the triangle, then

$$A = \sqrt{s(s - a)(s - b)(s - c)} \quad \text{where } s = \frac{a + b + c}{2}$$

a. Find the area of a triangle with sides of length 3, 4, and 5.

b. Find the area of a triangle with sides of length 1, 1, and 1.

c. Use Heron's formula to show that the area of an equilateral triangle of side length *x* is given by $(\sqrt{3}/4)x^2$.

d. What is the area of a triangle with sides of length 1, 2, and 5?

115. *Technote Passing* Suppose a friend, located in a room 80 feet above yours, 50 feet down the hall, and 35 feet across the hall, electronically transmits a note to you. Your receiver (cleverly disguised as a wristwatch) has a range of only 100 feet. Will you receive the message?

CHAPTER 1 REVIEW

EXERCISES 1–4   *List all the terms that describe the given number. Choose from natural, whole, integer, rational, irrational, real, and complex.*

1. -2.33

2. $\frac{12}{3}$

3. A number that when squared gives -11

4. π

EXERCISES 5–6   *Place the given list of numbers in ascending order.*

5. $1.5, \sqrt{2}, 1.4, \sqrt{3}, 2$

6. $\frac{2}{3}, 0.67, 0.66, \frac{2001}{3000}$

EXERCISES 7–10   *Write the set using interval notation and sketch the graph.*

7. $\{x \mid -5 \le x < -1\}$

8. $\{x \mid x > 3\}$

9. $\{x \mid x \le \frac{1}{2}\}$

10. $\{x \mid \pi \le x \le 2\pi\}$

EXERCISES 11–12   *Plot the given points, find the distance between them, and find the coordinates of the midpoint of the line segment connecting them.*

11. $(4, 5), (-2, -3)$

12. $(-1, -2), (-3, 7)$

EXERCISES 13–14   *Use a scatterplot to graph the given data. Then, use the graph to estimate the missing y-value.*

13.

x	0.7	1.2	1.9	2.5	3.2	3.7
y	2.1	1.5	1.0	1.2	?	2.6

14.

x	0.6	1.0	1.5	1.9	2.3	2.8	3.2	3.6	4.0
y	1.9	2.6	3.0	3.3	3.5	3.7	3.8	3.9	?

EXERCISES 15–18   *True or false.*

15. The operation of subtraction is commutative.

16. $3 \cdot 5 + 6 \div 2 = 18$

17. For all real numbers x, the distance between x and 2 is $x - 2$.

18. The degree of the product of two polynomials is the product of the degrees of the polynomials.

EXERCISES 19–22   *Perform the indicated operation and write the answer in the form a + bi.*

19. $(1 - 3i) + (4 + 2i)$

20. $(-2 + i) \cdot (5 - 2i)$

21. $\dfrac{1}{3 - i}$

22. $\dfrac{2 - i}{4 + 3i}$

EXERCISES 23–26 *Simplify the given expression. Write your answer without using negative exponents.*

23. $(-3)^4$

24. $(\frac{4}{5})^{-2}$

25. $\dfrac{(a^3b)^2}{a^4b^6}$

26. $\dfrac{x^{-2}y^3}{(xy^{-1})^{-3}}$

EXERCISES 27–30 *Convert the given number to scientific notation.*

27. 31,400,000

28. 0.00461

29. Eleven billion

30. One ten-thousandth

EXERCISES 31–32 *Evaluate the polynomial for the indicated value of the variable.*

31. $x^3 - 2x^2 - 3x$; $x = -2$

32. $-a^5 + 4a^2 - 4$; $a = -1$

EXERCISES 33–38 *Perform the indicated operation and simplify.*

33. $(2t^3 + 4t) - (3t^3 + t^2 - t)$

34. $(8x^2 - x + 3) + (2x^2 - 4x + 6)$

35. $(2y + 3)(y - 4)$

36. $(2 - w)(4 - 3w)$

37. $x(x - 3) + (x + 2)(x - 1)$

38. $(u^2 - 3u + 7)(u - 2)$

EXERCISES 39–46 *Factor the given polynomial.*

39. $9u^2 - 4$

40. $8x^3 + 125$

41. $t^2 + 4t - 32$

42. $2y^3x - 54x$

43. $9x^2 + 6x + 1$

44. $b^2 + 2ab + 8a + 4b$

45. $64s^6 - 1$

46. $2x^2 + 7x - 4$

EXERCISES 47–50 *Simplify the rational expression.*

47. $\dfrac{x + 3}{x^2 - 9}$

48. $\dfrac{v^2 - v - 2}{v^2 + v - 6}$

49. $\dfrac{t^3 - 1}{t^3 - t}$

50. $\dfrac{y^2x^2 + 4y^2}{x^2y^2 + 4y^2 + 4x^2 + 16}$

EXERCISES 51–56 *Perform the indicated operation and simplify.*

51. $\dfrac{3}{y + 2} - \dfrac{1}{y}$

52. $\dfrac{1}{x^2 + x} + \dfrac{1}{x^2 - x}$

53. $\dfrac{x^2}{x^2 - 9} \cdot \dfrac{3 - x}{x^2 + x}$

54. $\dfrac{v^3 + 8}{v^2 - 4} \div \dfrac{v^2 - 2v + 4}{v - 2}$

55. $\dfrac{\frac{1}{t} + \frac{1}{2}}{t + 2}$

56. $\dfrac{1 - \frac{1}{x}}{1 + \frac{1}{x}}$

EXERCISES 57–62 *Simplify the given radical expression.*

57. $\sqrt[3]{a^6b^4}$

58. $\sqrt{32u^3v^6}$

59. $\dfrac{9}{\sqrt{3}}$

60. $\dfrac{2}{\sqrt[3]{2}}$

61. $\dfrac{1}{x - \sqrt{2}}$

62. $\dfrac{x - 4}{\sqrt{x} - 2}$

EXERCISES 63–66 *Perform the indicated operation and simplify. Assume all variables represent positive numbers.*

63. $\sqrt{6s^5t} \cdot \sqrt{2st}$

64. $\dfrac{\sqrt[3]{54xy^2}}{\sqrt[3]{2xy}}$

65. $\sqrt{28} + 9\sqrt{7}$

66. $3\sqrt{2} - \sqrt{8}$

EXERCISES 67–70 *Convert from radical notation to exponential notation. Assume all variables represent positive numbers. Write your answer without using negative exponents.*

67. $\sqrt[4]{t^2}$

68. $\sqrt[3]{x^2y^4}$

69. $\dfrac{1}{\sqrt[3]{x}}$

70. $\dfrac{1}{\sqrt{u^3v^4}}$

EXERCISES 71–72 *Convert from exponential notation to radical notation.*

71. $x^{2/3}$

72. $t^{-5/2}$

EXERCISES 73–76 *Simplify the given exponential expression. Express the answer in exponential form without negative exponents.*

73. $(x^{1/3})^{1/4}$

74. $y^{2/5}y^{3/2}$

75. $\dfrac{a^{2/3}b^{4/3}}{a^{5/3}b^{-2/3}}$

76. $\left(\dfrac{xy^{-1/2}}{x^{1/4}y}\right)^2$

EXERCISES 77–84 *Parts a and b are connected: Part a involves an elementary concept, whereas part b involves related material from this chapter. First answer part a, and then use this result to answer part b.*

77. a. Add $\frac{1}{3} + \frac{1}{4}$.

    b. Add $\frac{1}{x} + \frac{1}{y}$.

78. a. Simplify $\sqrt{2^3 3^5}$.

    b. Simplify $\sqrt{x^3y^5}$.

79. a. Expand $(2x - 3y)^2$.

    b. It is given that $2x \geq 3y$. Simplify $\sqrt{4x^2 - 12xy + 9y^2}$.

80. a. Expand $3a(2a - 3)(a - 5)$.

    b. Factor $6a^3 - 39a^2 + 45a$.

81. a. Find the midpoint of the line segment connecting $(-2, 3)$ and $(2, 11)$.

    b. Find a point P such that the line segment connecting $(-2, 3)$ and $(0, 7)$ is half the length of the line segment connecting $(-2, 3)$ and P.

82. a. Compute the product of 2.8 and 100,000.

    b. Express 280,000 in scientific notation.

83. a. Simplify $2x(x + 1)(x - 1)$.

    b. Simplify $\dfrac{2x^3 - 2x}{x + 1}$.

84. a. If \$822.70 is deposited into an account paying 5% interest compounded annually, what will the balance be after 4 years?

    b. How much should be deposited into an account paying 5%

interest compounded annually so that the balance after 4 years is $500?

85. *Simple Interest*   Suppose that $1000 is deposited in an account paying 6% simple interest. What will the balance be after 8 years?

86. *Compound Interest*   If $2500 is deposited in an account paying 6.5% interest compounded annually, what will the balance be after 10 years?

87. *Volume of Pyramid*   The volume of a pyramid with height h and a square base with side length x is given by the monomial $\frac{1}{3}x^2h$. Find the volume of the Great Pyramid of Egypt given that the base has length 750 feet and the height is 450 feet. The dimensions of an Olympic-size swimming pool are approximately $164' \times 75' \times 6\frac{1}{2}'$. How many Olympic-size swimming pools could be drained into the pyramid?

The Great Pyramid at Giza near Cairo.

88. *Maximizing Area*   Consider a rectangle with dimensions l and w. Denote the perimeter of the rectangle by P, so that $P = 2l + 2w$, and let $d = l - w$, the difference between the dimensions.

a. Show that $\frac{1}{4}(P + 2d) = l$ and $\frac{1}{4}(P - 2d) = w$.

b. Use the expressions given in part a to find a formula for the area of the rectangle in terms of P and d.

c. Use your formula from part b to deduce that for a fixed perimeter P, the area is greatest when $d = 0$. For what kind of rectangle is $d = 0$?

89. *Investment Value*   Suppose an initial investment of D dollars is made in some enterprise that is expected to bring in periodic cash flows of $C_1, C_2, \ldots, C_n$ at the end of each of n years. If the interest rate on the investment is denoted by i, then the net present value of the investment at the end of the kth year is given by the expression

$$N_k = -D + \frac{C_1}{(1 + i)} + \frac{C_2}{(1 + i)^2} + \frac{C_3}{(1 + i)^3} + \cdots + \frac{C_k}{(1 + i)^k}$$

A negative value for N_k means the enterprise is not yet profitable, whereas a positive value for N_k means it is profitable in the sense that the rate of return i on the original investment has been exceeded. For an initial investment of $D = \$10,000$ and cash flows at the end of the first and second years of $C_1 = \$5000$ and $C_2 = \$7000$, find and simplify a rational expression for N_2 involving the variable i. Evaluate the expression for $i = 0.10$ (10%) and again for $i = 0.15$ (15%). For which of these interest rates is the enterprise profitable after 2 years?

CHAPTER 1 TEST

PROBLEMS 1–8　*Answer true or false.*

1. All rational numbers are real.

2. Some irrational numbers are not real.

3. π is a complex number.

4. $-\sqrt{16}$ is a real number.　*True*

5. For some real numbers x, $|x| = -x$.　*True*

6. The product of two polynomials is always a polynomial.　*True*

7. The quotient of two polynomials is always a polynomial.　*False*

8. The principal root of any positive number is positive.　*True*

PROBLEMS 9–14　*Give an example of each.*

9. A rational number

10. An irrational number

11. A number whose square is negative

12. A polynomial in two variables with degree 5

13. A perfect square trinomial

14. A binomial

PROBLEMS 15–20　*Simplify the given expression. Assume all variables represent positive numbers.*

15. $(-2x)^3$

16. $(u^{4/3}v^{-2/5})^{-1/2}$

17. $\left(\dfrac{a^{-3}b^4}{a^2b^{-1}}\right)^2$

18. $\sqrt{27x^5y^6}$

19. $\sqrt[4]{12a^3b^2}\,\sqrt[4]{20ab^6}$

20. $\dfrac{2}{\sqrt[3]{4}}$

PROBLEMS 21–24　*Perform the indicated operation and simplify whenever possible.*

21. $(3z^4 - 2z^3) - (5z^3 - z^2)$

22. $(2x + 3)(3x - 2)$

23. $\dfrac{1}{x^2 + 2x} + \dfrac{x}{x^2 - 4}$

24. $\dfrac{y + 2}{y^2 - 1} \cdot \dfrac{y^2 + 2y - 3}{y^2 + 2y}$

PROBLEMS 25–28　*Factor the given polynomial.*

25. $4x^2 - 9$

26. $t^2 + 2t - 15$

27. $u^3 + 8v^3$

28. $a^2 + ab - 2a - 2b$

29. The fact that $3(x - 2) = 3x - 6$ is a consequence of which property of real numbers?

30. Find an expression for the distance between $2x$ and -7.

31. If $5000 is deposited in an account paying 5.5% interest compounded annually, what will the balance be after 20 years?

32. A ball thrown upward from the ground with an initial velocity of 44 feet per second has height in feet after t seconds given by the polynomial $-16t^2 + 44t$. Find the height of the ball at times $t = 2$ and $t = 3$. What can you conclude about the height of the ball between 2 and 3 seconds?

Know these things (test)

2 EQUATIONS AND INEQUALITIES

Tropical rain forests, the home of perhaps as many as 10 million different species, are being deforested at alarming rates primarily by ranchers employing slash-and-burn techniques to clear land for grazing and by loggers extracting too much timber from the fragile ecosystem. In Exercise 66 of Section 7 we will see one instance of how mathematics can be used to model such phenomena.

SECTION 2.1   GRAPHS OF EQUATIONS

- ☐   **What doesn't your graphing calculator show that you may need to know?**
- ☐   **If 30 customers arrive per hour, how fast must customers be served so that the average time in line is 1 minute?**
- ☐   **How can a tiny island have a coastline more than a billion miles long?**
- ☐   **How can a graphing calculator be used to estimate the date on which the average temperature in Orlando, Florida, is highest?**

Many of the most profound mathematical and scientific results are relationships among quantities that can be expressed as simple algebraic equations. From Pythagoras' $c^2 = a^2 + b^2$ and Newton's $F = ma$ to Einstein's $E = mc^2$, the simplicity of these equations belies their significance: A handful of symbols is sufficient for the expression of history-altering ideas. Because of the central role played by equations in the application of mathematics to the real world, we will explore them in great depth in much of the remainder of this text.

We, as humans, do not have brains that come prewired to interpret mathematical equations. Indeed, it takes a great deal of experience and insight to glance at an algebraic equation and draw conclusions about the connections among its variables. By contrast, we tend to have great natural facility at interpreting visual input: We could probably recognize the face of a loved one in dim light at an odd angle reflected off a fun house mirror, for example. It is because of this ability to interpret visual stimuli that, wherever possible, we depict an equation graphically. To simplify things, we begin by restricting ourselves to equations involving only two variables, which we call x and y. We define the **graph of an equation in x and y** to be the set of points (x, y) in the coordinate plane such that x and y satisfy the equation. For example, the graph of the equation $y = 2x - 4$ is the set of points (x, y) in the coordinate plane such that x and y satisfy the equation.

EXAMPLE 1   ▨   Showing Points are on the Graph of an Equation

Show that the points $(0, -4)$ and $(3, 2)$ are on the graph of $y = 2x - 4$.

SOLUTION   Substituting 0 for x and -4 for y, we have $-4 = 2(0) - 4$, which is a true statement. Thus, $(0, -4)$ is on the graph of $y = 2x - 4$. Similarly, $2 = 2(3) - 4$ is a true statement, and so $(3, 2)$ is on the graph also.

The simplest technique for graphing an equation is **plotting points.** To do this, we choose a convenient value for one of the variables, and then use the equation to determine the value of the other variable. We then graph or "plot" the corresponding point. The process is repeated as necessary, and then the points are connected by drawing a smooth curve through the plotted points.

EXAMPLE 2   ▨   Graphing an Equation by Plotting Points

Sketch a graph of the equation $y = 2x - 4$.

SOLUTION   We choose a few convenient values of x, and compute the corresponding y-values.

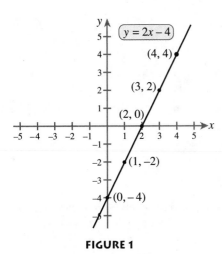

FIGURE 1

x	$y = 2x - 4$
0	-4
1	-2
2	0
3	2
4	4

Thus, the points $(0, -4)$, $(1, -2)$, $(2, 0)$, $(3, 2)$, and $(4, 4)$ are on the graph of $y = 2x - 4$. After plotting these points and "connecting the dots," we obtain the graph shown in Figure 1.

The points at which the graph of an equation crosses an axis are called the **intercepts.** More specifically, we make the following definitions.

Intercepts   □   The points where the graph of an equation in x and y crosses the x-axis are called the **x-intercepts,** and the points where the graph crosses the y-axis are called the **y-intercepts.** Note that an x-intercept is of the form $(a, 0)$. In practice, we often refer to the number a as the x-intercept. Similarly, a y-intercept is of the form $(0, b)$, and we often refer to the number b as the y-intercept.

In Example 2, $(2, 0)$ is the only x-intercept and $(0, -4)$ is the only y-intercept. If intercepts are not obvious from the graph, they can be found by using the fact that a point on the x-axis has y-coordinate 0, and a point on the y-axis has x-coordinate 0. Thus, to find any x-intercepts, we set $y = 0$ and find the corresponding x-values. Similarly, to find y-intercepts, we set $x = 0$ and find the corresponding y-values.

LINES AND CIRCLES

Notice that the graph of $y = 2x - 4$ in Figure 1 appears to be (and in fact is) a line. Had we known this information ahead of time, and known how to use it, much less work would have been required to obtain the graph. Indeed, one of the most useful tools for quickly plotting graphs is familiarity with certain special forms of equations. We consider two such forms here—lines and circles—and will see many others throughout the text.

Equations of the form $y = mx + b$ (like $y = 2x - 4$ or $y = -3x + 2$) are known as *linear equations* because their graphs are lines. We will establish this fact more firmly when we study linear equations in detail in Section 3.1; for now, we observe that if x and y are related by a linear equation, then the graph will be rising (or falling) at a constant rate. To see this, consider the effect on y of a unit increase in x. Since $y = mx + b$, increasing the x-value by 1 results in a change in the y-value of m units (increasing if m is positive, decreasing if m is negative). For example, in the case of the equation $y = 2x - 4$, every unit increase in x causes y to increase by 2 units, as illustrated by the following table.

x	$y = 2x - 4$
2	0
3	2
4	4
5	6
6	8

It follows that if x and y are related by a linear equation $y = mx + b$, then the graph will be rising (or falling) at a constant rate—like a line. As you already may have recognized, and as we will describe in detail in Section 3.1, m and b are the slope and y-intercept, respectively, of the line with equation $y = mx + b$.

Once an equation is recognized to be linear, its graph can be obtained simply by locating two points and connecting them with a straight line, as is demonstrated in the following example.

EXAMPLE 3  **Graphing a Linear Equation**

Sketch a graph of $y = -\frac{1}{2}x + 1$.

SOLUTION Since this equation is of the form $y = mx + b$, we locate two points and connect them with a straight line. Choosing convenient x-values—ones for which the corresponding y-values are easy to compute—we obtain the following table.

x	$y = -\frac{1}{2}x + 1$
0	1
2	0

Now we plot the points $(0, 1)$ and $(2, 0)$ and connect them with a line, as shown in Figure 2.

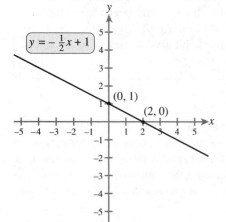

FIGURE 2

Like lines, circles can be easily plotted once their equations are in the proper form. We begin by defining a circle to be the set of points that are at an equal distance (the length of the radius) from a fixed point (the center). If we denote the center by (h, k) and the distance from the center to an arbitrary point on the circle (x, y) by r, we can apply the distance formula from Section 1.2 to obtain

$$r = \text{the distance from } (h, k) \text{ to } (x, y)$$
$$= \sqrt{(x - h)^2 + (y - k)^2}$$

Squaring both sides, we arrive at the standard equation of a circle.

Standard Equation of a Circle □

The standard form for the equation of a circle with radius r and center (h, k) is

$$(x - h)^2 + (y - k)^2 = r^2$$

EXAMPLE 4 ▦ **Finding the Equation of a Circle**

Find an equation of the circle with radius 2 and center $(3, -5)$.

SOLUTION We use the standard equation of a circle with $h = 3$, $k = -5$, and $r = 2$.

$$(x - 3)^2 + (y - (-5))^2 = 2^2$$
$$(x - 3)^2 + (y + 5)^2 = 4$$

EXAMPLE 5 ▦ **Graphing a Circle**

Graph the circle with equation $(x + 4)^2 + (y - 3)^2 = 16$.

SOLUTION We begin by rewriting our equation in standard form.

$$(x - (-4))^2 + (y - 3)^2 = 4^2$$

We then simply "read off" $h = -4$, $k = 3$, and $r = 4$. Thus, our circle is centered at $(-4, 3)$ and has radius 4. The graph is shown in Figure 3.

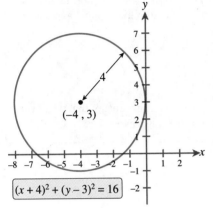

$(x + 4)^2 + (y - 3)^2 = 16$

FIGURE 3

Consider the equation $(x + 4)^2 + (y - 3)^2 = 16$ from Example 5. If we square the terms on the left-hand side, we obtain

$$(x + 4)^2 + (y - 3)^2 = 16$$
$$[x^2 + 2(4)x + 4^2] + [y^2 + 2(-3)y + (-3)^2] = 16$$
$$x^2 + 8x + 16 + y^2 - 6y + 9 = 16$$
$$x^2 + y^2 + 8x - 6y = -9$$

Now, in this form, the center and radius of the circle are difficult to identify. Thus, if we were given the task of graphing the equation $x^2 + y^2 + 8x - 6y = -9$, our first step would be to write it in standard form by somehow reversing the squaring out process just shown. So let's pretend for a moment that we don't already know the standard form of the equation, and see if we can recover it.

Our immediate goal is to write the left side as the sum of two perfect square trinomials—one involving x and the other involving y. To do this, we collect the x and y terms, leaving spaces for constant terms to be named later. This gives us

$$(x^2 + 8x +) + (y^2 - 6y +) = -9 + $$

Next, we must find a constant term ☐ such that $x^2 + 8x +$ ☐ is a perfect square. We know from our earlier work that 16 is the appropriate constant, but let's see if we can produce this result from scratch. From the formula $(x + b)^2 = x^2 + 2bx + b^2$, we see that the constant term is b^2, and since the coefficient of x is $2b$, we can identify b as one-half of the coefficient of x. It follows that the missing constant term can be computed by first dividing the coefficient of x by 2, and then squaring. In this case, we obtain

$$\left(\frac{8}{2}\right)^2 = 4^2 = 16$$

as expected. Applying this technique to the *y* terms as well, and adjusting the right side to reflect additions made to the left side, gives us

$$x^2 + 8x + \boxed{16} + y^2 - 6y + \left(\frac{-6}{2}\right)^2 = -9 + \boxed{16} + 9$$

$$(x + 4)^2 + (y - 3)^2 = 16$$

The process of adding an appropriate constant to create a perfect square trinomial is called **completing the square.**

Completing the Square on □
x² + Dx

To complete the square on an expression of the form $x^2 + Dx$, first divide the coefficient *D* of *x* by 2, then square this result, and add to the original expression to obtain

$$x^2 + Dx + \left(\frac{D}{2}\right)^2$$

Naturally, if the expression is part of an equation, we must ensure that whatever is added to one side of the equation is added to the other as well.

EXAMPLE 6   ■   Completing the Square to Graph a Circle

Graph the circle with equation $x^2 + y^2 - 2x + 4y - 20 = 0$.

SOLUTION   To find the center and radius, we must write the equation in standard form. To do this, we complete the square in both *x* and *y* as follows:

$(x - 1)^2 + (y + 2)^2 = 25$

$(1, -2)$

5

FIGURE 4

$$x^2 - 2x + y^2 + 4y = 20 \qquad \text{Grouping the } x \text{ and } y \text{ terms}$$

$$(x^2 - 2x + \quad) + (y^2 + 4y + \quad) = 20 \qquad \begin{array}{l}\text{Preparing to complete the} \\ \text{square in } x \text{ and } y\end{array}$$

$$(x^2 - 2x + \boxed{1}) + (y^2 + 4y + \quad) = 20 + 1 \qquad \begin{array}{l}\text{Adding } \left(\dfrac{-2}{2}\right)^2 = 1 \text{ to both} \\ \text{sides}\end{array}$$

$$(x^2 - 2x + \boxed{1}) + (y^2 + 4y + \boxed{4}) = 20 + 1 + 4 \qquad \begin{array}{l}\text{Adding } \left(\dfrac{4}{2}\right)^2 = 4 \text{ to both} \\ \text{sides}\end{array}$$

$$(x - 1)^2 + (y + 2)^2 = 25 \qquad \text{Expressing in standard form}$$

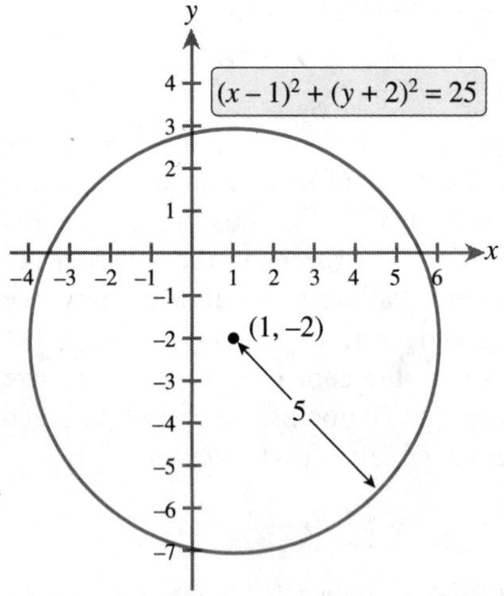

Thus, the circle is centered at $(1, -2)$, with radius 5. Its graph is shown in Figure 4.

CHOICE OF SCALE

The appearance of a graph depends greatly on the choice of **scale** for both the *x*- and *y*-axes. Shown in Figures 5–8 are four views of the graph of $y = x^3 - 0.03x^2 + 0.0002x$ using different scales. Notice that if we were to look only at Figures 5–7, we would conclude that the equation had only one *x*-intercept, whereas Figure 8 clearly shows that there are at least three *x*-intercepts.

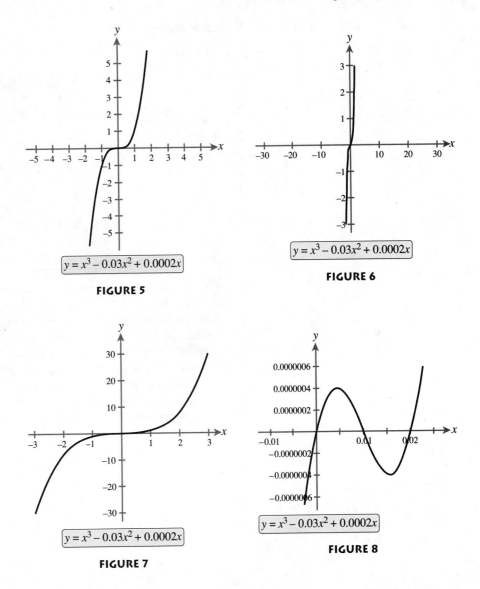

FIGURE 5

$$y = x^3 - 0.03x^2 + 0.0002x$$

FIGURE 6

$$y = x^3 - 0.03x^2 + 0.0002x$$

FIGURE 7

$$y = x^3 - 0.03x^2 + 0.0002x$$

FIGURE 8

$$y = x^3 - 0.03x^2 + 0.0002x$$

The scale that we select depends on which features of the graph interest us. For example, if we only need a rough idea of the shape of the graph for x-values between -3 and 3, then the view shown in Figure 7 is ideal. On the other hand, if we are primarily concerned with intercepts, then the scale used to generate Figure 8 is a better choice.

Often context—the meaning of the variables—will suggest an appropriate choice of scale. For example, if we are graphing an equation relating the average weight y (in pounds) of a child to his age x (in years), then, based on our knowledge of the ages of children and their body weights, we might choose our scales such that the view shows x-values (ages) from 0 to 16 and y-values (weights) from 0 to 160.

When graphing by plotting points, the key feature to be displayed is generally the data points themselves: We are the photographer and they are our subjects. And much like a photographer, we must choose our view so that all of our subjects are visible, yet are not bunched too tightly together.

E X A M P L E 7 ■ Choosing an Appropriate Scale

Graph the equation $y = 2560x^4 - 160x^2$ using the following table of values.

x	$y = 2560x^4 - 160x^2$
-0.4	39.9
-0.3	6.3
-0.2	-2.3
-0.1	-1.3
0	0
0.1	-1.3
0.2	-2.3
0.3	6.3
0.4	39.9

SOLUTION   We begin by noting that the x-values range from -0.4 to 0.4, and the y-values range from -2.3 to 39.9. If we choose the same scale on the x- and y-axes, we would obtain something like the graph in Figure 9. Note that at the given scale, all of the plotted points are within a narrow vertical strip; the points are practically on top of each other! It is virtually impossible to depict the graph using this scale. Instead, we should choose the scales on the x- and y-axes independently. The scale should be chosen so that the points of interest are spread out. The graph in Figure 10 shows the same points plotted on a set of coordinate axes with a different scale. Upon connecting the dots with a smooth curve, we obtain the sketch in Figure 11.

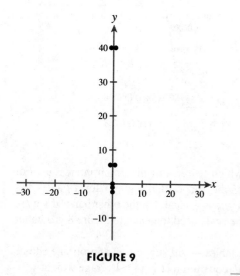

FIGURE 9

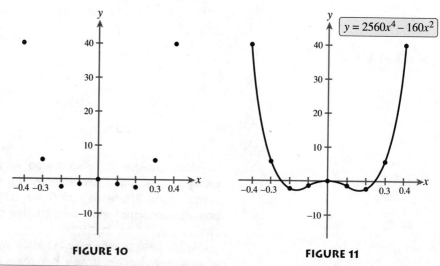

FIGURE 10

FIGURE 11

In the previous example, we plotted 9 points before we felt confident that our sketch captured the salient features of the graph. Depending on the complexity of the equation to be graphed, we might wish to plot 20 or more points. The graphing calculator relieves us of this tedium by instantly plotting dozens of points and connecting them in a reasonable fashion. Moreover, with a graphing calculator, we can easily adjust the view in mere seconds.

GRAPHING CALCULATORS

Graphing calculators are capable of plotting most any equation of the form $y = \square$, where $\square$ is an expression involving only the variable x. For example, to plot the graph of the equation $y = 3x^2 - 12x + 14$ with many models of graphing calculators, we would simply enter the expression `3X² - 12X + 14` as the `Y1` variable, and press the graph button. The resulting graph should look similar to that shown in Figure 12.

Many equations that are not of the form $y = \square$ can be plotted by first solving for y, and then graphing the resulting equation. For example, to graph $x^2 + 3y = 6$, we would first solve for y using algebraic techniques described later in this chapter, giving us $y = (6 - x^2)/3$, and then plot by setting the `Y1` variable to `(6 - X²)/3` and pressing the graph button. The resulting graph is shown in Figure 13.

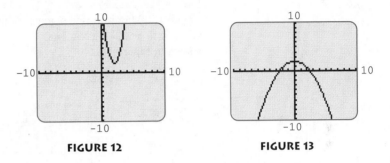

FIGURE 12   **FIGURE 13**

The graphing calculator is a lens through which we can view mathematical equations. To fully exploit the capabilities of this powerful instrument, we must learn how it is focused—that is, how to adjust the scale in such a way that the features of interest are revealed. The scale on a graphing calculator can be adjusted either by using the viewing window variables (see the Calculator Keys box titled ''Viewing Windows'') or by zooming (discussed in the next section). Regardless of the method we use, our goal is to produce a view of the graph in which the scale is both large enough to show all of the features of interest and yet small enough so that these features stand out. But in some cases this is an impossible task; it is a little like attempting to take a satellite photograph of North America that shows the Rio Grande River and the Continental Divide, as well as individual raccoons in your backyard. In some cases, the graph of an equation is best revealed by a series of views—sort of a graphical atlas—with each view demonstrating an important characteristic of the graph.

Producing useful graphs with a graphing calculator is as much art as it is science. In some instances, context—the real-world interpretation of the variables—dictates an appropriate initial viewing window. In other cases, our prior knowledge of what the graph *should* look like helps us select an initial viewing window. For example, we might recognize an equation as having a parabolic or linear graph, or we might be able to quickly determine an intercept. In practice, the most common method may be to simply begin with the standard viewing window and ''see what happens.'' In all cases, the process of finding an appropriate view is iterative: after selecting the viewing window, we plot the graph, and then, based on the appearance of this plot, we adjust the viewing window, repeating as necessary until we produce a satisfactory view. As we gain experience with this process, we'll eventually acquire a sort of graphical sixth sense that tells us which scale adjustments need to be made in order to improve the appearance of our graph.

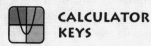

CALCULATOR KEYS

Viewing Windows

With all graphing calculators, the viewing window can be selected by simply indicating the range of *x*-values and the range of *y*-values that are to be shown. This is done by adjusting the values of the window variables Xmin, Xmax, Ymin, and Ymax, which indicate the minimum and maximum values of *x* and *y*, respectively, as suggested by Figure 14.

The first step to producing a graph is to choose appropriate initial settings for the window variables. A good place to start is with the calculator's default settings. Consider, for example, the equation $y = 3x^2 - 12x + 14$. When plotted with the standard viewing window on one popular brand of graphing calculator (Xmin=−10, Xmax=10, Ymin=−10, and Ymax=10), we obtain the graph shown in Figure 15. After noting the general shape of the graph, we refine the viewing window to Xmin=−5, Xmax=5, Ymin=0, and Ymax=20, which yields the graph shown in Figure 16.

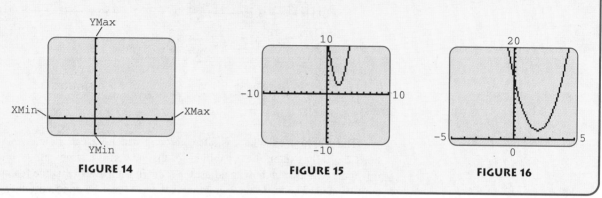

FIGURE 14   **FIGURE 15**   **FIGURE 16**

RULE OF THUMB   Initial values for the range settings are often clear from the context. For example, if *x* represents a percentage score on an exam, then set Xmin=0 and Xmax=100. If you are not sure what the initial values for the window variables should be, select your calculator's **default** settings. For example, the calculator graphics for this text were produced using a model with default settings of Xmin=−10, Xmax=10, Ymin=−10, and Ymax=10.

EXAMPLE 8   ■   Narrowing the Range of the Viewing Window

Use a graphing calculator to graph $y = x^3 - \frac{1}{2}x^2 - x + \frac{1}{2}$ with a view that shows the approximate locations of "humps" and intercepts.

SOLUTION   First we enter the expression $x^3 - \frac{1}{2}x^2 - x + \frac{1}{2}$ and set the window variables to the values Xmin=−10, Xmax=10, Ymin=−10, and Ymax=10. The plot is shown in Figure 17. Although this plot does give some idea of what the graph looks like, the intercepts and "humps" do not stand out. We can remedy this by selecting a smaller viewing window. We adjust the window variables so that Xmin=−2, Xmax=2, Ymin=−2, and Ymax=2. This yields the plot in Figure 18.

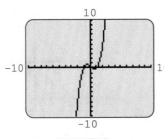

FIGURE 17

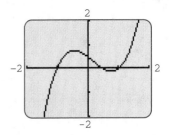

FIGURE 18

The coordinates of interesting points on the graph—such as the "humps" and intercepts from the previous example—can be approximated using the trace feature of a graphing calculator. We describe the general procedure below.

CALCULATOR KEYS

Tracing a Graph

When we select the trace feature on a graphing calculator, the cursor traces points on the graph within the viewing window, and the coordinates of the points on the graph are displayed. Figures 19 and 20 show graphs of $y = x^3 - \frac{1}{2}x^2 - x + \frac{1}{2}$ in which the trace feature has been enabled. In Figure 19, the cursor is located roughly at the point where the graph turns, and we see from the coordinates at the bottom of the viewing window that this point is approximately $(-0.43, 0.76)$. In Figure 20, we see that one of the x-intercepts has an x-coordinate of approximately 0.51. Note that the accuracy of these approximations depends on the viewing window chosen. We will discuss accuracy in greater detail in Section 2.2.

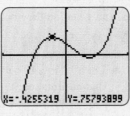

FIGURE 19

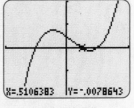

FIGURE 20

EXAMPLE 9   Identifying Points on a Graph

For the equation $y = x^3 - 8$, use a graphing calculator to estimate

a. the x-intercept to the nearest integer
b. the x-values of points on the graph having positive y-coordinates
c. the x-values of points on the graph having negative y-coordinates

SOLUTION
a. The graph of $y = x^3 - 8$ is shown in Figure 21. Using the trace feature, we estimate the x-intercept to be 2, as shown in Figure 22.

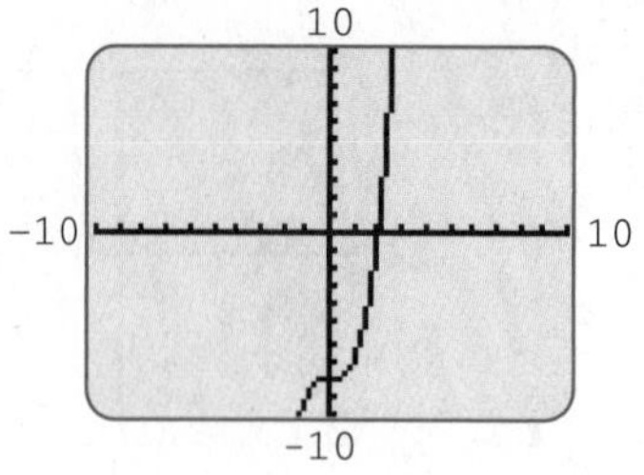

FIGURE 21

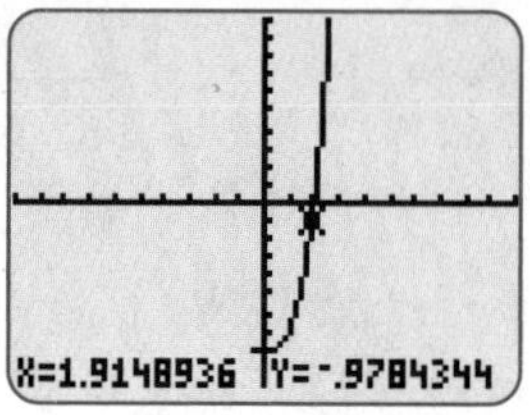

FIGURE 22

b. Here we are interested in points on the graph that lie *above* the x-axis. From Figure 22 we see that the graph is above the x-axis when x is greater than 2. Thus, for each $x > 2$, the corresponding y-coordinate is positive.

c. In this case, we are interested in points on the graph that lie *below* the x-axis. Since the graph is below the x-axis for x-values less than 2, the y-coordinates are negative for $x < 2$.

When an equation relates quantities from the real world, its graph can give us information about trends in the values of these quantities. The trace feature of a graphing calculator provides a convenient means for investigating such trends.

EXAMPLE 10   ▒   **Identifying Points on a Graph**

The average (Fahrenheit) temperature in Orlando, Florida, over the course of the year can be approximated by

$$y = 0.018x^4 - 0.45x^3 + 2.93x^2 - 1.5x + 61.5$$

where x is the number of months after January 1 (so $x = 0$ is January 1, $x = 1$ corresponds to February 1, and so forth). What are the lowest and highest average temperatures, and when do they occur?

SOLUTION   By setting $x = 0$, we determine that the y-intercept is 61.5, which suggests that the range of y-values should include 61.5. Moreover, since we are only interested in a 12-month period, the range in x-values should be limited to the interval [0, 12]. This information, together with our intuition about Florida temperatures, suggests that we try the range values Xmin=0, Xmax=12, Ymin=55, and Ymax=90. The result is shown in Figure 23. Using the trace feature, we see in Figure 24 that the minimum temperature is approximately 61.3° when $x = 0.26$ (the second week in January). Figure 25 shows that the maximum temperature is approximately 84.2° when $x = 6.3$ (the second week of July).

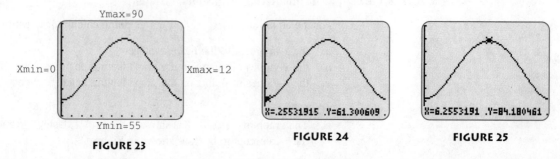

FIGURE 23

FIGURE 24

FIGURE 25

UNDERSTANDING AND MASTERY CHECKLISTS

CONCEPTS TO UNDERSTAND

- ☐ Graph of an equation
- ☐ Intercepts
- ☐ Graphs of lines
- ☐ Standard equation of a circle
- ☐ Completing the square
- ☐ Scale
- ☐ Viewing window
- ☐ Window variables
- ☐ Trace feature

SKILLS TO MASTER

- ☐ Sketch the graph of an equation by plotting points.
- ☐ Determine the intercepts of a graph.
- ☐ Graph a line given its equation.
- ☐ Find the equation of a circle given its radius and center.
- ☐ Complete the square to place the equation of a circle in standard form and then graph it.
- ☐ Choose an appropriate scale when plotting points.
- ☐ Select an appropriate viewing window by adjusting window variables.
- ☐ Identify key features of a graph by using the trace feature.

EXERCISES 2.1

EXERCISES 1–4   *Show that the given points are on the graph of the equation.*

1. $x^2 + y^2 = 100$; $(8, 6)$, $(0, -10)$, $(5\sqrt{2}, 5\sqrt{2})$

2. $x^2 + xy + y^2 = 7$; $(1, 2)$, $(-2, 3)$, $(\sqrt{7}, 0)$

3. $y = \dfrac{x}{x + 1}$; $(-2, 2)$, $(-\frac{4}{3}, 4)$, $(\sqrt{2}, 2 - \sqrt{2})$

4. $x = \sqrt{y^2 - 3y}$; $(0, 3)$, $(2, -1)$, $(\sqrt{10}, 5)$

EXERCISES 5–6   *Use the coordinates of the given points to construct a graph of the relevant portion of the equation. Be sure to choose an appropriate scale.*

5. $y = 1000(x^3 + 1)$

6. $y = x^3 - 3x^2$

x	y
−0.1	999
0.0	1000
0.1	1001
0.2	1008
0.3	1027

x	y
−1	−4
0	0
1	−2
2	−4
3	0
4	15

EXERCISES 7–12   *Sketch the graph of each of the following equations by plotting points. Identify any of the equations that are linear.*

7. $y = 2x - 4$

8. $y = x^2 - 4$

9. $y = -x^2 + 9$

10. $y = -3x + 6$

11. $y = \sqrt{x - 2}$

12. $y = \dfrac{10}{x^2 + 1}$

EXERCISES 13–24   *Find the center and radius of the circle and sketch its graph.*

13. $x^2 + y^2 = 25$

14. $x^2 + y^2 = 10$

15. $2x^2 + 2y^2 = 16$

16. $9 - x^2 = y^2$

17. $(x + 4)^2 + (y - 1)^2 = 4$

18. $(x - 3)^2 + (y + 1)^2 = 16$

19. $x^2 - 4x + y^2 + 6y + 13 = 4$

20. $x^2 + 2x + y^2 - 4y = -4$

21. $x^2 + y^2 - 8x + 2y - 15 = 0$

22. $x^2 + y^2 + 6x + 6y + 8 = 0$

23. $4x^2 + 12x + 4y^2 + 8y = 3$ (*Hint:* First divide through by the coefficient of x^2.)

24. $9x^2 + 9y^2 + 36x - 12y + 31 = 0$ (*Hint:* First divide through by the coefficient of x^2.)

EXERCISES 25–30   *Find the equation of the circle with the given properties.*

25. Center $(-1, 3)$ and radius 4

26. Center $(4, 0)$ and radius 2

27. Center $(4, -3)$ and passes through the origin

28. Center $(-3, -6)$ and passes through the point $(-3, 0)$

29. A diameter with endpoints $(-4, 2)$ and $(2, 8)$

30. A diameter with endpoints $(-3, 6)$ and $(5, -6)$

EXERCISES 31–36   *Find the graph that matches the given equation.*
Choose from i–vi.

37. $x^2 + y^2 = 4$

i.

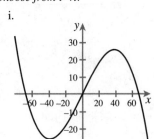

ii.

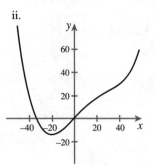

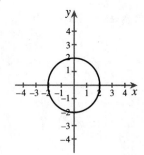

iii.

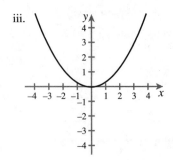

iv.

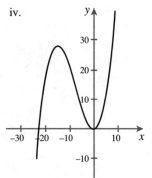

a.   b.

c.   d.

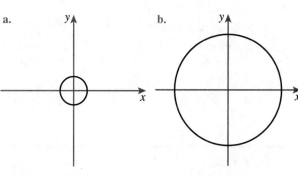

v.

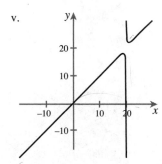

vi.

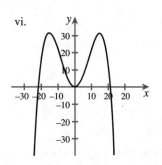

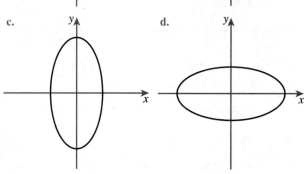

38. $y = x - 2$

31. $y = \dfrac{x^4}{108{,}000} - \dfrac{x^3}{1800} + x$ **32.** $y = x - \dfrac{x^3}{4500}$

33. $y = \dfrac{x^2}{3}$ **34.** $y = \dfrac{x^3}{60} + \dfrac{3x^2}{8}$

35. $y = \dfrac{9x^2}{32} - \dfrac{x^4}{1600}$ **36.** $y = x + \dfrac{1}{x - 20}$

EXERCISES 37–40   *Match each graph in parts a–d with a viewing window chosen from i–iv.*

i. Xmin=−2, Xmax=2, Ymin=−2, *and* Ymax=2

ii. Xmin=−8, Xmax=8, Ymin=−8, *and* Ymax=8

iii. Xmin=−2, Xmax=2, Ymin=−4, *and* Ymax=4

iv. Xmin=−4, Xmax=4, Ymin=−2, *and* Ymax=2

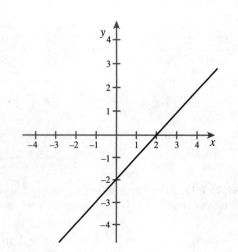

a.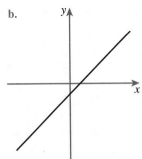

b.

c.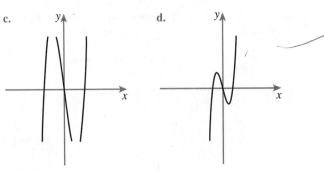

d.

c.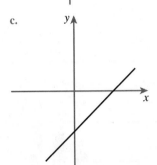

d.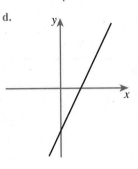

40.

39. $y = x^3 - 4x$

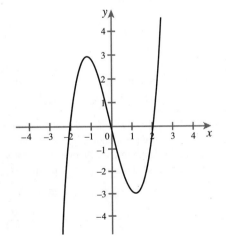

a.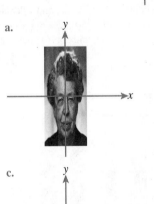

b.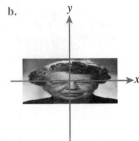

c.

d.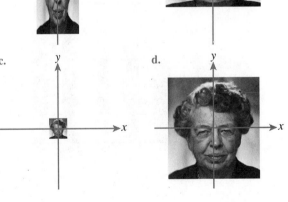

a.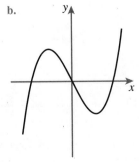

b.

EXERCISES 41–44   *For the given equation, use a graphing calculator to estimate the indicated value(s) to the nearest hundredth.*

a. the x-intercept(s)

b. the x-values of points on the graph having positive y-coordinates

c. the x-values of points on the graph having negative y-coordinates

41. $y = -\frac{1}{2}x^3 - 4$ **42.** $y = x^2 - 9$

43. $y = -x^2 - x + 6$ **44.** $y = x^3 - 4x$

EXERCISES 45–52   *Use a graphing calculator to estimate the indicated value(s) to the nearest hundredth.*

45. $y = \dfrac{x^3}{24} - \dfrac{x^2}{10} - \dfrac{22x}{5} + 12$; x- and y-intercepts

46. $y = \dfrac{x^4}{324} - \dfrac{17x^2}{36} + 15$; x- and y-intercepts

47. $y = \frac{1}{2}x^2 + 4x - 4$; coordinates of the lowest point on the graph

48. $y = |x - 8| + 13$; coordinates of the lowest point on the graph

49. $y = -\frac{1}{16}x^4 + \frac{1}{2}x^3 - \frac{1}{8}x^2 + x$; coordinates of the highest point on the graph

50. $y = \dfrac{12}{\sqrt{x^2 - 30x + 226}}$; coordinates of the highest point on the graph

51. $y = \dfrac{10}{x^2 - 16x + 64}$; x-coordinate(s) corresponding to $y = 2$

52. $y = -2x^2 - 16x - 20$; x-coordinate(s) corresponding to $y = 5$

■ APPLICATIONS

53. *U.S. Population*   The U.S. population for the years 1950–1990 can be approximated with the equation

$$y = 0.0002x^3 - 0.02x^2 + 3x + 151$$

where y represents the population (in millions) and x denotes the year (with $x = 0$ corresponding to 1950). Use an appropriate scale to plot the graph of the equation, and approximate the year when the population reached 190 million.

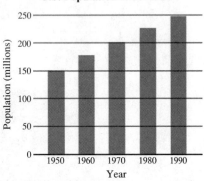

U.S. Population 1950–1990

Data source: U.S. Bureau of the Census.

54. *CO₂ Concentration*   The concentration of CO_2 in the atmosphere during the years 1960–1995 can be approximated with the equation

$$y = -0.0005x^3 + 0.04x^2 + 0.5x + 316$$

where y represents the concentration of CO_2 (measured in parts per million) and x denotes the year (with $x = 0$ corresponding to 1960). Use an appropriate scale to plot the graph of the equation, and approximate the year when the concentration first exceeded 330 parts per million.

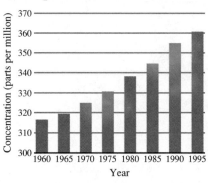

CO₂ Concentration 1960–1995

Data source: The Carbon Dioxide Information Analysis Center, U.S. Department of Energy.

55. *AIDS Cases*   The number of new AIDS cases reported during the years 1980–1996 can be approximated with the equation

$$y = -8.7x^4 + 177x^3 - 520x^2 + 1711x - 663$$

where y represents the number of new cases and x denotes the year (with $x = 0$ corresponding to 1980). Use an appropriate scale to plot the graph of the equation, and approximate the year when the number of new cases was largest.

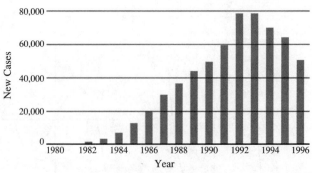

New AIDS Cases 1980–1996

Data source: U.S. Centers for Disease Control.

56. *California Population Density*   The population density of California for the years 1920–1990 can be approximated with either of the two equations

$$y = 0.0005x^3 - 0.04x^2 + 2.8x + 22$$

or

$$y = 0.00054x^3 + 0.072x^2 - 0.11x + 22$$

In both, y represents the density (people per square mile) and x the year (with $x = 0$ corresponding to 1920). For which years do these models agree?

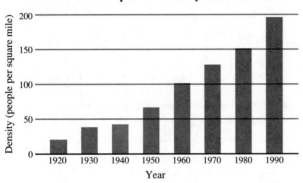

California Population Density 1920–1990

Data source: U.S. Bureau of the Census.

57. *Ferris Wheel*   An amusement park has a ferris wheel with a radius of 30 feet. The center of the wheel is 35 feet above the ground (see Figure 26).

a. Set up a coordinate system with the origin on the ground directly below the center of the wheel, and find an equation for the circular edge of the wheel.

b. If the ferris wheel becomes stuck with one passenger remaining near the top, and the distance from the point on the ground

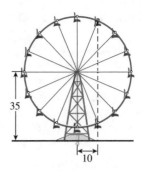

FIGURE 26

directly beneath the stranded passenger to the point on the ground directly beneath the center of the ferris wheel is 10 feet (see Figure 26), what is the length of the shortest ladder that will reach the person?

58. *Waiting Time*   A fast-food restaurant has determined that an acceptable average time for a customer to wait in line is 1 minute. A result from queuing theory suggests that the average length of time in minutes that a customer waits in line is given by $\dfrac{1}{x - y}$, where x is the number of customers served per hour and y is the number of customers who arrive per hour. Thus, the equation

$$1 = \frac{1}{x - y}$$

(or equivalently, $y = x - 1$) describes the relationship between x and y when the average waiting time is 1 minute.

a. Plot the graph of this equation and describe its shape.

b. During the busiest time of the day, customers arrive at a rate of 30 per hour. How many customers must be served per hour so that the time in line is still 1 minute? Illustrate this on your graph from part a.

▨ CONCEPTS AND CRITICAL THINKING

EXERCISES 59–62   *Answer true or false.*

59. The x-intercepts of the graph of an equation can be found by setting $x = 0$.

60. The y-intercepts of the graph of an equation can be found by setting $y = 0$.

61. The area of the viewing window is given by

$$\texttt{(Xmax-Xmin)(Ymax-Ymin)}$$

62. Essentially, graphing calculators sketch graphs by plotting many points.

EXERCISES 63–66   *Give an example of each.*

63. Four points lying on the graph of $x^2 + y^2 = 16$

64. A method by which the viewing rectangle can be adjusted

65. A reason why adjusting the viewing rectangle is important

66. An equation in x and y whose graph has no x-intercept

QUESTIONS FOR DISCUSSION OR ESSAY

67. Why is it usually difficult for you to graph by hand an equation that *cannot* be solved for either x or y? Why do you suppose a graphing calculator cannot graph such equations either?

68. When graphing an equation by plotting points, how do you know when you have plotted enough points? Also, how do you know what happens between the points that you have plotted?

69. A television image of a person is, in a sense, a graph. Of course, as with all graphs, the appearance depends on the choice of scale. It is often said that television "adds 10 pounds" to a subject; explain this effect.

PROJECTS FOR ENRICHMENT

70. *Hidden Intercepts*

    a. Graph the equation $y = 10x^4 \sin\left(\dfrac{0.1}{x}\right)$ with a graphing calculator. (The expression "sin" denotes the trigonometric function sine, but you needn't have any knowledge of trigonometry to solve this problem. Just use the [SIN] key on your calculator when entering the equation.) If you use range settings of Xmin=-10, Xmax=10, Ymin=-10, and Ymax=-10, your graph should look like the one shown in Figure 27.

    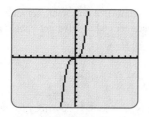

    FIGURE 27

    b. Use the graph you found in part a to estimate the number of x-intercepts of the equation.

    c. Adjust the viewing window to obtain a plot similar to the one shown in Figure 28. List the values of the window variables.

    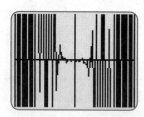

    FIGURE 28

    d. On the basis of the graph in part c, how many x-intercepts do you think there are?

    e. What lesson can be learned from the disparity of your answers to parts b and d?

71. *Relations and Fractal Curves*   In this section, we considered equations in the variables x and y. Such equations define relations between x and y. More formally, a **relation** is any set of ordered pairs (x, y). For example, the sets

$A = \{(9, -3), (4, -2), (1, -1), (0, 0), (1, 1), (4, 2), (9, 3)\}$
$B = \{(x, y) \mid y = x^2\}$
$C = \{(x, x^3) \mid x = 0, \pm1, \pm2, \ldots\}$
$D = \{(x, y) \mid x = 0, 1, 2, 3 \text{ and } y \geq x\}$
$E = \{(x, y) \mid x^2 + y^2 \leq 1\}$

are all relations. Thus, a relation can be a finite set of points, an infinite set of points defined by an equation, or a set of points arising from some other description. In any case, the graph of a relation R is the plot of the points in the set R.

a. Plot the graphs of the relations A–E given above.

It may not always be possible to describe a relation with concise set notation, as we did with the relations A–E. Consider the set of points obtained by the following process. Start with a line segment such as that shown in Figure 29. Divide the line segment into thirds, as shown in Figure 30. Take out the middle third of the segment and replace it with two line segments that form two sides of an equilateral triangle, as in Figure 31.

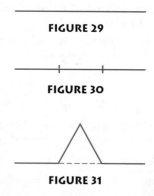

FIGURE 29

FIGURE 30

FIGURE 31

Notice that the total length of the new figure is $\frac{4}{3}$ times the length of original line segment. Now repeat this process on each of the four line segments. That is, take out the middle third of each line segment and replace it with two line segments, as shown in Figure 32. Another repetition of this process leads to the curve shown in Figure 33. If this process is repeated indefinitely, the result is a *fractal curve* similar to that shown in Figure 34. The set of points that form this curve is a relation.

FIGURE 32

FIGURE 33

FIGURE 34

An interesting fact about the resulting fractal curve is that it has infinite length. This is because the total length of the curve at each stage is $\frac{4}{3}$ times the length at the preceding stage, and the process is repeated infinitely many times.

b. Assuming the length of the line segment in Figure 29 is 1 unit, show that the area of the triangle in Figure 31 is $\sqrt{3}/36$.

c. Compute the total area of the regions in Figures 32 and 33 bounded above by the solid lines and below by the dashed line.

d. Even though the total area at each successive stage increases, the area of the region bounded by the fractal curve and the

dashed line is finite. In fact, the area is less than $\sqrt{3}/6$. Explain how we know this.

e. On the basis of what you've seen, explain how the area of an island could be finite even though the coastline has infinite length.

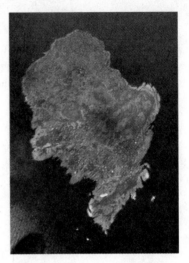

There is a sense in which a coastline, like that of this island in Lake Victoria, has infinite length.

SECTION 2.2   TECHNIQUES FOR SOLVING EQUATIONS

- ☐ **What logical error leads to both mathematical mistakes and ethnic and racial stereotypes?**
- ☐ **Are all infinities created equal?**
- ☐ **How is searching with a graphing calculator like exploring the skies with a viewfinder and a powerful telescope?**
- ☐ **How is equation solving like finding your way back to your hotel in a strange city?**
- ☐ **How can a graphing calculator be used to solve equations?**

SOLUTIONS AND SOLUTION SETS

A **solution** to an equation or inequality involving one variable is a value of the variable that makes the statement true. For example, consider the equation $3x + 5 = 11$. The number 2 is a solution of this equation since if we replace x with 2, we obtain the true statement $3 \cdot (2) + 5 = 11$.

EXAMPLE 1   ■   **Determining Whether a Given Value Is a Solution**

For each of the following equations or inequalities, determine whether or not the given value is a solution.

a. $3x^2 + 4x - 7 = 5x - 3$; $x = -1$

b. $\dfrac{1}{x} + \dfrac{x}{x + 2} = \dfrac{2}{x}$; $x = 2$

c. $\sqrt{x + 1} + x = 1$; $x = 3$

d. $3x + 1 < 6$; $x = 1$

SOLUTION   In each problem, we simply evaluate the left and right sides separately, then check to see whether the simplified forms match.

a. $x = -1$

$3x^2 + 4x - 7$	$5x - 3$
$3(-1)^2 + 4(-1) - 7$	$5(-1) - 3$
$3 - 4 - 7$	$-5 - 3$
-8	-8 ✓ Thus, -1 is a solution.

b. $x = 2$

$\dfrac{1}{x} + \dfrac{x}{x + 2}$	$\dfrac{2}{x}$
$\dfrac{1}{2} + \dfrac{2}{2 + 2}$	$\dfrac{2}{2}$
$\dfrac{1}{2} + \dfrac{2}{4}$	1
1	✓ Thus, 2 is a solution.

c. $x = 3$

$\sqrt{x + 1} + x$	1
$\sqrt{3 + 1} + 3$	1
$\sqrt{4} + 3$	1
5	× Thus, 3 is not a solution.

d.   Substituting 1 for x in the inequality $3x + 1 < 6$, we obtain $4 < 6$, which is true. Thus, 1 is a solution of $3x + 1 < 6$.

The collection of all solutions of an equation or inequality is called the **solution set**. In this chapter, we describe a variety of algebraic and graphical techniques for solving—that is, finding the solution sets of—equations and inequalities.

EQUIVALENCE   Two equations or inequalities are said to be **equivalent** if they have the same solution sets. For example, $3x = 3$ and $x = 1$ are equivalent because both have $\{1\}$ as their solution set. If the equations or inequalities have different solution sets, then they are said to be **nonequivalent.**

EXAMPLE 2   ■   **Determining Equivalence**

Classify each of the following as equivalent or nonequivalent.

a. $2x = 4$, $x = 2$ **b.** $3x + 5 = 17$, $3x = 12$

c. $x^2 = 9$, $x = 3$ **d.** $x + 1 < 2$, $x < 1$

SOLUTION

a. $2x = 4$ is equivalent to $x = 2$ since the solution set of both equations is $\{2\}$.

b. $3x + 5 = 17$ is equivalent to $3x = 12$ since the solution set for both equations is $\{4\}$.

c. $x^2 = 9$ is not equivalent to $x = 3$ since the solution set of $x^2 = 9$ is $\{-3, 3\}$, whereas the solution set of $x = 3$ is $\{3\}$.

d. $x + 1 < 2$ is equivalent to $x < 1$. The solution set of both inequalities is the set of all real numbers less than 1 or, in interval notation, $(-\infty, 1)$.

Our primary technique for solving equations is to apply arithmetic operations to generate successively simpler, but equivalent, equations. For example, to solve the equation $x + 5 = 8$, we would subtract 5 from both sides to produce the equivalent (but simpler) equation $x = 3$. The following table lists properties of equations that can be used to produce equivalent statements.

Properties of Equations ☐

Addition Property	$x = y$ is equivalent to $x + c = y + c$	You can add the same number to both sides.
Subtraction Property	$x = y$ is equivalent to $x - c = y - c$	You can subtract the same number from both sides.
Multiplication Property	$x = y$ is equivalent to $cx = cy \ (c \neq 0)$	You can multiply both sides by the same nonzero number.
Division Property	$x = y$ is equivalent to $\dfrac{x}{c} = \dfrac{y}{c} \ (c \neq 0)$	You can divide both sides by the same nonzero number.

E X A M P L E 3 ▨   **Using Properties of Equality to Solve Equations**

Indicate the property that is used at each numbered step in the solution of the following equation.

$$\frac{4x + 1}{3} = 5x - 7$$

Step 1 $4x + 1 = 15x - 21$

Step 2 $4x + 22 = 15x$

Step 3 $22 = 11x$

Step 4 $2 = x$

SOLUTION

Step 1   Here both sides were multiplied by 3; the *multiplication property* was used.

Step 2   21 was added to both sides; the *addition property* was used.

Step 3 $4x$ was subtracted from both sides; the *subtraction property* was used.

Step 4   Both sides were divided by 11; the *division property* was used.

Special care must be taken when using the division property. For example, if both sides of the equation $x^2 = x$ are divided by x, the equation $x = 1$ is obtained. But $x = 1$

is not equivalent to $x^2 = x$ since $x^2 = x$ has two solutions: 0 and 1. The problem with dividing by x is that if x assumes the value 0, we would be dividing by zero, and division by zero is forbidden. Thus, it is better to avoid division by variable expressions altogether.

> **WARNING!**   Avoid dividing both sides of an equation by an expression involving a variable. The division property applies only when we divide by expressions that cannot be zero.

APPROXIMATING SOLUTIONS GRAPHICALLY

Many equations and inequalities are extremely difficult—or even impossible—to solve by purely algebraic techniques. In such cases, we may still be able to *approximate* a solution with the aid of a graphing utility such as a graphing calculator or a computer. To do so, we must first understand the connection between graphs of equations and solutions.

Recall from Section 2.1 that the graph of an equation involving x and y is the set of all points (x, y) satisfying the equation. For example, the graph of the equation $y = x^2$ consists of all points of the form (x, y), where $y = x^2$. In other words, the graph of $y = x^2$ consists of points of the form (x, x^2). Thus, $(-2, 4)$, $(-1, 1)$, $(0, 0)$, $(1, 1)$, and $(2, 4)$ are all points on the graph of $y = x^2$. The graph of $y = x^2$ is shown in Figure 35.

We also saw in Section 2.1 that the x-intercepts of a graph are the points where the graph intersects the x-axis. In other words, an x-intercept of a graph is a point on the graph with y-coordinate 0. Thus, the graph of $y = x^2$, which is shown in Figure 35, has exactly one x-intercept, $(0, 0)$.

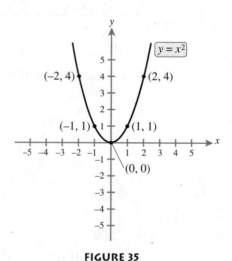

FIGURE 35

EXAMPLE 4   ■   Identifying x-intercepts

The graph of $y = \frac{1}{10}x^3 - \frac{1}{4}x^2 - \frac{21}{20}x + \frac{9}{5}$ is shown in Figure 36. Approximate the x-intercepts.

SOLUTION   The graph appears to intersect the x-axis at the points $(-3, 0)$, $(1.5, 0)$, and $(4, 0)$. Each of these points can be checked by substituting the x-coordinate into the original equation and simplifying to see that $y = 0$ is the corresponding y-coordinate. Thus, the x-intercepts are at $x = -3$, $x = 1.5$, and $x = 4$.

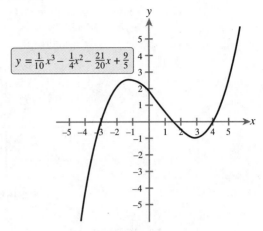

$$y = \frac{1}{10}x^3 - \frac{1}{4}x^2 - \frac{21}{20}x + \frac{9}{5}$$

FIGURE 36

A graphing calculator is an especially useful tool for locating intercepts. One approach is to simply plot the graph of the equation, and then estimate the x-intercepts using the trace feature. A second approach, one that is possible with many current models of graphing calculators, is to plot the graph and then instruct the calculator to automatically compute the x-intercepts. However, regardless of the degree to which we use the calculator, it is usually still necessary to adjust the view of the graph shown by the calculator so that the intercepts are clearly visible. In the box on the following page, we explain some of the features of graphing calculators that are invaluable for adjusting the viewing window and for finding intercepts.

Because even the "automatic" calculation of intercepts with a graphing calculator requires an appropriate view of the graph, we place most of our emphasis in the following examples on adjusting the viewing window.

EXAMPLE 5   ▓   **Finding Intercepts Using Zoom Boxes**

Approximate the x-intercepts of the equation $y = 1000x^3 - 15x^2 + 0.0002$.

SOLUTION   The plot of this equation with a viewing window defined by `Xmin=-10`, `Xmax=10`, `Ymin=-10`, and `Ymax=10` is given in Figure 37. Although it appears that the graph crosses the x-axis somewhere near the origin, it is impossible to tell *exactly* where the graph crosses, or even how many times it crosses.

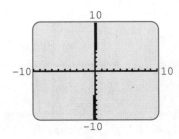

FIGURE 37

(Example 5 continues at the bottom of p. 121)

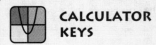

**CALCULATOR
KEYS**

Finding Intercepts

ZOOM IN   This provides us with a close-up view of a region of interest. By zooming in on an intercept repeatedly, we can approximate it with great accuracy. Figures 38 and 39 show the graph of $y = x^2 - 5$ before and after zooming in on the positive x-intercept.

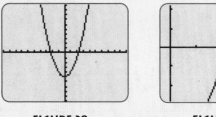
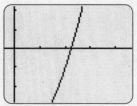

FIGURE 38   **FIGURE 39**

ZOOM OUT   Zooming out is the opposite of zooming in. When we zoom out, we obtain a broader view of the graph; it is as if we are viewing the graph from a greater distance. For example, Figure 40 shows a view of the graph of $y = 0.05x^2 - 8$ in which no intercept is visible. After zooming out (Figure 41), we see two x-intercepts.

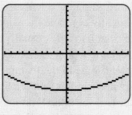
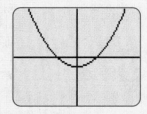

FIGURE 40   **FIGURE 41**

ZOOM BOX   With the zoom-box feature, we can draw a box around a region of interest, which then becomes our viewing window. Typically, the box is drawn by indicating the locations of diagonally opposite corners of the box. Figure 42 shows the graph of an equation that appears to have an x-intercept near $x = 4$. Figure 43 shows the zoom box that we have drawn around the region in which the x-intercept appears to lie. Figure 44 shows the viewing window that results from the zoom box of Figure 43. Note that there are actually three x-intercepts!

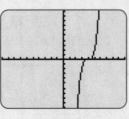
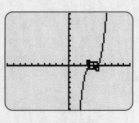
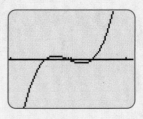

FIGURE 42   **FIGURE 43**   **FIGURE 44**

TRACE   The trace feature is especially useful for estimating intercepts. Figure 45 shows a graph of $y = x^2 - 5$ in which the trace feature has been enabled. The cursor has been moved just to the left of the x-intercept, and the x-coordinate of the cursor position is displayed as 2.1276596. In Figure 46, the cursor has been moved just to the right of the x-intercept, and the x-coordinate of the cursor is displayed as 2.3404255. Thus, the x-intercept lies between 2.1276596 and 2.3404255. For greater accuracy, we could zoom in on the intercept and repeat this process.

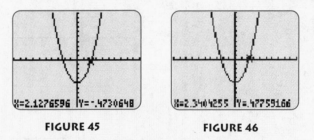

FIGURE 45   **FIGURE 46**

CALCULATE ZERO   Many graphing calculators are capable of "automatically" calculating intercepts (also called zeros or roots). With most models, the intercepts must be found one at a time by plotting the graph with a view that clearly shows an intercept, issuing the calculate-zero command, and then specifying a **left bound** (a number to the left of the intercept), a **right bound** (a number to the right of the intercept), and an **initial guess** (a rough approximation for the intercept, one that is located between the two bounds). The bounds and initial guess can usually be given either by positioning the trace cursor or by entering a number. Figures 47–50 illustrate the process for estimating the negative intercept of $y = x^2 - 5$. In general, the accuracy of the approximation depends on both the equation and the calculator being used. The calculator used for this illustration "guarantees" an approximation that is within 10^{-5} of the actual value.

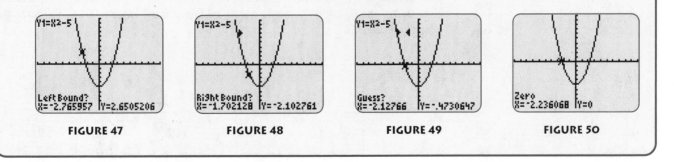

FIGURE 47   **FIGURE 48**   **FIGURE 49**   **FIGURE 50**

The sequence of plots shown in Figures 51–56 illustrate how zoom boxes can give us a better view of the graph near the x-intercepts. Notice that at each step, we construct a zoom box around the region where the graph appears to cross the x-axis.

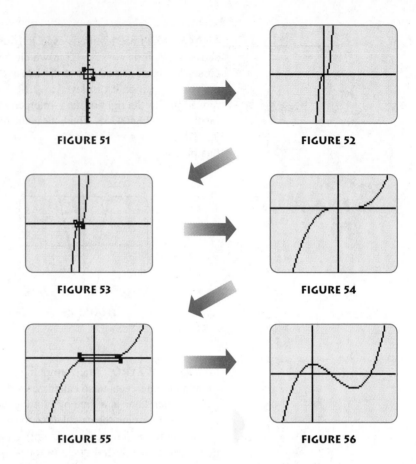

FIGURE 51

FIGURE 52

FIGURE 53

FIGURE 54

FIGURE 55

FIGURE 56

Now using the trace feature, we can obtain not only approximations for the x-coordinates of the intercepts, but also information concerning the accuracy of these approximations. For example, in Figure 57, the cursor has been moved just to the left of the first positive x-intercept, and the x-coordinate of the cursor position is displayed as 0.00404255. In Figure 58, the cursor has been moved just to the right of that same x-intercept (note that the y-coordinate has changed from positive to negative), and the x-coordinate of the cursor is displayed as 0.0043617. Thus, the x-intercept lies between 0.00404255 and 0.0043617, and so we are certain that the value 0.004 is correct to the nearest thousandth.

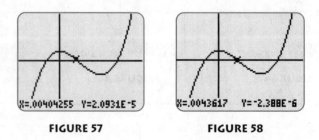

FIGURE 57

FIGURE 58

Repeating this process for the other two intercepts, we obtain -0.003 and 0.014, both of which are correct to the nearest thousandth. Thus, the three intercepts are -0.003, 0.004, and 0.014. Further accuracy can be achieved by zooming in near each of the intercepts. For example, by zooming in twice near the intercept 0.004 and using the trace feature (see Figures 59 and 60), we obtain the approximation 0.0043, which is correct to the nearest ten-thousandth.

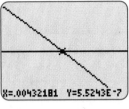
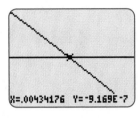

FIGURE 59   **FIGURE 60**

In the previous example, we approximated the values of x such that the y-coordinate, $y = 1000x^3 - 15x^2 + 0.0002$, was equal to 0. Thus, we found approximate solutions to the equation

$$1000x^3 - 15x^2 + 0.0002 = 0$$

This suggests the following strategy for solving equations graphically.

Steps for Solving Equations Graphically □

1. Algebraically rearrange the equation so that it is of the form □ = 0, where □ is an expression involving the variable.
2. Graph the equation $y = □$.
3. Find the x-intercepts of the graph. These are the solutions of the original equation. If there are no x-intercepts, then the equation has no real-valued solutions.

E X A M P L E 6 ▨ **Solving an Equation Graphically**

Solve $x^3 - 7x^2 = 14 - 17x$ graphically. Approximate to the nearest hundredth.

SOLUTION We begin by rewriting the equation so that it is of the form □ = 0.

$$x^3 - 7x^2 = 14 - 17x$$
$$x^3 - 7x^2 + 17x - 14 = 0 \qquad \text{Subtracting 14 and adding } 17x \text{ to both sides}$$

Next we use a graphing calculator to produce a graph of the equation $y = x^3 - 7x^2 + 17x - 14$. The graph in Figure 61 suggests that there is an x-intercept somewhere between 1 and 3. To be certain there is no unusual behavior near this intercept (as there was in Example 5), we zoom in by placing a zoom box around the region where the graph appears to cross the x-axis, as in Figure 62. The result is the viewing window shown in Figure 63.

At this point we are confident that there is only one intercept in the viewing window, and we use the automatic calculate-zero feature to obtain an approximation of 2. By zooming out repeatedly, we conclude that there are no other places where the graph crosses the x-axis. Thus, we have $x \approx 2$ as the only solution of the equation.

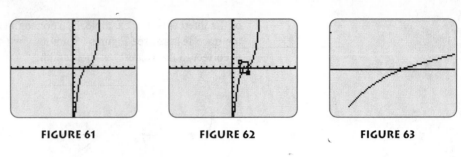

FIGURE 61   FIGURE 62   FIGURE 63

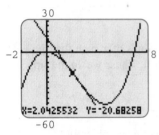

FIGURE 64

An alternative graphical technique can be used to solve equations like that of Example 6. If we graph $y = x^3 - 7x^2$ and $y = 14 - 17x$ on the same set of coordinate axes, then the point of intersection of the two graphs is a point (x, y) such that $y = x^3 - 7x^2$ *and* $y = 14 - 17x$. Thus, $x^3 - 7x^2 = 14 - 17x$. In other words, the x-coordinate of the point of intersection of the two graphs is a solution to the original equation. The coordinates of the point of intersection can be found either by using the zoom and trace features, as suggested by Figure 64, or by applying the calculate-intersect feature, which will be discussed in Section 2.8. Although it is generally easier to rearrange an equation so that it is of the form $\square = 0$ and then find its x-intercepts, as demonstrated in Example 6, there are occasions when it is more convenient to graph the left and right sides of the equation and then find any point(s) of intersection.

No matter what graphical technique is employed to solve an equation, it is important to recognize that the viewing window shows only a portion of the graph. Thus, it may be necessary to adjust the window variables, as we see in the following example.

E X A M P L E   7 ■ **Solving an Equation Graphically**

Approximate two solutions of $0 = \dfrac{x^2 - 5{,}000{,}000}{1000}$ to the nearest integer value.

SOLUTION We first plot the graph of $y = (x^2 - 5{,}000{,}000)/1000$. With the viewing window variables set to their default values, we obtain a view of the graph like the one in Figure 65. The graph is completely out of view! To remedy this, we take a bird's-eye view by zooming out repeatedly until a portion of the graph appears with two visible x-intercepts, as shown in Figure 66. We now use the calculate-zero feature to obtain the solutions $x \approx -2236$ and $x \approx 2236$.

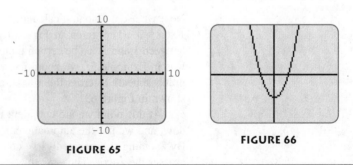

FIGURE 65

FIGURE 66

UNDERSTANDING AND MASTERY CHECKLISTS

CONCEPTS TO UNDERSTAND

☐ Solution set of an equation

☐ Equivalent statements

☐ Connection between x-intercepts and solutions

☐ Graphical approximations of solutions

☐ Zoom features of a graphing calculator

SKILLS TO MASTER

☐ Determine if a given number is a solution of an equation.

☐ Recognize operations resulting in equivalent equations.

☐ Find intercepts and other key features by zooming in and out and using the zoom-box feature.

☐ Approximate solutions to equations graphically.

EXERCISES 2.2

EXERCISES 1–20 *Determine whether or not the given value is in the solution set of the equation.*

1. $5x + 1 = 6$; $x = 1$

2. $3y - 1 = 6$; $y = \frac{7}{3}$

3. $4t - 3 = 2t + 7$; $t = 4$

4. $\dfrac{x + 3}{2} = \dfrac{4x + 1}{3}$; $x = -1$

5. $0.3s + 0.5 = 0.2(s - 1)$; $s = -7$

6. $0.4(u + 1) = 0.3(2u - 1)$; $u = 3.5$

7. $(3 \times 10^7)x + (2 \times 10^8) = 5 \times 10^8$; $x = 10$

8. $(2.3 \times 10^{-4})r + 0.3 = (1.1 \times 10^{-4})r$; $r = -2.5 \times 10^{-4}$

9. $z^2 + 5z = -6$; $z = -2$ 10. $x^2 - 9 = 0$; $x = -3$

11. $|3 - 4u| = |2u - 1|$; $u = 1$

12. $|2 - a^2| = a^2 - 2a + 4$; $a = 3$

13. $3t^3 + 4t^2 - t = 0$; $t = -1$

14. $(y + 3)(y^2 + 1) = 0$; $y = -3$

15. $x^2 = a$; $x = \pm\sqrt{a}$ $(a > 0)$

16. $as + b = c$; $s = \dfrac{c - b}{a}$, $a \ne 0$

17. $\sqrt{z^2 - 6z + 9} = \sqrt{2z} - 1$; $z = 2$

18. $\sqrt{(x + 2)^2} = x + 2$; $x = -4$

19. $\dfrac{6}{u - 2} = u - 1$; $u = 4$

20. $\dfrac{3x^2}{x + 1} = \dfrac{1}{2}$; $x = \dfrac{1}{2}$

EXERCISES 21–30 *Indicate whether or not each pair of equations is equivalent. Justify your answer.*

21. $x(x + 1) = 0$
     $x + 1 = 0$

22. $2x + 3 = 0$
     $2x = -3$

23. $x^2 = 9$
     $x = 3$

24. $-x = -1$
     $x = 1$

25. $2x + 1 = 4$
     $2x = 3$

26. $(x - 5)^2 = 16$
     $x - 5 = 4$

27. $\sqrt{x} = -1$
     $(\sqrt{x})^2 = (-1)^2$

28. $x = 4$
     $\dfrac{1}{x} = \dfrac{1}{4}$

29. $7x = 3$
     $x = \dfrac{3}{7}$

30. $(x - 5)(x + 2) = 0$
     $x + 2 = 0$

EXERCISES 31–34 *An equation is solved using the properties of equality. Indicate which property is used at each lettered step.*

31. $3x + 3 = 5$
   a. $3x = 2$
   b. $x = \dfrac{2}{3}$

32. $-6x + 3 = -15$
   a. $-6x = -18$
   b. $x = 3$

33. $\dfrac{2x + 3}{4} = \dfrac{3x - 1}{5}$
   a. $20\left(\dfrac{2x + 3}{4}\right) = 20\left(\dfrac{3x - 1}{5}\right)$
        $5(2x + 3) = 4(3x - 1)$
        $10x + 15 = 12x - 4$
   b. $-2x + 15 = -4$
   c. $-2x = -19$
   d. $x = \dfrac{19}{2}$

34. $(2x - 3)(3x + 7) = (6x + 1)(x - 4)$
        $6x^2 + 14x - 9x - 21 = 6x^2 - 24x + x - 4$
        $6x^2 + 5x - 21 = 6x^2 - 23x - 4$
   a. $5x - 21 = -23x - 4$
   b. $28x - 21 = -4$
   c. $28x = 17$
   d. $x = \dfrac{17}{28}$

EXERCISES 35–42   *An equation and the number of solutions are given. Use a graphing calculator to approximate the solutions to the nearest tenth.*

35. $x^3 - 6x^2 + 14x - 11 = 0$ (1 solution)

36. $x^3 - 4x^2 = 4x - 16$ (3 solutions)

37. $x^3 = 15x^2 + 88x + 100$ (3 solutions)

38. $x^3 - 2x + 1 = 0$ (3 solutions)

39. $\dfrac{7x}{x+3} = \dfrac{4x-1}{x-1}$ (2 solutions)

40. $\sqrt{3x+7} = \sqrt{x+2} + 1$ (2 solutions)

41. $x^3 - 6x^2 = \dfrac{198}{25} - \dfrac{299}{25}x$ (3 solutions)

42. $(x-10)^3 = \dfrac{x-10}{25}$ (3 solutions)

◼ APPLICATIONS

43. *Hang Time*   When a team mascot jumps off a 2-foot-high trampoline as part of a slam-dunking half-time show, her initial velocity is 20 feet per second. It can be shown that her height in feet t seconds after jumping is given by $h = -16t^2 + 20t + 2$. Find the mascot's hang time—that is, the length of time she is airborne.

44. *Breaking Even*   A retail outlet specializing in the sale of trail mix estimates that when x bags of the mix are sold, a profit of $x^3 - 6x^2 + 12x - 58$ dollars is earned. Estimate the number of bags that must be sold for the retailer to break even—that is, for the profit to be zero.

45. *Unknown Interest*   An investor deposits $500 into a savings account on January 1, 2000. Her only transaction is a deposit of $200 on January 1, 2002. The balance in the account as of January 1, 2005 is $1000. Assuming the bank pays interest compounded annually and that the interest rate has been constant, compute the interest rate. (*Hint*: Use the compound interest formula to find an expression for the accumulated value of each of the deposits. Set the sum equal to 1000 and solve.)

46. *Snowman Dimensions*   A snowman is constructed by placing three spherical snowballs on top of one another. The bottom sphere has a radius $\frac{1}{2}$ foot greater than that of the middle sphere, which, in turn, has a radius $\frac{1}{2}$ foot greater than that of the top sphere. The snowman sits in a barrel, which collects the water as the snowman melts. Based on the volume of water in the barrel, it is estimated that the snowman had a volume of 20 cubic feet. Find the radius of each of the spherical snowballs. (*Hint*: The volume of a sphere of radius r is $\frac{4}{3}\pi r^3$. If r represents the radius of the smallest snowball, then what are the radii of the other spheres?)

◼ CONCEPTS AND CRITICAL THINKING

EXERCISES 47–50   *Answer true or false.*

47. Some equations have no solutions.

48. Two equations are equivalent if and only if they have some solutions in common.

49. The equations $x(x-1) = (x+1)(x-1)$ and $x = x + 1$ are equivalent.

50. The x-intercepts of an equation of the form $y = \square$ correspond to solutions of $\square = 0$.

EXERCISES 51–54   *Give an example of each.*

51. An equation equivalent to $x = 4$

52. A property of equations

53. An equation with solution set $\{-1, 1\}$

54. A reason to avoid dividing both sides of an equation by an expression involving a variable

55. What's wrong with the following proof that $1 = 2$?

$$x = 0$$
$$x + x = 0 + x$$
$$2x = x$$
$$\frac{2x}{x} = \frac{x}{x}$$
$$2 = 1$$

56. A view of the graph of an equation is shown in Figure 67. Discuss the number of possible x-intercepts. In particular, is it possible for the graph to have two x-intercepts within the given viewing window? Could it have three x-intercepts? How about four or five?

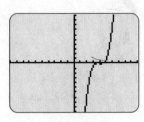

FIGURE 67

■ QUESTIONS FOR DISCUSSION OR ESSAY

57. Compare the process by which an intercept of the graph of an equation is located using a graphing calculator with the method that one would use to find a particular crater on the moon using a viewfinder (a small wide-angled telescope) and a powerful telescope. Explain why the powerful telescope alone would be useless.

58. Modern graphing calculators have capabilities that would seem extraordinary to algebra students of just 10 years ago. What additional features do you foresee for calculators 10 years from now? What about 25 years from now? How will these changes affect the way that algebra is taught?

59. An **identity** is an equation or inequality that is true for all values of the variables. When mathematics students are asked to *verify*

an identity, they are being asked to show that the given equation or inequality holds for all values of the variable for which the expression "makes sense." It is not uncommon for students to attempt to verify an identity by showing that it holds for some particular value of the variable. For example, some students, when asked to show that

$$(x + 1)^2 = x^2 + 2x + 1$$

would argue as follows, "Take $x = 2$. On the left-hand side you get 9 and on the right side you get 9, so the equation is true." What is wrong with this reasoning? Explain how this logical error (in nonmathematical contexts!) could lead to stereotypes and prejudice.

■ PROJECTS FOR ENRICHMENT

60. *The Case of the Wandering Amnesiac*   An amnesiac wanders into Inspector Magill's office, desperate for help in establishing his identity. The following exchange takes place:

Amnesiac: I have absolutely no idea who I am. I woke up in a strange hotel room in this city with no idea how I got there or why. You are my last hope.

Magill: Did you check for identification in your hotel room? Perhaps you have a driver's license or maybe your name would be in the hotel's register.

Amnesiac: I don't remember the name of the hotel, what room I was in, or even what street it was on. I was so panic-stricken that I simply wandered the streets randomly until I reached the police station. All I remember is that when I left my hotel room I turned right, passed five rooms and entered a stairwell. I walked down five flights of stairs, walked out of the lobby, and turned toward the setting sun. I walked 2 miles, turned left, walked 3 miles, took

another left, walked 100 yards, and turned right into the station.

Magill: You have a pretty good memory for an amnesiac! Come with me, I'll take you to your hotel room.

a. How did Inspector Magill find the hotel room?

b. The value 30 is obtained from a given number by the following process: The given number is multiplied by 3, and 9 is then subtracted from this result. This sum is then cubed. Finally, 4 is added to the result. Find the given number, and explain how this problem can be solved in the same manner that Inspector Magill solved the mystery of the missing hotel room.

61. *Cardinal Numbers*   Solution sets of equations vary tremendously in terms of their size, or **cardinality.** For example, the equation $x = 2$ has exactly one solution, $x = x + 1$ has no solutions, and the solution set of $x = x$ consists of all real numbers. We define the **cardinal number** of a finite set A to be the number of elements in A, and we denote this number by $n(A)$. Thus, for example, if $A = \{0, 1, -1\}$, then $n(A) = 3$. For each of parts a through d, find the cardinal number of the indicated set.

a. $\{0, 1, \ldots, 10\}$.

b. {The prime numbers less than 20} (A prime number is any natural number that has no factors other than 1 and itself. The first two prime numbers are 2 and 3.)

c. the solution set of $|x| = 4$—that is, the set of numbers whose distance from 0 is 4

d. the solution set of $|x| = -4$—that is, the set of numbers whose distance from 0 is -4

For infinite sets, cardinal numbers are not as easy to define. We must first define the notion of equivalence. Two sets A and B are said to be **equivalent** if their elements can be put in a one-to-one correspondence with each other. For example, the sets $\mathbb{N} = \{1, 2, 3, \ldots\}$ (the natural numbers) and $B = \{2, 4, 6, \ldots\}$ are equivalent because of the correspondence illustrated here:

$$
\begin{array}{ccc}
1, & 2, & 3, \ldots \\
\updownarrow & \updownarrow & \updownarrow \\
2, & 4, & 6, \ldots
\end{array}
$$

Any set that is equivalent to $\mathbb{N}$ is said to be an **infinite countable** set. The cardinal number of an infinite countable set is denoted with the symbol $\aleph_0$ (read "aleph nought"—aleph is the first letter of the Hebrew alphabet). The set of real numbers $\mathbb{R}$ is not equivalent to the set of natural numbers and so it is not countable. In some sense, there are many more real numbers than there are natural numbers. However—and this is a seemingly odd feature of equivalence—any interval of real numbers is equivalent to $\mathbb{R}$. As a consequence, there are just as many real numbers in the interval $(0, 1)$ as there are in $\mathbb{R}$ itself! The cardinality of any set that is equivalent to $\mathbb{R}$ is denoted by c. In parts e through h, determine the cardinality of the given set.

e. $\{1, 3, 5, \ldots\}$

f. $(-1, 1)$

g. The set of Fibonacci numbers 1, 1, 2, 3, 5, 8, 13, ..., where each number after the second is the sum of the previous two numbers.

h. The set of real numbers between 0.99998 and 0.99999

62. *Implications*   Statements are said to be *equivalent* if either both are true or both are false; that is, the first statement is true if and only if the second is true. For example, the statements "Tina works in the most populous city in California" and "Tina works in Los Angeles," are equivalent because the first statement is true if and only if the second is true. The mathematical symbol for equivalence is "$\Leftrightarrow$". If statements A and B are equivalent, then we write $A \Leftrightarrow B$.

If the truth of statement A implies the truth of statement B, but not necessarily vice versa, then we write $A \Rightarrow B$. Similarly, if the truth of B implies the truth of A, then we write $A \Leftarrow B$. For example, consider the following statements.

A: "I am a father."
B: "I am a parent."

Clearly all fathers are parents, but not all parents are fathers. Thus, $A \Rightarrow B$, but A and B are not equivalent.

For each of the following pairs of statements, indicate whether the first implies the second ($A \Rightarrow B$), the second implies the first ($A \Leftarrow B$), or both implications hold ($A \Leftrightarrow B$), in which case A and B are equivalent.

a. A: S is a square
   B: S is a rectangle.

b. A: Barbara is in Congress.
   B: Barbara is in the Senate.

c. A: Bob is older than Gene.
   B: Gene is younger than Bob.

d. A: $x + 3 = 5$
   B: $x = 2$

e. A: $x = 3$
   B: $x^2 = 9$

f. A: $\dfrac{x + 1}{x} = \dfrac{1}{x}$
   B: $x + 1 = 1$

g. A: $2x = x$
   B: $2 = 1$

SECTION 2.3   LINEAR EQUATIONS AND INEQUALITIES

☐ How can a room's dimensions be reconstructed from its perimeter?

☐ If a fly begins flying back and forth between two trains that are traveling toward each other on parallel tracks, what distance will the fly have traveled when the trains meet?

☐ If a pitcher on the back of a 60-mile-per-hour truck throws a 95-mile-per-hour fastball in the direction the truck is moving, how fast is the ball going from the perspective of someone on the ground?

☐ What property of equality is shared with human relationships?

LINEAR EQUATIONS

A **linear equation** in the variable x is any equation that can be put in the form $ax + b = 0$. Because the variable appears only to the first power, linear equations are quite easy to solve. For example, the linear equation $2x - 5 = 3$ can be solved as follows:

$$2x - 5 = 3 \qquad \text{Original equation}$$

$$(2x - 5) + 5 = 3 + 5 \qquad \text{Adding 5 to both sides}$$

$$2x = 8$$

$$\frac{2x}{2} = \frac{8}{2} \qquad \text{Dividing both sides by 2}$$

$$x = 4$$

We have used the algebraic properties discussed in Section 2.1 to show that the statement $x = 4$ is equivalent to the statement $2x - 5 = 3$. Thus, 4 is the only possible solution. As a simple check, we can substitute 4 for x in the original equation to get $2(4) - 5 = 8 - 5 = 3$.

Solving Linear Equations ☐

> A linear equation can be transformed to an equivalent equation of the form $x =$ ⌐a number⌐ by performing any combination of the following steps:
>
> **1.** Adding or subtracting the same expression on both sides of the equation
>
> **2.** Multiplying or dividing by the same (nonzero) number on both sides of the equation

> **RULE OF THUMB**   When solving linear equations, multiply out the parentheses and collect all the variable terms on one side of the equation and all the constant terms on the other side. It doesn't matter which side you choose for the variable, although the left side is more conventional.

E X A M P L E   1   ▦   **Solving a Linear Equation**

Solve the linear equation $2(t + 3) = 5t + 9$.

SOLUTION

$$2(t + 3) = 5t + 9$$

$$2t + 6 = 5t + 9 \qquad \text{Distributing}$$

$$2t + 6 - 5t = 5t + 9 - 5t \qquad \text{Subtracting } 5t \text{ from both sides}$$

$$-3t + 6 = 9$$

$$-3t + 6 - 6 = 9 - 6 \qquad \text{Subtracting 6 from both sides}$$

$$-3t = 3$$

$$\frac{-3t}{-3} = \frac{3}{-3} \qquad \text{Dividing both sides by } -3$$

$$t = -1$$

CHECK

$2(t + 3)$	$5t + 9$
$2(-1 + 3)$	$5(-1) + 9$
$2(2)$	$-5 + 9$
4	4 ✓

EXAMPLE 2   ■   **A Contradiction**

Solve the linear equation $\frac{1}{2}(4w + 3) = 2(w - \frac{2}{3})$. Verify the solution graphically.

SOLUTION

$$\frac{1}{2}(4w + 3) = 2\left(w - \frac{2}{3}\right)$$

$$2w + \frac{3}{2} = 2w - \frac{4}{3} \qquad \text{Distributing}$$

$$\frac{3}{2} = -\frac{4}{3} \qquad \text{Subtracting } 2w \text{ from both sides}$$

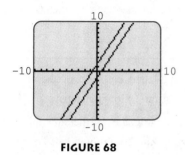

FIGURE 68

Since the resulting statement is a contradiction, there is no solution. In other words, the solution set is the empty set { }. Using a graphing calculator, we plot the graphs of $y = \frac{1}{2}(4w + 3)$ and $y = 2(w - \frac{2}{3})$ on the same set of axes, as shown in Figure 68. [Note that with most graphing calculators we would actually enter `Y1=1/2(4X+3)` and `Y2=2(X-2/3)`.] This view certainly suggests that the two graphs (which appear to be parallel lines) do not intersect, and hence there is no solution to $\frac{1}{2}(4w + 3) = 2(w - \frac{2}{3})$.

APPLICATIONS

An important step in using mathematics to solve real-life problems is the translation of verbal statements into mathematical statements. This step becomes much easier as you begin to recognize certain key words and phrases and their corresponding mathematical symbols. Consider, for example, the verbal statement "*the sum of 3 and a number is 7.*" If we begin by defining x to be the unknown number, then this statement translates into the algebraic statement $3 + x = 7$. Here the word *sum* becomes "+", and the word *is* becomes "=".

EXAMPLE 3   ■   **Translating Verbal Statements into Algebra**

Translate each of the following verbal statements into algebraic statements, and then solve for the variable.

a.   4 is 3 more than a number.
b.   Twice a number is 4 less than the number.
c.   The sum of three consecutive integers is 30.

SOLUTION

a.   Let y be the unknown number. Then

$$4 \text{ is } 3 \text{ more than } y$$
$$4 = y + 3$$
$$1 = y$$

b.   Let x be the unknown number. Then

$$\text{Twice } x \text{ is } 4 \text{ less than } x$$
$$2x = x - 4$$
$$x = -4$$

c. Let x be the smallest of the three integers. Then the other integers are $x + 1$ and $x + 2$. Thus, we have

$$x + (x + 1) + (x + 2) = 30$$
$$3x + 3 = 30$$
$$3x = 27$$
$$x = 9$$

The three integers are therefore 9, 10, and 11.

Linear equations often arise in applications from such diverse areas as geometry, physics, and finance, as we will see in the following examples.

EXAMPLE 4 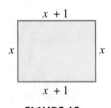 **Perimeter of a Room**

Bob is remodeling a rectangular room in his house. When he gets to the lumber yard, he realizes that he forgot to bring the room's measurements along. But he does remember that the length of the room is 1 foot more than the width and he needs 50 feet of baseboard to go around the edge of the room. What are the dimensions of the room?

SOLUTION The first step, and perhaps the most important one, is to read the problem carefully and make note of the information that is given. A sketch is often helpful if the problem is geometric.

- The room is shaped like a rectangle.
- The length is 1 foot more than the width.
- The perimeter of the room is 50 feet.

We next choose a variable to represent the quantity that is to be found, represent any other unknown quantities using that variable, and label the sketch accordingly, as shown in Figure 69.

$$x = \text{Width of the room}$$
$$x + 1 = \text{Length of the room}$$

Next, we use the information given in the problem to write an equation involving the variable. It is often helpful to first verbalize the equation. Because the perimeter is the sum of the side lengths, we have

$$\text{Sum of side lengths} = 50$$
$$x + (x + 1) + x + (x + 1) = 50$$

Now we can proceed to solve the equation.

$$x + (x + 1) + x + (x + 1) = 50$$
$$4x + 2 = 50$$
$$4x = 48$$
$$x = 12$$

After a solution has been found, it is important to reread the problem to see if the solution seems reasonable and to check the solution against the verbal statement of the

$x + 1$

x x

$x + 1$

FIGURE 69

problem. In this case, if the width is 12, the length must be $12 + 1 = 13$, and so the perimeter is $12 + 13 + 12 + 13 = 50$, as required. Thus, the dimensions are 12 feet by 13 feet.

The steps we followed to arrive at the solution to Example 4 are summarized as follows.

Problem-Solving Strategy ☐

> **1.** Read the problem carefully. Make note of any information. Draw a sketch if possible.
>
> **2.** Choose a variable to represent the unknown quantity. Represent other quantities in terms of the variable. Label the sketch.
>
> **3.** Verbalize an equation and then write it using the variable.
>
> **4.** Solve the equation.
>
> **5.** Check the answer.
>
> **6.** Answer the stated problem completely.

The second step (choosing a variable) may occasionally need to be modified by choosing two or more temporary variables to represent the different quantities in the problem. Eventually, however, you must select one of the variables as your main variable and rewrite each of the others in terms of the main variable. We will use this strategy in the next example.

EXAMPLE 5   ■   Distance, Rate, and Time

A passenger train leaves New York and travels toward Washington, D.C., at a rate of 50 miles per hour. One hour later, a supertrain leaves Washington, D.C., and travels toward New York on a parallel track at a rate of 200 miles per hour. If New York and Washington, D.C., are 225 miles apart, at what point will the two trains pass by each other?

SOLUTION   The information given in the problem is depicted in Figure 70, where d_1 and d_2 represent the distances traveled by the passenger train and the supertrain, respectively, at the point where the two trains meet.

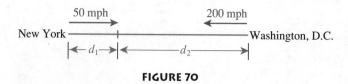

FIGURE 70

We assign the variable t for the time (in hours) that the passenger train has traveled. Since the supertrain left 1 hour later, it will have traveled $t - 1$ hours. Sometimes a table is a helpful way to organize all the information.

	Rate	Time	Distance
Passenger train	50	t	d_1
Supertrain	200	$t-1$	d_2

Since distance equals rate times time, we obtain the following two equations.

$$d_1 = 50t$$
$$d_2 = 200(t - 1)$$

But we also know that New York and Washington are 225 miles apart. This leads to the following equation.

Distance traveled by the passenger train + Distance traveled by the supertrain = 225

$$d_1 + d_2 = 225$$

By substituting for d_1 and d_2, we are able to solve for t.

$$50t + 200(t - 1) = 225$$
$$250t - 200 = 225$$
$$250t = 425$$
$$t = \frac{425}{250} = 1.7$$

In 1.7 hours, the passenger train will have traveled $50(1.7) = 85$ miles from New York, which is where the trains meet. As a check, the supertrain travels for 0.7 hour, which yields a distance of $200(0.7) = 140$ miles from Washington, D.C. The two distances total 225 miles, as required.

WARNING!   When assigning variables, we cannot use the same variable for two quantities that are different. In Example 5, it would have been incorrect to let d represent the distance traveled by *both* trains since the two distances are different. Also, variables must represent numbers. Thus, in Example 5 it would have been inappropriate to write "d_1 = New York."

Sometimes the only strategy needed for solving an application problem is to use a familiar formula. In the following example, we use the formula for simple interest.

EXAMPLE 6   ■   **Simple Interest**

Suppose a wealthy relative leaves you a sizable inheritance. You decide to save some of it for a new wardrobe for school next year. How much of the inheritance must be put into a bank paying simple interest at an annual rate of 5.25% in order to have $600 to spend on clothes after 1 year?

SOLUTION Recall the formula for working with simple interest: $B = P(1 + rt)$, where P is the principal, r is the annual interest rate, and t is the number of years. In this problem, we have $B = \$600$, $r = 0.0525$, and $t = 1$. Substituting these values into the formula leads to a linear equation.

$$600 = P(1 + 0.0525)$$
$$600 = 1.0525P$$
$$1.0525P = 600$$
$$P = \frac{600}{1.0525} \approx 570.07$$

So $570.07 is the amount that must be deposited.

Many common formulas involving several variables are linear in more than one variable. For example, the formula for converting Celsius temperatures to Fahrenheit temperatures ($F = \frac{9}{5}C + 32$) is linear in both C and F. Such formulas can be easily solved for any of the linear variables.

EXAMPLE 7   Converting to Celsius

Disc jockey Damon reports the weather each morning on CMTH, a radio station near the border of the United States and Canada. Unfortunately, his outdoor thermometer only reports the temperature in Fahrenheit, and his listeners have demanded that he give both Fahrenheit and Celsius readings. Use the formula $F = \frac{9}{5}C + 32$ to find a formula that Damon can use to convert Fahrenheit to Celsius.

SOLUTION   Starting with the formula $F = \frac{9}{5}C + 32$, we wish to solve for C in terms of F. In other words, we must isolate C on one side of the equation.

$$F = \frac{9}{5}C + 32$$
$$F - 32 = \frac{9}{5}C$$
$$\frac{F - 32}{9/5} = C$$
$$\frac{5}{9}(F - 32) = C$$

The formula $C = \frac{5}{9}(F - 32)$ will allow Damon to convert any Fahrenheit temperature to Celsius.

EXAMPLE 8   Rewriting an Equation to Solve it Graphically

Rewrite the equation $2x^2 - 5x - 2y = 4$ so that its graph can be plotted with a graphing calculator. Estimate the x-intercept(s) to the nearest hundredth.

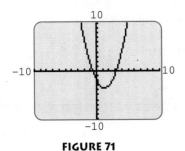

FIGURE 71

SOLUTION   We must solve the equation for y so that it is in the form $y = \square$.

$$2x^2 - 5x - 2y = 4$$
$$2x^2 - 5x - 4 = 2y$$
$$x^2 - \frac{5}{2}x - 2 = y$$

Now we plot the graph of $y = x^2 - \frac{5}{2}x - 2$ with a graphing calculator, as shown in Figure 71. By zooming in and using trace (or using the calculate-zero feature), we obtain $x \approx -0.64$ and $x \approx 3.14$ as the approximate x-intercepts.

LINEAR INEQUALITIES

The procedure for solving linear inequalities is virtually identical to that for solving linear equations. The only difference, although a significant one, occurs when both sides of the inequality are multiplied or divided by a *negative* number. In this case, the direction of the inequality must be reversed. This is perhaps easiest to see through an example. If we begin with the true statement $1 < 2$ and multiply both sides by -1 without reversing the inequality, we would obtain $-1 < -2$, which is false. However, if we reverse the inequality, we obtain the true statement $-1 > -2$.

We illustrate the procedure by solving $5 - 2x > 13$.

$$5 - 2x > 13 \qquad \text{Original inequality}$$
$$-2x > 8 \qquad \text{Subtracting 5 from both sides}$$
$$x < -4 \qquad \text{Dividing both sides by } -2 \text{ and reversing the inequality}$$

We applied algebraic properties to show that the statement $x < -4$ is equivalent to the original inequality. Thus, the solution is *the set of all real numbers x such that x is less than* -4. Using set-builder notation, we can write this as $\{x \mid x < -4\}$. In interval notation, the solution set is the interval $(-\infty, -4)$. We can also graph the solution on the number line, as shown in Figure 72.

FIGURE 72

Solving Linear Inequalities $\square$

> A linear inequality can be transformed into an equivalent inequality by performing any combination of the following steps:
>
> **1.** Adding or subtracting the same expression on both sides of the inequality
>
> **2.** Multiplying or dividing by the same **positive** number on both sides of the inequality
>
> **3.** Multiplying or dividing by the same **negative** number on both sides and **reversing the inequality**

EXAMPLE 9   ▨   **Solving a Linear Inequality**

Solve $4(t - 1) \geq t$ and graph its solution set.

SOLUTION

$$4(t - 1) \geq t$$
$$4t - 4 \geq t \qquad \text{Distributing}$$
$$4t \geq t + 4 \qquad \text{Adding 4 to both sides}$$
$$3t \geq 4 \qquad \text{Subtracting } t \text{ from both sides}$$
$$t \geq \frac{4}{3} \qquad \text{Dividing both sides by 3}$$

So the solution set is $[\frac{4}{3}, \infty)$ and the graph is shown in Figure 73.

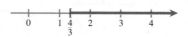

FIGURE 73

EXAMPLE 10   ▪   **Solving a Linear Inequality**

Solve $3t - 7 < 6t + 8$ and graph its solution set.

SOLUTION

$$3t - 7 < 6t + 8$$

$$3t < 6t + 15 \qquad \text{Adding 7 to both sides}$$

$$-3t < 15 \qquad \text{Subtracting } 6t \text{ from both sides}$$

$$t > -5 \qquad \text{Dividing both sides by } -3 \text{ and reversing the inequality}$$

So the solution set is $(-5, \infty)$ and the graph is shown in Figure 74.

FIGURE 74

UNDERSTANDING AND MASTERY CHECKLISTS

CONCEPTS TO UNDERSTAND

☐ Linear equations
☐ Translation of English into algebra
☐ Strategies for applied problem solving
☐ Distance, rate, and time
☐ Linear inequalities

SKILLS TO MASTER

☐ Identify and solve linear equations.
☐ Define appropriate variables when setting up applied problems.
☐ Convert quantitative relationships expressed in English to algebraic equations.
☐ Identify and solve linear inequalities.

EXERCISES 2.3

EXERCISES 1–24 *Solve the linear equation and check your answer.*

1. $x + 8 = 10$

2. $y - 17 = 34$

3. $3x - 4 = 2$

4. $-2x + 7 = 9$

5. $\frac{1}{2}z + 13 = \frac{5}{2}$

6. $-\frac{2}{3}w + 8 = 5$

7. $3x + 4 = 6x$

8. $-2x - 3 = 8x + 17$

9. $-2(m + 3) = 9$

10. $-3(n - 4) = 8$

11. $0.7(x - 0.4) = 0.2x$

12. $0.8(y + 0.4) = 0.6y$

13. $5(-2w + 9) = 3(w - 18) + 4$

14. $2z + 3(z - 6) = 5(z + 8)$

15. $(2.5 \times 10^9)q = 7.5 \times 10^{-5}$

16. $(-7.6 \times 10^{13})w + (7 \times 10^{12}) = 3.2 \times 10^{12}$

17. $-\frac{1}{4}(x + 0.35) = 0.2(x - 1) + 0.4$

18. $0.3(-2y - \frac{1}{2}) = \frac{1}{4}(0.3y - 7) + 9$

19. $-3(2p - 5) - 7(p + 1) = -13(p - 3) + 4$

20. $-9(w - 7) + 4(-2w + 5) + 6 = -3(3w - 8) + 40$

21. $(x + 2)(x - 3) = (x - 3)^2$

22. $(t - 1)(t - 4) = (t + 1)^2$

23. $(y - 1)^2 - (y + 1)^2 = 8$ **24.** $(x + 2)^2 - x^2 = x - 2$

EXERCISES 25—40 *Solve the inequality. Write your answer using interval notation and graph the solution.*

25. $x - 8 < 5$

26. $t + 3 \geq -1$

27. $23 - 4v < 27$

28. $5x - 3 \leq 22$

29. $2z + 7 \geq 5 - 6z$

30. $2y - 2 > 4y - 5$

31. $3(x - 1) \leq 17 - (8 - 3x)$

32. $6t - 2(t + 3) < 3(t + 11)$

33. $\frac{1}{4}w + 1 < w$

34. $2 - \frac{2}{5}x \leq 1$

35. $\frac{3}{4}s - 2 < \frac{5}{8}s - 3$

36. $\frac{1}{3}y - 1 \geq -2(-\frac{1}{6}y + \frac{1}{2})$

37. $2.3x - 9.1 < 16.4 - 5.7x$

38. $16.3 + 21.5t \geq 11.7t - 5.2$

39. $(z - 1)(z + 1) < (z - 3)(z + 1)$

40. $(x + 2)(x - 5) \leq (x - 1)(x - 3)$

EXERCISES 41–50 *Solve the formula for the indicated variable.*

41. $d = rt$ for t

42. $F = ma$ for m

43. $C = 2\pi r$ for r

44. $E = mc^2$ for m

45. $I = prt$ for t

46. $A = \frac{1}{2}bh$ for b

47. $P = 2w + 2l$ for l

48. $A = P(1 + r)$ for r

49. $F = G\dfrac{m_1 m_2}{r^2}$ for m_1

50. $s = -16t^2 + v_0 t + s_0$ for v_0

EXERCISES 51–56 *Solve the given equation for y. Use a graphing calculator to plot the graph of the resulting equation and estimate the x-intercept(s) to the nearest hundredth.*

51. $2x - 3y = 6$

52. $-4x - 5y = 20$

53. $4x^2 - 15 = 8x + 4y$

54. $3x^2 - 12x - 3y + 20 = 0$

55. $x^2y - x^2 - 3x + 4y = 0$

56. $y = x^3 - 2x^2y - 4x$

EXERCISES 57–62 *Translate each of the verbal statements into algebraic statements and then solve for the variable.*

57. 10 is 6 less than a number.

58. Twice a number is 5 more than the number.

59. The product of 4 and a number has the same value as the sum of 4 and that number.

60. One-third of a number has the same value as one less than that number.

61. One less than twice a number is less than 5.

62. Twice a number is at least 3 more than the number.

▇ APPLICATIONS

63. *Consecutive Integers*   The sum of two consecutive integers is 431. Find the two numbers.

64. *Consecutive Integers*   The sum of three consecutive integers is 363. Find the three numbers.

65. *Garden Dimensions*   A rectangular garden is 20 feet longer than it is wide. If the perimeter of the garden is 180 feet, find the length and width.

66. *Room Dimensions*   The width of a rectangular room is half its length. If the perimeter of the room is 51 feet, find the length and width.

67. *Exam Scores*   Juan scored 85 and 88 on the first two exams of a course. He must have an average of 90 to receive an A. What must he score on the third exam to get an A for the course?

68. *Bowling Scores.*   A world-record season bowling average of 245.63 was set by Doug Vergouven during the 1989–1990 season. (*Data source: The Guinness Book of World Records.*) If Doug were to bowl 230 and 240 in the first two games of a three-game series, what must he bowl in the third game to average 246 for the three-game series?

69. *Job Options*   Upon graduation from college, Meredith receives two job offers. One carries a combined annual salary and benefit package of $28,000, whereas the other has a benefit package of $6000, which represents 27% of the annual salary. Which job has the largest total salary and benefit package?

70. *Comparing Income*   Todd and Lisa have a combined monthly gross income of $3008. If Todd's salary is 12% less than Lisa's, how much does each earn?

71. *Saving for a Ring*   Jamal is just finishing his junior year of college and plans on proposing to his girlfriend after he graduates in 1 year. To be sure he has enough money to buy a $500 engagement ring, he wants to deposit some money in a bank account to pay for the ring 1 year later.

   a. What amount must he deposit in an account paying an annual interest rate of 6%?

   b. If he only has $425 to deposit, what annual interest rate must he find?

72. *Cost of Living Increase*   Tonya's contract specifies that she is to receive a cost-of-living increase in her salary each year that is equal to the rate of inflation for the previous year. The rate of inflation in 1998 was approximately 1.6% and Tonya's salary *after* the cost of living increase was $21,100. What was her salary before the increase?

73. *Radio Range*   A family is using a pickup truck and a rental van for an interstate move. In order to stay in contact during the trip, they have rented two-way radios with a range of 20 miles. The two vehicles start at the same time and travel in the same direction, the pickup truck at a constant speed of 55 miles per hour and the van at a constant speed of 45 miles per hour. How long before the two vehicles are beyond the range of the radios?

74. *Highway Rendezvous* Two groups of students from the same school are traveling in different cars to Florida during spring break. The first travels at an average speed of 55 miles per hour. The second leaves 30 minutes after the first and travels at an average speed of 65 miles per hour. How long will it take the second car to catch up to the first?

75. *Indy Racing* During the first 300 miles of a 500-mile race, Michael averaged 180 miles per hour. If Al averaged 178 miles per hour during the same time period, how far behind Michael was Al at the instant Michael passed the 300-mile mark? If Michael continues at the same rate, what speed would Al have to average for the remainder of the race in order to win?

76. *Paving Crews* Two paving crews are resurfacing 81 miles of freeway. The first crew starts at one end of the 81-mile stretch and completes 7 miles each week. The second crew starts 1 week later at the other end and completes 8 miles each week. How long will it take to complete the resurfacing? How much did each crew complete?

77. *Traveling Fly* Two trains are on parallel tracks headed toward each other, one at a rate of 6 miles per hour and the other at a rate of 4 miles per hour. At the instant the trains are 2 miles apart, a fly begins flying back and forth between the two trains at a rate of 10 feet per second. What total distance has the fly traveled by the time the trains meet?

78. *Calling Card Comparison* The following comparison of several popular calling card plans appeared in a 1998 *USA Today* advertisement.

	AT&T	MCI	Sprint
Service charge (per call)	0¢	99¢	30¢
Cost per minute	25¢	40¢	30¢

For each of the three plans, set up and solve a linear inequality to find the interval of possible times for a single call if the total bill for the call may not exceed $5.00. Assume that fractions of minutes are billed at the corresponding fraction of the per minute rate.

CONCEPTS AND CRITICAL THINKING

EXERCISES 79–82 *Answer true of false.*

79. If $-2x < 4$, then $x < -2$.

80. In interval notation, the solution set to $x < 3$ is $(-\infty, 3)$.

81. If x is the length of a rectangle with area A, then $\frac{A}{x}$ is the width of the rectangle.

82. The solution set to a linear inequality is either the empty set or a single interval.

EXERCISES 83–86 *Give an example of each.*

83. A nonlinear equation

84. A linear equation of the form $\square = 3$, having 2 as its only solution

85. A reason why "Let d = dimes" would not be an appropriate variable definition

86. A difference between the techniques for solving linear equations and inequalities

QUESTIONS FOR DISCUSSION OR ESSAY

87. In Example 5, we made the assumption that the rate is constant. What does your experience driving (or riding in) a car tell you about such an assumption? If the rate is not constant, what meaning does "rate" have in the equation distance = rate × time?

88. Checking your solutions is a good way to see if the solutions you found are correct. However, it does not provide a way to check if you found all of the solutions. Why is that? How do we know for sure that the solutions we found are the only ones?

89. The equations in Exercises 21–24 don't appear to be linear. Do they satisfy the definition of linear equations? Explain.

90. A majority of algebra students find word problems difficult. What makes them so difficult? Is it the case that word problems require a deeper level of understanding than standard problems? There are currently computer programs called *computer algebra systems* that can easily solve virtually all the standard problems in this text (and far more difficult problems as well), but there is no computer program currently available that will interpret and solve application problems. Why not?

■ PROJECTS FOR ENRICHMENT

91. *Relative Motion* Suppose baseball great Roger Clemens is traveling on the back of a flatbed truck going 60 mph when he throws a fastball, which under normal circumstances would have a speed of 95 mph. How fast will the ball travel, relative to an observer at the side of the road, if it is thrown

  a. in the direction the truck is moving?

  b. in the direction opposite that of the truck?

  c. If the observer runs out into the road immediately after the truck passes and if the ball is thrown toward the observer at the instant the truck is 60 feet in front of the observer, how long will it take the ball to reach the observer?

How fast will the ball travel relative to Roger Clemens if it is thrown

  d. in the direction the truck is moving?

  e. in the direction opposite that of the truck?

Now suppose the truck is going 110 mph when Roger throws the ball. Answer parts a–e again.

92. *Equivalence Relations* Equality in the set of real numbers is an example of an *equivalence relation* because it has the following three properties.

  I. Reflexive property: $a = a$ for all real numbers a.

  II. Symmetric property: If $a = b$, then $b = a$.

  III. Transitive property: If $a = b$ and $b = c$, then $a = c$.

  a. For each of the following, identify the property that is being used.

   i. If $15x = 3$, then $3 = 15x$.

   ii. If $x + 2 = 8$ and $8 = 2(x - 1)$, then $x + 2 = 2(x - 1)$.

   iii. $9y = 9y$

   iv. If $5t = s$ and $5t = r$, then $r = s$.

More generally, an **equivalence relation** is any relation that satisfies Properties I through III with ''='' replaced by that relation. Equality is only one of many relations that you have likely seen in the past. Other relations include $\equiv$ (congruence for triangles), $\approx$ (similarity for triangles), $<$, $\leq$, $>$, and $\geq$. However, not all of these relations are equivalence relations.

  b. Identify which of the preceding relations are equivalence relations. For those that are not, indicate which of the three properties do not hold.

Nonalgebraic relations can also be defined. For example, if A and B represent people, we can define the relation $A \sim B$ to mean such things as A ''looks like'' B, A ''likes'' B, A ''weighs the same as'' B, or A ''is married to'' B.

  c. Identify which of the preceding human relations are equivalence relations. For those that are not, indicate which of the three properties do not hold.

  d. Discuss three more examples of human relations. At least one must be an equivalence relation.

SECTION 2.4 **ABSOLUTE VALUE EQUATIONS AND INEQUALITIES**

> ☐ **How many years would it take a half-ton man to reach a weight of 200 pounds on a standard weight-loss program?**
>
> ☐ **How much variation in measured distance will improperly inflated car tires cause?**
>
> ☐ **How can distance be estimated with a bicycle tire?**

ABSOLUTE VALUE EQUATIONS

Equations involving absolute value, such as $|x| = 4$, can be solved by considering distances. According to the definition of absolute value, the equation $|x| = 4$ is equivalent to the statement ''the distance between x and 0 is 4.'' Since there are two values for x that are a distance 4 from 0 (namely, 4 and -4), the solution set for the equation $|x| = 4$ is $\{4, -4\}$. These observations generalize to the following rules for solving absolute value equations.

Solving Absolute Value Equations □

If $c > 0$, then $|u| = c$ if and only if $u = -c$ or $u = c$.

If $c < 0$, then $|u| = c$ has no solution since the absolute value of a number can never be negative.

EXAMPLE 1 ▦ Solving an Absolute Value Equation

Solve $|2t| = 10$.

SOLUTION Applying the rules just given, we obtain two equations to solve separately.

$$|2t| = 10$$
$$2t = -10 \quad \text{or} \quad 2t = 10$$
$$t = -5 \quad \text{or} \quad t = 5$$

CHECK

	$t = -5$			$t = 5$	
$|2t|$	10		$|2t|$	10	
$|2(-5)|$			$|2(5)|$		
$|-10|$			$|10|$		
10	✓		10	✓	

EXAMPLE 2 ▦ Solving an Absolute Value Equation

Solve $|3y + 4| = -2$.

SOLUTION Since the number on the right-hand side is negative, there is no solution. Thus, the solution set is the empty set.

EXAMPLE 3 ▦ Solving an Absolute Value Equation

Solve $\left|\frac{1}{2}x - 1\right| = \frac{1}{3}$.

SOLUTION We again apply the rules for solving absolute value equations to obtain two equations that can be solved separately.

$$\left|\frac{1}{2}x - 1\right| = \frac{1}{3}$$

$$\frac{1}{2}x - 1 = -\frac{1}{3} \quad \text{or} \quad \frac{1}{2}x - 1 = \frac{1}{3}$$

$$\frac{1}{2}x = \frac{2}{3} \quad \text{or} \quad \frac{1}{2}x = \frac{4}{3}$$

$$x = \frac{4}{3} \quad \text{or} \quad x = \frac{8}{3}$$

CHECK

$$x = \frac{4}{3} \qquad\qquad x = \frac{8}{3}$$

$$\left| \frac{1}{2} x - 1 \right| \qquad \frac{1}{3} \qquad\qquad \left| \frac{1}{2} x - 1 \right| \qquad \frac{1}{3}$$

$$\left| \frac{1}{2} \left(\frac{4}{3} \right) - 1 \right| \qquad\qquad\qquad \left| \frac{1}{2} \left(\frac{8}{3} \right) - 1 \right|$$

$$\left| \frac{2}{3} - 1 \right| \qquad\qquad\qquad\qquad \left| \frac{4}{3} - 1 \right|$$

$$\left| -\frac{1}{3} \right| \qquad\qquad\qquad\qquad\quad \left| \frac{1}{3} \right|$$

$$\frac{1}{3} \qquad \checkmark \qquad\qquad\qquad \frac{1}{3} \qquad \checkmark$$

Absolute value equations can be formed from statements involving distances. Given two numbers a and b, the distance between a and b can be found by computing $|a - b|$. Notice that it doesn't matter whether or not a is larger than b; either way, the absolute value gives the proper positive distance.

EXAMPLE 4   ▦   **Absolute Value and Distance**

Find all numbers x such that the distance between 5 and twice x is 9.

SOLUTION   The distance between 5 and twice x is given by $|5 - 2x|$. We set this equal to 9 and solve.

$$|5 - 2x| = 9$$

$$5 - 2x = -9 \qquad \text{or} \qquad 5 - 2x = 9$$

$$-2x = -14 \qquad \text{or} \qquad -2x = 4$$

$$x = 7 \qquad \text{or} \qquad x = -2$$

All of the absolute value equations that we have encountered so far have been relatively simple and were solved easily by algebraic means alone. More complex absolute value equations are usually easier to solve graphically, as in the following example.

EXAMPLE 5   ▦   **Solving an Absolute Value Equation Graphically**

Solve $|x + 2| = 5 - |x - 1|$ graphically.

SOLUTION   We begin by writing the equation in the form $\square = 0$.

$$|x + 2| = 5 - |x - 1|$$

$$|x + 2| + |x - 1| - 5 = 0$$

Next we use a graphing calculator to plot the graph of $y = |x + 2| + |x - 1| - 5$, as shown in Figure 75. Note that for many graphing calculators, the absolute value function is selected from a menu and has the form "`abs(`". Thus, for example, the expression $|x + 2|$ would look like "`abs(x+2)`" after it has been entered.

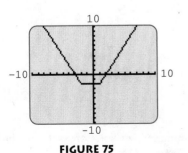

FIGURE 75

Using the trace feature, we find that the *x*-intercepts are 2 and -3, and it is easily verified that these satisfy the original equation. Thus, the equation has two solutions, 2 and -3.

COMPOUND INEQUALITIES

A compound inequality is formed by joining two linear inequalities with the words *and* or *or*. A simple example of a compound inequality is the statement

$$x > -3 \text{ and } x < 2$$

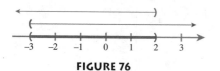

FIGURE 76

The solution set for this statement is the **intersection** (the portion common to both) of the intervals $(-3, \infty)$ and $(-\infty, 2)$—namely, the interval $(-3, 2)$. The graph of this solution set is shown in red in Figure 76. To solve more complicated compound inequalities using the word *and,* we simply solve each inequality separately and take the intersection of the solution sets.

EXAMPLE 6 ▦ Solving a Compound Inequality with And

Solve the compound inequality $6u - 8 \leq 16$ *and* $1 - 4u \leq -3$, and graph the solution set.

SOLUTION   We first solve each inequality separately.

$$
\begin{array}{ccc}
6u - 8 \leq 16 & \text{and} & 1 - 4u \leq -3 \\
6u \leq 24 & \text{and} & -4u \leq -4 \\
u \leq 4 & \text{and} & u \geq 1
\end{array}
$$

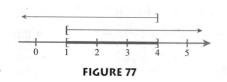

FIGURE 77

By intersecting the two sets $(-\infty, 4)$ and $[1, \infty)$, we obtain the solution set $[1, 4]$. The graph of this set is shown in red in Figure 77.

Compound inequalities formed with the word *or* are solved in a similar way, except instead of intersecting the two sets, we combine them; that is, we form their **union**. The union of two sets *A* and *B* is denoted $A \cup B$.

EXAMPLE 7 ▦ Solving a Compound Inequality with Or

Solve the compound inequality $\frac{1}{5}y + 4 < 3$ *or* $-5y - 3 \leq y$, and graph the solution set.

SOLUTION   First we solve each part separately.

$$
\begin{array}{ccc}
\dfrac{1}{5}y + 4 < 3 & \text{or} & -5y - 3 \leq y \\[2mm]
\dfrac{1}{5}y < -1 & \text{or} & -5y \leq y + 3 \\[2mm]
y < -5 & \text{or} & -6y \leq 3 \\[2mm]
y < -5 & \text{or} & y \geq -\dfrac{1}{2}
\end{array}
$$

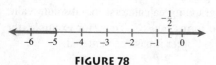

FIGURE 78

The solution is formed by taking the union of these two solutions sets to obtain $(-\infty, -5) \cup [-\frac{1}{2}, \infty)$, as shown in the graph in Figure 78.

> **RULE OF THUMB**   The solution of a compound inequality formed with the word *and* is found by solving each part separately and then taking the intersection of the two sets. The solution of a compound inequality formed with the word *or* is found by solving each part separately and then taking the union of the two sets.

Some compound inequalities are disguised as **double inequalities,** such as $-4 \leq 3x - 2 < 7$. This statement is equivalent to $-4 \leq 3x - 2$ *and* $3x - 2 < 7$ although it is more convenient to leave it as a double inequality so both inequalities can be solved together, as shown in Example 8.

EXAMPLE 8   ▧   Solving a Double Inequality

Solve the double inequality $-4 \leq 3x - 2 < 7$.

SOLUTION   We must isolate x as the middle term.

$$-4 \leq 3x - 2 < 7$$
$$-2 \leq 3x < 9 \qquad \text{Adding 2 to all ''sides''}$$
$$-\frac{2}{3} \leq x < 3 \qquad \text{Dividing all ''sides'' by 3}$$

FIGURE 79

The solution set is $[-\frac{2}{3}, 3)$, as shown in the graph in Figure 79.

Compound inequalities can often be used to solve application problems involving a range of solution values. You can spot such problems by the presence of such phrases as ''at least,'' ''at most,'' ''no more than,'' ''no less than,'' and so forth.

EXAMPLE 9   ▧   Weight Loss

A company is marketing a diet and exercise program that it claims will result in a weight loss of at least 1.5 pounds per week. For health reasons, they also warn that one should not lose more than 5 pounds per week. If Robert Earl Hughes (b. 1926, d. 1958) had started this program after he reached his world record weight of 1069 pounds, over what range of weeks would his weight have dropped to 203 pounds (his weight at age 6)?

SOLUTION   For each week, Robert would lose at least $\frac{3}{2}$ pounds and at most 5 pounds, and so we have

$$\text{Weight loss} \geq \frac{3}{2} \cdot (\text{Number of weeks}) \quad and \quad \text{Weight loss} \leq 5 \cdot (\text{Number of weeks})$$

Let x represent the number of weeks that Robert remained on the program. In dropping from 1069 to 203, he would have a weight loss of 866 pounds in x weeks. This leads to the inequalities $866 \geq \frac{3}{2}x$ and $866 \leq 5x$. Solving for x we obtain

$$\frac{2}{3} \cdot 866 \geq x \qquad \text{and} \qquad \frac{866}{5} \leq x$$

$$577\frac{1}{3} \geq x \qquad \text{and} \qquad 173\frac{1}{5} \leq x$$

Robert Hughes, shown just after he passed the 1000-pound mark.

If the company's claims are correct, it will take anywhere from $173\frac{1}{5}$ to $577\frac{1}{3}$ weeks for Robert to reach 203 pounds.

ABSOLUTE VALUE INEQUALITIES

Inequalities involving absolute value can be solved by rewriting them as compound inequalities, as we demonstrate in Table 1.

TABLE 1

Original inequality	Distance interpretation	Compound inequality	Interval and graph
$\lvert x \rvert < 3$	Numbers whose distance from 0 is less than 3	$-3 < x < 3$	$(-3, 3)$
$\lvert x \rvert > 3$	Numbers whose distance from 0 is greater than 3	$x < -3$ or $x > 3$	$(-\infty, -3) \cup (3, \infty)$

The inequalities in Table 1 lead us to the following general rules for dealing with absolute value inequalities.

Solving Absolute Value Inequalities ☐

If $c > 0$, then

Property 1. $\lvert u \rvert < c$ if and only if $-c < u < c$

Property 2. $\lvert u \rvert > c$ if and only if $u < -c$ or $u > c$

These properties are also valid if $<$ is replaced by $\leq$ and $>$ is replaced by $\geq$.

EXAMPLE 10 ▦ Solving an Absolute Value Inequality

Solve $\lvert 2x + 1 \rvert < 3$.

SOLUTION We first apply Property 1 and write $\lvert 2x + 1 \rvert < 3$ as the compound inequality $-3 < 2x + 1 < 3$. Next we solve for x.

$$-3 < 2x + 1 < 3$$
$$-4 < 2x < 2$$
$$-2 < x < 1$$

The solution is the interval $(-2, 1)$.

EXAMPLE 11 ▦ Solving an Absolute Value Inequality

Solve $\lvert 4 - y \rvert \leq -\frac{1}{2}$.

SOLUTION The rules for solving absolute value inequalities do not apply to $\lvert 4 - y \rvert \leq -\frac{1}{2}$ since the number on the right-hand side is negative. This inequality is a contradiction since absolute values are never negative, so the solution is the empty set.

EXAMPLE 12   ▨   **Solving an Absolute Value Inequality**

Solve $|1 - x| - 5 \geq -2$, and use a graphing calculator to confirm the solution.

SOLUTION   The inequality must be in the form $|u| \geq c$ before we apply Property 2.

$$|1 - x| - 5 \geq -2$$
$$|1 - x| \geq 3$$

According to Property 2, this is equivalent to the compound inequality $1 - x \leq -3$ *or* $1 - x \geq 3$. Solving each inequality separately we obtain

$$1 - x \leq -3 \quad \text{or} \quad 1 - x \geq 3$$
$$-x \leq -4 \quad \text{or} \quad -x \geq 2$$
$$x \geq 4 \quad \text{or} \quad x \leq -2$$

We form the union of these two intervals to obtain the solution $(-\infty, 2] \cup [4, \infty)$. To confirm this with a graphing calculator, we first place the inequality in the form $\square \geq 0$.

$$|1 - x| - 5 \geq -2$$
$$|1 - x| - 3 \geq 0$$

Next we plot the graph of $y = |1 - x| - 3$, as shown in Figure 80. Now if (x, y) is a point on the graph, then $y = |1 - x| - 3$, so that $|1 - x| - 3 \geq 0$ is equivalent to $y \geq 0$. Thus, the points on the graph that have x-coordinates satisfying $|1 - x| - 3 \geq 0$ have nonnegative y-coordinates (that is, the points are on or above the x-axis). Since the graph appears to be on or above the x-axis for $x \leq -2$ and also for $x \geq 4$ (the portions colored red in Figure 81), we conclude that the union of $(-\infty, -2]$ and $[4, \infty)$ is indeed the solution set of $|1 - x| - 3 \geq 0$.

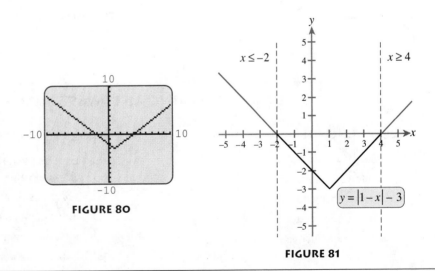

FIGURE 80

FIGURE 81

Absolute value inequalities are useful for expressing measurement errors. For example, if the length l of a metal pin is to be 4 centimeters with an error of no more than 0.01 centimeter, then we can write $|l - 4| \leq 0.01$ to indicate that l and 4 may not be more than 0.01 centimeter apart. By solving the inequality, we obtain the allowable range of l values.

$$|l - 4| \le 0.01$$
$$-0.01 \le l - 4 \le 0.01 \qquad \text{Property 1}$$
$$3.99 \le l \le 4.01 \qquad \text{Adding 4}$$

The answer could also have been expressed as 4 ± 0.01 centimeter, and could have been obtained without even considering absolute values. In the following example, however, the solution is not as obvious and would be difficult to obtain without using absolute values.

E X A M P L E 13   ▇   Bicycle Tires and Measurement Error

The circumference of a bicycle tire is to be 85 inches with an error of no more than 0.3 inch. Find the acceptable range of values for the radius of the tire.

SOLUTION   The circumference of a circle in terms of its radius is $C = 2\pi r$. Thus, if the circumference is to be 85 inches with a maximum error of ± 0.3 inch, we write $|2\pi r - 85| \le 0.3$ and proceed as follows:

$$|2\pi r - 85| \le 0.3$$
$$-0.3 \le 2\pi r - 85 \le 0.3 \qquad \text{Property 1}$$
$$84.7 \le 2\pi r \le 85.3 \qquad \text{Adding 85}$$
$$\frac{84.7}{2\pi} \le r \le \frac{85.3}{2\pi} \qquad \text{Dividing by } 2\pi$$
$$13.48 \le r \le 13.58 \qquad \text{Approximating}$$

So the radius r may be between 13.48 and 13.58 inches.

UNDERSTANDING AND MASTERY CHECKLISTS

CONCEPTS TO UNDERSTAND

☐ Absolute value as distance
☐ Absolute value equations
☐ Compound inequalities
☐ Union and intersection
☐ Absolute value inequalities

SKILLS TO MASTER

☐ Express the distance between real numbers using absolute value.
☐ Solve absolute value equations.
☐ Solve compound inequalities.
☐ Find the union or intersection of two sets of numbers.
☐ Express answers to compound inequalities as an interval or as the union of intervals.
☐ Solve absolute value inequalities, and express the answer using interval notation.

EXERCISES 2.4

EXERCISES 1–10 *Find the solution set for the absolute value equation and check your answer.*

1. $|x| = 6$
2. $|t| = 3$
3. $|3y| = \frac{5}{2}$
4. $\left|\frac{1}{2}x\right| = 3$
5. $|2u - 5| = 1$
6. $\left|1 + 3v\right| = \frac{1}{5}$
7. $\left|4 - \frac{1}{2}z\right| = -3$
8. $\left|\frac{2}{3}a + 4\right| = 16$
9. $|5.3x - 1.7| = 10.5$
10. $|104.6 - 99.1t| = -1347.2$

EXERCISES 11–16 *Write the verbal statement as an absolute value equation and then solve it.*

11. The distance between 5 and x is 7.
12. The distance between -3 and x is 2.
13. The distance between $-\frac{2}{3}$ and twice a number is $\frac{7}{3}$.
14. The distance between 1.5 and three times a number is 2.3.
15. The distance between 2 and half of a number is 5.
16. The distance between $\frac{1}{2}$ and a third of a number is 6.

EXERCISES 17–24 *Solve the compound inequality and graph the solution.*

17. $x + 3 > 1$ and $x - 2 < 4$
18. $\frac{1}{3}y \le -1$ or $6y + 1 \ge -5$
19. $5 \le 8t - 3 \le 9$
20. $3u - 2 \le 1$ or $4u + 1 > 9$
21. $-2x + 7 \le 1$ and $\frac{1}{2}x \le \frac{9}{4}$
22. $3w + 2 < -4$ and $-3w - 2 < 4$
23. $2z \ge 5$ or $\frac{1}{4}(z - 1) \ge 1$
24. $-3 < 1 - \frac{1}{2}x < 1$

EXERCISES 25–44 *Solve the absolute value inequality. Express your answer in interval notation.*

25. $|x| < 5$
26. $|u| \ge 2$
27. $|2w| < 6$
28. $\left|\frac{x}{7}\right| \le 1$
29. $\frac{1}{3}|z| > 4$
30. $|-3x| < 1$
31. $|s - 5| \le 2$
32. $|2u - 1| \ge \frac{5}{3}$
33. $|3x + 4| \ge \frac{1}{2}$
34. $|7t + 1| \le 2$
35. $|-2r + 1| < 8$
36. $|-8p - 3| \ge 2$
37. $\left|\frac{1}{2}x + \frac{3}{4}\right| > 4$
38. $|0.3w - 0.4| \le 0.1$
39. $|0.2y - 5| < -1.63$
40. $\left|\frac{4}{5}z + 9\right| > \frac{1}{3}$
41. $\left|\frac{3z - 2}{5}\right| \le 1$
42. $-2\left|\frac{4x - 3}{5}\right| > 1$
43. $|6t - 4| + 3 < 8$
44. $\frac{1}{3}|3y - 7| + 8 \le 10$

EXERCISES 45–48 *Write the verbal statement as an absolute value inequality and then solve it.*

45. The distance between x and 1 is less than 5.
46. The distance between t and -2 is at least 4.
47. The distance between twice a number and -3 is greater than 6.
48. The distance between -3 times a number and 4 is less than 2.

EXERCISES 49–52 *Use a graphing calculator to help you determine the answer.*

49. Plot the graph of $y = |3x - 2| - 5$. How is the solution set of $|3x - 2| = 5$ represented on the graph? What about the solution sets of $|3x - 2| > 5$ or $|3x - 2| < 5$?
50. Plot the graph of $y = |3x - 2| + 1$. How can you tell from this graph that $|3x - 2| < -1$ has no solutions, whereas the solution set of $|3x - 2| > -1$ is $(-\infty, \infty)$?
51. Find all solutions of $|x - 1| + |x - 2| = 3|x - 3|$.
52. Find all solutions of $|x| = |x - 2|$.

▮ APPLICATIONS

53. *Truck Rental* A moving truck can be rented for $19.95 per day plus $0.39 per mile. The total cost for one day can be expressed as $C = 0.39x + 19.95$, where x represents the number of miles traveled. What range of miles can be traveled if the total rental expense must be between $75 and $100?

54. *Rate of Return* An investment consultant guarantees that an initial investment of $10,000 will yield a balance between $15,000 and $18,000 after 5 years. Assuming simple interest, what range of interest rates is possible?

55. *Scale Accuracy* The manufacturer of a bathroom scale claims the scale is accurate to within $\frac{1}{2}$ pound. If Susan's true weight is 95 pounds, write an absolute value inequality that expresses the possible error in the weight W given by the scale.

56. *Copy Machine Cost* A copy machine has a purchase cost of $1200 and has an estimated page cost of 2¢ per page. What is the minimum number of pages that would have to be copied before the purchase of the copy machine would be more cost-effective than renting a machine at a cost of 5¢ per page?

57. *Tire Radius* If the rear tires of a certain car average 750 revolutions per minute, the distance traveled in 1 hour over the course of many trials varies between 59.80 and 60.69 miles. Find the range of values for the radius of the tires. (*Hint*: The distance traveled is the circumference of the tire, $2\pi r$, times the number of revolutions.)

58. *Scale Accuracy* A grocery store has a scale located in the produce department for customers to weigh the produce they buy.

A sign next to the scale warns that the scale may be off by as much as 2 ounces and so it should only be used to estimate the weight. If you buy a bag of apples that weighs 3.5 pounds (according to the inaccurate scale) and if the apples cost $0.79 per pound, how much variation in price could there be when you take it to the cash register, where there is an accurate scale?

59. *Temperature Conversion*   On a recent trip outside of the United States, Drew begins to feel ill. The only thermometer he can find gives readings in Celsius and is accurate to within 1° Celsius. If the thermometer gives a reading of 38° Celsius, what range of Fahrenheit temperatures does this indicate? The relationship between degrees Celsius C and degrees Fahrenheit F is given by $C = \frac{5}{9}(F - 32)$.

60. *Tire Spokes*   If a bicycle tire is to have a circumference of 85 inches, with a maximum possible error of 0.3 inch, the hub of the tire is 1.5 inches in diameter, and the rim and tire have a

combined radial thickness of 1 inch (see Figure 82), find the allowable range of values for the length of a spoke.

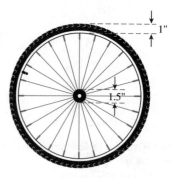

FIGURE 82

CONCEPTS AND CRITICAL THINKING

EXERCISES 61–64   *Answer true or false.*

61. For all real numbers x and y, $|x + y| = |x| + |y|$.

62. If x and y are both negative, then $|xy| = -xy$.

63. The solution set to an absolute value inequality may be the union of two disjoint intervals.

64. The distance between x and 2 is given by $|x - 2|$.

EXERCISES 65–68   *Give an example of each.*

65. An absolute value equation with empty solution set

66. A difference between the solution sets of inequalities of the form $|\,\square\,| < a$ and $|\,\square\,| > a$, where $a > 0$

67. An absolute value equation for which 0 is the only solution

68. An absolute value inequality having the empty set as its solution

69. When solving an absolute value inequality of the form $|u| > c$, the first step, according to Property 2, is to write the statement "$u > c$ or $u < -c$." It is not uncommon for beginning algebra students to ask at this juncture, "Where did the absolute value bars go?" How would you respond to this question?

70. Many students, when solving an absolute value inequality such as $|x + 3| > 2$, will write down $-2 > x + 3 > 2$ as their first step and then arrive at $-5 > x > -1$ as an answer. Even if you know nothing about absolute value, what is it about the answer that tells you something is wrong?

QUESTIONS FOR DISCUSSION OR ESSAY

71. In Example 9, we considered a diet and exercise program that claimed a weight loss of $1\frac{1}{2}$ pounds each week. This implies that weight loss will occur in a linear way, which is based on the assumption that calories will continue to be burned at the same rate. Why isn't this valid? What modification could be made to the weight-loss model?

72. In what sort of real-world contexts are absolute value inequalities likely to arise? Do you see a common thread in the applications in this exercise set?

PROJECTS FOR ENRICHMENT

73. *Estimating Distance with a Bicycle*   In this project, we will discover how an ordinary bicycle tire can be used as a measuring device. You will need a bicycle to perform this experiment.

    a. Measure the circumference of a tire to the nearest centimeter by marking the distance that the bicycle travels in one revolution of the tire. Perform this experiment five times.

    b. Explain why there is some discrepancy in the results of the five trials from part a.

    c. Compute the average of the five results from part a.

    d. Compute the difference between the largest and smallest results in part a. We will assume that the actual circumference deviates from the average obtained in part c by no more than this difference.

    e. Let C be the actual (unknown) circumference of the tire. Express the range of values of C as the solution set to an absolute value inequality and as an interval.

f. Suppose that you compute the distance traveled on a lengthy trip by counting revolutions of your tire. Using the result from part c, you estimate that the trip was 10 kilometers. Express the range of values of the actual trip length as the solution to an absolute value inequality and as an interval.

g. Take the bicycle to a track on which a 100-meter course is marked. Count the smallest integer number of revolutions required to travel the length of the track, then compute the distance by which the bicycle has overshot the starting point. For example, it might happen that 180 revolutions are required, with the bicycle ending up 1 meter past the finish line at the end of the 180th revolution.

h. Express the total distance traveled in part g in meters by adding the overshot distance to 100 meters.

i. Use the result from part h to estimate the circumference of the tire.

j. Assuming that the length of the track is 100 meters, with a maximum error of 1 centimeter, and that the error involved in computing the overshoot is negligible, find the maximum error in computing the circumference by this method. Now express the range of values of the circumference as the solution set to an absolute value inequality and as an interval.

k. Now suppose that, using the value of the circumference found in part i, you find that a certain trip was 10 kilometers long. Use the result of part j to express the range of values of the actual trip length as the solution set to an absolute value inequality and as an interval. What's the maximum error in computing the trip length by this method?

l. Compare and contrast the results from the two methods of computing the circumference.

m. What physical conditions could affect the accuracy of your results?

SECTION 2.5 **POLYNOMIAL EQUATIONS AND INEQUALITIES**

☐ **How many football fields could be covered with the material needed to construct a hot-air balloon capable of lifting the Statue of Liberty?**

☐ **How can you determine a bullet's muzzle velocity with only a stopwatch?**

☐ **If $u(u + 2) = 8$, why is it *not* true that either $u = 8$ or $u + 2 = 8$?**

☐ **How large would a snowball be if it consisted of all the water from the Great Lakes?**

☐ **If a missile is shot straight up at 4000 feet per second, for how long will it be out of the stratosphere?**

POLYNOMIAL EQUATIONS

Recall that a polynomial is an expression involving one or more variables raised to positive integer powers. In this section, we consider equations and inequalities involving polynomials with only one variable, such as those given here.

$$x^2 - x - 6 = 0$$
$$4x^3 = 16x$$
$$243u^5 + 32 = 0$$
$$x^2 - x - 2 < 0$$
$$-x^3 - x^2 \geq -6x$$

As in the previous two sections, we are interested in determining the solution sets of such equations and inequalities. The first technique we investigate in this section involves factoring and the following important property.

The Zero-Product Property   ☐ | For real numbers a and b, $ab = 0$ if and only if $a = 0$ or $b = 0$.

To illustrate the use of this property, consider the polynomial equation $x^2 - x - 6 = 0$. Equations such as this with 0 on the right-hand side are said to be in **standard form.** If we factor the left-hand side, we obtain $(x + 2)(x - 3) = 0$. Now applying the zero-product property, we know that either $x + 2 = 0$ or $x - 3 = 0$. Solving each of these, we obtain $x = -2$ or $x = 3$. Thus, the solution set for the original equation is $\{-2, 3\}$.

EXAMPLE 1 ◼ Solving a Polynomial Equation by Factoring

Solve $u^2 + 2u = 8$.

SOLUTION We first rewrite the equation in standard form (with a zero on the right-hand side) and then factor.

$$u^2 + 2u = 8$$
$$u^2 + 2u - 8 = 0 \qquad \text{Subtracting 8 from both sides}$$
$$(u + 4)(u - 2) = 0 \qquad \text{Factoring}$$

Now we apply the zero-product property.

$$u + 4 = 0 \qquad \text{or} \qquad u - 2 = 0$$
$$u = -4 \qquad \text{or} \qquad u = 2$$

CHECK

$u = -4$		$u = 2$	
$u^2 + 2u$	8	$u^2 + 2u$	8
$(-4)^2 + 2(-4)$		$(2)^2 + 2(2)$	
$16 - 8$		$4 + 4$	
8	✓	8	✓

So our solution set is $\{-4, 2\}$.

EXAMPLE 2 ◼ Solving a Polynomial Equation

Solve $4x^3 = 16x$.

SOLUTION

$$4x^3 = 16x$$
$$4x^3 - 16x = 0 \qquad \text{Writing in standard form}$$
$$4x(x^2 - 4) = 0 \qquad \text{Factoring out } 4x$$
$$4x(x + 2)(x - 2) = 0 \qquad \text{Factoring the difference of squares}$$
$$4x = 0 \quad \text{or} \quad x + 2 = 0 \quad \text{or} \quad x - 2 = 0 \qquad \text{Applying the zero-product property}$$
$$x = 0 \quad \text{or} \quad x = -2 \quad \text{or} \quad x = 2$$

CHECK

$x = 0$		$x = -2$		$x = 2$	
$4x^3$	$16x$	$4x^3$	$16x$	$4x^3$	$16x$
$4(0)^3$	$16(0)$	$4(-2)^3$	$16(-2)$	$4(2)^3$	$16(2)$
0	0 ✓	$4(-8)$	-32	32	32 ✓
		-32	✓		

The solution set is $\{-2, 0, 2\}$.

> **WARNING!**   In order to use the zero-product property to solve a polynomial equation, the equation must be in the standard form $\square = 0$. Thus, in Example 1, it would *not* be correct to simply factor $u^2 + 2u = 8$ as $u(u + 2) = 8$ and then claim that $u = 8$ or $u = 6$. See Exercise 109 for more discussion of this.

Solutions of equations of the form $p(x) = 0$ are called **zeros** of the polynomial $p(x)$. Thus, if $p(a) = 0$, then a is a zero of $p(x)$. Recall that $p(a)$ is obtained by replacing x with a in the polynomial $p(x)$.

EXAMPLE 3   ▨   **Verifying a Zero of a Polynomial**

Show that $t = -1$ is a zero for the polynomial $p(t) = 6t^3 + 7t^2 - t - 2$.

SOLUTION   We simply substitute -1 for t to obtain

$$p(-1) = 6(-1)^3 + 7(-1)^2 - (-1) - 2 = -6 + 7 + 1 - 2 = 0$$

Thus, -1 is a zero of the polynomial $p(t)$.

EXAMPLE 4   ▨   **Finding the Zeros of a Polynomial**

Find the zeros of the polynomial $p(x) = (x - 2)^3 - (x - 2)$.

SOLUTION   We solve the polynomial equation $(x - 2)^3 - (x - 2) = 0$ by factoring out the common binomial factor $(x - 2)$.

$$(x - 2)^3 - (x - 2) = 0$$
$$(x - 2)[(x - 2)^2 - 1] = 0$$
$$(x - 2)[(x^2 - 4x + 4) - 1] = 0$$
$$(x - 2)(x^2 - 4x + 3) = 0$$
$$(x - 2)(x - 3)(x - 1) = 0$$
$$x = 2, \, x = 3, \, x = 1$$

EXAMPLE 5   ▨   **Finding a Polynomial with a Given Zero**

Find the value for c in the polynomial $p(x) = x^2 + cx - 8$ if one of the zeros is -4.

SOLUTION   Since -4 is a zero, it must be true that $p(-4) = 0$. That is, when we substitute -4 into the polynomial, we must get 0.

$$p(-4) = 0$$
$$(-4)^2 + c(-4) - 8 = 0 \qquad \text{Substituting } -4 \text{ into the polynomial}$$
$$8 - 4c = 0$$
$$-4c = -8$$
$$c = 2$$

EXAMPLE 6   ■   Muzzle Velocity

Ammunition is often rated according to its muzzle velocity, the velocity at which the bullet leaves the gun. If a bullet is shot straight into the air with a muzzle velocity of v_0 feet per second, its height in feet after t seconds is given by the polynomial $h(t) = -16t^2 + v_0 t$. Find the muzzle velocity of a bullet that hits the ground $2\frac{1}{2}$ minutes after it is fired.

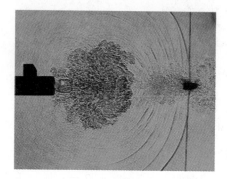

SOLUTION   Since the bullet hits the ground after $2\frac{1}{2}$ minutes (150 seconds), the value of the polynomial $h(t)$ must be zero when $t = 150$. Thus, $h(150) = 0$. Now we can proceed as in Example 5.

$$h(150) = 0$$
$$-16(150)^2 + v_0(150) = 0$$
$$-360{,}000 + 150v_0 = 0$$
$$150v_0 = 360{,}000$$
$$v_0 = 2400$$

So the muzzle velocity is 2400 feet per second.

EXAMPLE 7   ■   Approximating Solutions to a Polynomial Equation

Solve $x^5 - x^2 + 1 = 0$.

SOLUTION   Try as we might, we cannot factor $x^5 - x^2 + 1$ using only integer coefficients. It is thus impossible for us to use the zero-product property. Instead, we approximate any solutions of $x^5 - x^2 + 1 = 0$ using a graphing calculator. We begin by graphing $y = x^5 - x^2 + 1$, as shown in Figure 83. Next we zoom in on the x-intercept several times and then use the trace feature (or, alternatively, the calculate-zero feature) to estimate the coordinates of the x-intercept, as shown in Figure 84. Thus, an approximate value for a solution of $x^5 - x^2 + 1$ is -0.81.

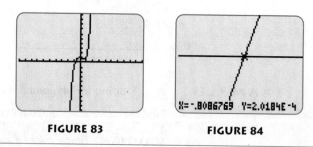

FIGURE 83   **FIGURE 84**

It is generally more difficult to find solutions for polynomial equations when the polynomial cannot be readily factored. In the previous example, we solved the equation graphically because the polynomial $x^5 - x^2 + 1$ didn't factor. Many (in some sense most) polynomial equations cannot be solved exactly by *any* algebraic technique, so the best we can do is approximate the solutions using either a graphing calculator or some other approximation technique (see Exercise 117).

EQUATIONS OF THE FORM $x^n - a = 0$

For now, and again in the next section, we consider some special types of polynomial equations that cannot be factored using integer coefficients and yet can be solved using basic algebraic techniques. The simplest type of polynomial equation that can be solved without factoring is one of the form $x^n - a = 0$. Since this equation can be rewritten in the form $x^n = a$, its solutions are nothing more than the nth roots of a. Recall from Section 1.8 that the number of nth roots of a depends on whether n is even or odd. Thus, we have the following rules for solving $x^n = a$.

Solving Equations of the Form $x^n = a$

1. If n is even and $a > 0$, then $x^n = a$ has two real solutions: $x = -\sqrt[n]{a}$ and $x = \sqrt[n]{a}$. We can condense this by writing $x = \pm\sqrt[n]{a}$.
2. If n is odd, then $x^n = a$ has one real solution: $x = \sqrt[n]{a}$.

EXAMPLE 8   Even nth Roots

Solve the equation $x^4 - 5 = 0$.

SOLUTION

$$x^4 - 5 = 0$$
$$x^4 = 5 \qquad \text{Rewriting in the form } x^n = a$$
$$x = \pm\sqrt[4]{5} \qquad \text{Since 4 is even, there are two 4th roots of 5.}$$

EXAMPLE 9   Odd nth Roots

Solve the equation $243u^5 + 32 = 0$.

SOLUTION

$$243u^5 + 32 = 0$$
$$243u^5 = -32$$
$$u^5 = -\frac{32}{243}$$
$$u = \sqrt[5]{-\frac{32}{243}} \qquad \text{Since 5 is odd, there is only one 5th root.}$$
$$u = -\frac{2}{3}$$

EXAMPLE 10   Air Balloons and the Statue of Liberty

A hot-air balloon capable of carrying three adults has a volume of approximately 3000 cubic yards. Estimate the number of football fields that could be covered with the material needed to construct a hot-air balloon capable of lifting the Statue of Liberty.

SOLUTION   For approximation purposes, we assume that the three passengers, fuel, and basket together weigh 1000 pounds. Thus, we assume that each pound of cargo

requires $\frac{3000}{1000} = 3$ cubic yards of hot air. Because the Statue of Liberty weighs 27,000 tons, or 54,000,000 pounds, the necessary volume for the air balloon is

$$54{,}000{,}000 \text{ pounds} \cdot 3 \text{ cubic yards/pound} = 162{,}000{,}000 \text{ cubic yards}$$

If we assume the balloon is spherical, the volume of the sphere in terms of its radius is $V = \frac{4}{3}\pi r^3$. To find the radius of the balloon, we must solve $162{,}000{,}000 = \frac{4}{3}\pi r^3$ for r.

$$162{,}000{,}000 = \frac{4}{3}\pi r^3$$

$$\frac{3 \cdot 162{,}000{,}000}{4\pi} = r^3$$

$$\sqrt[3]{\frac{3 \cdot 162{,}000{,}000}{4\pi}} = r$$

Evaluating with a calculator, we obtain $r \approx 338$ yards. The amount of material can now be found by using the formula for the surface area of a sphere, $S = 4\pi r^2$. Using $r = 338$ yards, we obtain $S = 4\pi(338)^2 \approx 1{,}440{,}000$ square yards. A football field measures 120 yards by $53\frac{1}{3}$ yards (including the end zones) for a total of approximately 6400 square yards. Thus, $1{,}440{,}000/6400 = 225$ football fields could be covered with the hot-air balloon material.

EXAMPLE 11   ◼   **Compound Interest**

A savings account deposit of \$1100 grows to \$1248.74 after 3 years. If interest is compounded annually, find the interest rate.

SOLUTION   Recall that the formula for interest compounded annually is $B = P(1 + r)^t$. We set $B = 1248.74$, $P = 1100$, and $t = 3$, and solve for the interest rate r.

$$1248.74 = 1100(1 + r)^3$$
$$1.13522 = (1 + r)^3 \qquad \text{Dividing both sides by 1100}$$
$$\sqrt[3]{1.13522} = 1 + r \qquad \text{Taking the cube root of both sides}$$
$$1.04318 = 1 + r$$
$$0.04318 = r \qquad \text{Subtracting 1 from both sides}$$

So the annual rate of interest is approximately 4.32%.

EXAMPLE 12   ◼   **Graphing a Circle Using a Graphing Calculator**

Graph the circle with equation $x^2 + y^2 = 9$.

SOLUTION   We begin by solving the equation for y.

$$x^2 + y^2 = 9$$
$$y^2 = 9 - x^2 \qquad \text{Subtracting } x^2 \text{ on both sides of the equation}$$
$$y = \pm\sqrt{9 - x^2} \qquad \text{Using square roots to solve for } y$$

Thus, we have two branches, $y = \sqrt{9 - x^2}$ and $y = -\sqrt{9 - x^2}$. We enter each of the branches and plot both graphs simultaneously, as shown in Figure 85. Note that the plot appears to be more elliptical than it does circular. This is because the calculator screen is wider than it is high, and so whenever the window variables Xmin and Xmax are given the same values as Ymin and Ymax, the units on the *x*-axis get spaced further apart than on the *y*-axis. With most curves, we would most likely never notice this phenomenon, but it is striking with circles. To an extent, we can compensate for this effect by changing the window variables so that the values for Xmin and Xmax are further apart than Ymin and Ymax. In Figure 86, we used Xmin=−15, Xmax=15, Ymin=−10, and Ymax=10. On some calculators, these are referred to as the ''square'' window settings. The ''square'' settings give us a more nearly circular graph.

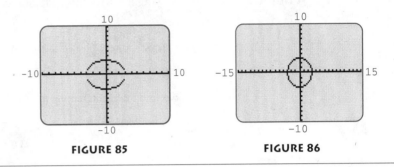

FIGURE 85   **FIGURE 86**

POLYNOMIAL INEQUALITIES

Consider the polynomial inequality $x^2 - x - 2 < 0$. If we let $y = x^2 - x - 2$, then we are interested in the *x*-values for which $y < 0$. Thus, the solutions of the inequality $x^2 - x - 2 < 0$ correspond to the *x*-values for which the graph of $y = x^2 - x - 2$ lies below the *x*-axis. Figure 87 shows the graph of $y = x^2 - x - 2$. Notice that the portion of the graph shown in red below the *x*-axis is between the two *x*-intercepts, which appear to be −1 and 2. Now to be precise, we should find exact values for the *x*-intercepts. We do this by solving the equation $x^2 - x - 2 = 0$.

$$x^2 - x - 2 = 0$$
$$(x - 2)(x + 1) = 0$$
$$x = 2, \ x = -1$$

Thus, the graph of $y = x^2 - x - 2$ is below the *x*-axis for *x*-values between −1 and 2, and so the solution set of $x^2 - x - 2 < 0$ is the interval $(-1, 2)$.

More generally, since the graph of a polynomial is an unbroken curve, the values of a polynomial $p(x)$ can only change sign at *x*-intercepts—that is, at zeros of $p(x)$. It follows that the solution set to a polynomial inequality of the form $p(x) > 0$ [or $p(x) < 0$, $p(x) \leq 0$, or $p(x) \geq 0$] consists of intervals, the endpoints of which are zeros of $p(x)$. This suggests the following strategy for solving polynomial inequalities.

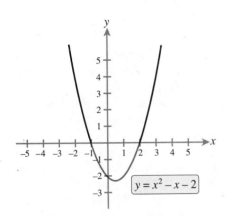

FIGURE 87

Strategy for Solving Polynomial Inequalities □

1. Write the inequality in **standard form** with 0 on the right-hand side and a polynomial $p(x)$ on the left-hand side [that is, $p(x) > 0$; $p(x) < 0$; $p(x) \geq 0$; or $p(x) \leq 0$].

2. Find the *x*-intercepts of the polynomial $p(x)$. In other words, solve the equation $p(x) = 0$. If the *x*-intercepts cannot be solved by algebraic means, estimate them graphically.

3. Determine the *x*-values for which the graph of $y = p(x)$ is above or below the *x*-axis (depending on *which* inequality symbol is being used—above for $>$ and below for $<$). Express the result using interval notation. If the inequality symbol is $\geq$ or $\leq$, the intervals will include their endpoints; otherwise not.

EXAMPLE 13 ◼ **Solving a Polynomial Inequality**

Solve $-x^3 - x^2 \geq -6x$.

SOLUTION We begin by writing the inequality in standard form.

$$-x^3 - x^2 + 6x \geq 0$$

Next we find the *x*-intercepts of $y = -x^3 - x^2 + 6x$.

$$-x^3 - x^2 + 6x = 0$$
$$-x(x^2 + x - 6) = 0$$
$$x(x + 3)(x - 2) = 0$$
$$x = 0, \ x = -3, \ x = 2$$

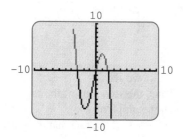

FIGURE 88

Then we use a graphing calculator to produce a graph of $y = -x^3 - x^2 + 6x$, as shown in Figure 88. Since the inequality symbol is $\geq$, we are interested in the *x*-values of the portion of the graph lying *on or above* the *x*-axis, shown in red in Figure 88. The graph is on or above the *x*-axis to the left of -3 and between 0 and 2. Thus, our solution is $(-\infty, -3] \cup [0, 2]$. Note that we include the endpoints because the inequality symbol is $\geq$ and not $>$.

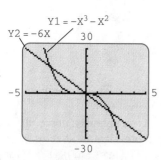

FIGURE 89

We recall that an alternative method for solving equations graphically is to graph the left and right sides of the equation on the same set of coordinate axes and find the *x*-coordinates of the points of intersection. Similarly, the inequality $-x^3 - x^2 \geq -6x$ from Example 13 can be solved by graphing $y = -x^3 - x^2$ and $y = -6x$ on the same set of coordinate axes, as in Figure 89, and identifying those intervals for which the graph of $y = -x^3 - x^2$ is on or above the graph of $y = -6x$.

Now, in general, we can approximate the *x*-coordinates of the points of intersection of the graphs, but in this instance we know from our work in Example 13 that the graphs cross exactly at $x = -3$, $x = 0$, and $x = 2$. (Why?) Because the graph of $y = -x^3 - x^2$ is on or above the graph of $y = -6x$ to the left of 3 and between 0 and 2 (including -3, 0, and 2), the solution set is $(-\infty, -3] \cup [0, 2]$.

EXAMPLE 14 ◼ **Solving a Polynomial Inequality**

Solve $x^5 + 3x^4 + 3x^3 + 5x^2 - 7 < 0$.

SOLUTION We begin by finding the *x*-intercepts of $y = x^5 + 3x^4 + 3x^3 + 5x^2 - 7$. We would like to solve the equation $x^5 + 3x^4 + 3x^3 + 5x^2 - 7 = 0$, but the left side doesn't factor, and we have no algebraic technique for solving this equation. We use the graph, shown in Figure 90, to *estimate* the *x*-intercepts. By using the calculate-zero feature or by zooming in on each of the *x*-intercepts and then using the trace feature, we produce the following approximations of the *x*-intercepts (accurate to the nearest

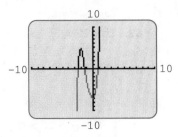

FIGURE 90

one-hundredth): -2.41, -1.33, and 0.83. Since the inequality symbol is $<$, we are interested in the x-values of the portion of the graph lying *below* the x-axis, which is red in Figure 90. The graph is below the x-axis to the left of -2.41 and between -1.33 and 0.83. Thus our solution is $(-\infty, -2.41) \cup (-1.33, 0.83)$.

> **WARNING!** When finding x-intercepts of unfamiliar polynomials with a graphing calculator, be careful not to omit any x-intercepts that are out of the viewing window. In Example 14, it would have been prudent to have zoomed out several times to make sure that there were no other x-intercepts.

UNDERSTANDING AND MASTERY CHECKLISTS

CONCEPTS TO UNDERSTAND

- ☐ Polynomials
- ☐ Zero-product property
- ☐ A polynomial equation in standard form
- ☐ Zeros of a polynomial
- ☐ Polynomial inequalities

SKILLS TO MASTER

- ☐ Recognize polynomial expressions and equations.
- ☐ Solve polynomial equations by factoring.
- ☐ Estimate zeros of a polynomial graphically.
- ☐ Solve polynomial equations of the form $x^n - a = 0$.
- ☐ Solve polynomial inequalities graphically.

EXERCISES 2.5

EXERCISES 1–30 *Find the real solution(s) of the given polynomial equation. Give exact values.*

1. $(x - 1)(x + 2) = 0$
2. $(a + 3)(a - 4)(a + 6) = 0$
3. $(2y + 1)(y + 3)(5y - 2) = 0$
4. $(3v + 2)(2v - 5) = 0$
5. $t^2 + 3t + 2 = 0$
6. $z^2 + 4z + 4 = 0$
7. $x^2 + 2x = 3$
8. $w^2 + 1 = 2w$
9. $(x - 2)(x^2 + 5x + 6) = 0$
10. $(t + 1)(t^2 - 9) = 0$
11. $(y + 1)(y + 4) = 4$
12. $x(x - 2) = 3$
13. $s^2 = 3s$
14. $5z = 2z^2$
15. $2x^2 + 5x - 3 = 0$
16. $3u^2 - 5u - 2 = 0$
17. $t^2 = 8$
18. $x^2 = 24$
19. $(y + 1)^2 = 3$
20. $(s - 2)^2 = 6$
21. $x^3 - 27 = 0$
22. $u^4 - 16 = 0$
23. $z^6 - 5 = 0$
24. $(x - 2)^5 + 4 = 0$
25. $(w + 3)^3 + w^3 = 0$
26. $(x^2 - 1)^3 - 27 = 0$
27. $(x + 2)(x^2 + 5x + 9) + x^2 + x - 2 = 0$
28. $t^2(t + 1) + t^2(t + 3) + (t - 2)(t + 1) + (t - 2)(t + 3) = 0$
29. $(2w + 1)^3 + 2(2w + 1)^2 + (2w + 1) = 0$
30. $(x + 1)^3 - (x + 1) = 0$

EXERCISES 31–38 *Use a graphing calculator to approximate the solution(s) to the nearest hundredth.*

31. $x^5 + x = 1$ (1 solution)
32. $x^2 = 2x^3 + 5$ (1 solution)
33. $x^3 - \frac{13}{3}x^2 - x + 15 = 0$ (2 solutions)
34. $\frac{1}{4}x^4 - \frac{25}{3}x^3 + 103x^2 = 560x - 1128$ (2 solutions)
35. $x^4 - 143x^2 = 100$ (2 solutions)
36. $0.01x^3 + 0.11x^2 - 2.7x + 2.8 = 0$ (3 solutions)
37. $x^3 + 7x^2 = 40x - 43.9$ (3 solutions)
38. $500x^3 - 5x = 100x^2 - 1$ (3 solutions)

EXERCISES 39–42 *Show that the given numbers are zeros of the polynomial.*

39. $p(x) = 2x^3 - x^2 - 5x - 2$; $-1, -\frac{1}{2}, 2$
40. $p(x) = 3x^3 + 8x^2 + 3x - 2$; $-1, -2, \frac{1}{3}$

41. $p(x) = x^3 - x^2 - 2x + 2; \ -\sqrt{2}, \sqrt{2}, 1$
42. $p(x) = x^3 - 2x^2 - 3x + 6, \ -\sqrt{3}, \sqrt{3}, 2$

EXERCISES 43–46 *Find the zeros of the given polynomial.*

43. $p(x) = x^3 - x^2 - 6x$
44. $p(z) = (z^2 - 4)(z^2 + 2z)$
45. $p(t) = 4t^3 - t$
46. $p(y) = 8y^4 + y$

EXERCISES 47–50 *Find the value(s) for c.*

47. One zero of $p(x) = x^2 + cx + 20$ is 5.
48. One zero of $p(x) = x^2 + x + c$ is -2.
49. One zero of $p(x) = cx^2 - 5x - 3$ is $-\frac{1}{2}$.
50. One zero of $p(x) = 3x^2 + cx + 6$ is $\frac{2}{3}$.
51. $p(x) = x^2 - cx + 4$ has only one zero.
52. $p(x) = cx^2 + 8x + c$ has only one zero.

EXERCISES 53–80 *Solve the polynomial inequality. Give exact answers wherever possible; when approximating, give answers accurate to the nearest hundredth.*

53. $(t + 3)(t - 1) \leq 0$
54. $(x - 2)(x + 1) > 0$
55. $(2y - 1)(y + 2) > 0$
56. $(s - 3)(4s + 2) \leq 0$
57. $x^2 + 4x \leq 5$
58. $t^2 + 6 > -5t$

59. $2x^2 + 3 > 7x$
60. $6w \geq w^2 + 8$
61. $u^2 \leq 8$
62. $x^2 \geq 25$
63. $x^3 < x^2 + 3$
64. $x^3 + 1 < 3x$
65. $x^4 + x > 1$
66. $x^4 < 1 - x^3$
67. $x(x + 1)(x - 3) > 0$
68. $(x - 1)(x + 2)(x + 4) \leq 0$
69. $(x - 5)^2(x + 1)^3 < 0$
70. $(t + 8)^2(t - 2)^4 > 0$
71. $x(x - 2)(x + 3) < -2$
72. $x^4 - 2x^3 + x + 1 \leq 0$
73. $(7x - 2)^4(2x + 3)^6 \leq 0$
74. $(4y + 1)^3(6y - 5)^3 \geq 0$
75. $3t^3 + 3t^2 > 0$
76. $2x^4 + 6x^3 \leq 0$
77. $2v^3 - 8v \geq 0$
78. $3y^4 - 27y^2 > 0$
79. $(x - 15)(x + 20)(x - 27) > -100$
80. $(x - 10)^2(x - 30) < -40$

EXERCISES 81–84 *Solve the given equation for y. Use a graphing calculator to plot the graph of the resulting equation and estimate the x-intercept(s) to the nearest hundredth.*

81. $y^2 + 3x - 4 = 0$
82. $x^2 + 4x + 4 + y^2 = 9$
83. $x^4 + 16y^4 = 9$
84. $2x^3 - 6x^2 + 6x - 2y^3 = 3$

■ APPLICATIONS

85. **Throwing a Ball** A ball thrown upward with an initial velocity of 160 feet per second has height $s = -16t^2 + 160t$ feet after t seconds. Solve the equation $-16t^2 + 160t = 0$ to find the time when it hits the ground.

86. **Falling Ball** A ball dropped from the top of the Sears Tower in Chicago has height $s = -16t^2 + 1454$ feet after t seconds. How long before the ball hits the ground?

87. **Compound Interest** A deposit of $5000 reaches a balance of $6288.60 after 4 years. If interest is compounded yearly, what is the annual percentage rate?

88. **Compound Interest** If you borrow $1000 from your father and pay him back $1100 after 2 years, what is the annual percentage rate, assuming interest is compounded yearly?

89. **Circumference of a Tree** A board foot is a unit of lumber measurement equal to 1 foot square by 1 inch thick. The number of board feet in a ponderosa pine can be approximated with the polynomial $B = 0.0015c^3$, where c is the circumference of the tree in inches at waist height. What is the circumference of a ponderosa pine with 2500 board feet?

90. **Radius of a Sphere** The volume of a sphere with radius r is $V = \frac{4}{3}\pi r^3$. If the volume of a sphere is 200 cubic centimeters, what is its radius?

91. **Great Lakes Snowball** Approximate the radius of a snowball that contains enough moisture to fill the Great Lakes. The total volume of water in the Great Lakes is approximately 5440 cubic miles and the amount of water in each cubic inch of loosely packed snow is approximately $\frac{1}{6}$ cubic inch.

92. **Dimensions of a Can** The volume of a cylinder with radius r and height h is $V = \pi r^2 h$. If a can of vegetables with a volume of 50 cubic inches is to have a height of twice the radius, find the dimensions of the can.

93. **Deer Population** The population of deer in a game preserve can be modeled by $P = 42t^2 + 36t + 1008$, where t is measured in years ($t = 0$ corresponds to 1998). In what year will the population be equal to 1620?

94. **Photograph Revenue** The more prints of a work a photographer sells, the less he can charge for each. Suppose that a certain photograph sells for $9900 when there is only one print, with a $100 reduction in price for each additional print produced.

    a. Find an expression for the price per print when x prints are made.

    b. Assuming all prints are sold, find an expression for the total revenue from the photograph.

    c. What can be said about the number of prints if the revenue exceeds $187,500?

95. **Height of a Projectile** A projectile is fired straight up with an initial velocity of 80 feet per second. Its height in feet after

t seconds is $s = -16t^2 + 80t$. During what time period will the projectile be above 96 feet?

96. *Area of a Corral*   A rancher has 200 feet of fencing with which to enclose a corral next to a stone wall (see Figure 91). If the width is x feet, the resulting area is given by $A = x(200 - 2x)$. For what range of x-values will the resulting area be at least 3200 square feet?

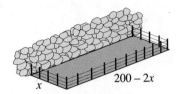

FIGURE 91

97. *Volume of a Box*   A rectangular package to be sent by the U.S. Postal Service can have a maximum combined length and girth (perimeter of the base) of 108 inches (see Figure 92). The volume of a box with a square base that exactly meets this restriction is $V = x^2(108 - 4x)$ cubic inches, where x is the width of the base in inches. Find all possible values of x for which the volume is positive.

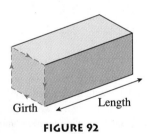

FIGURE 92

98. *Missile Flight*   If a missile is shot straight up at 4000 feet per second and we assume that air resistance is negligible (is it?), then the height of the missile in feet t seconds after blast-off is given by $h = -16t^2 + 4000t$. The stratosphere, the innermost portion of the atmosphere, extends approximately 30 miles from Earth's surface. For what period of time will the missile be out of the stratosphere? (There are 5280 feet in one mile.)

99. *Global Temperature*   The average global temperature during the years 1935–1995 can be modeled by the equation

$$y = 0.00003x^3 - 0.002x^2 + 0.03x + 61.5$$

where y represents the average temperature (measured in degrees Fahrenheit) and x denotes the year (with $x = 0$ corresponding to 1935). Assuming the equation continues to hold, approximate the year when the temperature first reaches 63°F.

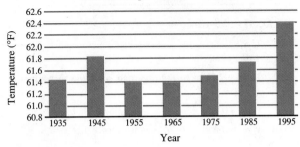

Data source: U.K. Meteorological Office.

100. *Intel Stock Performance*   The value of Intel stock from January 1996 through July 1998 can be approximated by

$$y = 21.7x^4 - 111x^3 + 162x^2 - 29.2x + 28.6$$

where x denotes the year ($x = 0$ corresponds to January 1996) and y is the value in dollars per share for that year. If the equation continues to hold, estimate the month in which Intel stock will be worth \$100.

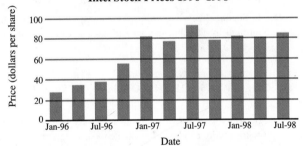

Data source: CSI.

One consequence of global warming would be widespread floods, such as the one shown here in Bangladesh.

■ CONCEPTS AND CRITICAL THINKING

EXERCISES 101–104   *Answer true or false.*

101. If $x(x + 2) = 4$, then we can conclude that either $x = 4$ or $x + 2 = 4$.

102. If $x(x + 2) = 0$, then we can conclude that either $x = 0$ or $x + 2 = 0$.

103. If $a > 0$ and $x^2 = a$, then it follows that $x = \sqrt{a}$.

104. If $a > 0$ and $x^3 = a$, then it follows that $x = \sqrt[3]{a}$.

EXERCISES 105–108   *Give an example of each.*

105. A polynomial having zeros of 1 and 2

106. A method other than the zero-product property that can be used to solve polynomial equations

107. A solution of an equation of the form $xp(x) = 0$, where $p(x)$ is a polynomial

108. A 1000th-degree polynomial equation having exactly 2 real solutions

109. Explain why $u(u + 2) = 0$ indicates that either $u = 0$ or $u + 2 = 0$, but $u(u + 2) = 8$ does *not* mean $u = 8$ or $u + 2 = 8$.

110. In Section 2.2, we mentioned the hazard of dividing equations through by variable factors and hence losing solutions. What solution would be lost if $x + 2$ was divided through on both sides of $x(x + 2) = x + 2$? Explain the correct way to solve this equation. Give another example of such an equation, other than any in the text, and discuss the proper way to solve it.

111. The polynomial equation $x^3 - 7x - 6 = 0$ factors as $(x - 3)(x + 1)(x + 2) = 0$ and hence has solutions $x = 3$, $x = -1$, and $x = -2$. Do you think this works in reverse? In other words, if you are given that $x = -1$, $x = 2$, and $x = 3$ are the solutions of $x^3 - 4x^2 + x + 6 = 0$, what can you conclude about how $x^3 - 4x^2 + x + 6$ can be factored? How do you think $2x^2 - 5x + 2$ should factor given that $2x^2 - 5x + 2 = 0$ has solutions $x = \frac{1}{2}$ and $x = 2$? Explain.

112. Consider the statement: A polynomial equation can have only as many solutions as the degree of the polynomial. Now solve the equation $x^3 - 8 = 0$. Does your solution in any way contradict the statement? Explain.

113. Use a graphing calculator to plot the graphs of $y = x^2$, and $y = x^4$. Then sketch $y = x^3$ and $y = x^5$. What do these graphs suggest about the shape of $y = x^n$? Use your results to explain why, if n is even, then $x^n = a$ has two solutions when $a > 0$ and no solutions when $a < 0$, whereas if n is odd, then $x^n = a$ has exactly one solution no matter what the value of a.

■ QUESTIONS FOR DISCUSSION OR ESSAY

114. In principle, the real solution set to any polynomial equation or inequality can be approximated using a graphing calculator. Is there any conceivable context in which exact solutions might be required? If not, is there any value in learning how to do so?

115. The primary graphical technique discussed for solving inequalities of the form $p(x) > q(x)$ is to first write the inequality in the form $p(x) - q(x) > 0$, and then graph the equation $y = p(x) - q(x)$ and estimate the intervals on which the graph lies above the x-axis. Alternatively, we could graph $y_1 = p(x)$ and $y_2 = q(x)$ in the same viewing window and estimate the intervals on which $y_1 > y_2$. Discuss the relative advantages and disadvantages of this alternative approach.

■ PROJECTS FOR ENRICHMENT

116. *The Effects of Shape on Surface Area*

a. Human beings have a density approximately that of water, which weighs about 62.4 pounds per cubic foot. Estimate the volume of a 200-pound man.

b. Estimate the radius of a ball that has the volume of a 200-pound man, and compute the surface area of such a ball.

c. For each of the following heights, estimate the radius and surface area of the cylinder such that the resulting volume is that of a 200-pound man.

    i. $h = 3$ feet
    ii. $h = 4$ feet

    iii. $h = 5$ feet
    iv. $h = 6$ feet
    v. $h = 7$ feet

d. Based on your results from parts b and c, what type of 200-pound man would most likely have the greatest surface area: a very short obese man, a muscular 6-footer, or a slender 7-footer?

e. Given that heat is dissipated through the skin, comment on the ideal bodily proportions for cold versus warm climates. What other factors besides surface area might affect the ability of the body to dissipate or retain heat?

© TSM/Viviane Moos, 1993

The short, stout build of the Inuit helps conserve body heat in the Arctic. Members of the Dinka tribe tend to be long-limbed and slender, an adaptation to the searing Sudanese heat.

117. *Locating Solutions*   We have seen that when polynomial equations cannot be solved *exactly* using factoring or some other algebraic method, it is possible to *approximate* the real solutions using a graphing calculator. There are also a number of nongraphical computational techniques for approximating solutions to equations. However, many of these techniques do require an initial guess for a solution, much as is necessary for the calculate-zero feature on a graphing calculator. How does one go about coming up with an initial guess without the use of a graph? One way is with a specialized form of the **Inter-mediate-Value Theorem,** abbreviated IVT, which we describe here. Consider a polynomial equation in the form $p(x) = 0$. If you can find two real numbers a and b such that $p(a)$ and $p(b)$ have opposite sign (that is, one is positive and the other is negative), then the IVT guarantees that there is a solution to the equation somewhere between a and b. Let's try it for the equation $x^2 - 2 = 0$. Consider the polynomial $p(x) = x^2 - 2$. Since $p(1) = -1$ is negative and $p(2) = 2$ is positive, the IVT says there is a solution to the equation between $x = 1$ and $x = 2$.

a. Use the IVT to show that there is another solution to $x^2 - 2 = 0$ between $x = -2$ and $x = -1$.

Notice that the IVT doesn't tell us what the solutions are. It just shows us intervals in which they lie. We would need to use other methods to actually find or approximate the solutions.

b. What are the actual solutions to the equation $x^2 - 2 = 0$? Show they are in the intervals we verified with the IVT.

c. Use the IVT to show that $x^3 + 5 = 0$ has a solution in the interval $(-2, -1)$. Then find the actual solution and confirm that it indeed lies in the interval $(-2, -1)$.

d. Use the IVT to show that $x^2 + 2x - 2 = 0$ has a solution in the interval $(0, 1)$ and another in the interval $(-3, -2)$. In Section 2.6 we'll see that the quadratic formula gives us the two solutions $x = -1 - \sqrt{3}$ and $x = -1 + \sqrt{3}$. Show these are in the intervals.

e. Use the IVT to show that $x^3 - 3x + 1 = 0$ has solutions in the intervals $(-2, -1)$, $(0, 1)$, and $(1, 2)$.

f. Describe a procedure for ''zeroing in'' on the solution of $x^3 - 3x + 1 = 0$ that lies in the interval $(0, 1)$.

SECTION 2.6   QUADRATIC EQUATIONS

☐ If there is a quadratic formula for solving second-degree polynomial equations, is there a cubic formula for solving third-degree equations? What about fourth-, fifth-, and higher-degree polynomial equations?

☐ How can you tell within 5 seconds whether or not the polynomial $3x^2 + 10x + 6$ factors, without ever attempting to factor it?

☐ A basketball player with a 48-inch vertical jump must leap 2 feet in order to dunk a basketball. What is the minimum time required for him to slam the basketball?

☐ What is the radius of a 1-foot-high paint can that holds just enough paint to cover its exterior?

As we discovered in Section 2.2, all first-degree polynomial (linear) equations are easily solved algebraically using properties of equations. In this section, we turn our attention to **quadratic equations,** polynomial equations of second degree that can be written in the form

$$ax^2 + bx + c = 0$$

Some quadratic equations can be solved by first factoring and then employing the zero-product property, but most are difficult or even impossible to factor. To solve such equations algebraically, we need to either *complete the square* or apply the *quadratic formula,* two specialized techniques for solving quadratic equations that are detailed in this section.

COMPLETING THE SQUARE

Recall from Section 2.5 that any equation of the form $\square^2 = d$ (with $d > 0$) is equivalent to $\square = \pm\sqrt{d}$. We refer to this fact as the **square root property.** This property can be used to solve some quadratic equations, as the following examples illustrate.

EXAMPLE 1 ■ Solving a Quadratic Equation of the Form $\square^2 = d$

Solve $(x - 13)^2 = 25$.

SOLUTION

$$
\begin{aligned}
(x - 13)^2 &= 25 \\
x - 13 &= \pm\sqrt{25} \qquad \text{Using the square root property} \\
x - 13 &= \pm 5 \\
x &= 13 \pm 5 \\
x &= 18, \; x = 8
\end{aligned}
$$

EXAMPLE 2 ■ Solving a Quadratic Equation of the Form $\square^2 = d$

Solve $x^2 + 6x + 9 = 10$.

SOLUTION

$$
\begin{aligned}
x^2 + 6x + 9 &= 10 \\
(x + 3)^2 &= 10 \qquad\qquad \text{Expressing the left side as} \\
& \qquad\qquad\qquad\quad \text{a perfect square} \\
x + 3 &= \pm\sqrt{10} \qquad \text{Using the square root property} \\
x &= -3 + \sqrt{10} \\
x &= -3 - \sqrt{10}, \; x = -3 + \sqrt{10}
\end{aligned}
$$

Note that in Example 2 we had to first express the left side as a perfect square. If the left side is not already a perfect square, then we can make it so by completing the square. Recall from Section 2.1 that to complete the square, we add an appropriate constant to both sides of the equation so that the left side becomes a perfect square. More specifically, to complete the square on $x^2 + bx$, we add $(b/2)^2$.

EXAMPLE 3 ▧ **Solving a Quadratic Equation by Completing the Square**

Solve $x^2 + 10x + 5 = 8$ for x by completing the square.

SOLUTION

$$x + 10x + 5 = 8$$

$$x^2 + 10x = 3 \qquad \text{Subtracting 5 from both sides to eliminate the constant term}$$

$$x^2 + 10x + \left(\frac{10}{2}\right)^2 = 3 + \left(\frac{10}{2}\right)^2 \qquad \text{Completing the square by adding the square of one-half the coefficient of } x \text{ to both sides}$$

$$x^2 + 10x + 25 = 3 + 25$$

$$(x + 5)^2 = 28 \qquad \text{Expressing the left side as a perfect square}$$

$$x + 5 = \pm\sqrt{28} \qquad \text{Using the square root property}$$

$$x = -5 \pm 2\sqrt{7}$$

If the coefficient of x^2 is not 1, then we simply divide both sides by this coefficient, as in the following example.

EXAMPLE 4 ▧ **Solving a Quadratic Equation by Completing the Square**

Solve $4x^2 - 24x + 11 = 0$.

SOLUTION

$$4x^2 - 24x + 11 = 0$$

$$4x^2 - 24x = -11 \qquad \text{Subtracting 11 from both sides to eliminate the constant term}$$

$$x^2 - 6x = -\frac{11}{4} \qquad \text{Dividing both sides by 4}$$

$$x^2 - 6x + \left(\frac{-6}{2}\right)^2 = -\frac{11}{4} + \left(\frac{-6}{2}\right)^2 \qquad \text{Completing the square by adding the square of one-half the coefficient of } x \text{ to both sides}$$

$$x^2 - 6x + 9 = -\frac{11}{4} + \frac{36}{4}$$

$$(x - 3)^2 = \frac{25}{4} \qquad \text{Expressing the left side as a perfect square}$$

$$x - 3 = \pm\frac{5}{2} \qquad \text{Using the square root property}$$

$$x = 3 \pm \frac{5}{2}$$

$$x = \frac{11}{2}, x = \frac{1}{2}$$

We summarize the technique of completing the square as follows.

Solving Quadratic Equations by Completing the Square ☐

1. Write the quadratic equation so that all the variable expressions are on the left side and the constant term is on the right.
2. Divide both sides of the equation by the coefficient of x^2.
3. Complete the square by adding the square of one-half the coefficient of x to both sides of the equation. (Divide the coefficient of x by 2, square the result, and then add to both sides of the equation.)
4. Express the left side as the square of a binomial.
5. Use the square root property and solve for x.

THE QUADRATIC FORMULA

Since completing the square is a somewhat tedious process, we will complete the square once and for all on the general quadratic equation $ax^2 + bx + c = 0$ and derive the **quadratic formula.**

$$ax^2 + bx + c = 0$$

$$ax^2 + bx = -c \qquad \text{Bringing the constant to the right side}$$

$$x^2 + \frac{b}{a}x = -\frac{c}{a} \qquad \text{Dividing both sides by } a, \text{ the coefficient of } x^2$$

$$x^2 + \frac{b}{a}x + \left(\frac{b}{2a}\right)^2 = -\frac{c}{a} + \left(\frac{b}{2a}\right)^2 \qquad \text{Adding the square of one-half the coefficient of } x \text{ to both sides}$$

$$\left(x + \frac{b}{2a}\right)^2 = \frac{b^2}{4a^2} - \frac{c}{a}\left(\frac{4a}{4a}\right) \qquad \text{Expressing the left side as the square of a binomial}$$

$$\left(x + \frac{b}{2a}\right)^2 = \frac{b^2 - 4ac}{4a^2} \qquad \text{Simplifying}$$

$$x + \frac{b}{2a} = \pm\sqrt{\frac{b^2 - 4ac}{4a^2}} \qquad \text{Using the square root property}$$

$$x + \frac{b}{2a} = \pm\frac{\sqrt{b^2 - 4ac}}{2a} \qquad \text{Simplifying}$$

$$x = \frac{-b \pm \sqrt{b^2 - 4ac}}{2a} \qquad \text{Solving for } x$$

Thus, we have obtained a solution to the general equation in terms of the coefficients a, b, and c.

The Quadratic Formula ☐

The solutions to the quadratic equation $ax^2 + bx + c = 0$ are given by

$$x = \frac{-b \pm \sqrt{b^2 - 4ac}}{2a}$$

E X A M P L E   5   ▦   **Solving a Quadratic Equation with the Quadratic Formula**

Solve $y^2 + 3y + 1 = 0$ with the quadratic formula.

SOLUTION   Here $a = 1$, $b = 3$, and $c = 1$. Substituting into the quadratic formula and simplifying we have

$$y = \frac{-b \pm \sqrt{b^2 - 4ac}}{2a}$$
$$= \frac{-3 \pm \sqrt{3^2 - 4 \cdot 1 \cdot 1}}{2 \cdot 1}$$
$$= \frac{-3 \pm \sqrt{5}}{2}$$

Thus, we obtain the two solutions $y = (-3 + \sqrt{5})/2 \approx -0.382$, and $y = (-3 - \sqrt{5})/2 \approx -2.618$.

E X A M P L E   6   ▦   **Solving a Quadratic Equation with the Quadratic Formula**

Solve $-10 + 2x^2 = 3x$ with the quadratic formula.

SOLUTION   We begin by writing the quadratic equation in standard form, with zero on the right side and the terms in descending order by power of x.

$$-10 + 2x^2 = 3x$$
$$2x^2 - 3x - 10 = 0$$

Now we can read off the coefficients, obtaining $a = 2$, $b = -3$, and $c = -10$. Substituting into the quadratic formula and simplifying, we obtain

$$x = \frac{-b \pm \sqrt{b^2 - 4ac}}{2a}$$
$$= \frac{-(-3) \pm \sqrt{(-3)^2 - 4 \cdot 2 \cdot (-10)}}{2 \cdot 2}$$
$$= \frac{3 \pm \sqrt{9 + 80}}{4}$$
$$= \frac{3 \pm \sqrt{89}}{4}$$

Thus, $x = (3 + \sqrt{89})/4 \approx 3.108$, or $x = (3 - \sqrt{89})/4 \approx -1.608$.

E X A M P L E   7   ▦   **An Application of the Quadratic Equation to an Equation of Motion**

Suppose that a ball is thrown upward from a height of 16 feet at 16 feet per second. How long will it take for the ball to hit the ground? (Assume that the height of the ball in feet after t seconds is given by $h = -16t^2 + v_0 t + h_0$, where v_0 is the initial velocity and h_0 is the initial height.)

SOLUTION   Using $v_0 = 16$ and $h_0 = 16$, we obtain $h = -16t^2 + 16t + 16$. The ball will hit the ground when $h = 0$, and so we must solve the quadratic equation $-16t^2 + 16t + 16 = 0$.

$$-16t^2 + 16t + 16 = 0$$

$$t^2 - t - 1 = 0 \qquad \text{Dividing through by } -16$$

$$t = \frac{-(-1) \pm \sqrt{(-1)^2 - 4(1)(-1)}}{2(1)} \qquad \begin{array}{l}\text{Applying the quadratic}\\\text{formula with } a = 1,\\ b = -1, \text{ and } c = -1\end{array}$$

$$= \frac{1 \pm \sqrt{5}}{2}$$

Thus, $t = \dfrac{1 + \sqrt{5}}{2} \approx 1.618$, or $t = \dfrac{1 - \sqrt{5}}{2} \approx -0.618$. Since the variable t represents a positive quantity, the desired solution is $t \approx 1.618$ seconds.

EXAMPLE 8   ▓   **Using the Quadratic Equation to Plot a Parabola**

Solve the equation $x = y^2 + 3y + 1$ for y. Use a graphing calculator to plot the graph(s) of the resulting equation(s) and estimate the intercept(s) to the nearest hundredth.

SOLUTION   We first rewrite the equation in standard quadratic form (in y) with 0 on the right-hand side.

$$x = y^2 + 3y + 1$$

$$y^2 + 3y + 1 - x = 0$$

Here $a = 1$, $b = 3$, and $c = 1 - x$. Substituting into the quadratic formula and simplifying, we obtain

$$y = \frac{-b \pm \sqrt{b^2 - 4ac}}{2a}$$

$$= \frac{-3 \pm \sqrt{3^2 - 4 \cdot 1 \cdot (1 - x)}}{2 \cdot 1}$$

$$= \frac{-3 \pm \sqrt{9 - 4 + 4x}}{2}$$

$$= \frac{-3 \pm \sqrt{5 + 4x}}{2}$$

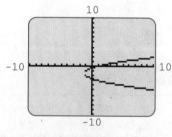

FIGURE 93

Now we plot graphs of the two equations $y = -\frac{3}{2} + \frac{1}{2}\sqrt{5 + 4x}$ and $y = -\frac{3}{2} - \frac{1}{2}\sqrt{5 + 4x}$. The result is shown in Figure 93. The intercepts could be estimated from these graphs using the trace feature, but all are easily found algebraically. To find the x-intercept, we set $y = 0$ in the original equation. This yields $x = 1$. To find the y-intercepts, we set $x = 0$ as follows:

$$y = \frac{-3 \pm \sqrt{5 + 4x}}{2}$$

$$= \frac{-3 \pm \sqrt{5 + 4 \cdot 0}}{2}$$

$$= \frac{-3 \pm \sqrt{5}}{2}$$

Thus, we have $y \approx -0.38$ and $y \approx -2.62$.

EXAMPLE 9   ■   Solving a Quadratic Equation with the Quadratic Formula

Solve $4z^2 - 12z + 9 = 0$ with the quadratic formula.

SOLUTION   Here $a = 4$, $b = -12$, and $c = 9$. Substituting into the quadratic formula and simplifying, we obtain

$$z = \frac{-b \pm \sqrt{b^2 - 4ac}}{2a}$$

$$= \frac{12 \pm \sqrt{(-12)^2 - 4 \cdot 4 \cdot 9}}{2 \cdot 4}$$

$$= \frac{12 \pm \sqrt{144 - 144}}{8}$$

$$= \frac{12}{8}$$

$$= \frac{3}{2}$$

Note that this quadratic equation has only one solution. This is because the expression under the radical, $b^2 - 4ac$, simplified to 0.

The quadratic formula can also be used to find nonreal complex solutions of quadratic equations. Recall that we define $i = \sqrt{-1}$, so that for positive real numbers a,

$$\sqrt{-a} = \sqrt{a} \cdot \sqrt{-1} = \sqrt{a}\, i$$

EXAMPLE 10   ■   Using the Quadratic Formula When There Are No Real Solutions

Use the quadratic formula to find all solutions of $2t^2 + 3t + 4 = 0$, and verify graphically.

SOLUTION   Here $a = 2$, $b = 3$, and $c = 4$. Substituting into the quadratic formula and simplifying we have

$$t = \frac{-b \pm \sqrt{b^2 - 4ac}}{2a}$$

$$= \frac{-3 \pm \sqrt{3^2 - 4 \cdot 2 \cdot 4}}{2 \cdot 2}$$

$$= \frac{-3 \pm \sqrt{-23}}{4}$$

$$= \frac{-3 \pm \sqrt{23}\, i}{4}$$

Note that this quadratic equation has no real solutions although it does have two nonreal complex solutions. The reason for this is that the expression under the radical, $b^2 - 4ac$, is negative. To investigate this outcome graphically, we plot the graph of $y = 2x^2 + 3x + 4$, as shown in Figure 94. Since the graph does not touch the x-axis, there are no x-intercepts and hence no real solutions to the equation. The precise nature of the complex solutions cannot be determined from the graph.

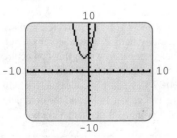

FIGURE 94

THE DISCRIMINANT

As we have seen in previous examples, quadratic equations may have either 2, 1, or 0 *real* solutions. Since the solutions to the quadratic equation $ax^2 + bx + c = 0$ are given by

$$x = \frac{-b \pm \sqrt{b^2 - 4ac}}{2a}$$

we see that the quantity inside the radical, $b^2 - 4ac$, will determine the number of real solutions. If $b^2 - 4ac$ is positive, then $\sqrt{b^2 - 4ac}$ is a positive real number, so that

$$\frac{-b + \sqrt{b^2 - 4ac}}{2a} \quad \text{and} \quad \frac{-b - \sqrt{b^2 - 4ac}}{2a}$$

are distinct real solutions. If $b^2 - 4ac$ is zero, then we have

$$x = \frac{-b \pm \sqrt{0}}{2a} = -\frac{b}{2a}$$

so that there is only one real solution. If $b^2 - 4ac$ is negative, then $\sqrt{b^2 - 4ac}$ isn't real, so that there are no real solutions. We define the **discriminant** of the quadratic equation $ax^2 + bx + c = 0$ to be the quantity $b^2 - 4ac$. We can use the discriminant to quickly determine the number of real solutions to a quadratic equation without actually finding the solutions themselves.

FERMAT'S LAST THEOREM

In this text, we are primarily interested in real-number solutions to equations. **Diophantine equations,** named after Diophantus of Alexandria, are equations for which only integer solutions are desired. For example, the equation $a^2 + b^2 = c^2$ has integer solutions consisting of Pythagorean triples, such as $a = 3$, $b = 4$, and $c = 5$. In fact, it is easily shown that $a^2 + b^2 = c^2$ has infinitely many integer solutions.

Sometime in the early 17th century, the amateur mathematician Pierre Fermat was reading a copy of *Arithmetic,* written by Diophantus himself, when he made what is undoubtedly the most significant marginal note of all time: "It is impossible to separate a cube into two cubes or a fourth power into two fourth powers or, in general, any power greater than the second into powers of like degree. I have discovered a truly marvelous demonstration, which this margin is too narrow to contain." Fermat was asserting that no equation of the form $a^3 + b^3 = c^3$, or $a^4 + b^4 = c^4$, or indeed $a^n + b^n = c^n$ for any $n > 2$ has integer solutions. In subsequent years, Fermat, in correspondence with the leading mathematicians of his time, proved that the equations $a^3 + b^3 = c^3$ and $a^4 + b^4 = c^4$ have no integer solutions, but his proof of the more general case referred

Andrew Wiles presenting his results

to in his marginal note was never found. His claim was dubbed "Fermat's last theorem."

Over the course of the next three centuries, entire branches of mathematics arose from ill-fated attempts at proving Fermat's last theorem. In spite of the best efforts of some of the greatest intellects the world has seen, the theorem stubbornly refused to yield a proof. The best that could be done was to establish the theorem for particular cases. By 1990, it had been shown that Fermat's last theorem holds for all equations $a^n + b^n = c^n$, where $n < 4,000,000$.

In June 1993, Andrew Wiles of Princeton University shocked the mathematical community when he announced that he had proven Fermat's last theorem. A brilliant mathematician with a distinguished record of research, Wiles had worked in relative seclusion for some 7 years on proving a conjecture, the truth of which would imply Fermat's last theorem. Upon close examination of the 200-page paper in which this conjecture is established, referees found several gaps in his proof, all but one of which was easily corrected. Wiles and his colleague Richard Taylor successfully bridged the last gap in 1994, and the complete proof has been verified by many prominent mathematicians.

Interpretation ☐
of the Discriminant

> Let $d = b^2 - 4ac$ be the discriminant of the quadratic equation
> $ax^2 + bx + c = 0$.
>
> **1.** If $d > 0$, then there are two real solutions.
> **2.** If $d = 0$, then there is one real solution.
> **3.** If $d < 0$, then there are no real solutions.

E X A M P L E   11 ▪ **Classifying with the Discriminant**

For each of the following quadratic equations, use the discriminant to find the number of real solutions.

a. $x^2 - 3x + 5 = 0$ **b.** $2x^2 + 3x - 1 = 0$ **c.** $9x^2 + 12x + 4 = 0$

SOLUTION
a. Here $d = (-3)^2 - 4 \cdot 1 \cdot 5 = 9 - 20 = -11$. Since $-11 < 0$, there are no real solutions.
b. Here $d = 3^2 - 4 \cdot 2 \cdot (-1) = 9 - (-8) = 17$. Since $17 > 0$, there are two real solutions.
c. Here $d = 12^2 - 4 \cdot 9 \cdot 4 = 144 - 144 = 0$. Thus, there is only one real solution.

EQUATIONS OF QUADRATIC TYPE

Recall that a quadratic equation in the variable x is of the form $ax^2 + bx + c = 0$ and can be solved by using the quadratic formula. Some equations, although not quadratic, are of **quadratic type**; that is, they can be written in the form $a\square^2 + b\square + c = 0$, where $\square$ is an expression involving the variable. Many equations of quadratic type can be solved using the quadratic formula.

Consider the equation $x^4 - 4x^2 + 1 = 0$. By rewriting the equation as $(x^2)^2 - 4(x^2) + 1 = 0$, we see that it is quadratic in the variable x^2. In fact, if we let $u = x^2$, then we may rewrite the equation as

$$u^2 - 4u + 1 = 0$$

Solving for u, we obtain

$$u = \frac{-(-4) \pm \sqrt{(-4)^2 - 4 \cdot 1 \cdot 1}}{2 \cdot 1} = \frac{4 \pm \sqrt{12}}{2} = 2 \pm \sqrt{3}$$

Now, using the fact that $u = x^2$, we have

$$x^2 = 2 + \sqrt{3} \quad \text{or} \quad x^2 = 2 - \sqrt{3}$$

from which it follows that

$$x = \pm\sqrt{2 + \sqrt{3}} \quad \text{or} \quad x = \pm\sqrt{2 - \sqrt{3}}$$
$$\approx \pm 1.932 \qquad\qquad \approx \pm 0.518$$

Thus, there are four solutions to $x^4 - 4x^2 + 1 = 0$.

EXAMPLE 12 ▨ Solving an Equation of Quadratic Type

Solve the following equation of quadratic type.

$$\left(\frac{x-3}{2}\right)^2 + 5\left(\frac{x-3}{2}\right) + 5 = 0$$

SOLUTION Here we let $u = \dfrac{x-3}{2}$ to obtain

$$u^2 + 5u + 5 = 0$$

Solving for u, we obtain

$$u = \frac{-5 \pm \sqrt{5^2 - 4 \cdot 1 \cdot 5}}{2}$$

$$= \frac{-5 \pm \sqrt{5}}{2}$$

Using the fact that $u = \dfrac{x-3}{2}$, we have

$$\frac{x-3}{2} = \frac{-5 \pm \sqrt{5}}{2}$$

$$x - 3 = -5 \pm \sqrt{5}$$

$$x = -2 \pm \sqrt{5}$$

UNDERSTANDING AND MASTERY CHECKLISTS

CONCEPTS TO UNDERSTAND

- ☐ Quadratic equations
- ☐ Completing the square
- ☐ The quadratic formula
- ☐ The discriminant
- ☐ Equations of quadratic type

SKILLS TO MASTER

- ☐ Recognize a quadratic equation.
- ☐ Solve a quadratic equation by completing the square.
- ☐ Solve a quadratic equation with the quadratic formula.
- ☐ Determine the number of real solutions of a quadratic equation by using the discriminant.
- ☐ Recognize and solve equations of quadratic type.

EXERCISES 2.6

EXERCISES 1–6 *Solve the given equation.*

1. $x^2 = 9$
2. $x^2 = 16$
3. $(x + 3)^2 = 16$
4. $(y - 2)^2 = 8$
5. $3(t - 1)^2 = 60$
6. $2(x + 6)^2 = 50$

EXERCISES 7–14 *Solve the given equation by completing the square.*

7. $x^2 + 5x + 6 = 0$
8. $x^2 - 4x - 21 = 0$
9. $x^2 - 2x - 3 = 0$
10. $x^2 + 4x + 1 = 0$

11. $y^2 + 3y + 1 = 0$

12. $s^2 - 5s + 2 = 0$

13. $x(x - 3) = 2$

14. $2x^2 + 6x = 8$

EXERCISES 15–36 *Solve using the quadratic formula.*

15. $x^2 - x - 6 = 0$

16. $x^2 - 4x - 21 = 0$

17. $6t^2 - t - 2 = 0$

18. $15y^2 - 4y - 4 = 0$

19. $x^2 + 6x = -9$

20. $y^2 - 14y = -49$

21. $u^2 - 6u + 3 = 0$

22. $x^2 + 4x + 2 = 0$

23. $2x^2 + 4x + 1 = 0$

24. $3t^2 - 5t - 5 = 0$

25. $x^2 + x = 1$

26. $y(2y + 7) = 1$

27. $x^2 - 1 = 0$

28. $y^2 + 1 = 0$

29. $w^2 - 4w + 5 = 0$

30. $2x^2 - 3x + 2 = 0$

31. $3y^2 - 2y + 1 = 0$

32. $(x + 1)^3 = (x - 1)^3 + 9x$

33. $2x^4 - 6x^2 + 1 = 0$

34. $v^4 - 2v^2 + 1 = 0$

35. $\left(\dfrac{w + 3}{5}\right)^2 + 3\left(\dfrac{w + 3}{5}\right) + 1 = 0$

36. $(5z + 4)^2 + 4(5z + 4) + 4 = 0$

EXERCISES 37–46 *Indicate the number of real solutions of the quadratic equation.*

37. $x^2 + 3x + 1 = 0$

38. $x^2 - 3x - 1 = 0$

39. $x^2 + 7 = 0$

40. $3y^2 - 5y + 3 = 0$

41. $2z^2 + 4z - 3 = 0$

42. $9w^2 + 12w + 4 = 0$

43. $4x^2 + 12x + 9 = 0$

44. $12x^2 - 11x - 36 = 0$

45. $6x^2 - 10x - 10 = 0$

46. $12s^2 + s - 6 = 0$

EXERCISES 47–50 *Solve for the indicated variable.*

47. $a^2 + xa + x = 0$ (Solve for a.)

48. $w^2 + pw + q = 0$ (Solve for w.)

49. $2x^2 + tx + t^2 = 0$ (Solve for t.)

50. $x^2 + 2xy + y^2 - 3 = 0$ (Solve for x.)

EXERCISES 51–54 *Solve the given equation for y. Use a graphing calculator to plot the graph(s) of the resulting equation(s) and estimate the intercept(s) to the nearest hundredth.*

51. $x = 2y^2 + 4y + 5$

52. $9x^2 + 4y^2 - 16y = 20$

53. $4x^2 - y^2 + 8x + 6y = 6$

54. $3x^2 - 6xy + 3y^2 + 2x - 7 = 0$

■ APPLICATIONS

55. *Dimensions of a Can* The height of a cylindrical soft drink can is 1 inch more than three times its radius. Its lateral surface area (the area of the "side" of the can, as opposed to the top or the bottom) is 30π square inches. Find the height and the radius of the can. (The lateral surface area L of a cylinder of radius r and height h is given by $L = 2\pi rh$.)

56. *Height of a Ball* The height of a ball (in feet) t seconds after it is thrown down from the top of the Sears Tower at 50 feet per second is given by $h = -16t^2 - 50t + 1454$. How long does it take for the ball to reach the ground?

57. *Vertical Jump* If a basketball player has a vertical jump of J inches, then it can be shown that her elevation in inches t seconds after jumping is given by $h = -192t^2 + 16\sqrt{3J}\, t$.

USC and 1996 Olympic basketball star Lisa Leslie dunking.

a. If a basketball player has a vertical jump of 24 inches, how long does it take her to reach an elevation of 1 foot?

b. Find the time required for a basketball player with a vertical jump of 48 inches to attain an elevation of 1 foot.

58. *Rectangle Dimensions* The length of a rectangle is 3 more than twice its width. Its area is 5. Find the dimensions of the rectangle.

59. *Self-Painting Can* Given that a certain paint covers approximately 400 square feet per gallon and that there are approximately 7.5 gallons per cubic foot, find the radius of a cylindrical paint can of height 1 foot that contains just enough paint to paint its own exterior. Interpret your answer. (*Hint*: The total surface area of a cylindrical can of height h and radius r is given by $A = 2\pi r^2 + 2\pi rh$. The volume is given by $V = \pi r^2 h$.)

60. *Profit* The price p (in dollars) of a product is related to the number of units sold, x, by the relationship $p = 11 - 0.01x$. Each unit costs \$4.00 to produce. If x units are sold, the *cost* of production is the total amount of money it costs to produce x units. The *revenue* generated by the product is the total amount of money obtained from sales of x units. The *profit* is the revenue minus the cost.

a. Explain why the cost to produce x units is $4x$, whereas the revenue from the sale of x units is $11x - 0.01x^2$.

b. How many units must be sold to obtain a revenue of \$1000?

c. What is the profit when the revenue is \$1000?

d. How many units must be sold to generate a profit of \$1000?

61. *Multiple Deposits*

a. If \$1000 is deposited in an account at the beginning of each year for 3 years and the account pays 5% interest compounded annually, what is the balance just after the third deposit is made?

b. If $1000 is deposited in an account at the beginning of each year for 3 years and the balance just after the third deposit is made is $4000, what is the interest rate?

62. *Multiple Deposits* Suppose $5 is deposited into an account on January 1, 1996; $15 on January 1, 1998; and $25 on January 1, 2000; and the balance in the account just after the last deposit is $60. Assuming that interest is compounded annually, find the interest rate.

CONCEPTS AND CRITICAL THINKING

EXERCISES 63–66 *Answer true or false.*

63. Some quadratic equations cannot be solved by factoring.

64. Some quadratic equations cannot be solved by completing the square.

65. Some quadratic equations cannot be solved by using the quadratic formula.

66. If $b^2 - 4ac > 0$, then the equation $ax^2 + bx + c = 0$ has 2 real solutions.

EXERCISES 67–70 *Give an example of each.*

67. A quadratic equation having 3 as its only solution.

68. A quadratic equation with no real solutions

69. A technique other than completing the square or the quadratic formula that can be used to solve some quadratic equations

70. An advantage of the quadratic formula over completing the square

QUESTIONS FOR DISCUSSION OR ESSAY

71. Some hand-held calculators can instantly solve any quadratic equation. One could argue that this eliminates the need for students to learn the quadratic formula. What do you think? If inexpensive calculators can instantly solve quadratic equations, should students be forced to learn the quadratic formula? For that matter, why should any of us learn to add, subtract, multiply or divide when $5 calculators can easily accomplish the same task?

72. A project in this exercise set is devoted to using a *cubic* formula, which gives a solution to cubic (third-degree) equations in terms of coefficients. There is also a *quartic* formula, which gives the solution to quartic (fourth-degree) equations in terms of the coefficients. Surprisingly, it was proven in the early 19th century by Abel that there is no quintic formula involving only the fundamental operations of arithmetic and radicals. A related result is that certain constructions with ruler and compass, such as trisecting an arbitrary angle and "squaring a circle" (constructing a square equal in area to that of a given circle) cannot be done. In spite of this, mathematicians (including the authors) regularly receive purported demonstrations of circle squaring or the trisection of an arbitrary angle from amateur mathematicians collectively referred to as "circle squarers." Why do you suppose

these circle squarers have such great difficulty accepting proofs that something (like squaring a circle) *cannot* be done? Are you willing to accept, based on the word of the established mathematical community alone, that there is not, nor can there ever be, a quintic formula for the solution of fifth-degree polynomial equations?

73. Describe the most efficient method for solving the equation $ax^2 + bx + c = 0$ in each of the following cases.

a. $a = 0$

b. $b = 0$

c. $c = 0$

74. In Example 8, we sketched the graph of $x = y^2 + 3y + 1$ by first solving the equation for y, obtaining $y = -\frac{3}{2} + \frac{1}{2}\sqrt{5 + 4x}$ and $y = -\frac{3}{2} - \frac{1}{2}\sqrt{5 + 4x}$, and then plotting both of these branches in the same viewing window. Suppose that we had instead begun by graphing $y = x^2 + 3x + 1$, the equation obtained by exchanging the roles of x and y. How is this graph related to that of $x = y^2 + 3y + 1$? In particular, how could we use the graph of $y = x^2 + 3x + 1$ as an aid in graphing $x = y^2 + 3y + 1$? What connection is there between the x- and y-intercepts of the two graphs?

PROJECTS FOR ENRICHMENT

75. *Cubic Formulas* A general **cubic equation** is of the form $ax^3 + bx^2 + cx + d = 0$ and has three complex solutions, either one or three of which are real. Just as the quadratic formula gives the solution to a quadratic equation in terms of its coefficients, the **cubic formula** gives the solution to a cubic equation in terms

of its coefficients. However, the cubic formula is considerably more difficult to use than the quadratic formula. The equation first must be placed in *normal form* and then in *reduced form* before the equation can be solved. Even after the equation is in reduced form, certain complications can arise (taking square roots of imaginary numbers, for example). In this project, we will deal only with "nice" examples in which there are no complications. Furthermore, we will find only one of the three solutions.

I. Finding the Reduced and Normal Forms

An equation of the form $x^3 + rx^2 + sx + t = 0$ is said to be in **normal form.** Note that the coefficient of the cube term is 1. A cubic equation can be written in normal form by dividing through by the coefficient of the cube term.

a. Express $5x^3 + 4x^2 - 10x + 9 = 0$ in normal form.

b. Express $-x^3 - 3x^2 - 4x - 7 = 0$ in normal form.

A cubic equation of the form $y^3 + py + q = 0$ is said to be in **reduced form.** Note that the coefficient of the cube term is 1 and there is no square term. To transform the normal equation $x^3 + rx^2 + sx + t = 0$ into reduced form, we make the substitution $x = y - \frac{r}{3}$.

c. Express $x^3 - 3x^2 + x - 6 = 0$ in reduced form.

d. Express $2x^3 - 6x^2 + 4x + 10 = 0$ in reduced form.

II. The Cubic Formula

A real solution to the equation $y^3 + py + q = 0$ (provided that $\frac{q^2}{4} + \frac{p^3}{27} > 0$), is given by

$$y = \sqrt[3]{-\frac{q}{2} + \sqrt{\frac{q^2}{4} + \frac{p^3}{27}}} + \sqrt[3]{-\frac{q}{2} - \sqrt{\frac{q^2}{4} + \frac{p^3}{27}}}$$

e. Find a solution to the equation $y^3 + 2y - 1 = 0$ using the cubic formula. Evaluate your answer using a calculator.

f. Find a solution of the equation $3x^3 - 6x^2 + 3x - 6 = 0$ by first writing it in normal form and then in reduced form. Solve the resulting equation for y and then find x. Evaluate your answer using a calculator and check it in the original equation.

76. *The Discriminant and Factoring Quadratic Equation*   It can be shown that for natural numbers n, $\sqrt{n}$ is irrational unless n is a perfect square. Thus, $\sqrt{2}$, $\sqrt{3}$, and $\sqrt{5}$ are irrational, whereas $\sqrt{4} = 2$ and $\sqrt{9} = 3$ are rational. It follows from this that if a, b, and c are all integers, then $\sqrt{b^2 - 4ac}$ will be irrational unless $b^2 - 4ac$ is a perfect square.

a. Assume a, b, and c are all integers and that $b^2 - 4ac = n^2$ for some positive integer n. Show that the quadratic equation $ax^2 + bx + c = 0$ has the solutions

$$x = \frac{-b + n}{2a} \quad \text{and} \quad x = \frac{-b - n}{2a}$$

b. If $b^2 - 4ac = n^2$, show that $ax^2 + bx + c$ factors as

$$\frac{1}{a}\left(ax + \frac{b - n}{2}\right)\left(ax + \frac{b + n}{2}\right)$$

Explain why each of the numbers $\frac{1}{a}$, a, $\frac{b - n}{2}$, and $\frac{b + n}{2}$ is rational.

c. It follows from part b that whenever the discriminant of a quadratic equation with integer coefficients is a perfect square, the quadratic can be factored using only rational coefficients. For each of the following quadratics, indicate whether or not it can be factored using rational coefficients. If it can be factored, use the formula from part b to produce a factorization.

   i. $x^2 + 8x + 144$ ii. $6x^2 - 3x - 2$

  iii. $12x^2 + 13x - 90$ iv. $3x^2 + 10x + 6$

77. *Investigating Equations of Motion*

a. Drop a coin from each of the following heights and measure the time required for the coin to strike the ground. For each height, perform four trials, and then compute the average time.

   i. 4 feet   ii. 6 feet

  iii. 8 feet   iv. 10 feet

If we assume that the only force acting on a projectile is a constant gravitational force, then when the projectile is released from an initial height of h with an initial upward velocity of v_0, the height after t seconds is given by $y = -\frac{1}{2}gt^2 + v_0t + h$, where the constant g represents the acceleration due to gravity. If we use feet for our unit of distance, seconds for our unit of time, and feet per second as our unit of velocity, then $g \approx 32$ ft/sec^2.

b. For each of the initial heights given in part a, use the equation of motion to compute the time required for the coin to drop.

c. Explain any discrepancies between your answers from part a and your answers from part b.

d. Since we are assuming that the gravitational force is constant, an object tossed into the air will slow down by 32 ft/sec each second. Thus, if v represents the velocity of the object in feet per second, then $v = v_0 - 32t$. Compute the velocity upon impact of each of the coins from part a.

e. Toss a coin into the air and observe how its velocity changes. At what point does the velocity appear to be zero?

f. Find a formula for the maximum height of an object in terms of its initial height and initial velocity. (*Hint:* Begin by using the result of part e and the formula from part d to find an expression for the time at which the maximum height occurs.)

g. Now use the result of part f to find a formula for the initial velocity in terms of the initial height and maximum height.

h. What is the initial upward velocity of an outfielder who leaps 3 feet into the air to prevent a home run?

i. How fast must a ball be thrown if it is to rise 1 mile in the air?

j. Throughout this project, we have assumed that gravity was the only force acting on the projectile and that the gravitational force was constant. What other forces might be relevant, and under what circumstances would the gravitational force vary?

SECTION 2.7 RATIONAL AND RADICAL EQUATIONS AND INEQUALITIES

☐ If the life of a section of the Amazon rain forest will be extended by 20 years if the fires set by ranchers are extinguished, and by 10 years if logging is banned, how long will the section of forest remain if neither cause of deforestation is eliminated?

☐ If a company of 3000 employees, 35% of whom are women, mandates that 60% of new hires be female, then how many employees must be added to achieve a 50–50 gender balance?

☐ If you drive 40 miles per hour for 50 miles and then cycle 20 miles per hour for another 50 miles, why isn't your average speed 30 miles per hour?

☐ Which two-man team could shingle a roof faster: one in which each worker is capable of shingling the roof alone in 12 hours, or one consisting of a foreman who can shingle a roof alone in 8 hours and an apprentice who would take 16 hours?

EQUATIONS INVOLVING RATIONAL EXPRESSIONS

Equations involving rational expressions like

$$\frac{1}{x} + \frac{2}{2+x} = \frac{2}{x}$$

can be converted to polynomial equations by **clearing fractions**—that is, by multiplying both sides by a common multiple of all denominators appearing in the equation. The resulting polynomial equation can then be solved using the standard techniques for solving polynomial equations (factoring, the quadratic formula, and so forth), producing one or more solution *candidates*. Since multiplication by a variable expression doesn't always yield an equivalent equation, these solution candidates must be checked by substitution into the original equation. In particular, if substitution of the solution candidate into the original equation gives rise to a division by zero, then, of course, the candidate is not a solution.

EXAMPLE 1 ■ An Equation Giving Rise to a Linear Equation

Solve $\dfrac{2}{x} + \dfrac{1}{2} = \dfrac{7}{6}$.

SOLUTION The least common denominator in this equation is $6x$. Multiplying both sides by $6x$ and simplifying we have

$$6x \cdot \left(\frac{2}{x} + \frac{1}{2}\right) = 6x \cdot \frac{7}{6}$$

$$6x \cdot \frac{2}{x} + 6x \cdot \frac{1}{2} = 7x$$

$$12 + 3x = 7x$$

$$-4x = -12$$

$$x = 3$$

CHECK $x = 3$

$$\dfrac{2}{x} + \dfrac{1}{2} \ \bigg| \ \dfrac{7}{6}$$

$$\dfrac{2}{3} + \dfrac{1}{2}$$

$$\dfrac{4+3}{6}$$

$$\dfrac{7}{6} \qquad\qquad \checkmark$$

EXAMPLE 2   ■   **An Equation with No Solution**

Solve $\dfrac{1}{x-1} + \dfrac{1}{x+1} = \dfrac{2}{x^2-1}$.

SOLUTION   We begin by factoring the denominator of the expression on the right.

$$\dfrac{1}{x-1} + \dfrac{1}{x+1} = \dfrac{2}{(x-1)(x+1)}$$

Next we multiply both sides by the least common denominator, $(x-1)(x+1)$.

$$(x-1)(x+1)\left(\dfrac{1}{x-1} + \dfrac{1}{x+1}\right) = (x-1)(x+1)\,\dfrac{2}{(x-1)(x+1)}$$

$$(x-1)(x+1)\,\dfrac{1}{x-1} + (x-1)(x+1)\,\dfrac{1}{x+1} = 2$$

$$(x+1) + (x-1) = 2$$

$$2x = 2$$

$$x = 1$$

CHECK   Substituting 1 for x in the original equation causes a division by zero. Thus, there is no solution.

EXAMPLE 3   ■   **An Equation Giving Rise to a Quadratic**

Solve $\dfrac{18}{x+2} + \dfrac{2}{x-3} = 5$.

SOLUTION

$$\dfrac{18}{x+2} + \dfrac{2}{x-3} = 5$$

$$(x+2)(x-3)\left(\dfrac{18}{x+2} + \dfrac{2}{x-3}\right) = (x+2)(x-3)5 \qquad \text{Multiplying by the LCD}$$

$$18(x-3) + 2(x+2) = 5(x^2 - x - 6) \qquad \text{Distributing}$$

$$20x - 50 = 5x^2 - 5x - 30$$

$$-5x^2 + 25x - 20 = 0$$

$$x^2 - 5x + 4 = 0 \qquad\qquad \text{Dividing through by } -5$$

$$(x-4)(x-1) = 0$$

$$x = 4, \ x = 1$$

CHECK

$$x = 4$$

$$\frac{18}{x + 2} + \frac{2}{x - 3} \quad\bigg|\quad 5$$

$$\frac{18}{4 + 2} + \frac{2}{4 - 3}$$

$$\frac{18}{6} + \frac{2}{1}$$

$$3 + 2$$

$$5 \qquad\qquad\qquad\bigg|\quad \checkmark$$

$$x = 1$$

$$\frac{18}{x + 2} + \frac{2}{x - 3} \quad\bigg|\quad 5$$

$$\frac{18}{1 + 2} + \frac{2}{1 - 3}$$

$$\frac{18}{3} + \frac{2}{-2}$$

$$6 - 1$$

$$5 \qquad\qquad\qquad\bigg|\quad \checkmark$$

As we have seen in the past, a graphing calculator can be used to confirm a solution to an equation found algebraically or, if exact answers are not needed, to provide approximations to solutions of equations that may be difficult (or even impossible) to solve algebraically. In the case of rational equations, there are some special challenges. Consider, for example, the equation from Example 3. To approximate the solutions graphically, we first rewrite the equation in the form

$$\frac{18}{x + 2} + \frac{2}{x - 3} - 5 = 0$$

and then plot the graph of

$$y = \frac{18}{x + 2} + \frac{2}{x - 3} - 5$$

as shown in Figure 95. Here we already encounter something unexpected. Instead of two *x*-intercepts corresponding to the two solutions found in Example 3, there appear to be four. The problem is with our calculator; in its default **connected mode,** it occasionally connects pieces of graphs of rational equations that should not be connected. The vertical lines that appear in Figure 95 at approximately $x = -2$ and $x = 3$ are a consequence of this phenomenon. There should not, in fact, be any points on the graph at $x = -2$ or $x = 3$ since the equation is not even defined at these values. Instead, there should be breaks in the graph at both $x = -2$ and $x = 3$. To more accurately depict such breaks, it's sometimes helpful to change the calculator to **dot mode** before plotting the graph of a rational equation. The graph shown in Figure 96 was done in dot mode. An even more accurate depiction is shown in Figure 97, where we have sketched in

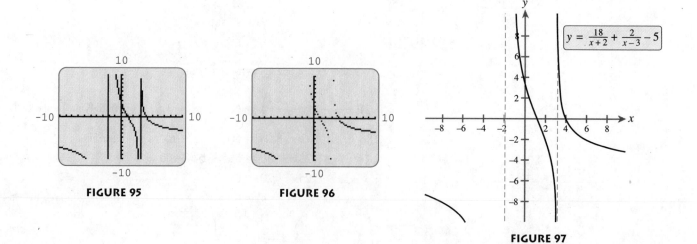

FIGURE 95

FIGURE 96

$$y = \frac{18}{x + 2} + \frac{2}{x - 3} - 5$$

FIGURE 97

dashed lines at $x = -2$ and $x = 3$ to clarify the breaks in the graph. Such lines, called **vertical asymptotes,** occur at x-values where rational equations (expressed in simplest terms) are undefined. We will study vertical asymptotes in more detail in Section 4.5; for now, we must simply be aware of their existence when we plot graphs of rational equations with a graphing calculator. Once the vertical asymptotes have been accounted for, we can estimate the "true" x-intercepts using the usual techniques of either zooming in and tracing or using the calculate-zero feature. In the case of the graph in Figure 95, we obtain estimates for the x-intercepts of $x = 1$ and $x = 4$, which agree with the solutions found in Example 3.

EXAMPLE 4   ▨   **Meeting a Personnel Goal—A Mixture Problem**

An international conglomerate currently has 3000 employees, only 30% of whom are women. The company personnel director has dictated that beginning immediately, 60% of the new hires are to be women until the company's target of 50% women is met. How many new employees must be hired for the company to meet its hiring goal? (Assume that no employees leave the company during the hiring period.)

SOLUTION   Let x be the total number of new employees to be hired. Since 30% of the current employees are women and 60% of the new hires are to be women, we have

$$\text{Total number of women} = \text{Current number of women}$$
$$+ \text{ Number of newly hired women}$$
$$= 30\% \cdot 3000 + 60\% \cdot x$$
$$= 900 + 0.6x$$

The total number of employees is just the sum of the original number of employees and the number of new employees. Thus, we have

$$\text{Total number of employees} = 3000 + x$$

Now the percentage of women in the company is given by the formula

$$\text{Percentage of women} = \frac{\text{Total number of women}}{\text{Total number of employees}} \cdot 100\%$$

or in this case

$$50\% = \frac{900 + 0.6x}{3000 + x} \cdot 100\%$$
$$0.5 = \frac{900 + 0.6x}{3000 + x}$$

Solving for x, we have

$$0.5(3000 + x) = 900 + 0.6x$$
$$1500 + 0.5x = 900 + 0.6x$$
$$-0.1x = -600$$
$$x = \frac{-600}{-0.1}$$
$$x = 6000$$

Thus, 6000 employees must be hired for the company to achieve its desired gender balance.

Real-world applications in which production (distance traveled, homes painted, lawns cut, units manufactured, and so forth) occurs at a constant rate often lead to rational equations. The fundamental formula relating output, rate of production, and time is given by

$$\text{Production} = \text{Rate} \times \text{Time}$$

a special case of which is the familiar distance = rate × time. Since solving for either rate or time gives a fractional expression (rate = production/time, time = production/rate), rational equations arise quite naturally in such contexts. Notoriously tricky problems in which a task is performed by two or more parties working together (children shoveling snow, a man and his daughter cutting grass, a painter and her apprentice painting a room, two pipes filling a pool, and so forth) can be solved quite easily by employing the following rule of thumb.

RULE OF THUMB   When solving work problems, find expressions for the **rate** at which the work is being accomplished. Use the assumption that when two parties work together, the rate at which work is performed is the **sum** of the rates for each party working alone.

EXAMPLE 5   ■   A Problem Involving Work

When a boy uses a toy shovel to help his mother shovel the driveway, the job takes 4 hours. If the boy were working alone, it would take him 4 times longer than it would his mother. How long would it take the boy's mother to shovel the driveway alone? How long would it take the boy?

SOLUTION   Let x be the time required for the boy's mother to shovel the driveway. Then $4x$ is the time required for the boy to shovel the driveway. The rates are then as follows:

$$\text{Mother's rate: } \frac{1 \text{ job}}{x \text{ hours}} \quad \text{Boy's rate: } \frac{1 \text{ job}}{4x \text{ hours}} \quad \text{Rate together: } \frac{1 \text{ job}}{4 \text{ hours}}$$

Since their rate together is the sum of their rates separately, we have

$$\frac{1}{x} + \frac{1}{4x} = \frac{1}{4}$$

$$4x\left(\frac{1}{x} + \frac{1}{4x}\right) = 4x \cdot \frac{1}{4} \qquad \text{Multiplying by the LCD}$$

$$4 + 1 = x$$

$$x = 5$$

Thus, it takes the boy's mother 5 hours, whereas the boy takes $4 \cdot 5 = 20$ hours to shovel the driveway.

INEQUALITIES INVOLVING RATIONAL EXPRESSIONS

Consider an inequality involving a rational expression of the form

$$\frac{P(x)}{Q(x)} > 0$$

where $P(x)$ and $Q(x)$ are polynomials. Since the sign of the quotient is determined by the sign of its numerator and denominator, and since polynomials like $P(x)$ and $Q(x)$

can only change signs at their zeros, it follows that the quotient $P(x)/Q(x)$ can change signs only at a zero of either $P(x)$ or $Q(x)$. For example, consider the inequality

$$\frac{x^2 - 9}{x^2 - 4} > 0$$

The zeros of the numerator are ± 3, and the zeros of the denominator are ± 2, so the quotient can only change signs at $-3, -2, 2,$ and 3. These four numbers divide the real number line into five intervals: $(-\infty, -3), (-3, -2), (-2, 2), (2, 3),$ and $(3, \infty)$—on each of which the quotient $(x^2 - 9)/(x^2 - 4)$ must be either entirely positive or entirely negative. Since we are solving $(x^2 - 9)/(x^2 - 4) > 0$, we must determine the intervals for which the quotient is positive. To help us with this task, we consider the graph of $y = (x^2 - 9)/(x^2 - 4)$, as shown in Figure 98 (connected mode), Figure 99 (dot mode), and Figure 100. Since we want to know where $y = (x^2 - 9)/(x^2 - 4)$ is positive, we look to see where the graph is *above* the x-axis. This occurs on the intervals $(-\infty, -3)$, $(-2, 2)$, and $(3, \infty)$. It follows that $(x^2 - 9)/(x^2 - 4) > 0$ on these intervals, and thus the solution set to our inequality is given by

$$(-\infty, -3) \cup (-2, 2) \cup (3, \infty)$$

Notice that two of the endpoints of these intervals—namely, $x = -2$ and $x = 2$—occur at breaks in the graph (that is, at vertical asymptotes). Since this will often be the case when solving rational inequalities, and since such breaks occur at x-values for which denominators in the rational expression are equal to zero, it is important that we do not clear denominators as we did with rational equations.

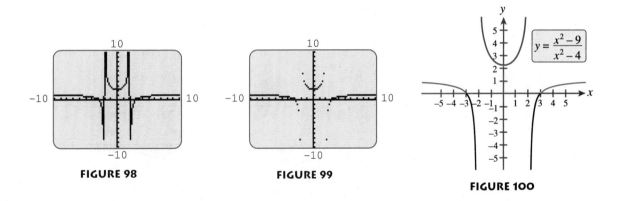

FIGURE 98   **FIGURE 99**   **FIGURE 100**

Our experience with the preceding example suggests the following strategy for solving rational inequalities.

Strategy for Solving ☐
Rational Inequalities

1.   Write the inequality in standard form with a rational expression

$$r(x) = \frac{P(x)}{Q(x)}$$

on the left side and 0 on the right [that is, $r(x) < 0$, $r(x) \le 0$, $r(x) > 0$, or $r(x) \ge 0$].   *(continued)*

(continued)

2. Solve $P(x) = 0$. The solutions are the x-intercepts of $y = r(x)$. If this equation cannot be solved algebraically, approximate the x-intercepts with a graphing calculator.

3. Solve $Q(x) = 0$. If there are any breaks in the graph, they will occur at these values. If the equation $Q(x) = 0$ cannot be solved algebraically, estimate the location of the breaks graphically. Be aware that your graphing calculator may attempt to ''fill-in'' the breaks, and so you may need to use your calculator's dot mode.

4. From the graph of $y = r(x)$, identify the portions of the graph that are either above or below the x-axis, depending on which inequality symbol is used. Express the solution set in interval notation using the results of steps 2 and 3 to determine the endpoints of the intervals.

EXAMPLE 6 ■ **An Inequality Involving Rational Functions.**

Solve $\dfrac{5}{2} - \dfrac{9}{x + 2} < \dfrac{1}{x - 3}$.

SOLUTION Subtracting $\dfrac{1}{x - 3}$ from both sides gives

$$\frac{5}{2} - \frac{9}{x + 2} - \frac{1}{x - 3} < 0$$

We now simplify the left side by converting to the common denominator $2(x + 2)(x - 3)$.

$$\frac{5(x + 2)(x - 3) - 9 \cdot 2 \cdot (x - 3) - 2(x + 2)}{2(x + 2)(x - 3)} < 0$$

$$\frac{5x^2 - 5x - 30 - 18x + 54 - 2x - 4}{2(x + 2)(x - 3)} < 0$$

$$\frac{5x^2 - 25x + 20}{2(x + 2)(x - 3)} < 0$$

$$\frac{5(x - 4)(x - 1)}{2(x + 2)(x - 3)} < 0$$

Setting the numerator to 0, we obtain $x = 4$ and $x = 1$. These are the x-intercepts. Setting the denominator to 0, we obtain $x = -2$ and $x = 3$. These are the locations of the breaks in the graph. We now use a graphing calculator to produce a graph of

$$y = \frac{5x^2 - 25x + 20}{2(x + 2)(x - 3)}$$

like the one shown in Figure 101 (connected mode) or Figure 102 (dot mode). We see that the graph lies below the x-axis (and hence, $y < 0$) for x between -2 and 1, and for x-values between 3 and 4. Thus the solution set is $(-2, 1) \cup (3, 4)$.

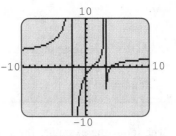

FIGURE 101

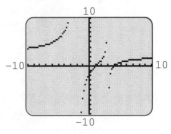

FIGURE 102

It should be noted that, in some contexts, an approximation of the solution to a rational inequality will be sufficient, and if so, much of the work shown in Example 6 can be avoided. In particular, to approximate the solution to the inequality

$$\frac{5}{2} - \frac{9}{x + 2} < \frac{1}{x - 3}$$

we need only plot the graph of

$$y = \frac{5}{2} - \frac{9}{x + 2} - \frac{1}{x - 3}$$

and estimate the intervals on which the graph appears to be below the *x*-axis.

EQUATIONS INVOLVING RADICALS

Our primary technique for solving equations involving radicals is to eliminate the radicals to produce a polynomial equation, in much the same way as we solved equations involving rational expressions by eliminating denominators.

Radicals are eliminated by first *isolating* the radical and then raising both sides of the equation to an appropriate power. If the equation still involves a radical expression, we may need to repeat the process. When the radicals have been eliminated, the resulting equation is solved using the standard techniques for polynomial equations, producing one or more solution candidates.

It is essential that we check our candidates in the original equation. To see why, consider the equation $\sqrt{x} = -1$. If we square both sides, we obtain $x = 1$. But clearly $x = 1$ is not a solution to the original equation. Thus, the equation $\sqrt{x} = -1$ has no real-number solution. The apparent contradiction of correct algebra producing incorrect results will be explored in Exercise 83.

We now summarize the technique for solving equations involving radical expressions.

Technique for Solving □
Equations Involving Radicals

1. ISOLATE the radical—if there is more than one radical, isolate one of them.
2. EXPONENTIATE—if the radical is a square root, square both sides; if the radical is a cube root, cube both sides; and so on.
3. REPEAT if necessary—if a radical remains after isolating and exponentiating, then isolate the remaining radical and exponentiate again.

(continued)

(continued)

4. SOLVE the resulting equation—typically, this is a polynomial equation.

5. CHECK your answer by substituting into the original equation. This is important: there may be extraneous solutions.

The following examples illustrate this technique for solving radical equations.

EXAMPLE 7 ■ **Solving an Equation Involving a Radical**

Find all real solutions of $\sqrt{3x+1} - 4 = 0$.

SOLUTION

$$\sqrt{3x+1} - 4 = 0$$
$$\sqrt{3x+1} = 4 \qquad \text{Isolating the radical}$$
$$(\sqrt{3x+1})^2 = 4^2 \qquad \text{Squaring both sides to eliminate the radical}$$
$$3x+1 = 16$$
$$3x = 15$$
$$x = 5$$

CHECK $x = 5$

$\sqrt{3x+1}$	4
$\sqrt{3\cdot 5+1}$	
$\sqrt{16}$	
4	✓

EXAMPLE 8 ■ **A Radical Equation with an Extraneous Solution**

Find all real solutions of $\sqrt{4x+17} - 2x = 1$.

SOLUTION

$$\sqrt{4x+17} = 2x+1 \qquad \text{Isolating the radical}$$
$$(\sqrt{4x+17})^2 = (2x+1)^2 \qquad \text{Squaring both sides to eliminate the radical}$$
$$4x+17 = 4x^2+4x+1 \qquad \text{Expanding}$$
$$-4x^2+16 = 0 \qquad \text{Collecting all expressions on the left side}$$
$$x^2-4 = 0 \qquad \text{Dividing both sides by } -4$$
$$(x-2)(x+2) = 0$$
$$x = 2, \ x = -2$$

CHECK

$x = 2$		$x = -2$	
$\sqrt{4x+17}-2x$	1	$\sqrt{4x+17}-2x$	1
$\sqrt{8+17}-2\cdot 2$		$\sqrt{-8+17}+4$	
$\sqrt{25}-4$		$\sqrt{9}+4$	
$5-4$		$3+4$	
1	✓	7	×

Thus, the only solution is $x = 2$.

EXAMPLE 9   ◼   **An Equation Involving Two Radicals**

Find all real solutions of $\sqrt{x + 2} - \sqrt{x - 3} = 1$.

SOLUTION

$$\sqrt{x + 2} - \sqrt{x - 3} = 1$$
$$\sqrt{x + 2} = 1 + \sqrt{x - 3} \qquad \text{Isolating a radical}$$
$$x + 2 = (1 + \sqrt{x - 3})^2 \qquad \text{Squaring both sides to eliminate the isolated radical}$$
$$x + 2 = 1 + 2\sqrt{x - 3} + (x - 3) \qquad \text{Expanding}$$
$$4 = 2\sqrt{x - 3} \qquad \text{Simplifying}$$
$$2 = \sqrt{x - 3} \qquad \text{Isolating the remaining radical}$$
$$4 = x - 3 \qquad \text{Squaring both sides to eliminate the radical}$$
$$x = 7$$

CHECK

$$
\begin{array}{c|c}
 & x = 7 \\
\sqrt{x + 2} - \sqrt{x - 3} & 1 \\
\sqrt{7 + 2} - \sqrt{7 - 3} & \\
\sqrt{9} - \sqrt{4} & \\
1 & \checkmark
\end{array}
$$

EXAMPLE 10   ◼   **An Equation Involving Cube Roots**

Solve $\sqrt[3]{x^2 + 1} = 2$.

SOLUTION   Cubing both sides, we obtain

$$(\sqrt[3]{x^2 + 1})^3 = 2^3$$
$$x^2 + 1 = 8$$
$$x^2 = 7$$
$$x = \pm\sqrt{7}$$

It is easily verified that $x = \sqrt{7}$ and $x = -\sqrt{7}$ are solutions.

Many problems involving distance give rise to equations involving radicals.

EXAMPLE 11   ◼   **A Triathlon**

The first two legs of a triathlon are set up in such a way that participants must swim from checkpoint A across a 4-mile-wide channel and then run to checkpoint B, 10 miles down the coast from checkpoint A, as shown in Figure 103. A certain participant can swim at an average speed of 3 miles per hour and can run at an average speed of 12 miles per hour. Given that her total time from checkpoint A to checkpoint B was 2 hours and 15 minutes, find the point where she came ashore. (Assume that she swims in a straight line.)

SOLUTION   Let x be the distance in miles from the point where the athlete came ashore to checkpoint B. Then $10 - x$ is the distance from the point directly across the channel from checkpoint A to the point where she came ashore. Using the Pythagorean Theorem, the swimming distance is

$$\sqrt{4^2 + (10 - x)^2} \quad \text{or} \quad \sqrt{x^2 - 20x + 116}$$

Refer to Figure 104 for a summary of these distances. Since the athlete swims at a rate of 3 miles per hour, the time spent swimming is

$$\frac{\text{Swimming distance}}{\text{Swimming rate}} = \frac{\sqrt{x^2 - 20x + 116}}{3}$$

Similarly, the time spent running is

$$\frac{\text{Running distance}}{\text{Running rate}} = \frac{x}{12}$$

Because the total time spent swimming and running was $2\frac{1}{4} = \frac{9}{4}$ hours, we must have

$$\text{Swimming time} + \text{Running time} = \frac{9}{4}$$

$$\frac{\sqrt{x^2 - 20x + 116}}{3} + \frac{x}{12} = \frac{9}{4}$$

Now we solve for x.

$$\frac{\sqrt{x^2 - 20x + 116}}{3} = \frac{9}{4} - \frac{x}{12} \qquad \text{Subtracting } \frac{x}{12} \text{ from both sides}$$

$$4\sqrt{x^2 - 20x + 116} = 27 - x \qquad \text{Multiplying both sides by 12}$$

$$16(x^2 - 20x + 116) = (27 - x)^2 \qquad \text{Squaring both sides}$$

$$16x^2 - 320x + 1856 = 729 - 54x + x^2$$

$$15x^2 - 266x + 1127 = 0$$

$$x = \frac{266 \pm \sqrt{266^2 - 4(15)(1127)}}{2(15)} \qquad \text{Using the quadratic formula}$$

$$x = 7, \ x = \frac{161}{15} \approx 10.7$$

Thus, the athlete landed at a point either 7 or 10.7 miles from checkpoint B.

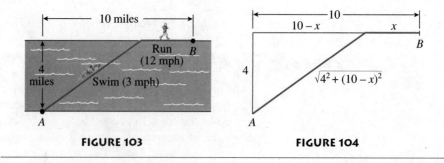

FIGURE 103   **FIGURE 104**

When algebraic techniques fail or if an exact answer isn't necessary, we can solve equations involving rational expressions graphically, as illustrated in the following example.

EXAMPLE 12   ▨   Approximating the Solution of a Radical Equation

Use a graphing calculator to approximate the smallest positive solution of $\sqrt{x + 8} - \sqrt{x - 1} = x$ to the nearest hundredth.

SOLUTION   We first rewrite the equation in standard form.

$$\sqrt{x + 8} - \sqrt{x - 1} - x = 0$$

Next we graph $y = \sqrt{x + 8} - \sqrt{x - 1} - x$ with a graphing calculator, as shown in Figure 105. Finally, we either zoom in on the x-intercept several times and use the trace feature, as shown in Figure 106, or we use the calculate-zero feature. In any case, the solution to the nearest hundredth is $x \approx 2.12$.

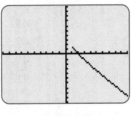

X=2.1233553 Y=-.0015178

FIGURE 105   **FIGURE 106**

UNDERSTANDING AND MASTERY CHECKLISTS

CONCEPTS TO UNDERSTAND

☐ Rational equations
☐ Clearing fractions
☐ Vertical asymptotes
☐ Work problems
☐ Rational inequalities
☐ Radical equations
☐ Significance of checking solutions

SKILLS TO MASTER

☐ Solve a rational equation by clearing fractions.
☐ Set up and solve work problems.
☐ Recognize vertical asymptotes.
☐ Solve rational inequalities graphically.
☐ Solve equations involving radicals.
☐ Systematically check solutions to rational and radical equations and inequalities.

EXERCISES 2.7

EXERCISES 1–24   *Solve the given equation.*

1. $2 = \dfrac{4}{x}$

2. $3 = \dfrac{2}{x - 1}$

3. $\dfrac{7}{y + 4} = \dfrac{3}{y - 4}$

4. $\dfrac{5}{y} = y$

5. $\dfrac{2}{2x + 1} - \dfrac{3}{4x + 1} = 0$

6. $\dfrac{x - 5}{3x + 2} = \dfrac{2x}{6x + 1}$

7. $\dfrac{2}{t} + \dfrac{3}{t^2 + t} = \dfrac{8}{t + 1}$

8. $\dfrac{2}{x - 1} = \dfrac{5}{x} + \dfrac{7}{x^2 - x}$

9. $\dfrac{s}{s^2 - 1} + \dfrac{2}{s + 1} = \dfrac{4}{s - 1}$

10. $\dfrac{2}{x + 3} - \dfrac{5x}{x^2 + 5x + 6} = \dfrac{3}{2x + 6}$

11. $\dfrac{3}{2x - 1} + \dfrac{x + 2}{6x^2 + x - 2} = \dfrac{2}{9x + 6}$

12. $\dfrac{x}{x + 1} = \dfrac{-6}{x - 1}$

13. $\dfrac{x}{x + 1} = \dfrac{3x - 3}{x + 5}$

14. $\dfrac{x}{x - 1} = \dfrac{3}{x^2 - x} - \dfrac{2}{x - 1}$

15. $\dfrac{x}{2x - 1} + \dfrac{3}{x} = \dfrac{2(x^2 + 1)}{2x^2 - x}$

16. $\dfrac{3x + 1}{2x - 1} + \dfrac{3x}{6x - 3} = \dfrac{13 + 8x}{x + 1}$

17. $\dfrac{1}{x + 2} + \dfrac{3}{x} = \dfrac{2x + 2}{x(x + 2)}$

18. $\dfrac{2 - 2x}{x^2 + 2x - 15} + \dfrac{4 - 3x}{x^2 - 3x} = \dfrac{5x - 1}{x^2 + 5x}$

19. $\dfrac{x + 2}{x - 1} + \dfrac{3x + 3}{x + 1} = \dfrac{2x^2 + 4x + 2}{x^2 - 1}$

20. $\dfrac{3x}{x - 3} + \dfrac{54}{x^2 - 9} = 2$

21. $\dfrac{7}{x + 5} = \dfrac{3}{x + 1} = -1$

22. $\dfrac{3}{x + 5} - 2 = \dfrac{x + 8}{x + 5}$

23. $\dfrac{x}{x + 1} + \dfrac{3}{x - 1} = \dfrac{6}{x^2 - 1}$

24. $\dfrac{1 + \dfrac{1}{x}}{1 - \dfrac{1}{x}} = 2$

EXERCISES 25–32   *Solve the given inequality.*

25. $\dfrac{x + 2}{x + 3} \le 2$

26. $\dfrac{5x}{x^2 - 9} \le 0$

27. $\dfrac{1}{x - 2} > \dfrac{4}{x + 1}$

28. $\dfrac{x + 5}{2x - 7} \ge -1$

29. $\dfrac{x}{x + 1} \le \dfrac{x - 2}{x + 3}$

30. $x + 6 \ge -\dfrac{20}{x - 6}$

31. $1 + \dfrac{1}{x} \ge \dfrac{6}{x^2}$

32. $\dfrac{1}{x^2 - 2x + 2} \ge 2$

EXERCISES 33–52   *Find the exact value(s) of the real solution(s) of the given equation.*

33. $\sqrt{x} = 3$

34. $\sqrt{s} = \sqrt{7}$

35. $\sqrt{x} = -2$

36. $\sqrt{x + 5} = 2$

37. $\sqrt{8x - 7} = 0$

38. $\sqrt{3y - 4} = -3$

39. $\sqrt{u + 1} = \sqrt{2u - 2}$

40. $\sqrt{3x + 4} = \sqrt{2x - 6}$

41. $\sqrt{4t + 1} = -\sqrt{4t + 1}$

42. $\sqrt{x - 3} - \sqrt{2} = 0$

43. $\sqrt{\dfrac{3}{r}} - \sqrt{\dfrac{r}{3}} = 0$

44. $\sqrt{5x - 3} - 1 = x$

45. $\sqrt{x + 5} - \sqrt{x} = 1$

46. $\sqrt{5x + 6} + \sqrt{4x + 1} = 7$

47. $\sqrt{x + 13} - x = 1$

48. $5 = a + \sqrt{a + 1}$

49. $2x + 2 = \sqrt{6x + 10}$

50. $z = 2 + \sqrt{2z - 4}$

51. $3 - 2\sqrt{y + 1} = 0$

52. $x - 3\sqrt{4 - x} = 0$

EXERCISES 53–58   *Use a graphing calculator to approximate any solution(s) to the nearest hundredth.*

53. $\dfrac{1}{x^2 + 1} = x$

54. $\dfrac{1}{x} = x - \dfrac{1}{x^2}$

55. $\dfrac{x + 1}{x^2 - 4} = 1 - x^2$

56. $\sqrt{x} = x^2 - 1$

57. $\sqrt{x} + \sqrt{x + 1} = x$

58. $\sqrt{4 - x} = x^2$

APPLICATIONS

59. Shadow Length   Suppose that a woman h feet tall is standing d feet from an l-foot-tall lamppost. It can be shown that if the length of the woman's shadow is s, then $\dfrac{l}{h} = \dfrac{s + d}{s}$.

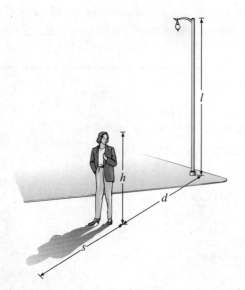

a. Solve this equation for s.

b. How long would the shadow of a woman be if the woman is 5 feet 6 inches and is standing 30 feet from a 40-foot-tall lamppost?

60. Gravitational Force   The gravitational force (in newtons) exerted by an object of mass m_1 kilograms on an object of mass m_2 kilograms with center of mass r meters distant from that of the first object is given by

$$F = G\,\dfrac{m_1 m_2}{r^2}$$

where $G = 6.673 \times 10^{-11}\,\text{N} \cdot \text{m}^2/\text{kg}^2$ is the universal gravitational constant.

a. Find the force of gravitational attraction between two objects with 100-kilogram masses that are 1 meter apart.

b. Solve the force equation for m_2.

61. Harmonic Mean   The harmonic mean of three numbers x, y, and z is defined by

$$h = \dfrac{3}{\dfrac{1}{x} + \dfrac{1}{y} + \dfrac{1}{z}}$$

a. Find the harmonic mean of 3, 4, and 10.

b. Solve the preceding equation for z.

c. Three numbers have a harmonic mean of 10.5. One of the numbers is 7 and another is 8. What is the third number?

62. *Electrical Resistance* When two resistors with resistances R_1 and R_2 are in parallel, the effective resistance R satisfies the equation

$$\frac{1}{R} = \frac{1}{R_1} + \frac{1}{R_2}$$

a. Solve this equation for R_1.

b. Suppose that two resistors in parallel have an effective resistance of 30 ohms. If the resistance of one is 75 ohms, what is the resistance of the other?

63. *River Current* Suppose that the record-holding racing boat *Miss Budweiser*, capable of traveling 140 miles per hour in still water, travels a total of 40 miles upstream and back.

a. If x denotes the speed of the current, find an expression for the time spent traveling upstream.

b. Find an expression for the time traveling downstream.

c. Find the speed of the current given that the total time was 35 minutes.

64. *Filling a Tank* Pipe A and pipe B working together require 2 hours less to fill a tank than pipe B does alone. If pipe B requires 1 hour less than pipe A to fill the tank, find the time required for each of the pipes to fill the tank.

65. *Roofing Estimate* A roofing contractor observes that his new employee, Brandon, works at half the rate of his foreman, Craig. Based on previous experience, the contractor knows that Craig can shingle a certain roof in 8 hours working alone. He concludes that if Brandon were allowed to work alone, he would take 16 hours to shingle the roof.

a. To determine the time required to shingle a roof when Craig and Brandon work together, the contractor simply assumes an average of 12 hours per roof per man. How long would it take the two to complete one roof using this assumption?

b. Find the actual time required to complete one roof, using the assumption that their rate working together is the sum of their rates working alone. (*Hint*: If x denotes the number of hours required when working together, then $1/x$ is their combined rate in roofs per hour.)

c. By how much will the contractor overestimate the time required for Craig and Brandon to shingle 20 roofs working together?

66. *Rain Forest Destruction* A section of the Amazon Rain Forest is being destroyed by two independent sources: fires set by ranch-

ers interested in creating ranges for their cattle and logging by timber companies. It is estimated that it would take fires (without logging) 10 years longer to destroy the section of forest than both fires and logging. It would take logging (without fires) 20 years longer than both fires and logging. How long will the section of forest stand if both logging and fires continue?

67. *Running Cable* Fiber-optic cable is to be run between two office buildings that are separated by a 20-yard roadway, as shown in Figure 107. The cable must start at point A, pass under the roadway, and then continue under the ground to point B. It costs $500 per yard to pass the cable under the roadway, and it costs $100 per yard to bury it under the ground the rest of the way to point B. If the total cost of laying the cable is $30,000 and if the distance from point B to the point directly opposite A is 100 yards, find the point where the cable reaches the other side of the road.

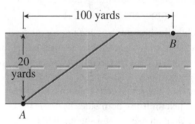

FIGURE 107

68. *Average Speed* Compute the average speed of a trip consisting of two legs: driving 40 miles per hour for 50 miles and then cycling 20 miles per hour for another 50 miles. Why isn't the average speed 30 miles per hour?

69. *Growth Rate* The rate of growth for a boy between the ages of 1 and 19 is approximated by

$$y = \frac{4}{x^2} - \frac{x}{2} + 9 + \frac{32}{3(x-14)^2 + 4}$$

where y is measured in centimeters per year and x in years.

a. Estimate the age at which the average rate of growth is 0 centimeters per year. What interpretation can be given to your answer?

b. Estimate the two ages at which the average rate of growth is 10 centimeters per year.

[*Data source*: Robert M. Malina and Claude Bouchard, *Growth, Maturation, and Physical Activity* (Champaign, IL: Human Kinetics Books, 1991), p. 52.]

■ **CONCEPTS AND CRITICAL THINKING**

EXERCISES 70–73 *Answer true or false.*

70. If $\dfrac{2x+2}{x-1} < 1$, then we can conclude that $2x + 2 < x - 1$.

71. If $\dfrac{2x+2}{x-1} = 1$, then we can conclude that $2x + 2 = x - 1$.

72. If $(2, 7)$ is a point on the graph of $y = p(x) - q(x)$, then we can conclude that 2 must be in the solution set of the inequality $p(x) > q(x)$.

73. Although it may be prudent to check solutions to polynomial equations, it is essential to check solutions to rational and radical equations.

EXERCISES 74–77 *Give an example of each.*

74. A radical equation having no real solution

75. A reason why checking is essential when solving radical equations

76. A rational equation for which checking is essential

77. A number that couldn't possibly be a solution of an equation of the following form, where $\square$ and $\triangle$ are expressions involving x:

$$\frac{\square}{x - 1} = \triangle$$

QUESTIONS FOR DISCUSSION OR ESSAY

78. Consider the following statement: Equations involving radical or rational expressions are solved by first eliminating the most distasteful part of the equation. What is meant by this? What would be the most "distasteful" part of an equation involving radicals? Of an equation involving a rational expression? How does this approach to problem-solving generalize to the solution of non-mathematical problems?

79. When doing algebra, it is not enough to know what can be done, one must also know what *cannot* be done. In this section, we have described in some detail how to solve inequalities involving rational expressions, such as $\frac{4}{x} < -1$. Many students find that even after doing several such problems the correct way, the temptation to "just multiply both sides by x" is overwhelming. What would the final answer be if a student had erroneously multiplied both sides by x and solved the resulting linear inequality? Why can't you "just multiply both sides by x"?

80. Many work problems (such as Example 5 of this section) are solved by making the assumption that the rate at which two people work together is the sum of their individual rates. Under what circumstances is this assumption valid? If a complete dental cleaning can be done in 30 minutes when *one* dentist is working, how long would it take 100,000 dentists working together to clean your teeth?

81. Rational equations could be solved in much the same way that we solved rational inequalities. That is, we could write a given rational equation in the form

$$\frac{P(x)}{Q(x)} = 0$$

and then set $P(x) = 0$. How does this approach differ from solving rational equations given in the text? What is the advantage to solving them the way we did?

PROJECTS FOR ENRICHMENT

82. *Minimizing Triathlon Time* In Example 11, we considered a triathlon in which an athlete starts from checkpoint *A*, swims across a 4-mile-wide channel, and then runs to checkpoint *B*, which is 10 miles down the coast from checkpoint *A*. Given a swimming rate of 3 miles per hour, a running rate of 12 miles per hour, and a total time of 2 hours and 15 minutes, we determined that the athlete must have landed either 7 or 10.7 miles from checkpoint *B*.

a. Does it seem reasonable that the time is the same for two different landing spots? Explain.

b. If we repeat the steps in Example 11 using a total time of 2 hours, we would arrive at the equation

$$\frac{\sqrt{x^2 - 20x + 116}}{3} + \frac{x}{12} = 2$$

where x denotes the distance from the point where the athlete came ashore to checkpoint *B*. Solve this equation for x. What does the answer indicate?

The expression

$$\frac{\sqrt{x^2 - 20x + 116}}{3} + \frac{x}{12}$$

gives the total time for the athlete to travel from *A* to *B*, assuming she comes ashore x miles from checkpoint *B*.

c. Evaluate the expression just given for $x = 0$, $x = 5$, and $x = 10$. Of these three landing points, which gives the athlete the best time?

d. Devise a strategy for determining where the athlete should land so that the total time is as small as possible.

83. *Equivalent Statements* When solving equations involving either rational expressions or radicals, it is essential to check solutions. Even if no mistake is made, an answer might not be a solution of the original equation. This is because the techniques used to solve such equations do not always yield equivalent equations.

When solving linear or quadratic equations, each step produces a statement that is equivalent to the previous statement. In this way, we can be sure that any solution of the last equation will be a solution of the first equation. This is not necessarily the case, however, with equations involving rational expressions or radicals. For example, the operation of squaring both sides of an equation will sometimes yield a nonequivalent equation: The old equation implies the new equation, but the new equation doesn't necessarily imply the old. It is for this reason that the answers to any equation involving radicals should be checked.

For each of the following operations, indicate whether or not it *always* produces an equivalent equation. If not, give an example that demonstrates this.

a. Adding a number to both sides of an equation

b. Multiplying both sides by zero

c. Multiplying both sides by an expression involving a variable

d. Dividing both sides by an expression involving a variable

e. Cubing both sides of an equation

f. Squaring both sides of an equation

SECTION 2.8   SYSTEMS OF EQUATIONS

- ☐ **How can chemical equations be balanced using systems of equations?**
- ☐ **How can hog lard and turkey breast be mixed to form a 90% fat-free mixture?**
- ☐ **If one runner defeats another by 10 meters in the first heat of a 50-meter race, and then for the second heat begins 10 meters behind the starting line to handicap himself, who will win the second heat?**

CLASSIFYING SYSTEMS OF EQUATIONS

We have solved a variety of equations—linear equations, quadratic equations, equations with rational expressions, equations involving radicals, and so on. In this section, we deal with **systems of equations,** which consist of more than one equation and involve more than one variable. Solving a system of equations entails finding values of each of the variables so that all of the equations are satisfied. For example, consider the following system of two equations in two unknowns.

$$2x + 3y = 8$$
$$6x - 3y = 0$$

It can be shown that $x = 1$, $y = 2$ is a solution of this system. In fact, the first equation is satisfied since $2 \cdot 1 + 3 \cdot 2 = 8$, and the second equation is satisfied since $6 \cdot 1 - 3 \cdot 2 = 0$. It is customary to express the solution as $(1, 2)$, which indicates that $x = 1$, $y = 2$ is a solution of the system.

Throughout this section, we will be referring to linear and nonlinear systems of equations. Linear equations involve only variables raised to the first power. For example, $2x + 3y = 5$ is linear, whereas $3x^2 + 4y^3 = 7$ is nonlinear. A **linear system of equations** is a system in which every equation is linear. A **nonlinear system of equations** includes at least one equation that is nonlinear. For example,

$$2x + 3y = 8$$
$$6x - 3y = 0$$

is a linear system of equations, whereas

$$x^2 + y^2 = 1$$
$$2x + 3y = 5$$

is a nonlinear system of equations.

EXAMPLE 1   ■   Verifying a Solution of a System of Nonlinear Equations

In parts a and b, check to see whether the given pair of numbers satisfies the system of equations

$$3x^2 - 5y = 7$$
$$2x + y = 10$$

a.   (2, 1)   **b.**   (3, 4)

SOLUTION

a.   Substituting $x = 2$ and $y = 1$ into the first equation we have

$$3(2)^2 - 5(1) = 12 - 5 = 7$$

Substitution into the second equation gives

$$2(2) + 1 = 5 \neq 10$$

Because the second equation doesn't hold, (2, 1) is not a solution.

b.   Substituting $x = 3$ and $y = 4$ into the first equation, we have

$$3(3)^2 - 5(4) = 27 - 20 = 7$$

Substitution into the second equation gives

$$2(3) + 4 = 6 + 4 = 10$$

Since (3, 4) satisfies both equations, it is a solution of the system.

SOLVING SYSTEMS GRAPHICALLY

Recall from Section 2.1 that the graph of an equation in x and y is the set of points (x, y) such that x and y satisfy the equation. It follows that if (x, y) is a solution to a system of equations, then the point (x, y) must lie on the graph of each of the equations that make up the system. Thus, the solution set of a system of two equations in x and y consists of the intersection points of their graphs. This suggests a technique for approximating the solution set to a system of equations: Graph each of the equations comprising the system, and then estimate the points of intersection of the resulting curves. We illustrate this method in the following examples.

EXAMPLE 2   ■   Solving a System Graphically

Solve the system

$$x - y = 4$$
$$x^3 + y = 7$$

SOLUTION   To plot the graphs, we must first solve each equation for y. This gives us

$$y = x - 4$$
$$y = 7 - x^3$$

Figure 108 shows the result of plotting these equations in the standard viewing window. The graphs appear to intersect near $(2, -2)$. After zooming in a few times, we obtain the view shown in Figure 109, from which we conclude that the coordinates of the point are approximated by $(2.07, -1.93)$. By zooming out repeatedly, we are convinced that there are no other points of intersection. Thus, the only solution to the system is $x \approx 2.07$, $y \approx -1.93$.

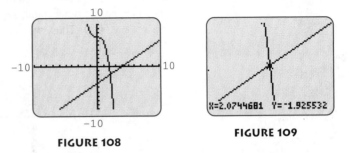

FIGURE 108

FIGURE 109

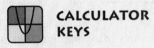

CALCULATOR
KEYS

Finding Intersection Points

Many graphing calculators are capable of automatically estimating the coordinates of the intersection points of two curves. The process typically involves plotting the two curves with a view that shows an intersection point, selecting the **calculate-intersect** command from a menu, identifying the two curves on which the point lies, and specifying an initial guess for the point. Figures 110–113 illustrate this process for estimating the coordinates of the point of intersection of the curves given in Example 2.

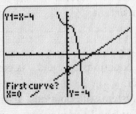

FIGURE 110

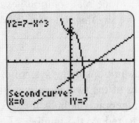

FIGURE 111

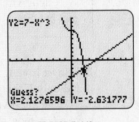

FIGURE 112

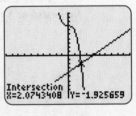

FIGURE 113

EXAMPLE 3   ■   **Solving a System Graphically**

Solve the system

$$2x - y = 7$$
$$x^2 + y^2 = 9$$

SOLUTION   Since the first equation is linear and the second is the equation of a circle, we are looking for the point(s) of intersection of a line and a circle. Because a line can cross a circle once, twice, or not at all, there may be 0, 1, or 2 solutions to this system. Solving the first equation for y gives us $y = 2x - 7$. The second equation gives us

$$x^2 + y^2 = 9$$
$$y^2 = 9 - x^2$$
$$y = \pm\sqrt{9 - x^2}$$

Figure 114 shows the graphs of $y = 2x - 7$, $y = \sqrt{9 - x^2}$, and $y = -\sqrt{9 - x^2}$ (note the use of the "square" window settings, as discussed in Example 12 of Section 2.5). With this view, it is difficult to tell if there is an intersection point. The view shown in Figure 115, on the other hand, suggests that the line $2x - y = 7$ does not intersect the circle $x^2 + y^2 = 9$. Thus, we conclude that the system has no solution.

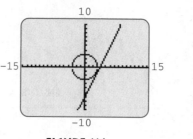

FIGURE 114

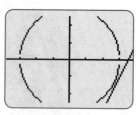

FIGURE 115

Some systems of equations have several or even infinitely many solutions. On the other hand, as the previous example illustrates, some systems have no solutions whatsoever. In this case, the equations of the system impose inconsistent conditions: It is impossible to satisfy both equations at the same time. More formally, we have the following definitions.

Consistent and □
Inconsistent Systems

> A system with at least one real solution is said to be **consistent.** If there are no real solutions, then we say that the system is **inconsistent.**

It follows from these definitions that a system of two equations in x and y is consistent if and only if the graphs of the two equations intersect.

There are several algebraic methods for solving systems of equations, all of which have a common goal: to reduce the number of equations and the number of variables in a step-by-step manner until there is only one equation in one unknown. For example, three equations in three variables are reduced to two equations in two variables, which are, in turn, reduced to one equation in one variable. The resulting equation is then solved for the lone remaining variable, and its value is used to find the values of the previously eliminated variables. In this section, we confine ourselves to systems of two equations in two unknowns.

SOLVING SYSTEMS OF EQUATIONS BY SUBSTITUTION

The most straightforward method (although not always the shortest) for solving systems of equations is **substitution:** solving one of the equations of a system for a variable and then substituting the resulting expression into the other equations that comprise the system. This technique is illustrated in the following example.

EXAMPLE 4   ■   Solving a System of Equations by Substitution

Solve the following system of equations.

$$(1) \qquad\qquad 3x + y = 10$$
$$(2) \qquad\qquad 2x - 3y = -8$$

SOLUTION   We begin by solving equation (1) for y.

$$3x + y = 10$$

(3)
$$y = 10 - 3x$$

Next we substitute $10 - 3x$ for y in equation (2).

$$2x - 3y = -8 \qquad \text{Equation (2)}$$
$$2x - 3(10 - 3x) = -8 \qquad \text{Substituting } 10 - 3x \text{ for } y$$
$$2x - 30 + 9x = -8$$
$$11x - 30 = -8$$
$$11x = -8 + 30$$
$$11x = 22$$
$$x = 2$$

Finally, we use equation (3) and the fact that $x = 2$ to find y.

$$y = 10 - 3x \qquad \text{Equation (3)}$$
$$y = 10 - 3(2) \qquad \text{Substituting 2 for } x$$
$$y = 4$$

Thus, the solution is given by $(x, y) = (2, 4)$. We now verify that $(2, 4)$ is indeed a solution of the system by substitution into the original system.

CHECK   Equation: $3x + y = 10$ Equation: $2x - 3y = -8$
Solution: $x = 2, y = 4$ Solution: $x = 2, y = 4$

$3x + y$	10	$2x - 3y$	-8
$3(2) + 4$		$2(2) - 3(4)$	
$6 + 4$		$4 - 12$	
10	✓	-8	✓

Note that in the previous example, we chose to solve the first equation for y. In general, our choice of which variable to solve for (and in which equation) is determined by convenience.

In all cases, solutions to systems of equations should be checked by substitution into the original system. For the remainder of the examples in this section, we shall omit the details of the verification process.

EXAMPLE 5   ▨   **Solving a System of Nonlinear Equations by Substitution**

Solve the following system of equations.

(1)
$$x - y = 1$$
(2)
$$x^2 + 2 = 6y$$

SOLUTION   In this case, we solve equation (1) for x and substitute into equation (2).

$$x - y = 1$$

(3)
$$x = y + 1$$

We now substitute $y + 1$ for x in equation (2).

$$x^2 + 2 = 6y \qquad \text{Equation (2)}$$
$$(y + 1)^2 + 2 = 6y \qquad \text{Substituting } y + 1 \text{ for } x$$
$$y^2 + 2y + 1 + 2 = 6y \qquad \text{Expanding}$$
$$y^2 - 4y + 3 = 0 \qquad \text{Collecting all terms on the left}$$
$$(y - 1)(y - 3) = 0 \qquad \text{Factoring}$$
$$y = 1, \; y = 3$$

Now we use equation (3), $x = y + 1$, to find the corresponding x-values.

$$\text{When } y = 1: \quad x = 1 + 1 = 2$$
$$\text{When } y = 3: \quad x = 3 + 1 = 4$$

Thus, the two solutions are $(2, 1)$ and $(4, 3)$. Note that in this case *both* solutions need to be verified.

EXAMPLE 6 ▇ **Determining the Weights of Disks Using Systems of Equations**

Suppose that you are at a bench-press contest and have just witnessed the lightweight champion press a bar weighted with 2 large disks and 1 small disk on each side, and the weight is announced as 275 pounds. Then, the eventual middleweight champion presses a bar with 3 large disks and 1 small disk on each side, and the weight is announced as 365 pounds. The heavyweight champion lifts a bar weighted with 4 large disks and 1 small disk, but you cannot hear the announced weight. Assuming the bar weighs 45 pounds, how much did the heavyweight lift?

SOLUTION We first organize the given information in a table.

TABLE 2

Weight class	Large disks (per side)	Small disks (per side)	Announced weight
Light	2	1	275
Middle	3	1	365
Heavy	4	1	?

Let x represent the weight of the larger disk, and y the weight of the smaller disk. Since the lightweight champion's bar had a total of 4 large disks and 2 small disks, and the bar weighs 45 pounds, we have

$$4x + 2y + 45 = 275, \text{ or}$$
(1)
$$4x + 2y = 230$$

Similarly, from the middleweight's lift we obtain

$$6x + 2y + 45 = 365, \text{ or}$$
(2)
$$6x + 2y = 320$$

Now solving equation (1) for y, we have

$$4x + 2y = 230$$
$$2y = 230 - 4x$$
(3) $$\qquad y = 115 - 2x$$

Substituting into equation (2) gives us

$$6x + 2(115 - 2x) = 320$$
$$2x + 230 = 320$$
$$2x = 90$$
$$x = 45$$

Now substituting 45 for x in equation (3), we obtain

$$y = 115 - 2(45)$$
$$= 25$$

Thus, the large disks weigh 45 pounds and the small disks weigh 25 pounds. Since the heavyweight raised a total of 8 large disks and 2 small disks (plus the bar), he lifted

$$8(45) + 2(25) + 45 = 455 \text{ pounds}$$

SOLVING SYSTEMS OF EQUATIONS BY ELIMINATION

A second method for solving systems of equations, **elimination,** consists of adding multiples of the equations that constitute the system in such a way that a variable is eliminated. For example, consider the following system of equations.

$$2x + y = 6$$
$$3x - y = 9$$

By the addition property of equations (if $a = b$ and $c = d$, then $a + c = b + d$), the sum of the left-hand sides of the equation equals the sum of the right-hand sides—that is,

$$
\begin{array}{r}
2x + y = 6 \\
+ \quad 3x - y = 9 \\
\hline
5x \qquad = 15
\end{array}
$$

This gives us $x = 3$, which can then be substituted into either of the original equations to find y. We substitute into the first equation:

$$2x + y = 6$$
$$2(3) + y = 6$$
$$y = 0$$

Thus, the solution is $(3, 0)$.

Note that by adding the two equations together, one of the variables (in this case, y) was eliminated. This gave us a simple equation involving only one variable. Of course, for an arbitrary system of two equations in two variables, there is no guarantee that simply adding the equations together will eliminate one of the variables. However, it is often the case that a variable will be eliminated upon addition if we first multiply each equation by an appropriate constant. The technique of elimination is illustrated in the following examples

E X A M P L E 7   ■   **Solving a System of Linear Equations by Elimination**

Solve the following system of equations by elimination.

(1) $$4x + 3y = 5$$
(2) $$3x + y = 4$$

SOLUTION   Adding these equations together eliminates neither x nor y. In fact, addition gives

$$
\begin{array}{r}
4x + 3y = 5 \\
+ \quad 3x + \ y = 4 \\
\hline
7x + 4y = 9
\end{array}
$$

and we're no better off than before. However, if we first multiply the second equation by -3, we obtain

$$(-3)(3x + y) = (-3)4$$
(3) $$-9x - 3y = -12$$

Since equation (3) includes the term $-3y$, and equation (1) includes the term $3y$, adding equations (1) and (3) eliminates y.

$$
\begin{array}{rl}
4x + 3y = 5 & \text{Equation (1)} \\
+ \quad -9x - 3y = -12 & \text{Equation (3)} \\
\hline
-5x \qquad = -7 &
\end{array}
$$

Thus $x = \frac{7}{5}$. We can now substitute into equation (2) to find y.

$$3x + y = 4$$
$$3\left(\frac{7}{5}\right) + y = 4$$
$$\frac{21}{5} + y = 4$$
$$y = 4 - \frac{21}{5}$$
$$= \frac{20}{5} - \frac{21}{5}$$
$$= -\frac{1}{5}$$

Thus, our solution is $(x, y) = (\frac{7}{5}, -\frac{1}{5})$.

In the previous example, we eliminated a variable by multiplying one of the equations of a system by a constant and then adding to the other equation. In the next example, it is more convenient to multiply *both* equations by a constant before addition. Our goal is the same as before: We would like to produce a system of two equations such that when the equations are added, one of the variables will be eliminated.

E X A M P L E 8 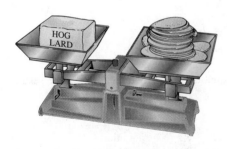 Solving a System of Linear Equations by Elimination

Solve the following system of equations by elimination.

$$3x - 5y = 5$$
$$4x - 7y = 6$$

SOLUTION Although either variable could be eliminated, we will eliminate x. To obtain equations in which the coefficients of x are opposites, we multiply the first equation by 4 and the second equation by -3.

Original equation	Multiply by . . .		New equation
$3x - 5y = 5$	4		$12x - 20y = 20$
$4x - 7y = 6$	-3	$+$	$-12x + 21y = -18$
			$y = 2$

Substituting 2 for y in $3x - 5y = 5$ gives

$$3x - 5(2) = 5$$
$$3x - 10 = 5$$
$$3x = 15$$
$$x = 5$$

Thus, our solution is $(x, y) = (5, 2)$

E X A M P L E 9 Solving a Mixture Problem Using Elimination

A foolish dieter, following the instructions of his nutritionist to the letter, is insistent that fat constitute exactly 10% of his diet. On a certain day, he must select from a menu consisting strictly of 92% fat-free turkey bologna and hog lard, which is virtually 100% fat. How much of each food should he consume if he wishes to eat 10 pounds of food?

SOLUTION Let x represent the number of pounds of turkey bologna eaten and y the number of pounds of hog lard. Then we have

Pounds of fat from turkey bologna
  + Pounds of fat from lard = Pounds of fat in mixture
$$8\% \cdot x + 100\% \cdot y = 10\% \cdot 10$$
$$0.08x + y = 1$$

On the other hand, we have

Pounds of turkey bologna + Pounds of lard = 10 pounds
$$x + y = 10$$

Thus, we must solve the system

(1) $$0.08x + y = 1$$
(2) $$x + y = 10$$

Multiplying equation (1) by -1 and adding to equation (2) gives us

$$-0.08x - y = -1$$
$$+\underline{x + y = 10}$$
$$0.92x = 9$$

Solving for x gives us $x = \dfrac{9}{0.92} \approx 9.8$. Substituting $x = 9.8$ into equation (2) gives

$$x + y = 10$$
$$9.8 + y = 10$$
$$y = 0.2$$

Thus, 9.8 pounds of turkey bologna mixed with 0.2 pound of lard yields 10 pounds of a mixture that is 90% fat-free.

So far we have illustrated the technique of elimination for solving linear systems of equations, but it can also be an effective technique for solving nonlinear systems, as illustrated in the following example.

EXAMPLE 10   ■   Solving a Nonlinear System of Equations

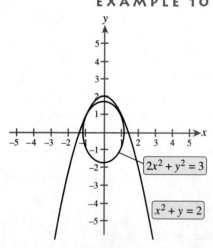

FIGURE 116

Find all points of intersection of the curves $x^2 + y = 2$ and $2x^2 + y^2 = 3$ shown in Figure 116.

SOLUTION   We begin by noting that intersection points of the graphs correspond to solutions of the system

(1) $ x^2 + y = 2$
(2) $ 2x^2 + y^2 = 3$

We eliminate the variable x by multiplying equation (1) by -2 and adding the result to equation (2).

Original equation	Multiply by . . .	New equation
$x^2 + y = 2$	-2	$-2x^2 - 2y = -4$
$2x^2 + y^2 = 3$	1	$+\ \ 2x^2 + y^2 = 3$
		$y^2 - 2y = -1$

Solving for y, we obtain

$$y^2 - 2y = -1$$
$$y^2 - 2y + 1 = 0$$
$$(y - 1)^2 = 0$$
$$y = 1$$

Substituting $y = 1$ into equation (1) gives us

$$x^2 + 1 = 2$$
$$x^2 = 1$$
$$x = \pm 1$$

Thus, the points of intersection are $(1, 1)$ and $(-1, 1)$, as shown in Figure 117.

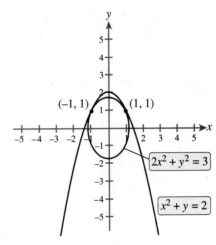

FIGURE 117

UNDERSTANDING AND MASTERY CHECKLISTS

CONCEPTS TO UNDERSTAND

- ☐ Linear and nonlinear systems of equations
- ☐ Intersection of graphs
- ☐ Consistent and inconsistent systems
- ☐ Substitution
- ☐ Elimination

SKILLS TO MASTER

- ☐ Classify systems of equations as linear or nonlinear.
- ☐ Solve systems of equations graphically.
- ☐ Classify systems of equations as consistent or inconsistent.
- ☐ Solve systems of equations by substitution.
- ☐ Solve systems of equations by elimination.
- ☐ Apply systems techniques in applied settings.

EXERCISES 2.8

EXERCISES 1–30 *Solve the system of equations. Some systems may be inconsistent.*

1. $2x + y = 10$
   $3x - y = 5$

2. $3s + t = 5$
   $2s - 2t = 14$

3. $-3x + y = 7$
   $3x + 4y = -2$

4. $2x + 3y = -7$
   $x + 5y = -14$

5. $2q + 3r = 3$
   $4q + 6r = 0$

6. $3x - 4y = 8$
   $9x - 12y = 2$

7. $3m - n = 2$
   $-4m + 2n = 2$

8. $3x + 5y = 2$
   $5x + y = 18$

9. $x - 5y = 5$
   $4x + 5y = 5$

10. $-2x - y = -2$
    $2x + y = -1$

11. $2z - 7w = -2$
$10z - 35w = -8$

12. $2x + 5y = 5$
$4x - 5y = 4$

13. $\dfrac{10}{3}a + 2b = -6$
$a + \dfrac{1}{5}b = 1$

14. $6a - 2b = \dfrac{7}{2}$
$3a + \dfrac{1}{2}b = \dfrac{5}{2}$

15. $2x + 3y = 9.8$
$3x - 5y = -8.1$

16. $2w - 5z = 12.9$
$3w + 4z = 4.4$

17. $0.5x - 0.2y = 1.36$
$1.5x + 0.4y = 5.78$

18. $1.2A + 4.6B = 18.5$
$-2.4A + 3.0B = 5.7$

19. $(2 \times 10^{-6})x + 10^{-6}y = 33$
$4x - y = 3 \times 10^7$

20. $(4 \times 10^4)k + (1.2 \times 10^4)m = 2 \times 10^4$
$(2 \times 10^4)k + (5 \times 10^4)m = -2.1 \times 10^5$

21. $x + y = 8$
$xy = 15$

22. $2x - 3y = 5$
$\dfrac{y}{x} = -1$

23. $2p_2 + 3q_2 = 30$
$p + q = 1$

24. $\dfrac{c}{d} - c = 4$
$c - 8d = 0$

25. $\dfrac{2}{a} + \dfrac{3}{b} = 0$
$\dfrac{4}{a} + \dfrac{12}{b} = -1$

26. $\dfrac{6}{p} + \dfrac{6}{q} = 33$
$-\dfrac{3}{p} + \dfrac{4}{q} = -6$

27. $x^2 + y^2 = 9$
$5x - y = -3$

28. $2(u + v)^3 - v = -11$
$(u + v)^3 - 3v = 7$

29. $x_4 + y_4 = 97$
$x_4 - y_4 = 65$

30. $\dfrac{x^2}{4} + \dfrac{y^2}{9} = 2$
$3x + y = 3$

EXERCISES 31–36 *Find the point(s) of intersection of the given pair of equations.*

31. $2x - y = 1; \ 3x + y = 9$

32. $x - 4y = 8; \ -3x - 2y = 4$

33. $y + x^2 = 4; \ y - x = 2$

34. $y = x^2 + 2x; \ y = -2x - 3$

35. $x^2 + y^2 = 2; \ x = y^2$

36. $x^2 + y^2 = 25; \ x + y = 1$

EXERCISES 37–42 *Use a graphing calculator to estimate the coordinates of the point(s) of intersection of the given pair of equations to the nearest hundredth.*

37. $y = x^3 - x + 2$
$y = x^2 + x + 1$

38. $y = x^2 - 3x + 7$
$y = 3x^2 - 11x + 15$

39. $y = 15 - x^2$
$y = 3\sqrt{4 - x^2} + 4$

40. $y = \sqrt{x^2 - 4x + 4}$
$y = x - 2$

41. $y = 3|x - 2|$
$y = 6 - 3|x - 4|$

42. $y = x$
$y = \dfrac{x^3 + x + 1}{x^2 + 1}$

APPLICATIONS

43. *Counting Coins* A street musician counts the coins in his hat and discovers that he has four more nickels than dimes, and that he has a total of 80¢ in nickels and dimes. How many nickels and how many dimes does the street musician have?

44. *Counting Coins* A budding numismatist (coin collector) has twice as many silver dollars as quarters; the total face value of the silver dollars and the quarters is $4.50. How many of each does she have?

45. *Low-Fat Mixture* A sundae is being made from 98% fat-free frozen yogurt and 80% fat-free dessert topping. How many ounces of each should be used if we wish to have a 10-ounce, 90% fat-free sundae?

46. *Achieving Racial Balance* In a certain school district, 40% of the students are African American, but only 10% of the 1000 teachers are. Since the student–teacher ratio is also unacceptably high, the school board decides that it will correct both problems by doubling the number of teachers, and hiring in such a way that the racial makeup of the resulting faculty reflects that of the student body. The personnel director determines that all of the new hires will be from two institutions: a state university of which 60% of the students are African American and a historically Black college of which 93.3% of the students are African American. How many new teachers should be hired from each of the schools?

47. *Age Comparison* Mark is 11 times older than his son Benjamin. In 27 years, Benjamin will be $\frac{1}{2}$ the age of his father. How old are they?

48. *Age Comparison* Two years ago, Juanita was 2.5 times older than her younger sister LaShonda. Ten years from now, Juanita will be 1.5 times older than LaShonda. How old are they now?

49. *Ticket Sales* For a certain Bob Dylan concert, two types of tickets were available: $30 tickets and $20 tickets. If a total of 13,000 tickets were sold and the gross receipts for the concert totaled $310,000, then how many of each kind of ticket were sold?

50. *Painting Rate*   It takes Susan 1 hour less to paint a room than it does David. Together, they take 2 hours. How long does it take each of them working alone?

51. *Raising a Whale*   Arnold and Franco are participating in a lift-a-thon known as "Raising the Whales" to raise money to protect the blue whale, of which there are, by some estimates, only 10,000 left in existence. Their goal is to lift a total weight equal to that of the largest blue whale ever captured, approximately 209 tons (*Data source: The Guinness Book of World Records*). If Arnold were lifting alone, he could "raise a whale" in 10 fewer hours than it would take Franco working alone. Together, they can complete the task in 15 hours. How long does it take each of them working alone?

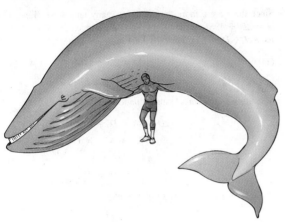

52. *Car Speed*   Car *A* leaves an intersection at 2:00 P.M. traveling north at a constant speed. Car *B* leaves the same intersection at 4:00 P.M. and travels east at a constant speed. At 6:00 P.M., the cars are 200 miles apart, and the total distance traveled by the cars is 280 miles. How fast is each car traveling?

53. *Sprint Times*   Two brothers are competing in a 50-meter race. The first time they race, the older brother wins by 10 meters. The older brother offers to start the next race 10 meters behind the starting line in order to make the race fair. Although both runners run at the same speed as before, the older brother still wins, this time by 0.5 second. What is the 50-meter time for both boys? (Assume each is running at a constant speed.)

54. *Cattle Speed*   Two ancient cows of equal speeds, each well versed in the rudiments of mathematics, are standing side by side

on a 120-foot railroad bridge that runs north-south. The cows hear the whistle of a train heading toward them from the north when the train is 1000 feet away from the bridge. The train is traveling at a speed of 53 feet per second. The impish bovines run at top speed in opposite directions to torment the engineer. Just as planned, they each are just grazed by the train as they clear the bridge. How fast can the cattle run, and how far were they initially from the north end of the bridge?

55. *Radar Search*   Two radar stations are located 50 miles apart along a coastline at the points marked *A* and *B* in Figure 118. A ship is in distress at a point *C*, which is 20 miles from *A* and 40 miles from *B*. Find the location of the ship relative to *A*. (*Hint*: Use point *A* as the origin of a coordinate system and find the points of intersection of the circle centered at *A* with radius 20 and the circle centered at *B* with radius 40.)

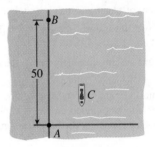

FIGURE 118

56. *Kidnap Caper*   A suitcase containing $250,000 in ransom money has just been picked up by a kidnapper. Little does he realize that his fate has just been sealed. Inspector Magill is on the case and a miniature radio transmitter is in the handle of the suitcase. Using two receivers that can detect the distance to the signal, Magill and an assistant are able to monitor the movement of the signal. When the signal finally stops moving, it is 2 miles away from Magill's position and 3 miles away from his assistant's position. If Magill is 4 miles west of his assistant, use intersecting circles to determine the location(s) where police should be sent.

■ CONCEPTS AND CRITICAL THINKING

EXERCISES 57–60   *Answer true or false.*

57. Replacing the first of two equations in a system by the sum of the two equations results in an equivalent system.

58. The system $x + y = 5$ is inconsistent
$$2x + 2y = 5$$

59. If even a single equation in a system is linear, then the entire system is considered to be linear.

60. If even a single equation in a system is nonlinear, then the entire system is considered to be nonlinear.

EXERCISES 61–64   *Give an example of each.*

61. A nonlinear system of equations

62. A linear system of equations with infinitely many solutions

63. A system of two linear equations for which the only solution is $x = 2$, $y = 3$.

64. An inconsistent system in which one of the equations is $2x + 3y = 5$

65. One of the most common mistakes made by beginning algebra students when solving applied problems is that of not carefully defining variables. Consider the following problem.

> There are four more than twice as many nickels as dimes. Together the nickels and dimes are worth $1.00. How many of each are there?

One key but often overlooked restriction is that any variable introduced should represent a number; many students make the mistake of defining variables to be inanimate objects, barnyard animals, and people. For example, when solving the above problem, it is not uncommon for students to begin this problem by writing something like, "Let N = nickels, and let D = dimes". The problem is that "nickels" is not a number. Thus, the student is forced to assume that N represents some number in some way associated with "nickels". But there is more than one number associated with "nickels"! Explain how this difficulty leads to the following incorrect solution.

Since there are 4 more than twice as many nickels as dimes, we have $N = 2D + 4$. Since the nickels and dimes together are worth 100¢, we have $N + D = 100$. Thus we must solve the system.

(1) $$N = 2D + 4$$
(2) $$N + D = 100$$

Substituting equation (1) into equation (2), we have

$$(2D + 4) + D = 100$$
$$3D + 4 = 100$$
$$3D = 96$$
$$D = 32$$

Thus, $N = 2D + 4 = 2 \cdot 32 + 4 = 68$.

■ QUESTIONS FOR DISCUSSION OR ESSAY

66. Explain the connection between the systems of equations in the first column and the corresponding set of statements in the second column.

	System	Statements
A.	$x + y = 2$ $2x + 2y = 4$	Edward is the father of Mark. Mark is the son of Edward.
B.	$x + y = 3$ $x + y = 5$	I am a crook. I am not a crook.
C.	$2x + 3y = 8$ $3x - y = 1$	The author was a Kennedy. The author served as president.

67. Some applied problems require natural number solutions. For example, in a problem involving the number of coins, it would be unacceptable to have a final answer such as "2.3 quarters, 4.7 dimes, and $-\sqrt{2}$ nickels." How do you suppose the authors construct systems of equations in such a way that the answers are guaranteed to be of the proper form? For example, can you construct a system of equations involving the variables x and y such that the solution is $x = 2$ and $y = -5$?

68. Explain how a system of three equations in three variables could be solved by generalizing the techniques of this section.

■ PROJECTS FOR ENRICHMENT

69. *Balancing Chemical Equations*   Chemical equations are a shorthand way of describing chemical changes. The left side of the chemical equation consists of the **reactants**, the chemicals that react with each other. The right side of the chemical equation consists of the **products**, the results of the chemical reaction. The left and right sides are separated by an arrow that is typically read as "yields." For example, the reaction consisting of hydrogen (in the form of hydrogen gas H_2) and oxygen (in the form of oxygen gas O_2) combining to form water would be represented as

$$H_2 + O_2 \rightarrow H_2O$$

Chemical explosions such as this one (created with 24 kilograms of gunpowder) result from the rapid release of gasses and heat in a chemical reaction.

This equation has not been balanced, however, and as such does not conform to the law of conservation of matter—there are two oxygen atoms on the left but only one on the right. The following chemical equation is a balanced version of this reaction.

$$2H_2 + O_2 \rightarrow 2H_2O$$

More generally, we must ensure that the number of atoms on each side of the yield sign are the same by affixing integer coefficients in front of the reactants and products. For example, consider the unbalanced chemical equation $Al_2O_3 + F_2 \rightarrow AlF_3 + O_2F_2$. To balance the equation, we begin by placing unknown coefficients in front of each of the reactants and products.

$$wAl_2O_3 + xF_2 \rightarrow yAlF_3 + zO_2F_2$$

Now, by equating the number of atoms of aluminum, fluorine, and oxygen, we obtain the following system of equations.

$2w = y$	Equating the number of aluminum atoms on each side
$2x = 3y + 2z$	Equating the number of fluorine atoms on each side
$3w = 2z$	Equating the number of oxygen atoms on each side

Unfortunately, there are not enough equations to determine the values of w, x, y, and z since there are four variables but only three equations. For the sake of convenience, we assume that $w = 1$. This eliminates one of the variables, giving us the following system of three equations in three unknowns.

(1) $2 = y$

(2) $2x = 3y + 2z$

(3) $3 = 2z$

From equation (1) we have $y = 2$. From equation (3) we have $z = \frac{3}{2}$. Substituting these values into equation (2) gives us

$$2x = 3(2) + 2\left(\frac{3}{2}\right)$$

$$2x = 6 + 3$$

$$x = \frac{9}{2}$$

Thus, our balanced chemical equation now reads

$$Al_2O_3 + \frac{9}{2} F_2 \rightarrow 2AlF_3 + \frac{3}{2} O_2F_2$$

However, we would like our coefficients to be integers. Since multiplying both sides by the same constant will give an equivalent balanced equation, we simply clear fractions by multiplying both sides by the least common denominator of all fractions appearing in the equation. In this case, we multiply by 2 to obtain

$$2Al_2O_3 + 9F_2 \rightarrow 4AlF_3 + 3O_2F_2$$

Balance the following chemical equations by setting up and solving a system of equations.

a. $Na + Cl_2 \rightarrow NaCl$

b. $Ag + H_2S + O_2 \rightarrow Ag_2S + H_2O$

c. $Cu + HNO_3 \rightarrow Cu(NO_3)_2 + H_2O + NO$

d. $Ca_3(PO_4)_2 + H_3PO_4 \rightarrow Ca(H_2PO_4)_2$

70. *Partial Fraction Decomposition*   Consider the following sum of rational expressions.

$$\frac{3}{x} - \frac{2}{x^2} + \frac{4}{x + 1}$$

We can easily add these terms together by finding a common denominator in the usual way. That is, since the least common denominator is $x^2(x + 1)$, we have

$$\frac{3(x)(x + 1) - 2(x + 1) + 4(x^2)}{x^2(x + 1)} = \frac{3x^2 + 3x - 2x - 2 + 4x^2}{x^2(x + 1)}$$

$$= \frac{7x^2 + x - 2}{x^2(x + 1)}$$

Partial fraction decomposition is the reverse of this procedure. For example, the partial fraction decomposition of

$$\frac{7x^2 + x - 2}{x^2(x + 1)} \quad \text{is} \quad \frac{3}{x} - \frac{2}{x^2} + \frac{4}{x + 1}$$

We confine ourselves to rational expressions of the form $P(x)/Q(x)$, where $P(x)$ and $Q(x)$ are polynomials with the degree of $P(x)$ less than the degree of $Q(x)$. We further assume that $Q(x)$ can be factored as the product of linear factors; that is, $Q(x)$ can be factored as $(x - a)^s(x - b)^t \cdots (x - c)^u$. It can be shown (it is called the Partial Fraction Decomposition Theorem) that with these restrictions, $P(x)/Q(x)$ can be written in the form

$$\frac{A_1}{x - a} + \frac{A_2}{(x - a)^2} + \cdots$$

$$+ \frac{A_s}{(x - a)^s} + \frac{B_1}{x - b} + \frac{B_2}{(x - b)^2} + \cdots$$

$$+ \frac{B_t}{(x - b)^t} + \cdots + \frac{C_1}{x - c} + \frac{C_2}{(x - c)^2} + \cdots + \frac{C_u}{(x - c)^u}$$

for some choice of the coefficients $A_1, A_2, \ldots, A_s, B_1, B_2, \ldots, B_t, C_1, C_2, \ldots, C_u$. This is not as confusing as it sounds. For example, according to the Partial Fraction Decomposition Theorem, the rational expression

$$\frac{3x^2 + 4x + 7}{x^3(x - 2)^2}$$

can be written in the form

$$\frac{A}{x} + \frac{B}{x^2} + \frac{C}{x^3} + \frac{D}{x - 2} + \frac{E}{(x - 2)^2}$$

with one term for each of the powers of x from one to three, and one term for each of the powers of $x - 2$ from one to two. We illustrate the method of finding the partial fraction decomposition with the rational expression

$$\frac{7x^2 + x - 2}{x^2(x + 1)}$$

since we already know what the answer should be. We begin by noting that, indeed, the degree of the numerator (2) is less than the degree of the denominator (3), and that the denominator is the product of linear factors. Next we use the Partial Fraction Decomposition Theorem to conclude that

$$\frac{7x^2 + x - 2}{x^2(x + 1)} = \frac{A}{x} + \frac{B}{x^2} + \frac{C}{x + 1}$$

for some choice of constants A, B, and C. Clearing all denominators by multiplying both sides through by $x^2(x + 1)$ gives us

$$7x^2 + x - 2 = Ax(x + 1) + B(x + 1) + Cx^2$$
$$7x^2 + x - 2 = Ax^2 + Ax + Bx + B + Cx^2$$

Collecting like terms on the right-hand side gives us

$$7x^2 + x - 2 = (A + C)x^2 + (A + B)x + B$$

Now, equating like powers of x on the left- and right-hand sides gives us the following system of equations.

(1) $\qquad\qquad 7 = A + C$

(2) $\qquad\qquad 1 = A + B$

(3) $\qquad\qquad -2 = B$

This system is especially easy to solve. From equation (3) we have $B = -2$. Substituting into equation (2), we get $1 = A + -2$, so that $A = 3$. Upon substituting $A = 3$ into equation (1), we get $C = 4$. Thus, as expected,

$$\frac{7x^2 + x - 2}{x^2(x + 1)} = \frac{3}{x} + \frac{-2}{x^2} + \frac{4}{x + 1}$$

Find the partial fraction decompositions for each of the following rational expressions.

a. $\dfrac{9x^2 + 4x + 1}{x(x + 1)^2}$
b. $\dfrac{x^3 + 8x^2 + 11x + 6}{x(x + 1)^3}$

c. $\dfrac{x^3 + 10x^2 - 12x + 5}{x^2(x - 1)^2}$

[handwritten: any problem here]

CHAPTER 2 REVIEW

EXERCISES 1–2 *Determine which of the given points are on the graph of the equation.*

1. $x^2y + 2xy^3 + x = -2$; $(-1, 1)$, $(2, -2)$, $(-2, 0)$

2. $\dfrac{2}{x + 1} = \dfrac{3}{\sqrt{y - 1}}$; $(\frac{2}{3}, 5)$, $(1, 10)$, $(-\frac{1}{3}, 2)$

EXERCISES 3–4 *Sketch the graph of each of the following equations by plotting points.*

3. $y = -2x^2 + 8$

4. $y = -\frac{1}{2}x + 3$

EXERCISES 5–10 *Find the center and radius of the circle and sketch its graph.*

5. $x^2 + y^2 = 16$

6. $x^2 + y^2 = \frac{1}{4}$

7. $(x - 2)^2 + (y + 1)^2 = 9$

8. $(x + 3)^2 + (y - 2)^2 = 25$

9. $x^2 + y^2 + 10x - 8y + 32 = 0$

10. $x^2 + y^2 - 2x - 6y + 6 = 0$

EXERCISES 11–14 *Find the equation of the circle with the given properties.*

11. Center $(2, 5)$ and radius 3

12. Center $(-2, 1)$ and radius 4

13. Center $(2, 4)$ and passes through $(-4, -4)$

14. The points $(5, 5)$ and $(-1, -3)$ are endpoints of a diameter.

EXERCISES 15–18 *Find the graph that matches the given equation. Choose from i–iv.*

i.

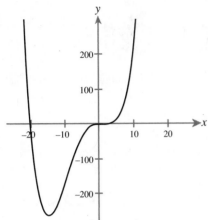

ii.

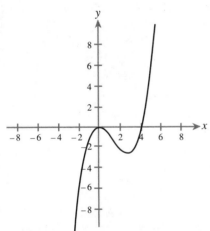

iii.

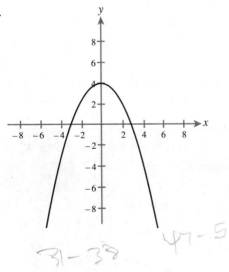

iv.

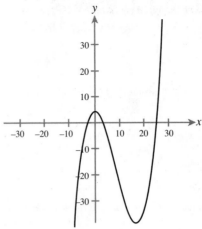

15. $y = \dfrac{x^3}{48} - \dfrac{x^2}{2} + 4$

16. $y = \dfrac{x^4}{80} + \dfrac{x^3}{5} - x^2$

17. $y = \dfrac{x^3}{4} - \dfrac{33x^2}{32}$

18. $y = 4 - 0.48x^2$

EXERCISES 19–22 *Use the trace feature of a graphing calculator to estimate the indicated value(s) to the nearest hundredth.*

19. $y = \dfrac{x^3}{8} - \dfrac{15x^2}{16} - \dfrac{129x}{32} + 5$; x- and y-intercepts

20. $y = \dfrac{1440}{x^2 + 36} - 2x - 32$; x- and y-intercepts

21. $y = \sqrt{-x^2 + 4x + 5}$; coordinates of the highest point on the graph

22. $y = x^2 - 3x + 1$; coordinates of the lowest point on the graph

EXERCISES 23–26 *Determine whether or not the given value is in the solution set of the equation or inequality.*

23. $x^5 + 2x - 3x^2 - 1 = 23$; $x = 2$

24. $3\sqrt{x - 7} - \dfrac{2x}{11} = 4$; $x = 11$

25. $4\,|\,\sqrt{5x + 4} - 7\,| + 16 = 0$; $x = 1$

26. $x^2 - 3x \le 4$; $x = -1$

EXERCISES 27–30 *Indicate whether or not each pair of equations or inequalities is equivalent.*

27. $x^2 + 2x = -1$, $x + 2 = -\dfrac{1}{x}$

28. $3x + 5 = 7$, $3x = 2$

29. $(x + 3)^2 = 4$, $x + 3 = 2$

30. $-x < 1$, $x < -1$

EXERCISES 31–34 *Solve the given equation with a graphing calculator. Give answers accurate to the nearest hundredth.*

31. $x^7 - x^3 + 1 = 0$

32. $4x^3 - 2x - 1 = 0$

33. $x^4 - 3x = 1$

34. $x^3 - 2x^2 = 5x - 6$

EXERCISES 35–38 *Solve the given equation.*

35. $3x + 4 = 9$

36. $\frac{2}{3}(x - 4) = \frac{1}{6}(x - 1)$

37. $2(x + 1) - 3(2x - 2) = 1$

38. $(x + 1)(x - 2) = (x - 3)^2$

EXERCISES 39–44 *Solve the inequality and graph the solution.*

39. $-3x - 4 < 8$

40. $2x + 1 \geq 5$

41. $\frac{1}{3}(x + 2) - \frac{1}{2}(x - 2) > 0$

42. $-2 < \frac{2x - 3}{4} \leq 6$

43. $2x - 1 \leq 5$ or $3 - 3x \leq 0$

44. $2x - 2 > -6$ and $5 - x \geq 1$

EXERCISES 45–46 *Solve for the indicated variable (assume all variables are nonzero).*

45. $ax + by = c$, for x

46. $y = mx - mx_1 + y_1$, for m

EXERCISES 47–50 *Find the solution set for each of the following equations and inequalities.*

47. $|2x - 5| = 3$

48. $|x^2 + 12x + 16| = -16$

49. $|-3x + 4| < 2$

50. $|2x + 4| > 1$

EXERCISES 51–54 *Write the verbal statement as an absolute value equation or inequality and then solve it.*

51. The distance between 3 times a number and 4 is 8.

52. The distance between -2 and half a number is 1.

53. The distance between twice a number and 4 is more than 6.

54. The distance between a number and -1 is no more than 2.

EXERCISES 55–58 *Find the real solution(s) of the given polynomial equation.*

55. $x(2x - 1)(x + 3) = 0$

56. $x^2 + 5x + 6 = 0$

57. $v^4 - 16 = 0$

58. $(x - 2)^3 - (x - 2) = 0$

EXERCISES 59–60 *Find the zeros of the given polynomial.*

59. $p(x) = 2x^2 + x - 6$

60. $p(x) = 3x^3 - 12x$

EXERCISES 61–64 *Solve the given polynomial equation or inequality. Find exact values whenever possible. If you must approximate, give answers correct to the nearest hundredth.*

61. $(s + 1)(s + 5) > 0$

62. $4t^3 > 4t^2$

63. $x^4 - 2x < 1$

64. $x^3 - 2x^2 \leq 5x - 6$

EXERCISES 65–70 *Use any method to find all complex solutions.*

65. $x^2 + x - 12 = 0$

66. $2x^2 - 4x = 9$

67. $w^2 + 4w + 2 = 0$

68. $x^2 + x + 2 = 0$

69. $6y^2 - y - 2 = 0$

70. $x^4 - 3x^2 = 4$

EXERCISES 71–74 *Indicate the number of real solutions for the given quadratic equation.*

71. $2x^2 + 3x - 4 = 0$

72. $2x^2 - 3x + 4 = 0$

73. $x^2 + 4\sqrt{5}x + 20 = 0$

74. $2x^2 - 5 = 0$

EXERCISES 75–80 *Solve the given rational equation or inequality.*

75. $\frac{2}{x} = \frac{3}{x + 1}$

76. $\frac{3x - 1}{2x} = \frac{3x + 4}{2x + 1}$

77. $\frac{x}{2x + 1} = \frac{x + 2}{5x}$

78. $\frac{x^2}{x + 1} = -3 + \frac{1}{x + 1}$

79. $\frac{3x - 9}{x - 1} > 0$

80. $\frac{2x^2 - 2}{x^2 - 16} < 0$

EXERCISES 81–84 *Solve the given equation involving radicals.*

81. $\sqrt{3x - 5} = 1$

82. $\sqrt{2x - 3} + 7 = 5$

83. $\sqrt{3a + 10} - a = 2$

84. $\sqrt{x - 1} + \sqrt{x + 4} = 5$

EXERCISES 85–90 *Solve the given equation for y. Use a graphing calculator to plot the graph of the resulting equation and estimate the x-intercept(s) to the nearest hundredth.*

85. $2x + 3y = 12$

86. $xy - x^2 + 4 = 0$

87. $y^4 - x = 5$

88. $y^3 + x = 2$

89. $x + 4y = y^2 + 5$

90. $x^2 - 4y = 5 - y^2$

EXERCISES 91–94 *Solve the given system of equations.*

91. $x + 5y = 13$
$2x + 3y = 12$

92. $3x - y = -7$
$4x + y = -7$

93. $2x - 4y = 5$
$-x + 2y = 6$

94. $x + 2y^2 = 6$
$x + y^2 = 5$

EXERCISES 95–96 *Find the point(s) of intersection of the given pair of equations.*

95. $2x + 3y = 6, 4x - y = 4$

96. $x^2 - y = 1, x^2 + y^2 - 2y = 7$

EXERCISES 97–98 *Use a graphing calculator to estimate the coordinates of the point(s) of intersection of the given pair of equations to the nearest hundredth.*

97. $y = 2x^3 - 3x^2 - 11x$
$y = x + 7$

98. $y = -2x^4 + 8x^3 - 8x^2 - 1$
$y = 2x^4 - 4x^2$

EXERCISES 99–108 *Parts a and b are connected: Part a involves a concept from earlier in the text, whereas part b involves related material from this chapter. First answer part a, and then use this result to answer part b.*

99. a. Compute $P(2)$, where $P(x) = 2x^3 - 6x^2 + 3x + 5$.

  b. Find a solution of $2x^3 - 6x^2 + 3x + 5 = 3$.

100. a. Factor $x^2 - x - 6$.

  b. Solve $\dfrac{x^2 - x - 6}{x^{14} + 7x^{12} + 6x^3 + 192.35x + 8} = 0$.

101. a. Simplify $\dfrac{\frac{2}{9}}{1-3}$.

   b. Find a solution of $\dfrac{2x^2}{1-\frac{1}{x}} = -\dfrac{1}{9}$.

102. a. Expand $(x+3)^2$.

   b. Solve $x^2 + 6x + 9 = 8$.

103. a. Evaluate $P(x) = (x+4)(x+5)(x+6)$ for $x = 1$.

   b. Find three consecutive integers whose product is 210.

104. a. Evaluate $Q(x) = 2x - 3$ for $x = -3$ and $x = 6$.

   b. Find a number c such that the solution set of $|2x - 3| = c$ is $\{-3, 6\}$.

105. a. Solve $x^2 + x - 2 = 0$.

   b. Solve $x^2 + x - 2 \le 0$.

106. a. Evaluate $6 \cdot 5 + (6 + 3) \cdot 10$.

   b. If a man has 15 coins in his pocket—some dimes and some nickels—and he has 3 more dimes than nickels, how many of each coin does he have?

107. a. Expand $(x-3)(x+3)(x+9)$.

   b. Find all solutions of $Z^3 + 9Z^2 - 9Z = 81$.

108. a. Evaluate $P(x) = (x+1)^2 + (x-2)^2$ for $x = 11$.

   b. Find the dimensions of a right triangle with hypotenuse 15 in which one side is 3 units longer than the other.

109. *Consecutive Integers*   Find two consecutive odd integers whose sum is 236.

110. *CD Price*   A compact disc (CD) costs $12.19 including 6% sales tax. What is the price of the CD before sales tax?

111. *Interception Risk*   Since many professional football players can jump and extend their hands to a height of 11 feet or more, a passed football is at some risk of interception whenever its height is 11 feet or below. If the quarterback releases the ball from a height of 6 feet with an upward velocity of 30 feet per second, then the height of the ball in feet t seconds after it is released is given by $h = -16t^2 + 30t + 6$. Find the time interval for which the football is at risk of being intercepted.

112. *Fortune 500*   A corporate CEO would like her company's total sales to surpass $625 million, which she estimates would be enough to be listed in the *Fortune 500*. If the company's revenue (in millions of dollars) for the sale of x million units of a product is given by $R = \frac{1}{3}x(100 - x)$, then for what interval of unit sales will the company attain her goal?

113. *Rectangle Dimensions*   The length of a rectangle is $\frac{1}{2}$ foot less than 6 times its width. Its area is 3 square feet. Find the dimensions of the rectangle.

114. *Car Wash Rate*   It takes Daniel 1 hour longer to wax the car than it does his brother Roberto. Together it takes them 50 minutes to wax the car. How long does it take each working separately?

115. *Triangle Dimensions*   What are the dimensions of a right triangle if one side is 3 inches less than twice the length of the other and the length of the hypotenuse is 51 inches?

116. *Counting Coins*   A woman has 11 coins in her pocket, all of which are either nickels or dimes. If the value of the coins is 75¢, how many of each type of coin does she have?

nere two.

CHAPTER 2 TEST

PROBLEMS 1–8 *Answer true or false.*

1. If $(x + 1)(x - 4) = 4$, then either $x = 3$ or $x = 6$.
2. If $(x + 1)(x - 4) = 0$, then either $x = -1$ or $x = 4$.
3. If $|x - 2| = 3$, then x is a distance of 3 units from 2.
4. If $|x| > 1$, then $-1 > x > 1$.
5. The equations $x^2 = 1$ and $x = 1$ are equivalent.
6. If $a^2 = b^2$, then $a = b$.
7. To solve $-1 < \dfrac{1}{x}$, we begin by multiplying both sides by x.
8. A solution of a system of equations in the variables x and y is any pair of numbers (x, y) that satisfy at least one of the equations.

PROBLEMS 9–12 *Give an example of each of the following.*

9. A quadratic equation with 3 as its only solution
10. A polynomial equation of degree 3 with solutions 1, 2, and 3
11. An inconsistent system of equations
12. An absolute value inequality with no solution

PROBLEMS 13–22 *Solve the given equation or inequality. Give exact answers wherever possible. When approximating, give answers accurate to the nearest hundredth.*

13. $-3t + 5 = 9t - 1$
14. $|2x + 4| = 7$
15. $|3 - 2y| > 1$
16. $u^3 - 4u = 0$
17. $x^4 - 2x^2 + x - 1 = 0$
18. $\dfrac{x}{x + 2} < 0$
19. $2x^2 - 4x = 8$
20. $x^2 - 5 \le 4x$
21. $\dfrac{2}{x + 4} = \dfrac{3}{x - 2}$
22. $\sqrt{2x + 1} + 1 = x$

23. The graph of an equation is shown in Figure 119, with a viewing rectangle defined by `Xmin=-10, Xmax=10, Ymin=-10, Ymax=10`. Match the viewing rectangles labeled i–iii with the range values in parts a–c.

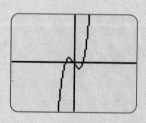

FIGURE 119

i.

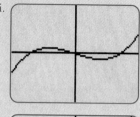

ii.

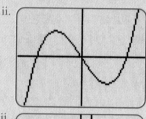

iii.

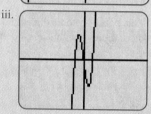

a. `Xmin=-2, Xmax=2, Ymin=-2, Ymax=2`
b. `Xmin=-2, Xmax=2, Ymin=-10, Ymax=10`
c. `Xmin=-10, Xmax=10, Ymin=-2, Ymax=2`

24. Solve the system of equations

$$3x - 2y = -1$$
$$2x + 3y = 21$$

25. The sum of three consecutive even integers is 306. Find the numbers.

26. A Habitat for Humanity crew can build a house in 3 days less time than it takes a local construction company. When both crews work together, a house is built every 2 days. How long does it take each crew working alone?

3 FUNCTIONS AND THEIR GRAPHS

The Philippine volcano Mount Pinatubo erupted on June 15, 1991, spewing millions of tons of sulfur dioxide and ash into Earth's atmosphere. Rising, the hot gas combined with water vapor to form a haze of sulfuric acid droplets, blocking sunlight and cooling Earth's atmosphere. Such naturally occurring phenomena complicate the task of measuring the effects of greenhouse gasses (such as carbon dioxide) on global warming. Atmospheric scientists believe that volcanic eruptions and emissions of chemicals known as CFCs from aerosol sprays and other sources contribute to the depletion of stratospheric ozone. In this chapter we will mathematically model both carbon dioxide and atmospheric ozone levels, and discuss the difficulty of modeling Earth's atmosphere.

SECTION 3.1 LINES

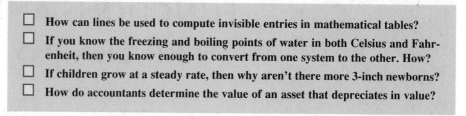

☐ How can lines be used to compute invisible entries in mathematical tables?

☐ If you know the freezing and boiling points of water in both Celsius and Fahrenheit, then you know enough to convert from one system to the other. How?

☐ If children grow at a steady rate, then why aren't there more 3-inch newborns?

☐ How do accountants determine the value of an asset that depreciates in value?

In this chapter, we introduce the single most important concept in college algebra and arguably in all of mathematics: *functions*. Besides being an elegant mathematical notion, functions serve as our primary medium for modeling the real world, providing us with a lens through which we can predict the future and unearth the past. So why, you might ask, would a chapter on functions begin with a section on lines? The answer is twofold. First, an important class of functions (linear functions—a topic discussed in Section 3.6) has graphs that are lines. Second, in much the same way that Earth appears flat from our vantage point on its surface, so too do most functions have graphs that appear to be lines after we have zoomed in a few times. Because of this "local linearity," we cannot fully understand functions without first mastering the more fundamental notion of lines.

SLOPE

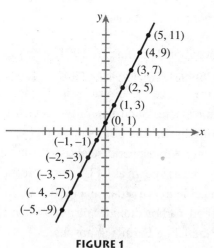

FIGURE 1

The **slope** of a nonvertical line is the number of units the line rises for every unit of horizontal change. For example, consider the line l shown in Figure 1. Note that in passing from $(1, 3)$ to $(2, 5)$, or from $(2, 5)$ to $(3, 7)$, the x-value is increased by 1 unit whereas the y-value is increased by 2 units. In fact, it doesn't matter which point on the line we start with: If the x-coordinate is increased by 1 unit, the y-coordinate will be increased by 2 units. Thus, the slope of l is 2.

Since the y-value is increased by 2 units for every 1 unit of change in the x-value, the change in y-coordinates and the change in x-coordinates will always be in a 2 to 1 ratio. For example, consider the points $(-2, -3)$ and $(4, 9)$. The difference in the x-coordinates is 6, and the difference in the y-coordinates is 12, exactly twice the difference in the x-coordinates.

We can define the slope of an arbitrary line as the ratio of the difference in the y-coordinates to the difference in the x-coordinates for any two points on the line. If the two points are (x_1, y_1) and (x_2, y_2), then the difference in the y-coordinates is $y_2 - y_1$, and the difference in the x-coordinates is $x_2 - x_1$, as shown in Figure 2. Thus, we define the slope, which is usually denoted by the letter m, as follows.

Definition of the Slope of a Line ☐

The slope of the line passing through the points (x_1, y_1) and (x_2, y_2) (where $x_1 \neq x_2$) is defined by

$$m = \frac{\text{Change in } y}{\text{Change in } x} = \frac{y_2 - y_1}{x_2 - x_1}$$

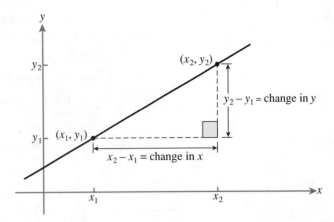

FIGURE 2

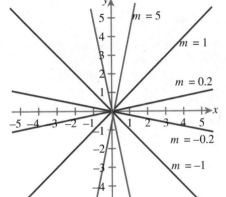

FIGURE 3

The reason for the restriction $x_1 \neq x_2$ is that if $x_1 = x_2$, then $x_2 - x_1 = 0$, so that the fraction $\dfrac{y_2 - y_1}{x_2 - x_1}$ would have a denominator of 0. The restriction $x_1 \neq x_2$ implies that we cannot define slope for *vertical* lines since all points on a vertical line have the same x-coordinate.

The slope of a line is a measure of its steepness; the steeper the line, the greater the slope. For example, a line with slope 1 will rise 1 unit for every unit that we move to the right, whereas a line with slope 5 will rise 5 units for every unit that we move to the right. Thus, a line with slope 5 is much steeper than a line with slope 1. If the slope of a line is negative, then moving 1 unit to the right will result in a *negative* change in the y-coordinate; that is, the line will descend. Figure 3 shows several lines, along with their slopes.

It is an essential point that for a given line, the quantity

$$m = \frac{y_2 - y_1}{x_2 - x_1}$$

is always the same; it doesn't matter which two points (x_1, y_1) and (x_2, y_2) are chosen. This is illustrated in Figure 4, in which it can be seen that triangle *PQR* is similar to (that is, has the same shape as) triangle *STU*. A consequence of this similarity is that corresponding sides are proportional, so that

$$\frac{QR}{PR} = \frac{TU}{SU}$$

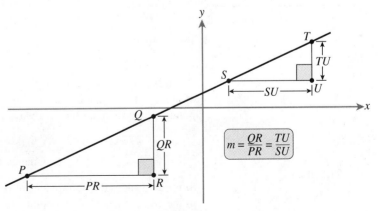

FIGURE 4

In other words, computing the slope with points P and Q would yield the same result as computing the slope with points S and T.

> **WARNING!** When computing the slope of a line given two points on the line, it doesn't matter which point is considered to be (x_1, y_1) and which is considered to be (x_2, y_2). It is essential, however, that the order of subtraction be consistent from numerator to denominator. In other words, we may compute slope as
>
> $$m = \frac{y_2 - y_1}{x_2 - x_1} \quad \text{or} \quad m = \frac{y_1 - y_2}{x_1 - x_2} \quad \text{but } not \quad m = \frac{y_2 - y_1}{x_1 - x_2} \quad \text{or} \quad m = \frac{y_1 - y_2}{x_2 - x_1}$$

EXAMPLE 1   Finding the Slope of a Line Given Two Points on the Line

The points $(-1, 3)$ and $(2, 4)$ are on the line l. Find the slope of l.

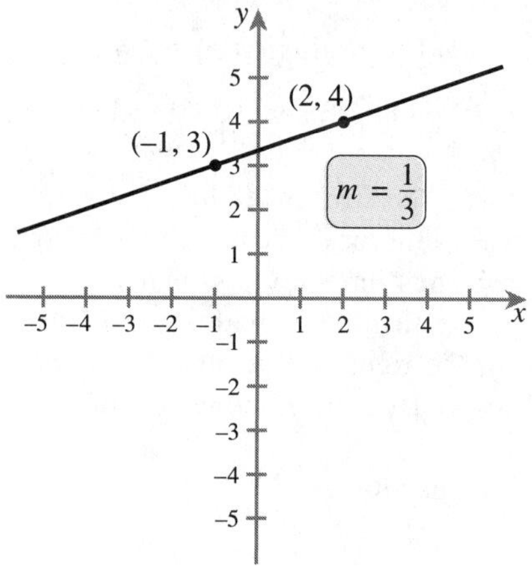

SOLUTION   We take (x_1, y_1) to be the point $(-1, 3)$ and (x_2, y_2) to be the point $(2, 4)$. Then

$$m = \frac{y_2 - y_1}{x_2 - x_1}$$

$$= \frac{4 - 3}{2 - (-1)}$$

$$= \frac{1}{3}$$

The graph of l is shown in Figure 5. Note that since m is positive, the line is rising. Because m is relatively small, the line rises gently.

FIGURE 5

EXAMPLE 2   Finding the Slope of a Horizontal Line

Find the slope of the horizontal line shown in Figure 6.

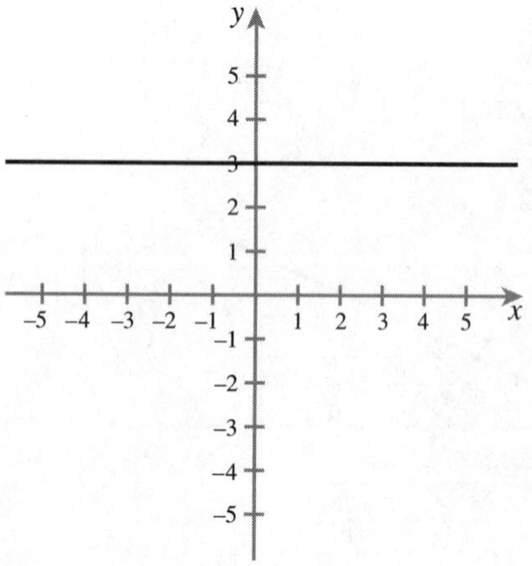

SOLUTION   We choose two points on the line, say $(2, 3)$ and $(5, 3)$. Then we have

$$m = \frac{3 - 3}{5 - 2} = \frac{0}{3} = 0$$

It isn't difficult to see that the slope of any horizontal line is zero. All y-coordinates of points on a horizontal line are the same, so the difference in the y-coordinates of two points on a horizontal line is always zero. Thus, the slope is

$$m = \frac{\text{Change in } y}{\text{Change in } x} = \frac{0}{\text{Change in } x} = 0$$

FIGURE 6

Similarly, since all x-coordinates of points on a vertical line are the same, the difference in the x-coordinates of two points on a vertical line is always zero. Thus, the slope of a vertical line is undefined. The relationship between the orientation of a line and it slope may be summarized as follows.

Relationship Between Slope   ☐
and Orientation

Description of line	Slope
Rising from left to right	Positive
Falling from left to right	Negative
Horizontal	Zero
Vertical	Not defined

PARALLEL AND PERPENDICULAR LINES

Since slope is a measure of the steepness of a line, it is not surprising that parallel lines have the same slope. Figure 7 shows three parallel lines, all with slope 2.

The relationship between the slopes of perpendicular lines is slightly more complicated. In Figure 8, a line l with slope m is shown, along with a line l' perpendicular to l. For convenience, we have assumed that the lines cross at the origin. Now the point $(1, m)$ is on line l and, using elementary geometry, it can be shown that the point $(-m, 1)$ is on line l'. To compute the slope of l' we will use the points $(0, 0)$ and $(-m, 1)$. If we let m' denote the slope of l', then we have

$$m' = \frac{1 - 0}{-m - 0} = -\frac{1}{m}$$

Thus, the slope of l' is the negative reciprocal of the slope of l.

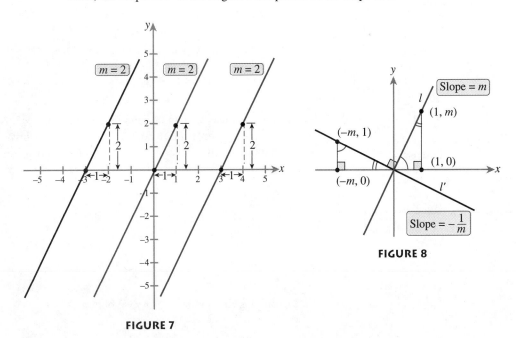

FIGURE 7

FIGURE 8

We summarize the relationships between parallel and perpendicular lines as follows.

Slopes of Parallel and ☐
Perpendicular Lines

1. Two nonvertical lines are parallel if and only if they have the same slope.

2. Two lines, neither of which is vertical, are perpendicular if and only if their slopes are negative reciprocals of one another—that is, if and only if the product of their slopes is -1.

E X A M P L E   3   ◼   **Finding the Slope of a Line Parallel to a Given Line**

Find the slope of a line parallel to the line passing through the points $(2, 3)$ and $(4, 7)$.

SOLUTION   Let l denote the line passing through $(2, 3)$ and $(4, 7)$. Then l has slope

$$m = \frac{7 - 3}{4 - 2} = \frac{4}{2} = 2$$

Since parallel lines have the same slope, a line parallel to l also has slope 2.

E X A M P L E   4   ◼   **Graphing a Line Passing Through a Given Point and Perpendicular to a Given Line**

Graph the line passing through the point $(3, 4)$ that is perpendicular to the line passing through $(-2, 5)$ and $(3, 6)$.

SOLUTION   Let l' denote the line passing through $(-2, 5)$ and $(3, 6)$, and let l denote the line perpendicular to l' and passing through $(3, 4)$. The slope of l' is given by

$$m' = \frac{6 - 5}{3 - (-2)} = \frac{1}{5}$$

Since l and l' are perpendicular to one another, l has slope

$$m = -\frac{1}{m'} = -\frac{1}{(1/5)} = -5$$

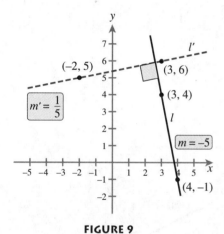

FIGURE 9

Thus, l is the line passing through $(3, 4)$ with slope -5. Now, beginning with the point $(3, 4)$, and moving 1 unit to the right and 5 units down, we obtain the point $(4, -1)$. To graph l, we simply connect the points $(3, 4)$ and $(4, -1)$, as shown in Figure 9.

EQUATIONS OF LINES

If (x_1, y_1) is a given point on a line with slope m (see Figure 10), then for an arbitrary point (x, y) on the line we have

$$\frac{y - y_1}{x - x_1} = m$$

Multiplying through by $x - x_1$, we obtain

$$y - y_1 = m(x - x_1)$$

This is called the **point-slope equation,** and it enables us to produce an equation of a line given the coordinates of a particular *point* on the line and the *slope* of the line.

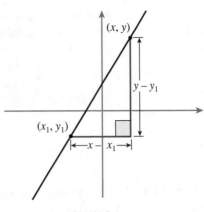

FIGURE 10

Point-Slope Equation of a Line ☐ | A line with slope m passing through the point (x_1, y_1) has equation $y - y_1 = m(x - x_1)$.

EXAMPLE 5 ▓ **Finding the Equation of a Line**

Find an equation of the line with slope -2 passing through the point $(1, -5)$. Graph the line with a graphing calculator.

SOLUTION   Using the point-slope equation with $m = -2$, $x_1 = 1$, and $y_1 = -5$, we obtain the following equation.

$$y - y_1 = m(x - x_1)$$
$$y - (-5) = -2(x - 1)$$
$$y + 5 = -2(x - 1)$$

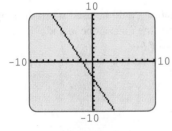

FIGURE 11

Solving for y, we obtain

$$y = -2(x - 1) - 5$$
$$= -2x - 3$$

The graph of $y = -2x - 3$ is shown in Figure 11.

The point-slope equation of a line can be used to find the equation of a line when either the slope or a point on the line is given indirectly. For example, if two points on the line are given (but not the slope), we can find the equation by first computing the slope, and then using either one of the points in the point-slope equation. The following examples illustrate how the point-slope equation can be used when either the slope or a point on the line is not given.

EXAMPLE 6 ▓ **Finding the Equation of a Line Given Two Points on the Line**

Find an equation of the line passing through the points $(-2, 6)$ and $(2, 7)$.

SOLUTION   We begin by finding the slope.

$$m = \frac{7 - 6}{2 - (-2)} = \frac{1}{4}$$

Now that we have the slope we can select either $(-2, 6)$ or $(2, 7)$ as the point (x_1, y_1). We choose $(-2, 6)$. From the point-slope equation, we have

$$y - y_1 = m(x - x_1)$$

$$y - 6 = \frac{1}{4}(x - (-2))$$

$$y - 6 = \frac{1}{4}(x + 2)$$

E X A M P L E   7   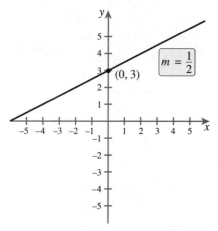   **Finding the Equation of a Line Given the Slope and the *y*-intercept**

Find an equation of the line with *y*-intercept 3 and slope $\frac{1}{2}$, the graph of which is shown in Figure 12.

SOLUTION   Since the *y*-intercept is 3, we know that the graph crosses the *y*-axis at the point $(0, 3)$. Thus, we have

$$y - y_1 = m(x - x_1)$$

$$y - 3 = \frac{1}{2}(x - 0)$$

$$y = \frac{1}{2}x + 3$$

FIGURE 12

In Example 7, we saw that the point-slope equation could be used when the slope and the *y*-intercept were given. More generally, if the line *l* has slope *m* and *y*-intercept *b*, then the point $(0, b)$ lies on *l*, and so the equation of *l* is

$$y - b = m(x - 0)$$
$$y - b = mx$$
$$y = mx + b$$

This is called the **slope-intercept equation.**

Slope-Intercept Equation   □ | A line with slope *m* and *y*-intercept *b* has equation $y = mx + b$.
of a Line

E X A M P L E   8   ▨   **Finding the Slope-Intercept Equation of a Line**

Find an equation of the line with slope -4 and *y*-intercept 6.

SOLUTION   We apply the slope-intercept equation with $m = -4$ and $b = 6$.

$$y = mx + b$$
$$y = -4x + 6$$

The slope-intercept equation of a line is often useful for extracting information about the line from its equation. For example, if an equation of a line is given, its slope can be determined by writing the equation in slope-intercept form. The slope-intercept equation is also especially useful for graphing. In particular, if an equation of a line is given and we wish to graph it using a graphing calculator, we must first place the equation in slope-intercept form. The following examples illustrate the utility of the slope-intercept form.

EXAMPLE 9 ▨ Finding the Slope Given the Equation of a Line

Find the slope and the y-intercept of the line with equation $5x + 4y = 8$.

SOLUTION We begin by writing the equation in slope-intercept form. To do this, we solve for y.

$$5x + 4y = 8$$
$$4y = -5x + 8$$
$$y = -\frac{5}{4}x + 2$$

Thus, the slope is $-\frac{5}{4}$ and the y-intercept is 2.

EXAMPLE 10 ▨ Finding the Equation of a Line Through a Given Point and Parallel to a Given Line

Find an equation of the line l that passes through $(-3, 5)$ and is parallel to the line with equation $2x + y = 10$.

SOLUTION We begin by finding the slope of the line with equation $2x + y = 10$. To do this, we write its equation in slope-intercept form.

$$2x + y = 10$$
$$y = -2x + 10$$

Thus, we can conclude that the line $2x + y = 10$ has slope -2. Since l is parallel to this line, it also has slope -2. Thus l has slope -2 and passes through the point $(-3, 5)$. Now that we know the slope and a point on the line, we can apply the point-slope equation.

$$y - y_1 = m(x - x_1)$$
$$y - 5 = -2(x - (-3))$$
$$y - 5 = -2(x + 3)$$

If desired, we can also express the equation in slope-intercept form as follows:

$$y - 5 = -2x - 6 \qquad \text{Distributing}$$
$$y = -2x - 1 \qquad \text{Solving for } y$$

EXAMPLE 11 ■ **Graphing a Line Given its Equation**

Graph the line with equation $2x - y = 4$.

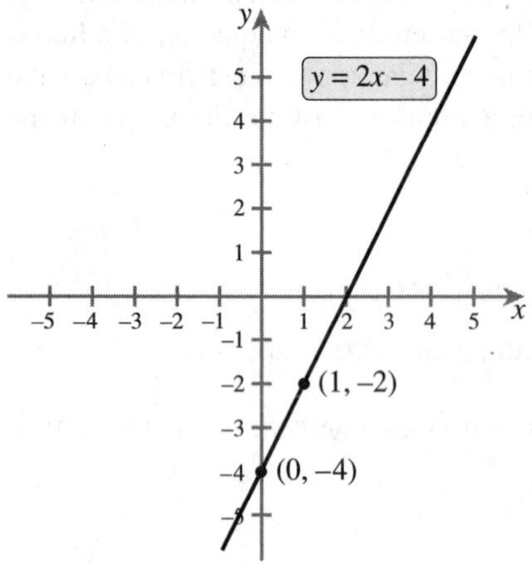

FIGURE 13

SOLUTION We begin by placing the equation in slope-intercept form.

$$\begin{aligned} 2x - y &= 4 \\ -y &= -2x + 4 \\ y &= 2x - 4 \end{aligned}$$

We can now conclude that the slope is 2 and the y-intercept is -4. To graph the line, we begin by plotting the y-intercept $(0, -4)$. Next we find one additional point by substituting a convenient value (such as 1) for x. With $x = 1$, we have

$$\begin{aligned} y &= 2x - 4 \\ &= 2 \cdot 1 - 4 \\ &= -2 \end{aligned}$$

Thus, $(1, -2)$ is a second point on the line. Finally, we simply ''connect the dots,'' as shown in Figure 13.

The slope-intercept equation can be used to derive the equation of a horizontal line. Since the slope of a horizontal line is zero, we have

$$\begin{aligned} y &= mx + b \\ &= 0x + b \\ &= b \end{aligned}$$

It's not necessary, however, to use the slope-intercept equation to produce the equation of a horizontal line. Points on a given horizontal line must have the same y-coordinates, but there is no restriction on their x-coordinates (x can be anything). Thus, the equation of a horizontal line is $y = C$, where C is a constant.

The equation of a vertical line can be found in a similar fashion. Since points on a vertical line have no restriction on their y-coordinates, but their x-coordinates must all be the same, the equation of a vertical line is $x = C$, where C is a constant.

The Equations of Vertical and Horizontal Lines □

> Horizontal lines have equations of the form $y = C$, where C is a constant.
>
> Vertical lines have equations of the form $x = C$, where C is a constant.

EXAMPLE 12 ■ **Finding the Equation of a Horizontal Line**

Find the equation of the horizontal line passing through the point $(2, 5)$.

SOLUTION Since the line is horizontal, we know that its equation is of the form $y = C$. Since the point $(2, 5)$ is on the line and has y-coordinate 5, C must be 5. Thus, the equation of the line is $y = 5$.

EXAMPLE 13 ■ **Finding the Equation of a Vertical Line**

Find the equation of the vertical line shown in Figure 14.

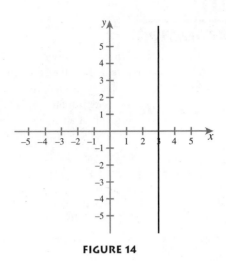

FIGURE 14

SOLUTION Every point on the line has x-coordinate 3. Thus, its equation is $x = 3$.

One minor deficiency of the point-slope and slope-intercept equations is that neither is sufficiently general to include every possible line. Both the point-slope equation, $y - y_1 = m(x - x_1)$, and slope-intercept equation, $y = mx + b$, involve m, the slope of the line. The difficulty is that vertical lines do not have a slope, so that neither equation is applicable for vertical lines. Primarily for this reason, we define the *standard equation of a line* as follows.

Standard Equation of a Line □

> An equation of the form $Ax + By = C$ is said to be a **standard equation of a line.** Every line has an equation of this form, and, conversely, every equation of this form is that of a line, provided A and B are not both 0. If the equation of a line is a standard equation, then it is said to be in **standard form.**

Note that even vertical lines can be written as equations in standard form. For example, the vertical line $x = 3$ from Example 13 is already in standard form, with $A = 1$, $B = 0$, and $C = 3$.

E X A M P L E 14 ■ **Finding a Standard Equation**

Find an equation in standard form for the line passing through $(2, 5)$ with slope -2.

SOLUTION Since we are given a point on the line, $(2, 5)$, and the slope of the line, -2, we use the point-slope equation.

$$y - 5 = -2(x - 2)$$

Now we rearrange to write the equation in standard form.

$$y - 5 = -2x + 4$$
$$2x + y = 9$$

LINEAR INTERPOLATION

It is often the case that we have no equation to express the precise relationship between two variables. Perhaps our only information is in the form of a limited number of data points. In order to use these data to make predictions, we must make certain assumptions regarding the (unknown) equation. **Linear interpolation** is the process of predicting values based on the assumption that the graph of an equation is linear "between" data points; **linear extrapolation,** on the other hand, is based on the assumption that the graph of an equation will extend in a linear fashion outside the range of the data points. These techniques are illustrated in the following example.

EXAMPLE 15 ▪ Linear Interpolation and Extrapolation

TABLE 1 Average weight for boys

Age (years)	Weight (pounds)
10	69
11	77
12	83
13	92
14	107

Data source: 1992 World Almanac

Table 1 gives the average weight for boys ages 10–14. Use linear interpolation or extrapolation to predict the average weight for a boy at each of the given ages.

a. 11 years, 3 months
b. 14 years, 6 months
c. 25 years

SOLUTION Let t represent the age in years and w the weight in pounds.

a. We interpolate by assuming the relationship between w and t is linear between the data points $(11, 77)$ and $(12, 83)$. We have

$$m = \frac{83 - 77}{12 - 11} = \frac{6}{1} = 6$$

Using the point-slope equation, we have

$$w - 77 = 6(t - 11), \text{ or}$$
$$w = 6t + 11$$

Now, 3 months = 0.25 year, so we are interested in $t = 11.25$. Thus

$$w = 6(11.25) + 11 = 78.5$$

This is illustrated in Figure 15.

b. Since $t = 14.5$ (14 years, 6 months) lies outside our given data, we use linear extrapolation. We find the equation of the line passing through the last two data points, $(13, 92)$ and $(14, 107)$, and we assume that this relationship holds for $t = 14.5$ as well. The slope of this line is given by

$$m = \frac{107 - 92}{14 - 13} = 15$$

Thus, the line has equation

$$w - 92 = 15(t - 13)$$
$$w = 15t - 103$$

When $t = 14.5$, we have $w = 15(14.5) - 103 = 114.5$.

c. Again $t = 25$ is outside the range of the given data, so we extrapolate from the last two data points. From part b, we know that the equation of the line joining these points is $w = 15t - 103$. Evaluating at $t = 25$ gives us

$$w = 15(25) - 103 = 272$$

Obviously, the average weight of 25-year-old "boys" is not 272 pounds: Extrapolation far from original data is not to be trusted.

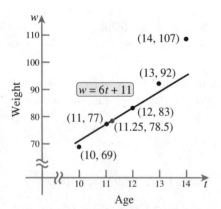

FIGURE 15

UNDERSTANDING AND MASTERY CHECKLISTS

CONCEPTS TO UNDERSTAND

- ☐ Slope of a line
- ☐ Parallel and perpendicular lines
- ☐ Point-slope equation of a line
- ☐ Slope-intercept equation of a line
- ☐ Standard equation of a line
- ☐ Linear interpolation and extrapolation

SKILLS TO MASTER

- ☐ Estimate the slope of a line given its graph.
- ☐ Compute the slope of a line through two points.
- ☐ Find the slope of a line given the slope of a line parallel or perpendicular to it.
- ☐ Find an equation of a line given a point on the line and the slope of the line.
- ☐ Find an equation of a line given two points on the line.
- ☐ Find the slope and *y*-intercept of a line given its equation.
- ☐ Place the equation of a line in standard form.
- ☐ Estimate values between or outside data points using linear interpolation or extrapolation.

EXERCISES 3.1

EXERCISES 1–12 *Find the slope of the line passing through the given pair of points.*

1. $(1, 3)$ and $(2, 7)$
2. $(3, 4)$ and $(-2, 4)$
3. $(-2, 6)$ and $(3, -5)$
4. $(3, -4)$ and $(-7, -1)$
5. $(0, 32)$ and $(100, 212)$
6. $(1492, 1)$ and $(1999, 1000)$
7. $(\frac{1}{2}, 2)$ and $(-1, 3\frac{1}{2})$
8. $(\frac{2}{3}, -3)$ and $(-\frac{3}{5}, -2)$
9. $(2, 3.5)$ and $(4.6, 7.9)$
10. $(1.02, 100)$ and $(0.98, 101)$
11. $(35, 1.5 \times 10^7)$ and $(40, 2.0 \times 10^7)$
12. $(2.3 \times 10^{-2}, 3.5)$ and $(4.5 \times 10^{-2}, 5.7)$

EXERCISES 13–16 *The graph of a line is given. Estimate the slope of the line.*

13.

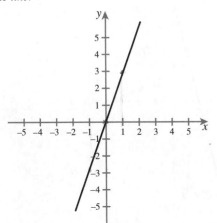

14.

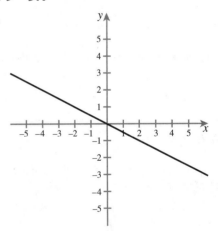

15.

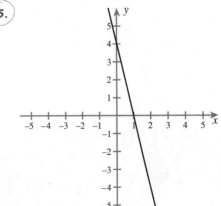

16.

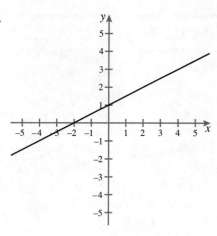

EXERCISES 17–20   *Find the slope of the line l with the given property.*

17. The line l is parallel to a line with slope -3.

18. The line l is perpendicular to the line passing through the points $(2, 7)$ and $(-1, 9)$.

19. The line l is perpendicular to a line with slope $\frac{3}{5}$.

20. The line l is parallel to the line passing through the points $(3, 4)$ and $(-2, 6)$.

EXERCISES 21–22   *Use the given information to determine the slopes of the lines l_1, l_2, l_3, and l_4.*

21. The lines l_1, l_2, l_3, and l_4 have slopes of $\frac{2}{3}$, -2, 0, and 3, but not necessarily in that order.

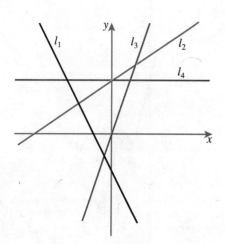

22. The lines l_1, l_2, l_3, and l_4 have slopes of 1, 4, $-\frac{1}{2}$, and -3, but not necessarily in that order.

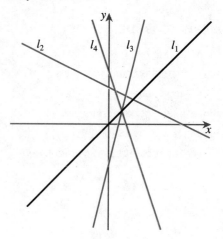

EXERCISES 23–24   *Find the slope of each of the lines l_1, l_2, and l_3.*

23.

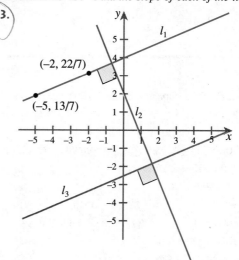

24.

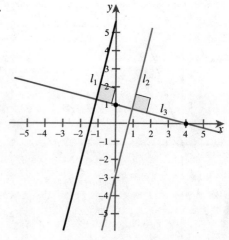

EXERCISES 25–34 *Find an equation of the line with the given slope m passing through the indicated point P. Express your final answer in slope-intercept form, if possible.*

25. $m = 1, P = (2, 4)$ **26.** $m = 3, P = (-1, 5)$

27. $m = -1, P = (2, 8)$ **28.** $m = 0, P = (0, 1)$

29. m is undefined, $P = (3, 7)$ **30.** $m = 7, P = (-1, -2)$

31. $m = 5, P = (\frac{1}{2}, \frac{2}{3})$ **32.** $m = -\frac{2}{5}, P = (-6, 2)$

33. $m = 0, P = (2, 1)$ **34.** m is undefined, $P = (-2, 1)$

EXERCISES 35–46 *Find an equation of the line passing through the indicated pair of points. Express your final answer in slope-intercept form, if possible.*

35. $(1, 4)$ and $(-2, 3)$ **36.** $(2, 2)$ and $(4, 7)$

37. $(2, 0)$ and $(0, 4)$ **38.** $(-3, 0)$ and $(0, 2)$

39. $(3, -14)$ and $(3, 7)$ **40.** $(19, 5)$ and $(23, 5)$

41. $(1.2, 3.4)$ and $(4.6, 5.9)$ **42.** $(\frac{1}{2}, \frac{3}{5})$ and $(-\frac{1}{4}, \frac{2}{5})$

43. $(3, \frac{7}{8})$ and $(-2, \frac{1}{8})$

44. $(1, 0.000002)$ and $(1, \text{your weight in kilograms})$

45. $(\pi, 2 \times 10^{100})$ and $(\pi, 2 \times 10^{-10})$

46. $(1999, 1.2 \times 10^9)$ and $(2001, 1.4 \times 10^9)$

EXERCISES 47–52 *Find the slope and the y-intercept of the line with the given equation. Then graph the line.*

47. $2x - 3y = 8$ **48.** $3x + 6y = 9$

49. $x = 3y + 4$ **50.** $4 - 3y = 0$

51. $y - 3 = -2(x + 3)$ **52.** $y + 2 = 3(x - 7)$

EXERCISES 53–60 *Find an equation of the line with the given properties. Express your final answer in slope-intercept form, if possible.*

53. Slope: $\frac{2}{3}$ **54.** Slope: -4
y-intercept: 2 y-intercept: -3

55. Parallel to: $y = 4x + 2$
y-intercept: 5

56. Parallel to: The line containing $(0, 3)$ and $(4, 0)$
Point on line: $(-2, 5)$

57. Perpendicular to: $x - 3y = 8$
Point on line: $(2, -4)$

58. Perpendicular to: $4x + 8y = 7$
Point on line: $(-2, -2)$

59. Parallel to: $x = 6$ **60.** Perpendicular to: $x = 19$
Point on line: $(3, 8)$ y-intercept: 7

EXERCISES 61–64 *Write the given equation in standard form.*

61. $y = 3x + 4$ **62.** $y - 8 = -2(x - 3)$

63. $y - \frac{2}{3} = 2(x - \frac{1}{2})$ **64.** $x = 2y + 8$

EXERCISES 65–68 *Find the point of intersection of the given lines.*

65. $y = 3x + 5$ and $y = -2x - 8$

66. $2x - 3y = 6$ and $3x + 4y = 12$

67. $x = 2$ and $y = 5$

68. $y = m_1 x + b_1$ and $y = m_2 x + b_2$. (Under what circumstances do these lines not intersect?)

EXERCISES 69–72 *Use slope to verify the given statement.*

69. The points $(1, 0)$, $(7, 4)$, $(15, 9)$ are not collinear (that is, do not lie on a line).

70. The points $(-2, -9)$, $(2, -1)$, and $(4, 3)$ are collinear (that is, lie on a line).

71. The points $(2, 3)$, $(3, 5)$, $(6, 3)$, and $(7, 5)$ form a parallelogram (that is, a four-sided figure having opposite sides that are parallel).

72. The points $(-1, 1)$, $(1, -1)$, $(5, 3)$, and $(3, 5)$ form a rectangle.

APPLICATIONS

73. *Mail-Order Cost* It costs $1000 in overhead per month to run a mail-order business. In addition, each customer order costs $5.00 to produce. Express the total monthly cost C in terms of N, the number of orders received.

74. *Temperature Conversion* The relationship between degrees Fahrenheit (F) and Celsius (C) is linear. At atmospheric pressure, water freezes at 32°F and 0°C, whereas water boils at 212°F and 100°C.

  a. Find an equation relating the variables F and C.

  b. Room temperature is usually taken to be 72°F. What is room temperature in degrees Celsius?

  c. For decades, the average human body temperature was assumed to be 98.6°F (this figure has since been adjusted to 98.2°F). What is 98.6°F in degrees Celsius?

  d. The temperature at the surface of the sun can reach 5500°C. Express this in degrees Fahrenheit.

The surface of the sun during a solar flare.

  e. Absolute zero is the lowest possible temperature. On the Celsius scale, absolute zero is $-273°$. What is absolute zero in degrees Fahrenheit?

75. *Growth Rate*   Suppose that a child were to grow at a constant rate from conception to maturity. Thus, if h represents the height (or length) of the child in inches and t represents the number of years *since conception,* then h and t are related linearly. Assume that at $t = 0$, $h = 0$. Assume further that the child is fully grown at 16 years old, at which time she is $5'6''$ tall.

   a. Find an equation relating h and t.

   b. Use the result of part a to estimate the length of the baby at birth.

   c. Estimate the height of the child at age 12.

76. *Sales Commission*   Each month a salesman earns a $2000 salary plus a 10% commission on all sales over $10,000. Let P be the salesman's pay and S be his monthly sales.

   a. Write an equation relating S and P (assuming $S > 10,000$).

   b. Graph the equation of part a.

   c. Use the result of part a to find the salesman's total pay in a month when he sold $25,000.

77. *Machinery Depreciation*   When computing the value of a piece of machinery for tax purposes, it is common practice to use linear depreciation. With linear depreciation, we assume that the value decreases at a constant rate so that value V and time T are related linearly. Suppose that in 1999 a crane was purchased for $200,000. After 10 years, the crane will have a salvage value of only $20,000. Use linear depreciation to estimate the value of the crane in 2003.

78. *Computer Depreciation*   Suppose that a computer system purchased new in 1992 for $2500 is worth $1500 in 1995. Assuming linear depreciation—that is, assuming the value of the computer system and its age are related linearly—estimate the value of the computer in 1997.

79. *Population Interpolation*   The U.S. population was approximately 227 million in 1980 and 249 million in 1990.

   a. Use linear interpolation to estimate the population in 1987.

   b. Use linear extrapolation to predict the population in 2000 (the next census year).

80. *GDP Interpolation*   The U.S. Gross Domestic Product (GDP) was approximately $3780 billion in 1980 and $4900 billion in 1990 (both in 1987 dollars).

   a. Use linear interpolation to estimate the GDP in 1988.

   b. Use linear extrapolation to predict the GDP in 2002.

■ CONCEPTS AND CRITICAL THINKING

EXERCISES 81–86   *Answer true or false.*

81. Slope isn't defined for horizontal lines.

82. A vertical line has a slope of 0.

83. If the y-coordinates of points on a line increase as the x-coordinates decrease, then the line has negative slope.

84. Parallel lines have the same slope.

85. A line with slope 3 is steeper than a line with slope -4.

86. Two lines having the same slope and a point of intersection are identical.

EXERCISES 87–90   *Give an example of each.*

87. An equation of a line that has undefined slope

88. An equation of a horizontal line

89. A line that does not intersect $2x + 7y = 11$

90. A line perpendicular to $3x - y = 5$

91. Explain why it doesn't matter whether one uses

$$\frac{y_2 - y_1}{x_2 - x_1} \quad \text{or} \quad \frac{y_1 - y_2}{x_1 - x_2}$$

to compute the slope.

■ QUESTIONS FOR DISCUSSION OR ESSAY

92. A line is an idealized, abstract construction. But do lines really exist? To what extent does a sketch of a line (whether done by hand or with a graphing calculator) differ from the line itself? Are rays of light lines? More generally, what connections are there between the abstract spaces of mathematics (like the coordinate plane and the number line) and physical space?

93. It can easily be shown that if the lines l_1 and l_2 have equations $a_1x + b_1y = c_1$ and $a_2x + b_2y = c_2$, respectively, then either l_1 and l_2 are the same line, intersect in a single point, or are parallel. However, as you might recall from high school geometry, lines may also be **skew**—nonintersecting and yet not parallel either. How do you explain this apparent contradiction?

94. One of the most common abuses of mathematics is to assume, with very little evidence, that some particular quantity and time are related linearly. In essence, a variable will vary linearly with time if it changes by the same amount for each unit of time—that is, if its rate of change is *constant*. Which of the following variables do you think will have a relationship with time that is approximately linear? Give reasons to support your answers.

   a. The world's population

   b. The national debt

   c. The price of a 12-ounce can of Coke

   d. The world record time for the 100-meter run

   e. The number of floating-point operations per second of the world's fastest computer

   f. The planet Earth's total reserves of crude oil

   g. The number of AIDS patients in the United States

   h. The total number of solar eclipses of which there is a human record

PROJECTS FOR ENRICHMENT

95. *Distance Between a Line and a Point* In this project, we derive a formula for the distance between a line *l* and a point *P* not on *l*. We assume that *l* is nonvertical. Suppose then that *l* has equation $y = mx + b$ and that *P* has coordinates (c, d), as shown in Figure 16. Clearly the shortest path between *P* and *l* is along *l'*, the line perpendicular to *l* through *P*. Then if *Q* is the point of intersection of *l* and *l'*, the distance between *P* and *l* is given by *PQ*.

   a. What is the slope of *l'*?

   b. What is the equation of *l'*?

   c. Find the coordinates of *Q*.

   d. Find an expression for the distance between *l* and *P*. Your answer should involve only the variables *m*, *b*, *c*, and *d*.

   e. Use the formula obtained in part d to find the distance between the line with equation $y = 4x + 7$ and the point $(2, 5)$.

   f. Use the formula obtained in part d to find the distance between the parallel lines $y = mx + b_1$ and $y = mx + b_2$. [*Hint:* Consider the point $(0, b_1)$ on the first line.]

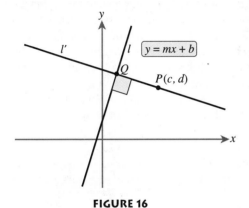

FIGURE 16

96. *Interpolating Logarithms* In this project, we learn how to make linear interpolations from tabular data. In particular, we estimate logarithms the way our parents did: through tabular interpolation. Although this method has little value today as a method for computing logarithms since $7 calculators can perform the computations instantly, the technique of tabular interpolation itself is still useful in other contexts. Logarithms will be studied extensively in Chapter 5.

Table 2 gives the logarithms of the numbers 1.1, 1.2, . . . , 9.8, 9.9 to four decimal places of accuracy. For example, suppose that you wish to find the logarithm of 3.5. You simply read off the entry in the third row and the fifth column—namely, 0.5441. Now suppose that you wish to find the logarithm of a number that falls *between* two numbers in the table. For example, suppose you want the logarithm of 2.33. According to Table 2, the logarithm of 2.3 is 0.3617 and the logarithm of 2.4 is 0.3802, so it seems reasonable to assume that the logarithm of 2.33 is somewhere between 0.3617 and 0.3802. To estimate the logarithm of 2.33 by the method of *linear interpolation,* we simply assume that the graph of the equation $y = \log x$ ("log x" stands for "logarithm of x") is linear between 2.3 and 2.4

   a.   **i.** Find the equation of the line that passes through the points $(2.3, 0.3617)$ and $(2.4, 0.3802)$.

       ii. Now use this equation to estimate the value of *y* when $x = 2.33$.

      iii. Estimate the logarithm of 8.86.

   b.   **i.** Find the equation of the line that passes through (a, b) and $(a + 0.1, c)$.

       ii. Suppose that *x* is between *a* and $a + 0.1$, and that *b* and *c* represent the logarithms of *a* and $a + 0.1$, respectively. Use the result of part i to estimate the logarithm of *x*.

      iii. Now use the table and the formula obtained in part ii to estimate the logarithm of 7.74.

TABLE 2

N	0	1	2	3	4	5	6	7	8	9
1	.0000	.0414	.0792	.1139	.1461	.1761	.2041	.2304	.2553	.2788
2	.3010	.3222	.3424	.3617	.3802	.3979	.4150	.4314	.4472	.4624
3	.4771	.4914	.5051	.5185	.5315	.5441	.5563	.5682	.5798	.5911
4	.6021	.6128	.6232	.6335	.6435	.6532	.6628	.6721	.6812	.6902
5	.6990	.7076	.7160	.7243	.7324	.7404	.7482	.7559	.7634	.7709
6	.7782	.7853	.7924	.7993	.8062	.8129	.8195	.8261	.8325	.8388
7	.8451	.8513	.8573	.8633	.8692	.8751	.8808	.8865	.8921	.8976
8	.9031	.9085	.9138	.9191	.9243	.9294	.9345	.9395	.9445	.9494
9	.9542	.9590	.9683	.9685	.9731	.9777	.9823	.9868	.9912	.9956

SECTION 3.2   FUNCTIONS

☐ **What will the volume of the sphere of influence of Martin Luther King's "I Have A Dream" speech be at the turn of the century?**

☐ **How fast must a cyclist ride in order to generate enough power to replace a small power plant?**

☐ **Does the association of a U.S. president to his vice president constitute a function?**

☐ **What type of business software is designed specifically for the rapid computation of thousands of function values?**

☐ **Why is it that no matter what expression you enter into your graphing calculator, its graph will not be a circle?**

It is impossible to overstate the significance of functions—both as a foundational mathematical concept and as a tool for describing relationships between real-world variables. The language of functions permeates every branch of the physical and social sciences: IQ scores are a function of nature and nurture, pressure is a function of temperature and volume, consumer demand is a function of price, population is a function of food supply, and computational power is a function of processor speed. We will see that whenever the value of one variable determines the value of another, this correspondence defines a function. In fact, many of the equations involving two variables that we have seen throughout the first two chapters of this text define functional relationships. Specifically, every time we solved for y in an equation involving both x and y, we were expressing y as a function of x. It is thus in a very real sense that this section is largely a formalization of a concept with which we already are quite familiar.

DEFINITION OF FUNCTION

Correspondences between sets of objects are very common in mathematics as well as in everyday life. Table 3 gives some examples of such correspondences.

TABLE 3   Correspondences between sets

To each human . . .	there corresponds . . .	a biological mother.
To each license plate number in a given state . . .	there corresponds . . .	a vehicle.
To each package weight . . .	there corresponds . . .	a first-class Postal Service delivery rate.
To each positive real number . . .	there corresponds . . .	the square of that number.

Each of the examples in Table 3 has the following form.

To each element of a set D . . .   there corresponds . . .   an element of a set R.

More specifically, to each element of a set D there corresponds *exactly* one element of a set R. Correspondences with this property are called *functions*.

Definition of a Function   □

> A **function** from a set D to a set R is a correspondence or rule that assigns to each element x of D exactly one element y of R. The set D is called the **domain** of the function and the elements x of D are the **input values.** The elements y of R that correspond to the input values are the **output values.** The set of all possible output values is called the **range** of the function.

EXAMPLE 1   ▓   Examples of Functions

Verify that each of the examples given in Table 3 are functions and identify the domain and range for each.

SOLUTION

a.   Since each human has exactly one biological mother, this correspondence is a function. The domain is the set of all humans. The range is the set of all mothers.

b.   You can imagine the resulting chaos if the same license number was assigned to two different vehicles. So it is by necessity that to each license number their corresponds exactly one vehicle, and thus this correspondence is a function. The domain is the set of all current license numbers, and the range is the set of all licensed vehicles.

c.   For any given package weight, there is only one first-class postal rate, and so this correspondence is a function. The domain is the set of all allowable package weights (up to 70 pounds for the Postal Service). The range is the set of all first-class package rates.

d.   Each positive real number has exactly one square. The domain is the set of all real numbers. Since the square of a real number can never be negative, the range is the set of nonnegative real numbers.

EXAMPLE 2   ▓   An Example of a Nonfunction

Explain why the correspondence between the set D of all telephone numbers and the set R of all telephones is not a function.

SOLUTION   It is not uncommon for several telephones to be connected to the same line and thus have the same number. Thus, there are elements of the set D that are paired with more than one element of R, and so the correspondence is not a function.

EQUATIONS AND FUNCTIONS

Functions involving sets of real numbers are often defined by an equation in two variables. By convention, we typically let the variable x represent the input and y the output. For example, the function described earlier that assigns to each positive real number the square of that number can be defined by the equation $y = x^2$. This equation specifies that to each input value x there corresponds the output value $y = x^2$, and so the equation defines a function. More generally, an equation of the form

$$y = \square$$

where $\square$ is an expression involving x, defines y as a function of x, with **independent variable** x and **dependent variable** y. It is important to note, however, that not all equations in x and y define y as a function of x. In order for y to be a function of x, there must correspond exactly one y for each x.

EXAMPLE 3   ■   **Testing an Equation to See if It Defines a Function**

Determine whether or not the equation $y - x^3 = 0$ defines y as a function of x. Find all y-values corresponding to the x-values given in the following table.

x	y
-2	
0	
1	
3	

SOLUTION   Solving for y, we have $y = x^3$. This shows that for each value of x there is exactly one value for y. Thus, the equation does define y as a function of x. We complete the table as follows.

x	$y = x^3$
-2	$(-2)^3 = -8$
0	$0^3 = 0$
1	$1^3 = 1$
3	$3^3 = 27$

EXAMPLE 4   ■   **An Equation That Does Not Define y as a Function of x**

Show that the equation $y^2 - x = 0$ does not define y as a function of x.

SOLUTION
We first solve for y.

$$y^2 - x = 0$$
$$y^2 = x$$
$$y = \pm\sqrt{x}$$

This indicates that for each positive value of x there are *two* values for y—namely, $y = \sqrt{x}$ and $y = -\sqrt{x}$. Thus, this equation does *not* define y as a function of x. Note, however, that since the equation can be written as $x = y^2$, with x on the left and an expression involving only y on the right, $y^2 - x = 0$ *does* define x as a function of y.

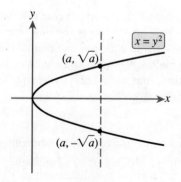

FIGURE 17

The graph of the equation $y^2 - x = 0$ from Example 4 shows us very clearly why it does not define the variable y as a function of x. By rewriting the equation as $x = y^2$ and plotting points, we obtain the graph shown in Figure 17. Notice that for any positive value of a, both of the points $(a, \sqrt{a})$ and $(a, -\sqrt{a})$ are on the graph: There are two distinct y-values corresponding to the same x-value. Graphically, this means that the vertical line $x = a$ intersects the graph twice. This is not allowed if the equation is to define y as a function of x since such a function must pair each x with exactly one y, which suggests the following test.

The Vertical Line Test ☐ | An equation defines y as a function of x if and only if no vertical line crosses the graph of the equation more than once.

EXAMPLE 5 ▓ Using the Vertical Line Test

Use the vertical line test to identify which of the following equations define y as a function of x.

a. $x^2 + y^2 = 4$

b. $y = (x + 1)^3 - 2$

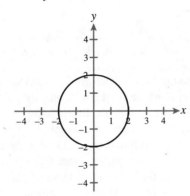

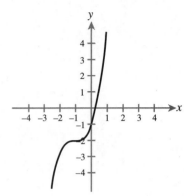

SOLUTION

a. $x^2 + y^2 = 4$ does not define y as a function of x since some vertical lines intersect the graph in two points.

b. $y = (x + 1)^3 - 2$ does define y as a function of x since no vertical line intersects the graph in more than one point.

FUNCTION NOTATION

As we observed earlier, an equation of the form $y = \square$, where $\square$ is an expression involving x, defines a function. In many circumstances, particularly when there is more than one function under discussion, it is convenient to attach names to functions. For example, the functions defined by the equations $y = \sqrt{x}$ and $y = x^2$ might be referred to as f and g, respectively. The output value of a function f associated with an input value x is denoted by $f(x)$, read "f of x." Thus, instead of writing $y = \sqrt{x}$, we write $f(x) = \sqrt{x}$. Functions expressed in this way are said to be written in **function notation.**

EXAMPLE 6 ▓ Using Function Notation

Write each of the following functions using function notation.

a. h is the function that assigns to each nonzero real number its reciprocal.

b. g is the function that assigns to each real number t the number $t^3 - t$.

c. P is the function that assigns to each real number one less than its square.

SOLUTION

a. The reciprocal of a number x is the number $1/x$. Thus $h(x) = 1/x$.

b. $g(t) = t^3 - t$

c. One less than the square of a real number x is the number $x^2 - 1$. Thus $P(x) = x^2 - 1$.

E X A M P L E 7   Constructing a Function from a Formula

Express the area of a circle as a function of its circumference.

SOLUTION   The area of a circle in terms of its radius is $A = \pi r^2$. The circumference of a circle in terms of its radius is $C = 2\pi r$. Solving the latter equation for r, we obtain $r = C/(2\pi)$. Substituting for r in the area formula gives

$$A = \pi r^2 = \pi \left(\frac{C}{2\pi}\right)^2 = \pi \frac{C^2}{4\pi^2} = \frac{C^2}{4\pi}$$

Thus $A(C) = \dfrac{C^2}{4\pi}$.

Example 7, in which we constructed a function named A with C as the independent variable, illustrates that not all functions are named f, and that the independent variable need not be x. In fact, the independent variable simply serves as a placeholder, and thus could be replaced by any symbol. For example, the function $f(x) = x^2 - 3x$ is exactly the same as the function $f(\square) = (\square)^2 - 3(\square)$: With either definition, the output is obtained by subtracting three times the input from the square of the input. Thus, in order to find the value of the function $f(x) = x^2 - 3x$ when $x = -1$, we simply replace all occurrences of x with -1 to obtain

$$f(-1) = (-1)^2 - 3(-1) = 1 + 3 = 4$$

> **WARNING!**   The function notation $f(-1)$ represents the output value of the function f when $x = -1$. It does not mean we are to multiply f by -1.

E X A M P L E 8   Evaluating Functions

Let $g(t) = t^3 - t$. Find the following function values.

a. $g(-2)$ **b.** $g(\text{⚭})$ **c.** $g(x + 1)$

SOLUTION
a. $g(-2) = (-2)^3 - (-2) = -8 + 2 = -6$
b. $g(\text{⚭}) = (\text{⚭})^3 - \text{⚭}$
c. $g(x + 1) = (x + 1)^3 - (x + 1)$
$\qquad\qquad = (x^3 + 3x^2 + 3x + 1) - (x + 1)$
$\qquad\qquad = x^3 + 3x^2 + 3x + 1 - x - 1$
$\qquad\qquad = x^3 + 3x^2 + 2x$

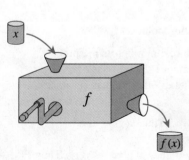

FIGURE 18

The process of evaluating a function can be illustrated using a "function machine," such nas the one suggested in Figure 18. When an input value x is fed into the machine, the machine operates on the number and produces an output value $f(x)$. If no restrictions are specifically placed on the input values, the domain of a function is assumed to be the set of all valid real number input values. The range is then the set of all resulting output values.

EXAMPLE 9 ▨ **Finding Domain**

Find the domain of $g(t) = \dfrac{t}{t^2 - 1}$.

SOLUTION All input values of t are valid except those that lead to a zero in the denominator. To determine where this occurs, we set $t^2 - 1 = 0$ and solve for t.

$$t^2 - 1 = 0$$
$$(t + 1)(t - 1) = 0$$
$$t = -1, t = 1$$

So the domain of g is the set of all real numbers t except -1 and 1. This set can be written in set-builder notation as $\{t \mid t \neq -1, t \neq 1\}$. In interval notation, we write $(-\infty, -1) \cup (-1, 1) \cup (1, \infty)$.

EXAMPLE 10 ▨ **Finding Domain**

Find the domain of $h(x) = \sqrt{2x - 3}$.

SOLUTION Since we are only considering values of x for which $h(x)$ is real, the valid input values of x are those for which $2x - 3 \geq 0$. Solving this linear inequality for x yields $x \geq \frac{3}{2}$. Thus, the domain of h is $[\frac{3}{2}, \infty)$.

GRAPHS OF FUNCTIONS

The **graph of a function f** is the set of points (x, y) such that $y = f(x)$. In other words, the graph of a function f is simply the graph of the equation $y = f(x)$. Thus, the techniques developed for graphing equations can also be applied to graphing functions.

The graph of a function depicts the correspondence between input values in the domain and output values in the range. Indeed, if (x, y) is a point on the graph of a function f, then y is the value of the function at x, as shown in Figure 19.

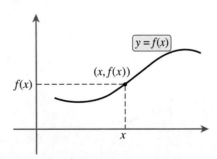

FIGURE 19

EXAMPLE 11   ▨   Graphing a Function

Sketch the graph of the function $f(x) = \frac{1}{2}x - 3$.

SOLUTION   Replacing $f(x)$ with y, we obtain $y = \frac{1}{2}x - 3$, a straight line with y-intercept -3 and slope $\frac{1}{2}$. The graph is shown in Figure 20.

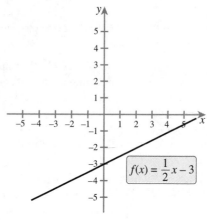

FIGURE 20

EXAMPLE 12   ▨   Graphing a Function with a Restricted Domain

Sketch the graph of the function $f(x) = \frac{1}{x}$.

SOLUTION   We first note that 0 is not in the domain of f. So there is no point on the graph with x-coordinate 0. Next, we find and plot some points satisfying $y = \frac{1}{x}$, as shown in Figure 21.

x	$y = \frac{1}{x}$	x	$y = \frac{1}{x}$
1	1	$\frac{1}{2}$	2
2	$\frac{1}{2}$	$\frac{1}{3}$	3
3	$\frac{1}{3}$	$-\frac{1}{2}$	-2
-1	-1	$-\frac{1}{3}$	-3
-2	$-\frac{1}{2}$		
-3	$-\frac{1}{3}$		

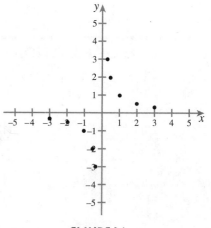

FIGURE 21

Notice that as the x-values get larger (in magnitude), the corresponding y-values get smaller, and so the points get closer and closer to the x-axis. Moreover, as the x-values get smaller, the y-values get larger, shooting up or down along the y-axis. We use these observations to complete the graph, as shown in Figure 22.

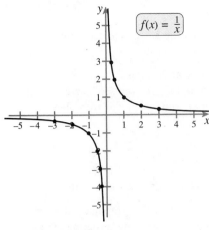

$f(x) = \frac{1}{x}$

FIGURE 22

As we noted, there is no point on the graph with x-coordinate 0. As a consequence, the graph does not cross the line $x = 0$, but rather approaches it arbitrarily closely. More precisely, the graph rises (or falls) without bound as x gets close to 0. For this reason, the line $x = 0$ (the y-axis) is called a *vertical asymptote*. The line $y = 0$ (the x-axis), on the other hand, is a *horizontal asymptote* since the graph approaches it arbitrarily closely as x goes toward ∞ and $-\infty$. We will study vertical and horizontal asymptotes in detail in Section 4.5.

Graphs of functions are easily plotted using a graphing calculator. In fact, a graphing calculator is really just a superhuman point-plotter. It chooses regularly spaced input values in the interval determined by `Xmin` and `Xmax` and then computes the corresponding output values according to the given formula. As long as the output values are in the interval specified by `Ymin` and `Ymax`, the input–output pairs are plotted as points on the display.

EXAMPLE 13 ▨ **Determining Domain and Range Using a Graphing Calculator.**

Use a graphing calculator to plot the graph of the function $f(x) = \sqrt{x^2 - 16}$, and use the graph to determine the domain and range of f.

SOLUTION We enter the expression $y = \sqrt{x^2 - 16}$ and plot it with a graphing calculator, as shown in Figures 23 and 24. To find the domain, we are looking for the set of all possible x-coordinates of points on the graph. The view in Figure 23 suggests that no portion of the graph lies between $x = -4$ and $x = 4$. The two views together suggest that the graph extends infinitely to the right and left. Thus, we conclude that the domain is $(-\infty, -4] \cup [4, \infty)$. Note that the numbers -4 and 4 are included in both intervals because both values can be substituted into the function. The range consists of the set of all possible y-coordinates of points on the graph. Since the graph lies strictly above the x-axis, and appears to extend infinitely upward, we conclude that the y-coordinates are all positive (or zero) and approach infinity. In other words, we conclude that the range is $[0, \infty)$.

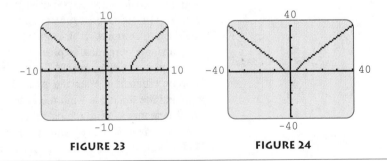

FIGURE 23   **FIGURE 24**

Functions are generally the method of choice for expressing relationships between variables in the real world—and even the not-so-real world—as the following example illustrates.

EXAMPLE 14 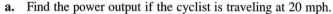 Computing the Power Output of a Cyclist

For a 170-pound cyclist on a 30-pound bicycle pedaling at v miles per hour, the power output in watts is given by $P(v) = 0.0178678v^3 + 2.01168v$.

a. Find the power output if the cyclist is traveling at 20 mph.

b. A small power plant will produce around 500,000,000 watts of power. Find the speed required for a single cyclist to generate enough power to replace a small power plant. Answer to the nearest mile per hour.

SOLUTION

a. Evaluating $P(20)$, we have

$$P(20) = 0.0178678(20)^3 + 2.01168(20) \approx 183.176 \text{ watts}$$

b. We plot the graph of $y = 0.0178678v^3 + 2.01168v$ with a graphing calculator and search for a point with y-coordinate 500,000,000. After some experimentation, we settle on the view shown in Figure 25. Next we zoom in until we can estimate the x-coordinate of the point to the nearest mile per hour, as shown in Figure 26. Thus, a cyclist capable of traveling at 3036 miles per hour would generate enough power to replace a small power plant.

The human-powered Gossamer Albatross in flight.

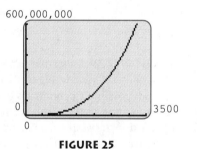

FIGURE 25

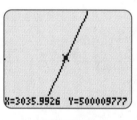

FIGURE 26

 CALCULATOR KEYS

Computing Function Values

A graphing calculator can be used in various ways to compute function values; the method you choose in a given context will likely depend on the number of values you wish to compute. If just one value is desired, it is generally easiest to directly enter a single numerical expression into the calculator, as would have been the case in part a of Example 14. However, if more than one value is desired, it may be more efficient to enter the actual function expression—just as you would when graphing the function—and then use your calculator's automatic **calculate-value** or **table** features. With the calculate-value feature, you would specify an x-value, and the corresponding y-value would be calculated and displayed on the graph. With the table feature, you would specify a range of x-values, and the corresponding y-values would be calculated and displayed in a table. Check your calculator manual for details on these and perhaps other features for computing function values.

UNDERSTANDING AND MASTERY CHECKLISTS

CONCEPTS TO UNDERSTAND

☐ Functions

☐ Domain and range of a function

☐ Vertical line test

☐ Function notation

☐ Graphs of functions

SKILLS TO MASTER

☐ Determine when a correspondence defines a function.

☐ Find the domain of a function.

☐ Use a graphing calculator to estimate the domain and range of a function.

☐ Given the graph of an equation, determine if the graph defines y as a function of x.

☐ Evaluate functions expressed with function notation.

☐ Express functional relationships using function notation.

☐ Sketch the graph of a function.

EXERCISES 3.2

EXERCISES 1–6   *Determine whether the correspondence from the set D to the set R defines a function.*

1.

D	−2	−1	0	1	2
R	4	1	0	1	4

2.

D	4	1	0	1	4
R	−2	−1	0	1	2

3.

U.S. Presidents and Their Vice Presidents	
President (D)	**Vice President (R)**
George Washington	John Adams
John Adams	Thomas Jefferson
Thomas Jefferson	Aaron Burr
Thomas Jefferson	George Clinton
James Madison	George Clinton
James Madison	Elbridge Gerry

4.

Car and Driver Top Ten for 1998	
Make (D)	**Model (R)**
Audi	A4
BMW	328i/M3
BMW	5-Series
Chevrolet	Corvette
Dodge	Intrepid
Honda	Accord
Honda	Prelude
Lexus	GS 300/400
Mazda	Miata
Porsche	Boxster

5.

Internet Growth	
Year (D)	**Internet Hosts (R)**
1991	376,000
1992	727,000
1993	1,313,000
1994	2,217,000
1995	4,852,000
1996	9,472,000
1997	16,146,000
1998	29,670,000

6.

Density	
Substance (D)	**Density (lb/ft³) (R)**
Hydrogen	0.006
Air	0.1
Styrofoam	6.2
Whale oil	48.0
Ice	57.4
Water	62.4
Aluminum	168.4
Lead	704.2
Gold	1203.8
Neutron star	4.4×10^{16}

14. $\dfrac{x^2}{16} + \dfrac{y^2}{9} = 1$

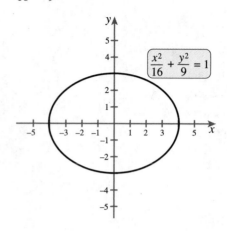

EXERCISES 7–12   *Determine whether the given equation defines y as a function of x.*

7. $3x + 4y = 12$

8. $y^2 - x^2 = 0$

9. $x^2 + y^2 = 1$

10. $x^2 + x - y = 0$

11. $0 = x - y^3$

12. $x = \sqrt{y}$

15. $x = y^3 - 4y$

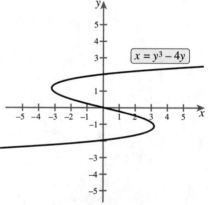

EXERCISES 13–18   *Use the vertical line test to determine whether the given equation defines y as a function of x.*

13. $y = -x^2 + 1$

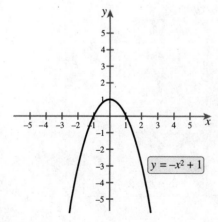

16. $x = y^3$

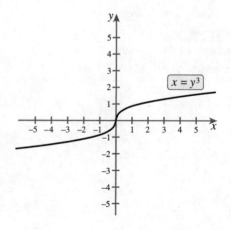

17. $xy = 4$

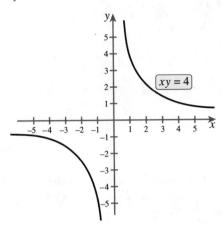

18. $|x| - |y| = 0$

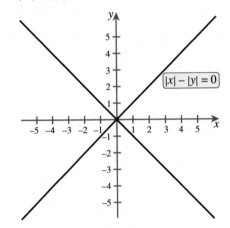

EXERCISES 19–28 *Evaluate the given function as indicated.*

19. $f(x) = x^2 + 3x + 2$
   a. $f(2)$
   b. $f(-3)$
   c. $f(0)$

20. $g(t) = t^3 - 4t$
   a. $g(1)$
   b. $g(-2)$
   c. $g(3)$

21. $f(x) = \dfrac{3x + 2}{4x - 5}$
   a. $f(4)$
   b. $f(5)$
   c. $f(-t)$

22. $h(y) = \dfrac{y^2 - 3}{y + 7}$
   a. $h(2)$
   b. $h(-6)$
   c. $h(x)$

23. $x(t) = 3t^2$
   a. $x(t^2)$
   b. $x(t - 1)$
   c. $x\left(\frac{5}{t}\right)$

24. $f(x) = \dfrac{2x + 2}{2x - 3}$
   a. $f(t)$
   b. $f(x - 1)$
   c. $f\left(\frac{x}{2}\right)$

25. $g(x) = (x + 1)^2$
   a. $g(x - 1)$
   b. $g(x^2)$
   c. $g\left(\dfrac{1 - a}{a}\right)$

26. $p(x) = \sqrt{9 - x^2}$
   a. $p(\square)$
   b. $p(x + 3)$
   c. $p(x - 3)$

27. $f(x) = \dfrac{1}{x - 1}$, $g(x) = x^2$
   a. $f(g(3))$
   b. $f(g(t))$
   c. $g(f(x + 1))$

28. $f(x) = x^2$, $g(y) = 2y + 1$
   a. $f(g(2))$
   b. $f(g(t))$
   c. $g(f(c^2))$

EXERCISES 29–32 *Express the given function using function notation.*

29. h assigns to every real number the absolute value of the number.

30. g assigns to every real number twice the cube of the number.

31. f assigns to every real number in its domain the reciprocal of 3 more than the number.

32. p assigns to each real number in its domain the reciprocal of the square root of the number.

EXERCISES 33–44 *Find the domain of the function.*

33. $f(x) = x^2$

34. $g(x) = \dfrac{4}{x}$

35. $h(x) = \dfrac{3}{x - 4}$

36. $r(s) = |s|$

37. $r(t) = \dfrac{5t^2 + 3t + 1}{t^2 - 9}$

38. $p(x) = \dfrac{x + 1}{x^2 - 5x + 6}$

39. $f(x) = \sqrt{x + 1}$

40. $g(t) = \sqrt{t - 3}$

41. $f(x) = \dfrac{\sqrt{x + 2}}{x^2 - 4x + 3}$

42. $g(y) = \dfrac{2}{\sqrt{2y - 6}}$

43. $f(x) = \dfrac{2}{|x| - 3}$

44. $g(x) = \dfrac{x}{x}$

EXERCISES 45–48 *Find a formula $f(x)$ for a function that satisfies the given correspondences between input and output values. There may be more than one correct answer.*

45.

x	0	1	2	3
$f(x)$	1	3	5	7

46.

x	0	1	2	3
$f(x)$	2	1	0	−1

47.

x	−1	0	1	2
$f(x)$	undefined	1	$\frac{1}{2}$	$\frac{1}{3}$

48.

x	2	5	10	17	26
f(x)	1	2	3	4	5

56.

x	$f(x) = -(x - 2)^2 + 4$
0	
1	
2	
3	
4	

EXERCISES 49–52   *Find a function f for which the given set is the domain. There may be more than one correct answer.*

49. $(-\infty, 7]$

50. All real numbers except -4 and 1

51. All real numbers except 0 and 3

52. $[-2, \infty)$

EXERCISES 53–56   *Complete the table and use the resulting points to sketch the graph of the function.*

53.

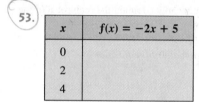

x	$f(x) = -2x + 5$
0	
2	
4	

54.

x	$f(x) = \frac{2}{3}x - 2$
-3	
0	
3	

55.

x	$f(x) = (x + 1)^2 - 3$
-3	
-2	
-1	
0	
1	

EXERCISES 57–60   *Use a graphing calculator to sketch the graph of the function and then use the graph to help identify or approximate the domain and range of the function.*

57.
 a. $f(x) = x^2$
 b. $f(x) = (x + 1)^2$
 c. $f(x) = x^2 + 1$
 d. $f(x) = \dfrac{1}{x^2 + 1}$

58.
 a. $f(x) = \sqrt{x}$
 b. $f(x) = \sqrt{x + 4}$
 c. $f(x) = \sqrt{x} + 4$
 d. $f(x) = \dfrac{1}{\sqrt{x + 4}}$

59.
 a. $f(x) = \sqrt{x^2 + 1}$
 b. $f(x) = x + \sqrt{x^2 + 1}$
 c. $f(x) = x\sqrt{x^2 + 1}$
 d. $f(x) = \dfrac{x}{\sqrt{x^2 + 1}}$

60.
 a. $f(x) = \sqrt{9 - x^2}$
 b. $f(x) = x + \sqrt{9 - x^2}$
 c. $f(x) = x\sqrt{9 - x^2}$
 d. $f(x) = \dfrac{x}{\sqrt{9 - x^2}}$

61.
 a. Use the quadratic formula to solve the equation $y = x^2 + 4x + 6$ for x.
 b. For what values of y does the equation of part a have a real-valued solution?
 c. What is the range of the function g defined by $g(x) = x^2 + 4x + 6$?

62.
 a. Find the domain of $f(x) = \dfrac{|x|}{x}$.
 b. Evaluate $f(2)$, $f(-3.76)$, and $f(10^{2345})$.
 c. Find the range of f.

■ APPLICATIONS

63. *Volume of a Sphere*   The volume of a sphere with radius r is given by $V = \frac{4}{3}\pi r^3$. Express the volume of a sphere as a function of its diameter.

64. *Area of a Circle*   The area of a circle with radius r is given by $A = \pi r^2$. Express the area of a circle as a function of its diameter.

65. *Area of a Square*   A square is inscribed in a circle. Express the area of the square as a function of the radius of the circle.

66. *Volume of a Cube*   A cube is inscribed in a sphere. Express the volume of the cube as a function of the radius of the sphere.

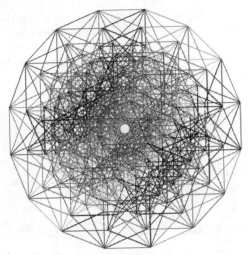

Just as a 2-dimensional cube (a square) has volume s^2, and a 3-dimensional cube has volume s^3, an 8-dimensional hypercube as depicted here has volume s^8.

67. *Area of a Triangle*   A line passing through the point $(2, 1)$ has intercepts $(a, 0)$ and $(0, b)$ that form the vertices of a right triangle, as shown in Figure 27. Express the area of the triangle as a function of a.

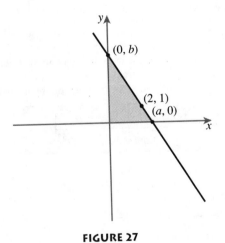

FIGURE 27

68. *Area of a Triangle*   Two perpendicular lines intersect at the point $(2, 3)$ (Figure 28). Their x-intercepts form the vertices of a right triangle, as shown. Express the area of the triangle as a function of the length of its hypotenuse.

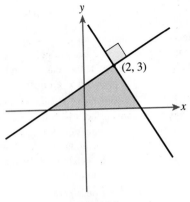

FIGURE 28

69. *Volume of a Box*   A box with a square base and no top is formed by cutting squares out of the corners of a piece of cardboard and then folding up the sides (Figure 29). If the cardboard measures $16'' \times 16''$ and the length of the edge of each cut-out square is denoted by x, express the volume of the resulting box as a function of x. What is the domain of this function? (*Hint:* Include in the domain only those values that make sense in the context of this problem.)

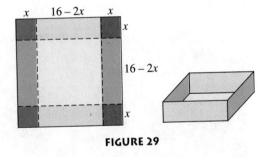

FIGURE 29

70. *Area of a Pasture*   A rancher wishes to enclose a rectangular pasture at the base of a cliff: 200 linear feet of fencing are available and no fencing is needed along the cliff side of the pasture. Express the area of the pasture as a function of x, the dimension perpendicular to the cliff face.

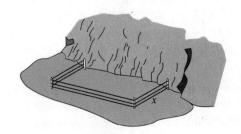

71. *Saving for a Vacation* The Reynolds estimate that in 4 years they will need $2000 for a family vacation to Disney World. They intend to shop around for the best interest rate they can find and then deposit the necessary principal so that it will grow to $2000 after 4 years. Use the formula for interest compounded annually to express the principal as a function of interest rate.

72. *Height of a Can* A research and development team for a company that produces canned foods is in the process of designing a can that must have a volume of 400 square centimeters. Use the formula $V = \pi r^2 h$ for the volume of a cylinder with radius r and height h to express the height of the can as a function of the radius. What is the domain of this function?

73. *Sphere of Influence* Electromagnetic waves—like those in radio and television transmissions, for example—travel at the speed of light, often in all directions. Although our radios pick up a signal almost instantaneously, the signal continues to travel through space at the speed of light. Thus, for example, Dr. Martin Luther King's famous 1963 "I have a dream" speech is even now reaching new portions of the galaxy. The radio waves from the broadcast began traveling through space at the speed of light in all directions from the point of origin. Since light travels at a rate of approximately 5.9×10^{12} miles per year, the distance the waves would have traveled after t years would be given by $r = (5.9 \times 10^{12})t$. Moreover, when the waves are a distance r from the point of origin, the volume of space (ignoring the fact that some of the signal will be blocked by Earth itself) through which they would have traveled would be given by $V = \frac{4}{3}\pi r^3$. Express

Dr. Martin Luther King delivers his "I have a dream" speech.

the volume V as a function of t, and compute the volume of space through which the speech will have traveled in the year 2000.

74. *Radius of an Oil Spill* An oil tanker develops a leak and begins spilling oil at a rate of 10,000,000 cubic centimeters per minute. The volume of oil that has spilled after t minutes is thus given by $V = 10,000,000t$ cubic centimeters. Assuming that the resulting oil spill has a circular shape with radius r and a constant thickness h, the volume can also be expressed as $V = \pi r^2 h$. If $h = 1$ centimeter, express the radius as a function of time and compute the radius of the spill after 2 hours.

In spite of containment efforts, oil spills continue to cause immeasurable damage to the environment.

75. *Subsidized Housing* An apartment complex receives a government subsidy and, in return, must base its monthly rent on family size and monthly income. Selected income and rent amounts for a family of four are given in the following table. Plot the points and construct a function $R(x)$ that gives the monthly rent for a family of four with an income of x dollars per month.

Income ($)	800	950	1150	1400
Rent ($)	100	160	240	340

76. *Cab Fare* Selected cab fares are given in the following table. Plot the points and construct a function $F(x)$ that gives as output values the fare that would be charged for any given distance x.

Distance (mi)	1	3	7	13
Fare ($)	1.50	2.50	4.50	7.50

■ CONCEPTS AND CRITICAL THINKING

EXERCISES 77–80 *Answer true or false.*

77. The domain of a function is the set of all output values.

78. No circle is the graph of $y = f(x)$, where f is a function.

79. If there is a vertical line that crosses the graph of an equation once, then the graph is the graph of a function.

80. For all functions f and real numbers x and y, $f(x + y) = f(x) + f(y)$.

EXERCISES 81–84 *Give an example of each.*

81. A function f with range $[5, \infty)$

82. A graph that does not define y as a function of x

83. A function whose domain and range are both $[0, \infty)$

84. A correspondence between two sets of people D and R that is not a function

85. Suppose that the range of $g(x)$ is $(-\infty, 4]$ and the domain of $g(x)$ is $(-\infty, \infty)$. Find the domain of $\dfrac{1}{g(x) - 5}$.

86. Define the function $f(x)$ by $f(x) = \dfrac{1}{q(x)}$, where q is another function. There is one number that is guaranteed *not* to be an element of the range of f, no matter how q is defined. What number is it?

87. Rational expressions are simplified by dividing out equal factors of the numerator and denominator. Thus, for example,

$$\frac{x^2 - 1}{x + 1} = \frac{(x + 1)(x - 1)}{x + 1} = x - 1$$

However, if we define

$$f(x) = \frac{x^2 - 1}{x + 1} \quad \text{and} \quad g(x) = x - 1$$

these two functions are not the same. Explain why this is so. In what sense is $\frac{x}{x} \neq 1$?

QUESTIONS FOR DISCUSSION OR ESSAY

88. A graphing calculator is capable of graphing equations of the form $y = \square$, where $\square$ is an expression involving x. As we have seen in this section, such equations define y as a function of x, and their graphs pass the vertical line test. Explain how a graphing calculator can be used as an aid in graphing an equation such as $x^2 + y^2 = 1$, even though its graph does not pass the vertical line test and so does not define y as a function of x.

89. In our discussion on equations and functions, we were careful to say that an equation "defines y as a function of x" rather than just saying that the equation was a function. What does it mean to define x as a function of y? Give an example of an equation that defines x as a function of y but that does not define y as a function of x. Now explain why it is not meaningful to say that an equation either is or is not a function.

90. In Example 14, we saw that a cyclist would have to travel at a rate of 3036 mph in order to produce enough power to run a small power plant. If we agree that 20 mph is a "normal" speed, the cyclist would have to travel 152 times faster than normal. If, instead of one cyclist traveling 152 times faster than normal, we had many cyclists traveling at normal speed, how many cyclists does your intuition tell you it should take to run the plant? Now compute how many it would take by using the facts from Example 14 that a cyclist traveling at 20 miles per hour can produce approximately 183 watts of power and a small power plant produces around 500,000,000 watts. What seems surprising about the answer? Can you explain why there is such a disparity between your intuitive answer and the actual number?

PROJECTS FOR ENRICHMENT

91. *Functions of Several Variables*   Many of the correspondences between real-world quantities require the use of more than one variable. Consider, for example, the correspondence that assigns to each rectangle of length l and width w an area A equal to lw. The equation $A = lw$ defines the area as a function of two variables that we can write as $A(l, w) = lw$. In this case, each pair of values (l, w) is associated with *exactly one* area A. For example, a rectangle with length $l = 8$ and width $w = 5$ yields an area of $A(8, 5) = 8 \cdot 5 = 40$, and no other area is possible.

In general, we define a **function f of two variables** to be a rule or correspondence that assigns to a pair of numbers (x, y) exactly one number $f(x, y)$. The **domain** of a function of two variables is the set of all pairs (x, y) that can be used as input values. The **range** is the set of all corresponding output values. The domain of the function $f(x, y) = x^2 + y^2$ is the set of all pairs (x, y) such that x and y are real numbers. The range is the set of all positive real numbers.

a. Evaluate the following functions for the indicated values.

   i. $d(r, t) = rt$; $d(55, 4)$, $d(20, \frac{1}{4})$

   ii. $f(x, y) = x^2 + 2xy + y^2$; $f(2, 3)$, $f(-1, 5)$

   iii. $h(u, v) = 1/(u - v)$; $h(10, 8)$, $h(0.3, 0.4)$

   iv. $g(x, y) = \sqrt{1 - x^2 - y^2}$; $g(0, 1)$, $g(\frac{1}{2}, \frac{1}{4})$

b. Describe the domain of each of the functions in part a.

Functions of three or more variables are also possible. Consider the correspondence defined by the equation $B = P(1 + r)^t$, where B is the balance in an account after a principal of P dollars is deposited for t years at an annual rate of interest r. So B is a function of the three variables P, t, and r, and we write $B(P, r, t) = P(1 + r)^t$. Notice here that each triple of values (P, r, t) leads to *exactly one* balance B. For example, if $P = 1000$, $t = 4$, and $r = 0.05$, then

$$B(1000, 4, 0.05) = 1000(1 + 0.05)^4 \approx 1215.51$$

and this is the only balance possible for these values of P, t, and r.

c. Give a precise definition for a function of three variables and also for its domain and range.

d. Evaluate the following functions for the indicated values.

 i. $V(l, w, h) = lwh$; $V(4, 6, 3)$, $V(5, \frac{3}{2}, 8)$

 ii. $g(x, y, z) = \sqrt{x^2 + y^2 + z^2}$; $g(0, -4, 3)$, $g(-2, 1, 5)$

 iii. $f(u, v, w) = \dfrac{uv + vw + uw}{uvw}$; $f(1, 2, 3)$, $f(-4, 3, \frac{1}{3})$

 iv. $x(a, b, c) = \dfrac{-b + \sqrt{b^2 - 4ac}}{2a}$; $x(1, -7, -10)$, $x(2, 0, -9)$

e. Describe the domain of each of the functions in part d.

92. *Spreadsheet Functions* A spreadsheet is well suited for studying functions. In this project, we use a spreadsheet to investigate the definition of a function, the domain and range of a function, and the graph of a function.

a. Consider the spreadsheet fragment shown in Figure 30. Use the definition of function given in this section to explain why these data define a function. We refer to such functions as **numerical functions.** What is the domain of this numerical function? What is the range?

	A	B
1	**Input**	**Output**
2	0	1
3	1	2
4	2	5
5	3	10
6	4	17
7	5	26
8		

FIGURE 30

b. Set up a worksheet of your own with the labels **Input** and **Output** in cells A1 and B1, respectively. Enter the input values **0** through **3** in cells A2 through A5. Next, enter the formula **=sqrt(4-a2^2)** in cell B2. Figure 31 shows what you should see on your screen just before you press the Enter key after typing the formula. Copy the formula from cell B2 to cells B3 through B5. Your screen should look similar to Figure 32. Write out, using standard function notation, the symbolic definition of the function $f(x)$ whose output values agree with those given in the output column of Figure 32. What does the entry **#NUM!** in cell B5 suggest about the input value 3?

	A	B
1	**Input**	**Output**
2	0	=sqrt(4-a2^2)
3	1	
4	2	
5	3	

FIGURE 31

	A	B
1	**Input**	**Output**
2	0	2
3	1	1.732050808
4	2	0
5	3	**#NUM!**

FIGURE 32

c. Use your spreadsheet's "fill" feature to enter the numbers $-3.0, -2.9, -2.8, \ldots, 2.9, 3.0$ in the cell range A2:A62. Copy the formula in cell B2 to the cell range B3:B62. Based on the numerical data, what appears to be the domain of the function $f(x)$ you identified in part a? What is the range? How do the domain and range of the symbolic function $f(x)$ differ from the domain and range of the numerical function defined by the input–output columns of your spreadsheet?

d. Use your spreadsheet to produce input–output tables for the following functions. Use the given starting, stopping, and increment values for the input column. In each case, estimate the domain and range of the symbolic function $f(x)$.

Function	Start	Stop	Increment
$f(x) = x\sqrt{4 - x^2}$	-3	3	0.1
$f(x) = \sqrt{-x^4 + 10x^2 - 9}$	-5	5	0.2
$f(x) = \dfrac{1}{\sqrt{25x + 1}}$	-1	3	0.1

e. Use algebraic techniques to find the domain and range of the function $f(x) = \dfrac{1}{\sqrt{25x + 1}}$ from part d. What does your answer tell you about the spreadsheet method for investigating domain and range?

Graphs produced by spreadsheets also define functions. Figures 33–35 show 3 different spreadsheet graphs of the same set of data.

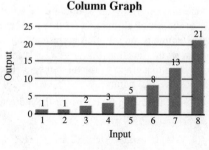

FIGURE 33

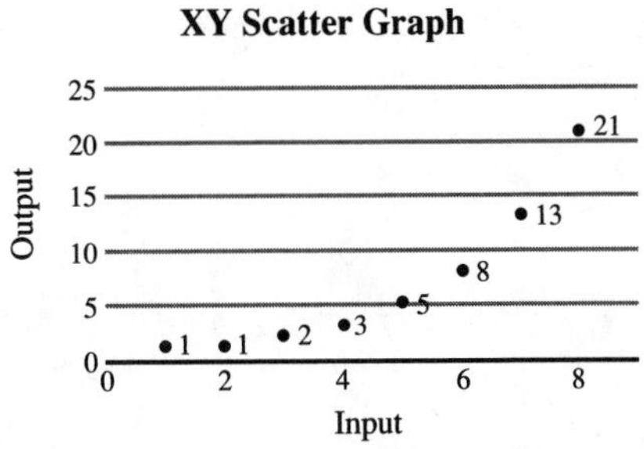

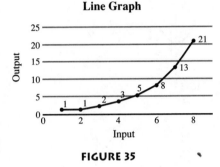

XY Scatter Graph

FIGURE 34

Line Graph

FIGURE 35

f. Use the definition of function given in this section to explain why these graphs define functions. We refer to such functions as **graphical functions.** Identify the domain and range of the graphical functions shown in Figures 33–35.

g. Use your spreadsheet's graphing feature to plot a column (or bar) graph, an xy (scatter) graph, and a line graph of the function from part c. Which type of graph most closely resembles a typical mathematical graph of a function?

SECTION 3.3 GRAPHS OF FUNCTIONS

☐ **Why does a soft drink can have the shape it does?**

☐ **What are the optimum dimensions of a box to be sent by mail?**

☐ **Why are functions satisfying $f(-x) = -f(x)$ said to be *odd*?**

☐ **How can a still photograph of a stretch of highway be used to estimate the speed at which the cars are traveling?**

☐ **How can the graph of a function be used to ensure that a bicycle racer receives a smooth handoff from a support vehicle?**

ZEROS OF A FUNCTION

We define a **zero** of a function to be any input value that yields zero as the output value. More formally, a zero of a function f is any number c such that $f(c) = 0$. It follows that if c is a zero of f, then $(c, 0)$ is an x-intercept of the graph of f. Thus, we can find the zeros of a function f algebraically by solving the equation $f(x) = 0$, or graphically by identifying the x-intercepts of the graph of $y = f(x)$.

EXAMPLE 1   **Finding Zeros Algebraically**

Find the zeros of the function $f(x) = x^2 - 4$ and illustrate them graphically.

SOLUTION   To find the zeros, we set $f(x) = 0$ and solve for x.

$$f(x) = 0$$
$$x^2 - 4 = 0$$
$$(x + 2)(x - 2) = 0$$
$$x = -2, \; x = 2$$

In Figure 36, we see that the zeros of f, -2 and 2, correspond to the x-intercepts of the graph of $y = x^2 - 4$.

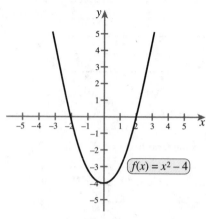

FIGURE 36

EXAMPLE 2 ▩ **Finding Zeros Graphically**

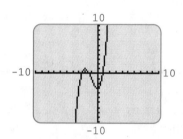

FIGURE 37

Use a graphing calculator to estimate the three zeros of $f(x) = x^3 + 3x^2 - 3$ to the nearest hundredth.

SOLUTION The graph of $y = x^3 + 3x^2 - 3$ shown in Figure 37 suggests three zeros (x-intercepts). As we noted in Section 2.5—and will show more formally in Chapter 4—a polynomial of degree 3 can have at most 3 zeros. Thus, we can be confident there are no other zeros outside of this view. We estimate each zero using the calculate-zero feature, obtaining $x \approx -2.53$, $x \approx -1.35$, and $x \approx 0.88$.

TRANSLATIONS

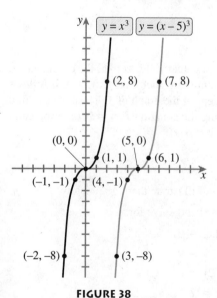

FIGURE 38

Many graphs are easily sketched by recognizing that they are merely shifts or **translations** of familiar graphs. For example, consider the graphs of $y = x^3$ and $y = (x - 5)^3$ shown in Figure 38. It is evident from the graphs and also from the following tables that the y-values are the same whenever the x-values for $y = (x - 5)^3$ are 5 greater than those for $y = x^3$. The net effect is that the graph of $y = (x - 5)^3$ has the same shape as the graph of $y = x^3$, but it is translated (shifted) 5 units to the right.

x	$y = x^3$		x	$y = (x - 5)^3$
-2	-8		3	-8
-1	-1		4	-1
0	0		5	0
1	1		6	1
2	8		7	8

More generally, replacing x with $x - h$ in an equation results in a horizontal shift of the graph by h units (to the right if $h > 0$, to the left if $h < 0$). Thus, we conclude that the graph of $y = f(x - h)$ is a horizontal shift by h units of the graph of $y = f(x)$. Similarly, if we replace y by $y - k$ in an equation, then the graph will be shifted k units vertically (up if $k > 0$, down if $k < 0$). Consequently, the graph of $y - k = f(x)$ or,

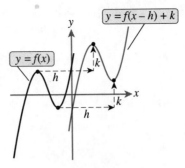

FIGURE 39

equivalently, $y = f(x) + k$ is a vertical shift by k units of the graph of $y = f(x)$. These observations are illustrated in Figure 39 (for $h, k > 0$), and are summarized as follows.

Translations of Functions □ | The graph of $y = f(x - h) + k$ is a translation h units horizontally and k units vertically of the graph of $y = f(x)$. If h is positive, then the translation is to the right; if h is negative, then the translation is to the left. If k is positive, then the translation is upward; if k is negative, then the translation is downward.

EXAMPLE 3 ■ **Graphing Translated Functions**

Use the graph of $f(x) = \sqrt{x}$ to sketch the graph of the given function.

a. $q(x) = \sqrt{x} + 3$ **b.** $r(x) = \sqrt{x + 2}$ **c.** $s(x) = \sqrt{x - 1} - 2$

SOLUTION

a. Since

$$q(x) = \sqrt{x} + 3$$
$$= f(x) + 3$$
$$= f(x - 0) + 3$$

we have $h = 0$ and $k = 3$. Thus, the graph of q is obtained by translating the graph of f 3 units upward, as shown in Figure 40.

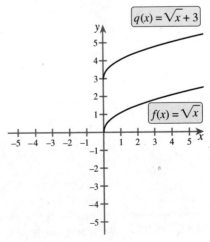

FIGURE 40

b. Since $r(x) = \sqrt{x + 2} = f(x + 2)$, we have $h = -2$, and $k = 0$. Thus, the graph of r is obtained by translating the graph of f 2 units to the left, as shown in Figure 41.

c. Here $s(x) = f(x - 1) - 2$, so that $h = 1$ and $k = -2$. Thus, the graph of s is obtained by translating the graph of f 1 unit to the right and 2 units down, as shown in Figure 42.

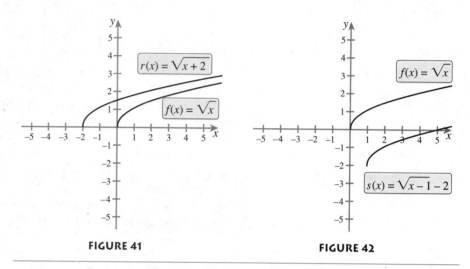

FIGURE 41   **FIGURE 42**

E X A M P L E   4   ■   **Constructing a Translated Function**

The graph of $f(x) = x^3$ is shown in Figure 43, together with the graph of a translated function $h(x)$. Find $h(x)$.

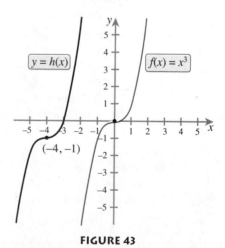

FIGURE 43

SOLUTION   The graph of h is that of f shifted 4 units to the left and 1 unit down. Thus, $h(x) = f(x + 4) - 1 = (x + 4)^3 - 1$.

REFLECTIONS   As Figure 44 illustrates, the **reflection about the x-axis** of the point (x, y) is the point $(x, -y)$. Thus, if we substitute $-y$ for y in an equation, the new graph will be a reflection of the old graph about the *x*-axis. Similarly, the **reflection about the y-axis** of the point

(x, y) is the point $(-x, y)$, and so substitution of $-x$ for x in an equation causes a reflection about the y-axis. **Reflection about the origin** is defined by reflecting about both the x-axis and the y-axis. Thus, the reflection about the origin of the point (x, y) is the point $(-x, -y)$, and it follows that if we substitute $-x$ for x and $-y$ for y in an equation, the new graph will be the reflection of the old graph about the origin.

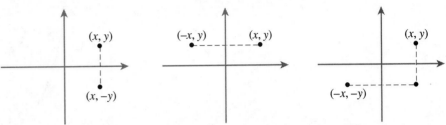

Reflection about the *x*-axis   **Reflection about the *y*-axis**   **Reflection about the origin**

FIGURE 44

When these principles are applied to the graph of a function $y = f(x)$, we obtain the following rules for reflections.

Reflections of Functions ☐

Function	Graph
$h(x) = f(-x)$	Reflection about the y-axis of the graph of f
$h(x) = -f(x)$	Reflection about the x-axis of the graph of f
$h(x) = -f(-x)$	Reflection about the origin of the graph of f

EXAMPLE 5 ▓ **Graphing Reflected Functions**

Use the graph of $f(x) = x^2 - 2x$, shown in Figure 45, to sketch the graphs of the following functions.

a. $h(x) = (-x)^2 - 2(-x) = x^2 + 2x$
b. $h(x) = -(x^2 - 2x) = -x^2 + 2x$

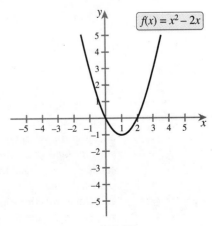

$f(x) = x^2 - 2x$

FIGURE 45

SOLUTION

a.   We note that $h(x) = f(-x)$ since $f(-x) = (-x)^2 - 2(-x) = x^2 + 2x$. Thus, the graph of h is the reflection of the graph of f about the y-axis, as shown in Figure 46.

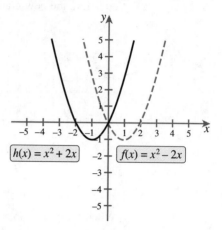

FIGURE 46

b.   Here $h(x) = -f(x)$ since $-f(x) = -(x^2 - 2x) = -x^2 + 2x$. Thus, the graph of h is the reflection of the graph of f about the x-axis, as shown in Figure 47.

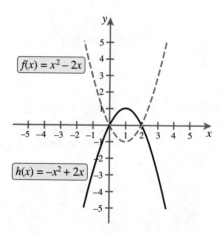

FIGURE 47

SYMMETRY

If the graph of an equation remains the same after reflecting about the y-axis, then the graph is said to be **symmetric with respect to the y-axis**. To test the graph of an equation for symmetry with respect to the y-axis, we simply replace x with $-x$ in the equation and check to see whether or not the resulting equation is equivalent to the original one. Thus, for the graph of a function f to be symmetric with respect to the y-axis, the equations $y = f(x)$ and $y = f(-x)$ must be equivalent—that is, $f(-x) = f(x)$. Such functions are said to be **even**. Similarly, functions that are **symmetric with respect to the origin**—that is, functions whose graphs remain the same after reflecting about the origin—can be shown to satisfy the equation $f(-x) = -f(x)$ and are said to be **odd.** We summarize these facts as follows.

Even and Odd Functions   □

Type of function	Graphical property	Algebraic property
Even	Symmetry with respect to the y-axis	$f(-x) = f(x)$ for each x in the domain of f
Odd	Symmetry with respect to the origin	$f(-x) = -f(x)$ for each x in the domain of f

EXAMPLE 6   ■   Identifying Even and Odd Functions

Use graphical as well as algebraic tests to determine whether the following functions are even or odd.

  a. $f(x) = |x|$ **b.** $f(x) = -x^2 + 6x$ **c.** $f(x) = x^3 - 3x$

SOLUTION

Function	Graph	Algebraic Test	Conclusion
a. $f(x) = |x|$	The graph appears to be symmetric with respect to the y-axis.	$f(-x) = |-x| = |x| = f(x)$	Even
b. $f(x) = -x^2 + 6x$	The graph is not symmetric with respect to either the y-axis or the origin.	$f(-x) = -(-x)^2 + 6(-x)$ $= -x^2 - 6x$ This is not equal to $f(x)$ or $-f(x)$.	Neither even nor odd
c. $f(x) = x^3 - 3x$	The graph appears to be symmetric with respect to the origin.	$f(-x) = (-x)^3 - 3(-x)$ $= -x^3 + 3x$ $= -f(x)$	Odd

> **RULE OF THUMB**   Notice that the function $f(x) = x^3 - 3x$ from part c of
> Example 6 includes only odd powers of x and was determined to be an odd
> function. This was not a coincidence. It can be shown (see Exercise 70) that a
> polynomial is even if all the powers of x are even, and it is odd if all the powers
> of x are odd.

EXAMPLE 7   ■   **Completing the Graph of a Function**

Complete the graph of the function f with the given property.

a. f is even   **b.** f is odd.

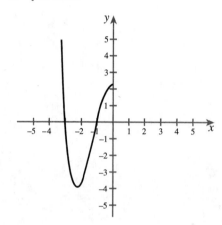   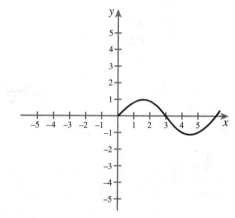

SOLUTION

a.   Since an even function is symmetric with respect to the y-axis, we simply reflect
the graph about the y-axis, as shown in Figure 48.

b.   Since an odd function is symmetric with respect to the origin, we first reflect the
graph about the x-axis to obtain the dashed portion shown in Figure 49, and then
we reflect the *dashed* portion across the y-axis to obtain the complete graph of f
in Figure 50.

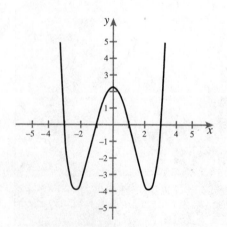

FIGURE 48

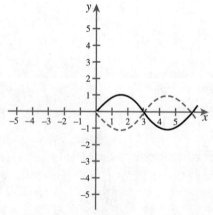

FIGURE 49

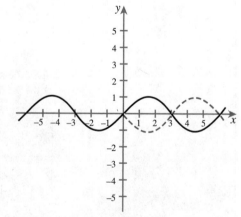

FIGURE 50

INCREASING AND DECREASING FUNCTIONS

In addition to showing symmetry, graphs can be used to unveil many other important characteristics of functions. Consider, for example, the function $f(x) = x^3 - 3x^2$, graphed in Figure 51. Notice that as we move from left to right, the graph rises until

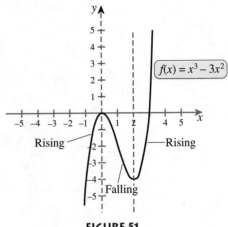

FIGURE 51

$x = 0$, falls in the interval from $x = 0$ to $x = 2$, and then rises to the right of $x = 2$. We say that the function f is *increasing* in the interval $(-\infty, 0)$, *decreasing* in the interval $(0, 2)$, and *increasing* in the interval $(2, \infty)$. The points $(0, 0)$ and $(2, -4)$, where the graph of f changes direction, are called *turning points*. More generally, we make the following definitions.

Increasing, Decreasing, and Turning Points □

A function f is said to be **increasing** or **decreasing** on an interval I according to the following properties.

Type of behavior	Graphical property	Algebraic property
Increasing	As we move from left to right in the interval I, the graph rises.	For any x_1 and x_2 in the interval I with $x_1 < x_2$, we have $f(x_1) < f(x_2)$.
Decreasing	As we move from left to right in the interval I, the graph falls.	For any x_1 and x_2 in the interval I with $x_1 < x_2$, we have $f(x_1) > f(x_2)$.

The points where f changes from increasing to decreasing or decreasing to increasing are called **turning points.**

In the case where a function changes from increasing to decreasing, we hit a high point or peak of the graph. Although such a peak might not be the highest point on the entire graph, it is certainly the highest point nearby and so the function value is called a *local maximum*. Similarly, at a point where the function changes from decreasing to increasing, the graph bottoms out at a *local minimum*. We summarize these definitions as follows.

Local Maximum and Minimum □

The value of f at $x = c$ is called **local maximum** if it is the largest value of f near c. The value of f at $x = c$ is called a **local minimum** if it is the smallest value of f near c. In Figure 52 we have identified points where a function f has local maximum or minimum values.

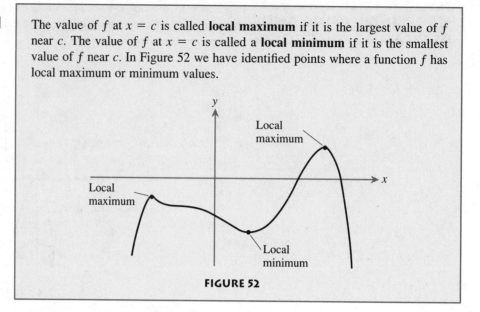

FIGURE 52

E X A M P L E 8 ▨ **Using a Graphing Calculator to Find Turning Points**

Use a graphing calculator to approximate the locations of the turning points for the function $f(x) = \frac{1}{8}x^3 - x^2 + 2$. Also, indicate the intervals on which f is increasing and the intervals on which it is decreasing.

SOLUTION   The graph of $y = \frac{1}{8}x^3 - x^2 + 2$, shown in Figure 53, has two turning points: a local maximum at $x = 0$ and a local minimum at a point in the fourth quadrant. Since $f(0) = 2$, there is a local maximum of 2 at $x = 0$. Using the trace feature, as shown in Figure 54, we find that f has a local minimum of approximately -7.48 near $x = 5.37$.

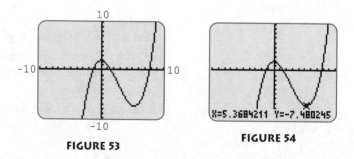

FIGURE 53

FIGURE 54

Since the graph of f is rising for x less than 0 and also for values of x greater than 5.37, f is increasing on $(-\infty, 0) \cup (5.37, \infty)$. Since the graph of f is falling between $x = 0$ and $x = 5.37$, f is decreasing on $(0, 5.37)$.

In Example 8, we were able to quickly obtain rough approximations for the co-ordinates of turning points using the trace feature of a graphing calculator. As one might guess, more accurate approximations can be found by zooming in. Alternatively, we can accurately estimate the coordinates of turning points in a more direct and efficient manner using a built-in feature of our graphing calculator.

CALCULATOR KEYS

Turning Points

Most graphing calculators are capable of "automatically" calculating the coordinates of turning points. With one popular model, for example, a turning point can be found by plotting the graph with a view that clearly shows the turning point, selecting the **calculate-minimum** or **calculate-maximum** feature (depending on whether the turning point corresponds to a local minimum or maximum), and specifying a **left bound** (a point to the left of the turning point), a **right bound** (a point to the right of the turning point), and an **initial guess** (a rough approximation for the turning point, one that is located between the two bounds). The bounds and initial guess can be given either by positioning the trace cursor or by entering a number for the x-coordinate. Figures 55–58 illustrate the use of the calculate-minimum feature for estimating the coordinates of one turning point of $f(x) = \frac{1}{8}x^3 - x^2 + 2$ from Example 8.

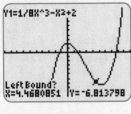

FIGURE 55

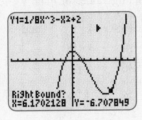

FIGURE 56

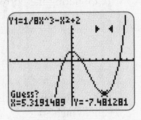

FIGURE 57

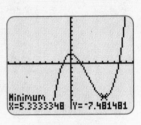

FIGURE 58

E X A M P L E 9 ■ **Maximizing Volume with a Graphing Calculator**

A rectangular package to be sent by the Postal Service can have a maximum combined length and girth (perimeter of the base) of 108 inches (see Figure 59). What are the dimensions of a box with a square base and with the largest possible volume that exactly meets this restriction?

SOLUTION The volume of a box with square base and length l is

$$V = x^2 l$$

To maximize volume, we let the sum of the length and girth be the largest allowable value. Thus, $l + 4x = 108$, or

$$l = 108 - 4x$$

Substituting $l = 108 - 4x$ into $V = x^2 l$, we obtain

$$V = x^2(108 - 4x)$$

Since neither x nor l can be negative, we have $x \geq 0$ and $108 - 4x \geq 0$, from which it follows that $x \leq 27$. Thus, we are interested in the value of x between 0 and 27 that makes $V = x^2(108 - 4x)$ as large as possible. We set the window variables xmin and xmax to 0 and 27, respectively. After some experimentation, we settle on values for ymin and ymax of 0 and 12,000, respectively. A plot of $y = x^2(108 - 4x)$ with this viewing window is shown in Figure 60. If only a rough approximation is required, we

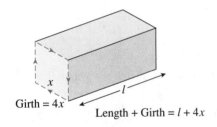

Girth = $4x$

Length + Girth = $l + 4x$

FIGURE 59

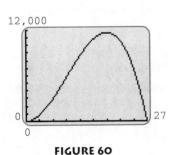

FIGURE 60

can simply use the trace feature, as shown in Figure 61. But if greater accuracy is desired, we would use the calculate-maximum feature, as shown in Figure 62. In either case, our work suggests that the maximum occurs at approximately $x = 18$.

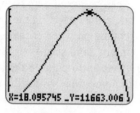

FIGURE 61

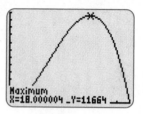

FIGURE 62

Computing the corresponding values of l and V, we find that

$$l = 108 - 4x \qquad \text{and} \qquad V = x^2 l$$
$$= 108 - 4 \cdot 18 \qquad\qquad\qquad = 18^2 \cdot 36$$
$$= 36 \qquad\qquad\qquad\qquad = 11{,}664$$

Thus, the box should be $18'' \times 18'' \times 36''$, which will result in a volume of 11,664 cubic inches.

UNDERSTANDING AND MASTERY CHECKLISTS

CONCEPTS TO UNDERSTAND

- ☐ Zeros of functions
- ☐ Translations
- ☐ Reflections about the axes and the origin
- ☐ Symmetry
- ☐ Even and odd functions
- ☐ Increasing and decreasing functions
- ☐ Turning points
- ☐ Local maximum and minimum

SKILLS TO MASTER

- ☐ Approximate zeros graphically.
- ☐ Given the graph of a function, sketch the graph of its translation.
- ☐ Given the graph of a function, sketch the graph of its reflection.
- ☐ Given a formula for $f(x)$ and the graphs of f and its translation g, find an equation for $g(x)$.
- ☐ Given a formula for $f(x)$ and the graphs of f and its reflection g, find an equation for $g(x)$.
- ☐ Given a function, determine the symmetries of its graph.
- ☐ Use symmetry as an aid in graphing a function.
- ☐ Determine whether a function is even, odd, or neither.
- ☐ Use a graphing calculator to approximate the intervals on which a graph is increasing or decreasing, and find its turning points.

EXERCISES 3.3

EXERCISES 1–10   *Find the exact values of all the real zeros of the given function.*

1. $f(x) = 3x + 1$

2. $p(s) = 4s + 6$

3. $g(x) = x^2 - 2x - 8$

4. $x(t) = t^2 - 9$

5. $h(x) = |2x + 5|$

6. $f(x) = \sqrt{3x - 9}$

7. $f(x) = \dfrac{x^2 - 4}{x^{345} + 345x^{23} - 172{,}896}$

8. $g(x) = |2x - 8| - 2$

9. $f(a) = 2 - \dfrac{6}{a}$

10. $h(y) = 9 - \sqrt{y}$

EXERCISES 11–16   *Use a graphing calculator to estimate the zeros of the given function to the nearest hundredth.*

11. $f(x) = 2x^3 - 3x^2 - 20x + 2$

12. $h(x) = x^3 - 2x^2 - 14x + 30$

13. $f(x) = \dfrac{15}{x^2 + 1} - 5x - 14$

14. $g(x) = x + 8 - \dfrac{5x + 88}{x^2 + 1}$

15. $f(x) = \sqrt{2x + 5} - \sqrt{x^2 + 1}$

16. $g(x) = \sqrt{4x + 6} - x - 2$

EXERCISES 17–20   *A function f and its graph are given. For each translated or reflected graph, determine the corresponding function g.*

17. $f(x) = |x|$

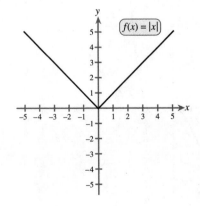

a.

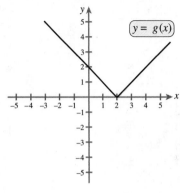

b.

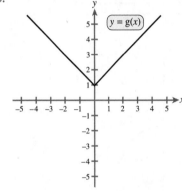

c.

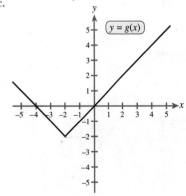

18. $f(x) = \sqrt{9 - x^2}$

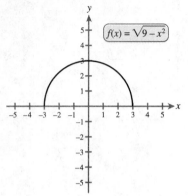

$f(x) = \sqrt{9 - x^2}$

a.

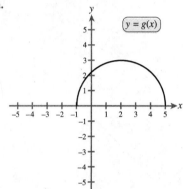

$y = g(x)$

b.

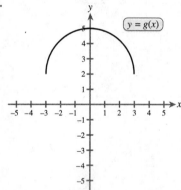

$y = g(x)$

c.

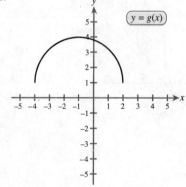

$y = g(x)$

19. $f(x) = \frac{1}{4}x^4 - \frac{1}{3}x^3 - x^2$

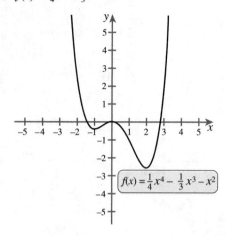

$f(x) = \frac{1}{4}x^4 - \frac{1}{3}x^3 - x^2$

a.

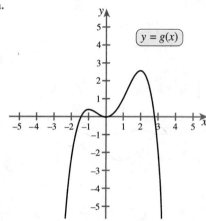

$y = g(x)$

b.

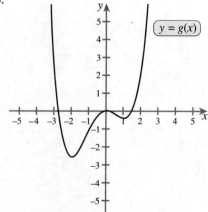

$y = g(x)$

20. $f(x) = x + 1 + \dfrac{6}{x^2 + 1}$

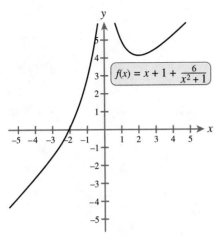

$$f(x) = x + 1 + \dfrac{6}{x^2 + 1}$$

a.

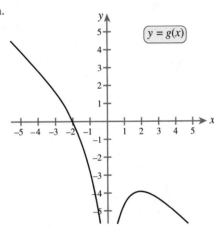

$y = g(x)$

b.

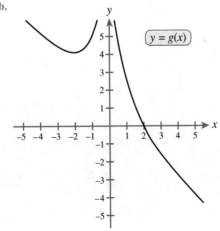

$y = g(x)$

EXERCISES 21–24 *Use the graph of the function f to sketch the graph of the function h.*

21.

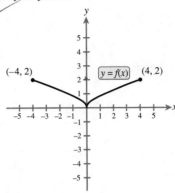

$(-4, 2)$ $y = f(x)$ $(4, 2)$

a. $h(x) = f(x - 2)$
b. $h(x) = f(x) + 3$
c. $h(x) = f(x + 1) - 4$
d. $h(x) = f(-x)$
e. $h(x) = -f(x) + 2$
f. $h(x) = -f(x - 1)$

22.

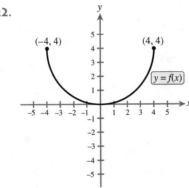

$(-4, 4)$ $(4, 4)$ $y = f(x)$

a. $h(x) = f(x) - 2$
b. $h(x) = f(x + 3)$
c. $h(x) = f(x + 1) - 4$
d. $h(x) = f(-x)$
e. $h(x) = -f(x) + 2$
f. $h(x) = -f(x - 1)$

23.

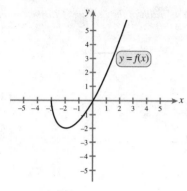

a. $h(x) = f(x - 3) + 2$

b. $h(x) = f(-x) - 1$

c. $h(x) = -f(x + 1)$

d. $h(x) = f(-x + 2) - 3$

24.

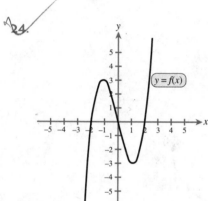

a. $h(x) = f(x + 2) - 3$

b. $h(x) = f(-x) - 1$

c. $h(x) = -f(x + 1)$

d. $h(x) = -f(x - 2) + 3$

EXERCISES 25–30   *Determine whether the given function is even, odd, or neither.*

25. $f(x) = x^4 - 4x^2$

26. $f(x) = 2x^3 - 8x + 1$

27. $f(x) = x\sqrt{x^2 + 1}$

28. $f(x) = \dfrac{1}{x^2 + 1}$

29. $f(x) = x^4 - \dfrac{1}{12}x$

30. $f(x) = |x + 2| - |x - 2|$

EXERCISES 31–34   *Complete the graph of the function by assuming that the function is (a) even and (b) odd.*

31.

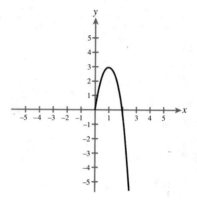

32.

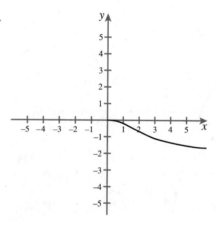

33.

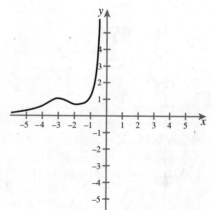

34.

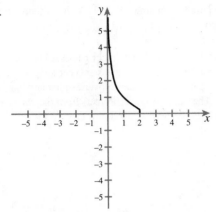

EXERCISES 35–42 *Use a graphing calculator to approximate the intervals on which the given function is increasing, the intervals on which it is decreasing, and the coordinates of any turning points. Classify each turning point as a local maximum or local minimum.*

35. $f(x) = x^2 - 4x + 5$ **36.** $g(x) = -x^2 - 6x - 5$

37. $h(x) = 2x^3 - 3x^2 - 6x + 5$

38. $p(x) = x^3 - 4x^2 + x + 1$

39. $q(x) = \dfrac{5}{x^2 - 4x + 5}$ **40.** $r(x) = \dfrac{-3}{x^2 - 6x + 10}$

41. $f(x) = x^4 - 3x^2 + x$ **42.** $g(x) = -2x^4 + 5x^2 + 7$

■ APPLICATIONS

43. *Translating a Population Function* Suppose a certain state's deer population can be approximated with the function $P(x) = x^2 + 14x + 100$, where x is the year ($x = 0$ corresponds to 1985) and $P(x)$ is the population (in hundreds) for that year. Use a horizontal shift to find a function $Q(x)$ that gives the same population approximations but for which $x = 0$ corresponds to 1990.

44. *Translating a Speed Record Function* The 1-mile automobile speed records for the years 1906–1939 can be approximated with the function $f(x) = 0.29x^2 - 2.7x + 134$, where x is the year ($x = 0$ corresponds to 1906) and $f(x)$ is the speed record in miles per hour. Use a horizontal shift to find a function $g(x)$ that gives the same speed approximations but for which $x = 0$ corresponds to 1910.

45. *Translating a Height Function* The height of a ball thrown from ground level with an initial velocity of 80 feet per second is given by $f(t) = -16t^2 + 80t$, where $f(t)$ is the height in feet and t is the time in seconds. Find a function $g(t)$ that gives the height of the ball at time t if it is thrown with the same initial velocity from the top of a 20-foot building. How do the graphs of f and g compare?

46. *Translating a Cost Function* A clothing company has daily production costs given by $C(x) = 20{,}000 - 12x + 0.05x^2$, where $C(x)$ is the total cost in dollars and x is the number of units produced. Because of a new labor contract, the daily cost is expected to rise $2000, independent of the number of units produced. Find the new cost function. How does its graph compare to the original function?

47. *Maximum Profit* An electronics company has determined that the cost for producing x security systems per month is $C(x) = 50x + 4050$ and the revenue from the sale of x systems per month is $R(x) = 200x - 0.5x^2$. How many systems must be sold per

month to break even (that is, for the profit to be zero)? Determine the number of systems that must be sold each month for the profit to be as large as possible.

48. *Maximum Height* A ball is thrown up into the air from the top of a 112-foot building with an initial velocity of 96 feet per second. Its height (in feet) after t seconds is given by $s(t) = -16t^2 + 96t + 112$. Find the maximum height of the ball. When does the ball hit the ground?

49. *Maximum Area* A point $P(x, y)$ lies on the line $y = -\frac{3}{4}x + 3$, as shown in Figure 63. Find the location of P so that the area of the shaded rectangle is as large as possible. (*Hint:* First find a function that expresses the area in terms of x.)

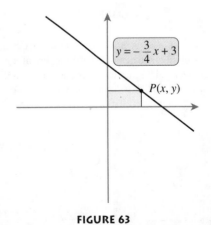

$y = -\dfrac{3}{4}x + 3$

$P(x, y)$

FIGURE 63

50. *Maximum Area* A line passing through the point (2, 1) has intercepts $(a, 0)$ and $(0, b)$. Find the values of a and b so that the

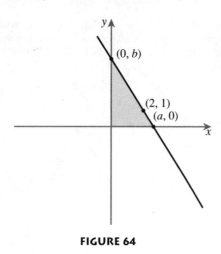

FIGURE 64

area of the shaded triangle in Figure 64 is as small as possible. (*Hint:* First find a function that expresses the area in terms of *a*.)

51. *Maximum Area*   A rancher wishes to enclose a rectangular pasture at the base of a cliff, as shown in Figure 65: 200 linear feet of fencing are available and no fencing is needed along the cliff side of the pasture. Find the dimensions of the pasture that will have the largest possible area. (*Hint:* First find a function that expresses the area in terms of *x*.)

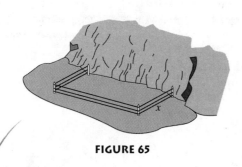

FIGURE 65

52. *Maximum Volume*   A box with a square base and no top is formed by cutting squares out of the corners of a piece of cardboard and then folding up the sides (see Figure 66). If the cardboard measures 16″ × 16″, find the dimensions of the box that

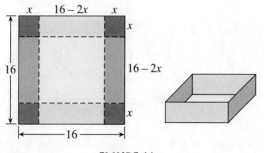

FIGURE 66

has the largest possible volume. (*Hint:* First find a function that expresses the volume in terms of *x*.)

53. *Water Bottle Handoff*   A bicycle racer passes a checkpoint at a constant rate of 33 ft/sec. At that instant, her support car accelerates from a dead stop in order to overtake her and hand her a water bottle. The car's distance (in feet) from the checkpoint after *t* seconds is given by the function $f(t) = \frac{4}{33}(33t^2 - t^3)$ for *t* from 0 to 33.

a. Plot the graph of *f* and find the car's maximum distance from the checkpoint.

b. Find a function *g*(*t*) that gives the racer's distance from the checkpoint after *t* seconds.

c. Determine when the racer and support car meet. This can be done algebraically, but you may wish to plot the graphs of *f* and *g* as well. How does this time compare to the time at which the support car is furthest from the checkpoint? Does this outcome seem reasonable? Explain.

d. Calculus can be used to show that the velocity of the support car as a function of *t* is given by $v(t) = \frac{4}{11}(22t - t^2)$, where *t* is again measured in seconds and the velocity is in feet per second. Find the velocity of the support car at the instant the racer and car meet. Is your answer a surprise? Why or why not?

54. *Traffic Flow*   A model for estimating traffic flow on a certain two-lane highway is given by the function $f(x) = -0.000111x^3 - 0.118x^2 + 65x$, where *x* denotes the number of vehicles per kilometer (traffic density) and $f(x)$ is the corresponding number of cars passing a given point on the road in 1 hour (traffic flow). For example, when the traffic density is 100 vehicles per kilometer, the traffic flow is $f(100) \approx 5209$ vehicles per hour.

a. Estimate the traffic flow when the traffic density is 150 vehicles per kilometer.

b. Estimate the traffic density when the traffic flow is measured to be 3000 vehicles per hour.

c. For what traffic density will traffic come to a stop? (*Hint:* For what value of *x* will $f(x) = 0$?)

d. For what traffic density is the traffic flow greatest? What is the greatest possible traffic flow?

[*Data source:* Neville D. Fowkes and John J. Mahony, *An Introduction to Mathematical Modeling* (Chichester, England: Wiley, 1994), p. 390.]

55. *Modeling CO₂ Concentration*   The atmospheric concentration of CO_2 has shown an overall upward trend in recent years, but it also oscillates quite predictably during the course of each year. For May 1994 through May 1996, CO_2 levels recorded at the Mauna Loa Observatory in Hawaii can be approximated by the function

$$f(x) = -35.9x^6 + 215.6x^5 - 473.3x^4$$
$$+ 455x^3 - 173x^2 + 13x + 361.45$$

where x denotes the year ($x = 0$ corresponds to May 1994) and $f(x)$ is the concentration of CO_2 in parts per million. Find the turning points of f. At approximately what time of the year is the concentration the lowest? At what time is it the highest?

CO₂ Concentration 1994–1996

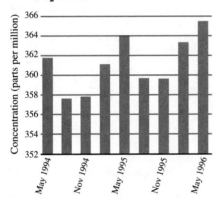

Data source: The Carbon Dioxide Information Analysis Center, U.S. Department of Energy.

56. *Global Temperature*   Long-term trends in average global temperature during the years 1985–1997 can be modeled with the function

$$f(x) = 0.0022x^3 - 0.041x^2 + 0.24x + 61.7$$

where $f(x)$ represents the average temperature (measured in degrees Fahrenheit) and x denotes the year (with $x = 0$ corresponding to 1985). According to this model, during which years was global temperature rising? When was it falling? This model doesn't reflect short-term fluctuations brought about by atmospheric events. Which of the following occurrences might be responsible for the fluctuations suggested by the data depicted in the bar chart?

Occurrence	Date
Chernobyl meltdown	April 26, 1986
Yellowstone fire	Summer 1988
Exxon Valdez oil spill	March 24, 1989
San Francisco earthquake	October 17, 1989
Mt. Pinatubo eruption	June 15, 1991
El Niño	1991–1995, 1997
Hurricane Andrew	August 1992
Peruvian tsunami	February 21, 1996

Global Temperature 1985–1997

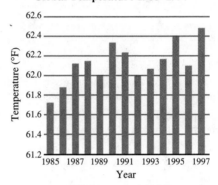

Data source: U.K. Meteorological Office.

57. *Modeling Gold Reserves*   Gold reserves in the United States for the years 1973–1989 can be modeled by the function

$$f(x) = -0.0026x^4 + 0.099x^3 - 1.14x^2 + 2.85x + 275$$

where x denotes the year ($x = 0$ corresponds to 1973) and $f(x)$ is the gold reserve for that year measured in millions of fine troy ounces. Find the intervals on which f is increasing and those on which it is decreasing. According to this model, for what year was the gold reserve highest?

U.S. Gold Reserves 1973–1989

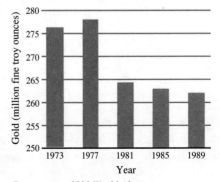

Data source: 1991 World Almanac.

58. *Modeling Waste Production*   The average number of pounds of waste produced each day by each person in the United States for the years 1980–1995 can be approximated with the polynomial function

$$f(x) = -0.0019x^3 + 0.04x^2 - 0.13x + 3.7$$

where x denotes the year ($x = 0$ corresponds to 1980) and $f(x)$ is the number of pounds of waste. Determine the time period(s) during which waste production was rising.

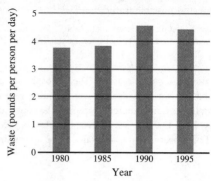

Data source: U.S. Environmental Protection Agency.

CONCEPTS AND CRITICAL THINKING

EXERCISES 59–62   *Answer true or false.*

59. If the graph of a function f lies entirely in the third quadrant, then 2 is in neither the domain nor the range of f.

60. If the graph of a function f crosses the y-axis once, then we can conclude that f has exactly one zero.

61. If a function f satisfies $f(-x) = f(x)$ for all x, then its graph will be symmetric with respect to the y-axis.

62. If h is defined by $h(x) = f(x - 2)$, then the graph of h will be a translation, 2 units to the left, of the graph of f.

EXERCISES 63–66   *Give an example of each.*

63. A function for which 2 and -2 are zeros

64. A function with a local maximum of 5 occurring at $x = 0$

65. An even function, the graph of which is a line

66. An odd function, the graph of which is a line

67. Suppose that f is an even function (so that the graph of f is symmetric with respect to the y-axis) and $g(x) = f(x - 3)$. What can be said about the graph of g? If f is odd, what can be said about the graph of g?

68. Show that if an odd function is defined at $x = 0$, then the graph of the function must pass through the origin. Explain why this fact is not contradicted by the given graph of an odd function.

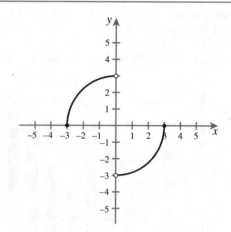

69. If n is even, show that $f(x) = x^n$ is an even function. If n is odd, show that f is an odd function.

70. Generalize Exercise 69 to show that if $f(x)$ is a polynomial, then f is even if all the powers of x are even, and f is odd if all the powers of x are odd.

71. An even function is symmetric with respect to the y-axis. An odd function is symmetric with respect to the origin. Why isn't there a term for a function that is symmetric with respect to the x-axis?

QUESTIONS FOR DISCUSSION OR ESSAY

72. We have observed that polynomial functions with all even powers of x are even and polynomial functions with all odd powers of x are odd. Why can't the functional notions of even and odd be defined this way instead of using the more abstract properties $f(-x) = f(x)$ and $f(-x) = -f(x)$?

73. To what extent are the graphical concepts of translation, reflection, and symmetry needed as tools for sketching graphs of functions? If you feel that they are no longer needed as graphing tools because of the graphing calculator, what purpose do they serve? If, on the other hand, you feel that they are still needed as graphing tools, when and how should they be used in conjunction with (or instead of) a graphing calculator?

74. When zooming in on turning points, it can happen that the graph will eventually appear as a horizontal line across the entire graphing calculator display. Experiment with zoom boxes of varying dimensions to see how this problem can be avoided. What is your conclusion as to the best dimensions of a zoom box for zooming in on turning points?

■ **PROJECTS FOR ENRICHMENT**

75. *Designing a Soft Drink Can*   An ordinary soft drink can has a volume of 355 cubic centimeters and, for the sake of simplicity, may be assumed to be a right circular cylinder. In this project, you will determine the dimensions of the cylindrical can that will provide the required volume with the least amount of aluminum.

   a. Write out the formula for computing the surface area of a cylindrical can in terms of its radius r and height h.

   b. Write out the formula for computing the volume of a cylindrical can in terms of its radius r and height h.

   c. Use the volume formula from part b and the fact that the volume must be 355 cubic centimeters to obtain an expression for h in terms of r. Substitute this expression into the surface area formula of part a. The result should be a function $S(r)$ that gives the surface area of a can of radius r with a volume of 355.

   d. Use a graphing calculator to approximate the value for r that gives the minimum value for $S(r)$. This is the radius of the can with minimum surface area. Find the height that corresponds to the radius you found.

   e. Approximate the actual radius and height of an ordinary soft drink can. How do the actual dimensions compare to the ones you found in part d? What explanation can you give for the discrepancy?

   f. Now suppose the top of the can is twice as thick as the sides and bottom. Find the dimensions of the can that will minimize the amount of aluminum in the can.

   g. Experiment with values for the thickness of the top (compared to the sides) until the dimensions of the can with a minimum amount of aluminum approximately coincide with the actual dimensions of a pop can. Does your value for the relative thickness of the top seem reasonable? Do you think this explains why the dimensions of a pop can are the way they are, or have other important factors still been ignored?

76. *Stretching and Shrinking a Function*   We have already seen how the graph of a function $y = f(x)$ is affected by the modifications $f(x + c)$, $f(x - c)$, $f(x) + c$, and $f(x) - c$. In this project, we investigate the effect of the modifications $cf(x)$ and $f(cx)$.

   a. Sketch the graph of $f(x) = x^2 + 1$.

   b. Sketch the graph of each of the following functions.

      i. $h(x) = 2(x^2 + 1)$ [that is, $h(x) = 2f(x)$]

      ii. $h(x) = 4(x^2 + 1)$

      iii. $h(x) = \frac{1}{2}(x^2 + 1)$

      iv. $h(x) = \frac{1}{4}(x^2 + 1)$

   c. Describe how each of the graphs in part b compare to the graph in part a. You may find it helpful to first consider what happens to several specific points on the graph of f. For example, the point $(2, 5)$ on the graph of f is "moved" away

from the x-axis to the point $(2, 10)$ on the graph in i of part b.

   d. Carefully write out a rule that describes how the graph of $h(x) = cf(x)$ can be obtained from the graph of a function f. You may find it convenient to consider two cases, one where $c > 1$ and the other where $0 < c < 1$. Explain how the case $c < 0$ can be dealt with by using your new rule together with a reflection.

   e. Try out your rule by sketching the graphs of $y = cf(x)$ for $c = 3$ and $c = \frac{1}{3}$ and the graph of the function f, shown in Figure 67.

   f. Sketch the graph of each of the following functions.

      i. $h(x) = (2x)^2 + 1$ [that is, $h(x) = f(2x)$]

      ii. $h(x) = (4x)^2 + 1$

      iii. $h(x) = (\frac{1}{2}x)^2 + 1$

      iv. $h(x) = (\frac{1}{4})^2 + 1$

   g. Describe how each of the graphs in part f compare to the graph in part a. You may find it helpful to first consider what happens to several specific points on the graph of f. For example, the point $(2, 5)$ on the graph of f is "moved" toward the y-axis to the point $(1, 5)$ on the graph in i of part f. In other words, the same y-coordinate is obtained from an x-coordinate that is half the distance to the y-axis.

   h. Carefully write out a rule that describes how the graph of $h(x) = f(cx)$ can be obtained from the graph of a function f. You may find it convenient to consider two cases, one where $c > 1$ and the other where $0 < c < 1$. Explain how the case $c < 0$ can be dealt with by using your new rule together with a reflection.

   i. Try out your rule by sketching the graphs of $y = f(cx)$ for $c = 3$ and $c = \frac{1}{3}$ and the graph of the function f shown in Figure 67.

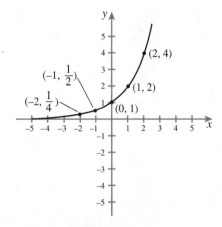

FIGURE 67

SECTION 3.4 **COMBINATIONS OF FUNCTIONS**

☐ How can the beating of a butterfly's wings in China create a tropical storm in Mexico?

☐ If the number of hot dogs consumed after t minutes by 135-pound hot-dog-eating champion Hirofumi Nakajima is $f(t)$, and his body weight after consuming n hot dogs is $g(n)$, then what does the function $g \circ f$ model?

☐ If f and g are functions and not just real numbers, what is meant by $f + g$, $f - g$, fg, or f/g?

☐ What do you get if you choose a number, square it, add 1, and then repeat this process (squaring and adding 1) 8 times?

ARITHMETIC COMBINATIONS OF FUNCTIONS

Just as numbers can be combined by the basic arithmetic operations of addition, subtraction, multiplication, and division, so too can functions. We can define the sum of two functions f and g by the formula $(f + g)(x) = f(x) + g(x)$. The operations of subtraction, multiplication, and division of functions are defined in a similar fashion, as follows.

☐ *Arithmetic Operations on Functions*

Operation	Definition
Addition	$(f + g)(x) = f(x) + g(x)$
Subtraction	$(f - g)(x) = f(x) - g(x)$
Multiplication	$(fg)(x) = f(x)g(x)$
Division	$\left(\dfrac{f}{g}\right)(x) = \dfrac{f(x)}{g(x)}$

EXAMPLE 1 ■ **Finding Combinations of Functions**

Let $f(x) = x^2 - 4$ and $g(x) = x + 2$. Compute each of the following.

a. $(f + g)(x)$ **b.** $(f - g)(x)$
c. $(fg)(x)$ **d.** $(f/g)(x)$

SOLUTION

a. $(f + g)(x) = f(x) + g(x)$
$\qquad\qquad = (x^2 - 4) + (x + 2)$
$\qquad\qquad = x^2 + x - 2$

b. $(f - g)(x) = f(x) - g(x)$
$\qquad\qquad = (x^2 - 4) - (x + 2)$
$\qquad\qquad = x^2 - x - 6$

c. $(fg)(x) = f(x)g(x)$
$$= (x^2 - 4)(x + 2)$$
$$= x^3 + 2x^2 - 4x - 8$$

d. $\left(\dfrac{f}{g}\right)(x) = \dfrac{f(x)}{g(x)}$
$$= \dfrac{x^2 - 4}{x + 2}$$
$$= \dfrac{(x - 2)(x + 2)}{x + 2}$$

It is tempting to simplify this expression to $x - 2$ by canceling the common factor $x + 2$. Note, however, that $x - 2$ is defined for all real numbers x, whereas

$$\frac{(x - 2)(x + 2)}{x + 2}$$

is undefined for $x = -2$. Thus, to be precise, we should express our answer as $(f/g)(x) = x - 2$ for $x \neq -2$.

EXAMPLE 2 ■ **Computing Combinations of Functions Defined Numerically**

Suppose that f and g are defined by Table 4. Compute $f + g$, fg, and f/g.

TABLE 4

x	$f(x)$	$g(x)$
0	−2	1
1	0	2
2	2	5
3	4	10
4	6	17

SOLUTION Since $(f + g)(x) = f(x) + g(x)$, we have, for example, $(f + g)(0) = f(0) + g(0) = -2 + 1 = -1$. In other words, the output values of $f + g$ are found simply by adding together the corresponding output values of f and g, as shown in Table 5. In a similar fashion, the output values of fg and f/g are found by multiplying or dividing the corresponding output values of f and g. See Table 5.

TABLE 5

x	$f(x)$	$g(x)$	$(f + g)(x)$	$(fg)(x)$	$(f/g)(x)$
0	−2	1	−1	$-2 \cdot 1 = -2$	$-2/1 = -2$
1	0	2	2	$0 \cdot 2 = 0$	$0/2 = 0$
2	2	5	7	$2 \cdot 5 = 10$	$2/5 = \frac{2}{5}$
3	4	10	14	$4 \cdot 10 = 40$	$4/10 = \frac{2}{5}$
4	6	17	23	$6 \cdot 17 = 102$	$6/17 = \frac{6}{17}$

In keeping with our convention of assuming that the domain (unless otherwise specified) is the set of real numbers for which the function "makes sense," the domain of $f + g$ is the set of real numbers x for which both $f(x)$ and $g(x)$ are defined. Thus, the domain of $f + g$ will consist of those real numbers that are in the domains of both f **and** g. Similarly, the domains of $f - g$ and fg consist of all real numbers that are contained in the domains of both f and g. However, since the expression $f(x)/g(x)$ is not defined when $g(x) = 0$, the domain of f/g consists of those real numbers x in the domains of both f and g such that $g(x) \neq 0$.

EXAMPLE 3 ■ Finding the Domains of Combinations of Functions

Let $f(x) = x - 2$, and $g(x) = \sqrt{x - 1}$. Find the domains of each of the following functions.

a. $(f + g)(x)$ **b.** $(f - g)(x)$

c. $(fg)(x)$ **d.** $(f/g)(x)$

e. $(g/f)(x)$

SOLUTION We begin by noting that $f(x) = x - 2$ is defined for all real numbers, whereas $g(x) = \sqrt{x - 1}$ is only defined for $x - 1 \geq 0$ or, equivalently, $x \geq 1$. Thus, the domain of f is $(-\infty, \infty)$ and the domain of g is $[1, \infty)$.

a–c. The domain of $f + g$, $f - g$ and fg is the set of real numbers x in the domains of both f and g. Thus, the domain of $f + g$, $f - g$, and fg is $[1, \infty)$.

d. As already noted, both f and g are defined on the interval $[1, \infty)$. However, since $g(1) = \sqrt{0} = 0$, 1 is excluded from the domain of f/g. Thus, the domain of f/g is $(1, \infty)$.

e. Since $f(2) = 2 - 2 = 0$, the number 2 is excluded from the domain of g/f. Thus, the domain of g/f consists of all real numbers in the interval $[1, \infty)$, with the exception of 2—that is, $[1, 2) \cup (2, \infty)$.

COMPOSITION OF FUNCTIONS

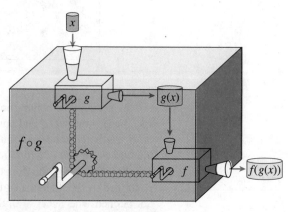

$f \circ g$

$x \xrightarrow{g} g(x) \xrightarrow{f} f(g(x))$

FIGURE 68

Functions can also be combined by *composing* or "stringing" them together—using the output of one function as the input of another. The **composition** of two functions f and g, written $f \circ g$, is defined by the formula $(f \circ g)(x) = f(g(x))$. The function $f \circ g$ takes an input value x and produces $f(g(x))$ as the corresponding output value, as shown in Figure 68.

We may also think of $f \circ g$ as the machine formed by first inputting x in the machine g, yielding the output $g(x)$, and then inputting $g(x)$ into the machine f, giving $f(g(x))$ as a final output, as shown in Figure 69. The function machine helps to illustrate what must occur in order for $(f \circ g)(x) = f(g(x))$ to be defined. First, $g(x)$ must be defined, so that x must be in the domain of g. Second, because we are using $g(x)$ as an input for f, $g(x)$ must be in the domain of f. Thus, we have the following formal definition of composition.

FIGURE 69

Definition of Composition □
of Functions

> The composition of the functions f and g, written $f \circ g$, is defined by
>
> $$(f \circ g)(x) = f(g(x))$$
>
> The domain of $f \circ g$ consists of those real numbers x in the domain of g such that $g(x)$ is in the domain of f.

E X A M P L E 4   ▧   **Computing the Composition of Two Functions**

Given that $f(x) = 2x^2 + 4x + 5$ and $g(x) = 2x + 1$, compute $f(g(x))$.

SOLUTION

$$
\begin{aligned}
(f \circ g)(x) &= f(g(x)) \\
&= f(2x + 1) &&\text{Replacing } g(x) \text{ with } 2x + 1 \\
&= 2(2x + 1)^2 + 4(2x + 1) + 5 &&\text{Substituting } 2x + 1 \text{ for } x \text{ in the formula for } f(x) \\
&= 2(4x^2 + 4x + 1) + 8x + 4 + 5 \\
&= 8x^2 + 16x + 11
\end{aligned}
$$

> **RULE OF THUMB**   When evaluating the composition of functions, begin on the inside and work your way out. That is, at each stage, evaluate the innermost function first.

E X A M P L E 5   ▧   **Computing Compositions of Functions**

Let

$$f(x) = \frac{x + 1}{x - 1} \quad \text{and} \quad g(x) = x^2$$

Compute each of the following compositions.

a. $(f \circ g)(x)$ **b.** $(g \circ f)(x)$

SOLUTION

a.
$$
\begin{aligned}
(f \circ g)(x) &= f(g(x)) \\
&= f(x^2) \\
&= \frac{x^2 + 1}{x^2 - 1}
\end{aligned}
$$

b.
$$
\begin{aligned}
(g \circ f)(x) &= g(f(x)) \\
&= g\left(\frac{x + 1}{x - 1}\right) \\
&= \left(\frac{x + 1}{x - 1}\right)^2 \\
&= \frac{(x + 1)^2}{(x - 1)^2}
\end{aligned}
$$

EXAMPLE 6   ■   Computing Compositions of Functions

Let $f(x) = \sqrt{x}$ and $g(x) = x^2 + 1$. Compute each of the following compositions and find their domains.

a. $(f \circ g)(x)$ **b.** $(g \circ f)(x)$

SOLUTION

a. $(f \circ g)(x) = f(g(x))$

$$= f(x^2 + 1)$$
$$= \sqrt{x^2 + 1}$$

Since $g(x)$ is defined for all real numbers and the expression under the radical, $x^2 + 1$, is always positive, the domain of $f \circ g$ is the set of all real numbers.

b. $(g \circ f)(x) = g(f(x))$

$$= g(\sqrt{x})$$
$$= (\sqrt{x})^2 + 1, \quad \text{for } x \geq 0$$
$$= x + 1, \quad \text{for } x \geq 0$$

Note that although $x + 1$ is defined for all real numbers x, $(g \circ f)(x)$ is not, since the inside function, $f(x) = \sqrt{x}$, is defined only for nonnegative x. Thus, the domain of $g \circ f$ is the set of all nonnegative real numbers.

> **WARNING!**   As the previous examples suggest, $f \circ g$ and $g \circ f$ are generally different functions. Thus, composition of functions is *noncommutative*—the order matters when composing functions.

EXAMPLE 7   ■   Computing Compositions of Functions Defined Numerically

Suppose that f and g are defined by Table 6. Compute $(f \circ g)(x)$ and $(g \circ f)(x)$ for $x = 0, 1, 2, 3, 4$.

SOLUTION   By definition, $(f \circ g)(x) = f(g(x))$. Thus, using the appropriate values from Table 6, we have

$$(f \circ g)(0) = f(g(0)) = f(1) = 0$$
$$(f \circ g)(1) = f(g(1)) = f(2) = 2$$

But

$$(f \circ g)(2) = f(g(2)) = f(5)$$

and $f(5)$ is not defined in Table 6. Thus, $(f \circ g)(2)$ is undefined. Similarly, $(f \circ g)(3)$ and $(f \circ g)(4)$ are undefined, as shown.

$$(f \circ g)(3) = f(g(3)) = f(10) = ?$$
$$(f \circ g)(4) = f(g(4)) = f(17) = ?$$

TABLE 6

x	$f(x)$	$g(x)$
0	−2	1
1	0	2
2	2	5
3	4	10
4	6	17

The computations for $g \circ f$ are much the same as those for $f \circ g$.

TABLE 7

x	$(f \circ g)(x)$	$(g \circ f)(x)$
0	0	undefined
1	2	1
2	undefined	5
3	undefined	17
4	undefined	undefined

$(g \circ f)(0) = g(f(0)) = g(-2) = ?$

$(g \circ f)(1) = g(f(1)) = g(0) = 1$

$(g \circ f)(2) = g(f(2)) = g(2) = 5$

$(g \circ f)(3) = g(f(3)) = g(4) = 17$

$(g \circ f)(4) = g(f(4)) = g(6) = ?$

Table 7 summarizes the values for $f \circ g$ and $g \circ f$.

DYNAMICAL SYSTEMS AND CHAOS THEORY

Suppose we enter a number into our calculator and repeatedly press the x^2 key. What will happen? The result depends on the value of the original number. If the original number is less than 1, then the result becomes smaller each time we press the x^2 key, so we get closer and closer to 0. On the other hand, if the original number is greater than 1, then the result becomes larger and larger as we repeatedly strike the x^2 key. This is an example of a **dynamical system:** A process is repeated over and over again, and the output from one step is the input for the next.

Our lives are tapestries of dynamical systems. The amount that we eat influences our metabolism, which affects the amount that we eat, which influences our metabolism. . . . Our personalities determine our friends, who influence our personalities, which determine our friends. . . . The amount that we sleep, our moods, our eating, drinking, and exercise habits are all bound together in an elaborate dynamical system. Every ecosystem is, in essence, a dynamical system. On every day in every acre of tropical rain forest, an intricate dance is played out among millions of competing, cooperating, and coexisting dynamical systems. Earth's atmosphere is an enormous dynamical system, the complexity of which has thus far vexed our most brilliant atmospheric scientists. In fact, one could argue that Earth's atmosphere is the second most complex dynamical system known, second only to what is perhaps the most complex structure in the known universe, the human brain. Our very thoughts are the product of the interactions and interconnections between literally billions of neurons. Synapses are being reinforced here and weakened there, leaving in their wake the stuff of which thoughts and fears, joy and sorrow are made.

In the last few decades, computers have provided us with a

A hurricane as photographed from a satellite.

lens through which dynamical systems can be viewed, the results of millions of iterations of certain processes played out on computer screens in a matter of minutes. These powerful instruments have illuminated largely unexplored mathematical territory and have given rise to the new science of **chaos,** which has caused nothing less than a paradigm shift in our understanding of the way in which dynamical systems unfold.

For an illustration of chaos, consider a billiard ball on a frictionless pool table with bumpers that absorb no energy. One would guess that the path of the billiard ball is completely determined by the initial velocity, position, and spin of the ball—the ball, without friction to slow it, will bounce around the table in a completely determined pattern. But in fact small, seemingly negligible factors will, after surprisingly few caroms, give rise to enormous variation. Surely one needn't account for the gravitational pull of Jupiter or the vibrations from a tree limb falling 20 miles away or the change in air pressure near the ball caused by the breeze from the hamster wheel at the house next door—but one must, if accuracy is desired after the first few caroms. Shockingly small changes in conditions can have enormous consequences.

This important aspect of chaos, extreme sensitivity to initial conditions, is often called **the butterfly effect.** The name arises from the degree to which atmospheric conditions can be chaotic; minor perturbations can become small eddies that can give rise to storms. It is said that the beating of a butterfly's wings at the right location at the right time can cause a tropical storm a thousand miles away a few days later. There is a certain degree of unpredictability in the weather that no computer, no matter how powerful, and that no measuring instruments, no matter how sensitive, will ever overcome.

UNDERSTANDING AND MASTERY CHECKLISTS

CONCEPTS TO UNDERSTAND

- ☐ Sum of functions
- ☐ Difference of functions
- ☐ Product of functions
- ☐ Quotient of functions
- ☐ Composition of functions

SKILLS TO MASTER

- ☐ Compute the sum, difference, product, and quotient of two given functions and find their respective domains.
- ☐ Evaluate arithmetic combinations of functions.
- ☐ Given two functions f and g, compute $(f \circ g)(x)$ and $(g \circ f)(x)$.
- ☐ Determine the domain of a function formed by composition.

EXERCISES   3.4

EXERCISES 1–12   *Find the following combinations. Specify the domain if it is anything other than all real numbers.*

a. $(f + g)(x)$

b. $(f - g)(x)$

c. $(fg)(x)$

d. $(f/g)(x)$

1. $f(x) = x, g(x) = 5$
2. $f(x) = 3x + 2, g(x) = x - 7$
3. $f(x) = x^2, g(x) = 3x + 1$
4. $f(x) = 2x + 5, g(x) = x^2 + 1$
5. $f(x) = x, g(x) = x$
6. $f(x) = \dfrac{2}{x}, g(x) = 3x$
7. $f(x) = 2x - 2, g(x) = \dfrac{2}{x + 5}$
8. $f(x) = \dfrac{1}{x}, g(x) = -\dfrac{1}{x}$
9. $f(x) = \sqrt{x - 2}, g(x) = x - 4$
10. $f(x) = 3x, g(x) = \sqrt{2x - 6}$

11.

x	$f(x)$	$g(x)$
0	7	0
1	0	4
2	-2	8

12.

x	$f(x)$	$g(x)$
1999	100	undefined
2000	0	10
2001	200	5

EXERCISES 13–30   *Compute both $(f \circ g)(x)$ and $(g \circ f)(x)$. Specify the domain if it is anything other than all real numbers.*

13. $f(x) = 2x + 3, g(x) = 4x - 5$
14. $f(x) = 3 - 2x, g(x) = \dfrac{3 - x}{2}$
15. $f(x) = x^2, g(x) = 2x + 7$
16. $f(x) = 3x - 4, g(x) = x^3$
17. $f(x) = \dfrac{3}{x}, g(x) = \dfrac{1}{3x}$
18. $f(x) = \dfrac{4}{x + 1}, g(x) = 2x + 4$
19. $f(x) = x + 1, g(x) = x^3 - 3x$
20. $f(x) = x^2 + 3x, g(x) = x - 5$
21. $f(x) = 7, g(x) = x^2 + 3x + 1$
22. $f(x) = \dfrac{1}{x}, g(x) = \dfrac{1}{x}$
23. $f(x) = \sqrt{x}, g(x) = x^4$
24. $f(x) = |x|, g(x) = 4x - 1$
25. $f(x) = 3x + 5, g(x) = \sqrt{x - 2}$
26. $f(x) = 2x + 3, g(x) = \sqrt{2x - 5}$
27. $f(x) = \dfrac{2}{x + 8}, g(x) = \sqrt[3]{\dfrac{2}{x}} - 8$
28. $f(x) = \dfrac{1}{x^2 - 9}, g(x) = 3$

29.

x	$f(x)$	$g(x)$
1	2	1
2	4	2
3	6	4
4	8	8

30.

x	f(x)	g(x)
0	10	5
5	15	10
10	20	15
15	10	0

EXERCISES 13–32 *Use the given graphs of f and g to complete the table.*

31.

x	f(x)	g(x)	(f ∘ g)(x)	(g ∘ f)(x)
0				
1				
2				

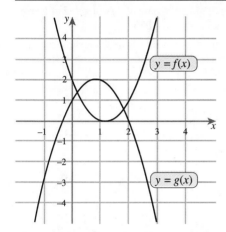

32.

x	f(x)	g(x)	(f ∘ g)(x)	(g ∘ f)(x)
0				
3				
6				
9				

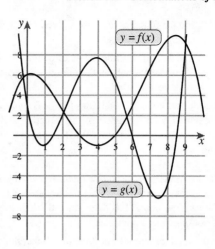

EXERCISES 33–38 *These exercises deal with the* **iterates** *of a function f, which are defined in the following way:*

$$f^1(x) = f(x)$$
$$f^2(x) = f(f(x))$$
$$f^3(x) = f(f(f(x)))$$
$$\vdots$$

Thus, the nth iterate of a function f, denoted $f^n(x)$, is found by composing f with itself n times. For example, if $f(x) = 3x$, then $f^2(x) = f(f(x)) = f(3x) = 3(3x) = 9x$.

33. If $f(x) = x + 5$, find $f^3(x)$.

34. If $f(x) = 2x + 1$, find $f^2(x)$.

35. If $f(x) = 2x$, find $f^4(x)$.

36. If $f(x) = x^2$, find $f^3(x)$.

37. If $f(x) = x^2$, then $f^n(a)$ can be found by entering a into a calculator and pressing the x^2 key (followed by the Enter key, if necessary) n times. Use this method to find $f^{10}(0.98)$ and $f^{10}(1.02)$. What happens to $f^n(0.98)$ and $f^n(1.02)$ as n gets larger?

38. If $f(x) = x^2 + 1$, then $f^n(a)$ can be found by entering a into a calculator, computing the square, adding 1, and then repeating the procedure n times. Compute $f^8(a)$ for the values of $a = -1$ and $a = 0.5$. What happens to $f^n(a)$ as n gets larger?

APPLICATIONS

39. *Refinery Production* An oil company owns two refineries. Over a 6-month period, the number of barrels of crude oil refined at each can be approximated with the functions $B_1(t) = 75t^2 - 450t + 1800$ and $B_2(t) = -85t^2 + 510t + 900$, where t is measured in months ($t = 1$ is the first month), and the function value for a given t is the production for that month, in thousands of barrels.

a. Graph both B_1 and B_2 and describe the production pattern for each refinery. What are the maximum and minimum production levels for each refinery for the 6-month period?

b. Write a function that gives the total production for the oil company over the 6-month period. Plot the graph of this function and describe the total production over this period. What are the maximum and minimum production levels?

40. *Company Sales* A company that markets office equipment has two salespeople, Pete and Tina. Over a 10-week period, Pete's sales can be approximated with the function $S_1(t) = -3t^2 + 36t + 92$, and Tina's sales can be approximated with $S_2(t) = 4t^2 - 40t + 200$. In both cases, t is measured in weeks ($t = 1$

is the first week), and the function value for a given t is the sales for that week, in hundreds of dollars.

a. Graph both S_1 and S_2 and describe the sales pattern for Pete and Tina. What are the maximum and minimum sales for both during the 10-week period?

b. Write a function that gives the total sales for the company for the 10-week period. Plot the graph of this function and describe the total sales over this period. What are the maximum and minimum sales?

41. *Hot Dog Contest* At the 1997 *Nathan's Famous Hot-Dog-Eating Contest,* 135-pound Hirofumi Nakajima defeated his arch rival, 6-foot-7-inch, 330-pound Ed "The Animal" Krachie, by consuming 24 hot dogs in 12 minutes.

a. Let $f(t)$ be the total number of hot dogs that Nakajima consumed after t minutes. Find an expression for $f(t)$. (Assume that Nakajima eats at a constant rate.)

b. Let $g(n)$ represent Nakajima's weight after consuming n hot dogs. Find an expression for $g(n)$. (You will need to estimate the weight of a hot dog and bun.)

c. Find expressions for $(f \circ g)(t)$ and $(g \circ f)(t)$.

d. Which of the functions, $f \circ g$ or $g \circ f$, is a real-world model? What does it model?

e. Nakajima's award-winning style consists of eating first the meat, and then the bread of each hot dog. What effect (if any) would this have on the model from part d?

42. *Homicides and Population Growth* The population of Indianapolis, Indiana, can be modeled by the function $g(t) = 0.04t + 7$, where t represents the year (with $t = 0$ corresponding to 1980), and $g(t)$ the population (in hundreds of thousands). On the other hand, an analysis of Bureau of Justice data reveals that the number of homicides in a city of a given population is approximately $f(P) = 25.25P + 15.998$, where P is the population of the city (in hundreds of thousands).

a. Find an expression for $(f \circ g)(t)$ and $(g \circ f)(P)$. Which of these two functions can be interpreted as a mathematical model? What does it model?

b. Use the result of part a to predict the number of homicides in Indianapolis in the year 2020.

c. The model used in part b rests on the assumption that population is the only factor that affects the homicide rate for a city. In reality, what other factors would come into play?

CONCEPTS AND CRITICAL THINKING

EXERCISES 43–46 *Answer true or false.*

43. $(f \circ g)(x) = f(x)g(x)$ for all real numbers x.

44. If $f(2) = g(2)$, then 2 is a zero of $f - g$.

45. If the domain of each of f and g is the set of all real numbers, then the domain of f/g is the set of all real numbers.

46. For any two functions f and g, $f \circ g = g \circ f$.

EXERCISES 47–50 *Give an example of each.*

47. A function f such that $f(x + c) = f(x)$, for all real numbers c

48. A pair of functions f and g such that $(f/g)(x) = x^2$

49. A pair of functions f and g such that $f \circ g = g \circ f$

50. A pair of functions f and g such that $(f \circ g)(x) = x$

51. Show that composition of functions is associative; that is, show that $f \circ (g \circ h) = (f \circ g) \circ h$.

52. It can be shown that

$$1 + x + x^2 + x^3 + \cdots + x^n = \frac{1 - x^{n+1}}{1 - x} \quad \text{for } x \neq 1$$

a. Find a formula for

$$1 + (x + 1) + (x + 1)^2 + (x + 1)^3 + \cdots + (x + 1)^n$$

b. Find a formula for $1 - x + x^2 - x^3 + \cdots + (-x)^n$.

53. Find a combination of rotations by 90° and reflections about the axes that yield a reflection about the line $y = x$.

54. Produce functions f and g such that $f \circ g$ is nowhere defined, even though g is defined everywhere. [*Hint:* Start by finding a function f that is defined only for $x > 0$. Then select a function g such that $g(x) < 0$.]

QUESTIONS FOR DISCUSSION OR ESSAY

55. If f and g are two linear functions, then which of the functions $f + g, f - g, fg, f/g, f \circ g$, and $g \circ f$ are guaranteed to be linear also? Justify your answer.

56. In what sense could an automobile manufacturing line be considered the composition of many functions? Give an example of

one of these factory functions and its domain and range. Find another multistep process that can be described in terms of composition of functions.

■ **PROJECTS FOR ENRICHMENT**

57. *Computing Iterates of a Function Graphically*   In Exercises 33–38, we defined the iterate $f^n(x)$ to be the function formed by composing f with itself n times. Suppose that for a given function f and a real number x_0, we wish to determine the behavior of $f^n(x_0)$ as n gets larger and larger. Does $f^n(x_0)$ jump around seemingly at random, does it converge to a single point, or does it bounce back and forth between two values? To answer such questions, we could simply compute $f(x_0)$, $f^2(x_0)$, $f^3(x_0)$, and so on until, if we are lucky, we discern a pattern. Computing $f^n(x_0)$ can be a time-consuming process, even with the aid of computers. There is, however, a way to use the graph of the function to estimate its iterates. For example, suppose that we are interested in computing iterates of the function $f(x) \doteq 2x(1 - x)$ with $x_0 = 0.2$. We begin by graphing the function f and also the line $y = x$. Next we locate $x_0 = 0.2$ on the x-axis and move vertically to locate the point $(0.2, f(0.2))$ on the graph. We then move horizontally until we hit the line $y = x$. We should now be at the point $(f(0.2), f(0.2))$. We then move vertically until we hit the graph of f; we are now at $(f(0.2), f(f(0.2)))$ or, more briefly, $(f(0.2), f^2(0.2))$. We again move horizontally until we hit the line $y = x$, which leaves us at $(f^2(0.2), f^2(0.2))$. We continue in this way—moving vertically until we hit the graph of the function, then moving horizontally until we hit the line $y = x$—until we have found the desired iterate. Figure 70 illustrates this procedure. It is evident from this figure that as n gets larger and larger, $f^n(0.2)$ approaches 0.5.

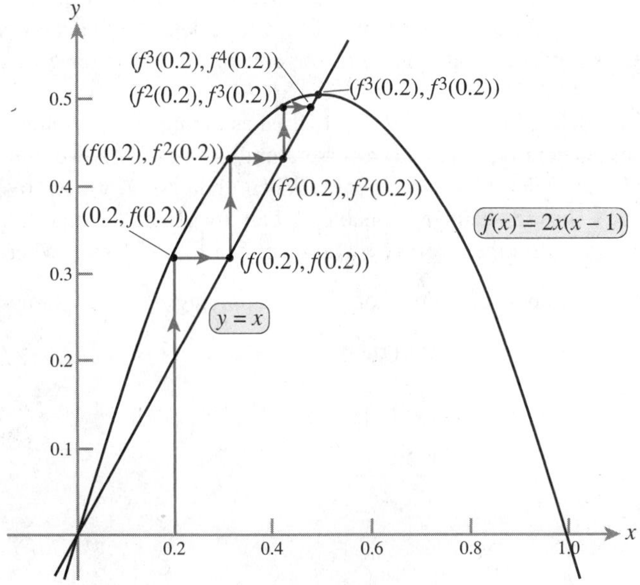

a. Let $f(x) = 2x(1 - x)$ as previously stated. Use a calculator to compute $f(0.2)$, $f^2(0.2)$, $f^3(0.2)$, and $f^4(0.2)$, confirming that these iterates indeed approach 0.5.

b. Use the graphical technique illustrated in Figure 70 to determine the first four iterates of f with $x_0 = 0.8$.

c. Now let $g(x) = 3.5x(1 - x)$. Use the graphical technique to determine the first 10 iterates of g with $x_0 = 0.3$. A sketch of the graph of g (along with the line $y = x$) is shown in Figure 71. Describe what happens to $g^n(x)$ as n gets larger and larger.

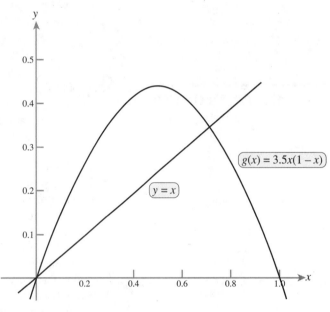

FIGURE 71

FIGURE 70

SECTION 3.5   INVERSES OF FUNCTIONS

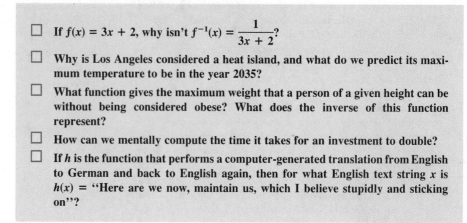

☐   If $f(x) = 3x + 2$, why isn't $f^{-1}(x) = \dfrac{1}{3x + 2}$?

☐   Why is Los Angeles considered a heat island, and what do we predict its maximum temperature to be in the year 2035?

☐   What function gives the maximum weight that a person of a given height can be without being considered obese? What does the inverse of this function represent?

☐   How can we mentally compute the time it takes for an investment to double?

☐   If h is the function that performs a computer-generated translation from English to German and back to English again, then for what English text string x is $h(x) = $ "Here are we now, maintain us, which I believe stupidly and sticking on"?

INVERSE FUNCTIONS

One could argue that a disturbing portion of our lives is spent performing actions that we ourselves then willingly undo. In the course of a single day, we dress and undress, walk up stairs that we soon will descend, drive 2 blocks north and 2 blocks south, stand up and sit down, crank up the volume and turn it down again. Actions and their opposites—tasks and their antitasks—much of what we do must later be undone if we wish to return to our original state. In the mathematical realm, the actions are functions: A function f takes a numerical input, processes it, and gives a numerical output. The inverse of f, denoted by f^{-1}, is a function that reverses this process, undoing the action of f. If f is a function that adds 2 to an input, then f^{-1} will subtract 2; if f multiplies the input by 8, then f^{-1} will divide by 8; and so forth. But just as there are actions in the real world that cannot be reversed, so too are there functions without inverses. In the remainder of this section, we formalize the notion of inverse, describe the class of functions that have inverses, and develop a method for computing the inverse of a function that has one.

Consider the function f defined by $f(x) = 2x + 1$. It takes the input, multiplies it by 2, and then adds 1. Now generally, when we reverse a multistep process, we undo actions in the reverse order in which they were done. (When undressing, we remove our shoes before our socks.) Thus, it seems reasonable that an inverse of f would first subtract 1, and then divide by 2. In other words, we are guessing that $f^{-1} = g$, where $g(x) = \dfrac{x - 1}{2}$. If g really does undo the action of f, then applying g to $f(x)$ should return us to x. In other words, we expect that $g(f(x)) = x$. In fact,

$$
\begin{aligned}
g(f(x)) &= g(2x + 1) \\
&= \frac{(2x + 1) - 1}{2} \\
&= \frac{2x}{2} \\
&= x
\end{aligned}
$$

As expected, g undid the action of f. Now let's see if f undoes the action of g by computing $f(g(x))$.

$$f(g(x)) = f\left(\frac{x-1}{2}\right)$$

$$= 2\left(\frac{x-1}{2}\right) + 1$$

$$= x - 1 + 1$$

$$= x$$

Our work here suggests the following, more formal definition of the inverse of a function.

Definition of f^{-1} □

> A function g satisfying $(g \circ f)(x) = x$ for all x in the domain of f and $(f \circ g)(x) = x$ for all x in the domain of g, is said to be the **inverse** of the function f. We write $g = f^{-1}$ or, equivalently, $f = g^{-1}$.

EXAMPLE 1 ▧ **Verification of Inverse Functions**

Show that the functions

$$f(x) = \frac{3x+2}{4} \qquad \text{and} \qquad g(x) = \frac{4x-2}{3}$$

are inverses of one another.

SOLUTION   We must show that $(f \circ g)(x) = (g \circ f)(x) = x$ for all x.

$(f \circ g)(x) = f(g(x))$

$$= f\left(\frac{4x-2}{3}\right)$$

$$= \frac{3\left(\frac{4x-2}{3}\right) + 2}{4}$$

$$= \frac{4x - 2 + 2}{4}$$

$$= \frac{4x}{4}$$

$$= x$$

$(g \circ f)(x) = g(f(x))$

$$= g\left(\frac{3x+2}{4}\right)$$

$$= \frac{4\left(\frac{3x+2}{4}\right) - 2}{3}$$

$$= \frac{3x + 2 - 2}{3}$$

$$= \frac{3x}{3}$$

$$= x$$

Thus, f and g are indeed inverses.

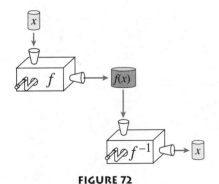

FIGURE 72

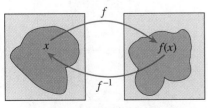

FIGURE 73

The inverse of a function is its opposite; it *undoes* what the function *does,* and vice versa, as illustrated in Figure 72. It is also useful to think of f and f^{-1} as mappings between sets, as depicted in Figure 73. Here we see that the domain of f is the range of f^{-1} and that the domain of f^{-1} is the range of f. In other words, the input values for f are the output values of f^{-1}, and vice versa.

The inverse of a function f can be viewed as that function which, when given an output of the function f, determines the input. However, it isn't always possible to determine the input to a function from the output, and for this reason not all functions have inverses. For example, consider the function $f(x) = x^2$. If it is given that the output of this function is 9, we cannot determine the input; it might have been either 3 or -3. Thus, the squaring process cannot be reversed: If the result of a squaring is a given positive number, then the original number cannot be specified with certainty. Thus, the function f doesn't have an inverse. In fact, the only functions that *do* have inverses are those for which distinct inputs yield distinct outputs. We define *one-to-one functions* to be such functions. More formally, we have the following definition.

Definition of ☐
One-to-One Functions

> If the function f has the property that $f(a) = f(b)$ only if $a = b$, then f is said to be **one-to-one,** or 1–1.

A function can easily be tested to see whether or not it is 1–1 by looking at its graph. If the function is 1–1, then for each y-value, there will be only one x-value. Thus, a line with equation $y = c$ (a horizontal line) will be crossed at most once by the graph of a 1–1 function. If a 1–1 function f were to cross $y = c$ two or more times, then there would be two or more distinct x-values a and b such that $f(a) = f(b) = c$, and f would, by definition, fail to be 1–1. This gives us the following test for 1–1 functions.

Horizontal Line Test ☐

> A function f is 1–1 if and only if no horizontal line crosses the graph of f more than once.

E X A M P L E 2 ▨   Testing to See Whether a Function is 1–1

Determine which of the following are the graphs of 1–1 functions.

a.
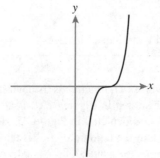

b.
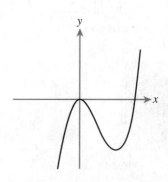

c.

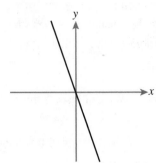

d.

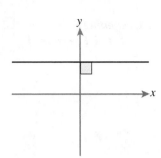

e.

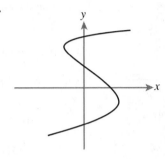

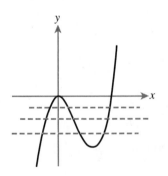

FIGURE 74

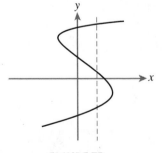

FIGURE 75

SOLUTION

a. No horizontal line crosses this graph more than once; the function is 1–1.

b. Several horizontal lines are shown in Figure 74 that cross the graph more than once; this function is not 1–1.

c. No horizontal line crosses this graph more than once; hence this function is 1–1.

d. The graph of this function is itself a horizontal line. Therefore, there is one horizontal line (the graph itself) that intersects the graph more than once. In fact, it intersects itself infinitely many times. This function is not 1–1.

e. Trick question! As we see in Figure 75, this graph fails the *vertical* line test, and so it is not even the graph of a function.

We have seen that if a function is 1–1, then it has an inverse. We now turn our attention to computing the inverse. Let f be a 1–1 function, and let $y = f(x)$. Since the inverse of a function returns the input to f when given the output, we can compute the inverse of f by solving the equation $y = f(x)$ for x. In fact, if we apply f^{-1} to both sides of the equation $y = f(x)$, we obtain

$$y = f(x)$$
$$f^{-1}(y) = f^{-1}(f(x)) \qquad \text{Applying } f^{-1} \text{ to both sides}$$
$$f^{-1}(y) = x \qquad \text{Simplifying by using the definition of } f^{-1}$$

Of course, once we know $f^{-1}(y)$, we can compute $f^{-1}(x)$ by simply substituting x for y in the formula for $f^{-1}(y)$. This suggests the following strategy for computing inverses.

Steps for Computing $f^{-1}(x)$ for   ☐
1–1 Functions f

> **1.** Solve the equation $y = f(x)$ for x to obtain $x = f^{-1}(y)$.
>
> **2.** Substitute x for y to find $f^{-1}(x)$.

The following examples illustrate this technique.

EXAMPLE 3   ▓   Computing the Inverse of a Linear Function

Find the inverse of the function $f(x) = 3x + 2$.

SOLUTION   We begin by solving the equation $y = 3x + 2$ for x.

$$y = 3x + 2$$

$$y - 2 = 3x \qquad \text{Subtracting 2 from both sides}$$

$$x = \frac{y - 2}{3} \qquad \text{Dividing both sides by 3 and exchanging the left and right sides}$$

Thus, we have

$$f^{-1}(y) = \frac{y - 2}{3}$$

Substituting x for y, we obtain

$$f^{-1}(x) = \frac{x - 2}{3}$$

EXAMPLE 4   ▓   Finding the Inverse of a Function Involving Radicals

Find $f^{-1}(x)$ if $f(x) = \sqrt[3]{2x - 7}$.

SOLUTION   We let $y = f(x)$ and solve for x in terms of y.

$$y = \sqrt[3]{2x - 7}$$

$$y^3 = (\sqrt[3]{2x - 7})^3 \qquad \text{Cubing both sides}$$

$$y^3 = 2x - 7$$

$$y^3 + 7 = 2x$$

$$2x = y^3 + 7$$

$$x = \frac{y^3 + 7}{2}$$

Thus, $f^{-1}(y) = \dfrac{y^3 + 7}{2}$, so that $f^{-1}(x) = \dfrac{x^3 + 7}{2}$.

GRAPHS OF INVERSE FUNCTIONS

Suppose that (a, b) is on the graph of $y = f(x)$. Then $b = f(a)$, from which it follows that $a = f^{-1}(b)$. Thus, (b, a) is a point on the graph of $y = f^{-1}(x)$. As Figure 76 suggests, the point (b, a) is the reflection of the point (a, b) about the line $y = x$. Thus, the graph of $y = f^{-1}(x)$ is the reflection of the graph of $y = f(x)$ about the line $y = x$.

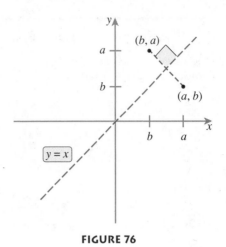

FIGURE 76

E X A M P L E 5   ▧   **Using the Graph of f to Graph f^{-1}**

The graph of a function f is given in Figure 77. Show that f^{-1} exists, and graph it.

SOLUTION   Since the graph passes the horizontal line test, f^{-1} exists. To graph f^{-1}, we simply reflect the graph of f about the line $y = x$, as shown in Figure 78.

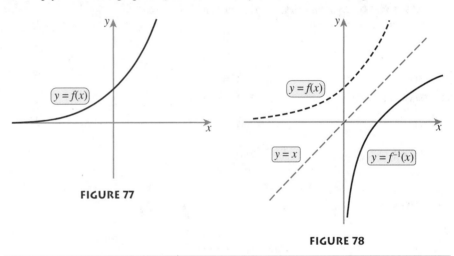

FIGURE 77

FIGURE 78

UNDERSTANDING AND MASTERY CHECKLISTS

CONCEPTS TO UNDERSTAND

☐ Inverse functions
☐ One-to-one functions
☐ Horizontal line test
☐ Graphs of inverse functions

SKILLS TO MASTER

☐ Verify or disprove that a given pair of functions are inverses of one another.
☐ Use the horizontal line test to determine whether a function is one-to-one.
☐ Compute the inverse of a simple function.
☐ Given the graph of a function, sketch its inverse.

EXERCISES 3.5

EXERCISES 1–10 *Determine whether the functions f and g are inverses of one another by evaluating $f(g(x))$ and $g(f(x))$.*

1. $f(x) = x + 2; g(x) = x - 2$

2. $f(x) = 3x; g(x) = \frac{x}{3}$

3. $f(x) = 2x + 1; g(x) = \frac{1}{2}x - 1$

4. $f(x) = \frac{1}{4}x - 3; g(x) = 4x + 12$

5. $f(x) = (x + 1)^3; g(x) = \sqrt[3]{x} - 1$

6. $f(x) = \sqrt[5]{x - 2}; g(x) = (x + 2)^5$

7. $f(x) = \dfrac{x + 3}{2x - 1}; g(x) = \dfrac{x + 3}{2x - 1}$

8. $f(x) = \dfrac{1}{x} - 4; g(x) = \dfrac{1}{x + 4}$

9. $f(x) = \sqrt{x - 6}; g(x) = x^2 + 6, x \geq 0$

10. $f(x) = 4 - x^2, x \geq 0; g(x) = \sqrt{x + 4}$

EXERCISES 11–20 *Determine whether the function is 1–1. Use a graphing calculator as necessary.*

11. $f(x) = -2x + 5$

12. $g(x) = \dfrac{x + 3}{4}$

13. $f(x) = x^2$

14. $f(x) = \sqrt[3]{x}$

15. $h(x) = x^3 + 2x + 1$

16. $f(x) = x^3 - x^2$

17. $g(x) = \dfrac{x^4}{12} - x^3$

18. $f(x) = x^5 + 1$

19. $h(x) = |x - 3|$

20. $f(x) = \sqrt{x + 2}$

EXERCISES 21–34 *Determine whether the function f is 1–1. If it is, find $f^{-1}(x)$.*

21. $f(x) = \frac{1}{3}x - 1$

22. $f(x) = \dfrac{-x + 3}{4}$

23. $f(x) = x^4$

24. $f(x) = x^3$

25. $f(x) = x^2 + 1, x \geq 0$

26. $f(x) = (x + 4)^2$

27. $f(x) = \sqrt{2x + 5}$

28. $f(x) = \sqrt{4 - x^2}$

29. $f(x) = \dfrac{x^2}{x^2 + 1}$

30. $f(x) = \dfrac{x}{x + 4}$

31.

x	$f(x)$
0	4
1	7
2	10
3	13
4	16

32.

x	$f(x)$
0	1
1	-1
2	-3
3	-1
4	1

33.

x	$f(x)$
-2	5
-1	1
0	0
1	1
3	10

34.

x	$f(x)$
-5	15
0	10
5	5
10	0
15	-5

EXERCISES 35–40 *Determine whether f^{-1} exists and, if so, sketch its graph.*

35.

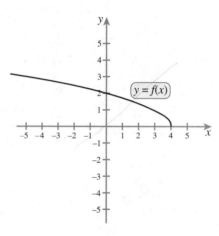

36.

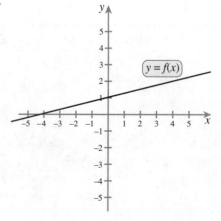

37.

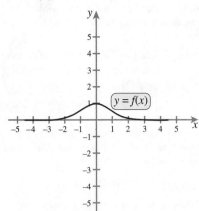

39.

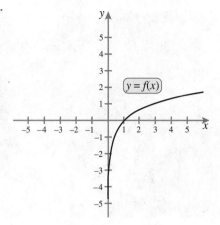

38.

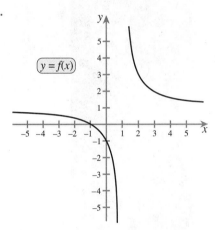

40.

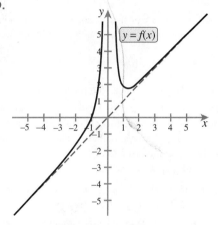

▦ APPLICATIONS

41. *Heat Islands*   Covered with heat-absorbing asphalt and stripped of cooling vegetation, large cities can be as much as 10°F warmer than the surrounding countryside. These heat islands wreak environmental and economic havoc by pumping up energy consumption and contributing to smog. As cities expand, the problem grows worse. In Los Angeles, for example, the yearly maximum temperature (computed as a 10-year running average) has risen from 98°F in 1935 to 104°F in 1985, an increase of 0.12°F per year.

a. Let $M(t)$ represent the yearly maximum temperature in Los Angeles in the year $1935 + t$. Find an expression for $M(t)$ by assuming that the yearly maximum temperature increases 0.12°F every year.

b. Use the result from part a to estimate the highest temperature in Los Angeles in 2035.

c. Find an expression for $M^{-1}(t)$.

d. Give an interpretation of $M^{-1}(t)$. In particular, what does $M^{-1}(120)$ represent?

42. *The Kelvin Scale*   The Kelvin temperature scale is commonly used in many of the sciences, including astronomy, chemistry, and physics. Essentially, the Kelvin temperature scale is a translated Celsius scale: A degree Kelvin is the same as a degree Celsius, with 0 Kelvin defined to be absolute zero (the coldest physically possible temperature). It can be shown that if K and F represent the temperature in degrees Kelvin and Fahrenheit, respectively, then

$$K = g(F) = \frac{5F + 2300}{9}$$

a. Find absolute zero in degrees Fahrenheit.

b. Interpret g^{-1}, and then find a formula for it.

c. According to data collected by the Mars Pathfinder mission, the temperature on Mars varies from 190 to 240 on the Kelvin

Martian landscape as photographed by the Mars Pathfinder.

scale. Find the corresponding temperature range in degrees Fahrenheit.

43. *Rule of 72* The time t (in years) it takes an investment to double if it is earning interest at an annual rate of $i\%$ is approximated by

$$t = f(i) = \frac{72}{i}$$

This formula is usually referred to as the rule of 72.

a. Some credit cards charge interest at rates of 18% and more. If no payments are made (which is generally not an option), find the time required for an initial charge of $400 to grow to $800, assuming an interest rate of 18%.

b. Interpret f^{-1}, and then find a formula for it.

c. Use your result from part b to estimate the interest rate for an investment that doubled in value after 8 years.

44. *Training Heart Rate* Tolerance to cardiovascular exertion depends on a variety of factors, including the subject's overall health and general level of aerobic fitness. Still, all other things being equal, younger people can tolerate higher pulse rates than can older people, so naturally the recommended training heart rate (THR) is adjusted for age. A general rule of thumb for the THR (in beats per minute) for a subject aged t years is given by

$$\text{THR} = h(t) = \frac{3(220 - t)}{4}$$

a. Find an appropriate training heart rate for a 20-year-old.

b. Interpret h^{-1}, and then find a formula for it.

c. Use your result from part b to estimate the age of a subject whose recommended training heart rate is 120 beats per minute.

d. Some health professionals consider formulas such as the one we have given for THR to be downright dangerous. Wherein lies the danger in applying this simple formula?

45. *Body Mass Index* Few mathematical models have created as much widespread controversy as those used to determine appropriate body weight ranges. One such model, based on the body mass index (BMI), dictates that the maximal healthy weight (in pounds) for a person h inches tall is given by

$$w = f(h) = 0.0355264h^2$$

a. When the 6′2″ Arnold Schwarzenegger competed as a bodybuilder, he sometimes weighed as much as 250 pounds. According to this formula, was he obese?

Arnold Schwarzenegger posing in competition.

b. What restriction must we place on the domain of f so that f will have an inverse? With this restriction in place, compute the inverse of f and give an interpretation.

c. When Walter Hudson died in 1991, he was estimated to have weighed 1400 pounds. Use your result from part b to estimate the minimum height for a healthy 1400-pounder.

d. List a few of the many factors not taken into account by our model f that are relevant to the determination of obesity.

CONCEPTS AND CRITICAL THINKING

EXERCISES 46–49 *Answer true or false.*

46. A function has an inverse if and only if it passes the vertical line test.

47. If $g(x) = \dfrac{1}{f(x)}$, then $g = f^{-1}$.

48. Only 1–1 functions have inverses.

49. If f is a 1–1 function, then so is $g(x) = f(x - 1)$.

EXERCISES 50–53 *Give an example of each.*

50. The graph of an equation that is not the graph of a function

51. The graph of a function that is not a 1–1 function

52. The graph of a 1–1 function having -3 as its only zero

53. A pair of functions f and g such that $f + g$ is 1–1, even though neither f nor g is 1–1

54. a. Simplify $f^{-1} \circ (f \circ g)$.

b. Suppose that f and $f \circ g$ are given. Explain how the result of part a could be used to find g.

c. Employ the technique you described in part b to find $g(x)$ if it is given that $(f \circ g)(x) = 6x + 21$ and $f(x) = 3x + 9$.

55. Define $(f \circ g \circ h)(x) = f(g(h(x)))$.

a. Show that $f^{-1} \circ f \circ p = p$, for any function p.

b. Show that $p \circ f \circ f^{-1} = p$, for any function p.

c. Suppose that the function f has *two* inverses, g and h. Use the result of part a to show that $g \circ f \circ h = h$.

d. Use the result of part b to show that $g \circ f \circ h = g$.

e. Use parts c and d to conclude that g and h are the same function and that a function can have at most one inverse.

56. An arbitrary linear function can be written as $f(x) = ax + b$. Compute f^{-1}.

57. Find a function f such that $f^{-1} = f$. (*Hint:* Start by looking for an operation that, when performed twice in a row, returns you to the original number.)

58. Which 1–1 functions have graphs that are symmetric with respect to the y-axis?

▨ QUESTIONS FOR DISCUSSION OR ESSAY

59. Discuss the analogy between functions and their inverses on the one hand, and numbers and their multiplicative inverses on the other.

60. What is the relationship between a function and its inverse with respect to the following properties: always positive, always negative, always increasing, always decreasing, 1–1, even, and odd? In other words, for example, if a function is always positive, then what can be said about its inverse?

61. The Web search engine AltaVista™ offers an online translation service. Given a textual document, it can translate from English to one of 5 languages, or it can translate from one of these languages to English. If we denote the AltaVista functions that translate from English to German and from German to English by f and g, respectively, then what are the domains and ranges of f and g? Would you expect f and g to be inverses of one another? Use the following facts to check.

• $(f \circ g)$("A rolling stone gathers no moss") = "A rolling rollenstein does not enter Moos."

• $(f \circ g)$("Row, row, row your boat gently down the stream") = "Series, series, rudders your boat easily down the current."

• $(f \circ g)$("Here we are now entertain us, I feel stupid and contagious") = "Here are we now, maintain us, which I believe stupidly and sticking on."

▨ PROJECTS FOR ENRICHMENT

62. *Solving Functional Equations*   If x, y, and z are real numbers, then the equation $xy = z$ can be solved for y by multiplying both sides by x^{-1}.

$$xy = z$$
$$x^{-1}xy = x^{-1}z$$
$$y = x^{-1}z$$

This strategy can be modified for solving certain **functional equations**, equations in which the unknown is a function. The basic idea is that instead of multiplying by the multiplicative inverse, we compose with the functional inverse. For example, if the functions f and h are known and f has an inverse, then the equation $f \circ g = h$ can be solved for g as follows:

$$f \circ g = h$$
$$f^{-1} \circ f \circ g = f^{-1} \circ h$$
$$g = f^{-1} \circ h$$

a. Find a function g such that $f \circ g = h$, where $f(x) = 2x - 4$ and $h(x) = \dfrac{2x - 10}{x - 2}$.

b. Find a function f such that $f \circ g = h$, where $g(x) = 2x - 1$ and $h(x) = \dfrac{10x - 2}{4x - 3}$.

c. Consider the functional equation $f \circ g = h$, with $f(x) = 3x + 4$, and $h(x) = \dfrac{7x + 1}{x + 1}$. The following is an *incorrect* solution of this functional equation.

$f \circ g = h$	The original equation
$f^{-1} \circ f \circ g = h \circ f^{-1}$	Composing with f^{-1} on both sides
$g = h \circ f^{-1}$	Simplifying
$g(x) = (h \circ f^{-1})(x)$	Evaluating both sides at x

$$g(x) = h(f^{-1}(x))$$

Using the definition of composition of functions

$$g(x) = h\left(\frac{x-4}{3}\right)$$

Substituting $\frac{x-4}{3}$ for $f^{-1}(x)$

$$= \frac{7\left(\frac{x-4}{3}\right) + 1}{\left(\frac{x-4}{3}\right) + 1}$$

Evaluating, using the fact that $h(x) = \frac{7x+1}{x+1}$

$$= \frac{7x - 25}{x - 1}$$

Simplifying

If we check our solution by evaluating $(f \circ g)(x)$, we find that

$$(f \circ g)(x) = f\left(\frac{7x - 25}{x - 1}\right)$$

$$= 3\left(\frac{7x - 25}{x - 1}\right) + 4$$

$$= \frac{25x - 79}{x - 1}$$

$$\neq h(x)$$

What went wrong?

SECTION 3.6 SELECTED FUNCTIONS

☐ **How long must one wait after drinking alcoholic beverages to ensure that the blood alcohol level falls below the legal limit?**

☐ **Will a baseball hit at 144 feet per second (about 98 miles per hour) at an angle of 60° hit the ceiling of the Houston Astrodome?**

☐ **How can a single function be used to compute federal income tax even when there are several tax brackets?**

☐ **How can you quickly mentally compute the day of the week for any date in any century?**

To fully exploit the power of functions for modeling the real world, it is not sufficient to merely master general concepts such as notation, graphs, combinations, and inverses. Instead, we must also gain familiarity with a few particularly useful classes of frequently occurring functions. In this section, we discuss linear, quadratic, and piecewise-defined functions; in the next two chapters, we will tackle polynomial, rational, exponential, and logarithmic functions. As we begin this functional odyssey, pay special attention to the key features that distinguish one species of function from another.

LINEAR FUNCTIONS

Recall that the graph of an equation of the form $y = mx + b$ is a line with slope m and y-intercept b. In fact, any nonvertical line can be put in this form, and it is for this reason that we define a **linear function** to be a function of the form $f(x) = mx + b$. In the case where $m = 0$, we obtain the **constant function** $f(x) = b$. Linear functions are certainly the easiest to study, and they are also among the most useful for applications.

EXAMPLE 1 ■ Linear Cost and Revenue Functions

A sidewalk hot dog vendor has a fixed daily cost of $60 for the rental of a cart and the vendor permit. The variable costs (for the hot dogs, buns, and condiments) average 30¢ per hot dog. Each hot dog sells for $1.50.

a. Construct a linear cost function that gives total cost for the day as a function of the number of hot dogs sold.

b. Construct a linear revenue function that gives the total revenue for the day as a function of the number of hot dogs sold.

c. How many hot dogs would have to be sold in a given day in order to break even? Illustrate this graphically.

SOLUTION

a. If x represents the number of hot dogs sold in one day, then

$$\text{Total cost} = \text{Fixed cost} + \text{Variable cost}$$
$$= 60 \qquad\quad + (\text{Cost per hot dog})(\text{Number of hot dogs})$$
$$= 60 \qquad\quad + (0.3)x$$

So the total cost function is $C(x) = 0.3x + 60$.

b. Again, if x represents the number of hot dogs sold in one day, then

$$\text{Total revenue} = (\text{Revenue per hot dog})(\text{Number of hot dogs})$$
$$= (1.50)x$$

So the total revenue function is $R(x) = 1.5x$.

c. To break even, the total revenue must equal the total cost. Thus,

$$R(x) = C(x)$$
$$1.5x = 0.3x + 60$$
$$1.2x = 60$$
$$x = 50$$

So 50 hot dogs must be sold in one day to break even that day. Graphically, 50 is the x-coordinate of the point of intersection of the graphs of R and C, as shown in Figures 79 and 80.

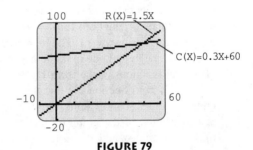

FIGURE 79   **FIGURE 80**

QUADRATIC FUNCTIONS

After linear functions, the next simplest class of functions consists of the **quadratic** functions—functions of the form $f(x) = ax^2 + bx + c$, where a, b, and c are constants and $a \neq 0$. This class of functions arises in many real-world contexts, including business (revenue and profit functions) and physics (equations of motion). Of course, we could simply use a graphing calculator to plot the graph of a quadratic function (and, in fact, we often do just that), but learning how to graph a quadratic by hand will help develop our intuition and "feel" for this important category of functions. This task is greatly simplified by the fact that, as we will soon see, there is a sense in which all quadratic functions descend from a common ancestor: $f(x) = x^2$. Thus, we turn our attention to the graph of this very special quadratic.

Since $x^2 \geq 0$ for all x, it follows that the smallest possible value of $f(x) = x^2$ is $f(0) = 0$. Furthermore, since f is even [that is, $f(-x) = f(x)$], the graph of f should be symmetric with respect to the *y*-axis. Indeed, the graph of *f* shown in Figure 81 reveals that $(0, 0)$ is the lowest point on the curve, and that the graph is symmetric about the *y*-axis. You might recognize the familiar shape of this graph as being that of a parabola. We will discuss parabolas in great detail later in the text; for now there are just two terms associated with parabolas that we need to know. First of all, the point where a parabola changes direction is called its **vertex.** Second, the line through the vertex dividing the parabola into two equal pieces is called the **axis of symmetry.** For the graph of $f(x) = x^2$, the vertex is the origin and the axis of symmetry is the *y*-axis, as shown in Figure 82.

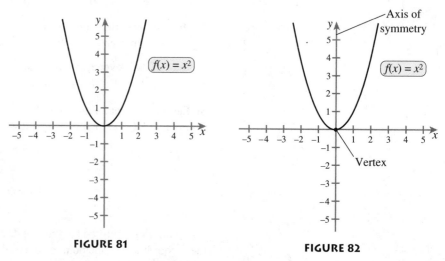

FIGURE 81   **FIGURE 82**

Like the graph of $y = x^2$, the graph of $f(x) = ax^2$, where $a \neq 0$, is a parabola with the origin as its vertex and the *y*-axis as its axis of symmetry. The sign of *a* determines the parabola's orientation: for $a > 0$, the parabola opens upward, and for $a < 0$, the parabola opens downward. The magnitude of *a* affects the "pointiness" of the graph: The larger the magnitude of *a*, the narrower the graph of *f*. Figure 83 shows the graphs of several functions of the form $f(x) = ax^2$.

Recall from Section 3.2 that the graph of $g(x) = f(x - h) + k$ is a translation *h* units horizontally and *k* units vertically of the graph of *f*. Thus, $g(x) = a(x - h)^2 + k$ is a translation of the graph of $f(x) = ax^2$, so that the graph of *g* is a parabola with (h, k) as its vertex and $x = h$ as its axis of symmetry. Furthermore, the parabola will open upward for $a > 0$ and downward for $a < 0$. These features are suggested by the graphs shown in Figures 84 and 85.

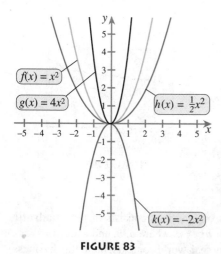

FIGURE 83

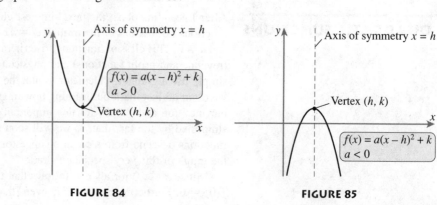

FIGURE 84   **FIGURE 85**

Now, for our final step, we note that by completing the square, any quadratic function $g(x) = ax^2 + bx + c$ can be written in the standard form $g(x) = a(x - h)^2 + k$. We summarize the key points of the preceding discussion as follows.

Quadratic Functions □

- A function of the form $f(x) = ax^2 + bx + c$ $(a \neq 0)$ is called a **quadratic function.** Its graph is a parabola.
- A quadratic function written as $f(x) = a(x - h)^2 + k$ $(a \neq 0)$ is said to be in standard form. Its graph is a parabola with vertex (h, k) and axis of symmetry $x = h$. The parabola opens upward for $a > 0$ and downward for $a < 0$.
- A quadratic function written in the form $f(x) = ax^2 + bx + c$ $(a \neq 0)$ can be written in standard form by completing the square.

E X A M P L E 2 ▒ **Graphing a Quadratic Function**

Sketch the graph of $f(x) = 2x^2 + 12x + 13$ and identify the vertex and axis of symmetry. Do not approximate.

SOLUTION   Since we are interested in exact results, we will graph the quadratic by hand. Our first step is to complete the square in order to write $f(x)$ in standard form. Note that since the coefficient of x^2 is 2, we begin by factoring 2 from both the x^2 and the x terms.

$$f(x) = 2x^2 + 12x + 13$$

$$= 2(x^2 + 6x + \quad) + 13$$

Factoring 2 from the square and first power terms; preparing to complete the square by leaving space for a constant term

$$= 2(x^2 + 6x + 9) + 13 - 2 \cdot 9$$

Adding $(\frac{6}{2})^2 = 9$ inside the parentheses to complete the square; compensating by subtracting $2 \cdot 9$ outside the parentheses

$$= 2(x + 3)^2 - 5$$

Simplifying

This is a quadratic in the standard form $f(x) = a(x - h)^2 + k$, with $a = 2$, $h = -3$, and $k = -5$. Thus, its graph opens upward (since $a > 0$), and it has $(-3, -5)$ as its vertex and $x = -3$ as its axis of symmetry. Plotting a few points and using symmetry, we obtain the sketch shown in Figure 86.

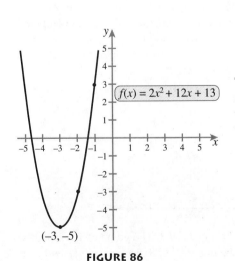

$f(x) = 2x^2 + 12x + 13$

$(-3, -5)$

FIGURE 86

x	y
-3	-5
-2	-3
-1	3

EXAMPLE 3 ▪ The Path of a Baseball

Houston Astrodome

The Astrodome in Houston, Texas, rises to a maximum height of 208 feet above the playing field. If a baseball is hit from a height of 3 feet with an initial velocity of 144 feet per second and an initial angle of 60°, and if we ignore air resistance, the height of the ball is approximated by $f(x) = -\frac{1}{324}x^2 + \sqrt{3}x + 3$, where x is the horizontal distance from home plate and all distances are in feet. Assuming the ball is hit fair and the playing field extends at least 300 feet from home plate, show that the ball will hit the dome ceiling somewhere above the field.

SOLUTION Because of the complexity of the coefficients of f, we use a graphing calculator to plot the graph of this quadratic function, as shown in Figure 87. By using the trace feature, we see that a height of more than 208 feet is obtained at an x-value of about 171 feet—which corresponds to a position within the playing field. It follows that, indeed, the ball will hit the ceiling of the dome somewhere above the playing field.

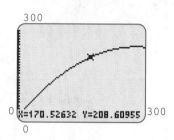

FIGURE 87

PIECEWISE-DEFINED FUNCTIONS

Occasionally, it is desirable to use two or more formulas to define a function. Consider, for example, the function f defined by

$$f(x) = \begin{cases} -x, & x \le -1 \\ 2x + 3, & x > -1 \end{cases}$$

This *piecewise* definition indicates that for an input value x less than or equal to -1, the output value is $-x$. For an input value x greater than -1, the output value is $2x + 3$. Thus, $f(-4) = -(-4) = 4$, whereas $f(1) = 2(1) + 3 = 5$. Some additional input–output pairs are given in Table 8. Notice that for each $x \le -1$, $f(x)$ is computed

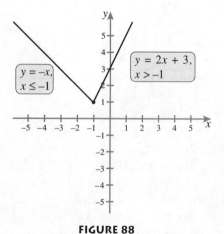

FIGURE 88

TABLE 8

x	Interval	Formula	$f(x)$
-3	$x \le -1$	$-x$	$-(-3) = 3$
-2	$x \le -1$	$-x$	$-(-2) = 2$
-1	$x \le -1$	$-x$	$-(-1) = 1$
0	$x > -1$	$2x + 3$	$2(0) + 3 = 3$
1	$x > -1$	$2x + 3$	$2(1) + 3 = 5$
2	$x > -1$	$2x + 3$	$2(2) + 3 = 7$

using $-x$, whereas for $x > 1$, $f(x)$ is computed using $2x + 3$. The graph of f is obtained by plotting the line $y = -x$ for $x \le -1$, and the line $y = 2x + 3$ for $x > -1$, as shown in Figure 88.

E X A M P L E 4 ▨ Evaluating and Graphing a Piecewise-Defined Function

Evaluate the function

$$f(x) = \begin{cases} x + 1, & x < 2 \\ x^2 - 4, & x \geq 2 \end{cases}$$

at the *x*-values given in the following table, and then sketch its graph.

x	$f(x)$
-1	
0	
1	
2	
3	

SOLUTION In order to compute a function value for a given *x,* we simply note whether $x < 2$ or $x \geq 2$ and use the corresponding "piece" of the function definition. For example, to compute $f(-1)$, we note that $-1 < 2$, and so we use the $x + 1$ piece of the function to obtain $f(-1) = (-1) + 1 = 0$. Similarly, since $3 \geq 2$, we compute $f(3)$ using the $x^2 - 4$ piece, obtaining $f(3) = (3)^2 - 4 = 5$. The remaining output values are computed in Table 9. The graph of *f* is obtained by plotting the line $y = x + 1$ for $x < 2$ and the parabola $y = x^2 - 4$ for $x \geq 2$, as shown in Figure 89. Notice that an open circle is used at the point $(2, 3)$ to indicate that the point is not included on the graph, and a closed circle is used at $(2, 0)$ to indicate that this point is included.

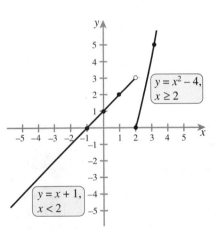

FIGURE 89

TABLE 9

x	Interval	Formula	$f(x)$
-1	$x < 2$	$x + 1$	$(-1) + 1 = 0$
0	$x < 2$	$x + 1$	$(0) + 1 = 1$
1	$x < 2$	$x + 1$	$(1) + 1 = 2$
2	$x \geq 2$	$x^2 - 4$	$(2)^2 - 4 = 0$
3	$x \geq 2$	$x^2 - 4$	$(3)^2 - 4 = 5$

> **WARNING!** A piecewise-defined function is *one* function with several pieces, not several different functions. Thus, for a given input value *x,* the corresponding output value $f(x)$ is computed using only one of the pieces.

Piecewise-defined functions can also be plotted with the aid of a graphing calculator. Although some graphing calculators have the ability to plot such functions directly, we can simply graph all of the expressions used to define the piecewise-defined function, while recognizing that for a given *x*-value only one of the graphs applies.

EXAMPLE 5 ■ **Graphing a Piecewise-Defined Function with a Graphing Calculator**

Graph the following function with the aid of a graphing calculator.

$$f(x) = \begin{cases} -x^3 + 6x^2 - 9x + 4, & x < 3 \\ x - 3, & x \geq 3 \end{cases}$$

SOLUTION   Unless our graphing calculator has the ability to plot piecewise-defined functions, we begin by plotting the graphs of both $y = -x^3 + 6x^2 - 9x + 4$ and $y = x - 3$ on the same screen, as shown in Figure 90. Now we recognize that only the portion of the graph of $y = -x^3 + 6x^2 - 9x + 4$ for which $x < 3$ (shown in red) is actually part of the graph of f; moreover, only the portion of the graph of $y = x - 3$ for which $x \geq 3$ (shown in blue) is part of the graph of f. Figure 91 shows the graph of f.

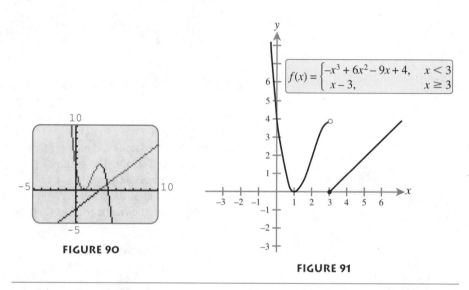

FIGURE 90

FIGURE 91

EXAMPLE 6 ■ **Constructing and Graphing a Piecewise-Defined Function**

A moving van rental company has two different rate schedules for a certain size of moving van. For vans used locally (a total distance of less than 100 miles), the rate is $40 plus 39¢ per mile. For long-distance rentals (a total distance of 100 miles of more), the rate is $200 plus 49¢ per mile for each mile over 400.

a. Construct a piecewise-defined function that can be used to compute the rate for any desired distance.
b. Plot the graph with the aid of a graphing calculator.
c. Determine how far one can move with $250.

SOLUTION

a. Let x represent the total distance traveled, and let $f(x)$ be the corresponding rental rate (in dollars). If $x < 100$, the rate is $40 + 0.39x$. If $100 \leq x \leq 400$, the rate is a fixed $200. If $x > 400$, the rate is $200 plus $0.49 for each mile over 400, or $200 + 0.49(x - 400)$. Thus, f is defined by

$$f(x) = \begin{cases} 40 + 0.39x, & 0 \leq x < 100 \\ 200, & 100 \leq x \leq 400 \\ 200 + 0.49(x - 400), & x > 400 \end{cases}$$

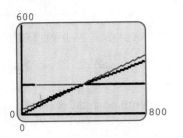

FIGURE 92

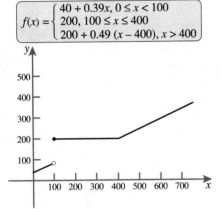

$$f(x) = \begin{cases} 40 + 0.39x, \, 0 \le x < 100 \\ 200, \, 100 \le x \le 400 \\ 200 + 0.49\,(x - 400), \, x > 400 \end{cases}$$

FIGURE 93

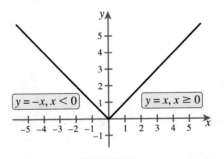

$$y = -x, x < 0 \qquad y = x, x \ge 0$$

FIGURE 94

b. We begin by graphing all three of the formulas that are used to define f, as shown in Figure 92. Note that only the red portion of each graph is actually part of the graph of f. In Figure 93, we show the graph of f.

c. From the graph we see that the y-coordinate attains the value 250 for an x-value larger than 400. Thus, we use the piece of f corresponding to $x > 400$, set $f(x)$ equal to 250, and solve for x.

$$200 + 0.49(x - 400) = 250$$
$$0.49(x - 400) = 50$$
$$x - 400 = \frac{50}{0.49}$$
$$x = \frac{50}{0.49} + 400$$

Thus, the total distance is $x = \dfrac{50}{0.49} + 400 \approx 502$ miles.

Two special examples of piecewise-defined functions are the absolute value function and the so-called step functions. Although the **absolute value function** $f(x) = |x|$ can be written as a single equation, it is equivalent to

$$f(x) = \begin{cases} -x, & x < 0 \\ x, & x \ge 0 \end{cases}$$

So the graph of $f(x) = |x|$ can be found by graphing $y = -x$ for $x < 0$ and $y = x$ for $x \ge 0$, as shown in Figure 94.

A **step function** is a piecewise-defined function taking only constant values over its pieces. For example, the function g defined by

$$g(x) = \begin{cases} 1, & x \le -3 \\ -1, & x > -3 \end{cases}$$

takes on constant values over each of its pieces, and is thus a step function. Its graph is shown in Figure 95.

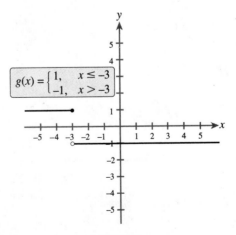

$$g(x) = \begin{cases} 1, & x \le -3 \\ -1, & x > -3 \end{cases}$$

FIGURE 95

Another example of a step function is the **floor function** $f(x) = \lfloor x \rfloor$, where $\lfloor x \rfloor$ is defined as follows.

The Floor Function ☐

> The **floor** of a real number x, denoted $\lfloor x \rfloor$, is the greatest integer less than or equal to x.

Applying this definition, we see that $\lfloor 2.6 \rfloor = 2, \lfloor 8 \rfloor = 8, \lfloor -4.1 \rfloor = -5$. Note in particular that the floor function does not simply round or truncate the decimal places. More generally, we observe that for every real number x between two consecutive integers n and $n + 1, \lfloor x \rfloor = n$. Thus, the graph of the floor function is constant between each pair of consecutive integers, as shown in Figure 96.

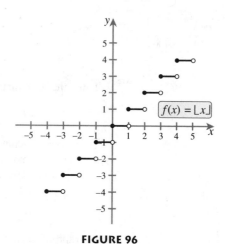

FIGURE 96

UNDERSTANDING AND MASTERY CHECKLISTS

CONCEPTS TO UNDERSTAND

☐ Linear functions
☐ Quadratic functions
☐ Parabolas
☐ Axis of symmetry of a parabola
☐ Vertex of a parabola
☐ Piecewise-defined functions
☐ Step functions
☐ The floor function

SKILLS TO MASTER

☐ Recognize a linear function and sketch its graph.
☐ Sketch the graph of a quadratic function.
☐ Find the vertex of a parabola.
☐ Evaluate and graph piecewise-defined functions.

EXERCISES 3.6

EXERCISES 1–12 *Identify the type of function that is given (piecewise-defined, quadratic, linear, or none of these).*

1. $f(x) = 3x^2 + 4$

2. $g(x) = 2x^3 + 5x + 7$

3. $h(x) = \dfrac{x^2 - 5x + 3}{x + 1}$

4. $p(x) = \sqrt{2x + 5}$

5. $q(x) = 2^x$

6. $H(x) = \begin{cases} 0, & x < 0 \\ 1, & x > 0 \end{cases}$

7. $f(x) = 3x + 9$

8. $g(x) = 2$

9. $f(t) = 4^{13}t + 7$

10. $r(s) = 2s^2 - 1$

11. $f(x) = \begin{cases} x, & x < 0 \\ x^2, & x > 0 \end{cases}$

12. $g(p) = 3p$

EXERCISES 13–16   *Plot the graph of the linear function and find the slope and y-intercept.*

13. $f(x) = 2x - 1$

14. $g(x) = -3x + 2$

15. $h(x) = \dfrac{-3x + 4}{4}$

16. $g(x) = \dfrac{x - 6}{2}$

EXERCISES 17–20   *Find the linear function having the given properties.*

17. The graph of h has slope $-\frac{1}{2}$ and $h(2) = 4$.

18. $f(-2) = 3$ and $f(2) = -3$.

19. The graph of g has y-intercept -4 and $g(3) = 0$.

20. The graph of f is the perpendicular bisector of the line segment connecting $(-2, 1)$ and $(4, 3)$.

EXERCISES 21–28   *Find the coordinates of the vertex of the graph of the given quadratic function.*

21. $f(x) = 3x^2 - 4$

22. $g(x) = x^2 + 4x + 9$

23. $f(x) = 2x^2 - 6x + 10$

24. $g(x) = ax^2 + c$

25. $m(x) = x^2 + 16x + 10$

26. $f(x) = 2^5 x^2 + 2^7 x$

27. $P(x) = 1 - 2x + x^2$

28. $h(x) = 3x + x^2 + 5$

EXERCISES 29–32   *Find the quadratic function having the given properties, and sketch its graph.*

29. The graph of f has vertex $(1, 3)$ and y-intercept 5.

30. The graph of h has vertex $(-2, -1)$ and $h(-1) = -3$.

31. The graph of g has axis of symmetry $x = 3$, $g(3) = -1$, and $g(2) = -2$.

32. The graph of f passes through the origin, $f(-1) = 0$, and $f(1) = 4$

EXERCISES 33–38   *Evaluate the piecewise-defined function as indicated and then sketch its graph.*

33. $f(x) = \begin{cases} -x - 4, & x \le 0 \\ 2x - 4, & x > 0 \end{cases}$

$f(-2)$, $f(0)$, $f(3)$

34. $g(x) = \begin{cases} 3x + 2, & x < 0 \\ -x + 2, & x \ge 0 \end{cases}$

$g(-4)$, $g(0)$, $g(1)$

35. $h(x) = \begin{cases} x^2 - 1, & x < 2 \\ \frac{1}{2}x + 1, & x \ge 2 \end{cases}$

$h(0)$, $h(2)$, $h(4)$

36. $g(x) = \begin{cases} -\frac{2}{3}x + 2, & x \le -3 \\ 13 - x^2, & x > -3 \end{cases}$

$g(-4)$, $g(-3)$, $g(0)$

37. $f(x) = \begin{cases} x, & x < -1 \\ 1, & -1 \le x \le 1 \\ x + 2, & x > 1 \end{cases}$

$f(-3)$, $f(-1)$, $f(4)$

38. $h(x) = \begin{cases} -x - 1, & x < -2 \\ x + 3, & -2 \le x \le 0 \\ 3, & x > 0 \end{cases}$

$h(-3)$, $h(0)$, $h(6)$

EXERCISES 39–42   *Use the graphs of f and g to sketch the graph of h.*

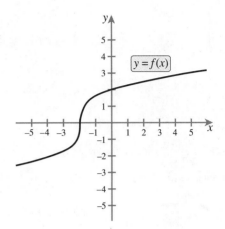

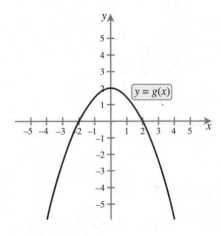

39. $h(x) = \begin{cases} f(x), & x \le -2 \\ g(x), & x > -2 \end{cases}$

40. $h(x) = \begin{cases} g(x), & x \le 0 \\ f(x), & x > 0 \end{cases}$

41. $h(x) = \begin{cases} g(x), & x \le -2 \\ x, & -2 < x < 0 \\ f(x), & x \ge 0 \end{cases}$

42. $h(x) = \begin{cases} f(x), & x < -2 \\ 0, & -2 \le x \le 2 \\ g(x), & x > 2 \end{cases}$

EXERCISES 43–46   *Use a graphing calculator to help you graph the piecewise-defined function.*

43. $f(x) = \begin{cases} -x^2 + 2x, & x \le 1 \\ (x - 2)^2, & x > 1 \end{cases}$

44. $g(x) = \begin{cases} -x^2 - 4x - 3, & x < -2 \\ x^2 + 4x + 5, & x \ge -2 \end{cases}$

45. $g(x) = \begin{cases} x^3 + 2x^2, & x < -1 \\ 1, & -1 \le x \le 1 \\ x^3 - 2x^2 + 2, & x > 1 \end{cases}$

46. $g(x) = \begin{cases} \dfrac{2}{x^2 + 1}, & x < -1 \\ -x, & -1 \le x \le 1 \\ \dfrac{-2}{x^2 + 1}, & x > 1 \end{cases}$

47. Let $f(x) = \lfloor x + 2 \rfloor$. Sketch the graph of f.

48. Let $h(x) = \lfloor x \rfloor - 2$. Sketch the graph of h.

49. For an interval $[a, b]$, define

$$f_{[a,b]}(x) = \begin{cases} 1, & a \le x \le b \\ 0, & \text{otherwise} \end{cases}$$

  a. Graph $f_{[0,1]}$.

  b. Graph $f_{[0,2]} + f_{[1,3]}$.

  c. Graph $f_{[0,1]} + f_{[1,2]} + f_{[2,3]} + \cdots$.

50. Let $f(x) = \dfrac{|x|}{x}$.

  a. Find the domain of f.

  b. Evaluate the following.

    i. $f(2)$ **ii.** $f(5)$

    iii. $f(-2)$ **iv.** $f(-1)$

  c. Find the range of f.

  d. Graph f.

51. Let $g(x) = ((x))$, where $((x))$ denotes the fractional part of x, so that, for example, $((3.456)) = 0.456$, $((2\frac{7}{8})) = \frac{7}{8}$, and so on.

  a. Evaluate the following

    i. $g(2.8)$ **ii.** $g(4.7)$

    iii. $g(\frac{23}{5})$

  b. Graph g for $x \ge 0$.

  c. Simplify $\lfloor x \rfloor + ((x))$ for $x \ge 0$.

52. The function "sgn," the signum function, is defined by

$$\text{sgn}(x) = \begin{cases} 1, & x > 0 \\ -1, & x < 0 \end{cases}$$

  a. Graph sgn(x).

  b. Let $f(x) = \text{sgn}(x - 2)$. Graph f.

■ APPLICATIONS

53. *Rental Profit* A landlord owns 50 apartments that can be rented on a monthly basis. He has fixed overhead costs of $5000 per month, plus $80 per month for every apartment that is rented. He has determined that all of the apartments will be rented if he charges $300 per month. However, for every $10 increase in rent, two fewer apartments will be rented.

  a. Find a linear cost function that expresses the landlord's total monthly cost as a function of the number of apartments rented.

  b. Find a linear demand function that expresses the rental price as a function of the number of apartments rented. [*Hint:* Find two points (x, p), where x is the number of units rented and p is the corresponding monthly rent, and then find the equation of the line passing through the two points.]

  c. Use the demand function from part b to find a revenue function that expresses the landlord's monthly revenue as a function of the number of apartments rented.

  d. Use the cost and revenue functions to find the profit function.

  e. How many apartments should be rented, and at what price, in order for the landlord to maximize his profit?

54. *Ticket Profit* A university has a football stadium that can hold 80,000 people. A recent analysis has determined that the cost to staff the stadium for a game is $50,000 plus 25¢ for each person in attendance. It has also been determined that all the seats will be filled if the average ticket price is $10, but 5000 fewer people will attend for each $1 increase in ticket price.

  a. Find a linear cost function that expresses the university's total cost to staff the stadium as a function of the number of people in attendance.

  b. Find a linear demand function that expresses the ticket price as a function of the number of people in attendance. [*Hint:* Find two points (x, p), where x is the number of people in attendance and p is the corresponding ticket price, and then find the equation of the line passing through the two points.]

  c. Use the demand function from part b to find a revenue function that expresses the university's revenue as a function of the number of people in attendance.

  d. Use the cost and revenue functions to find the profit function.

  e. How many people must attend, and at what ticket price, for the university to maximize profit?

55. *Path of a Baseball* If a baseball is hit from a height of 3 feet with an initial velocity of 100 feet per second and an initial angle of 45°, and if we ignore air resistance, the ball will follow the path given by the quadratic function $f(x) = -\frac{2}{625}x^2 + x + 3$, where x is the horizontal distance in feet from home plate and $f(x)$ is the corresponding height in feet above the playing field.

  a. Find the maximum height of the ball.

  b. Will the ball clear a 12-foot-high fence located 300 feet from home plate?

56. *Path of a Baseball* If a baseball is thrown from a height of 5 feet with an initial velocity of 80 feet per second and an initial angle of 30°, and if we ignore air resistance, the ball will follow the path given by the quadratic function

$$f(x) = -\frac{1}{300}x^2 + \frac{\sqrt{3}}{3}x + 5$$

where x is the horizontal distance in feet from where the ball is thrown and $f(x)$ is the corresponding height in feet.

a. Find the maximum height of the ball.

b. How far will the ball land from where it was thrown?

57. *Farm Acreage* The number of acres of U.S. farmland during the years 1930–1990 can be approximated with the quadratic function $f(x) = -0.21x^2 + 11.9x + 987$, where x is the year ($x = 0$ corresponds to 1930) and $f(x)$ is the total acreage in millions of acres. At what time was the number of acres highest? How many acres were there at that time?

U.S. Farm Acreage 1930–1990

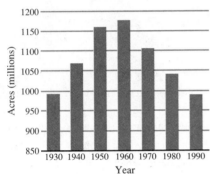

Data source: U.S. Department of Agriculture.

58. *Alcohol Consumption* The average number of gallons of alcoholic beverages consumed per year by each adult in the United States for the years 1970–1990 can be approximated with the quadratic function $f(x) = -0.053x^2 + 1.17x + 35.6$, where x is the year ($x = 0$ corresponds to 1970) and $f(x)$ is the average number of gallons consumed by each adult in that year. For what year was the alcohol consumption the greatest and what was the average rate of consumption at that time?

Alcoholic Beverage Consumption

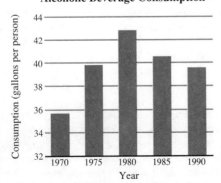

Data source: U.S. Department of Agriculture.

59. *Telephone Rate* Suppose the cost of a cell phone call from New York to Los Angeles is 95¢ for the first minute and 65¢ for each additional minute or fraction thereof. Then the cost of a phone call lasting x minutes can be computed using the function $C(x) = 0.95 - 0.65 \lfloor 1 - x \rfloor$. Plot the graph of this function. What is the longest phone call that can be made if the cost cannot exceed $5?

60. *Postal Rate* In 1999, the U.S. Postal Service rate for first-class mail was raised to 33¢ for the first ounce and 22¢ for each additional ounce or fraction thereof. The rate for an item weighing x ounces can be computed with the function $C(x) = 0.33 - 0.22 \lfloor 1 - x \rfloor$. Plot the graph of this function. What is the heaviest first-class item that can be sent for $3.00 or less?

61. *Tax Rates* Fred and Irma are trying to decide if they should file a joint tax return or if they should file separately. If they file separately, Fred has a taxable income of $40,000 and Irma has a taxable income of $50,000. If they file jointly, their combined taxable income is $85,000. The function S below computes taxes for separate returns and the function J computes taxes for joint returns. In both cases, x denotes the taxable income. Compute the total taxes they will pay if they file separately. How does this amount compare to the taxes they will pay if they file jointly?

$$S(x) = \begin{cases} 0.15x, & 0 \le x < 17,900 \\ 2685 + 0.28(x - 17,900), & 17,900 \le x < 43,250 \\ 9783 + 0.31(x - 43,250), & x \ge 43,250 \end{cases}$$

$$J(x) = \begin{cases} 0.15x, & 0 \le x < 35,800 \\ 5370 + 0.28(x - 35,800), & 35,800 \le x < 86,500 \\ 19,566 + 0.31(x - 86,500), & x \ge 86,500 \end{cases}$$

62. *Tax Rates* The tax rates for a single person filing a U.S. tax return in 1992 are given in Table 10. Construct a piecewise-defined function that expresses the total tax that must be paid as a function of taxable income.

TABLE 10   1992 Tax rates (single)

If taxable income is over ...	but not over ...	the tax is ...	of the amount over ...
$0	$21,450	15%	$0
$21,450	$51,900	$3,217.50 + 28%	$21,450
$51,900		$11,743.50 + 31%	$51,900

63. *Secret Code* A coded message from a terrorist organization is intercepted by an intelligence satellite. The message reads EFPSNKXEFL. As a leading authority on terrorist activity, Inspector Magill is called in to help break the code. He immediately recalls an incident from the previous day when a member of the same organization was arrested for the possession of an illegal assault rifle. Among other items found in his possession was a piece of paper upon which was written

$$f(x) = 3x + 5 - 26 \left\lfloor \frac{3x + 5}{26} \right\rfloor$$

Within minutes, Magill has broken the code and foiled a hijacking plan. What was the message? (*Hint:* First assign to each letter of the alphabet the number corresponding to its position. For example, A ↔ 1, B ↔ 2, ...)

64. *Christmas Break*   At a certain university, the spring semester begins on the first Monday after January 5. It follows (trust us!) that if Christmas falls on day n (with $n = 0$ being Sunday, $n = 1$ being Monday, and so on), then the January date on which the next spring semester will begin is given by

$$f(n) = 9 - n + 7 \left\lfloor \frac{n + 3}{7} \right\rfloor$$

Also, the day of the week on which Christmas falls in the year $1900 + x$ is given by

$$g(x) = 2 + x + \left\lfloor \frac{x}{4} \right\rfloor - 7 \left\lfloor \frac{2 + x + \left\lfloor \frac{x}{4} \right\rfloor}{7} \right\rfloor$$

a. On what day of the week did Christmas fall in 1975?

b. If Christmas falls on a Saturday, on what date does the next spring semester start?

c. What interpretation can be given to the function $f \circ g$? In particular, if the year is $1990 + x$ what does $(f \circ g)(x)$ represent?

d. Find an expression for $(f \circ g)(x)$. (It won't be pretty.)

e. Use your answer to part d to find the day on which the spring semester started for the 1998–1999 academic year.

CONCEPTS AND CRITICAL THINKING

EXERCISES 65–68   *Answer true or false.*

65. If x^2 appears in the formula defining a function, then the function is said to be quadratic.

66. The graph of a quadratic function is a parabola with a vertical axis of symmetry.

67. The floor function is an example of a step function.

68. A quadratic function cannot have both a maximum value and a minimum value.

EXERCISES 69–74   *Give an example of each.*

69. A piecewise-defined function that is linear on one piece and quadratic on another

70. A function whose graph is a parabola with vertex $(1, 3)$

71. Two linear functions, the graphs of which do not intersect

72. A quadratic function having no zeros

73. A quadratic function having one zero

74. A quadratic function having two zeros

QUESTIONS FOR DISCUSSION OR ESSAY

75. The function $f(x) = -\frac{1}{324}x^2 + \sqrt{3}x + 3$ in Example 3 takes into account the initial velocity, angle, and height of the ball, and also the force of gravity. What other factors might also affect the path of the ball? How do these factors compare in significance to the ones that were taken into account? How will these factors influence such things as the maximum height of the ball and the maximum distance the ball travels before it hits the ground (assuming it doesn't hit the ceiling first)?

76. How can you tell whether a graphical image is the plot of the graph of one piecewise-defined function, or of several functions?

77. Some quadratic functions have a local maximum, whereas others have a local minimum. How can you determine whether a given quadratic function has a local maximum or a local minimum without graphing the function?

78. Some people view taxes as payment for services rendered by the government: defending our borders, educating our children, protecting our environment, insuring our elderly and indigent, and so forth. Certainly the cost of a quart of milk is not dependent on one's income, nor is the cost of virtually any commodity available in the marketplace. Why then are taxes dependent on income? If we let i represent income and $T(i)$ represent the corresponding income tax, then what kind of function would T be if all individuals were taxed the same amount? What if all individuals were taxed at a flat rate? Some have suggested that anyone earning under a certain cutoff amount should not pay any tax, but that anyone earning above the cutoff should be taxed at a flat rate on the money they earn above the cutoff level. What kind of function would $T(i)$ be under these circumstances? What method of tax computation do you feel would be most fair?

■ **PROJECTS FOR ENRICHMENT**

79. *Driving Under the Influence*   Concern over the incidence of alcohol-related motor vehicle accidents has prompted lower limits on the legal level of alcohol in the bloodstream. Unfortunately, most people are unaware of the relationship between alcohol consumption and the level of alcohol in the bloodstream. Indeed, many are unaware that even small quantities of alcohol can raise the level beyond legal limits. In this project, we investigate the relationship between alcohol consumption and the alcohol level in the bloodstream, and we also see what bearing body weight has on the relationship.

Ethanol, or grain alcohol, is the form of alcohol that is found in beer, wine, and other liquors. It is eliminated from the body by the liver via an enzymatic process at the constant rate of 12 grams per hour. We assume that each serving of beer, wine, or other liquors contains 18 grams of ethanol. We further assume that the ethanol enters the bloodstream immediately after the drink is consumed.

a. Suppose one drink is consumed at time $t = 0$. Find a linear function $A_1(t)$ that gives the number of grams of ethanol remaining in the bloodstream after t hours. How long will it take for all the ethanol to be eliminated?

b. Now suppose that one drink is consumed at time $t = 0$ and a second is consumed 1 hour later, at time $t = 1$. Find a piecewise-defined function $A_2(t)$ that gives the number of grams of ethanol remaining in the bloodstream after t hours. Note that one piece will correspond to the interval $0 \le t < 1$ and that the second will correspond to $t > 1$. Plot the graph of A_2. How long will it take for all the ethanol to be eliminated?

c. Suppose that drinks are consumed at a regular rate of 2 per hour for 2 hours (that is, at times $t = 0$, $t = \frac{1}{2}$, $t = \frac{3}{2}$, $t = 2$). Find a piecewise-defined function $A_3(t)$ that gives the number of grams of ethanol remaining after t hours, and plot the graph of A_3.

d. Repeat part c to find a function A_4 for consumption at a rate of 3 drinks per hour.

If $A(t)$ denotes the number of grams of ethanol in the bloodstream at time t, the percentage levels $L(t)$ of blood serum ethanol at time t are approximated by

$$L(t) = \frac{A(t)}{310 \times (\text{Body weight in pounds})} \quad \text{for adult males}$$

and

$$L(t) = \frac{A(t)}{250 \times (\text{Body weight in pounds})} \quad \text{for adult females}$$

e. For each of the functions A_1–A_4 in parts a through d, find the functions $L_1(t)$ through $L_4(t)$ for an adult male weighing 160 pounds and an adult female weighing 130 pounds.

f. In many states, the legal limit is 0.1%. In other words, one is legally under the influence if $L(t) \ge 0.001$. For each of the functions $L_1(t)$ through $L_4(t)$ in part e, find the time interval during which the blood serum ethanol level is at or above 0.001. It may be helpful to sketch the graphs of the functions.

g. Discuss some of the limitations of this model for computing blood serum ethanol level.

80. *Day of the Week Computation*   To the list of human calculators—people capable of extraordinary mental feats such as rapid multiplication, extraction of cube roots, memorization of entire books—one addition should be made: you. You are about to learn a method that will enable you to compute the day of the week for any known date—the signing of the Declaration of Independence, the day you were born, next Christmas, Halloween in the year 2025, the day of your twenty-first birthday—quickly and without calculator or even pencil and paper.

In this project, we investigate a technique developed by mathematician John Horton Conway for computing the day of the week on which any given date falls in the standard, or Gregorian, calendar. It is based on the fact that in any given year, a certain collection of easily memorized dates (called reference dates) fall on the same day of the week—a day we will call Doomsday. Once Doomsday is known for a given year, the day on which any given date falls can easily be computed by comparison to a nearby reference date. Here are a few facts about the Gregorian calendar that we will need.

   i. There are 365 days in an ordinary year; thus, leap years have 366 days. The extra day is February 29.

   ii. Leap years occur every four years—whenever the last two digits form a number divisible by four. Thus 1840, 1984, and 1996 were all leap years. (These are the same years in which Summer Olympics and U.S. presidential elections are held.)

   iii. There is one exception to rule ii: Only every fourth century year—a year ending in "00"—is a leap year. Thus, 1700, 1800, and 1900 were not leap years, but 2000 is a leap year; 2100, 2200, and 2300 will not be leap years, but 2400 will be a leap year, and so forth.

a. Find the number of days separating each pair of dates in the tables given, and then explain why all these dates fall on the same day of the week.

Ordinary Year:

Dates	1/3, 2/28	2/28, 4/4	4/4, 5/9	5/9, 6/6
Days Apart				

Dates	6/6, 7/11	7/11, 8/8	8/8, 9/5	9/5, 10/10
Days Apart				

Dates	10/10, 11/7	11/7, 12/12
Days Apart		

Leap Year:

Dates	1/4, 2/29	2/29, 4/4	4/4, 5/9	5/9, 6/6
Days Apart				

Dates	6/6, 7/11	7/11, 8/8	8/8, 9/5	9/5, 10/10
Days Apart				

Dates	10/10, 11/7	11/7, 12/12
Days Apart		

We call the dates appearing in these tables *reference dates*. Since we will be using them extensively, let's see if we can develop a mnemonic device (memory aid) to help us remember them. First, we notice the following pattern: 4/4, 6/6, 8/8, 10/10, and 12/12 are all reference dates. Second, imagine that you have a 9 to 5 job at a 7-11 convenience store—a handy way of remembering the dates 9/5 and 7/11, and their transposes 5/9 and 11/7. We then note that no matter if we are in an ordinary or a leap year, the last day of February is a reference date. For convenience, we will think of this as the 0th day of March, or 3/0. Finally, we must remember that 1/3 is a reference date in an ordinary year, whereas 1/4 is a reference date in a leap year. Memorize the reference dates; all reference dates will fall on Doomsday for a given year.

b. At this point, we are able to find the day of the week on which any date falls provided that we know Doomsday for the year in question. For example, suppose that we are interested in finding the day on which Christmas falls in a year in which Doomsday is Saturday. Now December 12 is a reference day, so we know that December 12 is a Saturday. It follows that December 26 is a Saturday also and hence Christmas, December 25, is on a Friday. Use this technique to complete the following table.

Year	1776	1865	1998	2020
Doomsday	Thurs	Mon	Sat	
Valentine's Day (2/14)				
Independence Day (7/4)				Wed
Halloween (10/31)				
Christmas (12/25)				

c. Since 52 weeks is 364 days, but an ordinary year has 365 days and a leap year has 366 days, the day on which April 4 falls will advance one day in an ordinary year and 2 days in a leap year. Thus, Doomsday itself advances one day for an ordinary year and 2 days for a leap year. For example, if this year Doomsday is Monday, then next year it will be either Tuesday or Wednesday, depending on whether next year is an ordinary or leap year. Use this fact to complete the following table.

Year	1942	1943	1944	1945	1996	1997	1998
Doomsday		Sun					Sat

d. It is a fact that Doomsday 1900 was Wednesday. Thus, to figure Doomsday for any year in the 20th century, we begin with Wednesday and advance one day for each year and an additional day for each leap year. Thus, for example, to compute Doomsday for 1910, we advance 10 days—one for each year after 1900—and an additional 2 days to take into account the leap years of 1904 and 1908. Advancing 12 days from Wednesday brings us to a Monday. Thus, Doomsday 1910 was Monday. Use this technique to find Doomsday for each of the following years.

Year	1926	1953	1956	1978	1993
Doomsday					

e. Show that the number of days that one must advance from Wednesday to compute Doomsday for the year $1900 + x$ (with $0 \le x \le 99$) is given by $x + \lfloor \frac{x}{4} \rfloor$.

The formula given in part e often involves computations that are difficult to perform without pencil and paper. We now provide an alternative method that, although more complicated, involves smaller numbers and is thus ideal for mental computation. To find Doomsday in $1900 + x$, divide x by 12 to obtain a quotient q and a remainder r. Next, divide r by 4 to obtain a quotient s—ignore the remainder of this second division. Now add the numbers q, r, and s. For example, to compute Doomsday 1981, we divide 81 by 12 to get a quotient of 6 and a remainder of 9; dividing 9 by 4 gives us a quotient of 2. We add to obtain $6 + 9 + 4 = 17$; thus, Doomsday will fall 17 days after Wednesday—that is, on a Saturday. (An explanation for why this works is just beyond the scope of this text.) Try using this technique, without pencil and paper, to find the day of the week on which you were born.

f. To find Doomsdays for other centuries, we need only know Doomsday for the century year. Doomsdays for century years are as follows: 1700, Sunday; 1800, Friday; 1900, Wednesday; 2000, Tuesday. After this, the cycle repeats: Sunday, Friday, Wednesday, Tuesday, Sunday, Friday, Wednesday, Tuesday, and so forth. Use this information to complete the following table.

Event	Date	Day of the Week
Signing of the Declaration of Independence	7/4/1776	
Assassination of Abraham Lincoln	4/14/1865	
D-day	6/6/1944	
Neil Armstrong walks on the moon	7/20/1969	
Y2K eve	12/31/1999	

SECTION 3.7 # MODELING WITH FUNCTIONS AND VARIATION

☐ **What is the gravitational pull exerted on the moon by 600-pound Sumo wrestler Konishiki?**

☐ **If a man is 48 times as tall as a grasshopper is long, and a grasshopper can jump 30 inches, how far could a man-sized grasshopper jump?**

☐ **What will the total health-care expenditures be in the year 2003?**

☐ **How much is a graduate degree worth?**

☐ **What concentration of the ozone-depleting chemical CFC-11 will be present in the atmosphere in the year 2002?**

☐ **How are the gas mileage and weight of a car related?**

MATHEMATICAL MODELS

A **mathematical model** is a mathematical description of the behavior of some aspect of the real world. Mathematical models have been formulated for such diverse phenomena as a signal transmission in the human nervous system, the aerodynamics of a hummingbird, the spread of AIDS, deforestation of the Amazon Rain Forest, the jet stream, the path of Halley's comet, and the entire global economy. All the major theories of the physical sciences are, in essence, mathematical models. Examples include Einstein's general theory of relativity, which models gravity, and superstring theory, which is a model of the very fabric of space itself.

Real-world phenomena tend to be so extraordinarily complex and subject to so many random influences that any description, mathematical or otherwise, is necessarily incomplete. A mathematical model is an attempt to capture the salient features of a real-world phenomenon, but we sometimes find that in striving for simplicity, we have left out so many key features that our model is like a bad made-for-TV movie: overly simplistic and unrealistic. Nonetheless, the accuracy and utility of certain mathematical models is startling. For example, our current model of the motion of Earth, the sun, and the moon is so good that it correctly predicts the time of eclipses to within seconds, decades in advance. On the other hand, in spite of an astronomical amount of data collection, the use of the most powerful supercomputers in existence, and the efforts of some of the most brilliant minds on the planet, existing models of Earth's atmosphere are not good enough to reliably predict the weather even 3 or 4 days in advance!

You have already seen many examples of mathematical models in the text—from falling objects and stopping distance in previous chapters to CO_2 levels and alcohol consumption in this chapter. However, in these earlier examples, we were concerned primarily with how a model could be used to solve a real-world problem. In this section, we investigate not just how models are used to solve problems, but also how models are constructed.

VARIATION

Many of the most important mathematical models arising in the natural sciences can be conveniently described using the language of **variation.** For example, the force acting on an object *varies directly* as the acceleration that the object is undergoing; the strength of the gravitational pull between two bodies *varies inversely* as the square of the distance between them and *varies jointly* as the masses of the bodies. The terminology of variation is summarized in the following table.

Variation Terminology □

Type of variation	Equation	Terminology
Direct Variation	$y = kx$	"y varies directly as x," "y is proportional to x," or "y is directly proportional to x"
Inverse Variation	$y = \frac{k}{x}$	"y varies inversely as x" or "y is inversely proportional to x"
Joint Variation	$z = kxy$	"z varies jointly as x and y"

In all cases, the constant k is called the **constant of proportionality.**

If y varies directly as x, then we can conclude that $y = kx$ for some constant k. Note, however, that this information alone is not enough to completely specify the relationship between the two variables. Additional data is required to determine k, the constant of proportionality, as illustrated in the following examples.

E X A M P L E 1 ▨ **Computing the Constant of Proportionality**

It is given that y varies directly as x, and that $y = 3$ when $x = 2$. Find the constant of proportionality.

SOLUTION Since y varies directly as x, we may write $y = kx$ for some constant k. Because $y = 3$ when $x = 2$, we have $3 = k \cdot 2$, from which it follows that $k = \frac{3}{2}$. Therefore, $y = \frac{3}{2}x$.

E X A M P L E 2 ▨ **Computing the Constant of Proportionality**

Suppose that z varies jointly with x and y, and that $z = 40$ when $x = 2$ and $y = 5$. Find the constant of proportionality.

SOLUTION Since z varies jointly with x and y, we know that $z = kxy$ for some constant k. Substituting 2 for x, 5 for y, and 40 for z gives us

$$40 = k \cdot 2 \cdot 5$$
$$40 = 10k$$
$$k = 4$$

Therefore, $z = 4xy$.

Once the constant of proportionality is determined, the model is complete and can then be used to make estimates, as in the following examples.

E X A M P L E 3 ▨ **Estimating the Time of a Trip**

The travel time for a certain trip varies inversely with the average speed. If the trip takes 30 minutes at an average speed of 25 miles per hour, how long will it take at 100 miles per hour?

SOLUTION Let t represent the time for the trip in minutes and r the average speed in miles per hour. Then, since the time varies inversely with the rate, we have

$$t = \frac{k}{r}$$

Using the fact that $t = 30$ when $r = 25$ gives us

$$30 = \frac{k}{25}$$

and so

$$k = 25 \cdot 30 = 750$$

Thus, we have $t = 750/r$. When $r = 100$, this gives us $t = 750/100 = 7.5$ minutes.

EXAMPLE 4    **Estimating Body Weight as a Function of Height**

If all men were shaped similarly, then the weight of a man would be proportional to the cube of his height. Suppose that the weight of a man 5′10″ is 170 pounds. What would the weight of a similarly shaped 8-foot-tall man be?

SOLUTION   Let w represent the weight of a man in pounds and h his height in inches. Since weight is proportional to the cube of height, we have $w = kh^3$. Now 5′10″ is 70 inches, so that we have

$$170 = k \cdot 70^3$$

$$k = \frac{170}{70^3} \approx 0.0004956$$

Thus $w = 0.0004956h^3$. Since 8 feet is 96 inches, the weight of an 8-foot-tall man would be

$$w = 0.0004956 \cdot 96^3 \approx 438.5 \text{ pounds}$$

Robert Wadlow, who reached a world record height of 8′11″ shortly before his death at the age of 22, towers above actresses Maureen O'Sullivan and Ann Morris.

Often it is convenient to express the relationship between several variables as a combination of two or more of the standard types of variation, as in the following examples.

EXAMPLE 5    **Expressing the Relationship Among Several Variables**

It is given that w varies jointly as the square of x and the cube of y and inversely as z. Find w when $x = 1$, $y = 2$, and $z = 10$, given that $w = 9$ when $x = 2$, $y = 3$, and $z = 36$.

SOLUTION   Combining the variation statements, we have

$$w = k \frac{x^2 y^3}{z}$$

Now substituting $w = 9$, $x = 2$, $y = 3$, and $z = 36$, we can solve for k.

$$9 = k \frac{2^2 3^3}{36}$$

$$9 = 3k$$

$$k = 3$$

Thus,

$$w = 3\,\frac{x^2y^3}{z}$$

Finally, substituting $x = 1$, $y = 2$, and $z = 10$, we can determine the value of w.

$$w = 3 \cdot \frac{1^2 2^3}{10}$$

$$= 3 \cdot \frac{4}{5}$$

$$= \frac{12}{5}$$

EXAMPLE 6 ■ Estimating the Gravitational Pull of an Object

The gravitational pull between two objects varies jointly as their masses and inversely as the square of the distance between them. The weight of an object on Earth is, by definition, the gravitational pull between the object and Earth. Find the gravitational pull that 600-pound Sumo wrestler Konishiki exerts on the moon, which weighs 1.62064×10^{23} pounds and is 234,912 miles from Earth.

SOLUTION   We begin by finding the general equation of variation. Let d be the distance between two objects, let m_1 and m_2 be their masses (or weights), and let F be the force due to gravity. Since F varies jointly with m_1 and m_2 and inversely with the square of d, we have

$$F = k\,\frac{m_1 m_2}{d^2}$$

In order to find the constant k, we choose a scenario in which all of the quantities F, m_1, m_2, and d are known. Because the gravitational force between a 1-pound object on the surface of Earth is 1 pound, if we knew the weight of Earth and the distance from Earth's surface to its center, we would have enough information to evaluate k. After consulting an almanac (or the numerography inside the front cover of this text) we discover that Earth has a weight of approximately 1.3176×10^{25} pounds and that its radius is approximately 3963 miles. Thus, we have $F = 1$ pound, $d = 3963$ miles, $m_1 = 1$ pound, and $m_2 = 1.3176 \times 10^{25}$ pounds. Substituting into the equation of variation, we find

$$1 \text{ lb} = k \frac{1 \text{ lb} \cdot 1.3176 \times 10^{25} \text{ lb}}{(3963 \text{ mi})^2}$$

and solving for k, we obtain

$$k = \frac{(3963)^2 \text{ mi}^2}{1.3176 \times 10^{25} \text{ lb}} \approx 1.19197 \times 10^{-18} \frac{\text{mi}^2}{\text{lb}}$$

Now, to find the gravitational pull between Konishiki and the moon, we use this value of k and let $m_1 = 600$ pounds, $m_2 = $ the weight of the moon $\approx 1.62064 \times 10^{23}$ pounds, and $d = $ the distance from Earth to the moon $\approx 234{,}912$ miles. Thus, the force is given by

$$F \approx 1.19197 \times 10^{-18} \frac{\text{mi}^2}{\text{lb}} \cdot \frac{600 \text{ lb} \cdot 1.62064 \times 10^{23} \text{ lb}}{(234{,}912 \text{ mi})^2}$$

$$\approx 0.00210 \text{ lb}$$

DATA FITTING

Although some sophisticated mathematical models are constructed from underlying principles, many are formed simply by finding a function that fits a given set of data. Often the *form* of the function (for example, linear or quadratic) is either known or assumed, and the given data are then used to determine the function precisely. For example, suppose that we wish to find a linear function that is consistent with the following data.

x	y
2	5
3	7

One approach would be to find the slope of the line connecting $(2, 5)$ and $(3, 7)$ and then use the point-slope form for the equation of a line. However, we will use a method that is more general and can be applied in many other situations. We first note that since y is to be a linear function of x, we must have $y = ax + b$ for some choice of constants a and b. (In the context of data fitting, it is conventional to represent linear functions in the form $y = ax + b$ rather than $y = mx + b$.) Substituting 2 for x and 5 for y, we have

$$5 = a(2) + b$$

Now substituting 3 for x and 7 for y gives us

$$7 = a(3) + b$$

Thus, we must solve the following system.

(1) $\qquad\qquad\qquad\qquad 2a + b = 5$

(2) $\qquad\qquad\qquad\qquad 3a + b = 7$

Subtracting the first equation from the second gives us $a = 2$. Substituting 2 for a in equation (1) gives

$$2(2) + b = 5$$
$$4 + b = 5$$
$$b = 1$$

Thus, we have $y = 2x + 1$.

EXAMPLE 7 ■ Finding a Parabola Passing Through Two Points

Suppose that we are given that the variable y is a function of the variable x of the form $y = x^2 + bx + c$, and that we are provided with the following data.

x	y
2	4
4	8

Find values for b and c so that $y = x^2 + bx + c$ is consistent with the data.

SOLUTION   We use the two data points to form two equations in the unknowns b and c. Using the point $(2, 4)$, we set $x = 2$ and $y = 4$ to obtain

$$4 = 2^2 + b \cdot 2 + c$$
$$4 = 4 + 2b + c$$
$$2b + c = 0$$

Similarly, the point $(4, 8)$ gives us

$$8 = 4^2 + b \cdot 4 + c$$
$$8 = 16 + 4b + c$$
$$4b + c = -8$$

Thus, we must solve the following system of equations.

(1) $\qquad\qquad\qquad 2b + c = 0$

(2) $\qquad\qquad\qquad 4b + c = -8$

Subtracting the first equation from the second gives us $2b = -8$, or $b = -4$. Substituting -4 for b in equation (1) gives us $2(-4) + c = 0$, so that $c = 8$. Thus, we have $y = x^2 - 4x + 8$.

EXAMPLE 8 ■ Finding a Parabola Passing Through Three Points

It is known that s can be expressed as a quadratic function $f(t)$.

a. Find $f(t)$ and sketch its graph given the following data.

t	s
0	13
1	7
5	23

b. Find the value of t that minimizes s.

SOLUTION

a. An arbitrary quadratic function of t is of the form $f(t) = at^2 + bt + c$. We use the given data points to form a system of equations in the unknowns a, b, and c.

Using $(0, 13)$: $\qquad\qquad\qquad 13 = a \cdot 0^2 + b \cdot 0 + c$

$$c = 13$$

Using $(1, 7)$: $\qquad\qquad\qquad 7 = a \cdot 1^2 + b \cdot 1 + c$

$$a + b + c = 7$$

Using $(5, 23)$: $\qquad\qquad\qquad 23 = a \cdot 5^2 + b \cdot 5 + c$

$$25a + 5b + c = 23$$

Thus, we must solve the following system of equations.

(1) $\qquad\qquad\qquad\qquad\qquad c = 13$

(2) $\qquad\qquad\qquad\qquad\qquad a + b + c = 7$

(3) $\qquad\qquad\qquad\qquad\qquad 25a + 5b + c = 23$

Substituting $c = 13$ in equations (2) and (3) and simplifying gives us

(4) $\qquad\qquad\qquad\qquad\qquad a + b = -6$

(5) $\qquad\qquad\qquad\qquad\qquad 25a + 5b = 10$

Now, multiplying equation (4) by -5 and adding to equation (5) we have

$$-5a - 5b = 30$$
$$\underline{+ \quad 25a + 5b = 10}$$
$$20a = 40$$
$$a = 2$$

Substituting 2 for a in equation (4) gives $2 + b = -6$, and so $b = -8$. Hence, $a = 2$, $b = -8$, and $c = 13$, so that $f(t) = 2t^2 - 8t + 13$. The graph of $f(t)$ is shown in Figure 97.

b. Since $f(t) = 2t^2 - 8t + 13$ is a quadratic function, its graph is a parabola, and its minimum value will occur at the vertex of the parabola. We can find the vertex using the techniques of Section 3.6 or by approximating with a graphing calculator, as shown in Figure 98, which suggests that the vertex is the point $(2, 5)$. Thus, the minimum occurs at $t = 2$ and the minimum value is 5.

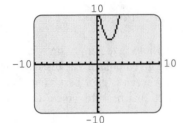

FIGURE 97

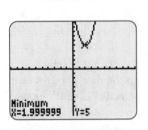

FIGURE 98

In the previous examples, the form of the model was given, and just enough data were provided to determine the model precisely. In practice, the data will not fit the model precisely; the model can only approximate the data. Our task then is to choose the model that most nearly approximates the given data. For example, suppose that we wish to model the data given in Table 11 with a linear function. Unless we are extremely fortunate, the given points will not lie on any one line. Therefore, we must choose the line that, in some sense, best approximates the given data.

TABLE 11

Year	1985	1986	1987	1988	1989	1990
U.S. health-care expenditure (in billions of dollars)	422.6	454.8	494.1	546.0	602.8	666.2

Data source: 1992 World Almanac.

If we let t be the number of years since 1985 (so that 1985 corresponds to $t = 0$, 1986 corresponds to $t = 1$, and so on), and let E be the total U.S. health-care expenditures in billions of dollars, then we obtain the graph of the data points (t, E) shown in Figure 99. Now it appears that no line passes through all of these points, so the best we can do is to find a line that "almost" passes through all of these points. Shown in Figure 100 is the line that appears to the authors to best fit the data; your eyes may tell you something different.

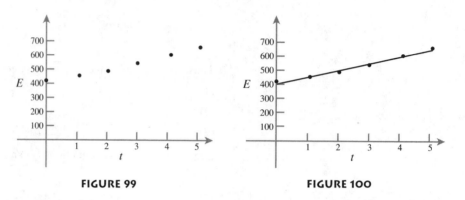

FIGURE 99　　　　　　　　　　**FIGURE 100**

The primary advantage of this method is its simplicity; we simply drew the line that seemed to come the closest to passing through the given points. The disadvantage is obvious; there is no guarantee that the line we have chosen is the "best" line. What we need is a mathematical method for selecting a line that best fits given data. One such method is called the **least-squares best fit.**

LEAST-SQUARES BEST FIT

Suppose we are given three data points (x_1, y_1), (x_2, y_2), and (x_3, y_3), as shown in Figure 101, and we would like to find the equation $y = ax + b$ of the line that in some sense most nearly fits the data. If the line were to pass through each of the three points, then the vertical distances d_1, d_2, and d_3 would all be zero. In fact, the distances d_1, d_2, and d_3 are a measure of the error involved in approximating the data points with the line. The line with the least-squares best fit is obtained by selecting a and b in such a way as to minimize the sum of the squares of the errors—that is, to minimize the quantity $d_1^2 + d_2^2 + d_3^2$. The following definition extends this idea to any number of data points.

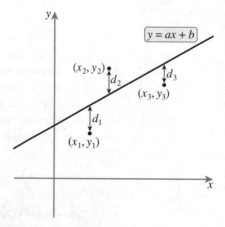

FIGURE 101

Least-Squares Best Fit

> If data points (x_1, y_1), (x_2, y_2), . . . , (x_n, y_n) are given, then the **line with the least-squares best fit,** or **regression line,** is that line for which the sum of the squares of the errors, $d_1^2 + d_2^2 + \cdots + d_n^2$, is as small as possible. The process of finding the coefficients a and b for the line $y = ax + b$ with the least-squares best fit is called **linear regression.**

Fortunately, graphing calculators can perform linear regressions for us. The general steps are as follows.

CALCULATOR KEYS

Linear Regression

The **linear regression** feature of a graphing calculator can be used to find a linear function of the form $y = ax + b$ that has the best least-squares fit to a collection of data points (x, y). There are several steps to the process.

Step 1. Clear any existing statistical data.

Step 2. Enter the points (x, y) as statistical data.

Step 3. Select the linear regression option from the list of available regression options.

After step 3 has been completed, the calculator will display values for a and b, the slope and y-intercept, respectively. Some calculators will also display a third number, r, the coefficient of correlation. Although a full discussion of the significance of r is beyond the scope of this text, we note that when r is close to either 1 or -1, the regression line is a good fit to the data. Note also that some calculators use $y = a + bx$ instead of $y = ax + b$ as the general form for the regression line. In this case, when the values for a and b are reported, a gives the y-intercept and b gives the slope. Be sure to check your calculator to see which form is used.

EXAMPLE 9   Performing Linear Regression with a Graphing Calculator

Find an equation of the form $y = ax + b$ for the line with the least-squares best fit to the points $(1, 5)$, $(2, 6)$, and $(6, 13)$.

SOLUTION   We begin by clearing any existing statistical data and entering the points $(1, 5)$, $(2, 6)$, and $(6, 13)$ as statistical data. Next we select the linear regression feature. The calculator should return a screen much like the one in Figure 102, which

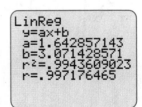

FIGURE 102

indicates that $a \approx 1.64$, $b \approx 3.07$, and $r \approx 0.997$. Thus, the regression line has equation $y = 1.64x + 3.07$. Since the value of r is very close to 1, the line is a good fit.

---

skip  **EXAMPLE 10** ■ **Finding the Least-Squares Best Fit**

a. Use the data in Table 12 to find a linear model with the least-squares best fit.

TABLE 12

Year	1985	1986	1987	1988	1989	1990
U.S. health-care expenditure (in billions of dollars)	422.6	454.8	494.1	546.0	602.8	666.2

Data source: 1992 World Almanac.

b. Use the model developed in part a to predict the total cost of health care in the year 2003.

SOLUTION

a. Let t represent the number of years after 1985, and $E(t)$ the total U.S. health-care expenditure (in billions of dollars) for year t. Then we wish to obtain a model of the form $E(t) = at + b$. We begin by clearing any existing statistical data from the calculator and then entering the following data points.

$$(0, 422.6), (1, 454.8), (2, 494.1), (3, 546.0), (4, 602.8), (5, 666.2)$$

Next we employ the linear regression feature to obtain $a \approx 48.97$, $b \approx 408.66$, and $r \approx 0.99$. We conclude that the linear model with the least-squares best fit to the given data has equation $E(t) = 48.97t + 408.66$, and because the coefficient of correlation r is nearly 1, the fit is a good one. Figure 103 shows a graph of the least-squares best fit line along with a plot of the data points themselves.

b. Recall that t represents the number of years after 1985. Thus, the year 2003 corresponds to $t = 18$. Substituting 18 for t in the linear model gives

$$E = 48.97t + 408.66 = 48.97 \cdot 18 + 408.66 \approx 1290.12$$

Thus, according to this model, the total of all health-care expenditures in the year 2003 will be approximately $1290 billion.

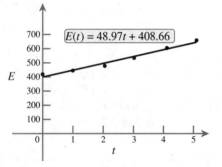

FIGURE 103

UNDERSTANDING AND MASTERY CHECKLISTS

CONCEPTS TO UNDERSTAND

- ☐ Mathematical models
- ☐ Variation: direct, joint, and inverse
- ☐ Constant of proportionality
- ☐ Data fitting
- ☐ Least-squares best fit
- ☐ Linear regression

SKILLS TO MASTER

- ☐ Determine the constant of proportionality.
- ☐ Solve a variation problem.
- ☐ Find a function of a given type whose graph passes through given data points.
- ☐ Find the equation of a line with the least-squares best fit to a collection of data points.
- ☐ Perform linear regression on a collection of data points and use the result to estimate missing data.

EXERCISES 3.7

EXERCISES 1–6 *Find the constant of proportionality.*

1. y varies directly as x, and $y = 8$ when $x = 2$.

2. z varies directly as y, and $z = 10$ when $y = 5$.

3. z varies inversely as x, and $z = 12$ when $x = 4$.

4. y varies inversely as t, and $y = 8$ when $t = \frac{1}{2}$.

5. w varies jointly as x and y, and $w = 15$ when $x = \frac{1}{2}$ and $y = 3$.

6. z varies jointly as s and t, and $z = 2$ when $s = 10$ and $t = 20$.

EXERCISES 7–12 *Solve the variation problem.*

7. w varies directly as x. Find w when $x = 4$, given that $w = 9$ when $x = 3$.

8. y varies directly as x. Find y when $x = 6$, given that $y = 10$ when $x = 3$.

9. q varies jointly as x and y. Find q when $x = 2$ and $y = 5$, given that $q = 90$ when $x = 3$ and $y = 5$.

10. A varies jointly as x and y. Find A when $x = 12$ and $y = 5$, given that $A = 18$ when $x = 3$ and $y = 10$.

11. V varies directly as the square of x and inversely as the cube of y. Find V when $x = 3$ and $y = 2$, given that $V = 2$ when $x = 4$ and $y = 3$.

12. W varies jointly as x and y and inversely as z. Find W when $x = 5$, $y = 3$, and $z = 1$, given that $W = 2$ when $x = 7$, $y = 2$, and $z = 4$.

EXERCISES 13–18 *Find the function of the given form whose graph passes through the indicated points.*

13. $f(x) = \dfrac{k}{x^2}; (-3, 2)$

14. $f(x) = ax + b; (2, 5), (4, 9)$

15. $f(x) = ax^2 + bx; (2, -4), (-1, 5)$

16. $g(x) = a(x + 2)(x - 3) + bx(x + 2) + cx(x - 3); (0, 1), (-2, 4), (3, 10)$

17. $h(x) = ax^2 + bx + c; (0, 5), (1, 3), (2, 3)$

18. $g(x) = ax^2 + bx + c; (-1, -1), (0, 4), (1, 5)$

EXERCISES 19–22 *Find the equation of the line with the least-square best fit.*

19. $(1, 3), (4, 4), (7, 6)$

20. $(-1, -4), (-2, -2), (-4, 0)$

21. $(-1, -6), (0, -5), (1, -4), (2, -1)$

22. $(1, 2), (2, 5), (3, 4), (4, 7)$

EXERCISES 23–26 *Find the line with the least-squares best fit to the given data points, and then use the equation of the line to predict the unknown y-value.*

23.

x	0	5	10	15	20	25
y	90	101	116	130	138	?

24.

x	0	4	8	12	16	20
y	35	45	56	69	78	?

25.

x	0	1	2	3	4	5
y	80	45	12	?	-50	-78

26.

x	0	2	4	6	8	10
y	200	130	?	-10	-60	-120

APPLICATIONS

27. *Board Feet* Suppose that the number of board feet of lumber in a ponderosa pine varies directly as the cube of its circumference at waist height. If a ponderosa pine with a circumference of 100 inches yields 1500 board feet of lumber, how much can be obtained from one with a circumference of 120 inches?

28. *Falling Ball* The distance traveled by a ball dropped from the top of a tall building varies directly as the square of the time since its release. If a ball has fallen 64 feet after 2 seconds, how far will it have fallen after 3 seconds?

29. *Kinetic Energy* The kinetic energy of a body in motion varies jointly as the mass of the object and the square of its velocity. If the kinetic energy of a 4000-pound vehicle traveling at a rate of 30 miles per hour is 200,000 joules, find the kinetic energy of the same vehicle traveling 60 miles per hour.

30. *Potential Energy* The potential energy of an object at rest near the surface of Earth varies jointly as its mass and height above the ground. If an object with a mass of 10 kilograms is 3 meters above the ground and has a potential energy of 294 joules, find the potential energy of an object with a mass of 15 kilograms that is 12 meters above the ground.

31. *Coulomb's Law* According to Coulomb's law, the electric force exerted between two charged particles at rest varies jointly as the charges of the particles and inversely as the square of the distance between the particles. Two electrons, each with a charge of 1.6×10^{-19} coulomb, are 1 centimeter apart and exert a force of 2.3×10^{-24} newton. How much force is exerted if the electrons are 2 centimeters apart?

32. *Grasshopper Jump* If a $1\frac{1}{2}$-inch-long grasshopper can jump 30 inches, and if jumping ability varies directly as length, how far could a 6-foot-tall human jump? Do you think jumping ability varies directly as length? Explain.

33. *Ant Strength* If an ant weighing 0.01 gram can lift an object weighing 0.2 gram, and if strength varies directly as body weight, how much could a 150-pound human lift? Do you think strength varies directly as body weight? Explain.

Leaf cutter ants

34. *Skidding Vehicles* The force needed to prevent a vehicle from skidding on a circular curve varies jointly as the weight of the car and the square of the speed and inversely as the radius of the curve. If it takes 4840 pounds of force to keep a certain car traveling at a certain speed from skidding on a curve, how much force will it take to prevent a truck that is twice as heavy but is traveling half as fast from skidding on the same curve?

35. *Powerlifting Records* The following table shows several weight classes of the men's world powerlifting bench-press records. Find the line with the least-squares best fit to the data, and estimate the bench-press record for the 100-kilogram weight class. The actual record is 261.5 kilograms, held by Ed Coan.

Weight class (kg)	52	60	75	90
Bench-press record (kg)	155	180	217.5	255

Data source: 1993 Guinness Book of World Records.

36. *Blood Serum Ethanol Level* Ethanol is a form of alcohol found in beer, wine, and other liquors. It is eliminated from the body by the liver, via an enzymatic process, at a constant rate. The following table shows the blood serum ethanol level for a 150-pound male taken at half-hour intervals following four 12-ounce beers. Find the line with the least-squares best fit to the data and predict when the level is below 0.001 (the legal limit for operating a motor vehicle in many states).

Time (minutes)	0	30	60	90
Blood serum ethanol ($\times 10^{-3}$)	1.56	1.42	1.29	1.15

37. *Ozone Depletion* Chlorofluorocarbons (CFCs) are a family of chemicals used as refrigerants and in the manufacture of numerous consumer products. The increasing concentration of CFCs in the environment has been shown to be a cause of the depletion of the ozone layer. The following table shows the rise in atmospheric concentration of one member of this chemical family, CFC-11, from 1980 to 1990. Find the line with the least-squares best fit to the data and predict the concentration of CFC-11 in 2002.

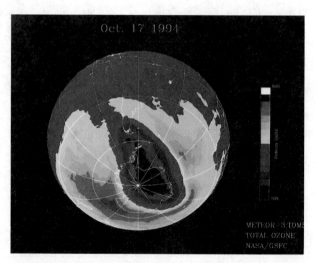

An elongated ozone hole covers much of Antarctica and the tip of South America.

Year	1980	1982	1984	1986	1988	1990
Concentration of CFC-11 (parts per trillion)	180	195	215	230	255	275

Data source: 1993 Environmental Almanac.

38. *Water Consumption* The use of bottled water in the United States has shown a steady increase in recent years. The following table shows the per capita (that is, per person) consumption for the years 1985–1990. Find the line with the least-squares best fit to the data and predict the per capita consumption in 2000.

Year	1985	1986	1987	1988	1989	1990
Bottled water consumption (gallons per person per year)	4.4	5.1	5.7	6.4	7.3	8.0

Data source: 1993 Environmental Almanac.

39. *Prison Population* The following table shows the number of inmates in federal and state prisons in 1980, 1985, and 1990. Find a quadratic model that expresses the number of prisoners as a function of the year, with 1980 corresponding to year 0. Predict the number of inmates in the year 2000.

Year	1980	1985	1990
Number of inmates (in thousands)	320	480	740

Data source: U.S. Bureau of Justice Statistics.

40. *Average Income* The average yearly earnings for a year-round full-time worker are related to education level. Use the following data from 1990 to find a quadratic model that expresses the average earnings as a function of the number of years of education beyond 8th grade. According to the model you found, what average earnings could be expected for someone with a master's degree (usually 17 years of education)? What educational level corresponds to the smallest average earnings? Does this seem reasonable?

Years of education beyond 8th grade	0	4	8
Average earnings	$16,000	$24,000	$36,000

Data source: U.S. Bureau of the Census.

CONCEPTS AND CRITICAL THINKING

EXERCISES 41–44 *Answer true or false.*

41. If y varies directly as x, then doubling x will cause y to double as well.

42. The line with the least-squares best fit to a set of points must pass through at least one of the points.

43. If y varies inversely to x, then doubling x will cause y to double as well.

44. The line with the least-squares best fit is the line for which the sum of the squares of the errors is as small as possible.

EXERCISES 45–48 *Give an example of each.*

45. A collection of three points for which the regression line passes through all three points

46. An equation involving two variables such that one variable varies directly with respect to the other

47. A collection of three data points and a quadratic model that fits the data points with absolutely no error

48. A type of variation

QUESTIONS FOR DISCUSSION OR ESSAY

49. Explain what is meant by the statement, "Three distinct nonlinear points completely determine a parabola." Is the statement true? Why or why not? What would happen if you tried to fit a function of the form $f(x) = ax^2 + bx + c$ to three points that lie on a straight line?

50. Discuss the primary hazards in the use of mathematical models for predicting real-world phenomena. What safeguards must be taken in order to minimize the risks?

51. The area of a circle varies directly as the square of the radius. If the radius of a circle is doubled, does this mean that the area will double? Explain, using an example to illustrate. In general, if a quantity y varies directly as the nth power of x, what happens to y if x is doubled? Justify your answer.

52. If the surface area of an object of a given shape varies directly as the square of its length and the volume varies directly as the cube of its length, then what can you say about the ratio of volume to surface area? How does this discussion relate to the rate at which small hamburgers grill compared to large ones? What about the rate at which large ice cubes melt compared to small ones?

53. If all men were shaped similarly, then body weight w would vary directly with the cube of height h, so that $w = kh^3$, as mentioned in Example 4. Clearly, men are not all of the same shape. But what about the shape of the "typical" 5-foot-tall man? Is it the same as that of the "typical" 6-foot-tall man or "typical" 7-foot-tall man? Using the units of pounds for weight and feet for height, estimate appropriate constants of proportionality for 5-footers, 6-footers, and 7-footers. Explain your results.

◼ **PROJECTS FOR ENRICHMENT**

54. *Transformed Least-Squares Fit*   We have seen how the least-squares method can be used to find the line of the form $y = ax + b$ that has the least-squares best fit to a set of data points. Here we consider a slight extension that will enable us to fit an equation of the form $y = a \cdot u(x) + b$, where u is some expression involving x. Consider, for example, the set of data points shown in Table 13 and graphed in Figure 104.

TABLE 13

x	0	1	2	3
y	-1.3	-0.8	0.7	3.2

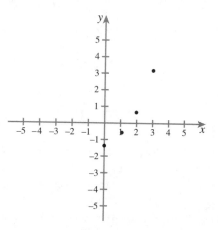

FIGURE 104

From the graph, we suspect that the data points lie on the graph of an equation of the form $y = ax^2 + b$. Notice that if we define $u = x^2$, this equation has the form $y = au + b$, which is linear in u (or x^2) and y. Thus, we may be successful in looking for a linear fit for the data points that are *transformed* in Table 14 by pairing each y with x^2. Indeed, the plot of x^2 versus y in Figure 105 does look linear. So we find the line with the least-squares best fit to the data in Table 14. We find $a = 0.5$ and $b = -1.3$. Our model is thus $y = 0.5x^2 - 1.3$, and it is not hard to check that this equation fits the original data points very well.

For each of the sets of data in parts a through d, find an equation of the given form that has the best transformed least-squares fit. Plot each equation along with the data points to show that the fit is good.

TABLE 14

x^2	0	1	4	9
y	-1.3	-0.8	0.7	3.2

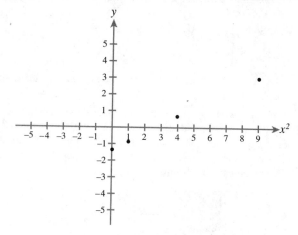

FIGURE 105

a. $y = ax^3 + b$

x	0	1	2	3
y	2.2	1.8	0.8	-2.4

b. $y = ax^4 + b$

x	1	$\sqrt{2}$	2	$\sqrt{5}$
y	-4.1	-3.9	-2.8	-1.8

c. $y = a\sqrt{x} + b$

x	1	2	3	4
y	11	13	14	16

d. $y = a\dfrac{1}{x} + b$

x	-3	-1	2	4
y	6	7	4	5

55. *Car Models*   Table 15 provides a compilation of data on ten different 1997 automobiles. In this project, we investigate relationships between the various characteristics given in the table. For example, we will see how gas mileage relates to engine size, power, and weight. You will need a graphing calculator or spreadsheet that is capable of performing linear regression (least-squares best fit). A spreadsheet may be useful for organizing and plotting the data.

a. Describe any general relationships you see in the table. For example, larger engine sizes tend to generate more horsepower.

TABLE 15

Make and Model	Engine Size (liters)	Power (hp)	Weight (lb)	Top Speed (mph)	Highway Mileage (mpg)
BMW 328i	2.8	190	3197	128	29
Cadillac Eldorado	4.6	275	3821	148	26
Chevrolet Camaro Z28	5.7	285	3442	149	25
Dodge Neon	2.0	132	2428	117	33
Ferrari 456 GT	5.5	436	3898	185	16
Ford Taurus LX	3.0	145	3326	112	29
Honda Civic LX	1.6	106	2387	116	35
Jaguar XJ6	4.0	245	4080	144	23
Mercedes Benz C36	3.6	276	3549	155	22
Toyota Camry LE	2.2	133	3020	116	30

b. Let y denote the top speed of a car and let x denote its engine size. Plot the set of points (x, y) given by Table 15. Do the points appear to lie on (or near) a single straight line? Find the least-squares best fit model of the form $y = ax + b$ for the engine size and top speed data in Table 15.

c. Repeat part b using power instead of engine size as the x variable. How do your results compare? Which appears to be the better predictor of top speed, power or engine size? (*Hint:* Compare the r values for the two models.)

d. Repeat part b using weight instead of engine size as the x variable. How do your results compare? Which of engine size, power, or weight appears to be the best predictor of top speed? Use your best predictor to estimate the top speed of a car with a 2.4-liter engine, capable of 141 hp, and weighing 2822 lb.

e. Develop and compare three linear models for predicting the highway gas mileage of a car as a function of engine size, power, and weight. Which of the three appears to be the best predictor? Use your best predictor to estimate the highway gas mileage of a car with a 2.4-liter engine, capable of 141 hp, and weighing 2822 lb.

So far, we have attempted to model the top speed and gas mileage of a car as a function of just one of the variables engine size, power, or weight. In actuality, all three of these variables, as well as perhaps others, play roles of varying importance in determining the top speed and gas mileage of a car.

f. Identify as many variables as you can think of that might affect top speed and gas mileage.

Using a technique called *multiple linear regression,* we can take more than one variable into account. For the following discussion, let s denote the top speed of a car, m denote the highway gas mileage, e denote the engine size, p denote the power, and w denote the weight. Applying multiple linear regression to the data in Table 15, we obtain the following two models:

$$s = 0.27p - 2.89e + 91.1$$
$$m = -0.004w - 0.05p + 1.15e + 47$$

g. Notice that the coefficient of w is very small in the gas mileage model and that w doesn't even appear in the top speed model. What does this suggest?

h. Find the top speed and highway gas mileage for a car with a 2.4-liter engine, capable of 141 hp, and weighing 2822 lb. A Mitsubishi Galant with these characteristics has a top speed of 130 mph and highway gas mileage of 28 mpg. How do these values compare to those predicted by the models? How might this information be used?

CHAPTER 3 REVIEW

EXERCISES 1–4 *Find the slope of the line satisfying the given properties.*

1. Passing through $(2, 5)$ and $(-1, 4)$

2. Passing through $(2, 5)$ with y-intercept 5

3. With equation $y = 4x - 7$

4. With equation $3x + 4y = 6$

EXERCISES 5–20 *Find the equation of the line satisfying the indicated properties. Express your answer in slope-intercept form* $y = mx + b$, *if possible.*

5. Slope 3 and passing through $(1, 5)$

6. Slope $-\frac{1}{2}$ and passing through $(-4, 8)$

7. Slope undefined and passing through $(2, \pi)$

8. Slope 0 and passing through $(5, -2)$

9. Passing through $(-1, 6)$ and $(3, -2)$

10. Passing through $(2, 4)$ and $(-2, 7)$

11. Passing through $(\frac{1}{2}, 3)$ and $(-2, \frac{3}{4})$

12. Passing through $(4.1, -1.2)$ and $(2.3, 6.6)$

13. With slope -2 and y-intercept 2

14. Slope $\frac{1}{3}$ and x-intercept -1

15. With y-intercept 5 and x-intercept 1

16. Slope undefined and y-intercept 0

17. Parallel to $y = -3x + 1$ and passing through $(7, 1)$

18. Passing through $(-2, 4)$ and perpendicular to the line containing $(3, -1)$ and $(5, -1)$

19. Perpendicular to $2x - 3y = 1$ and passing through $(-4, 2)$

20. Parallel to $5x + 2y = 10$ and passing through $(0, 0)$

EXERCISES 21–24 *Use the following table to determine whether the indicated correspondence from the set D to the set R defines a function.*

Chicago Bulls 1997–98 Roster		
Player	**Height**	**Weight**
Keith Booth	6-6	226
Randy Brown	6-2	191
Jud Buechler	6-6	228
Scott Burrell	6-7	218
Ron Harper	6-6	216
Michael Jordan	6-6	216
Steve Kerr	6-3	181
Toni Kukoc	6-11	232
Rusty LaRue	6-2	185
Luc Longley	7-2	292
Scottie Pippen	6-7	228
Dennis Rodman	6-6	220
Dickey Simpkins	6-10	264
Bill Wennington	7-0	277

21. $D =$ the set of players, $R =$ the set of heights

22. $D =$ the set of heights, $R =$ the set of weights

23. $D =$ the set of weights, $R =$ the set of heights

24. $D =$ the set of players, $R =$ the set of weights

EXERCISES 25–28 *Determine whether the given equation defines y as a function of x.*

25. $y - x^2 + 1 = 0$

26. $x - y^3 = 1$

27. $x^2 + y^2 = 5$

28. $x = y^4 - 2$

EXERCISES 29–34 *Evaluate the given function as indicated.*

29. $g(x) = 2x^2 - 3x + 1$
   a. $g(4)$
   b. $g(-3)$

30. $h(x) = \dfrac{x}{2x + 5}$
   a. $h(0)$
   b. $h(-5)$

31. $f(x) = \dfrac{x^2}{x^2 + 1}$
   a. $f(-x)$
   b. $f(x + 1)$

32. $g(x) = (x + 2)^2$
   a. $g(x - 2)$
   b. $g(\frac{1}{x})$

33. $h(x) = \begin{cases} 4 - x^2, & x \le 2 \\ -x + 4, & x > 2 \end{cases}$
   a. $h(0)$
   b. $h(5)$
   c. $h(a)$ if $a > 2$

34. $f(x) = \dfrac{|x - 3|}{x - 3}$
   a. $f(-2)$
   b. $f(10)$
   c. $f(c)$ if $c < 3$

EXERCISES 35–38 *Find the domain of the function.*

35. $f(x) = x^3 + 1$

36. $g(x) = \sqrt{3 - x}$

37. $h(t) = \dfrac{t}{\sqrt{2t + 1}}$

38. $f(x) = \dfrac{x^2}{4 - x^2}$

EXERCISES 39–42 *Use a graphing calculator to estimate the domain and range.*

39. $h(x) = \sqrt{x^2 - 4}$

40. $f(x) = \dfrac{2x^2}{x^2 + 1}$

41. $f(x) = \dfrac{|1 - x|}{1 - x}$

42. $g(x) = \dfrac{\sqrt{x + 1}}{\sqrt{x + 1}}$

EXERCISES 43–44 *A function f and its graph are given. For each translated or reflected graph, determine the corresponding function g.*

43. $f(x) = 2x^3 - 6x + 1$

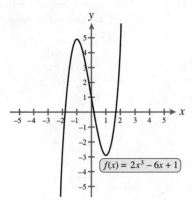

$f(x) = 2x^3 - 6x + 1$

44. $f(x) = x^3 - 3x^2 + 1$

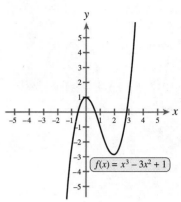

$f(x) = x^3 - 3x^2 + 1$

a.

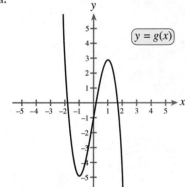

$y = g(x)$

b.

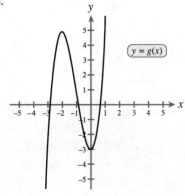

$y = g(x)$

a.

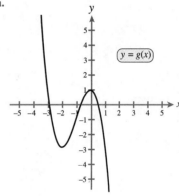

$y = g(x)$

b.

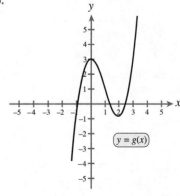

$y = g(x)$

EXERCISES 45–46 *Use the graph of f to sketch the graph of h.*

45.

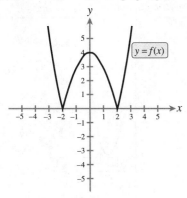

a. $h(x) = f(x + 3)$

b. $h(x) = f(x - 2) + 1$

c. $h(x) = -f(x) - 2$

46.

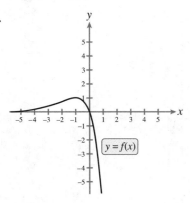

a. $h(x) = f(x) - 3$

b. $h(x) = f(x + 1) - 4$

c. $h(x) = f(-x) + 2$

EXERCISES 47–50 *Determine whether the given function is even, odd, or neither.*

47. $f(x) = x(4x^2 - 1)$

48. $g(x) = x^4 - x$

49. $f(x) = |x + 1|$

50. $h(x) = \sqrt{x^2 - 1}$

EXERCISES 51–54 *Find the exact values of all the real zeros of the given function.*

51. $h(x) = 4x - 9$

52. $f(x) = x^2 - x - \frac{5}{16}$

53. $g(x) = 3 - \frac{8}{x}$

54. $g(x) = |3x - 1| - 5$

EXERCISES 55–58 *Use a graphing calculator to estimate the following features of the given function to the nearest hundredth.*

a. *Zeros*

b. *Coordinates of the turning points (Classify as local maximum or minimum.)*

c. *Intervals on which the function is increasing or decreasing*

55. $f(x) = x^3 - 3x + 1$

56. $g(x) = 3x^4 + 4x^3 - 12x^2 + 2$

57. $h(x) = \dfrac{x^2 - 3}{2x - 4}$

58. $g(x) = |x^3 - x^2 + 1|$

EXERCISES 59–66 *Find the following combinations of the functions f and g. Specify the domain if it is anything other than the set of all real numbers.*

a. $(f + g)(x)$

b. $(fg)(x)$

c. $(f/g)(x)$

d. $(f \circ g)(x)$

e. $(g \circ f)(x)$

59. $f(x) = x^2; g(x) = 2x - 1$

60. $f(x) = 4x + 3; g(x) = \dfrac{2}{x}$

61. $f(x) = \dfrac{1}{x - 4}; g(x) = x^3$

62. $f(x) = \sqrt{x + 2}; g(x) = x^2 - 1$

63. $f(x) = 2x^2; g(x) = \sqrt{2x + 4}$

64. $f(x) = \dfrac{x}{x - 1}; g(x) = \dfrac{2}{x}$

65.

x	$f(x)$	$g(x)$
0	1	3
1	2	2
2	3	0
3	0	1

66.

x	$f(x)$	$g(x)$
−3	−2	−1
−2	0	2
−1	1	0
0	−3	3

EXERCISES 67–70 *Determine whether the given functions are inverses.*

67. $f(x) = x - 5; g(x) = x + 5$

68. $f(x) = 3x - 1; g(x) = \frac{x}{3} + 1$

69. $f(x) = x^3 - 4; g(x) = \sqrt[3]{x} + 4$

70. $f(x) = (x - 2)^2, x \geq 2; g(x) = \sqrt{x} + 2$

EXERCISES 71–78 *Determine whether the function is 1–1. If so, compute its inverse.*

71. $f(x) = 3x - 2$

72. $g(x) = x^2 + 5$

73. $h(x) = x^3 - x + 1$

74. $g(x) = (x + 2)^3 + 3$

75. $f(x) = \sqrt{x - 4}$

76. $h(x) = \dfrac{1}{x - 1}$

77.

x	$f(x)$
1	2
2	5
3	10
4	17
5	26

78.

x	$f(x)$
1	2
2	5
3	10
4	5
5	2

EXERCISES 79–80 *Use the graph of f to sketch the graph of f^{-1}.*

79.

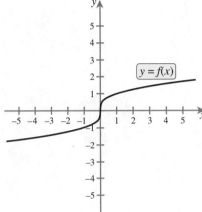

80.

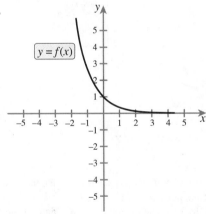

EXERCISES 81–86 *Identify the type of function that is given (linear, quadratic, piecewise-defined, or none of these), sketch its graph, and locate its intercepts. Use a graphing calculator as necessary.*

81. $h(x) = -(x - 3)^2 + 1$

82. $g(x) = \dfrac{1}{2}x - 4$

83. $f(x) = \dfrac{x + 2}{x - 2}$

84. $h(x) = \frac{1}{4}x^4 - 2x^2 + 2$

85. $g(x) = \begin{cases} 2x + 3, & x \le 1 \\ 5 - x^2, & x > 1 \end{cases}$

86. $f(x) = \begin{cases} x^2 + 4x, & x \le -2 \\ x, & x > -2 \end{cases}$

EXERCISES 87–90 *Sketch the graph of the given quadratic function and locate the exact coordinates of its vertex and intercepts.*

87. $f(x) = (x + 1)^2 - 4$

88. $f(x) = -2(x - 1)^2$

89. $g(x) = -x^2 + 8x - 17$

90. $h(x) = 3x^2 - 12x + 16$

EXERCISES 91–94 *Find the function having the given properties.*

91. f is linear, the graph of f has slope $-\frac{1}{2}$, and $f(0) = 3$

92. g is linear, the graph of g has y-intercept -2, and $g(1) = 5$

93. h is quadratic, the graph of h has vertex $(2, -1)$, and $h(4) = 3$

94. f is quadratic, the graph of f has axis of symmetry $x = -1$, $f(-1) = 4$, and $f(0) = 3$

EXERCISES 95–98 *Solve the variation problem.*

95. z varies directly as a. Find z when $a = 3$, given that $z = 20$ when $a = 4$.

96. y varies inversely as the square of x. Find y when $x = \frac{1}{4}$, given that $y = 3$ when $x = 3$.

97. T varies jointly as x and y. If $T = 1$ when $x = \frac{1}{2}$ and $y = \frac{1}{3}$, find T when $x = 2$ and $y = 3$.

98. P varies jointly as x and the square of y and inversely as the cube of z. If $P = 5$ when $x = 4$, $y = 5$, and $z = 2$, find P when $x = 2$, $y = 3$, and $z = 3$.

EXERCISES 99–100 *Find the function of the given form whose graph passes through the indicated points.*

99. $f(x) = ax^3 + b$; $(1, 2), (-1, 4)$

100. $h(x) = \sqrt{ax + b}$; $(2, 3), (10, 5)$

EXERCISES 101–102 *Find the equation of the line with the least-squares best fit to the given points.*

101. $(1, 2), (3, 3), (6, 5)$

102. $(-3, 5), (-1, 3), (2, 1), (5, -2)$

EXERCISES 103–104 *Find the line with the least-squares best fit to the given data points, and then use the equation of the line to predict the unknown y-value.*

103.

x	0	10	20	30	40	50
y	42	67	88	118	135	?

104.

x	0	3	6	9	12	15
y	205	175	150	?	90	65

EXERCISES 105–115 *Parts a and b are connected: Part a involves a concept from earlier in the text; part b involves related material from this chapter. First answer part a, and then use this result to answer part b.*

105. a. Solve $x^2 + 6 = -5x$.

    b. Find all zeros of $f(x) = x^2 + 5x + 6$.

106. a. Solve $x^2 + 3x < 1$.

    b. Find the domain of $f(x) = \sqrt{x^2 + 3x - 1}$.

107. a. Solve $y = \dfrac{1}{x + 2}$ for x.

    b. Find $f^{-1}(x)$ for the function $f(x) = \dfrac{1}{x + 2}$.

108. a. Expand $(x + 2)(x^2 + 1)$.

    b. Find all zeros of $g(t) = t^3 + 2t^2 + t + 2$.

109. a. Expand $(2x - 1)^3$.

    b. Find functions f and g such that $(f \circ g)(x) = 8x^3 - 12x^2 + 6x - 1$.

110. a. Solve $y(x - 5) = 2x - 7$ for x.

    b. Let $f(x) = \dfrac{2x - 7}{x - 5}$. Find $f^{-1}(x)$.

111. a. Simplify $\dfrac{3}{x + 5} - \dfrac{4}{x - 2}$.

    b. Find the domain and all zeros of $f - g$, where $f(x) = \dfrac{3}{x + 5}$ and $g(x) = \dfrac{4}{x - 2}$.

112. a. Solve the following equation for k.
$$9 = k\,\dfrac{4}{7}$$

    b. If y is directly proportional to x and $y = 9$ when $x = \frac{4}{7}$, find y when $x = 4$.

113. a. Simplify $(k\sqrt{x})^4$.

    b. If y is proportional to the square root of x, and z is proportional to the fourth power of y, how does z vary with x?

114. a. Graph $y = x^3 - x^2$ and $y = 2x - 1$ on the same set of coordinate axes.

    b. Estimate the zeros of $f(x) = x^3 - x^2 - 2x + 1$.

115. *Area of a Circle* Express the area of a circle as a function of its circumference.

116. *Area of a Rectangle* A point $P(x, y)$ lies on the parabola $y = 4 - x^2$, as shown in Figure 106. Express the area of the shaded rectangle as a function of x. Plot the graph of the function and estimate the value of x that will yield the largest possible area for the rectangle.

117. *Height of a Balloon* A hot-air balloon is rising vertically from a point 200 feet from an observer on the ground (Figure 107). Express the height h of the balloon as a function of its distance d from the observer. What domain of d values makes sense for the function h?

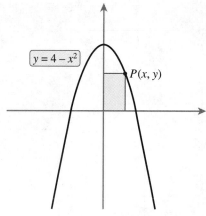

FIGURE 106

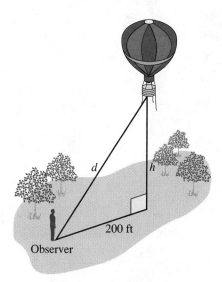

FIGURE 107

118. *Minimum Cost* A company has determined that the total production cost for manufacturing x units is $C(x) = 0.01x^3 - 4.5x^2 + 475x + 36{,}000$. Estimate the intervals on which C is increasing and the intervals on which it is decreasing. Estimate the number of units that must be manufactured to minimize the cost.

119. *Path of a Football* If a football is kicked from ground level with an initial velocity of 64 feet per second and an initial angle of 45°, and if we ignore air resistance, the ball will follow the path given by the quadratic function $f(x) = -\frac{1}{128}x^2 + x$, where x is the horizontal distance in feet from where the ball was kicked and $f(x)$ is the corresponding height of the ball in feet.

    a. Find the maximum height of the ball.

    b. Assuming the ball is kicked straight, will it clear a 10-foot-high goal post 30 yards away? What is the farthest distance from which the ball could be kicked and still clear the goal post?

120. *Overnight Rate* Suppose an overnight package delivery service charges $8.50 for the first pound and $2.00 for each additional pound or fraction thereof. Then the rate for a package weighing x pounds can be determined using the function $C(x) = 8.5 - 2\lfloor 1 - x \rfloor$. Plot the graph of this function. What is the heaviest package that can be sent if the cost cannot exceed $31?

121. *Volunteer Model* The percentage of the adult population doing volunteer work is related to educational level, as suggested by the following table. Find the line with the least-squares best fit to the data and predict the percentage of the population with 18 years of education who are involved in volunteer work.

Years of education	10	12	14	16
Percent doing volunteer work	8.3	18.8	28.1	38.4

Data source: U.S. Bureau of Labor.

122. *Accidental Death Rate* The death rate for work-related non-manufacturing accidental deaths has been on the decline in recent years. The following table shows the death rates for the years 1970, 1975, 1980, 1985, and 1990. Find the line with the least-squares best fit to the data and predict the death rate for the years 1995 and 2020. Are your answers reasonable? Explain.

Year	1970	1975	1980	1985	1990
Death rate (per 100,000)	21	17	15	12	10

Data source: 1991 Accident Facts, National Safety Council.

Construction of the Hoover Dam in the 1930s resulted in 114 accidental deaths.

123. *Compact Disc Sales* The following table shows the number of CDs sold during each of the years 1985, 1988, and 1989. Find a quadratic model that expresses the number of CDs sold as a function of the year, with 1985 corresponding to year 0. Predict the number of CDs sold in 1992.

Year	1985	1988	1989
CDs sold (millions)	23	150	207

Data source: Inside the Recording Industry: A Statistical Overview, Recording Industry Association of America.

CHAPTER 3 TEST

PROBLEMS 1–8 *Answer true or false.*

1. If a horizontal line intersects the graph of an equation in more than one point, the equation does not define y as a function of x.

2. Unless otherwise specified, we assume that the domain of a function f is the set of all input values such that the expression $f(x)$ is defined.

3. If $h(x)$ is defined by $h(x) = f(x - 2) + 3$, then the graph of h can be obtained by translating the graph of f 2 units to the right and 3 units upward.

4. An even function with domain all real numbers cannot be 1–1.

5. Only functions satisfying the horizontal line test are 1–1, and only 1–1 functions have inverse functions.

6. The graph of a piecewise-defined function is really just the graph of several different functions plotted on the same set of coordinate axes.

7. If y varies directly as x, then y is also a linear function of x.

8. Linear regression is the process of finding an equation of the line that passes through a given set of data points.

PROBLEMS 9–14 *Give an example of each.*

9. A linear function and its inverse

10. A quadratic function and the coordinates of the vertex of its graph

11. A piecewise-defined function that is linear over each of its three pieces

12. A set of three data points such that the regression line passes through all of the points

13. A function, the graph of which is symmetric with respect to the origin

14. A polynomial function that is neither even nor odd

PROBLEMS 15–16 *Find an equation of the line with the given properties.*

15. Passes through the points $(-2, 3)$ and $(-2, 7)$

16. Passes through the point $(-1, 4)$ and is perpendicular to the line with equation $4x + 3y = 12$

17. Determine whether the equation $(1/y) + x = 2$ defines y as a function of x. If it does, find the domain of the function.

18. Verify or disprove that

$$f(x) = \frac{3x + 1}{2} \quad \text{and} \quad g(x) = \frac{2x - 1}{3}$$

are inverses of one another.

19. Given that

$$f(x) = \begin{cases} 3x + 1, & x \le 2 \\ 4x - 5, & x > 2 \end{cases} \quad \text{and} \quad g(x) = x^2$$

compute each of the following quantities.

  a. $f(3)$

  b. $(f \circ g)(\sqrt{2})$

  c. $(g \circ f)(3)$

20. Find the domain of the function

$$f(x) = \frac{\sqrt{x - 2}}{x - 3}$$

21. Use the graph of f given in Figure 108 to produce the graph of $h(x)$, where $h(x) = f(x - 1) - 3$.

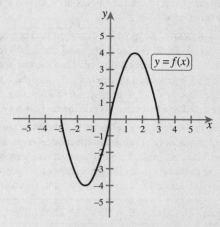

FIGURE 108

22. Find the coordinates of all zeros and turning points of $f(x) = x^3 - 6x^2$ to the nearest hundredth, and classify each turning point as a local maximum or local minimum. In addition, indicate the intervals on which f is increasing and the intervals on which it is decreasing.

23. Find the coordinates of the vertex of $f(x) = 3x^2 - 6x + 10$, and sketch the graph.

24. It is given that z varies inversely as x and directly as y. Find z when $x = 3$ and $y = 4$, given that $z = 5$ when $x = 2$ and $y = 3$.

25. It is given that $f(x) = ax^2 + bx$, and that $(2, -4)$ and $(-1, 5)$ are points on the graph of f. What is $f(3)$?

26. The *1994 Guinness Book of World Records* gives the following world records for women's track and field events:

- 100 meters: Delorez Florence Griffith Joyner (USA)   10.49 seconds

- 200 meters: Delorez Florence Griffith Joyner (USA)   21.34 seconds

- 400 meters: Marita Koch (East Germany)   47.60 seconds

Use linear regression to estimate the world record time for 800 meters.

4 POLYNOMIAL AND RATIONAL FUNCTIONS

This mountain range exists only in cyberspace; it is a computer-generated *fractal landscape*. Fractals are self-similar; that is, a small portion of a fractal appears to be a scaled down version of the whole. Compare the appearance of the mountain peak with the mountain itself and, in turn, the mountain with the entire range of mountains. Fractals are used to model many aspects of nature, including coastlines, clouds, ocean waves, and plants, all of which have self-similarity properties. In this chapter we will investigate polynomial functions, an essential tool for constructing computer-generated fractal images.

SECTION 4.1   POLYNOMIAL FUNCTIONS

□   **How have fertility rates fluctuated in the 20th century?**

□   **Why is it that polynomial functions can provide such startlingly accurate models of real-world phenomena for short time periods and yet are nearly useless for making long-term predictions?**

□   **How can polynomials be used to generate stunning fractal images?**

□   **How can we ever be certain that a graphing calculator is showing us all the important graphical features of a polynomial?**

From a strictly mathematical point of view, polynomials can be viewed as fundamental building blocks from which other important classes of functions can be constructed. And as we will discover, because of their simplicity, variety, and the smoothness of their graphs, polynomials are often used to construct mathematical models of real-world phenomena. In Section 3.6, we explored first- and second-degree polynomials and their graphs (lines and parabolas) in some detail. Much of this chapter, and this section in particular, is devoted to extending our analysis to higher-degree polynomials. We begin with some definitions.

In Chapter 1, a monomial was defined as an expression of the form ax^k (such as $2x^3$ or $5x^2$), where a is a real number and k is a nonnegative integer. We then defined a polynomial to be a sum of monomials, such as $2x^3 - 5x^2 + 1$, and identified the degree of a polynomial as being the highest power of the variable, so that, for example, the degree of $2x^3 - 5x^2 + 1$ is 3. A *polynomial function* is simply a function defined by a polynomial. More formally, we have the following definition.

Definition of a □
Polynomial Function

A **polynomial function of degree n** has the form

$$f(x) = a_n x^n + a_{n-1} x^{n-1} + \cdots + a_1 x + a_0$$

where n is a nonnegative integer and $a_n \neq 0$. The **coefficients** of $f(x)$ are the numbers $a_1, \ldots, a_n$

Polynomial functions of degree 1, such as $f(x) = \frac{1}{2}x - 3$, and polynomial functions of degree 2, such as $g(x) = 2x^2 + 4x + 5$, can be graphed by hand using the techniques for lines and parabolas discussed in Chapter 3. On the other hand, it is often quite tedious to sketch the graphs of polynomial functions of degree 3 or higher. For this reason, we often rely on a graphing calculator to help us with polynomial graphs. Still, as we have seen many times, a poor choice of viewing window can lead to misleading results, and so a basic understanding of the graphical features of polynomials is essential.

END BEHAVIOR OF POLYNOMIALS

The behavior of the graph of a function to the far right and to the far left (that is, for large positive and negative values of the independent variable) is called the **end behavior** of the function. Consider the polynomial function $f(x) = x^3$. For large positive values of x, $f(x)$ will be very large. For example, $f(1000) = 1000^3 = 1,000,000,000$. In fact, as x gets larger and larger, $f(x)$ becomes arbitrarily large. We say that $f(x)$

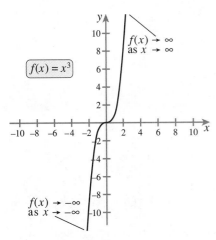

$f(x) = x^3$

$f(x) \to \infty$
as $x \to \infty$

$f(x) \to -\infty$
as $x \to -\infty$

FIGURE 1

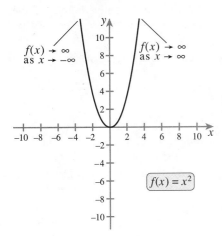

$f(x) \to \infty$
as $x \to -\infty$

$f(x) \to \infty$
as $x \to \infty$

$f(x) = x^2$

FIGURE 2

approaches infinity as x approaches infinity, and we write $f(x) \to \infty$ as $x \to \infty$. Similarly, for large negative values of x, $f(x)$ will be a very large negative number. For example, $f(-1000) = -1,000,000,000$. Thus, $f(x) \to -\infty$ as $x \to -\infty$. The graph of f provides compelling evidence of this end behavior. Notice how the graph soars upward as we move to the far right and plummets sharply as we move to the far left, as shown in Figure 1.

Let us now determine the end behavior of the polynomial function $f(x) = x^2$. When a large positive number is squared, the result is a very large positive number. For example, $1000^2 = 1,000,000$. If a large negative number is squared, the result is also a large positive number. For example, $(-1000)^2 = 1,000,000$. Thus, $f(x) \to \infty$ as $x \to \infty$ and $f(x) \to \infty$ as $x \to -\infty$. These facts can also be determined by noting that, since the graph of $y = x^2$ is a parabola opening upward, the function values are approaching infinity on the far right and the far left, as shown in Figure 2.

The difference in the end behaviors of x^2 and x^3 arises from the fact that x^2 is never negative, whereas x^3 *is* negative for $x < 0$. More generally, the end behavior of x^n depends on whether n is even or odd: For n odd, x^n behaves like x^3, whereas when n is even, x^n behaves like x^2.

Generalizing still further, the end behavior of any function of the form $f(x) = ax^n$ can be determined by considering the sign of ax^n. Although this isn't difficult to determine, we can save ourselves some effort by noting that the end behavior of ax^n will either be identical to or opposite that of x^n, according to whether a is positive or negative. Collecting these results gives us the following strategy for determining the end behavior of monomial functions.

RULE OF THUMB To determine the end behavior of $f(x) = ax^n$:
1. First determine the end behavior of x^n. Use the fact that x^n behaves like x^3 for n odd and like x^2 for n even. That is, for n even, the graph of x^n approaches ∞ on both ends, whereas for n odd, the graph approaches $-\infty$ on the left and ∞ on the right.
2. If $a > 0$, ax^n behaves like x^n; if $a < 0$, the end behavior of ax^n will be opposite that of x^n, in the sense that where the graph of one goes up, the other will go down.

EXAMPLE 1 ▧ **Determining the End Behavior of Monomials**

Determine the end behavior of each of the following functions and match each to its graph.

a. $f(x) = x^4$ **b.** $g(x) = x^7$
c. $h(x) = -2x^3$ **d.** $p(x) = -3x^2$

i.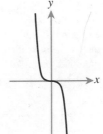

ii.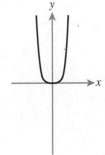

iii.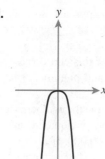

iv.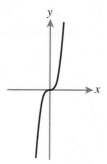

SOLUTION

a. Since the exponent 4 is even, $f(x) = x^4$ behaves like x^2. Thus, $f(x) \to \infty$ as $x \to \infty$ and also as $x \to -\infty$. The graph is (ii).

b. Since the exponent 7 is odd, $g(x) = x^7$ behaves like x^3. Thus, $g(x) \to \infty$ as $x \to \infty$, and $g(x) \to -\infty$ as $x \to -\infty$. The graph is (iv).

c. The graph of $y = x^3$ tends toward infinity on the right and negative infinity on the left. Multiplication by -2 reverses the end behavior. Thus, $h(x) = -2x^3 \to -\infty$ as $x \to \infty$, and $h(x) \to \infty$ as $x \to -\infty$. The graph is (i).

d. The exponent is even, but multiplication by -3 changes the sign. Thus, $p(x) \to -\infty$ as $x \to \infty$ and also as $x \to -\infty$. The graph is (iii).

Thus far, we have dealt only with *monomials*—that is, polynomials having only one term. For a more general polynomial, it is tempting to simply graph the function using a graphing calculator and then to note whether the graph is rising or falling on the far right and on the far left of the viewing window. The difficulty is in knowing how *wide* the viewing window should be, as the following example illustrates.

EXAMPLE 2 **Investigating End Behavior Graphically**

Investigate the end behavior of $f(x) = x^3 - 1,000,000x^2$ with a graphing calculator.

SOLUTION We have chosen a very naive approach: Begin with a standard viewing window and make adjustments until we are confident that we are, in fact, witnessing the "true" end behavior of the function. The main steps are shown in Table 1; considerations of space prevent us from showing all of the missteps that we might have taken along the way.

The final graph in Table 1 suggests that $f(x) \to \infty$ as $x \to \infty$, and $f(x) \to -\infty$ as $x \to -\infty$.

TABLE 1

Plot	Range	Problem	Solution
	Xmin=-10 Xmax=10 Ymin=-10 Ymax=10	The graph is invisible because of the choice of scale. In fact, the graph is "on top of" the y-axis.	Increase the range of y-values.
			(continued)

TABLE 1   *(continued)*

Plot	Range	Problem	Solution
	Xmin=-10 Xmax=10 Ymin=-1,000,000 Ymax=1,000,000	The range of *y*-values is still too small; the graph is cut off at the bottom. Also, the view is too narrow.	Widen the view by increasing the range of *x*-values. We will also make the range of *y*-values *huge*.
	Xmin=-1,000,000 Xmax=1,000,000 Ymin=-1E18 Ymax=1E18	The graph begins to turn up at the far right, but the view is too narrow to determine if this continues.	Widen the view. Again, we must increase our range of *y*-values to produce a usable plot.
	Xmin=-1E10 Xmax=1E10 Ymin=-1E30 Ymax=1E30		

Even after going through this extraordinary effort, we still have little confidence in our answer. Perhaps an even wider viewing window might show yet another turn in the graph. Again, we see the limitations of the graphing calculator. Fortunately, the end behavior of a polynomial can be determined without using a graphing calculator at all.

The end behavior of $f(x) = x^3 - 1,000,000x^2$ can be viewed as a contest between the monomials x^3 and $1,000,000x^2$. As x approaches infinity, x^3 approaches infinity, whereas $-1,000,000x^2$ approaches negative infinity. The end behavior of the sum of these two monomials—namely, $f(x)$—depends on which of them is dominant. As we will see in Exercise 54, the monomial of highest degree, which in this case is x^3, "wins." Thus, x^3 approaches infinity faster than $-1,000,000x^2$ approaches negative infinity. In fact, the end behavior of $f(x)$ is identical to that of x^3. More generally, the end behavior of a polynomial is determined by its term of highest degree.

End Behavior of Polynomials   ☐ | The end behavior of a polynomial is identical to that of its term of highest degree.

E X A M P L E   3   ▨   **Finding the End Behavior of a Polynomial**

Find the end behavior of each of the following polynomials.

a. $f(x) = x^4 - 2x^3 + 4x - 5$
b. $g(x) = 7 + 4x - 3x^2 - 2x^3$
c. $h(x) = -7x^6 - 4x^5 + 3x + 1$

SOLUTION

a. Since the term of highest degree is x^4, the end behavior of $f(x)$ will be identical to that of x^4 (see Figures 3 and 4). Thus, $f(x) \to \infty$ as $x \to \infty$ and $f(x) \to \infty$ as $x \to -\infty$.

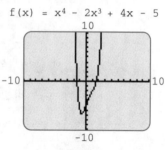

FIGURE 3

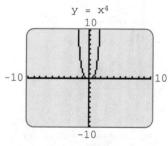

FIGURE 4

b. The term of highest degree is $-2x^3$, so the end behavior of $g(x)$ will be identical to that of $-2x^3$ (see Figures 5 and 6). Thus, $g(x) \to -\infty$ as $x \to \infty$ and $g(x) \to \infty$ as $x \to -\infty$.

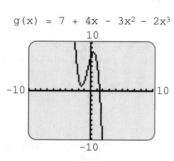

FIGURE 5

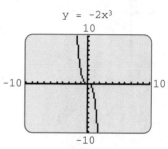

FIGURE 6

c. The term of highest degree is $-7x^6$, so the end behavior of $h(x)$ is identical to that of $-7x^6$ (see Figures 7 and 8). Since $-7x^6 \to -\infty$ as $x \to \infty$ and $-7x^6 \to -\infty$ as $x \to -\infty$, the same is true of $h(x)$. Thus, $h(x) \to -\infty$ as $x \to \infty$ and $h(x) \to -\infty$ as $x \to -\infty$.

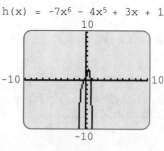

FIGURE 7

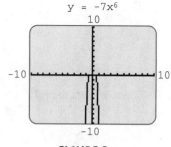

FIGURE 8

ZEROS OF POLYNOMIALS

Recall from Section 3.3 that a zero of a function f is a number c such that $f(c) = 0$. In other words, the zeros of a function are simply those input values that produce zero as an output value. We also saw in Section 3.3 that the real zeros of a function coincide with its x-intercepts. Indeed, if c is a real number for which $f(c) = 0$, then $(c, 0)$ is a point on both the graph of f and also the x-axis, and is thus an x-intercept. The zeros, and hence the x-intercepts, of a polynomial function can often be found by factoring, as in the following example.

EXAMPLE 4 ■ **Finding Zeros of a Polynomial Function**

Find the real zeros of $f(x) = 2x^3 + 6x^2 - 20x$ and locate the x-intercepts of the graph of f.

SOLUTION We first factor $f(x)$ as completely as possible.

$$f(x) = 2x^3 + 6x^2 - 20x$$
$$= 2x(x^2 + 3x - 10) \qquad \text{Factoring out the monomial } 2x$$
$$= 2x(x + 5)(x - 2) \qquad \text{Factoring}$$

Setting $2x(x + 5)(x - 2) = 0$ and solving gives us $x = 0$, $x = -5$, or $x = 2$. Thus, these are the only zeros of f, and they correspond to the x-intercepts $(0, 0)$, $(-5, 0)$, and $(2, 0)$. The graph of f is shown in Figure 9.

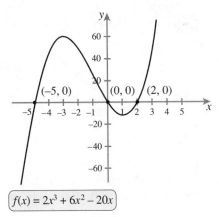

$f(x) = 2x^3 + 6x^2 - 20x$

FIGURE 9

Notice that the zeros of the polynomial function f in Example 4 were found by setting the factors of $f(x)$ equal to zero. This suggests an important connection between the zeros and factors of a polynomial. We will investigate this connection more thoroughly in upcoming sections—for now, we simply note the following.

The Number of Zeros of   □
a Polynomial

> If f is a polynomial function with real coefficients and degree n ($n \geq 1$), then f has at most n real zeros. Moreover, if n is odd, then f must have at least one real zero.

EXAMPLE 5 ■ **Finding the Zeros of a Polynomial with a Graphing Calculator**

Estimate the zeros of $f(x) = 0.05x^4 + 0.75x^3 - 0.2x^2 - 3x + 2$ to the nearest hundredth using a graphing calculator.

SOLUTION The graph shown in Figure 10 suggests that f has three zeros near the origin. By zooming in and using the trace feature, or by using the calculate-zero feature, we estimate the zeros, as shown in Figures 11–13, to be $x \approx -2.32$, $x \approx 0.73$, and $x \approx 1.57$.

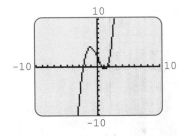

FIGURE 10

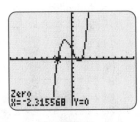

FIGURE 11

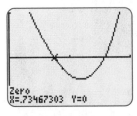

FIGURE 12

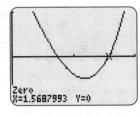

FIGURE 13

Since the term of highest degree is $0.05x^4$, we know $f(x) \to \infty$ as $x \to -\infty$, and so the graph of f must turn upward to the left of the viewing window. After zooming out, a fourth zero emerges to the left of the origin, as shown in Figure 14. We zoom in near this zero and obtain $x \approx -14.99$, as in Figure 15.

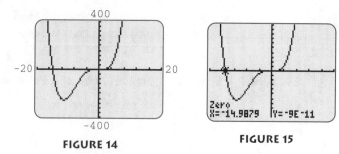

FIGURE 14   **FIGURE 15**

Thus, the zeros of f are approximately -14.99, -2.32, 0.73, and 1.57. Since a fourth-degree polynomial has at most four real zeros, there are no others.

The main reason we study zeros of polynomials is that they arise as solutions of polynomial equations. For example, the solutions of the polynomial equation $2x^3 + 6x^2 = 20x$ are the same as the solutions of $2x^3 + 6x^2 - 20x = 0$. Thus, solving $2x^3 + 6x^2 = 20x$ is equivalent to finding the zeros of the polynomial $f(x) = 2x^3 + 6x^2 - 20x$. We saw in Example 4 that f has exactly three zeros—$x = -5$, $x = 0$, and $x = 2$—so these are the only solutions of the polynomial equation. In general, if $g(x)$ and $h(x)$ are polynomials, then the equation $g(x) = h(x)$ can be solved by finding the zeros of the polynomial $f(x) = g(x) - h(x)$.

E X A M P L E   6   **An Application Involving Zeros**

The Arab oil embargo of the 1970s led to gasoline rationing, long lines, and inflated prices.

Unleaded gas prices were sampled at the end of five-year periods beginning in 1977, as depicted in Figure 16. Based on this limited data set, the average price of unleaded gas in the United States for the years 1977–1997 can be modeled by the polynomial

$$g(x) = -0.014x^4 + 0.625x^3 - 8.9x^2 + 43.4x + 65.6$$

where x is the year ($x = 0$ corresponds to 1977) and $g(x)$ is the average price (in cents) for a gallon of unleaded gas in that year. According to this model, for what years between 1977 and 1997 was the average price of a gallon of unleaded gas $1.00?

SOLUTION   It appears from Figure 16 that there were three times between 1977 and 1997 when the average price was $1.00. Since we are interested in the times

Unleaded Gas Prices 1977–1997

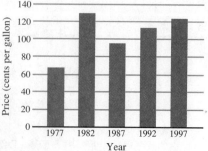

Data source: U.S. Energy Information Administration.

FIGURE 16

at which the price was \$1.00 and $g(x)$ gives the price in cents, we set $g(x) = 100$ to obtain

$$-0.014x^4 + 0.625x^3 - 8.9x^2 + 43.4x + 65.6 = 100$$

or equivalently,

$$-0.014x^4 + 0.625x^3 - 8.9x^2 + 43.4x - 34.4 = 0$$

Thus, we wish to find the zeros of the polynomial

$$f(x) = -0.014x^4 + 0.625x^3 - 8.9x^2 + 43.4x - 34.4$$

We plot the graph of f as shown in Figure 17 and use the trace or calculate-zero feature to approximate the x-intercepts, giving us $x \approx 0.97$, $x \approx 8.84$, and $x \approx 13.17$. By rounding to the nearest integer, we see that these values correspond to the years 1978, 1986, and 1990. Note that there is apparently another zero to the right of the viewing window, but we ignore this value since it corresponds to a year beyond 1997.

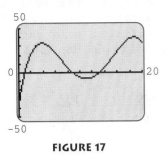

FIGURE 17

TURNING POINTS OF POLYNOMIALS

If we were to show a small child the graphs of polynomial functions and ask her to describe them, she would no doubt characterize them in terms of the number and size of the "humps"; in essence, the child would be focusing on the *turning points*—the points at which the graph changes from increasing to decreasing or decreasing to

MORPHING

When you see the faces of men of different races blend from one to another in a commercial for Schick razors, Michael Jackson transform into a panther in a music video, or the T-2000 robot metamorphose from liquid metal to human in the movie *Terminator II*, you are witnessing *morphing*, a spectacular graphic image manipulation technique. The most striking aspect of a morphing sequence is its fluidity; one image melts into another so seamlessly that our eyes tell us that the transformation, no matter how logic-defying, is actually taking place.

Although morphing has existed only since the 1980s, it is now commonplace in several visual media. Films, television commercials, television series, and music videos have employed morphing techniques to create startling special effects, and now software is available that enables those with personal computers to create their own morphing sequences.

Let us consider the steps required to create a morph sequence from an original to a final image. First, the original and final images are digitized; that is, the images are encoded as numerical data, in much the same way as images and music are stored on compact discs. Then the animator defines a correspondence between points on the initial and final images—for example, a point on the nose

A visual effect created with the aid of morphing.

of a tiger might correspond to a point on the beak of an eagle; a certain point on the hindquarters of the tiger might correspond to a point on the wing of the eagle; and so forth. Next the images are blended from one to the other, step by step, in such a way that the intermediate steps seem plausible. For example, when morphing a tiger into an eagle, we don't want to see a cavity form in the middle of the tiger's body that is then gradually filled in by feathers, nor do we want to see a paw *abruptly* change into a talon, or see a creature with whiskers on one side of its face and feathers on the other. The production of a smooth, plausible morph requires a combination of sophisticated mathematics, computational power, and often the artistic touch and intuition of the animator.

Mathematically, morphing is essentially an interpolation problem: a set of data points is given and intermediate values are to be determined. There are many ways in which data points can be "connected," and so there are many ways in which the morphing sequence can be generated. One of the most popular methods of interpolating is to use polynomial functions. Because polynomial functions are "smooth" (their graphs have no breaks or sharp changes of direction), the resulting morphing sequence likewise appears smooth.

increasing. Not only are turning points the most visually prominent aspect of the graph of a polynomial function, they also play a key role in the application of polynomial models to the real world.

Surprisingly, there is a close connection between the zeros of a polynomial and its turning points. To see this, suppose c_1 and c_2 are zeros of a polynomial function f, and suppose further that f has no other zeros between c_1 and c_2. Then the graph of f must cross the x-axis at $x = c_1$ and $x = c_2$ but nowhere between them. Although the precise shape of f between c_1 and c_2 is impossible to determine from this limited information, we show two possibilities in Figures 18 and 19. Notice that in either case, f has a

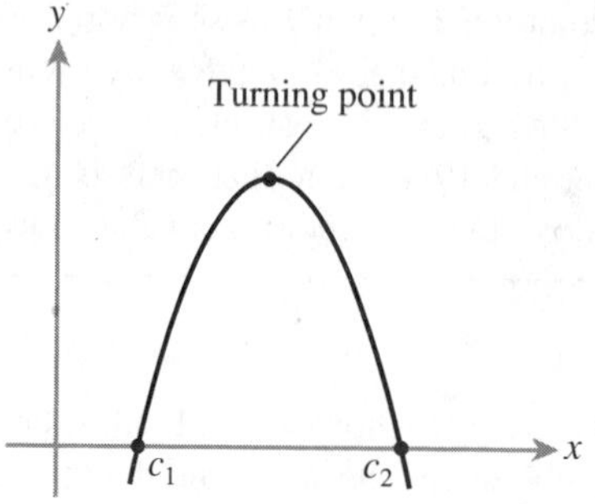

FIGURE 18

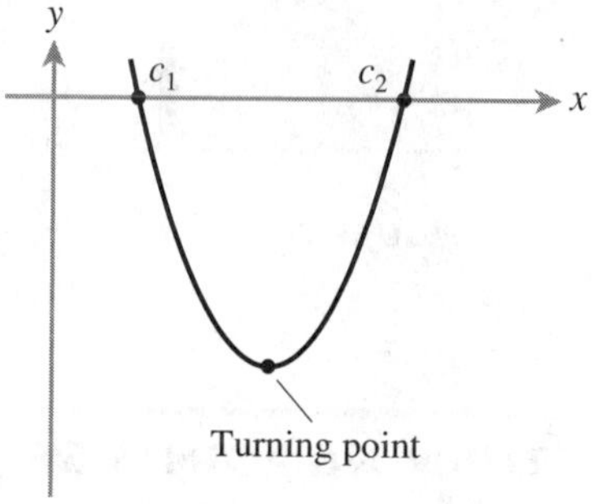

FIGURE 19

turning point somewhere between c_1 and c_2. In fact, this is always the case: A polynomial always has at least one turning point between adjacent zeros. So if f has m distinct real zeros, then f must have at least $m - 1$ turning points, one between each pair of adjacent zeros. It can also be shown, using techniques from calculus, that a polynomial of degree n has at most $n - 1$ turning points. These facts are summarized as follows.

The Number of Turning Points of a Polynomial

> Suppose f is a polynomial of degree n with m distinct real zeros, where $1 \le m \le n$. Then f has at least $m - 1$ and at most $n - 1$ turning points. If the degree n is even, then f has at least one turning point.

EXAMPLE 7 **Finding the Turning Points of a Polynomial with a Graphing Calculator**

Use a graphing calculator to estimate the coordinates of *all* the turning points of $f(x) = \frac{1}{4}x^4 - 7x^3 + 10x^2$.

SOLUTION Two turning points are visible in the graph of f shown in Figure 20. After zooming in and using the trace feature, or alternatively using the calculate-maximum/minimum feature, we estimate these points to be (0, 0) and (1, 3.25). See Figure 21 for the latter.

Now the graph of f is decreasing on the far right of the viewing window. However, because $f(x)$ will behave like its term of highest degree, $\frac{1}{4}x^4$, we know that $f(x) \to \infty$ as $x \to \infty$. Thus, the graph must turn upward somewhere to the right of the viewing window. After a process of trial and error, we obtain a view of the graph (Figure 22) that shows the third turning point. We estimate its coordinates to be approximately $(20, -12{,}000)$, as shown in Figure 23.

Since f has degree 4, there can be at most $4 - 1 = 3$ turning points. Thus, we have found *all* of the turning points of f, namely (0, 0), (1, 3.25), and $(20, -12{,}000)$.

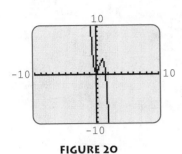

FIGURE 20

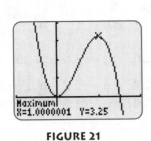

FIGURE 21

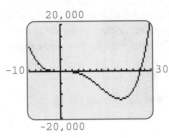

FIGURE 22

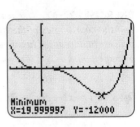

FIGURE 23

UNDERSTANDING AND MASTERY CHECKLISTS

CONCEPTS TO UNDERSTAND

☐ Polynomial function
☐ End behavior
☐ Zero
☐ Turning point

SKILLS TO MASTER

☐ Identify a polynomial function and its degree.
☐ Determine the end behavior of a polynomial function.
☐ Find the zeros of a polynomial function.
☐ Find the turning points of a polynomial function.

EXERCISES 4.1

EXERCISES 1–4 *Determine the end behavior of the given polynomial and match it with its graph.*

1. $f(x) = x^6$

2. $g(x) = -x^5$

3. $p(x) = -2x^4$

4. $h(x) = 4x^3$

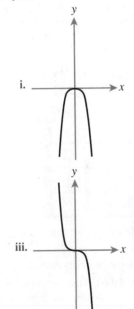

EXERCISES 5–12 *Determine the end behavior of the given polynomial. Verify your results with a graphing calculator.*

5. $f(x) = x^2 - 3x + 7$

6. $h(x) = -2x^3 + 6x^2 + 3x - 1$

7. $p(x) = -4x^4 + 7x - 1$

8. $g(x) = -3x^2 + 4x^4$

9. $h(x) = 10,000x^3 - 0.001x^5$

10. $f(x) = 2x^5 + 3x^2 - 100$

11. $g(x) = 10^7 + x^3$

12. $h(x) = -0.000001x^2 + 1,000,000x$

EXERCISES 13–16 *Indicate*

a. *the maximum number of real zeros of the given polynomial, and*

b. *the maximum number of turning points.*

13. $P(x) = mx + b$ **14.** $P(x) = ax^2 + bx + c$

15. $P(x) = ax^5 + bx^4 + cx^3 + dx^2 + ex + f$

16. $P(x) = ax^6 + bx^5 + cx^4 + dx^3 + ex^2 + fx + g$

EXERCISES 17–22 *Factor the polynomial and find its zeros.*

17. $f(x) = x^2 - 2x - 8$ **18.** $f(x) = x^2 + 8x + 15$

19. $f(x) = 3x^3 - 3x$ **20.** $f(x) = 4x^3 - 16x^2 + 16x$

21. $f(x) = -x^4 - 12x^3 - 36x^2$

22. $f(x) = x^4 - 25x^2$

EXERCISES 23–32 *Use a graphing calculator to estimate the zeros and the coordinates of the turning points of the polynomial to the nearest hundredth. Classify each turning point as a local maximum or minimum.*

23. $f(x) = x^3 - 6x^2 + 5x + 13$

24. $g(x) = x^3 + 9x^2 + 20x - 1$

25. $h(x) = \dfrac{x^3}{8} + 2x^2 + 2x + 3$

26. $g(x) = -\dfrac{x^3}{4} + 5x^2 - \dfrac{x}{4} + 2$

27. $p(x) = -\dfrac{x^4}{4} + 3x^3 + \dfrac{x^2}{2} - 9x$

28. $f(x) = \dfrac{x^4}{4} + 4x^3 + 10x^2 + 1$

29. $h(x) = 0.1x^3 - 0.7x^2 - 5.6x - 4$

30. $f(x) = 0.05x^3 + 0.55x^2 - 5.8x + 8$

31. $g(x) = 0.1x^5 - 0.01x^6 - 1$

32. $q(x) = 4x - 0.001x^4 + 9$

▨ APPLICATIONS

33. *AIDS Deaths* The number of AIDS deaths reported in the United States for each of the years 1983–1996 can be approximated with the polynomial function

$$f(x) = -72x^3 + 1170x^2 - 275x + 2640$$

where x is the year ($x = 0$ corresponds to 1983) and $f(x)$ is the number of deaths reported in that year. According to this model, when did the number of deaths peak and in what year after 1990 did the number of deaths drop to 10,000? Do you think this estimate is valid? Explain why this polynomial model works only for a limited time period by considering the end behavior of $f(x)$. If $f(1)$ represents the number of AIDS deaths in 1984 (from January 1 through December 31), then what would $f(1.5)$ represent?

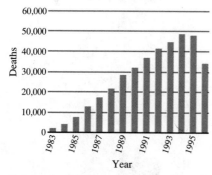

AIDS Deaths 1983–1996

Data source: U.S. Centers for Disease Control.

34. *Family Size* The average number of people per family unit in the United States for the years 1940–1990 can be modeled by the polynomial function

$$f(x) = 0.000002x^4 - 0.0002x^3 + 0.006x^2 - 0.06x + 3.76$$

where x is the year ($x = 0$ corresponds to 1940) and $f(x)$ is the average number of people per family unit during that year. Does f have any turning points? If so, approximate the coordinates. Can you give any explanations for the trends described by the function? Does it appear as though the model is valid for the year 2000? What graphical feature of polynomials can help explain why this polynomial model works only for a limited time period?

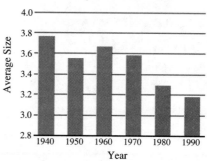

Family Size 1940–1990

Data source: U.S. Bureau of the Census.

35. *Paper Waste* The percentage of paper waste remaining after recycling during the years 1960–1995 can be modeled by the polynomial function

$$f(x) = 0.00016x^4 - 0.01x^3 + 0.22x^2 - 1.5x + 18$$

where x is the year ($x = 0$ corresponds to 1960) and $f(x)$ is the percentage of waste remaining in that year. Approximate the coordinates of any turning points. During which years between 1960 and 1995 was recycling increasing?

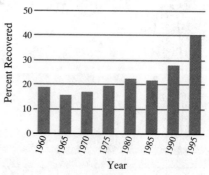

Paper Waste Recycling 1960–1995

Data source: U.S. Environmental Protection Agency.

36. *Fertility Rates* The fertility rate in the United States (the number of live births per 1000 women of childbearing age) for the years 1930–1990 can be modeled by the polynomial function

$$f(x) = 0.00011x^4 - 0.013x^3 + 0.44x^2 - 3.6x + 87$$

where x is the year ($x = 0$ corresponds to 1930) and $f(x)$ is the fertility rate for that year. Approximate the coordinates of any turning points. Why does this model appear to be of limited utility for years after 1990?

Fertility Rates 1930–1990

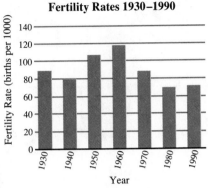

Data source: U.S. Bureau of the Census.

37. *Death Row Numbers* The number of U.S. prisoners awaiting execution on death row for the years 1990 to 1995 can be modeled either by

$$f(t) = -1.2t^3 + 16.1t^2 + 90.9t + 2349.3$$

or

$$g(t) = 7.2t^2 + 107.2t + 2345.8$$

where t represents the year (with $t = 0$ corresponding to 1990).

a. Describe the behaviors of the graphs of both f and g as $t \to \infty$.

b. Which of the two models is most likely to provide the best fit for the next century? Why?

c. When (if ever) do the two models agree?

d. Use a graphing calculator to estimate the maximum amount by which the models differ for the years 1990–1993. [*Hint:* Consider the function $|f(x) - g(x)|$.]

■ CONCEPTS AND CRITICAL THINKING

EXERCISES 38–43 *Answer true or false.*

38. If f is a fourth-degree polynomial and $f(x) \to \infty$ as $x \to \infty$, then $f(x) \to -\infty$ as $x \to -\infty$.

39. If f is a fifth-degree polynomial and $f(x) \to \infty$ as $x \to \infty$, then $f(x) \to -\infty$ as $x \to -\infty$.

40. Between any two zeros of a polynomial there must be at least one turning point.

41. Between any two turning points of a polynomial there must be at least one zero.

42. Only the term of highest degree has any effect on the graph of a polynomial.

43. Only the term of highest degree has any effect on the end behavior of the graph of a polynomial.

EXERCISES 44–47 *Give an example of each.*

44. A fourth-degree polynomial with no zeros

45. A cubic polynomial with no turning points

46. A cubic polynomial with two turning points

47. A cubic polynomial $f(x)$ such that $f(x) \to \infty$ as $x \to -\infty$

48. It is known that a polynomial f has four real zeros: -3, 1, 5, and 7. What can be said about the number and location of the turning points of f? What can be said about the degree of f; is it necessarily 4? Explain.

49. Use end behavior to explain why all polynomials of odd degree have at least one real zero.

50. Use your knowledge of the end behavior of polynomials to explain why a fourth-degree polynomial can have one or three turning points, but never two. More generally, what can you say about the possible number of turning points for polynomials of degree n, where n is even? What about polynomials of degree n, where n is odd?

■ QUESTIONS FOR DISCUSSION OR ESSAY

51. When zooming in to find a turning point, it often happens that the view shown by the calculator is "too flat"; that is, the entire graph appears horizontal, and thus it is impossible to pin down the x-coordinate of the turning point. Explain how this problem can be remedied either using a zoom box or adjusting the window variables.

52. In the real world, many quantities tend to level off over time. For example, the level of radioactivity present in a person exposed to radiation approaches zero as time goes on. Give three examples of real-world phenomena that tend to level off, and explain why polynomials don't make good models for such phenomena.

53. In Example 6, we considered a polynomial

$$g(x) = -0.014x^4 + 0.625x^3 - 8.9x^2 + 43.4x + 65.6$$

that approximated the average price of unleaded gas in the U.S. for the years 1977–1997. For example, $g(5) \approx 129.5$ approximates the average price for 1982 (in cents). What interpretation can be given to the values of $g(x)$ if x is not an integer? For example, how should we interpret $g(5.5) \approx 126.2$?

■ **PROJECTS FOR ENRICHMENT**

54. *Relating the Magnitudes of Monomials with Their Degrees*   In this project, we explore the relationship between the degree of a monomial and its magnitude. Let $f(x) = 0.0001x^3$ and $g(x) = 1000x^2$.

    a. Compute $f(10)$ and $g(10)$ and then $f(100)$ and $g(100)$. Explain why $g(x)$ is larger than $f(x)$ for these relatively small values of x.

    b. Find a value of x (other than 0) for which $f(x) = g(x)$.

    c. For what values of x will $f(x)$ be greater than $g(x)$?

    d. Explain why $[f(x) - g(x)] \to \infty$ as $x \to \infty$.

    e. Suppose that $P(x) = ax^n$ and $Q(x) = bx^{n+1}$, where both a and b are positive. For what values of x is $Q(x) > P(x)$?

    f. Explain why the end behavior of a polynomial is determined by the term of highest degree.

55. *Complex-Valued Polynomials and the Mandelbrot Set*   Until now, we have considered only *real-valued* functions with domains in the set of real numbers. In other words, we assumed that every function had only real numbers as input and output values. But we can also consider functions that allow nonreal input values and yield complex output values that may or may not be real. For example, suppose f is defined as $f(x) = x^2$, and we allow both real and nonreal input values. Thus, for example, $f(i) = i^2 = -1$ and $f(1 + i) = (1 + i)^2 = 1 + 2i + i^2 = 2i$. We say that f is a **complex-valued** function or, since f is a polynomial, a complex-valued polynomial. We can also consider functions that have nonreal numbers in their definitions—$f(x) = x^2 + ix$, for example. Here we compute $f(i) = i^2 + i^2 = -2$ and $f(1 + i) = (1 + i)^2 + i(1 + i) = -1 + 3i$.

    a. Evaluate each of the following functions at $x = i$.

       i. $f(x) = x^2 - x$

       ii. $f(x) = x^3 + 2x^2 - 3i$

       iii. $f(x) = 3x^4 - 2x^2 + x$

       iv. $f(x) = x^5 - 2ix^4 + 3x^2 - 5ix + 1$

    b. Evaluate each of the following functions at $x = 1 + i$ and $x = a + bi$.

       i. $f(x) = x^2 + 1$ ii. $f(x) = x^2 + i$

       iii. $f(x) = x^2 + (1 + i)$ iv. $f(x) = x^2 + x$

    The complex-valued polynomial $f(x) = x^2 + c$ generates a surprisingly intricate *fractal* image called the **Mandelbrot set**. To see how this happens, we must first consider the geometric representation of a complex number in the **complex plane**. The complex plane looks very much like the rectangular coordinate plane, but instead of the *x*-axis, we have the real axis, and instead of the *y*-axis, we have the imaginary axis. A complex number $a + bi$ is represented on the complex plane as the point with coordinates (a, b). The complex plane is shown in Figure 24 with several complex numbers plotted.

    The Mandelbrot set is defined to be the set of numbers c in the complex plane for which the sequence c, $f(c)$, $f(f(c))$,

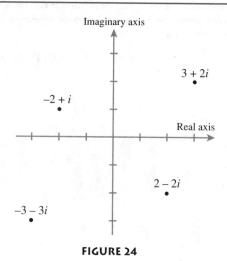

FIGURE 24

$f(f(f(c)))$, ... is *bounded* in the sense that if we were to plot these numbers in the complex plane, they would all be within a fixed distance of the origin. Notice that each number in the sequence (starting with the second) is found by using the previous number as the input value for f. We show below how to compute a few terms of the sequence for two different values of c.

$c = -2, f(x) = x^2 - 2$	$c = 1 + i, f(x) = x^2 + 1 + i$
$f(-2) = 2$	$f(1 + i) = 1 + 3i$
$f(2) = 2$	$f(1 + 3i) = -7 + 7i$
$f(2) = 2$	$f(-7 + 7i) = 1 - 97i$
$\vdots$	$\vdots$

Now the sequence -2, 2, 2, ... is bounded, and so -2 is a member of the Mandelbrot set. However, the sequence $1 + i$, $1 + 3i$, $-7 + 7i$, $1 - 97i$, ... is not bounded; in fact, each successive point is more than twice as far from the origin as the previous point. Thus, $1 + i$ is not in the Mandelbrot set.

c. Determine whether the following numbers are in the Mandelbrot set.

    i. -1 ii. 2

    iii. i iv. $\frac{1}{4}i$

If it were possible to plot all the points in the complex plane that belong to the Mandelbrot set, we would obtain a graph similar to the one shown in Figure 25. A more interesting ''graph'' can be obtained if we assign colors to the points that are not in the Mandelbrot set. This is done on the basis of how quickly the sequence c, $f(c)$, $f(f(c))$, ... moves away from the origin. One such image is shown in Figure 26. One of the fascinating properties of the Mandelbrot set (the black region) is that it contains an infinite number of miniature copies of itself. These can be found by zooming in near the boundary of the set. Also near the boundary is an endless collection of stunning images, two of which are shown in Figures 27 and 28.

FIGURE 26

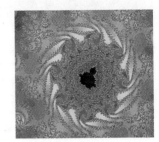

FIGURE 27

FIGURE 25

FIGURE 28

SECTION 4.2   DIVISION OF POLYNOMIALS

- [ ] **How can you divide two polynomials without writing down a single variable?**
- [ ] **How could you evaluate a 17th-degree polynomial function for any given value of x using only a $\boxed{\times}$ key and a $\boxed{+}$ key 17 times each?**
- [ ] **If every person with a flu virus passes it on to 3 other people, how many people will have had the flu 20 days after the first person became infected?**

DIVIDING POLYNOMIALS

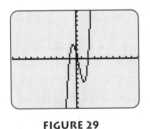

FIGURE 29

We have seen that the exact values of the zeros of a polynomial can be found by factoring. But what if a polynomial cannot be easily factored? Consider, for example, the polynomial $f(x) = 3x^3 - 4x^2 - 5x + 2$. It is not at all clear how to factor this polynomial, if indeed it can be factored. However, using a graphing calculator, we see in Figure 29 that there are three zeros. Two of these zeros, $x = -1$ and $x = 2$, can be determined from the graph and verified by substitution into f. The third zero can at least be approximated using either the trace or calculate-zero feature. However, if we wish to find it exactly, we must take a different approach. Intuition suggests that since $x = -1$ and $x = 2$ are zeros, $f(x)$ should have factors $(x + 1)$ and $(x - 2)$ (we will see why this is the case in Section 4.3). In other words, $f(x) = (x + 1)(x - 2)q(x)$ for some polynomial $q(x)$. As we see in the following example, this information is sufficient to find the remaining zero.

EXAMPLE 1   ▨   **Factoring a Polynomial Using Long Division**

Given that $(x + 1)$ and $(x - 2)$ are factors of $f(x) = 3x^3 - 4x^2 - 5x + 2$, factor $f(x)$ completely and determine the zeros.

SOLUTION   Since $(x + 1)$ and $(x - 2)$ are both factors of $f(x)$, the product of these two factors, $(x + 1)(x - 2) = x^2 - x - 2$, is also a factor. Thus we have

$$3x^3 - 4x^2 - 5x + 2 = (x^2 - x - 2)q(x)$$

for some polynomial $q(x)$. Now to find $q(x)$ we divide both sides of the equation by $x^2 - x - 2$ to obtain

$$\frac{3x^3 - 4x^2 - 5x + 2}{x^2 - x - 2} = q(x)$$

We can simplify this expression for $q(x)$ by performing **long division,** a process for dividing two polynomials that is similar to ordinary long division of natural numbers. As the first step, we choose a monomial that, when multiplied by the leading term of the denominator, yields the leading term of the numerator. We select $3x$ since $3x \cdot x^2 = 3x^3$. The monomial $3x$ is then multiplied through the denominator, $x^2 - x - 2$, as illustrated.

$$
\begin{array}{r}
3x \phantom{{} - 5x + 2} \\
x^2 - x - 2 \overline{)3x^3 - 4x^2 - 5x + 2} \\
3x^3 - 3x^2 - 6x \phantom{{} + 2} \\
\hline
-x^2 + x + 2
\end{array}
$$

We select $3x$ since $3x \cdot x^2 = 3x^3$.

Multiplying $x^2 - x - 2$ by $3x$ to obtain $3x^3 - 3x^2 - 6x$

Subtracting the second line from the first

Next, we choose a monomial that, when multiplied by the leading term of the denominator, produces the leading term of the remainder just found, $-x^2 + x + 2$. In this case, we select -1 and continue as follows.

$$
\begin{array}{r}
3x - 1 \phantom{{} + 2} \\
x^2 - x - 2 \overline{)3x^3 - 4x^2 - 5x + 2} \\
3x^3 - 3x^2 - 6x \phantom{{} + 2} \\
\hline
-x^2 + x + 2 \\
-x^2 + x + 2 \\
\hline
0
\end{array}
$$

We select -1 since $-1 \cdot x^2 = -x^2$.

Multiplying $x^2 - x - 2$ by -1 to obtain $-x^2 + x + 2$

Subtracting

The completed division shows that

$$\frac{3x^3 - 4x^2 - 5x + 2}{x^2 - x - 2} = 3x - 1$$

and so

$$
\begin{aligned}
3x^3 - 4x^2 - 5x + 2 &= (x^2 - x - 2)(3x - 1) \\
&= (x + 1)(x - 2)(3x - 1)
\end{aligned}
$$

Setting the factors of $f(x)$ equal to 0, we find that the three zeros of f are $x = -1$, $x = 2$, and $x = \frac{1}{3}$.

RULE OF THUMB   When performing long division, write the terms of each polynomial in descending order. For example, when computing

$$\frac{3x - 4x^2 + 1 - 2x^3}{1 + 3x}$$

begin by rewriting as

$$\frac{-2x^3 - 4x^2 + 3x + 1}{3x + 1}$$

In Example 1, $x^2 - x - 2$ divided evenly into $3x^3 - 4x^2 - 5x + 2$ because $x^2 - x - 2$ was a factor of $3x^3 - 4x^2 - 5x + 2$. In general, it is always possible to simplify the quotient of a polynomial and another of lesser degree, even if the second is not a factor of the first. This observation is formalized in the **division algorithm.**

Division Algorithm ☐

Suppose $f(x)$ and $d(x)$ are polynomials, $d(x) \neq 0$, and the degree of $d(x)$ is less than or equal to the degree of $f(x)$. Then there are unique polynomials $q(x)$ and $r(x)$ such that

$$f(x) = d(x) \cdot q(x) + r(x) \quad \text{or} \quad \frac{f(x)}{d(x)} = q(x) + \frac{r(x)}{d(x)}$$

where either $r(x) = 0$ or the degree of $r(x)$ is less than the degree of $d(x)$. The polynomials $f(x)$, $d(x)$, $q(x)$, and $r(x)$ are referred to, respectively, as the **dividend, divisor, quotient,** and **remainder.**

EXAMPLE 2 ▓ **Using Long Division to Find a Quotient and Remainder**

Use long division to find the quotient and remainder when dividing $f(x) = x^3 - 5x^2 + 20$ by $d(x) = x - 3$. In other words, find $q(x)$ and $r(x)$ so that $x^3 - 5x^2 + 20 = (x - 3)q(x) + r(x)$.

SOLUTION Since there is no x term in the polynomial $x^3 - 5x^2 + 20$, we insert $0x$ to ensure that terms with like powers line up. We divide as follows:

$$
\begin{array}{r}
x^2 - 2x - 6 \\
x - 3 \overline{\smash{)}\, x^3 - 5x^2 + 0x + 20} \\
\underline{x^3 - 3x^2} \\
-2x^2 + 0x \\
\underline{-2x^2 + 6x} \\
-6x + 20 \\
\underline{-6x + 18} \\
2
\end{array}
$$

Thus, $q(x) = x^2 - 2x - 6$ and $r(x) = 2$. It follows that

$$x^3 - 5x^2 + 20 = (x - 3)(x^2 - 2x - 6) + 2$$

Equivalently, we can write

$$\underbrace{\frac{\overbrace{x^3 - 5x^2 + 20}^{\text{Dividend}}}{\underbrace{x - 3}_{\text{Divisor}}}} = \overbrace{x^2 - 2x - 6}^{\text{Quotient}} + \frac{\overbrace{2}^{\text{Remainder}}}{x - 3}$$

SYNTHETIC DIVISION

A special process called **synthetic division** has been developed for performing divisions in the case where the divisor is of the form $x - c$, as in Example 2. Essentially, synthetic division is a compact and efficient bookkeeping method for keeping track of the steps performed in long division. By eliminating unnecessary symbols and space, the entire process can be performed on three lines without ever writing a single variable. Although not an essential tool, synthetic division is quite easily learned and also an excellent example of an efficient mathematical *algorithm*—or step-by-step procedure.

Let us begin our derivation of this procedure by finding shortcuts in the work we performed for Example 2. The following table shows several successive simplifications.

Remove all *x*'s	Remove unnecessary numbers and "+" signs	Condense by bringing numbers up	Combine top line with bottom line
$\begin{array}{r} 1 \ -2 \ -6 \\ \hline -3)1 \ -5 \ +0 \ +20 \\ \underline{1 \ -3} \\ -2 \ +0 \\ \underline{-2 \ +6} \\ -6 \ +20 \\ \underline{-6 \ +18} \\ 2 \end{array}$	$\begin{array}{r} 1 \ -2 \ -6 \\ \hline -3)1 \ -5 \ \ 0 \ \ 20 \\ \underline{-3} \\ -2 \\ \underline{6} \\ -6 \\ \underline{18} \\ 2 \end{array}$	$\begin{array}{r} 1 \ -2 \ -6 \\ \hline -3)1 \ -5 \ \ 0 \ \ 20 \\ -3 \ \ 6 \ \ 18 \\ \hline -2 \ -6 \ \ 2 \end{array}$	$\begin{array}{r} -3\rfloor 1 \ -5 \ \ 0 \ \ 20 \\ \underline{-3 \ \ 6 \ \ 18} \\ 1 \ -2 \ -6 \ \ 2 \end{array}$

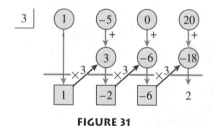

FIGURE 30

FIGURE 31

We can further simplify the procedure if we omit the "−" from the divisor −3. This changes the signs of the numbers in the second row and, as a consequence, allows us to add down the columns instead of subtracting. The final version of this **synthetic division** procedure is shown in Figure 30. To summarize, the numbers circled in the top row of Figure 31 are the coefficients of the dividend $x^3 - 5x^2 + 20$. The 3 on the far left is from the divisor $x - 3$. The 1 in the bottom row is carried down from the top row. Each boxed number is used to obtain the number to its right by following these steps.

1. Multiply the boxed number by 3 and place the product in the circle above and to the right.

2. Add the two circled numbers to obtain the boxed number in the bottom row.

The resulting boxed numbers in the bottom row are the coefficients of the quotient $x^2 - 2x - 6$, and the unboxed 2 is the remainder. Note that the degree of the quotient $x^2 - 2x - 6$ is 1 less than the degree of the dividend $x^3 - 5x^2 + 20$.

EXAMPLE 3 ■ **Using Synthetic Division to Find a Quotient and Remainder**

Use synthetic division to find the quotient and remainder when $3x^4 - 12x^2 + 8x + 4$ is divided by $x + 2$.

SOLUTION Since the divisor is $x + 2 = x - (-2)$, we place −2 to the far left. The top row is formed using the coefficients of the dividend $3x^4 - 12x^2 + 8x + 4$, including a 0 for the coefficient of the missing x^3 term. After bringing the leading coefficient 3 down to the bottom row, we proceed as in Figure 31, multiplying by −2 and then adding at each stage.

$$\begin{array}{r} -2\rfloor 3 \ \ \ \ 0 \ -12 \ \ \ 8 \ \ \ \ 4 \\ \underline{-6 \ \ \ 12 \ \ \ 0 \ -16} \\ 3 \ -6 \ \ \ \ 0 \ \ \ 8 \ -12 \end{array}$$

Since the divisor $x + 2$ has degree 1, and the dividend $3x^4 - 12x^2 + 8x + 4$ has degree 4, the quotient will have degree 3. Using the first four numbers in the bottom row as the coefficients, we obtain the quotient $3x^3 - 6x^2 + 0x + 8$. The remainder is -12. Thus,

$$3x^4 - 12x^2 + 8x + 4 = (x + 2)(3x^3 - 6x^2 + 8) - 12$$

EXAMPLE 4   ▇   **Using Synthetic Division to Factor a Polynomial**

Given that $(x + 4)$ and $(x - 1)$ are factors of the polynomial $f(x) = 4x^4 + 16x^3 - 3x^2 - 13x - 4$, use synthetic division to factor $f(x)$ and then find the zeros. Verify using a graphing calculator.

SOLUTION   We first use synthetic division to divide $4x^4 + 16x^3 - 3x^2 - 13x - 4$ by $x + 4$.

$$
\begin{array}{r|rrrrr}
-4 & 4 & 16 & -3 & -13 & -4 \\
   &   & -16 & 0 & 12 & 4 \\
\hline
   & 4 & 0 & -3 & -1 & 0
\end{array}
$$

Since the remainder is 0, we have confirmed that $x + 4$ is indeed a factor of $f(x)$. The first four numbers in the bottom row are the coefficients of the quotient $4x^3 - 3x - 1$, and so $f(x) = (x + 4)(4x^3 - 3x - 1)$. Now if $x - 1$ is a factor of $f(x)$, it must also be a factor of $4x^3 - 3x - 1$. Thus, we use synthetic division to divide $x - 1$ into $4x^3 - 3x - 1$.

$$
\begin{array}{r|rrrr}
1 & 4 & 0 & -3 & -1 \\
  &   & 4 & 4 & 1 \\
\hline
  & 4 & 4 & 1 & 0
\end{array}
$$

Since the remainder is 0, $x - 1$ is indeed a factor. The first three numbers in the bottom row are the coefficients of $4x^2 + 4x + 1$. Because $4x^2 + 4x + 1 = (2x + 1)^2$, it follows that

$$4x^4 + 16x^3 - 3x^2 - 13x - 4 = (x + 4)(x - 1)(2x + 1)^2$$

Thus, setting each of the factors of $f(x)$ equal to zero, we determine that the zeros of f are $x = -4$, $x = 1$, and $x = -\frac{1}{2}$. These values are consistent with the x-intercepts shown in Figure 32.

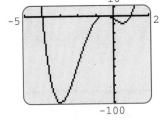

FIGURE 32

Notice that in all of the synthetic division examples, the remainder turned out to be zero or some other constant. The division algorithm guarantees that this will always be the case. Indeed, since the divisor $x - c$ is a polynomial of degree 1, and since the remainder's degree must be less than that of the divisor, the remainder must have degree zero and so must be a constant. In light of this, the division algorithm for a polynomial $f(x)$ and divisor $x - c$ gives us

$$f(x) = (x - c)q(x) + k$$

Now, if we set $x = c$, we obtain $f(c) = 0 \cdot q(c) + k = k$. In other words, $f(c)$ is the remainder obtained by dividing $f(x)$ by $x - c$. This fact is known as the **Remainder Theorem.**

Remainder Theorem ☐	If a polynomial $f(x)$ is divided by $x - c$, the remainder is $f(c)$.

E X A M P L E 5 ▮ Evaluating a Polynomial with the Remainder Theorem

Use the Remainder Theorem to evaluate $f(x) = 2x^3 - 5x^2 + x - 4$ at $x = -3$.

SOLUTION We use synthetic division to find the remainder when $2x^3 - 5x^2 + x - 4$ is divided by $x + 3 = x - (-3)$.

$$
\begin{array}{r|rrrr}
-3 & 2 & -5 & 1 & -4 \\
   &   & -6 & 33 & -102 \\
\hline
   & 2 & -11 & 34 & -106
\end{array}
$$

Thus, $f(-3) = -106$.

UNDERSTANDING AND MASTERY CHECKLISTS

CONCEPTS TO UNDERSTAND

☐ Long division
☐ Division algorithm
☐ Synthetic division
☐ Remainder Theorem

SKILLS TO MASTER

☐ Use long division to find a quotient and remainder when dividing one polynomial by another.

☐ Use synthetic division to find a quotient and remainder when dividing a polynomial by $x - c$.

☐ Divide a given factor into a polynomial to find the other factors.

☐ Use synthetic division and the Remainder Theorem to find a function value.

E X E R C I S E S 4.2

EXERCISES 1–12 *Use long division to find the quotient $q(x)$ and remainder $r(x)$ when $f(x)$ is divided by $d(x)$.*

1. $f(x) = 2x^2 - 5x - 12;\ d(x) = x - 4$
2. $f(x) = 3x^2 + x - 10;\ d(x) = x + 2$
3. $f(x) = 4x^3 - 3x^2 + 20x - 15;\ d(x) = x^2 + 5$
4. $f(x) = 2x^3 + 4x^2 - 1;\ d(x) = x^2 + 2x - 1$
5. $f(x) = 2x^4 - 7x^3 - 6x + 9;\ d(x) = 2x^2 - x + 3$
6. $f(x) = 9x^4 - 3x^3 + 2x - 4;\ d(x) = 3x^2 - 2$
7. $f(x) = x^2 + 1;\ d(x) = x + 1$
8. $f(x) = 2x^2 - x + 4;\ d(x) = x - 2$
9. $f(x) = 3x^3 - 4x + 2;\ d(x) = x^2 - 3$
10. $f(x) = x^4;\ d(x) = x^3 + 1$
11. $f(x) = x^5;\ d(x) = x^3 + x + 1$
12. $f(x) = 3x^4 + x^2 - 1;\ d(x) = x^2 - x + 3$

EXERCISES 13–24 *Use synthetic division to find the quotient $q(x)$ and remainder $r(x)$ when $f(x)$ is divided by $d(x)$.*

13. $f(x) = 2x^2 + 9x - 18;\ d(x) = x + 6$
14. $f(x) = 5x^2 - 13x - 6;\ d(x) = x - 3$
15. $f(x) = x^3 - 2x^2 - 2x + 4;\ d(x) = x - 2$
16. $f(x) = x^3 + 3x^2 - 9x + 5;\ d(x) = x + 5$
17. $f(x) = 3x^4 - 8x^3 - 10x + 3;\ d(x) = x - 3$
18. $f(x) = 2x^4 - 14x^3 + 5;\ d(x) = x - 7$
19. $f(x) = x^2 - 2;\ d(x) = x + 1$
20. $f(x) = x^2 + 2x;\ d(x) = x - 2$

21. $f(x) = x^3 + 6x^2 - 16x + 2$; $d(x) = x + 8$

22. $f(x) = 1 - 2x^3$; $d(x) = x - 4$

23. $f(x) = 1 + x^3 - 4x^4$; $d(x) = x - \frac{1}{3}$

24. $f(x) = 2x^4 + x^3 - 4x^2 + 1$; $d(x) = x + \frac{1}{2}$

EXERCISES 25–32   *Use the given factor(s) of the polynomial to find the remaining factors, and then identify the zeros. Verify the zeros graphically.*

25. $x + 2$; $f(x) = x^3 + 6x^2 + 3x - 10$

26. $x - 4$; $f(x) = x^3 - 3x^2 - 10x + 24$

27. $x - 3$; $p(x) = 2x^3 - 9x^2 + 7x + 6$

28. $x + 1$; $h(x) = 3x^3 + 10x^2 + x - 6$

29. $x + 4$ and $x - 1$; $g(x) = x^4 + 9x^3 + 22x^2 - 32$

30. $x - 6$ and $x + 3$; $f(x) = x^4 - 2x^3 - 27x^2 + 108$

31. $2x + 1$ and $x - 8$; $h(x) = 6x^4 - 47x^3 - 13x^2 + 38x + 16$

32. $5x - 1$ and $x - 1$; $p(x) = 10x^4 + 13x^3 - 43x^2 + 23x - 3$

EXERCISES 33–38   *Use synthetic division and the Remainder Theorem to find the indicated function values.*

33. $f(x) = 2x^3 - x^2 - 4x + 6$

   a. $f(3)$ **b.** $f(-2)$ **c.** $f(\frac{1}{2})$

34. $g(x) = 3x^3 + 7x^2 - 18x + 4$

   a. $g(-4)$ **b.** $g(2)$ **c.** $g(-\frac{1}{3})$

35. $h(x) = 3x^4 - 2x^3 - 4x + 3$

   a. $h(-1)$ **b.** $h(5)$ **c.** $h(-2.1)$

36. $h(x) = -x^4 + 3x^2 + 5x - 10$

   a. $h(3)$ **b.** $h(-1)$ **c.** $h(1.4)$

37. $g(x) = \frac{1}{2}x^4 - \frac{5}{3}x^3 + \frac{2}{3}x^2 - \frac{9}{2}x$

   a. $g(4)$ **b.** $g(-2)$

38. $f(x) = \frac{2}{5}x^4 - \frac{21}{2}x^2 + \frac{5}{2}x + \frac{3}{4}$

   a. $f(5)$ **b.** $f(-5)$

■ CONCEPTS AND CRITICAL THINKING

EXERCISES 39–42   *Answer true or false.*

39. When one polynomial is divided by another, the degree of the remainder is always greater than that of the quotient.

40. If the remainder obtained when dividing the polynomial $f(x)$ by $x + 3$ is 8, then $f(3) = 8$.

41. If the remainder obtained when dividing the polynomial $f(x)$ by $x - 3$ is 8, then $f(3) = 8$.

42. The most efficient method for computing the quotient and re-mainder when dividing a fifth-degree polynomial by a fourth-degree polynomial is synthetic division.

EXERCISES 43–46   *Give an example of each.*

43. A third-degree polynomial having $x + 2$ as a factor

44. A polynomial $P(x)$ such that when $P(x)$ is divided by $x - 1$ the remainder is 3

45. A fourth-degree polynomial having $x + 1$ and $x - 2$ as factors

46. An application of synthetic division

■ QUESTIONS FOR DISCUSSION OR ESSAY

47. Explain how the trace feature on your graphing calculator could be used to evaluate a polynomial function $f(x)$ at specific values for x. Discuss the advantages and disadvantages of this tech-nique, as compared to the synthetic division process described in this section. Does your calculator have any built-in feature for evaluating functions? If so, how does it compare to using the trace feature or synthetic division?

48. Compare and contrast long division of polynomials with long division of real numbers.

■ PROJECTS FOR ENRICHMENT

49. *Factors of $f(x) = x^n - a^n$ and Sums of the Form $1 + a + a^2 + \cdots + a^n$*   In this project, we use synthetic division to inves-tigate the factors of $x^n - a^n$, and then see how this helps us find sums of the form $1 + a + a^2 + \cdots + a^n$. To begin, we consider the polynomial $g(x) = x^2 - a^2$. Using the difference of squares formula, we obtain $g(x) = (x - a)(x + a)$. Next we consider

$h(x) = x^3 - a^3$. The difference of cubes formula gives us $h(x) = (x - a)(x^2 + ax + a^2)$. Notice that both $g(x)$ and $h(x)$ have $x - a$ as a factor.

   a. Complete the following synthetic division to divide $f(x) = x^4 - a^4$ by $x - a$, and then write out the factored form of $f(x)$.

$$\underline{a\,|}\ \ 1 \quad 0 \quad 0 \quad 0 \quad -a^4$$
$$a \quad a^2$$
$$\overline{\ 1 \quad a}$$

b. Use synthetic division to divide $f(x) = x^5 - a^5$ by $x - a$, and write out the factored form of $f(x)$.

c. Complete the following synthetic division to divide $f(x) = x^n - a^n$ by $x - a$. Write out the factored form of $f(x)$.

$$\overbrace{}^{n-1\ \text{zeros}}$$
$$\underline{a\,|}\ \ 1 \quad 0 \quad 0 \quad \cdots \quad 0 \quad -a^n$$
$$a \quad a^2$$
$$\overline{\ 1 \quad a}$$

d. By setting $x = 1$ in the factored form of $f(x)$ from part c, show that

$$1 - a^n = (1 - a)(1 + a + a^2 + \cdots + a^{n-1})$$

and thus

(1) $$\frac{1 - a^n}{1 - a} = 1 + a + a^2 + \cdots + a^{n-1}$$

Equation (1) enables us to find sums of the form $1 + a + a^2 + a^3 + \cdots + a^n$. Indeed, if we replace n with $n + 1$ in equation (1), we obtain

(2) $$1 + a + a^2 + \cdots + a^n = \frac{1 - a^{n+1}}{1 - a}$$

Now suppose we wish to find the sum $1 + 3 + 3^2 + 3^3 + \cdots + 3^6$. In equation (2), we set $a = 3$ and $n = 6$ to obtain

$$1 + 3 + 3^2 + 3^3 + \cdots + 3^6 = \frac{1 - 3^{6+1}}{1 - 3}$$
$$= \frac{1 - 2187}{-2}$$
$$= 1093$$

e. Use equation (2) to find the following sums.

   i. $1 + 4 + 4^2 + 4^3 + \cdots + 4^8$

   ii. $1 + \frac{1}{2} + \left(\frac{1}{2}\right)^2 + \cdots + \left(\frac{1}{2}\right)^{10}$

   iii. $1 + 6 + 36 + 216 + \cdots + 10{,}077{,}696$

   iv. $1 + \frac{2}{3} + \frac{4}{9} + \cdots + \frac{4096}{531{,}441}$

f. A job pays \$1 in the first month, \$2 in the second month, \$4 in the third month, and so on. How much will the job pay over the course of 5 years?

g. Suppose a highly contagious flu virus is brought into the United States by a returning tourist. To model the spread of this virus, we will assume that the flu is passed to three other people on the first day, and it continues to spread in such a way that each person who gets the flu on a given day passes it to three other people the next day. If this pattern continues, how many people will have had the flu after 10 days? After 20 days?

h. Is the assumption in part g realistic? Why or why not? What factors affect the rate at which a contagious disease spreads? Is it likely that an entire population will contract a contagious disease? Explain.

50. *Synthetic Division on a Spreadsheet* A spreadsheet is an ideal tool for performing synthetic division. Suppose, for example, we wish to divide $f(x) = 2x^5 - 12x^4 - 13x^3 + 5$ by $x - 7$. We begin by setting up the first row of the synthetic division in cells A1 through G1, as shown in Figure 33. Some of the formulas needed for rows 2 and 3 are also shown in Figure 33. The remaining formulas for rows 2 and 3 can be entered simply by copying from cell C2 to cells D2 through G2, and from cell C3 to cells D3 through G3. The coefficients of the resulting quotient and the remainder will be computed and displayed in cells B3 through G3. In this case, the quotient is $q(x) = 2x^4 + 2x^3 + x^2 + 7x + 49$ and the remainder is 348 (see Figure 34). Higher-degree polynomials can be handled by copying the formulas to more columns.

	A	B	C	D	E	F	G
1	7	2	−12	−13	0	0	5
2			=A1*B3				
3		=B1	=C1+C2				

FIGURE 33

	A	B	C	D	E	F	G
1	7	2	−12	−13	0	0	5
2			14	14	7	49	343
3		2	2	1	7	49	348

FIGURE 34

a. Use synthetic division to find the quotient $q(x)$ and remainder $r(x)$ when $f(x)$ is divided by $d(x)$.

   i. $f(x) = 2x^5 - 19x^4 + 25x^3 + 2x^2 - 86x + 87$; $q(x) = x - 8$

   ii. $f(x) = 3x^6 + 33x^5 - 5x^4 - 43x^3 + 130x^2 - 7x + 177$; $q(x) = x + 11$

   iii. $f(x) = x^8 - 10$; $q(x) = x - 2$

b. Use synthetic division and the Remainder Theorem to find the indicated function value.

   i. $f(x) = x^5 + 3x^2 - 5x + 1$; $f(5)$

   ii. $f(x) = 3x^6 - 10x^4 + 2x^2 + 5$; $f(-3)$

   iii. $f(x) = x^8 + x^7 + x^6 + \cdots + x + 1$; $f(2)$

SECTION 4.3   ZEROS AND FACTORS OF POLYNOMIALS

> ☐ How can a polynomial whose graph never touches the *x*-axis have six zeros?
>
> ☐ How can a graphing calculator be used as an aid in factoring polynomials?
>
> ☐ Is there a cubic formula for solving third-degree polynomial equations much like the quadratic formula for solving quadratic equations? If so, why haven't either of the authors memorized it?
>
> ☐ How can the zeros of a polynomial be used to model bungee jumping?

THE FACTOR THEOREM

We have already seen evidence of a close connection between the factors and zeros of polynomials. In Section 4.1, for example, we used the factors of a polynomial to find the exact values of its zeros. Now we complete the connection by showing that the zeros of a polynomial can be used to find its factors. Suppose that $x = c$ is a zero of a polynomial f. Then $f(c) = 0$. But according to the Remainder Theorem, $f(c)$ is the value of the remainder when $f(x)$ is divided by $x - c$. Thus, $f(x) = (x - c)q(x) + 0$, which shows that $x - c$ is a factor of $f(x)$. We have just proven the Factor Theorem.

Factor Theorem   ☐

> A polynomial function f has a zero c if and only if $x - c$ is a factor of $f(x)$. In other words, $f(c) = 0$ if and only if $f(x) = (x - c)q(x)$ for some nonzero polynomial $q(x)$.

EXAMPLE 1   ▨   Using Known Zeros of a Polynomial to Find the Remaining Zeros

The polynomial function $f(x) = 5x^4 + 8x^3 - 29x^2 - 20x + 12$ has among its zeros $x = -3$ and $x = 2$. Find the remaining zeros.

SOLUTION   Since f has zeros $x = -3$ and $x = 2$, the Factor Theorem tells us that $f(x)$ has factors $x + 3$ and $x - 2$. Thus, we use synthetic division to divide $f(x)$ by $x + 3$, and then apply synthetic division again to divide the quotient by $x - 2$. [Alternatively, we could use long division to divide $f(x)$ by $(x + 3)(x - 2) = x^2 + x - 6$.]

$$
\begin{array}{r|rrrrr}
-3 & 5 & 8 & -29 & -20 & 12 \\
   &   & -15 & 21 & 24 & -12 \\
\hline
   & 5 & -7 & -8 & 4 & 0
\end{array}
\qquad
\begin{aligned}
&\frac{5x^4 + 8x^3 - 29x^2 - 20x + 12}{x + 3} \\
&= 5x^3 - 7x^2 - 8x + 4
\end{aligned}
$$

$$
\begin{array}{r|rrrr}
2 & 5 & -7 & -8 & 4 \\
  &   & 10 & 6 & -4 \\
\hline
  & 5 & 3 & -2 & 0
\end{array}
\qquad
\frac{5x^3 - 7x^2 - 8x + 4}{x - 2} = 5x^2 + 3x - 2
$$

From the last line of the division by $x - 2$, we see that $5x^2 + 3x - 2$ is a factor of $f(x)$. But $5x^2 + 3x - 2 = (x + 1)(5x - 2)$, and so we have

$$
\begin{aligned}
5x^4 + 8x^3 - 29x^2 - 20x + 12 &= (x + 3)(5x^3 - 7x^2 - 8x + 4) \\
&= (x + 3)(x - 2)(5x^2 + 3x - 2) \\
&= (x + 3)(x - 2)(x + 1)(5x - 2)
\end{aligned}
$$

Setting each factor equal to zero, we determine that the zeros of f are -3, 2, -1, and $\frac{2}{5}$.

EXAMPLE 2   ■   **Finding a Polynomial with Given Zeros**

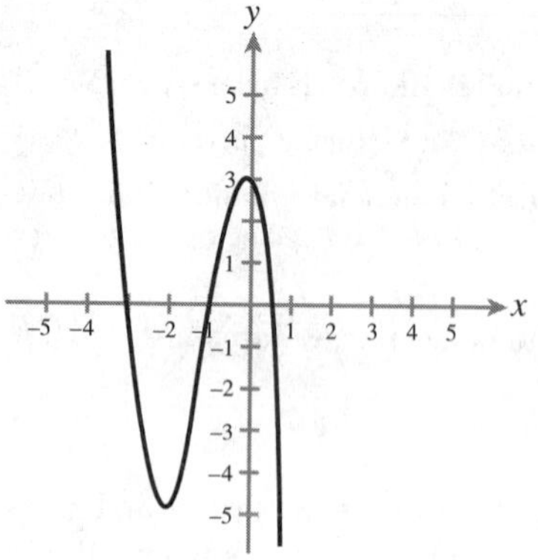

Find a possible third-degree polynomial function f for the graph shown in Figure 35. Verify with a graphing calculator.

SOLUTION   The function graphed in Figure 35 appears to have zeros of -3, -1, and $\frac{1}{2}$. According to the Factor Theorem, each zero c corresponds to a factor of the form $x - c$. So $f(x)$ must have factors $x + 3$, $x + 1$, and $x - \frac{1}{2}$. Because the product of these three factors will be a third-degree polynomial, we know that $f(x)$ cannot have any other factors involving x. But $f(x)$ could also have a constant factor, which we will denote by a. So f must have the form

$$f(x) = a\left(x - \frac{1}{2}\right)(x + 3)(x + 1)$$

To find the value of a, we note that the graph appears to have a y-intercept of $(0, 3)$, and thus we assume that $f(0) = 3$. This gives us

$$f(0) = 3$$

$$a\left(0 - \frac{1}{2}\right)(0 + 3)(0 + 1) = 3 \qquad \text{Setting } x = 0 \text{ in } f(x)$$

$$a\left(-\frac{3}{2}\right) = 3 \qquad \text{Multiplying out}$$

$$a = 3\left(-\frac{2}{3}\right) = -2$$

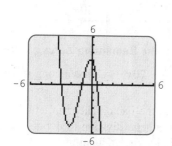

FIGURE 35

Finally, we multiply out the factors of $f(x)$ to obtain the standard polynomial form.

$$f(x) = -2\left(x - \frac{1}{2}\right)(x + 3)(x + 1)$$

$$= (-2x + 1)(x + 3)(x + 1) \qquad \text{Multiplying } -2 \text{ through } \left(x - \frac{1}{2}\right)$$

$$= (-2x + 1)(x^2 + 4x + 3) \qquad \text{Multiplying } (x + 3)(x + 1)$$

$$= (-2x^3 - 8x^2 - 6x) + (x^2 + 4x + 3) \qquad \text{Distributing } (-2x + 1)$$

$$= -2x^3 - 7x^2 - 2x + 3$$

FIGURE 36

The graph of f shown in Figure 36 looks very much like that shown in Figure 35.

In the preceding example, we used the fact that the product of three factors of the form $x - c$ is a third-degree polynomial. More generally, the product of n factors of this form is an nth-degree polynomial. Conversely, an nth-degree polynomial can have no more than n such factors and so can have no more than n real zeros. However, it can happen that an nth-degree polynomial will have fewer than n real zeros. For example, the fifth-degree polynomial $f(x) = x^5 - 2x^4 + x^3$, which factors as $f(x) = x^3(x - 1)^2$, has as its only zeros $x = 0$ and $x = 1$. Since the zero $x = 0$ arises from x^3, a factor of degree 3, it is said to be a zero of multiplicity 3. Likewise, $x = 1$ is a zero of multiplicity 2 since it arises from the factor $(x - 1)^2$.

Multiple Zeros   □

> If $(x - c)^k$ is a factor of a polynomial $f(x)$, but $(x - c)^{k+1}$ is not a factor, then c is called a **zero of multiplicity k**.

The multiplicity of a real zero of a polynomial affects the shape of the graph of the polynomial near the zero. Consider the polynomials $f(x) = (x - 1)^2$ and

$g(x) = (x - 1)^3$, shown in Figures 37 and 38. Both have $x = 1$ as a zero, but the multiplicity is 2 for f and 3 for g. When we compare the graphs of f and g, we see that the graph of f touches but does not cross the x-axis, whereas the graph of g passes through the x-axis. To see why, let's consider the signs of $f(x)$ and $g(x)$ for x-values just to the left and just to the right of 1.

x	$f(x) = (x - 1)^2$ (even multiplicity)	$g(x) = (x - 1)^3$ (odd multiplicity)
0.9	$(0.9 - 1)^2 = 0.01$; positive	$(0.9 - 1)^3 = -0.001$; negative
1.1	$(1.1 - 1)^2 = 0.01$; positive	$(1.1 - 1)^3 = 0.001$; positive

Note that the sign of $f(x)$ is the same on either side of 1, whereas the sign changes for $g(x)$. More generally, if a polynomial function h has a zero c with even multiplicity, then the sign of $h(x)$ will be the same on either side of c, and so the graph of h will touch but not cross the x-axis at c. On the other hand, if the multiplicity of c is odd, the sign of $h(x)$ will change at c, and the graph will pass through the x-axis at c.

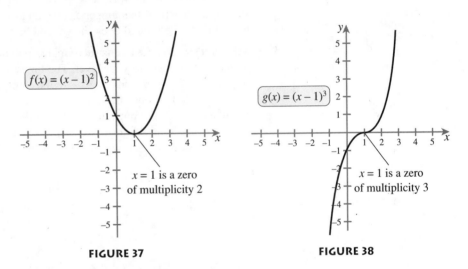

FIGURE 37   **FIGURE 38**

EXAMPLE 3   ■ **Finding a Polynomial with Multiple Zeros**

Find the third-degree polynomial $f(x)$ with the graph shown in Figure 39. Verify with a graphing calculator.

SOLUTION   Since the graph turns at $x = -2$, but passes through the x-axis at $x = 1$, we conclude that $x = -2$ is a zero of even multiplicity, whereas $x - 1$ is a zero of odd multiplicity. Because $f(x)$ is a third-degree polynomial, it follows that

$$f(x) = a(x + 2)^2(x - 1)$$

for some constant a. To find the value of a, we note that f has y-intercept $(0, 4)$, and so $f(0) = 4$.

$$f(0) = 4$$
$$a(0 + 2)^2(0 - 1) = 4 \qquad \text{Setting } x = 0 \text{ in } f(x)$$
$$a(-4) = 4 \qquad \text{Simplifying}$$
$$a = -1$$

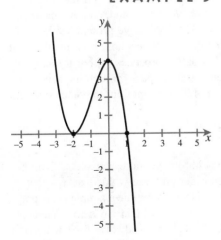

FIGURE 39

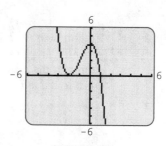

FIGURE 40

Thus, $f(x) = -1(x + 2)^2(x - 1)$. The graph of f shown in Figure 40 looks very much like the graph in Figure 39.

E X A M P L E 4   **Modeling Bungee Jumping with a Cubic Polynomial**

Suppose you are standing on a building 25 feet above the ground. A bungee jumper leaps from a platform 64 feet above you. She passes your level after 2 seconds, again after 6 seconds on the way back up, and after 9 seconds on the way back down. Assuming the height of the jumper can be roughly approximated with a cubic polynomial, how close to the ground does she get and how high does she rebound?

SOLUTION   We first construct a cubic polynomial $h(t)$ that gives the height of the jumper with respect to your level. The times at which the jumper passes your level—namely, $t = 2$, $t = 6$, and $t = 9$—are the zeros of the function h since the height (with respect to you) at those times is 0. Thus, we know that $h(t)$ must have factors $t - 2$, $t - 6$, and $t - 9$. We must also allow for the fact that $h(t)$ may have a constant factor, and so we settle on the form $h(t) = a(t - 2)(t - 6)(t - 9)$. Because the initial height of the jumper is 64 feet, we have $h(0) = 64$. This enables us to solve for a.

$$h(0) = 64$$
$$a(0 - 2)(0 - 6)(0 - 9) = 64$$
$$-108a = 64$$
$$a = -\frac{64}{108} = -\frac{16}{27}$$

Thus, $h(t) = -\frac{16}{27}(t - 2)(t - 6)(t - 9)$. The graph of h is shown in Figure 41. Using the trace or calculate-maximum/minimum feature, we determine that the minimum height is 12.3 feet below your level, or 12.7 feet above the ground, and the maximum height after the rebound is 7.5 feet above you, or 32.5 feet above ground level.

However accurate this model might be for the first 9 seconds, it is invalid soon after. The cubic polynomial $h(t)$ approaches $-\infty$ as t approaches ∞, but it seems unlikely that the bungee jumper will puncture Earth's crust, pass through its mantle and molten core, knife through its crystalline core, and then be shot out the other side into space.

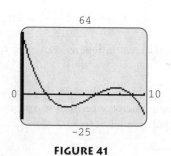

FIGURE 41

COMPLEX ZEROS

We have seen many examples of polynomials with one or more zeros. Are there examples of polynomials with no zeros? If we consider only real-valued zeros—that is, those that correspond to x-intercepts—the answer is yes. A simple example is the polynomial $f(x) = x^2 + 1$. It is clear from the graph in Figure 42 that f has no x-intercepts and hence has no real zeros. However, if we set $x^2 + 1 = 0$ and solve for x within the complex number system, we see that $x = \pm i$. Thus, the imaginary numbers i and $-i$

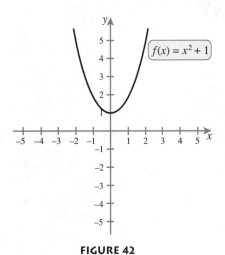

FIGURE 42

are both zeros of f. As it turns out, if we allow complex-valued zeros, every polynomial of degree 1 or higher will have at least one zero. This was first proven in 1799 by the famous German mathematician Carl Friedrich Gauss (he was 22 at the time). Because of its importance to algebra, his result is called the **Fundamental Theorem of Algebra.**

Fundamental Theorem of Algebra   ☐   If $f(x)$ is a polynomial of degree 1 or higher, then f has at least one zero in the complex number system.

We can extend the Fundamental Theorem of Algebra to show that a polynomial of degree n must have exactly n zeros (counting multiplicities). To do this, we first show that a polynomial of degree n can be factored into n factors of the form $x - c$. Suppose $f(x)$ is a polynomial of degree n, with $n \geq 1$. Then, by the Fundamental Theorem of Algebra, f has a zero; call it c_1. By the Factor Theorem, this means $f(x)$ can be factored as

$$f(x) = (x - c_1)q_1(x)$$

where $q_1(x)$ is a polynomial of degree $n - 1$. Applying the Fundamental Theorem of Algebra to the polynomial $q_1(x)$, we know that it too must have a zero; call it c_2. By the Factor Theorem, $q_1(x) = (x - c_2)q_2(x)$ for some polynomial $q_2(x)$ of degree $n - 2$. Combined with the earlier factorization, we now have

$$f(x) = (x - c_1)(x - c_2)q_2(x)$$

If we continue this process, we eventually arrive at a polynomial factor $q_n(x)$ of degree 0 (a constant), and here the process stops. The result is a *complete* factorization of $f(x)$ into linear factors.

Linear Factorization Theorem   ☐   If $f(x)$ is a polynomial of degree n, with $n \geq 1$, then $f(x)$ can be expressed as a product of linear factors in the following way:

$$f(x) = a(x - c_1)(x - c_2) \cdots (x - c_n)$$

where $c_1, c_2, \ldots, c_n$ are complex numbers and a is the leading coefficient of $f(x)$.

An immediate consequence of the Linear Factorization Theorem is that an *n*th-degree polynomial has *n* zeros (counting multiplicities), each of which corresponds to a factor of the form $x - c$. However, neither the Fundamental Theorem of Algebra nor the Linear Factorization Theorem says anything about how to find the zeros or factors. For this, we must still depend on techniques described earlier or on techniques that we will see in the next section.

EXAMPLE 5   ▨   **Finding the Complex Zeros of a Polynomial**

Given that 4 is a zero of $f(x) = x^3 - 4x^2 + 4x - 16$, find all the complex zeros of f and write $f(x)$ as a product of linear factors.

SOLUTION   Since 4 is a zero of f, $x - 4$ must be a factor or $f(x)$. We apply synthetic division to find a second factor.

$$
\begin{array}{r|rrrr}
4 & 1 & -4 & 4 & -16 \\
  &   & 4 & 0 & 16 \\
\hline
  & 1 & 0 & 4 & 0
\end{array}
\qquad
\frac{x^3 - 4x^2 + 4x - 16}{x - 4} = x^2 + 4
$$

So $f(x) = (x - 4)(x^2 + 4)$. Now the zeros that correspond to $x^2 + 4$ can be found as follows:

$$x^2 + 4 = 0$$
$$x^2 = -4$$
$$x = \pm\sqrt{-4} = \pm 2i$$

Thus, the zeros of f and $x = 4$, $x = 2i$, and $x = -2i$, and we have

$$f(x) = (x - 4)(x - 2i)(x + 2i)$$

EXAMPLE 6   ▨   **Finding the Complex Zeros of a Polynomial**

Find all the complex zeros of the polynomial $f(x) = x^5 - 2x^4 - x^3 + 4x^2 - 2x - 4$ and express the polynomial as a product of linear factors.

SOLUTION   We first plot the graph of f to see if any integer zeros can be easily obtained. It appears from the graph in Figure 43 that $x = -1$ and $x = 2$ are zeros of f. It also appears that $x = -1$ has even multiplicity and so we deduce that $f(x)$ has factors $(x + 1)^2(x - 2)$. To verify these factors, and also to find any remaining factors, we divide $f(x)$ by $(x + 1)^2(x - 2)$. Although this division could be done by performing synthetic division three times in succession (once for each of the factors), we choose to divide in one step, using long division. Now $(x + 1)^2(x - 2) = x^3 - 3x - 2$, so we divide $f(x)$ by $x^3 - 3x - 2$ as follows:

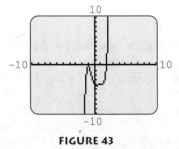

FIGURE 43

$$
\require{enclose}
\begin{array}{r}
x^2 - 2x + 2 \\
x^3 - 3x - 2 \enclose{longdiv}{x^5 - 2x^4 - x^3 + 4x^2 - 2x - 4} \\
\end{array}
$$

$$
\begin{array}{r}
x^5 -3x^3 - 2x^2 \\
\hline
-2x^4 + 2x^3 + 6x^2 - 2x \\
-2x^4 + 6x^2 + 4x \\
\hline
2x^3 - 6x - 4 \\
2x^3 - 6x - 4 \\
\hline
0
\end{array}
$$

Thus,

$$f(x) = (x^3 - 3x - 2)(x^2 - 2x + 2)$$
$$= (x + 1)^2(x - 2)(x^2 - 2x + 2)$$

Since $x^2 - 2x + 2$ cannot be factored using integer coefficients, we find its zeros using the quadratic formula.

$$x^2 - 2x + 2 = 0$$
$$x = \frac{-(-2) \pm \sqrt{(-2)^2 - 4(1)(2)}}{2(1)} = \frac{2 \pm \sqrt{-4}}{2} = \frac{2 \pm 2i}{2} = 1 \pm i$$

Thus, $x^2 - 2x + 2 = [x - (1 + i)][x - (1 - i)]$. Putting all the factors together, we have

$$f(x) = (x + 1)^2(x - 2)[x - (1 + i)][x - (1 - i)]$$

The five zeros of f are $x = -1$ (multiplicity 2), $x = 2$, $x = 1 + i$, and $x = 1 - i$.

It is no coincidence that the two nonreal zeros in Example 6 ($x = 1 + i$ and $x = 1 - i$) differ only in the operators $+$ and $-$. Pairs of complex numbers of the form $a + bi$ and $a - bi$ are called **complex conjugates.** If a polynomial has real coefficients, all nonreal zeros will appear as complex conjugate pairs. In other words, if a polynomial $f(x)$ with real coefficients has a complex zero $a + bi$, then it must also have a zero $a - bi$. (See Exercise 63 for an outline of the proof.)

EXAMPLE 7   ▨   **Finding a Polynomial with Known Complex Zeros**

Find a polynomial $f(x)$ of least degree with real coefficients and zeros $x = -3$ and $x = 2 - 4i$.

SOLUTION   The polynomial must have a second nonreal zero $x = 2 + 4i$. The two factors corresponding to the nonreal zeros $2 - 4i$ and $2 + 4i$ are $x - (2 - 4i)$ and $x - (2 + 4i)$. Multiplying these out yields a quadratic factor with real coefficients, as shown.

$$[x - (2 - 4i)][x - (2 + 4i)] = x^2 - (2 - 4i)x - (2 + 4i)x + (2 - 4i)(2 + 4i)$$
$$= x^2 - 4x + [2^2 - (4i)^2]$$
$$= x^2 - 4x + 4 + 16$$
$$= x^2 - 4x + 20$$

The factor corresponding to the real zero -3 is $x + 3$, and so we know that $f(x) = (x^2 - 4x + 20)(x + 3)$. Multiplying out gives $f(x) = x^3 - x^2 + 8x + 60$, a polynomial of least degree with real coefficients and the desired zeros.

Note that if we did not require the polynomial in Example 7 to have real coefficients, a polynomial with least degree and zeros $x = -3$ and $x = 2 - 4i$ would be the second-degree polynomial

$$f(x) = (x + 3)[x - (2 - 4i)] = x^2 + (1 + 4i)x + (-6 + 12i)$$

Complex zeros need not come in conjugate pairs if the coefficients of the polynomial are allowed to be nonreal.

UNDERSTANDING AND MASTERY CHECKLISTS

CONCEPTS TO UNDERSTAND

- ☐ The Factor Theorem
- ☐ Multiplicity of a zero
- ☐ The Fundamental Theorem of Algebra
- ☐ The Linear Factorization Theorem

SKILLS TO MASTER

- ☐ Use given zeros of a polynomial to find the remaining zeros.
- ☐ Find a polynomial of a specified degree and with given zeros.
- ☐ Find the complex zeros of a polynomial.

EXERCISES 4.3

EXERCISES 1–6 *Use the given zero(s) of f to factor f(x), and then find the remaining zeros.*

1. -3; $f(x) = x^3 + 3x^2 - 4x - 12$
2. 1; $f(x) = x^3 + 5x^2 + 3x - 9$
3. 5; $f(x) = 2x^3 - 7x^2 - 14x - 5$
4. -6; $f(x) = 3x^3 + 10x^2 - 44x + 24$
5. $-\frac{1}{3}, -4$; $f(x) = 3x^4 + 7x^3 - 19x^2 + 5x + 4$
6. $\frac{3}{2}, -1$; $f(x) = 2x^4 - x^3 - 5x^2 + x + 3$

EXERCISES 7–12 *Use a graphing calculator to identify the integer zeros of f, and then factor f(x) completely to find the remaining zeros.*

7. $f(x) = 5x^3 + 3x^2 - 12x + 4$
8. $f(x) = 3x^3 - 10x^2 - x + 12$
9. $f(x) = 6x^3 - 13x^2 + x + 2$
10. $f(x) = 6x^3 + 13x^2 + 4x - 3$
11. $f(x) = x^4 - x^3 - 5x^2 + 3x + 6$
12. $f(x) = x^4 + x^3 - 7x^2 - 5x + 10$

EXERCISES 13–20 *Find a polynomial function f that satisfies the given properties.*

13. Degree 2; zeros -3 and 4
14. Degree 3; zeros -1, 2, and 4
15. Degree 3; zeros -2, 1, and 5; $f(0) = 5$
16. Degree 2; zeros -4 and -1; $f(0) = 8$
17. Degree 4; zeros 0 and 2 (both with multiplicity 2); $f(-1) = 3$
18. Degree 4; zeros -1 and 1 (both with multiplicity 2); $f(2) = 1$
19. Degree 5; zeros -1 (multiplicity 3) and $\frac{1}{2}$ (multiplicity 2)
20. Degree 5; zeros 0 (multiplicity 1), $-\frac{2}{3}$ (multiplicity 2), and 1 (multiplicity 2)

EXERCISES 21–26 *Find the polynomial with the given degree and graph. Check your answer with a graphing calculator.*

21. Degree 2

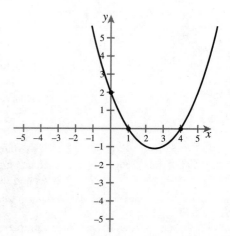

22. Degree 2

23. Degree 3

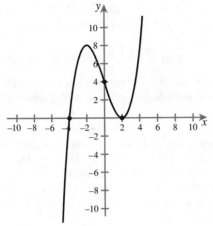

24. Degree 3

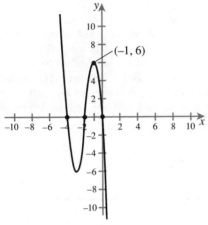

25. Degree 4

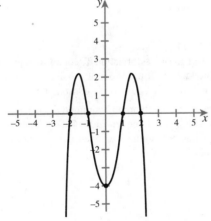

26. Degree 4

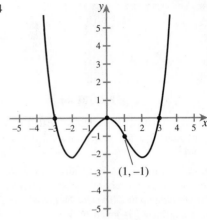

EXERCISES 27–30   *Use the given zero of f to factor f(x), and then find the remaining zeros.*

27. $-2i$; $f(x) = x^3 - 3x^2 + 4x - 12$

28. i; $f(x) = x^3 + 2x^2 + x + 2$

29. $1 + i$; $f(x) = x^4 - 5x^3 + 4x^2 + 2x - 8$

30. $2 - i$; $f(x) = x^4 - 3x^3 - 5x^2 + 29x - 30$

EXERCISES 31–38   *Find all the complex zeros of f and write f(x) as a product of linear factors. Use a graphing calculator as necessary.*

31. $f(x) = x^2 + 1$ **32.** $f(x) = x^2 - 4x + 13$

33. $f(x) = x^3 - 2x^2 + 5x$ **34.** $f(x) = x^3 + 2x^2 + 4x$

35. $f(x) = x^4 + 3x^2 - 4$ **36.** $f(x) = x^4 - 4x^3 + 5x^2$

37. $f(x) = x^5 - 5x^4 + 9x^3 - 5x^2$

38. $f(x) = x^5 - x^4 - x + 1$

EXERCISES 39–44   *Find a polynomial of least degree with real coefficients having the given zeros.*

39. -2 and $2i$ **40.** 0 and $2 + i$

41. i and $1 - i$ **42.** $-1, 3,$ and $-4i$

43. $0, \pm\sqrt{2},$ and $-2 + 2i$ **44.** $1, -1 + 3i, 1 - 3i$

■ **APPLICATIONS**

45. *Dimensions of a Box*   A box with no top is formed by cutting squares out of the corners of a rectangular piece of cardboard and then folding up the sides (see Figure 44). The volume of the box in cubic inches is given by $V(x) = 4x^3 - 72x^2 + 320x$, where x denotes the length of the side of a cut-out square (in inches). Find the zeros of V and use them to determine the

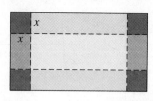 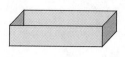

FIGURE 44

domain of V (that is, the values for x that make sense in this problem) and also the dimensions of the original piece of cardboard.

46. *Dimensions of a Box*   A box with no top is formed by cutting squares out of the corners of a rectangular piece of cardboard and then folding up the sides (see Figure 44). The volume of the box in cubic inches is given by $V(x) = 4x^3 - 48x^2 + 135x$, where x denotes the length of the side of a cut-out square (in inches). Find the zeros of V and use them to determine the domain of V (that is, the values for x that make sense in this problem) and also the dimensions of the original piece of cardboard.

47. *Breaking Even*   A company that manufactures bicycle frames has determined that the monthly profit (in dollars) from the sale of x frames is given by

$$P(x) = -0.005x^3 + 1.5x^2 + 50x - 15,000$$

Find the number of frames that must be sold per month for the profit to be zero. For what x-values will the profit be positive?

48. *Breaking Even*   A company that produces copy machines has a monthly profit (in dollars) from the sale of x copy machines given by

$$P(x) = -0.1x^3 + 20.5x^2 - 75x - 67,500$$

Find the number of copy machines that must be sold per month for the profit to be zero. For what x-values will the profit be positive?

49. *Height of a Ball*   A ball is projected upward from 64 feet below ground level. It passes ground level after 1 second and again after 4 seconds on its way back down. It is known that the height of the ball with respect to ground level is given by a quadratic function $s(t)$, where t denotes the number of seconds after the object has been projected upward. Find $s(t)$. What is the maximum height of the ball?

50. *Maximum Profit*   A roofing company has determined that because of fixed operating costs, they will lose $10,000 in a given month if no roofs are completed. Moreover, because of the company's variable cost structure, they will break even (that is, have zero profit) if they complete either 10 roofs or 30 roofs each month. Find a quadratic function $P(x)$ that expresses the company's monthly profit in terms of the number of roofs completed. How many roofs should be completed to maximize profit?

CONCEPTS AND CRITICAL THINKING

EXERCISES 51–54   *Answer true or false.*

51. If -4 is a zero of a polynomial $f(x)$, then $x - 4$ is a factor of $f(x)$.

52. All polynomials of degree 1 or greater have at least one real zero.

53. All nth-degree polynomials have exactly n distinct complex zeros.

54. If $2 + 3i$ is a zero of a polynomial $f(x)$ with real coefficients, then $2 - 3i$ is a zero of $f(x)$ also.

EXERCISES 55–58   *Give an example of each.*

55. A second-degree polynomial with no real zeros

56. A third-degree polynomial with exactly two real zeros

57. A fourth-degree polynomial with exactly two x-intercepts

58. A cubic polynomial $f(x)$ such that $x^2 + 5x + 6$ divides evenly into $f(x)$

59. Use the Linear Factorization Theorem to help you explain why the product of all the zeros of a polynomial with leading coefficient 1 must be equal to the constant term (or its opposite).

60. Explain why a polynomial with odd degree and real coefficients must have at least one real zero. Can anything be said along this line about polynomials of even degree with real coefficients? Explain.

QUESTIONS FOR DISCUSSION OR ESSAY

61. The Factor Theorem states that there is a one-to-one correspondence between the zeros of a polynomial and its linear factors. In other words, each zero c of a polynomial $f(x)$ corresponds to a factor $x - c$, and vice versa. This does *not* mean, however, that there is only one polynomial of a given degree that corresponds to a given set of zeros. Explain why this is so. It may be helpful to think of how two different polynomials of the same degree can be constructed so that they have the same zeros. If we know all the zeros of f, what additional information is required to determine $f(x)$?

62. Many physical applications involve cyclical or oscillating behavior. For example, the motion of a pendulum, the path of a weight attached to a spring, alternating current, and average monthly temperatures all exhibit cyclical behavior. Discuss the limitations in using polynomials to model cyclical phenomena.

PROJECTS FOR ENRICHMENT

63. *Complex Conjugates*   The complex conjugate of a number $a + bi$ is $a - bi$. In general, the conjugate of a complex number z is denoted by $\bar{z}$. Thus, if $z = a + bi$, then $\bar{z} = a - bi$. In this project, we investigate some of the properties of complex conjugates. Throughout, we let $z = a + bi$ and $w = c + di$.

a. Show that $z + \bar{z}$ is a real number

b. Show that $z\bar{z}$ is a real number.

c. Show that $\overline{z + w} = \bar{z} + \bar{w}$.

d. Show that $\overline{z \cdot w} = \bar{z} \cdot \bar{w}$.

e. Use part d to argue that $\overline{z^n} = (\bar{z})^n$.

f. If a is a real number, show that $\bar{a} = a$.

g. If $f(x) = a_n x^n + a_{n-1}x^{n-1} + \cdots + a_1 x + a_0$ is a polynomial with real coefficients, show that $f(\bar{z}) = \overline{f(z)}$.

h. If $f(x) = a_n x^n + a_{n-1}x^{n-1} + \cdots + a_1 x + a_0$ is a polynomial with real coefficients and z is a zero of f, show that $\bar{z}$ is also a zero of f.

64. *Cardan's Formula*   The zeros of a general quadratic polynomial $f(x) = ax^2 + bx + c$ are easily found with the quadratic formula. To find exact values for the zeros of the general cubic polynomial $f(x) = ax^3 + bx^2 + cx + d$, where $a \neq 0$, we require **Cardan's formula,** a formula that gives the solution to cubic equations in terms of their coefficients.

a. A cubic equation is said to be in **reduced form** if the square term has coefficient 0 and the cubed term has coefficient 1; that is, if it is of the form $z^3 + pz + q = 0$. Show that the equation $ax^3 + bx^2 + cx + d = 0$ can be transformed into a cubic equation in reduced form by making the substitution $x = z - \frac{b}{3a}$ and then dividing through by a.

b. Write the cubic equation $x^3 - 9x^2 + 9x + 62 = 0$ in reduced form.

Cardan's formula can be applied only to cubic equations in reduced form. To express Cardan's formula concisely, we define the following three variables.

$$s = \sqrt{\frac{q^2}{4} + \frac{p^3}{27}}, \quad u = \sqrt[3]{\frac{-q}{2} + s}, \quad \text{and} \quad v = \sqrt[3]{\frac{-q}{2} - s}$$

According to Cardan's formula, the three solutions to the equation $z^3 + pz + q = 0$ are given by

$$z_1 = u + v, \quad z_2 = \frac{-(u + v) + (u - v)\sqrt{3}i}{2}$$

and

$$z_3 = \frac{-(u + v) - (u - v)\sqrt{3}i}{2}$$

c. Use Cardan's formula to find the zeros of each of the following polynomials.

  i. $f(x) = x^3 + 63x - 316$

  ii. $g(x) = x^3 - 27x - 54$

  iii. $h(x) = x^3 - 9x^2 + 9x + 62$ (*Hint:* First write in the reduced form $z^3 + pz + q = 0$ and solve the resulting equation for z. Then use the relationship $x = z - \frac{b}{3a}$ to find the corresponding values of x.)

d. Use Cardan's formula in conjunction with the Factor Theorem to find factorizations of each of the polynomials f, g, and h defined in part c.

SECTION 4.4   REAL ZEROS OF POLYNOMIALS

☐ Why is it that the polynomial $P(x) = x^{20} - $ (your age in seconds)$x^{19} + $ (your weight in kilograms)$x^5 + 2x + 2.7$ cannot possibly cross the positive x-axis exactly once?

☐ You are told that a certain polynomial has an x-intercept somewhere between 1 and 10. How could you evaluate the function at 20 points, and, based on this information, determine the x-intercept to within 3 decimal places of accuracy?

☐ Why is it that any polynomial $P(x)$ with integer coefficients, leading coefficient 1, and satisfying $P(0) = 140$ has at most 2 rational zeros?

☐ How can it be that for a given loan and repayment schedule, there are two possible interest rates?

We have already seen that graphing calculators are invaluable aids for investigating the real zeros of a polynomial. However, we have also seen that a graphing calculator has its limitations. If the zeros are not integers, we may be limited to finding approximations.

More importantly, if the calculator is not used carefully, it can give misleading information about the number of zeros or the multiplicity of a zero. Two examples of these limitations are given in Table 1.

TABLE 1 Limitations of finding zeros of polynomials graphically

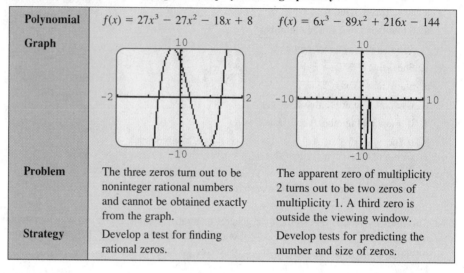

Polynomial	$f(x) = 27x^3 - 27x^2 - 18x + 8$	$f(x) = 6x^3 - 89x^2 + 216x - 144$
Graph		
Problem	The three zeros turn out to be noninteger rational numbers and cannot be obtained exactly from the graph.	The apparent zero of multiplicity 2 turns out to be two zeros of multiplicity 1. A third zero is outside the viewing window.
Strategy	Develop a test for finding rational zeros.	Develop tests for predicting the number and size of zeros.

THE RATIONAL ZERO THEOREM

Suppose that we are interested in finding only the rational zeros of a quadratic polynomial $f(x) = ax^2 + bx + c$, where the coefficients a, b, and c are integers. Now, because any rational number can be written as a fraction p/q in lowest terms, we will search for zeros of the form p/q, where p and q have no common factors other than 1. Since $f(p/q) = 0$, we have

$$a\left(\frac{p}{q}\right)^2 + b\left(\frac{p}{q}\right) + c = 0 \qquad \text{Definition of a zero}$$

$$ap^2 + bpq + cq^2 = 0 \qquad \text{Multiplying through by } q^2$$

$$ap^2 + bpq = -cq^2 \qquad \text{Subtracting } cq^2 \text{ on both sides}$$

$$p(ap + bq) = -cq^2 \qquad \text{Factoring out } p \text{ on the left}$$

The last step shows that p is a factor of the left-hand side, and so p must be a factor of the right-hand side as well. But since p and q have no common factors other than 1, p must be a factor of c. In a similar fashion, we can show that q must be a factor of a. This narrows our search considerably: The only possible rational zeros of $f(x) = ax^2 + bx + c$ are those of the form p/q, where p divides into the constant term c, and q divides into the leading coefficient a. The Rational Zero Theorem is a generalization of these results for polynomials of nth degree.

Rational Zero Theorem □

> If the polynomial $f(x) = a_n x^n + a_{n-1} x^{n-1} + \cdots + a_1 x + a_0$ has integer coefficients, then every rational zero of f has the form p/q, where p and q have no common factors other than 1, p is a factor of the constant term a_0, and q is a factor of the leading coefficient a_n.

According to the Rational Zero Theorem, if we find all the factors of the constant term a_0 and the leading coefficient a_n, and form a list of the rational numbers of the form

$$\frac{\text{Factor of } a_0}{\text{Factor of } a_n}$$

then all the rational zeros of the polynomial will be in the list. Trial and error (perhaps with the assistance of a graphing calculator) will lead us to the actual rational zeros.

· E X A M P L E 1   ▨   **Finding Rational Zeros**

Use the Rational Zero Theorem to find the rational zeros of

$$f(x) = 27x^3 - 27x^2 - 18x + 8$$

SOLUTION   First we list all the factors of the constant term and leading coefficient.

Factors of the constant term 8: $\pm 1, \pm 2, \pm 4, \pm 8$

Factors of the leading coefficient 27: $\pm 1, \pm 3, \pm 9, \pm 27$

Next we form all quotients of the factors of 8 divided by the factors of 27.

Possible rational zeros: $\pm 1, \pm\dfrac{1}{3}, \pm\dfrac{1}{9}, \pm\dfrac{1}{27}$

$$\pm 2, \pm\dfrac{2}{3}, \pm\dfrac{2}{9}, \pm\dfrac{2}{27}$$

$$\pm 4, \pm\dfrac{4}{3}, \pm\dfrac{4}{9}, \pm\dfrac{4}{27}$$

$$\pm 8, \pm\dfrac{8}{3}, \pm\dfrac{8}{9}, \pm\dfrac{8}{27}$$

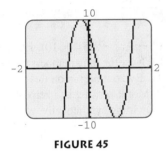

FIGURE 45

Since there are so many possible zeros, naive trial and error will likely be quite inefficient. Instead, we use information from the graph of f to help narrow down the list. From the graph in Figure 45, we can rule out integer zeros. Moreover, the middle zero appears to be approximately $\frac{1}{3}$, which is in our list of possible zeros. Thus, we begin by testing $\frac{1}{3}$ with synthetic division.

$$\begin{array}{r|rrrr} \tfrac{1}{3} & 27 & -27 & -18 & 8 \\ & & 9 & -6 & -8 \\ \hline & 27 & -18 & -24 & 0 \end{array}$$

The remainder 0 tells us that $x = \frac{1}{3}$ is indeed a zero. This fact, together with the first three numbers in the bottom row, indicates that two of the factors of $f(x)$ are $x - \frac{1}{3}$ and $27x^2 - 18x - 24$. Furthermore,

$$27x^2 - 18x - 24 = 3(9x^2 - 6x - 8) = 3(3x + 2)(3x - 4)$$

Thus, the other two zeros occur when $3(3x + 2)(3x - 4) = 0$; namely, when $x = -\frac{2}{3}$ and $x = \frac{4}{3}$. So the three zeros of f are $-\frac{2}{3}, \frac{1}{3}$, and $\frac{4}{3}$.

WARNING!   The Rational Zero Theorem can be applied only to polynomials with integer coefficients. Moreover, many polynomials with integer coefficients have no rational zeros.

EXAMPLE 2   ■   **Finding the Zeros of a Polynomial**

Find all the zeros of $f(x) = x^4 + \frac{3}{5}x^3 - \frac{27}{5}x^2 - 3x + 2$ and write $f(x)$ as a product of linear factors.

SOLUTION   Notice that f has noninteger coefficients and so we cannot apply the Rational Zero Theorem to f. However, the zeros of f coincide with the solutions of $x^4 + \frac{3}{5}x^3 - \frac{27}{5}x^2 - 3x + 2 = 0$, and we can clear the fractions in this equation by multiplying through by 5 to obtain $5x^4 + 3x^3 - 27x^2 - 15x + 10 = 0$. Thus, we will apply the Rational Zero Theorem to the polynomial function $g(x) = 5x^4 + 3x^3 - 27x^2 - 15x + 10$ (note that f and g are *different* polynomials—they just have the same zeros).

Factors of the constant term 10:	$\pm 1, \pm 2, \pm 5, \pm 10$
Factors of the leading coefficient 5:	$\pm 1, \pm 5$
Possible rational zeros:	$\pm 1, \pm\frac{1}{5}, \pm 2, \pm\frac{2}{5}, \pm 5, \pm 10$

These are the only possible rational zeros of g and so also of f. To narrow down the list, we plot the graph of g. (Note that we could have chosen to use f, but g is somewhat easier to work with since it has integer coefficients.) It appears in Figure 46 that g has a zero at $x = -1$, and we can easily test this using synthetic division.

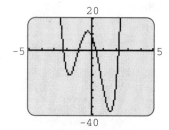

FIGURE 46

$$
\begin{array}{r|rrrr}
-1 & 5 & 3 & -27 & -15 & 10 \\
   &   & -5 & 2 & 25 & -10 \\
\hline
   & 5 & -2 & -25 & 10 & 0
\end{array}
$$

So $x = -1$ is a zero and $g(x)$ factors as $(x + 1)(5x^3 - 2x^2 - 25x + 10)$. From the graph, we see that there is another zero between 0 and 1. Now the only numbers between 0 and 1 in our list of possible rational zeros are $\frac{1}{5}$ and $\frac{2}{5}$. By zooming in on this positive zero, as shown in Figure 47, we can rule out $\frac{1}{5} = 0.2$. Moreover, if $\frac{2}{5}$ is a zero of $g(x)$, then it must also be a zero of $5x^3 - 2x^2 - 25x + 10$. Dividing synthetically, we obtain

X=.39893617 Y=.0360215

FIGURE 47

$$
\begin{array}{r|rrrr}
\frac{2}{5} & 5 & -2 & -25 & 10 \\
            &   & 2 & 0 & -10 \\
\hline
            & 5 & 0 & -25 & 0
\end{array}
$$

So $\frac{2}{5}$ is indeed a zero. The synthetic division for $\frac{2}{5}$ also tells us that another factor of $g(x)$ is $5x^2 - 25 = 5(x^2 - 5)$. This factor has two irrational zeros—namely, $x = \pm\sqrt{5}$. Thus, the four zeros of g (and f also) are $x = -1$, $x = \frac{2}{5}$, $x = -\sqrt{5}$, and $x = \sqrt{5}$. This gives us

$$f(x) = a(x + 1)\left(x - \frac{2}{5}\right)(x + \sqrt{5})(x - \sqrt{5})$$

Since $f(x) = x^4 + \frac{3}{5}x^3 - \frac{27}{5}x^2 - 3x + 2$ has a leading coefficient of 1, it follows that $a = 1$, which gives us

$$f(x) = (x + 1)\left(x - \frac{2}{5}\right)(x + \sqrt{5})(x - \sqrt{5})$$

DESCARTES' RULE OF SIGNS

Our next test gives us information about the number of real zeros of a polynomial. In Section 4.3, we saw that a polynomial of degree n will have n zeros. However, because of possible multiplicities and complex zeros, an nth-degree polynomial may have fewer than n distinct real zeros. **Descartes' Rule of Signs** gives us more information about the number of real zeros by considering the number of times that successive coefficients of the polynomial change from positive to negative or negative to positive. These changes are referred to as **variations in sign.** The following table gives several examples.

Polynomial	Coefficient signs	Variations in sign
$x^3 + 2x + 5$	$+ + +$	0
$2x^3 - x^2 - 1$	$+ - -$	1
$5x^3 - 3x^2 + x + 4$	$+ - + +$	2

Descartes' Rule of Signs ☐

Let $f(x) = a_n x^n + a_{n-1} x^{n-1} + \cdots + a_1 x + a_0$ be a polynomial with real coefficients.

1. The number of positive real zeros (counting multiplicities) is either equal to the number of variations in sign of $f(x)$ or is less than that number by an even integer.
2. The number of negative real zeros (counting multiplicities) is either equal to the number of variations in sign of $f(-x)$ or is less than that number by an even integer.

E X A M P L E 3 ▧ Using Descartes' Rule of Signs

Apply Descartes' Rule of Signs to the polynomial $f(x) = 2x^3 - x^2 - 1$.

SOLUTION Since $f(x)$ has only one variation in sign, f must have exactly one positive real zero. To test for negative zeros, we first simplify $f(-x)$.

$$f(-x) = 2(-x)^3 - (-x)^2 - 1 = -2x^3 - x^2 - 1$$

Since $f(-x)$ has no variations in sign, we conclude that f has no negative zeros. The graph of f in Figure 48 shows one zero near 1. Because of Descartes' Rule of Signs, we can be certain that there are no other real zeros. (Note that if we had just used the graph and not Descartes' Rule of Signs, we could not rule out the presence of zeros outside the viewing window.)

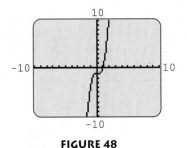

FIGURE 48

Since Descartes' Rule of Signs does not "pin down" the number of zeros unless there is only one variation in sign or none at all, it is usually best to complement the information given by Descartes' Rule of Signs by graphing the polynomial with a graphing calculator.

EXAMPLE 4 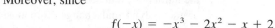 **Determining the Number of Real Zeros of a Polynomial**

Apply Descartes' Rule of Signs to $f(x) = x^3 - 2x^2 + x + 2$.

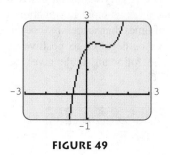

FIGURE 49

SOLUTION Since $f(x)$ has two variations in sign, f has either two or no positive real zeros. Moreover, since

$$f(-x) = -x^3 - 2x^2 - x + 2$$

has only one variation in sign, f has one negative zero. From the graph of f in Figure 49, we see one negative zero, no positive zeros, and two turning points. Because f is a cubic polynomial, and cubic polynomials have at most two turning points, f has no additional turning points. Thus, the graph of f will continue to rise as x gets larger, and f has no additional real zeros.

In the previous example, Descartes' Rule of Signs was superfluous. All the pertinent information could be obtained directly from the graph. This is not always the case, as the following examples illustrate.

EXAMPLE 5 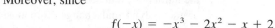 **Finding the Zeros of a Polynomial**

Find the zeros of $f(x) = 6x^3 - 89x^2 + 216x - 144$.

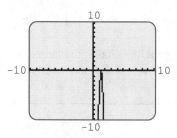

FIGURE 50

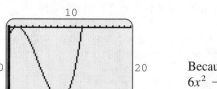

FIGURE 51

SOLUTION The graph of f is shown in Figure 50. At first glance it appears that f has only one zero of multiplicity 2—or perhaps 2 distinct zeros very close together—somewhere in the interval $[1, 2]$. However, since $f(x)$ has three variations in sign, Descartes' Rule of Signs tells us that f has either one or three positive zeros (counting multiplicities), and so a single positive zero of multiplicity 2 is impossible, as is the presence of exactly 2 positive zeros very close together. Thus, there must be a zero to the right of the viewing window. After some experimentation, we settle on the view shown in Figure 51. This suggests another zero at $x = 12$, which can be confirmed by synthetic division.

$$\begin{array}{r|rrrr} 12 & 6 & -89 & 216 & -144 \\ & & 72 & -204 & 144 \\ \hline & 6 & -17 & 12 & 0 \end{array}$$

Because the remainder is 0, we know that $f(x) = (x - 12)(6x^2 - 17x + 12)$. Moreover, $6x^2 - 17x + 12 = (2x - 3)(3x - 4)$ and so f has two more zeros at $x = \frac{3}{2}$ and $x = \frac{4}{3}$

It is not uncommon for a polynomial to have no rational zeros, and in this case it is very difficult—perhaps impossible—to obtain exact values for the zeros. Fortunately, we can approximate irrational zeros with a graphing calculator.

EXAMPLE 6 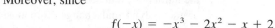 **Finding the Zeros of a Polynomial**

Show that $f(x) = x^3 + x - 1$ has exactly one real zero, which is irrational, and use a graphing calculator to approximate it to the nearest hundredth.

SOLUTION One real zero is readily apparent from the x-intercept of the graph of f shown in Figure 52. However, f could have as many as three real zeros because it is a

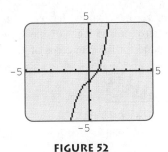

FIGURE 52

polynomial of degree 3. To investigate the number of real zeros, we apply Descartes' Rule of Signs. Since $f(x)$ has only one variation in sign, f must have one positive zero. Since $f(-x) = -x^3 - x - 1$ has no variations in sign, f has no negative zeros. Thus, f has precisely one real zero. To see if a rational value can be found, we apply the Rational Zero Theorem.

Factors of the constant term -1:	± 1
Factors of the leading coefficient 1:	± 1
Possible rational zeros:	± 1

But $f(-1) = -3$ and $f(1) = 1$, so neither 1 nor -1 is a zero. Thus, f has no rational zeros. To approximate the irrational zero, we can either zoom in and use the trace feature, or apply the calculate-zero feature, to obtain $x \approx 0.68$.

UPPER AND LOWER BOUNDS TEST

Our final test helps us determine upper and lower bounds for the zeros of a polynomial. We say that a real number b is an **upper bound** for the real zeros of a polynomial if none of the zeros is greater than b. Similarly, a number b is a **lower bound** if none of the zeros is less than b. Suppose we suspect that a number b is an upper (or lower) bound for the real zeros of a polynomial f. How can we determine this for sure? The graph of f can be misleading because of the possibility that a zero lies outside the viewing window. Instead, we can apply the following test, the proof of which is beyond the scope of this text.

Upper and Lower Bound Test □

> Let $f(x) = a_n x^n + a_{n-1}x^{n-1} + \cdots + a_1 x + a_0$ be a polynomial with real coefficients and positive leading coefficient a_n
>
> **1.** If $b > 0$ and each number in the last row of the synthetic division of $f(x)$ by $x - b$ is either positive or zero, then b is an upper bound for the real zeros of f.
>
> **2.** If $b < 0$ and the numbers in the last row of the synthetic division of $f(x)$ by $x - b$ alternate in sign (with 0 counting either as positive or negative), then b is a lower bound for the real zeros of f.

E X A M P L E 7 ▓ **Determining the Number of Zeros of a Polynomial**

Use the Upper and Lower Bounds Test to help you determine the number of zeros of the function $f(x) = x^6 - 4x^4 + 6x^2 - 4$.

SOLUTION The graph of f in Figure 53 shows two zeros. However, since f has degree 6, there could be as many as six zeros. Moreover, all we can tell from Descartes' Rule of Signs is that, since $f(x)$ and $f(-x)$ both have three variations in sign, f could have either one or three positive zeros and either one or three negative zeros.

Because the graph suggests that there are no zeros greater than 2, we apply the Upper and Lower Bounds Test to see if $x = 2$ is an upper bound for the real zeros of f.

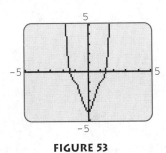

FIGURE 53

$$\begin{array}{r|rrrrrrr} 2 & 1 & 0 & -4 & 0 & 6 & 0 & -4 \\ & & 2 & 4 & 0 & 0 & 12 & 24 \\ \hline & 1 & 2 & 0 & 0 & 6 & 12 & 20 \end{array}$$

Since all the numbers in the last row are positive or zero, we know that $x = 2$ is an upper bound on the real zeros of f—there are no zeros to the right of the viewing window. Next we test to see if $x = -2$ is a lower bound for the real zeros of f.

$$
\begin{array}{r|rrrrrrr}
-2 & 1 & 0 & -4 & 0 & 6 & 0 & -4 \\
   &   & -2 & 4 & 0 & 0 & -12 & 24 \\
\hline
   & 1 & -2 & 0 & 0 & 6 & -12 & 20 \\
   & + & - & + & - & + & - & +
\end{array}
$$

Because the numbers in the last row alternate in sign (note how 0 is first counted as positive and then as negative), $x = -2$ is indeed a lower bound. Thus, f has only the two real zeros shown in Figure 53.

UNDERSTANDING AND MASTERY CHECKLISTS

CONCEPTS TO UNDERSTAND

- ☐ The Rational Zero Theorem
- ☐ Descartes' Rule of Signs
- ☐ Upper and Lower Bounds Test

SKILLS TO MASTER

- ☐ Use the Rational Zero Theorem to find the rational zeros of a polynomial.
- ☐ Apply Descartes' Rule of Signs to a polynomial to obtain information about the number of zeros.
- ☐ Apply the Upper and Lower Bounds Test to a polynomial to obtain information about its zeros.

EXERCISES 4.4

EXERCISES 1–10 *Use the Rational Zero Theorem to list all the possible rational zeros of the function. Use synthetic division and/or a graphing calculator to help you determine which are actually zeros.*

1. $f(x) = x^4 - 12x^2 + 27$
2. $f(x) = x^3 + 3x^2 - 2x - 6$
3. $g(x) = x^3 - 3x^2 - 4x + 12$
4. $h(x) = x^4 - 17x^2 + 16$
5. $f(x) = 4x^4 - 4x^3 - 9x^2 + x + 2$
6. $h(x) = 9x^3 + 9x^2 - 16x + 4$
7. $g(x) = 2x^5 + 3x^4 - 2x - 3$
8. $h(x) = 25x^5 - 4x^3 + 25x^2 - 4$
9. $f(x) = x^3 - \frac{9}{2}x^2 + \frac{1}{2}x + 6$
10. $g(x) = x^4 - \frac{10}{3}x^3 - 12x^2 + \frac{58}{3}x - 5$

EXERCISES 11–20 *Apply Descartes' Rule of Signs to the given function. Then use a graphing calculator to determine the precise number of positive and negative zeros.*

11. $f(x) = x^3 + 4$
12. $h(x) = x^4 - 3x^2 - 1$
13. $g(x) = 2x^4 + x^2 + 3$
14. $g(x) = 5x^3 - x^2 - 1$
15. $f(x) = x^3 - 4x^2 + 7x - 2$
16. $h(x) = 3x^3 + x^2 + 2x + 6$
17. $f(x) = x^5 - 10x^4 - 11x^3 - 5$
18. $g(x) = x^4 + 11x^3 - 12x^2 + 3$
19. $h(x) = x^3 - x^2 - \frac{101}{100}x + \frac{99}{100}$
20. $f(x) = x^3 + \frac{1}{100}x^2 - \frac{301}{100}x - \frac{101}{50}$

EXERCISES 21–26 *Use the Upper and Lower Bounds Test to confirm the given bounds for the zeros of the function.*

21. $f(x) = x^4 - 5x^3 - 11x^2 + 33x - 18$
    Lower: -4; Upper: 7
22. $g(x) = x^4 + 3x^3 - 27x^2 + 3x - 28$
    Lower: -8; Upper: 5
23. $h(x) = x^4 - 10x^3 - 5$
    Lower: -1; Upper: 11
24. $g(x) = x^4 + 11x^3 - 12x^2 + 6$
    Lower: -13; Upper: 1
25. $f(x) = x^4 - 62x^3 + 962x^2 - 62x + 960$
    Lower: -1; Upper: 62
26. $h(x) = x^4 - 3024x^2 - 3024$
    Lower: -56; Upper: 55

EXERCISES 27–36 *Find the exact values of the real zeros of the function. Use any of the tools described in this chapter.*

27. $h(x) = x^3 + 6x^2 - x - 6$

28. $g(x) = x^3 - 13x - 12$

29. $f(x) = x^3 - 14x^2 + 25x - 12$

30. $g(x) = x^3 + 11x^2 + 2x + 22$

31. $h(x) = x^3 + \frac{7}{3}x^2 - \frac{23}{12}x + \frac{1}{4}$

32. $g(x) = x^3 - \frac{7}{12}x^2 - \frac{7}{8}x - \frac{1}{6}$

33. $h(x) = 8x^3 + x^2 - 16x - 2$

34. $f(x) = 4x^3 - 9x^2 - 6x + 2$

35. $g(x) = x^4 + x^3 - 120x^2 - 121x - 121$

36. $h(x) = x^4 + 26x^3 + 170x^2 + 26x + 169$

EXERCISES 37–42 *Find the exact values of the real solutions of the polynomial equation.*

37. $x^3 + x^2 - 4x - 4 = 0$

38. $x^3 + 3x^2 + 2x + 6 = 0$

39. $x^4 - 27 = 6x^2$

40. $2x^4 - 21x^2 - 5 = 5x^3 + 19x$

41. $8x^5 - 8x^3 = 1 - x^2$

42. $x^5 + 18x = 11x^3$

EXERCISES 43–48 *Use the Rational Zero Theorem to show that the function has no rational zeros. Use a graphing calculator to approximate any irrational zeros to the nearest hundredth.*

43. $f(x) = x^3 + 3x + 1$

44. $g(x) = x^3 - x^2 - 2$

45. $h(x) = x^4 + 2x - 2$

46. $g(x) = x^4 - x^3 - 1$

47. $h(x) = x^5 - 10x^4 - 11x^3 - 5$

48. $f(x) = x^4 + 11x^3 - 12x^2 + 3$

■ APPLICATIONS

---

49. *Dimensions of a Box*   A box with a square base is to be constructed so that its height is 1 inch more than twice its base length. Find the dimensions of the box if the volume must be 9 cubic inches.

50. *Dimensions of a Box*   A box with a square base is to be constructed so that its height is 1 inch less than three times its base length. Find the dimensions of the box if the volume must be 100 cubic inches.

51. *Makeshift Box*   A box with no top is formed by cutting squares out of the corners of a 10″ × 5″ rectangular piece of cardboard and then folding up the sides (see Figure 54). What size square must be cut out of each corner if the resulting box is to have a volume of 18 cubic inches?

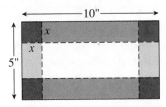
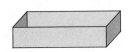

FIGURE 54

52. *Postal Regulations*   A rectangular package to be sent by the U.S. Postal Service can have a maximum combined length and girth (perimeter of the base) of 108 inches (see Figure 55). Find the dimensions of a box with a square base and volume 10,800 cubic inches if the combined length and girth is exactly 108 inches.

53. *Poverty Percentage*   The percentage of U.S. families below the poverty level for the years 1960–1995 can be modeled by the polynomial

$$p(x) = -0.0012x^3 + 0.078x^2 - 1.45x + 17.6$$

where x is the year (with $x = 0$ corresponding to 1960) and $p(x)$ is the percentage of families below the poverty level in that year.

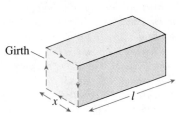

Girth = $4x$
Length + girth = $l + 4x$

FIGURE 55

According to this model, in what year(s) were 10.3% of U.S. families below the poverty level?

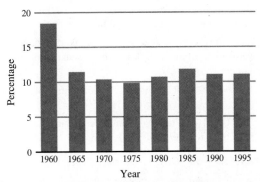

U.S. Families Below Poverty Level 1960–1995

Data source: U.S. Bureau of the Census.

54. *Sunday Accidents*   Of the total number of auto accidents that will occur on a given Sunday, the percentage that will have occurred by hour x (with $x = 0$ corresponding to midnight) can be modeled by the polynomial function

$$p(x) = -0.00182x^4 + 0.0884x^3 - 1.274x^2 + 8.951x$$

for *x* between 0 and 24. According to this model, by what time(s) had half of the Sunday accidents occurred? During what Sunday hour do the greatest number of accidents occur?

Sunday Accidents

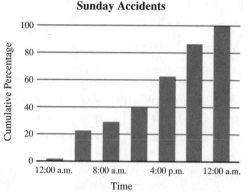

Data source: *1988 Accident Facts,* National Safety Council.

EXERCISES 55–58 *The **present value** of a payment in the amount P made t years from now is defined to be $P(1 + r)^{-t}$, which is the amount that, if invested now, would grow to P in t years. For convenience, we define $v = 1/(1 + r)$, so that the present value of a payment P made t years from now is Pv^t. The present value of a series of payments is simply the sum of the present values of each of the payments. To compute the interest rate(s) corresponding to a loan repayment schedule, we equate the present value of all payments made to that of all distributions received.*

55. *Installment Loan*   A loan of $3000 is received at time 0, and 3 payments of $1100 are made at times 1, 2, and 3.

   a. Find an expression (involving *v*) for the present value of the payments.

b. Set the present value of the payments equal to $3000 (the present value of the loan distribution) and apply Descartes' Rule of Signs to show that there is a unique value for *v*, and hence a unique interest rate *r*.

c. Use a graphing calculator to approximate the value of *v*, and use $v = 1/(1 + r)$ to find the corresponding interest rate *r*.

56. *Mortgage Loan*   A mortgage loan of *L* dollars is to be paid off in *n* monthly installments of *p* dollars (assume $np > L$). Use Descartes' Rule of Signs to show that there is a unique positive value for *v* and, hence, a unique positive interest rate.

57. *Line-of-Credit Loan*   A line-of-credit loan is initiated with a distribution of $1000 at time 0, payments at times 1 and 2, a second distribution at time 3, and a final payment at time 4.

   a. Denote the unspecified payments by p_1, p_2, and p_3, and find an expression (involving *v*) for the present value of the payments.

   b. Denote the unspecified distribution by *d*, and find an expression (involving *v*) for the present value of the distributions.

   c. Equate the expressions found in parts a and b and apply Descartes' Rule of Signs to determine the number of possible interest rates.

   d. Use a graphing calculator to approximate the interest rate(s) if the payments were all $400 and the second receipt was $500.

58. *Line-of-Credit Loan*   A line-of-credit loan is initiated with a payment of $25 at time 0, a distribution of $50 at time 1, and a payment of $26 at time 3. Show that two different positive interest rates result.

CONCEPTS AND CRITICAL THINKING

EXERCISES 59–62 *Answer true or false.*

59. A polynomial of the form $f(x) = x^3 + ax^2 + bx + 1$ (where *a* and *b* are integers) can have at most 2 rational zeros.

60. According to Descartes' Rule of Signs, the number of positive zeros of a polynomial $f(x)$ is equal to the number of variations in sign of $f(x)$.

61. It is possible that $\frac{1}{3}$ is a zero of a polynomial of the form $f(x) = x^4 + ax^2 + x + 3$, where *a* is an integer.

62. The Upper and Lower Bounds Test enables us to narrow our search for real zeros of a polynomial.

EXERCISES 63–66 *Give an example of each.*

63. A polynomial that, according to Descartes' Rule of Signs, has either 1, 3, or 5 positive zeros

64. A fourth-degree polynomial that, according to Descartes' Rule of Signs, has either 2 or 0 negative zeros

65. A fifth-degree polynomial that, according to the Rational Zero Theorem, has ±1 as its only possible rational zeros

66. A fourth-degree polynomial with five terms that, according to Descartes' Rule of Signs, has no positive zeros

67. Apply Descartes' Rule of Signs to $P(x) = x^{20} - $ (your age in seconds)$x^{19} + $ (your weight in kilograms)$x^5 + 2x + 2.7$.

68. Explain why any polynomial $P(x)$ with integer coefficients, leading coefficient 1, and satisfying $P(0) = 149$ has at most 2 rational zeros.

QUESTIONS FOR DISCUSSION OR ESSAY

69. According to the Rational Zero Theorem, if *p/q* is a zero of a polynomial $f(x)$, then *p* must be a factor of the constant term and

q must be a factor of the leading coefficient. What can be said if the constant term is zero? What if the leading coefficient is 1?

Explain how the Rational Zero Theorem can be used to show that $\sqrt{2}$ is an irrational number.

70. For what types of polynomials is Descartes' Rule of Signs most useful? Give some examples to help support your claim.

71. In light of all the other procedures we've seen for obtaining information about the zeros of a polynomial, discuss the usefulness of the Upper and Lower Bounds Test.

▣ PROJECTS FOR ENRICHMENT

72. *Proof of the Rational Zero Theorem*

    a. Prior to stating the Rational Zero Theorem, we showed that if p/q is a zero of $f(x) = ax^2 + bx + c$ and p and q have no common factors, then p must be a factor of c. Show that q must be a factor of a as well.

    b. Generalize the technique used in the text for quadratic polynomials to show that if p and q have no common factors and p/q is a zero of

$$f(x) = a_n x^n + a_{n-1} x^{n-1} + \cdots + a_1 x + a_0$$

    then p is a factor of a_0.

    c. Generalize the technique used in part a to show that if p and q have no common factors and p/q is a zero of

$$f(x) = a_n x^n + a_{n-1} x^{n-1} + \cdots + a_1 x + a_0$$

    then q is a factor of a_n.

73. *The Bisection Method* There often is an easy way to show that a polynomial f has a zero between two numbers a and b. We simply compute $f(a)$ and $f(b)$ and check to see if one output value is negative and the other positive. If so, then there is a zero between a and b. For example, we can be certain that $f(x) = x^3 + x^2 - 4$ has a zero between 1 and 2 since $f(1) = -2 < 0$ and $f(2) = 8 > 0$.

    a. Show that the given polynomial has a zero between a and b.

        i. $g(x) = x^3 + x - 3;\ a = 1,\ b = 2$

        ii. $f(x) = -x^3 + x^2 - 2x + 9;\ a = 2,\ b = 3$

        iii. $h(x) = x^4 - 9x^2 + x + 4;\ a = 2,\ b = 3$

        iv. $g(x) = x^4 - 8x + 2;\ a = 0,\ b = 1$

        v. $h(x) = x^5 + 4x^2 + 3;\ a = -2,\ b = -1$

        vi. $f(x) = x^5 + 2x^3 + 1;\ a = -1,\ b = 0$

The Bisection Method gets its name from the fact that we repeatedly *bisect* (cut in half) intervals $[a, b]$ in which we know there is a zero. More specifically, we perform the following steps:

    1. Find two numbers a and b so that $f(a)$ and $f(b)$ have opposite signs.

    2. Compute $c = (a + b)/2$, the number halfway between a and b.

    3. Test to see if the zero lies between a and c or between c and b. That is, compute $f(c)$ and compare its sign to that of $f(a)$ and $f(b)$. If $f(a)$ and $f(c)$ have opposite signs, then a zero is between a and c. If $f(c)$ and $f(b)$ have opposite signs, then the zero is between c and b.

    4. Rename a and b so they denote the two numbers (either the old a and c or the old c and b) between which the zero lies.

    5. Repeat steps 2, 3, and 4 until the zero is approximated to the desired accuracy.

Consider the function $f(x) = x^3 + x^2 - 4$. As we saw earlier, f has a zero between $a = 1$ and $b = 2$. Thus, we set $c = 1.5$. Since $f(1) = -2$, $f(1.5) = 1.625$, and $f(2) = 8$, we know there is a zero between 1 and 1.5. So we set $a = 1$ and $b = 1.5$ and repeat the procedure.

    b. Continue the process to estimate the zero of $f(x) = x^3 + x^2 - 4$ that lies between 1 and 2. Stop when a and b agree to two decimal places.

    c. You are told that a certain polynomial function has an x-intercept somewhere between 1 and 100. Explain how you could evaluate the function at 20 points and, based on this information, determine the x-intercept to within 3 decimal places of accuracy.

    d. *For students with programming experience* The Bisection Method is ideally suited for a computer or programmable calculator. The following pseudocode suggests how the program might be written. Note that the program requests values for a and b, and also for the desired accuracy e. When the desired accuracy is reached, the program displays the approximate zero.

INPUT a	Input the left endpoint.
INPUT b	Input the right endpoint.
IF $f(a)*f(b) > 0$ THEN	If $f(a)$ and $f(b)$ don't have opposite
OUTPUT "$f(a)$ and $f(b)$	signs, there may not be a zero between
must have opposite sign"	a and b, so we should stop.
STOP	
ENDIF	
INPUT e	Input the desired accuracy.
WHILE $b - a > e$ DO	The program stops when $b - a \le e$.
$(a + b)/2 \to c$	Let c be the point halfway between a and b.
IF $f(a)*f(c) < 0$ THEN	If $f(a)$ and $f(c)$ are opposite in sign:
$c \to b$	let c be the new right endpoint;
ELSE	otherwise:
$c \to a$	let c be the new left endpoint.
ENDIF	
ENDWHILE	
OUTPUT $(a + b)/2$	The approximate zero is $(a + b)/2$.

Write a program for your graphing calculator that implements the bisection method. Test the program on the function $f(x) = x^3 + x^2 - 4$ using $a = 1$, $b = 2$, and $e = 0.0001$. You should get 1.3146 as an approximate answer. Once the program is working correctly, use it to approximate the zeros of the functions in part a to within $e = 0.000001$.

SECTION 4.5   RATIONAL FUNCTIONS

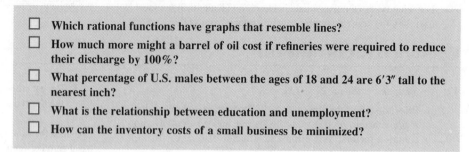

☐ Which rational functions have graphs that resemble lines?

☐ How much more might a barrel of oil cost if refineries were required to reduce their discharge by 100%?

☐ What percentage of U.S. males between the ages of 18 and 24 are 6′3″ tall to the nearest inch?

☐ What is the relationship between education and unemployment?

☐ How can the inventory costs of a small business be minimized?

A **rational function** is a function of the form

$$f(x) = \frac{p(x)}{q(x)}$$

where p and q are polynomials. Examples include

$$g(x) = \frac{2x + 3}{x^2 - 4} \quad \text{and} \quad h(x) = \frac{2}{x - 5}$$

as well as all polynomial functions, such as $f(x) = x^2 + 3x$, where the denominator is assumed to be 1. In this section, we only consider rational functions *in lowest terms*— that is, rational functions $f(x) = p(x)/q(x)$ for which $p(x)$ and $q(x)$ have no common factors. See Exercise 62 for a discussion of rational functions that are not in lowest terms.

DOMAIN AND ZEROS

Since a rational function f is constructed from polynomials p and q, it is not surprising that key features of f—its domain, x-intercepts, end behavior, and so forth—prove to be closely tied to the behaviors of these polynomials. In particular, since $f(x) = p(x)/q(x)$, $f(x)$ is defined only where both $p(x)$ and $q(x)$ are defined and where the denominator $q(x)$ is nonzero. Thus, because polynomials are defined for all real numbers, the domain of a rational function is found by excluding the zeros of the polynomial in the denominator.

Domain of a Rational Function ☐

> The domain of a rational function $f(x) = p(x)/q(x)$ consists of all real numbers x such that $q(x) \neq 0$.

We've seen that zeros of the denominator q determine the domain of f. On the other hand, since a fraction is only zero when its numerator is zero, the zeros of f correspond to the zeros of its numerator.

Zeros of a Rational Function ☐

> If $f(x) = p(x)/q(x)$ is a rational function in lowest terms, then $f(x) = 0$ if and only if $p(x) = 0$.

EXAMPLE 1   ■   **Finding the Domain and Zeros of a Rational Function**

Find the domain and zeros of the rational function $f(x) = \dfrac{3x + 6}{x + 1}$.

SOLUTION   The values that must be excluded from the domain can be found by locating the zeros of the denominator:

$$x + 1 = 0$$
$$x = -1$$

Thus, the only zero of the denominator is $x = -1$, and so the domain of f is the set of all real numbers except -1. To find the zeros of f, we simply find the zeros of the numerator.

$$3x + 6 = 0$$
$$3x = -6$$
$$x = -2$$

Thus, -2 is the only zero of f.

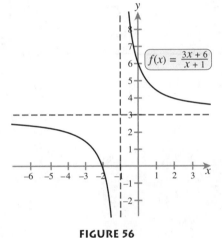

$f(x) = \dfrac{3x + 6}{x + 1}$

FIGURE 56

As we saw in Section 4.1, graphs of polynomial functions are always smooth, unbroken curves that, in the long run, either rise to infinity or sink to negative infinity. With rational functions, there is far more variation. The graph of a rational function can abruptly break, forming a steep, nearly vertical cliff as it approaches a *vertical asymptote*. Its end behavior might be like that of a polynomial—approaching positive or negative infinity on the far left and far right—or, alternatively, its graph might level off, gradually approaching a *horizontal asymptote*. For example, consider the graph of the function $f(x) = (3x + 6)/(x + 1)$ from Example 1, as shown in Figure 56. In particular, notice the break in the graph across the line $x = -1$ (this is a vertical asymptote). Notice also that the graph of f levels off along the horizontal line $y = 3$ as x approaches $\pm\infty$ (this is a horizontal asymptote). The remainder of this section is devoted to investigating these features of rational functions in more detail.

VERTICAL ASYMPTOTES

Let's take a closer look at the breaks that sometimes occur in the graphs of rational functions. In general, the graph of a rational function will be broken at an x-value for which the function is not defined. In other words, there will be a break in the graph of the rational function $f(x) = p(x)/q(x)$ at every zero of q. For example, consider again the function

$$f(x) = \frac{3x + 6}{x + 1}$$

from Example 1. Since the denominator has a zero of -1, it follows that the graph of f will break at $x = -1$. Let's investigate the behavior of $f(x)$ near this x-value.

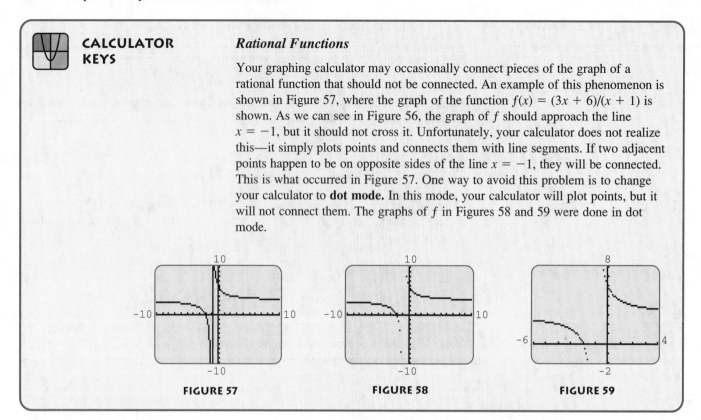

CALCULATOR KEYS

Rational Functions

Your graphing calculator may occasionally connect pieces of the graph of a rational function that should not be connected. An example of this phenomenon is shown in Figure 57, where the graph of the function $f(x) = (3x + 6)/(x + 1)$ is shown. As we can see in Figure 56, the graph of f should approach the line $x = -1$, but it should not cross it. Unfortunately, your calculator does not realize this—it simply plots points and connects them with line segments. If two adjacent points happen to be on opposite sides of the line $x = -1$, they will be connected. This is what occurred in Figure 57. One way to avoid this problem is to change your calculator to **dot mode.** In this mode, your calculator will plot points, but it will not connect them. The graphs of f in Figures 58 and 59 were done in dot mode.

FIGURE 57   **FIGURE 58**   **FIGURE 59**

We begin by forming a table of function values $f(x)$ for x near -1.

x approaches -1 from the right:

x	-0.9	-0.99	-0.999
$f(x) = \dfrac{3x + 6}{x + 1}$	33	303	3003

x approaches -1 from the left:

x	-1.1	-1.01	-1.001
$f(x) = \dfrac{3x + 6}{x + 1}$	-27	-297	-2997

Our work suggests that as x approaches -1 from the right, $f(x)$ approaches ∞, whereas as x approaches -1 from the left, $f(x)$ approaches $-\infty$. It follows that the graph of f will become nearly vertical near -1, shooting up on one side and down on the other, as we have already seen in Figures 56–59. Because the graph approaches the vertical line $x = -1$, we say that the line $x = -1$ is a *vertical asymptote* for the graph of f. When sketching the graphs of rational functions by hand, it is customary to depict vertical asymptotes using dashed lines.

Definition of Vertical Asymptotes ☐

The line $x = a$ is a **vertical asymptote** for the graph of a rational function f if $f(x)$ grows without bound (approaches ∞ or $-\infty$) as x approaches a from the right and from the left. If $f(x) = p(x)/q(x)$ is a rational function in lowest terms, then the line $x = a$ is a vertical asymptote for the graph of f if and only if $q(a) = 0$.

EXAMPLE 2 ■ **Finding Vertical Asymptotes**

Find any vertical asymptotes for the graph of $f(x) = \dfrac{6}{x^2 - 4}$.

SOLUTION We begin by finding the x-values for which f is not defined. Setting the denominator equal to zero gives us

$$x^2 - 4 = 0$$
$$(x + 2)(x - 2) = 0$$
$$x = -2, x = 2$$

Thus, the graph of f will have vertical asymptotes at $x = -2$ and $x = 2$. Figure 60 shows the graph of f as depicted by a graphing calculator. Using this graph as a guide, we've sketched the graph of f shown in Figure 61.

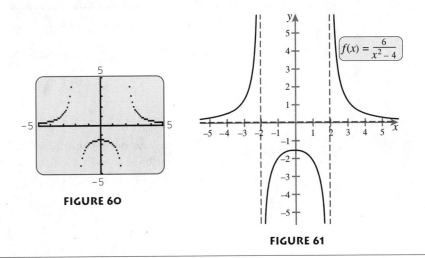

FIGURE 60

FIGURE 61

HORIZONTAL ASYMPTOTES

As we have already seen, the graph of a rational function may eventually level off, approaching a horizontal line called a horizontal asymptote. More precisely, we have the following definition.

Definition of a Horizontal Asymptote ☐

The line $y = b$ is a **horizontal asymptote** for the graph of a rational function f if $f(x)$ approaches b as x approaches ∞ or $-\infty$.

To approximate a horizontal asymptote graphically, we select a viewing window that reveals the end behavior of the graph. Then, if the graph appears to be horizontal, we can use the trace feature to estimate the limiting value b. We illustrate this technique with the function from Example 1.

EXAMPLE 3   ■   **Finding a Horizontal Asymptote Graphically**

Use a graphing calculator to estimate the horizontal asymptote (if any) of

$$f(x) = \frac{3x + 6}{x + 1}$$

SOLUTION   Figure 62 shows the graph of f with the standard viewing window. To view the end behavior of f, we set the window variables to Xmin=−500, Xmax=500, Ymin=−10, and Ymax=10, obtaining the view shown in Figure 63. As we trace the graph from left to right (Figure 64), it appears that the function values are approaching 3. Tracing from right to left, we see that $f(x)$ approaches 3 as well. Based on this evidence, it appears that $y = 3$ is a horizontal asymptote for the graph of f.

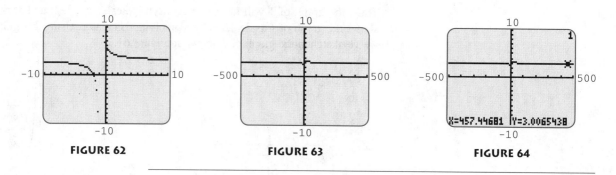

FIGURE 62   FIGURE 63   FIGURE 64

We now use our knowledge of the end behavior of polynomials to develop an *algebraic* method for finding horizontal asymptotes of rational functions. Since a rational function is the quotient of two polynomials, it can be written in the form

$$f(x) = \frac{a_n x^n + a_{n-1} x^{n-1} + \cdots + a_1 x + a_0}{b_m x^m + b_{m-1} x^{m-1} + \cdots + b_1 x + b_0}$$

where neither a_n nor b_m is 0. Because polynomials, in the long run, tend to behave like their terms of highest degree, we would expect that the end behavior of $f(x)$ could be found by replacing the numerator and denominator by their terms of highest degree. In fact, $f(x)$ does indeed behave like the quotient

$$\frac{a_n x^n}{b_m x^m}$$

Thus, for example, to determine the end behavior of

$$f(x) = \frac{x^2 + 1}{2x^3 - 3x + 4}$$

we need only consider the end behavior of the much simpler function

$$\frac{x^2}{2x^3} = \frac{1}{2x}$$

Now as x gets large, the denominator does also, and the quotient tends toward 0. Thus, $f(x) \to 0$ as $x \to \infty$ and as $x \to -\infty$. It follows that the line $y = 0$ is a horizontal asymptote of the graph of $f(x)$. With hindsight, this result was predictable: Since the degree of the denominator is greater than that of the numerator, the denominator grows much more rapidly than the numerator, and thus their quotient tends toward 0. More generally, the end behavior of a rational function depends on the relative degrees of numerator and denominator, as follows.

Locating Horizontal Asymptotes   □

The end behavior of the rational function

$$f(x) = \frac{a_n x^n + a_{n-1} x^{n-1} + \cdots + a_1 x + a_0}{b_m x^m + b_{m-1} x^{m-1} + \cdots + b_1 x + b_0}$$

is the same as that of

$$\frac{a_n x^n}{b_m x^m}$$

There are 3 possible cases, depending on the relatives sizes of n and m.

Case	$\dfrac{a_n x^n}{b_m x^m}$ simplifies to	End behavior
1. $n > m$	$\dfrac{a_n x^{n-m}}{b_m}$ (a monomial)	The quotient approaches $\pm\infty$, and so there are **no horizontal asymptotes.**
2. $n = m$	$\dfrac{a_n}{b_n}$	$y = \dfrac{a_n}{b_n}$ is the horizontal asymptote.
3. $n < m$	$\dfrac{a_n}{b_m x^{m-n}}$	The denominator approaches $\pm\infty$, so the quotient approaches 0. Thus, $y = \mathbf{0}$ (the x-axis) is the horizontal asymptote.

EXAMPLE 4   ▨   **Locating Vertical and Horizontal Asymptotes**

Locate any vertical and horizontal asymptotes and sketch the graph of

$$f(x) = \frac{x^2 - 2x + 2}{2x^2 - 4x}$$

SOLUTION $f(x)$ is undefined where $2x^2 - 4x = 0$. Solving for x we have

$$2x^2 - 4x = 0$$
$$2x(x - 2) = 0$$
$$x = 0, \ x = 2$$

Thus, there are vertical asymptotes at both $x = 0$ and $x = 2$. To determine the end behavior of f, we simply ignore all but the leading terms in the numerator and denominator to obtain $\dfrac{x^2}{2x^2} = \dfrac{1}{2}$, so that $y = \dfrac{1}{2}$ is the horizontal asymptote. (This is Case 2 described in the preceding box.) One view of the graph is shown in Figure 65. The graph in Figure 66 includes the asymptotes.

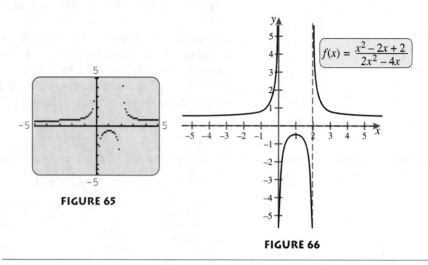

FIGURE 65

$$f(x) = \frac{x^2 - 2x + 2}{2x^2 - 4x}$$

FIGURE 66

E X A M P L E   5   ▦   Locating Vertical and Horizontal Asymptotes

Locate any vertical and horizontal asymptotes and sketch the graph of

$$f(x) = \frac{80x}{16x^2 + 1}$$

SOLUTION   The graph of f has no vertical asymptotes since the denominator $16x^2 + 1$ has no real zeros. The horizontal asymptote is the x-axis because the degree of the numerator is less than the degree of the denominator (Case 3). One view of the graph is shown in Figure 67. At first glance, it appears that the y-axis is a vertical asymptote. However, we know that is not the case since f has no vertical asymptotes. To see more clearly what is happening near the y-axis, we adjust the scale on the x-axis; the result is shown in Figure 68.

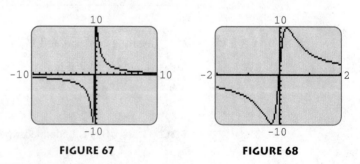

FIGURE 67

FIGURE 68

E X A M P L E   6   ▦   A Rational Function with No Vertical or Horizontal Asymptotes

Show that $f(x) = \dfrac{x^4}{x^2 - 2x + 2}$ has no vertical or horizontal asymptotes. Describe the end behavior of f.

SOLUTION   To check for vertical asymptotes, we find the zeros of the denominator by setting $x^2 - 2x + 2 = 0$ and using the quadratic formula.

$$x = \frac{-(-2) \pm \sqrt{(-2)^2 - 4(1)(2)}}{2(1)} = \frac{2 \pm \sqrt{-4}}{2}$$

Since the only zeros of the denominator are nonreal, the graph of f has no vertical asymptotes. There is no horizontal asymptote either because the degree of the numerator is larger than the degree of the denominator (Case 1). To determine the end behavior, we ignore all but the terms of highest degree in the numerator and denominator to obtain $x^4/x^2 = x^2$. Since $x^2 \to \infty$ as $x \to \pm\infty$, $f(x) \to \infty$ also as $x \to \pm\infty$. The graph of f is shown in Figure 69. Notice the similarity to the graph of $y = x^2$ shown in Figure 70.

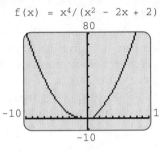

FIGURE 69

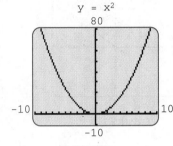

FIGURE 70

EXAMPLE 7   ■   **Graphing an Unknown Rational Function**

Suppose f is a rational function with vertical asymptote $x = -3$ and horizontal asymptote $y = 0$. Moreover, suppose $f(x) > 0$ for $x \neq -3$. Sketch a possible graph of f.

SOLUTION   Since $x = -3$ is a vertical asymptote, and since $f(x)$ is always positive, we know that $f(x)$ approaches ∞ as x approaches -3 from the left and the right. Because $y = 0$ is a horizontal asymptote, $f(x)$ must approach 0 (the x-axis) as x approaches ∞ or $-\infty$. Moreover, since $f(x)$ is always positive, we know that the graph must approach the x-axis from above. A *possible* sketch for f is shown in Figure 71.

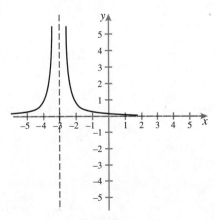

FIGURE 71

INCLINED ASYMPTOTES

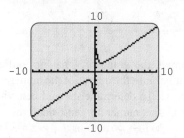

FIGURE 72

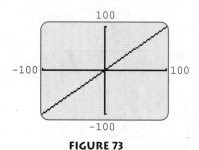

FIGURE 73

We have seen that the end behavior of a rational function is like that of a polynomial whenever the degree of the numerator exceeds the degree of the denominator. More specifically, such a rational function has no horizontal asymptote, and the function values approach either positive or negative infinity as $x \to \infty$ and as $x \to -\infty$. We now consider the particular case in which the degree of the numerator is exactly one more than the degree of the denominator. Consider, for example, the function

$$f(x) = \frac{x^2 + 1}{x}$$

The graph of f shown in Figure 72 certainly suggests that there is no horizontal asymptote. To investigate the end behavior further, we zoom out to obtain the view shown in Figure 73. Here we see that not only does f have end behavior like that of a polynomial, but the graph of f actually behaves like that of a linear (first-degree) polynomial. Indeed, the graph of f appears linear with a sufficiently large viewing window.

To see more clearly what's happening, let us investigate f algebraically. Dividing the numerator by x gives us $f(x) = x + 1/x$. Now, for very large positive or negative values of x, $1/x$ is a very small number and, hence, $f(x) = x + 1/x \approx x$. Thus, we would expect that in the long run, the graph of f would approach the graph of $y = x$. Since the line $y = x$ is neither vertical nor horizontal but rather at an incline, we say that $y = x$ is an **inclined asymptote** for the function f. Figure 74 shows the graph of f with its inclined asymptote $y = x$ depicted as a dashed line.

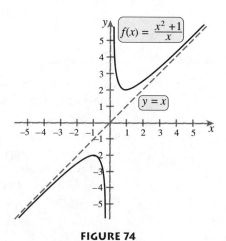

FIGURE 74

More generally, we find inclined asymptotes as follows.

Inclined Asymptotes □

Suppose $f(x) = p(x)/q(x)$ is a rational function for which the degree of $p(x)$ is *exactly* 1 *more* than the degree of $q(x)$. Then we can use division to rewrite f in the form

$$f(x) = ax + b + \frac{r(x)}{q(x)}$$

where the degree of $r(x)$ is less than the degree of $q(x)$. The line $y = ax + b$ is an inclined asymptote for f.

E X A M P L E   8   ■   Locating Vertical and Inclined Asymptotes

Find the vertical and inclined asymptotes for

$$f(x) = \frac{x^2 + 5x + 7}{x + 2}$$

and sketch the graph.

SOLUTION   Setting the denominator equal to 0 and solving, we see that there is a vertical asymptote at $x = -2$. To find the inclined asymptote, we must first divide $x^2 + 5x + 7$ by $x + 2$. We use synthetic division.

$$
\begin{array}{r|rrr}
-2 & 1 & 5 & 7 \\
   &   & -2 & -6 \\
\hline
   & 1 & 3 & 1
\end{array}
$$

From the last row, we see that the quotient is $x + 3$ and the remainder is 1. Thus

$$\frac{x^2 + 5x + 7}{x + 2} = x + 3 + \frac{1}{x + 2}$$

and so

$$f(x) = x + 3 + \frac{1}{x + 2}$$

This shows that $y = x + 3$ is an inclined asymptote. Using a graphing calculator, we obtain the view shown in Figure 75. The graph in Figure 76 includes the asymptotes.

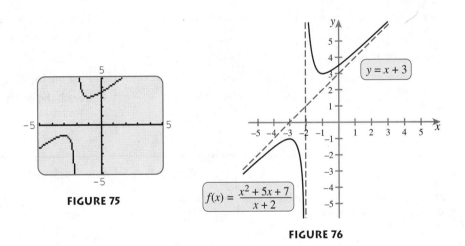

FIGURE 75

FIGURE 76

E X A M P L E   9   ■   Minimizing Inventory Cost

An appliance retailer sells 500 microwave ovens each year. By averaging ordering costs and storage costs, the retailer has determined that if x microwaves are ordered at a time, the yearly inventory cost will be

$$C(x) = 5x + 40{,}000 + \frac{4500}{x}$$

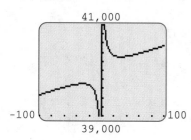

FIGURE 77

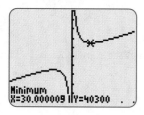

FIGURE 78

a. Estimate the number of microwaves that should be ordered each time so that the cost is as small as possible. How often should each order be placed?

b. Find the inclined asymptote of C and interpret its slope.

SOLUTION

a. We want to choose a viewing window so that a minimum value is shown. After some experimentation, we settle on the view shown in Figure 77. Because we are only concerned with positive values of x, we look for the lowest point on the graph of C to the right of the y-axis. Using the trace or calculate-minimum feature, we estimate the minimum cost to be $40,300 per year when 30 microwaves are ordered each time (see Figure 78). If 500 microwaves are sold each year and 30 must be ordered each time to minimize inventory cost, an order must be placed $\frac{500}{30} \approx 17$ times each year, or every 22 days.

b. By ignoring the rational term $4500/x$, we see that $y = 5x + 40,000$ is the inclined asymptote for the graph of C. Thus, we know that for very large values of x, the graph of $C(x)$ will look very much like the graph of the line $y = 5x + 40,000$. Since the slope of this line is 5, we conclude that for very large values of x, the y-coordinates on the graph of C will increase by approximately 5 units for every unit increase in x. In other words, for very large order sizes, the inventory cost will increase by approximately $5.00 for every additional unit ordered.

UNDERSTANDING AND MASTERY CHECKLISTS

CONCEPTS TO UNDERSTAND

☐ Domain and zeros of a rational function

☐ Vertical asymptotes

☐ Horizontal asymptotes

☐ Inclined asymptotes

SKILLS TO MASTER

☐ Find the domain and zeros of a rational function.

☐ Find the vertical asymptote(s) of a rational function.

☐ Find the horizontal asymptote of a rational function.

☐ Find the inclined asymptote of a rational function.

☐ Sketch the graph of a rational function.

E X E R C I S E S   4.5

EXERCISES 1–8 *Find the domain and zeros of the rational function and match it with its graph.*

1. $f(x) = \dfrac{1}{x - 2}$

2. $f(x) = \dfrac{x}{x - 1}$

3. $f(x) = \dfrac{x + 1}{x}$

4. $f(x) = \dfrac{x - 2}{x - 1}$

5. $f(x) = \dfrac{1}{x + 2}$

6. $f(x) = \dfrac{1}{x^2 + x - 2}$

7. $f(x) = \dfrac{x^2 - 4}{x^2 - 1}$

8. $f(x) = \dfrac{x + 1}{x + 2}$

i.

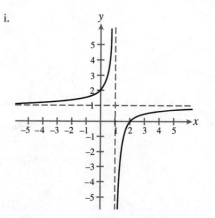

ii.

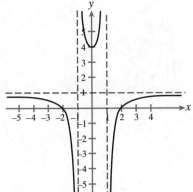

vi.

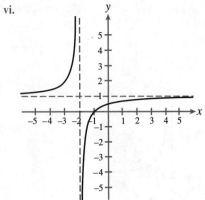

iii.

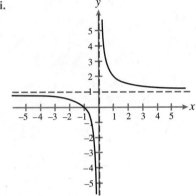

vii.

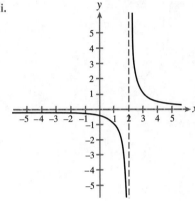

iv.

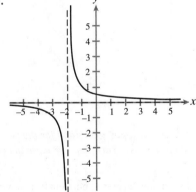

viii.

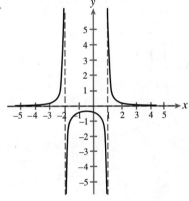

v.

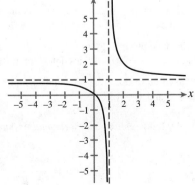

EXERCISES 9–12   *Use the graph of the function f to sketch the graph of the function g.*

9.

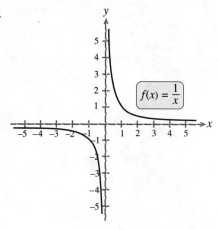

$f(x) = \dfrac{1}{x}$

a. $g(x) = \dfrac{1}{x - 3}$

b. $g(x) = \dfrac{1}{x} + 2$

c. $g(x) = -\dfrac{1}{x}$

10.

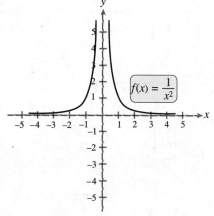

$f(x) = \dfrac{1}{x^2}$

a. $g(x) = \dfrac{1}{x^2} - 3$

b. $g(x) = \dfrac{1}{(x + 2)^2}$

c. $g(x) = -\dfrac{1}{x^2}$

11.

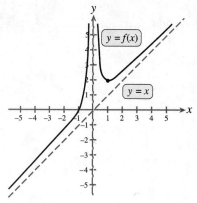

$y = f(x)$

$y = x$

a. $g(x) = f(x - 1)$

b. $g(x) = f(x) - 2$

c. $g(x) = -f(x)$

12.

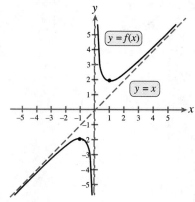

$y = f(x)$

$y = x$

a. $g(x) = f(x) + 2$

b. $g(x) = f(x + 1)$

c. $g(x) = -f(x)$

EXERCISES 13–16   *Construct a possible graph for the rational function with the given properties. There may be more than one correct answer.*

13. f has a vertical asymptote at $x = 4$, horizontal asymptote $y = 0$, and $f(x) < 0$ for all $x \ne 4$.

14. h has a vertical asymptote at $x = -2$, horizontal asymptote $y = 0$, $h(x) > 0$ if $x < -2$, and $h(x) < 0$ if $x > -2$.

15. g has a vertical asymptote at $x = -1$, horizontal asymptote $y = 2$, an x-intercept at $(-\frac{3}{2}, 0)$, and a y-intercept at $(0, 3)$.

16. h has a vertical asymptote at $x = 2$, horizontal asymptote $y = -3$, and zeros at $x = 0$ and $x = 4$.

EXERCISES 17–30   *Find all vertical and horizontal asymptotes and sketch the graph.*

17. $f(x) = \dfrac{1}{x + 3}$ **18.** $f(x) = \dfrac{1}{x - 2}$

19. $h(x) = \dfrac{x}{x + 2}$

20. $g(x) = \dfrac{x - 3}{x + 4}$

21. $f(x) = \dfrac{3x}{x - 2}$

22. $g(x) = \dfrac{x + 1}{3 - 2x}$

23. $g(x) = \dfrac{x}{x^2 - 9}$

24. $f(x) = \dfrac{x^2}{x^2 - 9}$

25. $g(x) = \dfrac{45x^2}{9x^2 + 1}$

26. $f(x) = \dfrac{1}{x^2 + 9}$

27. $h(x) = \dfrac{1}{(x + 1)^2}$

28. $f(x) = \dfrac{-x}{(x - 3)^2}$

29. $f(x) = \dfrac{x^2}{(x - 2)^2}$

30. $g(x) = \dfrac{(x - 2)^2}{x^2}$

EXERCISES 31–36   *Find the vertical and inclined asymptotes and sketch the graph.*

31. $g(x) = 2x + \dfrac{1}{x}$

32. $h(x) = -x + \dfrac{1}{x}$

33. $f(x) = \dfrac{x^2}{x - 1}$

34. $h(x) = \dfrac{x^2 + 1}{x + 1}$

35. $g(x) = \dfrac{-3x^2 - 6x + 1}{x + 2}$

36. $f(x) = \dfrac{4x^2 - 4x + 2}{x - 1}$

EXERCISES 37–46   *Find all vertical, horizontal, and inclined asymptotes and sketch the graph.*

37. $f(x) = \dfrac{2}{x^2 + 1}$

38. $h(x) = \dfrac{8}{x^2 - 2x + 3}$

39. $h(x) = \dfrac{12x + 24}{x - 15}$

40. $h(x) = \dfrac{-2}{x - 3}$

41. $g(x) = \dfrac{x^2 - 4x + 7}{x - 3}$

42. $f(x) = \dfrac{1 - 16x}{x + 12}$

43. $h(x) = \dfrac{-1}{x^2 - 9}$

44. $h(x) = \dfrac{2x^2 + 9x + 5}{x + 4}$

45. $f(x) = \dfrac{x^3}{x - 3}$

46. $g(x) = \dfrac{2x^4}{x^2 - 1}$

■ APPLICATIONS

47. *Pollution Control*   Based on data from a 1973 study on the cost of reducing oil refinery discharge, the increase in refinery costs can be modeled by the function

$$f(x) = \frac{x}{1000(100 - x)}$$

where x is the desired percent reduction in discharge and $f(x)$ is the corresponding additional cost in dollars per barrel crude. Identify any vertical or horizontal asymptotes and sketch the graph of f. What percent reduction is possible with an increase of $0.003 per barrel crude? According to this model, is a 100% reduction attainable? Explain. [*Data source:* Maynard Hufschmidt et al., *Environment, Natural Systems, and Development* (Baltimore: The John Hopkins University Press, 1983)].

48. *Nerve Excitation*   The minimum voltage needed to excite a nerve fiber is related to the time during which the current flows. According to data from one source, this relationship can be modeled by the function

$$f(x) = \frac{17}{x + 0.113} + 23.2$$

where x is the time during which current flows (in milliseconds) and $f(x)$ is the minimum current needed to excite the nerve fiber (in millivolts). Identify any vertical or horizontal asymptotes and sketch the graph of f. How long must the current flow to excite a nerve fiber with a voltage of 40 millivolts? [*Data source:* Douglas Riggs, *The Mathematical Approach to Physiological Problems* (Cambridge: M.I.T. Press, 1970)].

49. *Unemployment Rate*   The percentage rate of unemployment in the United States in 1992 for people with x years of education can be modeled by the function

$$f(x) = \frac{72,900}{100x^2 + 729}$$

Identify any vertical or horizontal asymptotes and sketch the graph of f. What level of education corresponds to an unemployment rate of 4.9%? What happens to the unemployment rate as the level of education increases? According to this model, is there an education level that corresponds to an unemployment rate of 0%? Explain. (*Data source:* U.S. Bureau of Labor Statistics).

50. *Height Distribution*   The percentage of U.S. males between the ages of 18 and 24 years who are within a half inch of a given height can be modeled by the function

$$f(x) = \frac{256}{(2x - 139)^2 + 16}$$

where x is the height (in inches). Identify any vertical or horizontal asymptotes and sketch the graph of f. What height cor-

7'6" NBA center Shawn Bradley

responds to the highest percentage? What interpretation could be given to this height? (*Data source:* U.S. National Center for Health Statistics).

51. *Bicycle Average Cost*   A company that manufactures bicycles has fixed costs of $100,000 and variable costs of $100 per bicycle. The total cost of producing x bicycles is thus $C(x) = 100,000 + 100x$. The average cost per bicycle is found by dividing the total cost by the number of bicycles produced. Thus, the average cost function is

$$\overline{C}(x) = \frac{100,000 + 100x}{x}$$

Sketch the graph of the average cost function. What is the horizontal asymptote and what information can be obtained from it?

52. *Skate Average Cost*   A company that manufactures inline skates has fixed costs of $80,000 and variable costs of $50 per pair of skates, so that the total cost of producing x pairs of skates is thus $C(x) = 80,000 + 50x$. The average cost per pair of skates is found by dividing the total cost by the number of pairs of skates produced. Thus, the average cost function is

$$\overline{C}(x) = \frac{80,000 + 50x}{x}$$

Sketch the graph of the average cost function. What is the horizontal asymptote and what information can be obtained from it?

53. *Basketball Inventory Cost*   A sporting goods retailer sells 580 basketballs each year. By averaging ordering costs and storage costs, the retailer has determined that if x balls are ordered at a time, the yearly inventory cost will be

$$C(x) = \frac{3}{2}x + 600 + \frac{5046}{x}$$

a. Estimate the number of balls that should be ordered each time so that the cost is as small as possible. How often should each order be placed?

b. Find the inclined asymptote of C and interpret its slope.

54. *Television Inventory Cost*   A retail appliance store sells 1500 television sets per year. By averaging ordering costs and storage costs, the retailer has determined that if x televisions are ordered at a time, the yearly inventory cost will be

$$C(x) = 5x + 30,000 + \frac{28,200}{x}$$

a. Estimate the number of televisions that should be ordered each time so that the cost is as small as possible. How often should each order be placed?

b. Find the inclined asymptote of C and interpret its slope.

CONCEPTS AND CRITICAL THINKING

EXERCISES 55–58   *Answer true or false.*

55. The graphs of all rational functions have vertical asymptotes.

56. The graph of a rational function may cross its horizontal asymptote but does not cross any of its vertical asymptotes.

57. The zeros of a rational function in lowest terms are the zeros of its numerator.

58. The vertical asymptotes of a rational function can be found by setting the numerator equal to zero.

EXERCISES 59–60   *Give an example of each.*

59. A rational function with horizontal asymptote $y = 3$ and vertical asymptotes $x = 1$ and $x = -1$

60. A rational function having neither vertical nor horizontal asymptotes

QUESTIONS FOR DISCUSSION OR ESSAY

61. In exercises 51 and 52, we considered average cost functions of the form

$$\overline{C}(x) = \frac{a + bx}{x}$$

Could a function of this form have a minimum value? In other words, is it possible to find a production level x for which the average cost is as small as possible? Explain. Is this situation

consistent with what you would expect for a company's average cost? Why or why not?

62. We have seen that if $f(x) = p(x)/q(x)$ is in simplest terms with $q(a) = 0$, then the line $x = a$ is a vertical asymptote for the graph of f. What happens if f is not in simplest terms? To help you answer this question, consider some examples:

$$f(x) = \frac{x^2 - 4}{x - 2}, \quad f(x) = \frac{x}{x^3 - x}, \quad \text{and} \quad f(x) = \frac{x^2 - 2x + 1}{x^2 + x - 2}$$

If $f(x)$ is not in simplest terms and $q(a) = 0$, is it necessarily the case that $x = a$ is a vertical asymptote? If not an asymptote, what happens at $x = a$?

63. Rational functions of the form

$$f(x) = \frac{ax + b}{cx + d}, \quad c \neq 0$$

can be graphed quite easily without the use of a graphing calculator. Consider, for example, the function

$$f(x) = \frac{2x + 3}{x + 1}$$

The vertical and horizontal asymptotes divide the plane into four regions, which we have numbered in Figure 79. It can be shown that the graph must lie either entirely in regions 1 and 3 or entirely in regions 2 and 4. Why do you think this is? Once the regions have been determined, a sketch of the graph is easily completed, as shown in Figure 80.

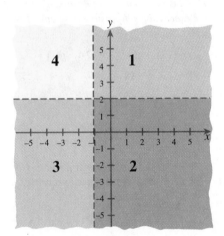

FIGURE 79

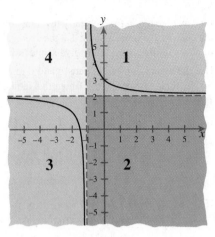

FIGURE 80

Describe a procedure for locating the horizontal and vertical asymptotes of the graph of a function $f(x) = \dfrac{ax + b}{cx + d}, c \neq 0$, and for deciding in which pair of regions to place the graph. Illustrate your procedure with an example.

64. In Example 7, we sketched the graph of a function f with a vertical asymptote at $x = -3$, a horizontal asymptote at $y = 0$, and with $f(x) > 0$ for $x \neq -3$. Explain why there could be more than one correct answer for this example, and give an example of another possible solution. Do you think one answer could be "more right" than another? Why or why not? What additional information would have to be given to narrow down the choice of possibilities? It has been said that one of the strengths of mathematics over some other disciplines is that an answer is either right or wrong. There is no gray area. In your experience with mathematics, how often have you found this to be the case? Do you think this "all or nothing" phenomenon is more or less likely to be the case when mathematics is applied to the real world? Explain.

■ **PROJECTS FOR ENRICHMENT**

---

65. *Minimizing Inventory Cost*   A shoe retailer wishes to minimize the costs incurred in handling a certain style of shoe that must be ordered periodically and kept in stock as they are sold to customers. Since the expenses involved in ordering and storing the shoes depend on how often orders are made and how long items are stored, the retailer wishes to minimize the *total cost per year*. We consider three separate costs that are involved in the total cost to the retailer.

    i. A *fixed cost* of $30 *per order* that is independent of the amount ordered. This cost includes such things as record keeping and other paperwork, employees' time, and so on.

    ii. A *purchase cost* of $15 *per pair* (throughout, "pair" will mean "pair of shoes").

    iii. An *inventory holding cost* of $2 *per pair per year*. This cost covers the expenses of keeping the shoes in the store or warehouse.

Let x denote the number of pairs ordered at a time and assume that the shoes are sold at a constant rate of 1200 pairs per year. Because shortages are not tolerable, the time between two consecutive orders depends only on the number ordered and the rate they are sold.

a. Find an expression for the number of reorders per year.

b. Find an expression for the total cost per order (not including inventory cost).

c. Use the expressions from parts a and b to find an expression for the yearly ordering costs.

d. Find an expression for the yearly inventory cost. Since x pairs are ordered at a time, assume that the average number of pairs of shoes in inventory at any given time is $x/2$.

e. Use the expressions from parts c and d to obtain a function for the total cost per year. Denote this total cost function by $T(x)$.

f. Use a graphing calculator to plot the graph of $T(x)$ and approximate the value for x that yields the minimum cost. This represents the number of shoes that should be ordered at a time. How often must this quantity of shoes be ordered?

g. What would change in this scenario if the retailer can order shoes only in quantities of 18?

CHAPTER 4 REVIEW

EXERCISES 1–6   *Determine the end behavior of the given polynomial. Verify your results with a graphing calculator.*

1. $g(x) = -3x^6$ 2. $h(x) = 2x^5$

3. $f(x) = 4x^3 - x^2 + 9x + 3$

4. $g(x) = -x^4 + 5x^2 - x + 8$

5. $h(x) = x^3 + 100 - 0.00005x^4$

6. $f(x) = (2 \times 10^4)x^2 + (2 \times 10^{-4})x^3$

EXERCISES 7–10   *Use a graphing calculator to estimate the zeros and the coordinates of any turning points of the given polynomial to the nearest hundredth. Classify each turning point as a local maximum or minimum.*

7. $f(x) = \dfrac{x^3}{3} - 3x^2 - 3$ 8. $g(x) = \dfrac{x^4}{9} + x^2 - 4x + 2$

9. $g(x) = 0.1x^4 + x^3 - 4x^2 + x - 4$

10. $h(x) = 0.01x^5 - 0.15x^4 - 0.4x^3 + 2x^2$

EXERCISES 11–14   *Use long division to find the quotient $q(x)$ and remainder $r(x)$ when $f(x)$ is divided by $d(x)$.*

11. $f(x) = 2x^2 + 3x - 2; d(x) = x + 2$

12. $f(x) = 9x^3 - 3x^2 + 7x + 1; d(x) = 3x - 2$

13. $f(x) = x^4 - 5x^2 + 2; d(x) = x^2 + 4x$

14. $f(x) = x^5 + 2x^3 - x^2 - 2; d(x) = x^3 - 1$

EXERCISES 15–18   *Use synthetic division to find the quotient $q(x)$ and remainder $r(x)$ when $f(x)$ is divided by $d(x)$.*

15. $f(x) = 3x^3 - 17x^2 + 22x - 8; d(x) = x - 4$

16. $f(x) = 4x^3 + 8x^2 - 9x - 13; d(x) = x + 2$

17. $f(x) = -2x^4 + 4x^2 + x - 3; d(x) = x + 3$

18. $f(x) = 32x^5 - 1; d(x) = x - \frac{1}{2}$

EXERCISES 19–22   *Use the Remainder Theorem to find the indicated function values.*

19. $f(x) = 3x^3 + 11x^2 + 2x - 6$

     a. $f(\frac{3}{2})$ b. $f(-2)$

20. $g(x) = 5x^4 - 6x^3 - 32x^2 + 7$

     a. $g(-3)$ b. $g(3.2)$

21. $h(x) = -2x^4 + 5x^3 + 3x$

     a. $h(5)$ b. $h(\frac{1}{2})$

22. $f(x) = \frac{1}{3}x^5 + 2x^4 + \frac{1}{2}x + 3$

     a. $f(-6)$ b. $f(2)$

EXERCISES 23–28   *Find all the real zeros of the given polynomial function. Give exact values if possible. Use any of the tools described in this chapter.*

23. $f(x) = 2x^2 + 5x - 3$ 24. $g(x) = 4x^3 - 4x^2 - 24x$

25. $h(x) = x^4 - 11x^2 + 18$

26. $g(x) = 4x^3 + 5x^2 - 18x + 9$

27. $h(x) = 6x^3 + 7x^2 - 1$

28. $f(x) = x^4 - 2x^3 - 5x^2 + 8x + 4$

EXERCISES 29–36   *Find a polynomial function f of least degree with real coefficients satisfying the given properties.*

29. Zeros -2, 3, and 5

30. Zeros -3, 0, and 4; $f(1) = 10$

31. Zeros 1 and -1 (both with multiplicity 2); $f(0) = 4$

32. Zeros $\frac{1}{2}$ (multiplicity 2) and 0 (multiplicity 3)

33. Graph appears as follows

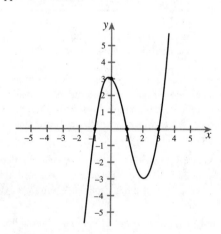

34. Graph appears as follows

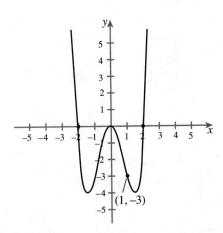

(1, −3)

35. Zeros −3 and $4i$

36. Zeros 0, 2, and $1 - 2i$

EXERCISES 37–40 *Write the polynomial function as a product of linear factors and identify all the complex zeros. Use any of the tools described in this chapter.*

37. $f(x) = x^2 - 4x + 5$

38. $g(x) = x^3 + 3x^2 + 4x + 12$

39. $h(x) = x^4 + 7x^2 - 144$

40. $f(x) = x^5 - 4x^4 + 10x^3 - 12x^2 + 5x$

EXERCISES 41–44 *Use the Rational Zero Theorem to list all the possible rational zeros of the polynomial. Then use synthetic division and/or a graphing calculator to help you determine which are actually zeros.*

41. $f(x) = x^3 - 2x^2 - 5x + 6$

42. $h(x) = 3x^3 - 10x^2 + 11x - 4$

43. $g(x) = 4x^4 - 8x^3 - x^2 + 8x - 3$

44. $h(x) = 9x^5 - x^3 - 9x^2 + 1$

EXERCISES 45–48 *Apply Descartes' Rule of Signs to the given function. Then use a graphing calculator to determine the precise number of positive and negative zeros.*

45. $h(x) = x^3 + 2x + 1$

46. $f(x) = x^4 - 4x^3 - x + 5$

47. $g(x) = -x^4 - 5x^3 + 2x^2 + 8x$

48. $h(x) = 2x^5 - 6x^3 + 3x^2 - 1$

EXERCISES 49–50 *Use the Upper and Lower Bounds Test to confirm the given bounds for the zeros of the function.*

49. $f(x) = x^4 - 2x^3 - 49x^2 - 2x + 48$
Lower: −7; Upper: 9

50. $g(x) = x^4 + 5x^3 - 299x^2 + 5x - 300$
Lower: −21; Upper: 16

EXERCISES 51–54 *Find the exact real solutions of the given polynomial equation.*

51. $x^3 + 4x + 12 = 7x^2$

52. $4x^3 + 56x^2 = x + 14$

53. $x^3 - \frac{37}{12}x^2 - \frac{7}{2}x - \frac{2}{3} = 0$

54. $3x^4 - 125x^2 + 19x = -57x^3 + 42$

EXERCISES 55–58 *Show that the function has no rational zeros. Use a graphing calculator to approximate all irrational zeros to the nearest hundredth.*

55. $f(x) = x^3 - 5x - 1$

56. $g(x) = x^4 + 2x^3 - x^2 - 7$

57. $g(x) = 0.05x^4 + x^3 - x^2 + x - 3$

58. $h(x) = x^5 - 15x^4 - x + 5$

EXERCISES 59–62 *Find the domain and zeros of the rational function and match it with its graph.*

59. $f(x) = \dfrac{1}{x - 3}$

60. $f(x) = \dfrac{x + 2}{x - 3}$

61. $f(x) = \dfrac{x + 2}{x^2 - 9}$

62. $f(x) = \dfrac{1}{x + 3}$

i.

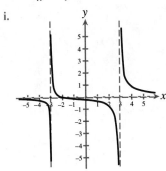

ii.

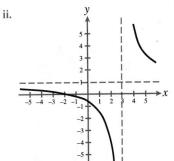

iii.

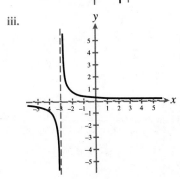

iv.

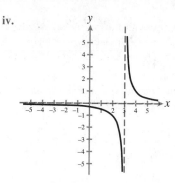

EXERCISES 63–72   *Find all vertical and horizontal asymptotes and sketch the graph.*

63. $f(x) = \dfrac{1}{x + 2}$

64. $h(x) = \dfrac{x + 3}{x - 1}$

65. $g(x) = \dfrac{2x + 5}{3x - 1}$

66. $h(x) = \dfrac{-1}{x^2 - 4}$

67. $f(x) = \dfrac{3 - x}{x^2}$

68. $g(x) = \dfrac{x}{x^2 + 5}$

69. $g(x) = \dfrac{2x^2}{x^2 + x + 2}$

70. $h(x) = \dfrac{x^4}{x + 2}$

71. $f(x) = \dfrac{2}{(1 - x)^2}$

72. $h(x) = \dfrac{x + 1}{(x - 3)^2}$

EXERCISES 73–76   *Find the vertical and inclined asymptotes and sketch the graph.*

73. $h(x) = -2x - \dfrac{1}{x}$

74. $g(x) = \dfrac{x^2 + 2x}{x - 2}$

75. $h(x) = \dfrac{3x^2 + 8x - 9}{x + 3}$

76. $f(x) = \dfrac{x^2 - 6x + 14}{2x - 8}$

EXERCISES 77–86   *Parts a and b are connected: Part a involves an elementary concept, whereas part b involves related material from this chapter. First answer part a, and then use this result to answer part b.*

77. **a.** Add $x - 3 + \dfrac{2}{x - 1}$.

   b. Find the inclined asymptote of $f(x) = \dfrac{x^2 - 4x + 5}{x - 1}$.

78. **a.** Expand $(x + 2)(x - 2)(x - 1)$.

   b. Find the zeros of $f(x) = x^4 - x^3 - 4x^2 + 4x$.

79. **a.** Solve $x^2 + 3x - 1 = 0$.

   b. Find all zeros of $f(x) = x^3 + 3x^2 - x$.

80. **a.** Factor $x^3 - 5x^2 - 6x$.

   b. Find all vertical asymptotes of $\dfrac{x + 1}{x^3 - 5x^2 + 6x}$.

81. **a.** Simplify $(x - 1)(x + 5) + 9$.

   b. Simplify $\dfrac{x^2 + 4x + 4}{x - 1}$.

82. **a.** Find all complex solutions of $x^2 + 2x + 5 = 0$.

   b. Express $f(x) = x^2 + 2x + 5$ as a product of linear factors.

83. **a.** If a function f has the x- and y-axes as its horizontal and vertical asymptotes, respectively, what are the asymptotes for $h(x) = f(x - 3) - 2$?

   b. Find a function with vertical asymptote $x = 3$ and horizontal asymptote $y = -2$.

84. **a.** Compute $f(-x)$ for $f(x) = x^4 + 4x^2 + 2$.

   b. Show that f has no real zeros.

85. **a.** Compute $(2 - 3i) + (2 + 3i)$ and $(2 - 3i)(2 + 3i)$.

   b. Find a polynomial with zeros $2 \pm 3i$.

86. **a.** Find $f(0)$, $f(1)$, and $f(2)$ for $f(x) = x^3 - 2x^2 - x + 2$.

   b. Find an interval of length 1 in which the function f has a turning point.

87. *Maximum Profit*   A company has determined that its profits can be modeled by a function of the form $f(x) = ax^2 + bx + c$, where x is the number of units sold and $f(x)$ is the corresponding profit. Find f given that the profit is zero for $x = 1000$ and $x = 5000$ and, because of fixed costs, the company loses \$15,000 if no units are sold. How many units must be sold in order to maximize profit? What is the largest possible profit?

88. *Oscillating Spring*   A spring is hung from a fixed point and an object is attached to its free end, as shown in Figure 81. If no external force is applied to the object, it will remain at rest in its *equilibrium* position. If the object is pulled down and then released, its position relative to the equilibrium position can be modeled for a brief period of time with a polynomial function of the form $f(t) = at^4 + bt^2 + c$, where t is the time after the object is released and $f(t)$ is the directed distance away from the equilibrium position at that time (a negative value indicates that the object is above the equilibrium position). Find f if the object is pulled 6 centimeters below the equilibrium position and then passes the equilibrium position at $t = 0.11$ and again at $t = 0.27$. (*Hint:* Since f is an even function, its graph is symmetric about the y-axis.) Where is the object at $t = 0.18$ second? At what point does the model fail to provide reasonable results?

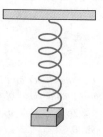

FIGURE 81

89. *Alaskan Temperature*   The average monthly temperature in Juneau, Alaska, over a 12-month period can be approximated with the function

$$f(x) = 0.0134x^4 - 0.33x^3 + 2.07x^2 - 0.456x - 5.04$$

where x denotes the month ($x = 0$ corresponds to January) and $f(x)$ is the average Celsius temperature for that month. Approximate the zeros of f in the interval $[0, 12)$. What do these zeros represent?

90. *Mathematics Degrees*   The number of bachelor's degrees conferred in mathematics during the years 1970–1995 can be approximated with the polynomial function

$$f(x) = -4.6x^3 + 215x^2 - 3080x + 27{,}500$$

where x denotes the year ($x = 0$ corresponds to 1970) and $f(x)$ is the number of degrees for that year. Estimate the year(s) in which 14,500 bachelor's degrees in mathematics were conferred. Approximate the coordinates of any turning points.

Mathematics Degrees 1970–1995

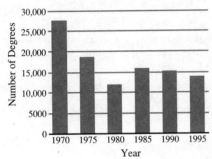

Data source: U.S. Department of Education.

91. *Calculator Cost*   A company that manufactures calculators has determined that the total cost for producing x calculators is given by $C(x) = 15{,}000 + 20x$. The average cost per calculator is thus given by

$$\overline{C}(x) = \frac{15{,}000 + 20x}{x}$$

Sketch the graph of $\overline{C}$. As the number of calculators increases, what value is the average cost approaching?

CHAPTER 4 TEST

PROBLEMS 1–8 *Answer true or false.*

1. A polynomial of degree 3 with real coefficients must have at least one real zero.

2. The graph of a polynomial function of degree 4 can have at most three turning points.

3. According to Descartes' Rule of Signs, if a polynomial function has three variations in sign, then it must have three positive zeros.

4. There is exactly one polynomial of degree 3 with zeros 1, -2, and 4.

5. If a polynomial of degree 2 divides evenly into one of degree 5, the quotient will have degree 3.

6. If $2 + 3i$ is a zero of a polynomial with real coefficients, then so is $2 - 3i$.

7. A rational function must have a vertical asymptote.

8. A rational function can have at most one horizontal asymptote.

PROBLEMS 9–14 *Give an example of each.*

9. A polynomial function f for which $f(x) \to -\infty$ as $x \to \infty$

10. A polynomial function of least degree and with zeros $-1, 3$, and 4

11. A polynomial function of least degree with real coefficients and with zeros 2 and $-3i$

12. A polynomial function of degree 4 with real coefficients and with no real zeros

13. A rational function with vertical asymptote $x = -1$ and horizontal asymptote $y = 1$

14. A rational function with vertical asymptote $x = 2$ and inclined asymptote $y = x - 3$

PROBLEMS 15–16 *Factor the given polynomial and find the exact values of all the zeros.*

15. $f(x) = 2x^4 - 7x^3 - 4x^2$

16. $f(x) = 6x^3 + x^2 - 19x + 6$

17. Find the quotient and remainder when $f(x) = 3x^4 - 12x^2 + 5x + 14$ is divided by $d(x) = x + 2$.

18. Given that $f(x) = 5x^4 - 2x^3 + 45x^2 - 18x$ has $3i$ as one of its zeros, find the remaining zeros and express $f(x)$ as a product of linear factors.

19. Use the Rational Zero Theorem to list all the possible rational zeros of $f(x) = 2x^3 + x^2 - 12x + 9$. Use synthetic division and/or a graphing calculator to help you determine which are actually zeros.

20. Apply Descartes' Rule of Signs to $f(x) = 2x^4 - 6x^3 + x^2 - 8x - 7$. Then use a graphing calculator to determine the precise number of positive and negative zeros.

21. Show that $f(x) = x^3 - x - 2$ has no rational zeros, and use a graphing calculator to estimate any irrational zeros to the nearest hundredth.

22. Find any vertical or horizontal asymptotes for the function

$$f(x) = \frac{x + 1}{2x - 3}$$

and sketch its graph.

23. U.S. homicide rates from 1985 to 1997 can be modeled by the function $f(x) = -0.006x^3 + 0.054x^2 + 0.13x + 8$, where x is the year ($x = 0$ corresponds to 1985) and $f(x)$ is the number of homicides per 100,000 people.

   a. Estimate the year(s) in which the homicide rate was 8.7 per 100,000.

   b. Approximately when did homicide rates peak? What was the rate at that time?

   c. According to this model, during what period did the crime rate drop?

5 EXPONENTIAL AND LOGARITHMIC FUNCTIONS

As the world population has exploded, so too has humankind's influence on Earth. This satellite photograph of North America at night shows the extent of our influence: where there is light, there is humanity. And there is light from New York to Los Angeles, from Montreal to Mexico City. Population growth profoundly affects the consumption of natural resources, environmental quality, and the entire global economy, and thus predicting future population has become increasingly important. In this chapter we will develop mathematical models for predicting the growth of human and other populations.

SECTION 5.1   **EXPONENTIAL FUNCTIONS**

☐ **If a certain population of bacteria grows exponentially, how long would it take for it to fill a space the size of our solar system?**

☐ **How can a doctor estimate the concentration of a medication in the bloodstream?**

☐ **Can a bank compound interest every instant of every day without going bankrupt?**

☐ **In 1965, Intel founder and techno-visionary Gordon Moore predicted that the number of transistors on a microprocessor would double every 18–24 months. How accurate has he been so far?**

From the population of a city to the growth of a single cell, from the value of a retirement account to the price of a personal computer, from drug concentration in the body to drug use in society, there are countless contexts in which mathematical models are used to predict how a given quantity will vary over time. One of the first steps in developing mathematical models of this type is to consider the growth rate of the quantity. For example, the average global temperature in 1980 was 61.9° Fahrenheit, and, since then, it has increased at an average rate of 0.029° per year. Thus, t years after 1980, we would expect that the average global temperature would have risen $0.029t$ degree. It follows that in the year $1980 + t$, the average global temperature (in degrees Fahrenheit) can be estimated by the linear function

$$G(t) = 61.9 + 0.029t$$

In this section and much of this chapter, we model phenomena that grow not at a constant rate (like global temperature), but rather at a *constant percentage rate*. For example,

- During the 1990s the cost of living increased by about 3% per year.

- The amount of radioactive carbon 14 in the Dead Sea Scrolls is decreasing by about 0.012% per year.

- The number of Internet hosts grew by about 90% per year in the 1990s.

- The world population in 2000 is 6 billion and is growing at about 1.3% per year.

If a quantity $Q = Q(t)$ grows at a constant percentage rate, then $Q(t)$ is what is known as an *exponential function*. Among the many real-world phenomena that can be described using exponential functions are population growth, the spread of disease and information, radioactive decay, concentrations of drugs in the body, sales patterns, and compound interest. The first of these—population growth—will motivate our definition of an exponential function.

THE EXPONENTIAL FUNCTION WITH BASE A

Suppose a biologist has been conducting a carefully controlled experiment on the growth of a bacteria culture. Unfortunately, a lab assistant misplaced all the data except that which appears in Table 1. To avoid the expense of starting the experiment over, the biologist decides to construct a function that estimates the number of bacteria at any given time.

The pattern in Table 1 suggests that the bacteria population is doubling (growing by 100%) every hour. In fact, the values in the second column are successive powers of 2. After 1 hour, the population (in thousands) is $2^1 = 2$; after 2 hours it is $2^2 = 4$;

TABLE 1

Time	Number of bacteria (in thousands)
12:00 noon	1
1:00 P.M.	2
2:00 P.M.	4
3:00 P.M.	8
4:00 P.M.	16

after 3 hours it is $2^3 = 8$; and so on. In general, if we let t represent the number of hours after 12:00 noon, then the number of bacteria after t hours is 2^t thousand. Thus, the function that the biologist is seeking would appear to be the *exponential function*

$$f(t) = 2^t$$

where t denotes the number of hours after 12:00 noon, and the population $f(t)$ is measured in thousands. Note that the function f can be used to estimate the bacteria population at any time. Indeed, since 12:00 noon represents $t = 0$, times before or after 12:00 noon can be treated as negative or positive values of t, respectively. For example, 9 A.M. is 3 hours before 12:00 noon, and so we set $t = -3$ to obtain

$$f(-3) = 2^{-3} = \frac{1}{2^3} = \frac{1}{8} = 0.125$$

which corresponds to 125 bacteria. The population at 6:00 P.M. ($t = 6$) is given by

$$f(6) = 2^6 = 64$$

which corresponds to 64,000 bacteria. For fractions of an hour, we use fractional exponents. Thus, for example, the bacteria population at 2:30 P.M. ($t = \frac{5}{2}$) is given by

$$f\left(\frac{5}{2}\right) = 2^{5/2} = (\sqrt{2})^5 \approx 5.66$$

which corresponds to 5660 bacteria. Irrational input values can be justified using methods from calculus. We simply assume that expressions such as $2^{\sqrt{2}}$ are valid and, should the need arise, can be approximated using the power key on a calculator.

EXAMPLE 1 ▪ Graphing an Exponential Function

Sketch the graph of the function $f(x) = 2^x$, and determine its domain and range.

SOLUTION We begin by selecting a few convenient x-values and computing the corresponding function values. For example, with $x = -3$, $f(-3) = 2^{-3} = \frac{1}{8}$. Thus, $(-3, \frac{1}{8})$ is a point on the graph. Other values are shown in Table 2. After plotting these points and connecting them with a smooth curve, we obtain the graph shown in Figure 1.

TABLE 2

x	$f(x) = 2^x$
-3	$\frac{1}{8}$
-2	$\frac{1}{4}$
-1	$\frac{1}{2}$
0	1
1	2
2	4
3	8

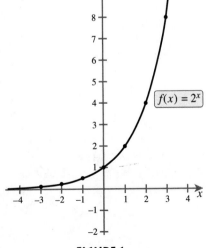

FIGURE 1

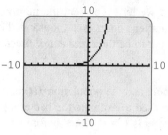

FIGURE 2

Alternatively, we could use a graphing calculator to obtain a graph similar to that shown in Figure 2. Recall that the domain of a function is the set of all possible input values (*x*-coordinates on the graph) and the range is the set of all possible output values (*y*-coordinates on the graph). Thus, our graph suggests that the domain of *f* is the set of all real numbers, and the range of *f* is the set of positive real numbers, or $(0, \infty)$.

The function $f(x) = 2^x$ is more properly called the *exponential function with base 2*. For arbitrary bases, we give the following definition.

Definition of the Exponential Function Base a

The function $f(x) = a^x$, for $a > 0$ and $a \neq 1$, is the **exponential function with base *a*.**

When $a > 1$, the graph of the exponential function $f(x) = a^x$ rises sharply as *x* gets larger, as shown in Figure 2, for example. However, when $a < 1$, the graph falls as *x* gets larger, as illustrated by the following example.

EXAMPLE 2 Graphing an Exponential Function with a Base Less Than 1

Sketch the graph of the exponential function $g(x) = (\frac{1}{2})^x$, and determine its domain and range.

SOLUTION Once again, we begin by selecting a few convenient *x*-values and computing the corresponding function values. For example, with $x = -3$, $g(-3) = (\frac{1}{2})^{-3} = 2^3 = 8$. Thus, $(-3, 8)$ is a point on the graph. Other values are shown in Table 3. After plotting these points and connecting them with a smooth curve, we obtain the graph shown in Figure 3.

TABLE 3

x	$g(x) = (\frac{1}{2})^x$
-3	8
-2	4
-1	2
0	1
1	$\frac{1}{2}$
2	$\frac{1}{4}$
3	$\frac{1}{8}$

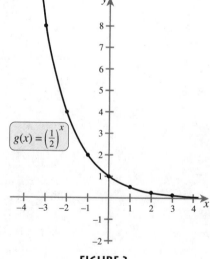

$g(x) = \left(\frac{1}{2}\right)^x$

FIGURE 3

On the basis of the graph, we again conclude that the domain is the set of all real numbers, and the range is the set $(0, \infty)$.

The graph of $g(x) = (\frac{1}{2})^x$ in Figure 3 appears to be a mirror image of the graph of $f(x) = 2^x$ from Figure 1. To confirm this, recall that the graph of a function g is the reflection about the y-axis of the graph of a function f if and only if $f(-x) = g(x)$. In this case, we have

$$f(x) = 2^x$$
$$f(-x) = 2^{-x}$$
$$= \frac{1}{2^x} = \left(\frac{1}{2}\right)^x = g(x)$$

Thus, the graphs of f and g are indeed reflections of one another about the y-axis. More generally, we have the following property.

***The Reflective Property of
Exponential Functions***   ☐

> For $a > 0$ and $a \neq 1$, the graph of $g(x) = (\frac{1}{a})^x$ is the reflection about the y-axis of the graph of $f(x) = a^x$

Figure 4 illustrates the reflective property of exponential functions.

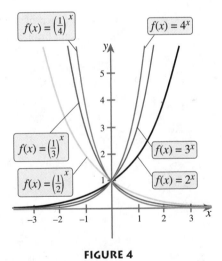

FIGURE 4

Several important properties of exponential functions are illustrated by the examples in Figure 4. First, we note that each exponential function f satisfies $f(x) > 0$ for all real numbers x. Thus, the domain of f is the set of all real numbers and the range is the set of all *positive* real numbers. Note also that since $a^0 = 1$, all exponential functions have y-intercept $(0, 1)$. Also, when the base a is greater than 1 [as with $f(x) = 2^x$, $f(x) = 3^x$ or $f(x) = 4^x$], the graph rises from left to right, so that f is an increasing function. On the other hand, when the base is less than 1 [as with $f(x) = (\frac{1}{2})^x$, $f(x) = (\frac{1}{3})^x$, $f(x) = (\frac{1}{4})^x$], the graph falls from left to right, and so f is a decreasing function. Finally, graphs of exponential functions pass the horizontal line test and so exponential functions are 1–1. These properties of exponential functions are summarized as follows.

Properties of □
Exponential Functions

Let $f(x) = a^x$, $a > 0$, $a \neq 1$.

1. The domain of f is the set of all real numbers, and the range of f is the set of all positive real numbers.

2. f has y-intercept $(0, 1)$ (that is, $a^0 = 1$).

3. f is increasing if $a > 1$ and decreasing if $0 < a < 1$.

4. f is 1–1 (that is, the graph of f passes the horizontal line test).

Exponential functions arise frequently in medical sciences such as pharmacology. In the next example, we see how drug levels can be modeled with exponential functions.

EXAMPLE 3 ▨ **Medication in the Body**

A 4-milligram dose of a certain medication is eliminated from the body in such a way that the quantity remaining at a given time is $\frac{2}{3}$ of the amount that was there 1 hour earlier. Thus, after t hours, the amount remaining in the body is given by the function

$$g(t) = 4\left(\frac{2}{3}\right)^t$$

a. How much of the medication remains after 3 hours? After $5\frac{1}{2}$ hours?
b. At what time will the amount in the body be half the original quantity?

SOLUTION

a. After 3 hours, the amount of medication remaining is given by

$$g(3) = 4\left(\frac{2}{3}\right)^3 = 4 \cdot \frac{8}{27} \approx 1.185 \text{ mg}$$

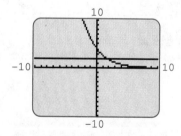

After $5\frac{1}{2}$ hours, the amount is

$$g\left(\frac{11}{2}\right) = 4\left(\frac{2}{3}\right)^{\frac{11}{2}} = 4\left(\frac{2}{3}\right)^{5.5} \approx 0.430 \text{ mg}$$

FIGURE 5

b. Since 4 milligrams of medication are in the body initially, we are interested in the time at which 2 milligrams of the substance remain. Thus, we are interested in the value of t for which

$$4\left(\frac{2}{3}\right)^t = 2$$

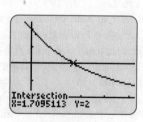

The solution of this equation corresponds to the x-coordinate of the point of intersection of the graphs of $y = 2$ and $y = 4(\frac{2}{3})^t$ (see Figure 5). Zooming in on the point of intersection and using the calculate-intersect feature gives us a t-value of approximately 1.71, as shown in Figure 6. Thus, after 1.71 hours (roughly 1 hour and 43 minutes), half the original dose of medication remains in the body.

FIGURE 6

The 1–1 property of exponential functions is often useful for solving simple equations involving exponents. It can be stated in the following way.

The 1–1 Property of
Exponential Functions □ | For all real numbers r and s, if $a^r = a^s$, then $r = s$.

EXAMPLE 4 ▒ **Solving an Exponential Equation**

Use the 1–1 property of exponential functions to solve the exponential equation $\dfrac{1}{4^x} = 64$.

SOLUTION First we must rewrite both sides of the equation as a power of 4—that is, in the form $4^{\square}$. This is possible since $\dfrac{1}{4^x}$ can be written as 4^{-x} and 64 can be written as 4^3. We proceed as follows:

$$\frac{1}{4^x} = 64$$

$$4^{-x} = 4^3 \qquad \text{Using the identity } \frac{1}{a^n} = a^{-n} \text{ and rewriting 64 as } 4^3$$

$$-x = 3 \qquad \text{Applying the 1–1 property}$$

$$x = -3$$

Notice that an exponential identity was needed in the preceding example when we claimed that $\dfrac{1}{4^x} = 4^{-x}$. Using techniques from calculus, the familiar exponential identities that we have already seen for integer and rational exponents can be shown to be valid for any real number exponent. For convenience, we summarize these identities again.

Exponential Identities □ | Let a and b be positive real numbers and let r and s be real numbers.

1. $a^r a^s = a^{r+s}$ **5.** $\left(\dfrac{a}{b}\right)^r = \dfrac{a^r}{b^r}$

2. $\dfrac{a^r}{a^s} = a^{r-s}$ **6.** $a^0 = 1$

3. $(a^r)^s = a^{rs}$ **7.** $a^{-r} = \dfrac{1}{a^r}$

4. $(ab)^r = a^r b^r$

E X A M P L E 5 ▪ Solving an Exponential Equation

Solve the exponential equation $3^x \cdot 3^{x+1} = \frac{1}{27}$.

SOLUTION

$$3^x \cdot 3^{x+1} = \frac{1}{27}$$

$$3^{x+x+1} = \frac{1}{27} \qquad \text{Applying the exponential identity } a^r a^s = a^{r+s}$$

$$3^{2x+1} = \frac{1}{3^3} \qquad \text{Rewriting 27 as a power of 3}$$

$$3^{2x+1} = 3^{-3} \qquad \text{Using the exponential identity } \frac{1}{a^n} = a^{-n}$$

$$2x + 1 = -3 \qquad \text{Applying the 1–1 property}$$

$$2x = -4$$

$$x = -2$$

It should be noted that the technique shown in Examples 4 and 5 is practical only for exponential equations that can easily be written in the form

$$a^{\square} = a^{\circ}$$

The method is not useful for such seemingly simple equations as $2^x = 25$. In this case, it is not possible to rewrite 25 as an integer power of 2. We will develop an alternative approach using logarithms in Section 5.4.

THE NATURAL EXPONENTIAL FUNCTION

The frequency with which the number e appears in mathematical formulas from diverse branches of mathematics (number theory, probability, statistics, and calculus, for example) is rivaled only by that of π. Like π, e is an irrational number and so its decimal expansion neither repeats nor terminates. Its decimal expansion begins

$$e = 2.71828\ldots$$

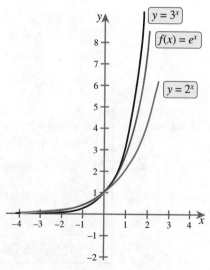

FIGURE 7

Perhaps the most important function in all of mathematics is the exponential function with base e, $f(x) = e^x$, also known as the **natural exponential function.** Since e is a number between 2 and 3, the graph of the natural exponential function $f(x) = e^x$ lies between the graphs of $y = 2^x$ and $y = 3^x$, as shown in Figure 7.

E X A M P L E   6   ■   **Using a Translation to Graph an Exponential Function**

Use the graph of $f(x) = e^x$ to sketch the graph of $g(x) = e^{x+3}$.

SOLUTION   Since $g(x)$ can be obtained by replacing x with $x + 3$ in the function $f(x) = e^x$, the graph of g can be found by translating the graph of f to the left 3 units, as shown in Figure 8. Note that the y-intercept $(0, 1)$ on the graph of $y = e^x$ is trans-d 3 units to the left to obtain the corresponding point $(-3, 1)$ on the graph of $g(x) = e^{x+3}$.

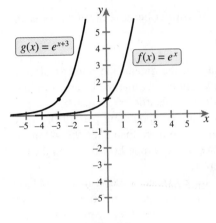

FIGURE 8

The number e arises in a natural way from the study of **compound interest.** In Section 1.4, we computed *annually compounded interest* using the formula $B = P(1 + r)^t$, where B is the balance in an account after t years when a principal of P dollars is deposited at an annual rate of interest r compounded annually. A more general approach (see Exercise 52) would lead to the following formula.

Compound Interest Formula   ☐

If a principal of P dollars is deposited in an account paying an annual rate of interest r compounded n times per year, then the balance B in the account after t years is given by

$$B = P\left(1 + \frac{r}{n}\right)^{nt}$$

E X A M P L E   7   ■   **Computing Compound Interest**

There is an apocryphal tale of a young George Washington throwing a silver dollar across the Potomac River. Suppose that instead of throwing a dollar, he had deposited it in 1776 in a bank that paid 6% interest. How much would that dollar be worth in the year 2000 if interest were compounded

a.   annually?   **b.**   quarterly?
c.   daily?   **d.**   every minute?

SOLUTION

a.   Letting $P = 1$, $r = 0.06$, $t = 224$, and $n = 1$ in the compound interest formula, we obtain

$$B = P\left(1 + \frac{r}{n}\right)^{nt} = 1\left(1 + \frac{0.06}{1}\right)^{1(224)} = (1.06)^{224} \approx 466{,}137.26$$

b.   Using $n = 4$, the number of quarters in a year, we obtain

$$B = P\left(1 + \frac{r}{n}\right)^{nt} = 1\left(1 + \frac{0.06}{4}\right)^{4(224)} = (1.015)^{896} \approx 621{,}689.96$$

c. With $n = 365$, we obtain

$$B = 1\left(1 + \frac{0.06}{365}\right)^{365(224)} \approx 686{,}180.14$$

d. Letting $n = 365 \cdot 24 \cdot 60 = 525{,}600$, the number of minutes in a year, we obtain

$$B = 1\left(1 + \frac{0.06}{525{,}600}\right)^{525{,}600(224)} \approx 686{,}937.94$$

Note that the answer your calculator gives may differ because of round-off error.

From the preceding example, it is evident that the greater the number of compounding periods, the greater the balance. One might expect that the balance could be made as large as desired simply by compounding with sufficient frequency. Surprisingly, though, there is a limit; no matter how many compounding periods are used, George Washington's dollar could never accumulate to more than \$686,938.47 in 224 years. To see where this amount comes from, we first examine how the natural exponential function is connected to compound interest.

Suppose you were lucky enough to find a bank that paid 100% interest and you deposited a dollar for one year. By setting $P = 1$, $r = 1.00$, and $t = 1$ in the compound interest formula, and by considering larger and larger values of n (the number of compounding periods), we obtain the following table of (approximate) values.

n	$B = \left(1 + \dfrac{1}{n}\right)^n$
1	2
10	2.59374
100	2.70481
1000	2.71692
10,000	2.71815
100,000	2.71827
1,000,000	2.71828

Notice that as the number of compounding periods increases, the balance gets closer to the number e. Actually, this is no coincidence since the number e can be *defined* as the limiting value of $[1 + (\frac{1}{n})]^n$. So if the bank compounded interest *continuously,* your dollar would increase in value to exactly e dollars after 1 year. For arbitrary values of P, r, and t, we have the following general formula for **continuously compounded interest.**

Continuously □
Compounded Interest

> If a principal of P dollars is deposited in an account paying an annual rate of interest r compounded continuously, then the balance B in the account after t years is given by
>
> $$B = Pe^{rt}$$

EXAMPLE 8   ▨   Computing Continuously Compounded Interest

Suppose that George Washington deposited his dollar in 1776 in a bank that compounded interest continuously at a rate of 6%. How much would it be worth in the year 2000?

SOLUTION   By applying the formula for continuously compounded interest with $P = 1$, $r = 0.06$, and $t = 224$, the balance would be

$$B = 1e^{(0.06)224} = e^{13.44} \approx 686{,}938.47$$

Thus, the dollar would be worth \$686,938.47 in the year 2000.

UNDERSTANDING AND MASTERY CHECKLISTS

CONCEPTS TO UNDERSTAND

☐ Exponential function with base a
☐ The 1–1 property of exponential functions
☐ Exponential identities
☐ The natural exponential function
☐ Compound interest
☐ Continuously compounded interest

SKILLS TO MASTER

☐ Sketch the graph of an exponential function.
☐ Use the 1–1 property of exponential functions to solve exponential equations.
☐ Use the compound interest formula to find the balance in an account.
☐ Use the continuously compounded interest formula to find the balance in an account.

EXERCISES 5.1

EXERCISES 1–2 *Sketch the graph of f and use translations or reflections to sketch the graph of g.*

1. $f(x) = 3^x$
   a. $g(x) = -3^x$ b. $g(x) = 3^{-x}$
   c. $g(x) = 3^x + 2$ d. $g(x) = 3^{x+2}$

2. $f(x) = \left(\frac{1}{4}\right)^x$
   a. $g(x) = -\left(\frac{1}{4}\right)^x$ b. $g(x) = \left(\frac{1}{4}\right)^{-x}$
   c. $g(x) = \left(\frac{1}{4}\right)^x - 3$ d. $g(x) = \left(\frac{1}{4}\right)^{x-2}$

EXERCISES 3–6 *Determine the value of a for which* $f(x) = a^x$ *has the indicated graph.*

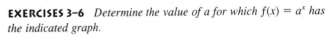
3.

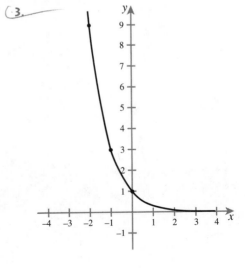

4.

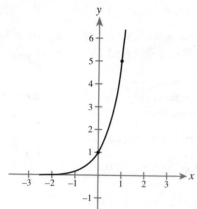

EXERCISES 7–10   *Use the given graph of* $f(x) = e^x$ *to sketch the graph of* $g(x)$.

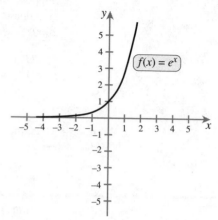

$f(x) = e^x$

7. $g(x) = e^{-x}$ **8.** $g(x) = -e^x$

9. $g(x) = e^{x-2}$ **10.** $g(x) = e^x + 3$

EXERCISES 11–14   *Use a graphing calculator to help you solve the given equation. Approximate all answers to the nearest hundredth.*

11. $4e^{0.1t} = 13$ **12.** $2e^{-0.2t} = 1$

13. $1200e^{-0.05t} = 100$ **14.** $8000e^{0.07t} = 9000$

EXERCISES 15–24   *Use the 1–1 property of the exponential function to solve the given equation.*

15. $3^x = 81$ **16.** $4^{-x} = 64$

17. $2^{3x-4} = 32$ **18.** $5^{1-x} = 125$

19. $\dfrac{1}{3^x} = 27$ **20.** $2^{x-1} = \frac{1}{64}$

21. $4^{x-3} = 16 \cdot 4^{2x}$ **22.** $3^{x+2} = \dfrac{81}{3^x}$

23. $2^{x^2+x} = 4$ **24.** $5^{x^2-5x} = \frac{1}{625}$

EXERCISES 25–26   *Determine the balance B if a principal of P dollars is deposited in an account for t years at an annual rate of interest r compounded n times per year. Give answers to the nearest penny.*

25. $P = \$1000$, $t = 10$ years, $r = 6\%$

    a. $n = 1$

    b. $n = 4$

    c. $n = 12$

    d. $n = 365$

    e. Compounded continuously

26. $P = \$2500$, $t = 5$, $r = 8\%$

    a. $n = 1$

    b. $n = 4$

    c. $n = 12$

    d. $n = 365$

    e. Compounded continuously

5.

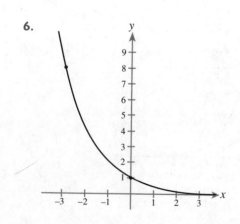

6.

■ APPLICATIONS

27. *Population Growth*   A breeder who supplies pet stores with gerbils has determined that a certain gerbil population can be approximated over a 4-month period with the function

$$f(t) = 20\left(\frac{3}{2}\right)^t$$

where t is measured in months. How many gerbils were present initially? Create a table showing the gerbil population at the end of each of the first 4 months. Assuming the pattern continues beyond 4 months, predict how many gerbils will be present after 6 months and again after 5 years.

28. *Population Growth*   An exterminating company has developed a new chemical-free method for exterminating cockroaches. They claim that a cockroach population with initial size 10,000 will number $f(t) = (10,000)2^{-t}$ after t days. Plot this function. Predict how many cockroaches will remain after 1 week. Estimate the day on which the last cockroach will die.

29. *Compound Interest*   On the occasion of the birth of their first grandchild, two proud grandparents deposit $5000 in a certificate of deposit (CD) in the child's name. The CD pays an annual interest rate of 8%, and it matures in 18 years, just in time for the child to begin college. Determine the amount that will have accumulated after 18 years if interest is compounded quarterly. What if interest is compounded continuously?

30. *Compound Interest*   The grandparents in Exercise 29 wish to determine whether the CD they purchased will pay for the first year of their grandchild's college education. Assuming that 1 year at their alma mater currently costs $15,000 and the amount is likely to increase an average of 5% per year, use the formula for interest compounded annually to estimate what 1 year of college will cost in 18 years.

31. *Energy Consumption*   Energy consumption in the United States during the years 1920–1995 can be approximated using the exponential function $Q(t) = 17.3 \cdot 1.02^t$, where t denotes the year ($t = 0$ corresponds to 1920) and $Q(t)$ is the energy consumed in that year (in quadrillions of Btu's). Estimate the energy consumption in the years 1940 and 1995, and use these values to determine the percent increase from 1940 to 1995. What does this model predict for energy consumption in 2010?

Top: A field of windmills generates electricity near Altamont, California. Bottom: A heliostat field focuses solar energy on a central receiving tower.

32. *Computer Transistors*   Moore's law, formulated in 1965 by Gordon Moore, cofounder of Intel, states that the number of transistors that can be put on a computer processor chip will approximately double every 18–24 months. In 1971, Intel introduced the 4004 processor with 2300 transistors. According to Moore's law, the number of transistors t years after 1971 can be approximated by $N = 2300 \cdot 2^{t/2}$. Use this function to estimate the number of transistors on a processor in 1997. The actual number of transistors on the Pentium II processor introduced by Intel in 1997 was 7.5 million. How well does your estimate compare to this value? What does Moore's law predict for the number of transistors in the year 2001?

Intel Processors 1971–1997

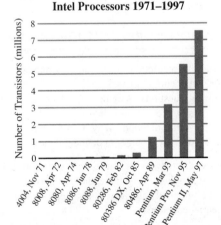

Data source: Intel.

33. *Population Equation*   At the beginning of the section, we considered a bacteria population that was modeled by the function $f(t) = 2^t$, where t is in hours ($t = 0$ corresponds to 12:00 noon)

Energy Consumption 1920–1995

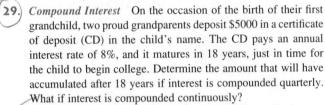

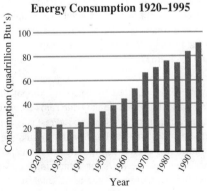

Data source: U.S. Department of Energy.

and $f(t)$ is the population in thousands. Estimate the time at which the population reaches 25,000.

34. Intersection Points   Use a graphing calculator to estimate the point(s) of intersection for the graphs of $y = 2^x$ and $y = x^2$. [*Hint:* It may be simpler to approximate the *three* zeros of $h(x) = 2^x - x^2$ than to use the calculate-intersect feature.]

35. Density Function   The function $f(x) = (1/\sqrt{2\pi})e^{-x^2/2}$ is known as the *standard normal probability density function,* and it is important in the study of probability and statistics. Use a graphing calculator to determine the maximum value of f. What happens to the value of $f(x)$ as x approaches infinity or negative infinity?

36. Medication Concentration   A certain medication is ingested and begins spreading throughout the bloodstream. The concentration (in nanograms per milliliter) of the medication in the bloodstream after t hours is given by $Q(t) = 4te^{-0.8t}$. Use a graphing calculator to plot this function. Estimate the maximum concentration of the medication and the time at which it occurs. What happens to the concentration as time increases beyond this point?

37. Contagious Virus   A group of 100 students returns from a trip to Europe, where they were exposed to a highly contagious virus. The virus begins spreading among the other 900 students in their school in such a way that t days after their return, the number of students with the virus is modeled by

$$N(t) = \frac{100{,}000}{100 + 900e^{-0.15t}}$$

a. How many students have the virus after 2 days? After 5 days?

b. Use a graphing calculator to help you plot the graph of $N(t)$ for $t > 0$.

c. Predict the number of days it will take for 800 of the school's 1000 students to be infected.

d. What happens to $N(t)$ as t approaches infinity?

38. Knowledge Retention   For a certain high school with a graduating class of 200 students, the number of classmates' names that a typical student can remember t years after graduation is modeled by $R(t) = 80 + 120e^{-0.75t}$.

a. How many students will a graduate remember 1 year after graduation? 5 years after graduation?

b. Use a graphing calculator to estimate the number of years after graduation until a graduate will remember less than half of her classmates.

c. According to this model, how many classmates will a graduate never forget?

CONCEPTS AND CRITICAL THINKING

EXERCISES 39–42   *Answer true or false.*

39. The graphs of $f(x) = 2^x$ and $g(x) = (\frac{1}{2})^x$ are reflections of one another about the y-axis.

40. Simple interest can be modeled with an exponential function.

41. All exponential functions are 1–1.

42. For some exponential functions f, $f(x) \to 0$ as $x \to \infty$.

EXERCISES 43–46   *Give an example of each.*

43. An exponential function that is decreasing over its entire domain

44. Two exponential functions such that the graph of the second is a translation, 6 units to the left, of the graph of the first

45. A point on the graph of every exponential function of the form $f(x) = a^x$

46. A number that is not in the range of any exponential function

QUESTIONS FOR DISCUSSION OR ESSAY

47. Discuss the restrictions placed on the value of a in the definition of the exponential function $f(x) = a^x$. In particular, consider the following questions: Why does the definition exclude the value $a = 1$? Isn't $f(x) = 1^x$ a valid function? Why isn't it considered an exponential function? Why must we have $a > 0$? What difficulties do we encounter in defining the function $f(x) = (-2)^x$?

48. Discuss the limitations of the technique used in Example 4 for solving simple exponential equations. Describe a procedure for using a graphing calculator to find approximate solutions to exponential equations.

49. Although many quantities grow exponentially over relatively short periods of time, exponential models can yield bizarre predictions for longer time periods. Consider, for example, the exponential function $f(t) = 2^t$, which was used to model the bac-

teria population at the beginning of this section. If the growth were to continue at this exponential pace, show that the number of bacteria at the end of 1 week would be 3.74×10^{50}. Assuming "average" size bacteria, this would be enough bacteria to fill a sphere with diameter that of our solar system. What is wrong with the model? What would really happen to the bacteria population over time?

50. A careful comparison of banking practices would reveal that many methods are used by banks to compound interest. Even among banks that claim to have the same compounding periods, there are differences in when the interest is actually paid into an account. Contact several local banks and inquire about their interest rates and compounding schemes for a specific type of account. Discuss some ways for comparing banks so that you can choose the one that pays "the best" interest.

PROJECTS FOR ENRICHMENT

51. *Savings Plans* We have seen how the balance in an account can be computed using the compound interest formula. Here we consider the effect of compound interest on deposits made at equally spaced time intervals throughout the year. To that end, suppose that an amount A is deposited k times each year in an account earning an annual rate of interest r compounded n times per year (we assume the account is opened with the first deposit of the amount A and that the deposits are equally spaced from that point on). After t years, the balance in the account will have accumulated to

(1) $$B = \dfrac{A\left[\left(1 + \dfrac{r}{n}\right)^{nt} - 1\right]}{1 - \left(1 + \dfrac{r}{n}\right)^{-n/k}}$$

a. Use equation (1) to find the balance after 10 years with monthly deposits of $100 in an account paying 5% interest compounded quarterly.

b. Use equation (1) to find the balance for the following savings plans.

   i. Monthly deposits of $50 in an account paying 6% interest compounded monthly for 20 years

   ii. Quarterly deposits of $150 in an account paying 6% interest compounded monthly for 20 years

   Note that for each plan, the same total is deposited each year. Which plan yields the larger balance? Explain.

c. Solve equation (1) for A. This will yield a formula for finding the deposit amount that will yield a balance B after t years in an account where the annual rate of interest r is compounded n times per year and deposits are made k times per year.

d. Find the deposit amount necessary to achieve the following savings goals.

   i. A balance of $50,000 after 20 years with an interest rate of 5% compounded daily and deposits made monthly

   ii. A balance of $1,000,000 after 40 years with an interest rate of 8% compounded monthly and deposits made monthly

   For both of parts i and ii use the standard compound interest formula to determine the amount of a single deposit that would be necessary to achieve the indicated savings goals.

e. Use a graphing calculator to find the number of years that would be necessary to achieve the following savings goals.

   i. A balance of $10,000 in an account with an interest rate of 5.5% compounded yearly and $50 deposits made monthly

   ii. A balance of $100,000 in an account with an interest rate of 7.25% compounded monthly and $50 deposits made weekly

52. *Deriving Compound Interest Formulas* The formula for compound interest can be derived from the formula for simple interest. We consider two specific cases before tackling the general case. First, suppose that a principal of P dollars is deposited in an account paying an annual rate of interest r compounded annually. Then, from the simple interest formula, the interest earned after the first year is given by $I = Pr(1)$. Thus, the balance at the end of the first year is $B = P + Pr = P(1 + r)$.

a. Apply the simple interest formula to the new balance $P(1 + r)$ to obtain an expression for the interest earned during the second year. Then add this interest to the balance $P(1 + r)$ and simplify to obtain the expression $P(1 + r)^2$ for the balance after 2 years.

b. Repeat part a for the third year to show that the balance after 3 years is $P(1 + r)^3$.

Generalizing this procedure, we obtain a balance of $P(1 + r)^t$ after t years for interest compounded annually. Now suppose interest is compounded quarterly, or 4 times per year. This means that the annual rate of interest r must be divided equally among the 4 quarters. In other words, the quarterly rate of interest is $r/4$.

c. Use the simple interest formula to write an expression for the interest earned on a principal P during the first quarter.

d. Add the interest from part c to the original principal P and simplify to obtain an expression for the balance after the first quarter. It should be $P(1 + r/4)$.

e. Repeat parts c and d several times to show that the balance is $P(1 + r/4)^4$ at the end of the first year and $P(1 + r/4)^8$ at the end of the second year.

Thus, with interest compounded 4 times per year, the principal after t years is $P(1 + r/4)^{4t}$. Finally, suppose interest is compounded n times per year. From the pattern that has been established, it seems clear that the principal after t years is $P(1 + r/n)^{nt}$.

f. What does the quantity r/n represent? What about nt?

53. *Continuously Compounded Interest* Using the fact that $(1 + 1/m)^m$ approaches e as m approaches infinity, we can derive the formula for continuously compounded interest from the compound interest formula $B = P(1 + r/n)^{nt}$.

a. Show that the formula for compound interest can be rewritten as

$$B = P\left[\left(1 + \dfrac{1}{n/r}\right)^{n/r}\right]^{rt}$$

b. Let $m = n/r$. What happens to the value of m if r remains fixed but n approaches infinity?

c. Substituting $m = n/r$ into the formula in part a, we obtain

$$B = P\left[\left(1 + \frac{1}{m}\right)^m\right]^{rt}$$

Using your observation from part b, argue that as the number of compounding periods approaches infinity, the balance gets closer in value to Pe^{rt}.

d. Surprisingly, continuous compounding does not offer a significant advantage over annual compounding at the same rate.

To see this, suppose you were to find a bank that pays 5% interest compounded continuously. If you deposit P dollars, the balance after t years would be $Pe^{0.05t}$. Now suppose a different bank pays 5.13% interest compounded annually. If you deposit P dollars there, the balance after t years would be $P(1 + 0.0513)^t$. In which of these banks should you invest your money?

e. Find a value for r for which quarterly compounding at an annual rate r will be equivalent to continuous compounding at an annual rate of 5%.

SECTION 5.2   LOGARITHMIC FUNCTIONS

☐ According to Newton, how long does it take a 185° turkey to cool to 100°?

☐ What makes acid rain acidic?

☐ How could you cross a river with an unlimited supply of pizza boxes, a parachute, and a mathematician?

☐ How many decibels is the song of the blue whale, a sound with more than 6 quintillion times the energy of the faintest sound perceptible to the human ear?

☐ How can a pathologist use body temperature to estimate time of death?

THE LOGARITHM FUNCTION WITH BASE A

As Figures 9 and 10 suggest, and as we observed in the previous section, the exponential function $f(x) = a^x$ (with $a > 0$ and $a \neq 1$) passes the horizontal line test and is thus 1–1: If $a^x = a^y$, then $x = y$.

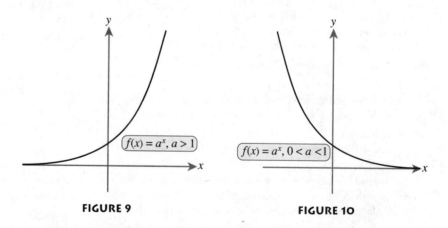

$f(x) = a^x, a > 1$

$f(x) = a^x, 0 < a < 1$

FIGURE 9

FIGURE 10

The 1–1 property of exponents was used in Section 5.1 to solve exponential equations like $2^x = 8$. In this case, for instance, the 1–1 property tells us that $2^x = 8$ can have at most one solution, and since $2^3 = 8$, we can conclude that 3 is the *only* solution of $2^x = 8$.

Now, let us consider this equation in a slightly different light. If we let $f(x) = 2^x$, then our equation could be written as $f(x) = 8$. From a functional point of view, we have been given the output (namely, 8) of the function and are asked to find the corresponding input. In essence, we are "undoing" the base 2 exponential function: We're given the result of an exponentiation, and we're asked to produce the original number. Recall from Section 3.5 that the function which undoes f in this sense is called the *inverse of f* and is denoted by f^{-1}. Using the language of inverse functions, then, solving $f(x) = 8$ is equivalent to computing $f^{-1}(8)$. For convenience, we repeat the definition of inverse function.

Definition of f^{-1} ☐

A function g satisfying $(g \circ f)(x) = x$ for all x in the domain of f and $(f \circ g)(x) = x$ for all x in the domain of g is said to be the **inverse** of the function f. We write $g = f^{-1}$ or, equivalently, $f = g^{-1}$.

As we discussed in Section 3.5, only 1–1 functions have inverses. It follows that because exponential functions are 1–1, $f(x) = a^x$ has an inverse. We define *logarithms* to be the inverse functions of exponentials. Intuitively, we think of logarithmic functions as undoing exponential functions in much the same way that subtracting 7 undoes adding 7, that dividing by 3 undoes multiplication by 3, that taking a cube root undoes cubing a number, and so forth. More precisely, we have the following definition.

Definition of the Logarithm
Function with Base a ☐

The function $g(x) = \log_a x$ (read, "log base a of x") for $a > 0$ and $a \neq 1$, the **logarithm function with base a,** is the inverse of the exponential function $f(x) = a^x$. Equivalently, $y = \log_a x$ if and only if $a^y = x$. Thus, we may think of $\log_a x$ as the power to which a must be raised in order to obtain x.

Since logarithmic functions are inverses of exponential functions, their graphs are reflections of one another about the line $y = x$. The graphs of a few exponential functions and the logarithmic functions that are their inverses are shown in Figure 11.

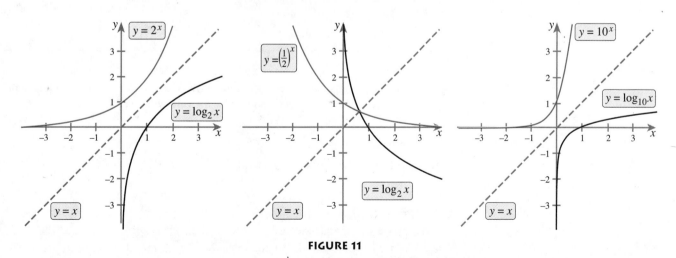

FIGURE 11

Several properties of logarithmic functions are suggested by the graphs in Figure 11, each of which can be verified from the definition of logarithms. For example, the graphs suggest that all logarithmic functions $g(x) = \log_a x$ have an x-intercept of 1, or that $\log_a 1 = 0$. Since $a^0 = 1$, $\log_a 1$ is indeed 0. This and other properties of logarithmic functions are summarized as follows.

Properties of □
Logarithmic Functions

Let $g(x) = \log_a x$, $a > 0$, $a \neq 1$.

1. The domain of g is $(0, \infty)$, and the range of g is $(-\infty, \infty)$.
2. The y-axis is a vertical asymptote of the graph of g.
3. g has x-intercept 1 (that is, $\log_a 1 = 0$).
4. g is increasing if $a > 1$ and decreasing if $0 < a < 1$.
5. g is 1–1 (that is, if $\log_a x = \log_a y$, then $x = y$).

Two additional properties of logarithms can be obtained directly from the definition of logarithmic functions as inverses of exponential functions. Since $f(x) = a^x$ and $g(x) = \log_a x$ are inverses, each "undoes" the other; that is, $f(g(x)) = x$ and $g(f(x)) = x$. This gives us

$$f(g(x)) = x \qquad g(f(x)) = x$$
$$f(\log_a x) = x \qquad g(a^x) = x$$
$$a^{\log_a x} = x \qquad \log_a a^x = x$$

The first of these properties is merely a restatement of the fact that $\log_a x$ is the power to which a must be raised in order to get x. The second property gives us the self-evident statement that "x is the power to which a must be raised in order to get a^x."

Inverse Identities □

Let $a > 0$ and $a \neq 1$.

Property	**Example**
1. $a^{\log_a x} = x$ for all $x > 0$	$2^{\log_2 5} = 5$
2. $\log_a(a^x) = x$ for all x	$\log_3(3^{-8}) = -8$

Some logarithms can be computed directly from the definition or, equivalently, using the inverse identities, as illustrated by the following example.

EXAMPLE 1 ▪ **Computing Logarithms**

Compute each of the following quantities.

a. $\log_2 8$ b. $\log_{1/3} 9$

c. $\log_{123.456789} 1$ d. $\log_4(-2)$

SOLUTION

a. We need to find the power to which 2 must be raised in order to get 8. Since $2^3 = 8$, $\log_2 8 = 3$.

b. We are interested in finding the power to which $\frac{1}{3}$ must be raised in order to get 9. Since $\left(\frac{1}{3}\right)^{-2} = 9$, $\log_{1/3} 9 = -2$.

c. To what power must 123.456789 be raised in order to get 1? Since $123.456789^0 = 1$, $\log_{123.456789} 1 = 0$. As previously stated, $\log_a 1 = 0$ for all possible bases a.

d. Here we are interested in solutions to $4^y = -2$. But $4^y > 0$ for all values of y, so this equation has no solution. In other words, $\log_4(-2)$ is not defined. More generally, the domain of logarithmic functions consists of all positive numbers. Thus, $\log_a x$ is defined only for $x > 0$.

Most logarithms cannot be evaluated using the technique illustrated in Example 1. Consider, for example $\log_2 5$. Here we wish to find the power to which 2 must be raised in order to get 5. Since $2^2 = 4$ and $2^3 = 8$ we would expect that the desired power is somewhere between 2 and 3. In the following example, we consider a graphical approach to estimating logarithms such as $\log_2 5$. A more direct approach is considered in Section 5.3.

EXAMPLE 2   ■   **Estimating a Logarithm Graphically**

Use a graphing calculator to estimate $\log_2 5$.

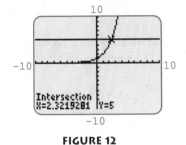

FIGURE 12

SOLUTION   Since $\log_2 5$ is that power to which 2 must be raised in order to get 5, we are looking for the value of x for which $2^x = 5$. Thus, we plot the graph of $y = 2^x$ and $y = 5$, and use the calculate-intersect feature to estimate the x-coordinate of the point of intersection, as shown in Figure 12. We obtain $x \approx 2.32$.

From the definition of logarithm, $y = \log_a x$ if and only if $a^y = x$. In other words, for every logarithmic equation, there is a corresponding exponential equation, and vice versa. This equivalence will prove to be a useful tool both for solving equations involving logarithms and for deriving logarithmic identities.

Equivalence of Exponential and Logarithmic Equations   ☐

> The logarithmic equation $y = \log_a x$ and the exponential equation $a^y = x$ are equivalent.

The following examples illustrate the equivalence of exponential and logarithmic equations.

EXAMPLE 3   ■   **Writing Logarithmic Equations as Exponential Equations**

Write an equivalent exponential equation for each of the following logarithmic equations.

a. $\log_{10} 1000 = 3$ **b.** $\log_x 100 = 2.1$ **c.** $\log_3(x^2) = 12$

SOLUTION
a. $10^3 = 1000$
b. $x^{2.1} = 100$
c. $3^{12} = x^2$

> **RULE OF THUMB** When converting between exponential equations and logarithmic equations, it is helpful to keep the following rules of thumb in mind.
>
> ■ The base of the logarithm is the base of the exponent (that is, the quantity being raised to the power).
>
> ■ The value of the logarithm is the exponent.

EXAMPLE 4 ▨ **Writing Exponential Equations as Logarithmic Equations**

For each exponential equation, write an equivalent logarithmic equation.

a. $0.5^{-2} = 4$ **b.** $e^0 = 1$ **c.** $1.015^{12t} = 1{,}000{,}000$

SOLUTION
a. $\log_{0.5} 4 = -2$
b. $\log_e 1 = 0$
c. $\log_{1.015} 1{,}000{,}000 = 12t$

The correspondence between exponential and logarithmic equations is also useful for finding inverses of functions involving logarithms or exponents, as we see in the following example.

EXAMPLE 5 ▨ **Finding the Inverse of a Function Involving Logarithms**

Compute the inverse of $f(x) = 4 \log_2(x + 1)$.

SOLUTION Recall from Section 3.5 that if $y = f(x)$ defines a 1–1 function, then $x = f^{-1}(y)$. Thus, to find $f^{-1}(y)$, we solve the equation $y = f(x)$ for x.

$$y = 4 \log_2(x + 1) \qquad \text{Replacing } f(x) \text{ with } y$$

$$\frac{y}{4} = \log_2(x + 1)$$

$$2^{y/4} = x + 1 \qquad \text{Converting to exponential form}$$

$$x = 2^{y/4} - 1 \qquad \text{Solving for } x$$

Thus, we have $f^{-1}(y) = 2^{y/4} - 1$. Substituting x in place of y, we obtain

$$f^{-1}(x) = 2^{x/4} - 1$$

NATURAL AND COMMON LOGARITHMS

The most commonly used logarithmic bases—by far—are 10 and e. The relative prominence of base 10 logarithms, also known as **common logarithms,** derives from the fact that 10 is the base for our number system. Although their role in computation (see Exercises 83 and 84 in Section 5.3) is mostly of historical significance in the posttransistor era, common logarithms survive in many scientific contexts, including decibel ratings for loudness, the Richter scale for earthquakes, and the pH level for measuring acidity. If we were to encounter intelligent life from a distant galaxy, we would be shocked if we found that common logarithms enjoyed any special status there. By contrast, one could argue that base e, or **natural logarithms,** would eventually emerge as the logarithm of choice in any civilization. Arising naturally in such seemingly unrelated areas as continually compounded interest and radioactive decay, natural logarithms (and their base e) seem part of the very fabric of our universe.

Special notation has developed for expressing common and natural logarithms. In the case of common logarithms the base is omitted, whereas with natural logarithms, the symbol "ln" is used.

Notation for Common and   ☐
Natural Logarithms

> $\log x = \log_{10} x$. Base 10 logarithms are called **common logarithms.**
>
> $\ln x = \log_e x$. Base e logarithms are called **natural logarithms.**

All of the properties that apply to general logarithm functions apply to common and natural logarithms as well. For this reason, the following properties should not be viewed as new formulas, but rather as particular instances of properties that we have already encountered.

Properties of Common and   ☐
Natural Logarithms

> ■ $10^{\log x} = x$ and $e^{\ln x} = x$
>
> ■ $\log(10^x) = x$ and $\ln(e^x) = x$
>
> ■ $\log 1 = \ln 1 = 0$
>
> ■   The domain of both $f(x) = \log x$ and $g(x) = \ln x$ is the set of positive real numbers.

E X A M P L E   6   ■   **Computing Common and Natural Logarithms**

Compute each of the following quantities, using a calculator as necessary.

a. $\log 1000$ **b.** $\ln \frac{1}{e}$ **c.** $\log 2.4 \times 10^8$

SOLUTION

a.   Because the base isn't given, it is understood to be 10. Thus, we must find the power to which 10 is raised in order to get 1000. Since $10^3 = 1000$, $\log 1000 = 3$.

b.   Here the base is e. We must find the power to which e is raised in order to get $1/e$. Since $e^{-1} = 1/e$, $\ln(1/e) = -1$.

c.   This one requires a calculator. However, even without a calculator, we can make a rough estimate. Since $\log(10^x) = x$, it follows that, for example, $\log(10^8) = 8$ and $\log(10^9) = 9$. Thus, since 2.4×10^8 is between 10^8 and 10^9, its common logarithm is somewhere between 8 and 9. Using the common logarithm key on our calculator, we find that $\log 2.4 \times 10^8 \approx 8.38$.

Although computing common logarithms with a calculator is a fairly simple task, there is considerable value in having an intuitive sense of the approximate size of logarithms, particularly in applied contexts, as a quick check of the reasonability of our results. By generalizing the results of part c of Example 6, we obtain the following approximation strategy.

Approximating Common
Logarithms Using
Scientific Notation

> If a number x is expressed in scientific notation as $x = a \times 10^r$, where $1 \le a < 10$, and r is an integer, then
>
> $$r \le \log x < r + 1$$
>
> In other words, to find a rough estimate of $\log x$, first write x in scientific notation as a number between 1 and 10 times 10 raised to a power. We can then conclude that $\log x$ is between the power and one more than the power.

EXAMPLE 7   Graphing a Function Involving the Natural Logarithm

Sketch the graph of $f(x) = \ln x$ and use it to sketch the graph of $g(x) = \ln x - 2$.

SOLUTION   Of course we could plot this with a graphing calculator, but let's see if we can produce a rough sketch of these graphs by hand. Since $f(x) = \ln x$ is the inverse of the natural exponential function e^x, the graph of f is the reflection of the graph of $y = e^x$ about the line $y = x$, as shown in Figure 13. Note that the graph of f has the y-axis as a vertical asymptote, an x-intercept of 1, and is defined only for positive values of x. Moreover, since the base e is greater than 1, the graph of f increases from left to right. Now $g(x) = \ln x - 2 = f(x) - 2$. So the graph of g can be obtained by translating the graph of f down 2 units. The result is shown in Figure 13.

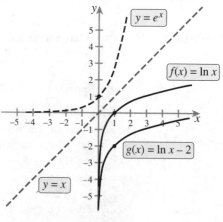

FIGURE 13

Computations involving natural and common logarithms can be performed directly with most calculators, including graphing calculators. Consequently, graphs of functions involving common and natural logarithms can be obtained with relative ease using a graphing calculator. In the following example, we use a graphing calculator to explore Newton's Law of Cooling, a mathematical model of temperature change.

EXAMPLE 8   ▓   **An Application Involving Natural Logarithms**

According to Newton's Law of Cooling, the time it takes for an object to cool from an initial temperature T_0 to a temperature T, given that the temperature of the surrounding air is a constant C, is given by

$$t = k \ln\left(\frac{T_0 - C}{T - C}\right)$$

where k is a constant. Suppose a turkey is taken out of the oven and is placed in air having a constant temperature of 70°F. If the turkey initially has an internal temperature of 185°F and $k = 12.5$ (for time measured in minutes), find the time required for the turkey to cool to the following temperatures.

a. $T = 160°F$ **b.** $T = 100°F$ **c.** $T = 80°F$

What happens to the time as T gets closer to 70°F? According to this model, will the turkey ever cool to 70°F? Use a graphing calculator to sketch the graph of t as a function of T, and then identify the domain.

SOLUTION   We first set $C = 70$, $T_0 = 185$, and $k = 12.5$ to obtain

$$t = 12.5 \ln\left(\frac{185 - 70}{T - 70}\right) = 12.5 \ln\left(\frac{115}{T - 70}\right)$$

Now we substitute the given values for T:

a. $T = 160$: $t = 12.5 \ln\left(\dfrac{115}{160 - 70}\right) \approx 3.1$ minutes

b. $T = 100$: $t = 12.5 \ln\left(\dfrac{115}{100 - 70}\right) \approx 16.8$ minutes

c. $T = 80$: $t = 12.5 \ln\left(\dfrac{115}{80 - 70}\right) \approx 30.5$ minutes

As T gets closer to 70, t increases without bound. The value $T = 70$ would yield an undefined expression inside the logarithm. Thus, according to the model, the turkey will never reach 70°F. In practice, of course, we suspect that it will. We conclude that the model fails to give realistic results for values of T near 70. This is illustrated by the graph of

$$t = 12.5 \ln\left(\frac{115}{T - 70}\right)$$

shown in Figure 14. Here we see that t (the y-coordinate) approaches infinity as T (the x-coordinate) approaches 70 from the right. The domain of the function is $(70, \infty)$.

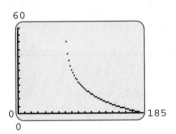

FIGURE 14

Logarithmic scales are often used when measuring quantities that vary over a large range of possible values. Examples include the decibel scale for measuring the loudness of sound and the Richter scale for measuring the intensity of earthquakes. In the case of the decibel scale, we establish a baseline by assigning an intensity I_0 to the faintest sound perceptible to most humans. A sound with intensity I is then assigned a decibel rating of

$$D = 10 \cdot \log\left(\frac{I}{I_0}\right)$$

EXAMPLE 9 ■ Measuring the Loudness of Sound

A blue whale flirts with onlookers.

The sound of a blue whale (which can be heard up to 530 miles away) can reach an intensity of 6.3×10^{18} times that of I_0. Find the number of decibels of this sound.

SOLUTION   We set $I = (6.3 \times 10^{18})I_0$ and compute D.

$$
\begin{aligned}
D &= 10 \cdot \log\left(\frac{I}{I_0}\right) \\
&= 10 \cdot \log\left(\frac{6.3 \times 10^{18} I_0}{I_0}\right) \\
&= 10 \cdot \log(6.3 \times 10^{18}) \\
&\approx 10 \cdot 18.8 \\
&= 188
\end{aligned}
$$

Thus, the sound of a blue whale can reach 188 decibels.

UNDERSTANDING AND MASTERY CHECKLISTS

CONCEPTS TO UNDERSTAND

☐ Logarithm function with base a

☐ Inverse properties of logarithms and exponents

☐ Natural logarithms

☐ Common logarithms

SKILLS TO MASTER

☐ Sketch the graph of a logarithmic function.

☐ Evaluate a logarithm by applying the definition.

☐ Estimate a logarithm graphically.

☐ Compute a natural logarithm using a calculator.

☐ Compute a common logarithm using a calculator.

☐ Find the inverse of a function involving logarithms or exponents.

EXERCISES 5.2

EXERCISES 1–8 *Evaluate the expression using a calculator. Give answers accurate to 5 decimal places.*

1. $\ln 15.2$
2. $\log 110$
3. $\log(\frac{2}{3})$
4. $\ln\sqrt{21}$
5. $\log(\sqrt{3})$
6. $\ln(\frac{17}{5})$
7. $(\ln 0.41)^3$
8. $\log(3.52 \times 10^{47})$

EXERCISES 9–24 *Evaluate the expression without the use of a calculator.*

9. $\log_2 16$
10. $\log_3 \frac{1}{9}$
11. $\log_3 0$
12. $\log_{23} 1$
13. $\ln e^2$
14. $\log 10{,}000$
15. $\log_6 \frac{1}{216}$
16. $\log_7 343$
17. $\log_{1/5} 25$
18. $\log_{3/4}(\frac{4}{3})^{100}$
19. $\log_{0.1} 1000$
20. $\log_{\sqrt{2}} 16$
21. $\log(10^{100})$
22. $\ln \sqrt{e}$
23. $\ln(e^4 \cdot e^3)$
24. $\log(10^4 \cdot 10^{-3})$

EXERCISES 25–30 *Use a graphing calculator to estimate the value of the expression to the nearest hundredth.*

25. $\log_5 37$
26. $\log_2 12$
27. $\log_2 50$
28. $\log_{100} 8$
29. $\log_{1/2} 108$
30. $\log_{1/3} 11$

EXERCISES 31–38 *Write an exponential equation equivalent to the given logarithmic equation.*

31. $\log_2 \frac{1}{4} = -2$
32. $\log_{1/3} 9 = -2$
33. $\log 1000 = 3$
34. $\ln 7 = x$

35. $\ln(x + 1) = 2$

36. $\log x = 4$

37. $\log_x 10 = 3$

38. $\ln(3x + 1) = x$

EXERCISES 39–46 *Write a logarithmic equation equivalent to the given exponential equation.*

39. $3^2 = 9$

40. $10^{-1} = \frac{1}{10}$

41. $\left(\frac{1}{2}\right)^{-3} = 8$

42. $\left(\frac{1}{4}\right)^2 = \frac{1}{16}$

43. $e^x = 5$

44. $e^{3x} = \frac{1}{2}$

45. $e^{-0.013t} = \frac{1}{2}$

46. $1.06^t = 2$

EXERCISES 47–52 *Use the given graph of $f(x) = \ln x$ to graph $g(x)$, and then find the domain of g. Verify your result with a graphing calculator.*

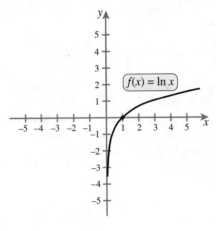

$f(x) = \ln x$

47. $g(x) = -\ln x$

48. $g(x) = \ln(-x)$

49. $g(x) = \ln(x - 3)$

50. $g(x) = 2 + \ln x$

51. $g(x) = \ln(-x) - 1$

52. $g(x) = -\ln(x + 2)$

EXERCISES 53–58 *Compute the inverse of each function, and graph both the function and its inverse on the same set of coordinate axes.*

53. $f(x) = \log_3 x$

54. $f(x) = 2 \ln x$

55. $f(x) = e^x - 5$

56. $f(x) = \log(x - 4)$

57. $f(x) = \log(1000x)$

58. $f(x) = 5^{x-3}$

EXERCISES 59–62 *Use a graphing calculator to help you sketch the graph of the given function, and find its domain.*

59. $f(x) = x \ln x$

60. $f(x) = (x^2 - 1)\ln x$

61. $f(x) = \dfrac{x}{\ln(x + 1)}$

62. $f(x) = \dfrac{\ln x}{x - 1}$

■ **APPLICATIONS**

EXERCISES 63–66 *The following exercises deal with Newton's Law of Cooling, which can be expressed using the formula*

$$t = k \ln\left(\frac{T_0 - C}{T - C}\right)$$

Here t is the time it will take for an object to cool to a temperature T, T_0 is the initial temperature, C is the temperature of the surrounding air, and k is a constant.

63. *Coffee Temperature* Find the time required for a hot cup of coffee with an initial temperature of 190°F to cool to a temperature of 80°F if the temperature of the surrounding air is 70°F and $k = 20$ (for time measured in minutes).

64. *Ice Tea Temperature* A freshly brewed pitcher of tea, currently at a temperature of 175°F, is placed in a refrigerator that has a constant temperature of 40°F. Find the time required for the tea to reach a temperature of 45°F given that $k = 50$ (for time measured in minutes).

65. *Body Temperature* Inspector Magill is called to the scene of a murder. The victim is lying in a room that has a constant temperature of 75°F. The victim's body temperature is currently 81.2°F, down from the normal body temperature of 98.6°F. After considering the height and weight of the victim, Magill arrives

at a value of $k = 10$ (for time measured in hours). How long has the victim been dead?

66. *Body Temperature Revisited* After further consideration (see Exercise 65), Inspector Magill decides that he is not sufficiently confident in his k value to fix the time of death. To determine k, he notes that the victim's body temperature at 3:00 P.M. Wednesday is 81.2°F and that by 4:00 A.M. Thursday it has dropped to 77.0°F. Assuming that the body was kept in a 75°F room, what value does Magill find for k, and what was the time of death?

EXERCISES 67–70 *The following exercises deal with the loudness of sound, which is measured using* **decibels.** *The faintest sound perceptible to most individuals is assigned an intensity of I_0, and a sound of intensity I has a decibel rating of*

$$D = 10 \cdot \log \frac{I}{I_0}$$

Find the decibel rating for the given sound.

67. A whisper with $I = 110 I_0$

68. A quiet conversation with $I = 9000 I_0$

69. A circular saw with intensity 10 billion times that of I_0

70. An auto horn with intensity 100 billion times that of I_0

EXERCISES 71–74   *The following exercises deal with the acidity of an aqueous solution, which is dependent on the solution's hydrogen ion concentration. Since these concentrations may be very small, it is convenient to measure acidity using the formula*

$$pH = -\log[H^+]$$

where $[H^+]$ is the hydrogen ion concentration in moles per liter. In general, the smaller the pH, the more acidic the solution. Find the pH for the given solution.

71. Orange juice with $[H^+] = 2.8 \times 10^{-4}$

72. Milk with $[H^+] = 3.97 \times 10^{-7}$

73. Beer with $[H^+] = 3.16 \times 10^{-5}$

74. Lemon juice with $[H^+] = 6.3 \times 10^{-3}$

75. *Mile-thick Paper*   How many times must a 0.003-inch-thick piece of paper be folded to obtain a thickness of at least 1 mile?

76. *Doubling Your Money*   Suppose that you have invested $1 into First Fantasy Bank's Double-Your-Money Account, in which your balance doubles every year. Use the graph of $y = 2^x$ to estimate the number of years required to obtain a balance of $1,000,000,000 (one billion dollars). Then express your answer in terms of logarithms.

77. *Typing Speed*   A learning model suggests that the time in weeks that it will take to learn to type w words per minute (wpm) is given by

$$t = 5 \ln\left(\frac{100}{100 - w}\right)$$

a. How many weeks will it take to learn to type 60 wpm? 80 wpm?

b. What happens to the time as w approaches 100 wpm? According to this model, can 100 wpm be reached? Explain.

c. Use a graphing calculator to help you sketch the graph of t as a function of w. What is the domain of this function?

78. *Sales Growth*   A model for the sales of a new product suggests that the time in months needed for the sales to reach S units is given by

$$t = 4 \ln\left(\frac{50,000}{50,000 - S}\right)$$

a. How many months will it take to reach sales of 15,000 units? 40,000 units?

b. What happens to t as S approaches 50,000? According to this model, can 50,000 units be sold? Explain.

c. Use a graphing calculator to help you sketch the graph of t as a function of S. What is the domain of this function?

CONCEPTS AND CRITICAL THINKING

EXERCISES 79–84   *Answer true or false.*

79. The natural logarithm function and the exponential function with base e are inverses of one another.

80. For any $a > 0$, $a \neq 1$, $\log_a 0 = 1$.

81. The logarithm of a negative number is negative.

82. If $a > b$, then $\log a > \log b$.

83. $y = \log_a x$ is equivalent to $a^y = x$.

84. The domain of the natural logarithm function consists of all real numbers.

EXERCISES 85–89   *Give an example of each.*

85. A number having a negative common logarithm

86. A number whose natural and common logarithms are the same

87. A real-world context in which common logarithms are employed

88. A characteristic shared by all logarithmic functions, but possessed by no exponential function

89. A function $f(x)$ satisfying $\ln x < f(x) < e^x$ for all $x > 0$

90. Graph the function

$$f(x) = \frac{\ln x}{x}$$

What value does $f(x)$ approach as x becomes larger and larger?

91. Graph the function

$$f(x) = \log\left(\frac{x + 1}{2x - 8}\right)$$

Find the domain of f and any vertical or horizontal asymptotes.

QUESTIONS FOR DISCUSSION OR ESSAY

92. John Napier of Scotland invented logarithms in the 16th century as a labor-saving device for doing computations. But *who* in 16th-century Europe would have had need for them? What computationally intensive activities might have occupied scientists and mathematicians of this era?

93. Although Napier is generally credited with the invention of logarithms, the Swiss watchmaker Burgi discovered logarithms independently at about the same time. The mathematical literature is filled with such instances. For example, Newton and Leibniz invented calculus independently at about the same time. Are such

occurrences simply coincidences? What explains the tendency for mathematicians, working independently, to make major discoveries simultaneously?

94. In Exercises 71–74, we considered the acidity of a solution as measured by the pH level. If normal rain is considered to have

High levels of sulfur in industrial emissions are a primary cause of acid rain.

a pH of approximately 5.6, and a recent rainfall is determined to have a hydrogen ion concentration of 6.3×10^{-4}, is the recent rain more or less acidic than normal rain? Which has the larger hydrogen ion concentration? If the primary causes of acid rain are sulfur dioxide and nitrogen oxides, do these compounds increase or decrease the hydrogen ion concentration of rain? In general, describe the effect on the pH level if the hydrogen ion concentration increases or decreases.

95. In Exercises 65 and 66, Inspector Magill applied Newton's Law of Cooling to estimate the time of death. What characteristics of a body (in addition to height and weight) would influence the cooling rate? Does Newton's Law of Cooling take into account all means by which heat is transferred? If so, of what relevance is the windchill factor, and how do you explain microwave ovens?

PROJECTS FOR ENRICHMENT

96. *The Leaning Tower of Pizza*   It can be shown that (in principle) pizza boxes can be stacked on a table with each box overhanging the box below it so that the top box extends as far as you please past the edge of the table—without toppling! Unfortunately, producing a large overhang requires a tremendous number of boxes. It can be shown that the greatest possible overhang with n ($n > 1$) 12-inch boxes is given by

$$6 \cdot \left(1 + \frac{1}{2} + \frac{1}{3} + \frac{1}{4} + \frac{1}{5} + \cdots + \frac{1}{n-1} \right) \text{ inches}$$

a. Compute the greatest overhang possible with 6 pizza boxes.

b. It can be shown that

$$1 + \frac{1}{2} + \frac{1}{3} + \frac{1}{4} + \frac{1}{5} + \cdots + \frac{1}{n-1} - \ln(n-1) \approx \epsilon$$

where ϵ is a constant known as Euler's number. Estimate Euler's number by computing

$$1 + \frac{1}{2} + \frac{1}{3} + \frac{1}{4} + \frac{1}{5} + \cdots + \frac{1}{n-1} - \ln(n-1)$$

for $n = 10$.

c. Find an approximation for the maximum overhang possible with n boxes that involves Euler's number and natural logarithms. (*Hint:* If

$$S = 1 + \frac{1}{2} + \frac{1}{3} + \frac{1}{4} + \frac{1}{5} + \cdots + \frac{1}{n-1}$$

and $S - \ln(n-1) \approx \epsilon$, then what can be said of S?)

d. Using the approximation for the maximum possible overhang, derived in part c, compute the number of pizza boxes required to obtain an overhang of 10 feet.

e. Suppose that a pizza box is 2 inches thick. How high would the stack of pizza boxes in part d be? (Express your answer in light-years.)

f. Explain how a very narrow river could be crossed using only a parachute and pizza boxes.

g. In this project, we've ignored many aspects of reality in order to obtain our result. Describe some of the factors that would complicate the unusual river-crossing technique described in part f.

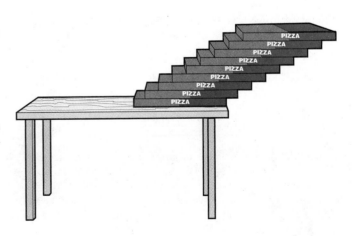

SECTION 5.3 LOGARITHMIC IDENTITIES AND EQUATIONS

- ☐ How powerful would the "bombquake" be that would result from simultaneously detonating all of the world's nuclear warheads?
- ☐ How can multiplication and division be performed simply by sliding pieces of paper?
- ☐ How many jamboxes would it take to cause a fatal musical overdose?
- ☐ If the entire population of China simultaneously jumped a foot off the ground, how strong would the resulting earthquake be?

LOGARITHMIC IDENTITIES

As we have already seen, exponential and logarithmic functions are inextricably intertwined. Indeed, logarithms depend on exponentials for their very definition. In the previous section, we exploited this connection to recast exponential equations as logarithmic equations and vice versa, so that, for example, the exponential equation $y = a^x$ could be rewritten as the equivalent logarithmic equation $x = \log_a y$. In fact, any statement written in the language of exponentials can be translated into an equivalent statement involving logarithms. In this section, we use three familiar exponential identities to derive three new logarithmic identities. Then, armed with our graphing calculator and our newly enlarged collection of logarithmic identities, we tackle logarithmic equations.

We begin our development of logarithmic identities by considering the exponential identity

$$a^b a^c = a^{b+c}$$

If we let $u = a^b$ and $v = a^c$, so that $b = \log_a u$ and $c = \log_a v$, then we have

$$uv = a^{b+c}$$

The corresponding logarithmic equation is

$$\log_a(uv) = b + c$$

Thus, we have the logarithmic identity

$$\log_a(uv) = \log_a u + \log_a v$$

Similarly, the exponential identities

$$\frac{a^b}{a^c} = a^{b-c} \qquad \text{and} \qquad (a^b)^n = a^{bn}$$

yield the logarithmic identities

$$\log_a\left(\frac{u}{v}\right) = \log_a u - \log_a v \qquad \text{and} \qquad \log_a(u^n) = n \log u$$

respectively.

The following is a summary of logarithmic identities. For the sake of completeness, we have included the inverse identities developed in the previous section, and have restated each of the rules for common and natural logarithms as well.

Logarithmic Identities ☐

Let a, u, and v be positive real numbers with $a \neq 1$, and let n be any real number.

Arbitrary logarithms	**Common logarithms**	**Natural logarithms**
1. $\log_a(uv) = \log_a u + \log_a v$	$\log(uv) = \log u + \log v$	$\ln(uv) = \ln u + \ln v$
2. $\log_a\left(\frac{u}{v}\right) = \log_a u - \log_a v$	$\log\left(\frac{u}{v}\right) = \log u - \log v$	$\ln\left(\frac{u}{v}\right) = \ln u - \ln v$
3. $\log_a(u^n) = n \log_a u$	$\log(u^n) = n \log u$	$\ln(u^n) = n \ln u$
4. $\log_a a^u = u$	$\log a^u = u$	$\ln a^u = u$
5. $a^{\log_a u} = u$	$10^{\log u} = u$	$e^{\ln u} = u$

The identities just listed are relatively simple, and, indeed, most students commit them to memory relatively quickly. Strangely, the difficulty lies not in the correct application of these five identities, but with resisting the overwhelming temptation to manufacture similar-looking—but false—formulas. For example, after learning the first logarithmic identity, beginning students are often lured into believing that $\log_a(uv) = \log_a(u + v)$ or that $\log_a u \log_a v = \log_a u + \log_a v$, neither of which is true!

WARNING!   Be aware of the temptation to use seemingly plausible (but bogus) formulas that, on the surface, resemble the logarithmic identities.

The following two examples illustrate the use of logarithmic identities for eliminating products, quotients, and powers within logarithmic expressions.

EXAMPLE 1 ▓ **Expanding with Logarithmic Identities**

Expand the following expression as a sum, difference, or multiple of logarithms.

$$\log\left(\frac{x\sqrt{y}}{z}\right)$$

SOLUTION

$$\log\left(\frac{x\sqrt{y}}{z}\right) = \log(x\sqrt{y}) - \log z \qquad \text{Identity 2}$$

$$= \log x + \log(y^{1/2}) - \log z \qquad \text{Identity 1}$$

$$= \log x + \frac{1}{2}\log y - \log z \qquad \text{Identity 3}$$

EXAMPLE 2 ▓ **Expanding with Logarithmic Identities**

Expand the following expression as a sum, difference, or multiple of logarithms.

$$\ln\left(\frac{3y^2}{x^5}\right)^3$$

SOLUTION

$$\ln\left(\frac{3y^2}{x^5}\right)^3 = 3\ln\left(\frac{3y^2}{x^5}\right) \qquad \text{Identity 3}$$

$$= 3[\ln(3y^2) - \ln(x^5)] \qquad \text{Identity 2}$$

$$= 3[\ln 3 + \ln(y^2) - \ln(x^5)] \qquad \text{Identity 1}$$

$$= 3(\ln 3 + 2\ln y - 5\ln x) \qquad \text{Identity 3}$$

$$= 3\ln 3 + 6\ln y - 15\ln x$$

Logarithmic identities can be useful for simplifying complex expressions involving logarithms. As we will see shortly, the simplification of complicated expressions involving logarithms is often the first step in solving logarithmic equations.

EXAMPLE 3 ■ **Simplifying with Logarithmic Identities**

Express $3\log_a(x^2) + 4\log_a x$ as a single logarithm.

SOLUTION

$$3\log_a(x^2) + 4\log_a x = \log_a(x^2)^3 + \log_a(x^4) \qquad \text{Identity 3}$$

$$= \log_a(x^6 \cdot x^4) \qquad \text{Identity 1}$$

$$= \log_a(x^{10})$$

EXAMPLE 4 ■ **Simplifying with Logarithmic Identities**

Express $2\ln x - \ln y + 6\ln z$ as a single logarithm.

SOLUTION

$$2\ln x - \ln y + 6\ln z = \ln(x^2) - \ln y + \ln z^6 \qquad \text{Identity 3}$$

$$= \ln\left(\frac{x^2}{y}\right) + \ln(z^6) \qquad \text{Identity 2}$$

$$= \ln\left(\frac{x^2 z^6}{y}\right) \qquad \text{Identity 1}$$

LOGARITHMIC EQUATIONS

Equations involving one or more logarithmic expressions are called **logarithmic equations.** Many logarithmic equations can be solved by converting to exponential equations. For example, the equation

$$\ln x = 2$$

is equivalent to

$$x = e^2$$

This solution could also be found by exponentiating both sides of the original equation, as follows:

$$\ln x = 2$$

$$e^{\ln x} = e^2 \qquad \text{Exponentiating both sides}$$

$$x = e^2 \qquad \text{Using the property } a^{\log_a x} = x$$

In general, any equation of the form $\log_a \square = c$ can be solved for $\square$, either by converting to exponential form or by exponentiating both sides.

EXAMPLE 5   ■   Solving a Logarithmic Equation

Solve $\log_2(x + 1) = 5$.

SOLUTION   We convert from the base 2 logarithmic form to the base 2 exponential form.

$$\log_2(x + 1) = 5$$
$$x + 1 = 2^5$$
$$x = 32 - 1 = 31$$

Another useful technique for solving logarithmic equations is to apply the 1–1 property of logarithms. This property was first observed in Section 5.2; we restate it here for convenience.

The 1–1 Property of Logarithms   □ | For all positive real numbers u and v, if $\log_a u = \log_a v$, then $u = v$.

EXAMPLE 6   ■   Using the 1–1 Property of Logarithms to Solve an Equation

Solve $\log(x^2 + 2x - 5) = \log(x + 1)$.

SOLUTION

$$\log(x^2 + 2x - 5) = \log(x + 1)$$
$$x^2 + 2x - 5 = x + 1 \qquad \text{Using the 1–1 property of logarithms}$$
$$x^2 + x - 6 = 0$$
$$(x - 2)(x + 3) = 0 \qquad \text{Factoring}$$
$$x = 2, x = -3$$

CHECK

$x = 2$		$x = -3$	
$\log(x^2 + 2x - 5)$	$\log(x + 1)$	$\log(x^2 + 2x - 5)$	$\log(x + 1)$
$\log[(2)^2 + 2(2) - 5]$	$\log(2 + 1)$	$\log[(-3)^2 + 2(-3) - 5]$	$\log[(-3) + 1]$
$\log(4 + 4 - 5)$	$\log 3$	$\log(9 - 6 - 5)$	$\log(-2)$
$\log 3$	✓		Not defined

Thus, the only solution is $x = 2$.

WARNING!   Since logarithms are defined only for positive real numbers, it is important to check all solutions to logarithmic equations by substituting them into the original equation.

Some logarithmic equations must be algebraically rearranged using logarithmic identities before they can be solved, as illustrated in the following example.

EXAMPLE 7 ▨ Solving an Equation Involving Natural Logarithms

Solve $\ln 2 + 3 \ln(x - 1) = 2 \ln 4$.

SOLUTION We begin by expressing each side in the form $\ln \square$.

$$\ln 2 + 3 \ln(x - 1) = 2 \ln 4$$

$$\ln 2 + \ln(x - 1)^3 = \ln(4^2) \qquad \text{Using } n \ln u = \ln(u^n)$$

$$\ln[2(x - 1)^3] = \ln 16 \qquad \text{Using } \ln(uv) = \ln u + \ln v$$

$$2(x - 1)^3 = 16 \qquad \text{Using the 1–1 property of logarithms}$$

$$(x - 1)^3 = 8$$

$$x - 1 = 2 \qquad \text{Taking the cube root of both sides}$$

$$x = 3$$

CHECK $x = 3$

$\ln 2 + 3 \ln(x - 1)$	$2 \ln 4$
$\ln 2 + 3 \ln(3 - 1)$	$\ln(4^2)$
$\ln 2 + 3 \ln 2$	$\ln 16$
$4 \ln 2$	
$\ln(2^4)$	
$\ln 16$	✓

EXAMPLE 8 ▨ Comparing Sound Intensity

The decibel rating D of a sound with intensity I is given by

$$D = 10 \cdot \log \frac{I}{I_0}$$

where I_0 is the intensity of the faintest sound perceptible to the human ear. How much more intense is the 140-decibel sound of a jet taking off than that of a 120-decibel jackhammer?

SOLUTION For comparison purposes, we set $I_0 = 1$. The relative intensity of the jet at takeoff can be found as follows:

$$140 = 10 \log I$$

$$14 = \log I$$

$$I = 10^{14} \qquad \text{Converting to exponential form}$$

For the jackhammer, we have

$$120 = 10 \log I$$

$$12 = \log I$$

$$I = 10^{12} \qquad \text{Converting to exponential form}$$

Now $10^{14} = 100 \cdot 10^{12}$, and so the jet is 100 times more intense than the jackhammer.

EXAMPLE 9 ▓ **The Nuclear Earthquake**

According to the mathematician John Paulos, the TNT equivalent of all the nuclear warheads on Earth is 25,000 megatons. Based on empirical evidence, the energy E (in ergs) of an earthquake is related to the Richter scale reading R by the equation $\log E = 11.8 + 1.5R$.

a. If all the warheads were simultaneously detonated, how powerful would the resulting "bombquake" be? (A megaton is one million tons, and a ton of TNT carries about 3.4×10^{16} ergs of energy.)

b. How many megatons of TNT would it take to create an earthquake of magnitude 7 on the Richter scale?

A loaded ICBM missile is an ominous reminder of the nuclear threat.

SOLUTION

a. We first compute the energy involved in such an explosion.

$$E = (25{,}000 \text{ megatons of TNT}) \times \frac{10^6 \text{ tons of TNT}}{1 \text{ megaton of TNT}} \times \frac{3.4 \times 10^{16} \text{ ergs}}{1 \text{ ton of TNT}}$$

$$= 8.5 \times 10^{26} \text{ ergs}$$

Using the equation $\log E = 11.8 + 1.5R$, we have

$$\log(8.5 \times 10^{26}) = 11.8 + 1.5R$$
$$26.929 = 11.8 + 1.5R$$
$$1.5R = 15.129$$
$$R = 10.086$$

Thus, the world's supply of nuclear warheads contains an energy equal to that of an earthquake of magnitude 10.086 on the Richter scale! This is greater than that of any recorded earthquake.

b. We set $R = 7$ and solve for E.

$$\log E = 11.8 + 1.5R$$
$$\log E = 11.8 + 1.5 \cdot 7$$
$$\log E = 22.3$$
$$E = 10^{22.3} \approx 2.0 \times 10^{22} \text{ ergs}$$

Converting to megatons of TNT we find

$$2.0 \times 10^{22} \text{ ergs} \times \frac{1 \text{ ton of TNT}}{3.4 \times 10^{16} \text{ ergs}} \times \frac{1 \text{ megaton of TNT}}{10^6 \text{ tons of TNT}}$$

$$\approx 0.59 \text{ megaton of TNT}$$

APPROXIMATIONS INVOLVING LOGARITHMS

In Section 5.2, we considered a technique for estimating logarithms graphically. Here we consider a more algebraic approach, one that uses the logarithmic identities discussed in this section.

EXAMPLE 10 ▓ **Approximating Logarithmic Expressions**

Estimate $\log_2 5$.

SOLUTION   We begin by setting $x = \log_2 5$ and then converting to exponential form.

$$x = \log_2 5$$

$$2^x = 5 \qquad \text{Converting to exponents}$$

$$\ln 2^x = \ln 5 \qquad \text{Taking the natural logarithm of both sides}$$

$$x \ln 2 = \ln 5 \qquad \text{Using the identity } \ln u^n = n \ln u$$

$$x = \frac{\ln 5}{\ln 2} \qquad \text{Solving for } x$$

Finally, using a calculator, we obtain $x \approx 2.3219$.

The process used in Example 10 can be used to rewrite any logarithmic expression of the form $\log_a u$ as an expression involving only natural logarithms, common logarithms, or even logarithms to an arbitrary base b. In so doing, we obtain the *change-of-base formula*. This formula, though useful, need not be memorized; it is always possible to change the base by using the procedure illustrated in Example 10.

The Change-of-Base Formula   □

> Let a and b be positive real numbers with $a \neq 1$ and $b \neq 1$. Then, for $u > 0$,
>
> $$\log_a u = \frac{\log_b u}{\log_b a} = \frac{\log u}{\log a} = \frac{\ln u}{\ln a}$$

EXAMPLE 11   ▨   **A Graph Involving the Base 2 Logarithm Function**

Estimate (to the nearest hundredth) the coordinates of any turning points of the function $f(x) = \log_2 x - x$.

SOLUTION   We assume that our graphing calculator does not compute base 2 logarithms directly. Thus, we must convert to either common or natural logarithms. We will use natural logarithms. From the change-of-base formula, we obtain

$$f(x) = \log_2 x - x$$

$$= \frac{\ln x}{\ln 2} - x$$

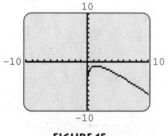

FIGURE 15

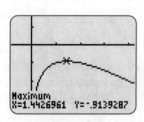

FIGURE 16

Plotting f in the standard viewing window, as in Figure 15, we see a single turning point (a local maximum). After zooming out a few times to convince ourselves that there are no others, we can find the coordinates of this turning point by either zooming in and tracing or by using the calculate-maximum feature. Figure 16 shows the view after zooming in and applying the calculate-maximum feature. The turning point has approximate coordinates $(1.44, -0.91)$.

The need for approximation also arises in situations where logarithmic equations cannot be solved algebraically. As we have seen many times in the past, approximate solutions of equations can be found by first writing the equation in the form $f(x) = 0$, then graphing the function f on a graphing calculator, and finally estimating the zeros with the trace or calculate-zero feature. However, if an equation involves nonpolynomial terms (such as $\ln x$ or $\log x$), it can be difficult to tell how many solutions the equation has. Because of this, and because of our knowledge of polynomials, it is often helpful to first put the equation in the form $g(x) = p(x)$, where $p(x)$ is a polynomial, and then plot the graphs of g and p to see how many times they intersect. Solution values can be approximated either by determining the intersection points of g and p, or by finding the zeros of f.

EXAMPLE 12   ▧   **Determining the Number of Solutions of a Logarithmic Equation**

Determine the number of solutions of the equation $\frac{1}{18}x^3 + 4x + 1 = x^2 + \ln x$.

SOLUTION We first regroup terms so that all the polynomial terms are on one side of the equation.

$$\frac{1}{18}x^3 + 4x + 1 = x^2 + \ln x$$

$$\frac{1}{18}x^3 - x^2 + 4x + 1 = \ln x$$

Next we plot the functions $p(x) = \frac{1}{18}x^3 - x^2 + 4x + 1$ and $g(x) = \ln x$, as shown in Figure 17. Because the solutions to $p(x) = g(x)$ correspond to the x-coordinates of the points of intersection of the graphs of p and g, we attempt to locate and approximate any intersection points of the two curves. One intersection point can be seen near the point $(6, 2)$. Since $p(x)$ is a third-degree polynomial with a positive leading coefficient, we know that its graph must turn around and start increasing to the right of the viewing window. Thus, a second intersection point is certain. With the slightly modified viewing window shown in Figure 18, we see that there is indeed a second intersection point. Since p can have no other turning points, there can be no other intersection points. Thus, there are exactly two solutions. We could approximate these solutions either by zooming in on the points of intersection or by using the calculate-intersect feature.

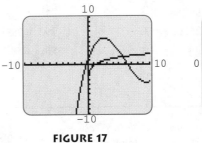

FIGURE 17

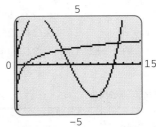

FIGURE 18

UNDERSTANDING AND MASTERY CHECKLISTS

CONCEPTS TO UNDERSTAND

- ☐ Logarithmic identities
- ☐ Logarithmic equations
- ☐ The 1–1 property of logarithms
- ☐ The change-of-base formula

SKILLS TO MASTER

- ☐ Use the logarithmic identities to rewrite expressions involving logarithms.
- ☐ Solve equations involving logarithms.
- ☐ Approximate a logarithm base a by changing to natural or common logarithms.

EXERCISES 5.3

EXERCISES 1–8 *Use logarithmic identities to expand each expression, rewriting it as a sum, difference, or multiple of logarithms.*

1. $\ln(x^2 y)$

2. $\log_2(xy\sqrt{z})$

3. $\log_5\left(\dfrac{x^3 y^4}{z^{-3}}\right)$

4. $\log_4\left(\dfrac{15\sqrt{q}}{\sqrt{r}}\right)$

5. $\ln\left(\dfrac{e}{7x^3}\right)$

6. $\ln\left(\dfrac{xy^2}{\sqrt{z}}\right)^3$

7. $\log\sqrt{\dfrac{y\sqrt{x}}{z^2}}$

8. $\log_a \sqrt{x^3 + y^3}$

EXERCISES 9–18 *Use logarithmic identities to write each expression as a single logarithm.*

9. $\log x - 2\log y$

10. $2\ln x + \frac{1}{2}\ln y$

11. $3\ln x + 4\ln y - \ln z$

12. $\frac{1}{2}\log(2x - y) + \frac{1}{2}\log(x + 2y)$

13. $\frac{1}{3}\ln(x - z) - \frac{2}{3}\ln(x + z)$

14. $4[\log_5(x^2 - y^2) - \log_5(x + y)]$

15. $2\log 4 - \log 6 + 3\log 2$

16. $\log_3 8 - 4\log_3(x^2)$

17. $2\ln(x + 3) - \ln(x^2 + 5x + 6)$

18. $\log(x + y) - \frac{1}{2}\log(x^2 + 2xy + y^2)$

EXERCISES 19–44 *Solve the logarithmic equation.*

19. $\frac{1}{3}\log_3 x = 1$

20. $2\log_4 x = 32$

21. $\log_x 4 = 2$

22. $\log_x \frac{1}{9} = 2$

23. $3\log_5 x = 2$

24. $2\log_2 x = 3$

25. $\log_4(2x + 1) = -2$

26. $\log_2(1 - 3x) = -1$

27. $\log_3(27^x) = -1$

28. $\log_5(5^x) = 2$

29. $2\log_3(x - 1) + \log_3 9 = 2$

30. $\log(2x + 3) = 2\log x$

31. $\ln(x + 7) - \ln(x + 2) = \ln(x + 1)$

32. $\log_2(3 - x) + \log_2(2 - x) = \log_2(1 - x)$

33. $\log(2x^2 + 11x + 5) - \log(2x + 1) = 2$

34. $\log_a(x + 1) + \log_a(x - 2) = \log_a(x + 6)$

35. $\log_2(27x^3 - 1) - \log_2(9x^2 + 3x + 1) = 1$
(*Hint:* Factor $27x^3 - 1$.)

36. $\log_{1/4} x + \log_{1/4}(3x - 5) = -1$

37. $\log_a(x + 7) = \log_a(x + 1)$

38. $\log_5(x^7) = 7\log_5 x$

39. $\log\sqrt{x^2 + 1999} = 3$

40. $\log\sqrt[3]{x^2 - 15x} = \frac{2}{3}$

41. $\ln(x^2) = 2\ln x$

42. $(\ln x)^2 = 2\ln x$

43. $\ln(\ln x) = 1$

44. $\log[\log(x + 1)] = \log 2$

EXERCISES 45–50 *Use the technique of Example 10 to estimate the logarithm to 5 decimal places.*

45. $\log_3 16$

46. $\log_4 25$

47. $\log_{12} 0.341$

48. $\log_{23} 1250$

49. $\log_\pi 10$

50. $\log_\pi e$

EXERCISES 51–56 *Approximate the solutions of the given equation to the nearest hundredth.*

51. $x - 2 = \ln x$

52. $\ln(x + 1) = x^2 - 1$

53. $\log_5 x = \frac{3}{2}x - 2$

54. $3 = x + \log_3 x$

55. $0.288x^2 = \ln(1 - x) + 0.002x^4 + 1$

56. $1.5x^2 = 0.0625x^3 + 5x + \ln x$

▪ APPLICATIONS

EXERCISES 57–60 *The following exercises deal with the decibel rating D of a sound with intensity I as given by $D = 10 \cdot \log(I/I_0)$, where I_0 is the faintest sound perceptible to the human ear.*

57. *Hearing Loss* The threshold of pain for most individuals is 120 decibels, whereas permanent hearing damage can be caused by even brief exposure to a sound rated at 160 decibels. How many times more intense is a sound of 160 decibels than one of 120 decibels?

58. *Rock Concert* According to *The Guinness Book of World Records*, the sound level at the mixing tower during Iron Maiden's set at the "Monsters of Rock" Festival reached a record 124 decibels. The loudest shout recorded by the *Guinness Book* was one of 113 decibels. How many times more intense was the sound produced by Iron Maiden than that of the champion shouter?

Iron Maiden in concert.

59. *Mass Choir* Suppose that 1000 people sing a note at precisely the same volume and pitch at exactly the same moment. Assuming that the collective sound is 1000 times more intense than that of an individual singer, how much louder is the sound of the group than that of an individual?

60. *Fatal Jambox*   It is said that a 200-decibel sound can be deadly. How many times more intense is a 200-decibel sound than a 100-decibel jambox? Does it follow that this number of jamboxes would result in a fatal musical overdose?

EXERCISES 61–64 *The following exercises deal with the equation* $\log E = 11.8 + 1.5R$, *which relates the energy E (in ergs) of an earthquake to the Richter scale reading R.*

61. *TNT Earthquake*   If 1 ton of TNT carries 3.4×10^{16} ergs of energy, how many tons of TNT would it take to equal the energy output of an earthquake registering 5 on the Richter scale?

62. *Automobile Earthquake*   The total number of miles driven by cars and light trucks in the United States in 1995 was approximately 2.2×10^{12}. If we assume that each vehicle was being driven at 40 miles per hour and had a 200-horsepower engine, then each expended about 1.34×10^{14} ergs per mile. What magnitude earthquake has the same energy output as the total energy used by vehicles in 1995?

63. *Comparing Earthquakes*   The earthquake that shook Armenia on December 7, 1988, registered 6.8 on the Richter scale. The

San Francisco earthquake of April 18, 1906, registered 8.3 on the Richter scale. How many times more powerful, as measured by energy output, was the San Francisco earthquake than that of Armenia?

64. *Jumping Earthquake*   It has been hypothesized that if each Chinese man, woman, and child were to jump at precisely the same instant, an earthquake of magnitude 4.3 on the Richter scale would result. Assuming that the energy of the ground movement is proportional to the number of people jumping and that there are 1.1 billion people in China and 6 billion people in the world, of what magnitude would the earthquake be that would result if all the people on Earth were to jump simultaneously?

CONCEPTS AND CRITICAL THINKING

EXERCISES 65–72 *Answer true only if the given equation is an identity for all positive x and y; otherwise answer false.*

65. $\log xy = (\log x)(\log y)$

66. $\log(x + y) = \log x + \log y$

67. $\log xy = \log x + \log y$

68. $y \ln x = \ln yx$

69. $y \ln x = \ln x^y$

70. $\log_2 \dfrac{x}{y} = \dfrac{\log_2 x}{\log_2 y}$

71. $\log_2 \dfrac{x}{y} = \log_2 x - \log_2 y$

72. $\log_2 \dfrac{x}{y} = \log_2 (x - y)$

EXERCISES 73–76 *Give an example of each.*

73. A reason why solution candidates to logarithmic equations must be checked

74. A number greater than 20 whose logarithm can be computed if the logarithm of 3 is known

75. Two numbers x and a such that $\log_a x = 2$

76. A logarithmic equation having no solution

77. Graph the following functions on the same coordinate system.

$$f(x) = \log x$$
$$g(x) = \log(x \cdot 10)$$
$$h(x) = \log(x \cdot 0.1)$$

What relationship do you see among the graphs of these functions? What property of logarithms explains this relationship?

78. Graph $y = \ln(e^3 x)$. Now find a number k so that the graph of $y = k + \ln x$ is the same. What properties of logarithms explains this?

QUESTIONS FOR DISCUSSION OR ESSAY

79. As was mentioned in the text, many students struggle with logarithms because there are so many seemingly plausible identities that are simply not true. For example, it is not true that for all x and y, $\log_a x \cdot \log_a y = \log_a(x + y)$. What makes this nonrule so tempting to accept as true? What other tempting nonrules in-

volving logarithms can you develop? How would one go about showing that a nonrule (such as the one given) is false?

80. In Exercise 64, we made the assumption that the ground movement arising from a number of people jumping at the same moment would be proportional to the number of people jumping. Is this a reasonable assumption? What other factors besides the number of people jumping would affect the resulting seismic disturbance?

81. As we have seen, the decibel rating of a sound with intensity I is computed using the formula $D = 10 \log(I/I_0)$, where I_0 is the intensity of the faintest sound that is perceptible to the human ear. Sound intensity itself is measured in watts per square centimeter. For example, I_0 is by common agreement assigned the value 10^{-16} watts per square centimeter, and the intensity of a whisper is approximately 10^{-13} watts per square centimeter. Why do you think the decibel is preferred as a measure of the loudness of a sound, rather than the intensity? More generally, under what circumstances do you think logarithms would be useful in formulas that define measurement scales?

82. In this section, we saw how a calculator could be used to compute the logarithm to any base of any number. Prior to the advent of the handheld calculator, such tasks were typically accomplished using a process that required extensive logarithm tables (see Exercise 84 for an example of such a table). This is just one example of how calculators have changed the way in which mathematics is taught. What are some other changes that have occurred? (A discussion with your parents or an older student may help with this question.) What are some changes that you think have yet to occur, but are inevitable? Give some examples of basic skills that you think should never be replaced by a calculator.

▪ PROJECTS FOR ENRICHMENT

83. *Constructing a Slide Rule*   Slide rules were widely used before inexpensive electronic calculators became available. In this project, we construct a simple slide rule and use it to perform multiplications.

    Begin by copying Figure 19 and separating the copied version into two pieces by cutting along the dashed line. Now secure the lower piece so that it will not move.

    To multiply two numbers between 1 and 10 whose product is less than 10, say 2.2 and 3.1, simply slide the top piece over the bottom so that the 1 at the left of the top piece is directly above 2.2 on the bottom piece. Next find 3.1 on the top piece and read off the number directly below it (≈ 6.8) on the bottom piece. If the numbers are such that their product exceeds 10, say 4.2 and 5.3, place the 1 at the *right-hand end* of the top piece directly above 4.2; then find 5.3 on the top piece and read off the number directly below it (≈ 2.2). The answer is then obtained by multiplying 2.2 by 10 to give 22.

    In general, when using slide rules, we must keep track of powers of ten. For this reason, it is helpful to first convert the numbers that we are multiplying to scientific notation. For example, to compute 120×4100, we first rewrite the product as $(1.2 \times 10^2)(4.1 \times 10^3)$, then multiply 1.2 and 4.1 with the slide rule to obtain 4.9, next multiply 10^2 and 10^3 to obtain 10^5, and finally write our answer as 4.9×10^5.

    a. Compute the following products using your makeshift slide rule.

    i. 2.5×6.8 ii. 0.0024×39
    iii. 350×180

    A slide rule like that depicted in Figure 19 could be formed in the following way. First, take a piece of ordinary graph paper, and mark a number line from 0 to 1. Next, compute the common logarithms of the integers from 1 to 10. Then, for each integer, make a tick mark at the position corresponding to its logarithm, and label the tick mark with the corresponding integer. For example, the number 2 was written at the position 0.301, since $\log 2 \approx 0.301$. For increased accuracy, smaller tick marks could be made at locations corresponding to the logarithms of the multiples of 0.1—that is, the numbers $1.1, 1.2, \ldots, 9.9$. Thus, the location of a number on the slide rule is determined by its logarithm.

    b. Find the ratio of the distance between 1 and 2 to the distance between 2 and 4 on the slide rule.

    c. What number on the slide rule is the same distance from 9 as 1 is from 2?

    d. Explain the process by which numbers are multiplied with a slide rule.

84. *Logarithm Tables*   John Napier developed logarithms as a labor-saving device for doing arithmetic calculations. Today, although logarithmic functions are extremely important in modeling real-world phenomena, logarithms are rarely used as a computational aid for performing arithmetic because inexpensive

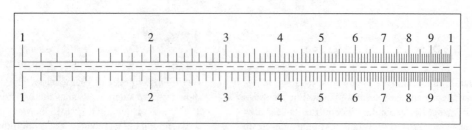

FIGURE 19

TABLE 4

N	0	1	2	3	4	5	6	7	8	9
1	.0000	.0414	.0729	.1139	.1461	.1761	.2041	.2304	.2553	.2788
2	.3010	.3222	.3424	.3617	.3802	.3979	.4150	.4314	.4472	.4624
3	.4771	.4914	.5051	.5185	.5315	.5441	.5563	.5682	.5798	.5911
4	.6021	.6128	.6232	.6335	.6435	.6532	.6628	.6721	.6812	.6902
5	.6990	.7076	.7160	.7243	.7324	.7404	.7482	.7559	.7634	.7709
6	.7782	.7853	.7924	.7993	.8062	.8129	.8195	.8261	.8325	.8388
7	.8451	.8513	.8573	.8633	.8692	.8751	.8808	.8865	.8921	.8976
8	.9031	.9085	.9138	.9191	.9243	.9294	.9345	.9395	.9445	.9494
9	.9542	.9590	.9683	.9685	.9731	.9777	.9823	.9868	.9912	.9956

electronic calculators are readily available. In this project, we investigate the methods by which our predecessors performed lengthy arithmetic computations.

I. Tables

Until fairly recently, logarithms were calculated by hand (using complicated formulas from calculus), and then tabulated in lengthy logarithm tables. Table 4 is an abbreviated version of a common (base 10) logarithm table.

To find the logarithm of a number between 1 and 10, say 3.9, we read off the entry in row 3 and column 9—namely, 0.5911. Thus, $\log 3.9 \approx 0.5911$. Logarithms of numbers larger than 9.9 or smaller than 1.0 can be found by converting to scientific notation and using the properties of logarithms. For example, we compute $\log 230$ as follows:

$$\log 230 = \log(2.3 \times 10^2)$$
$$= \log 2.3 + \log(10^2)$$
$$= 0.3617 + 2$$
$$= 2.3617$$

a. Approximate each the following logarithms using Table 4.

   i. $\log 6700$ **ii.** $\log 0.0023$

  iii. $\log 15{,}000{,}000$

To perform calculations using logarithms, we must be able to compute **antilogarithms.** The antilogarithm of x is simply the number 10^x. If the number x is between 0 and 1, then the antilogarithm of x (10^x) can be found by locating x in Table 4, and then reading off the number whose logarithm is x. For example, to compute $10^{0.1139}$, we locate 0.1139 in the table, finding that 1.3 is the number whose logarithm is 0.1139. Thus, $10^{0.1139} \approx 1.3$. If the number x does not appear in the table, then, as an estimate, simply choose the nearest number in the table. For example, to compute $10^{0.52}$, we locate the entry in the table nearest to 0.52, namely 0.5185, and find that $10^{0.52} \approx 10^{0.5185} \approx 3.3$. Antilogarithms of numbers larger than 1 or smaller than 0 can be found using the properties of exponents. For example, we compute $10^{4.75}$ and $10^{-2.8}$ as follows:

$$10^{4.75} = 10^4 10^{0.75} \qquad\qquad 10^{-2.8} = 10^{-3} \cdot 10^{0.2}$$
$$\approx 10^4 \cdot 10^{0.7482} \qquad\quad \approx 10^{-3} \cdot 10^{0.2041}$$
$$\approx 10^4 \cdot 5.6 \qquad\qquad\quad \approx 10^{-3} \cdot 1.6$$
$$= 56{,}000 \qquad\qquad\qquad \approx 0.0016$$

b. Approximate each of the following antilogarithms using the logarithms in Table 4.

   i. $10^{3.89}$ **ii.** $10^{-4.05}$

II. Multiplication and Division

The product or quotient of any two numbers can be estimated using the properties of logarithms and a logarithm table. For example, the identity $x \cdot y = 10^{\log(xy)} = 10^{\log x + \log y}$ can be used to compute 348×5700 as follows:

$$348 \times 5700 = 10^{\log(348 \times 5700)}$$
$$= 10^{\log 348 + \log 5700}$$
$$\approx 10^{2.5441 + 3.7559} \qquad \text{Using the table to estimate } \log 348 \text{ and } \log 5700$$
$$= 10^{6.3}$$
$$= 10^{0.3} 10^6$$
$$\approx 2 \cdot 10^6 \qquad\qquad \text{Using the table to estimate the antilogarithm of } 0.3$$
$$= 2{,}000{,}000$$

c. Perform each of the following operations using only addition, subtraction, and Table 4.

   i. $79{,}000 \times 62.5$ **ii.** 0.0003×426.7

d. Show that for any two positive numbers x and y, the quotient x/y is equal to the antilogarithm of the difference in their logarithms. Use this fact to compute $426/0.032$.

e. Explain how one could evaluate $3200^{2.3}$ using only a logarithm table, addition, subtraction, and multiplication. (*Hint:* Start with the fact that $3200^{2.3} = 10^{\log(3200^{2.3})}$, and then use a logarithmic identity.)

SECTION 5.4   EXPONENTIAL EQUATIONS AND APPLICATIONS

☐   Could the Shroud of Turin be 2000 years old?

☐   Which would you rather have: \$1000 in an account paying 10% interest or \$2000 in an account paying 5% interest?

☐   If eliminating all sources of contamination drops the pollution level of a lake by 15% after 5 years, then how long will it take for the pollution level to drop by 50%?

☐   How can a coroner estimate the initial dose of a heart medication taken 24 hours before a fatal heart attack, even if later doses were given?

EXPONENTIAL EQUATIONS

Equations such as $2^x = 2^3$, $3^x = 50$, and $2^{2x-1} = 3^x$ in which the variable appears in an exponent are called **exponential equations.** Some exponential equations can be solved using algebraic techniques we have already seen. For example, to solve $2^x = 2^3$, we use the fact that exponential functions are 1–1 to obtain $x = 3$. Until now, we have solved equations like $3^x = 50$ and $2^{2x-1} = 3^x$ graphically, a surefire—though sometimes lengthy—method. Our focus in this section is on the development of algebraic techniques for solving exponential equations, techniques that are, in many instances, both more efficient and more accurate than graphical techniques. Still, we should not lose sight of the fact that solutions to exponential equations can be approximated graphically. Moreover, the graphing calculator is an excellent tool for confirming results obtained algebraically.

The most straightforward technique for solving exponential equations is also the most effective: taking the logarithm of both sides. The central idea is to eliminate exponents using the identity $\log_a u^n = n \log_a u$. For example, consider the equation $3^x = 50$. Taking common logarithms of both sides gives us:

$$3^x = 50$$
$$\log 3^x = \log 50$$
$$x \log 3 = \log 50$$
$$x = \frac{\log 50}{\log 3} \approx 3.56088$$

Note that taking the logarithm of both sides has the effect of "bringing the variable out of the exponent." Since exponential equations are precisely those equations that have a variable in the exponent, taking the logarithm of both sides is a good way to begin solving many such equations. Moreover, by using either common or natural logarithms, we can easily obtain a decimal approximation with a calculator.

EXAMPLE 1   ▓   Solving an Exponential Equation

Solve $7^{3x} = 15$.

SOLUTION

$$7^{3x} = 15$$
$$\log 7^{3x} = \log 15 \qquad \text{Taking the common logarithm of both sides}$$
$$3x \log 7 = \log 15 \qquad \text{Using the identity } \log u^n = n \log u \text{ on both sides}$$

$$(3 \log 7)x = \log 15 \qquad \text{Rearranging on the left side}$$

$$x = \frac{\log 15}{3 \log 7} \qquad \text{Solving for } x$$

$$x \approx 0.46389 \qquad \text{Using a calculator}$$

EXAMPLE 2 ■ **Solving an Exponential Equation**

Solve $2^{2x-1} = 3^x$. Check the result graphically.

$2x \ln 2 = x \ln 3$

SOLUTION

$$2^{2x-1} = 3^x$$

$$\ln(2^{2x-1}) = \ln(3^x) \qquad \text{Taking the natural logarithm of both sides}$$

$$(2x - 1)\ln 2 = x \ln 3 \qquad \text{Using the logarithmic identity } \ln u^n = n \ln u$$

$$(2 \ln 2)x - \ln 2 = (\ln 3)x \qquad \text{Multiplying out on the left side}$$

$$(2 \ln 2)x - (\ln 3)x = \ln 2 \qquad \text{Rearranging so that all variable terms are on one side}$$

$$(2 \ln 2 - \ln 3)x = \ln 2 \qquad \text{Factoring out } x$$

$$x = \frac{\ln 2}{2 \ln 2 - \ln 3} \qquad \text{Solving for } x$$

$$\approx 2.40942 \qquad \text{Approximating using a calculator}$$

To check this solution graphically, we plot the graph of $y = 2^{2x-1} - 3^x$ (see Figure 20) and approximate the x-intercept, obtaining $x \approx 2.40942$ (see Figure 21).

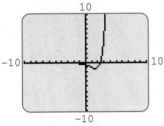

FIGURE 20

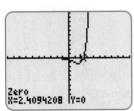

FIGURE 21

When solving exponential equations such as $2^{2x-1} = 3^x$ from Example 2, linear equations like $(2x - 1)\ln 2 = x \ln 3$ often arise. Such equations are easily solved if it is kept in mind that "ln 2" and "ln 3" are nothing more than real numbers like 5 or 7. In fact, we solved the equation $(2x - 1)\ln 2 = x \ln 3$ in exactly same way as we would the equation $(2x - 1) \cdot 5 = x \cdot 7$. Specifically, we eliminated the parentheses using the distributive property, collected all terms containing an x on one side, and then solved for x. The following parallel solutions illustrate these similarities.

$$(2x - 1) \cdot 5 = x \cdot 7 \qquad\qquad (2x - 1) \cdot \ln 2 = x \cdot \ln 3$$

$$10x - 5 = 7x \qquad\qquad (2 \ln 2)x - \ln 2 = (\ln 3)x$$

$$10x - 7x = 5 \qquad\qquad (2 \ln 2)x - (\ln 3)x = \ln 2$$

$$(10 - 7)x = 5 \qquad\qquad (2 \ln 2 - \ln 3)x = \ln 2$$

$$x = \frac{5}{3} \qquad\qquad x = \frac{\ln 2}{2 \ln 2 - \ln 3}$$

A similar process is used in the following example.

EXAMPLE 3 ■ **Compound Interest**

Suppose that $1000 is deposited into an account paying 10% interest, and on the same day $2000 is deposited into an account paying 5% interest. How long will it take for the balance in the 10% account to equal the balance in the 5% account? Assume that interest is compounded annually.

SOLUTION   Let t be the number of years required for the balances to be equal. Using the formula for compound interest, $B = P(1 + r)^t$, the balance in the 5% account is given by $B = 2000 \cdot 1.05^t$, whereas the balance in the 10% account is given by $B = 1000 \cdot 1.1^t$. Equating the balances we have

$$\text{Balance on \$2000 at 5\%} = \text{Balance on \$1000 at 10\%}$$

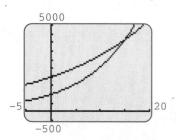

5000

−5   20

−500

FIGURE 22

$$2000 \cdot 1.05^t = 1000 \cdot 1.1^t$$

$$2 \cdot 1.05^t = 1.1^t \qquad \text{Dividing both sides by 1000}$$

$$\log(2 \cdot 1.05^t) = \log(1.1^t) \qquad \text{Taking the common logarithm of both sides}$$

$$\log 2 + \log(1.05^t) = \log(1.1^t) \qquad \text{Using } \log uv = \log u + \log v$$

$$\log 2 + t \log 1.05 = t \log 1.1 \qquad \text{Using } \log u^n = n \log u$$

$$t \log 1.05 - t \log 1.1 = -\log 2 \qquad \text{Rearranging}$$

$$t(\log 1.05 - \log 1.1) = -\log 2 \qquad \text{Factoring out the variable}$$

$$t = \frac{-\log 2}{\log 1.05 - \log 1.1} \qquad \text{Solving for the variable}$$

$$\approx 14.9 \qquad \text{Approximating using a calculator}$$

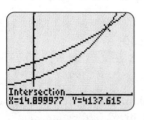

Intersection
X=14.899977 Y=4137.615

FIGURE 23

Thus, it will take approximately 14.9 years for the balances to be equal. We can check this solution by finding the intersection of the graphs of $y = 2000 \cdot 1.05^x$ and $y = 1000 \cdot 1.1^x$ as in Figure 22. In Figure 23, we see that the point of intersection is approximately $(14.9, 4138)$, which confirms our solution of $t \approx 14.9$.

Many exponential equations either cannot be solved exactly using the methods discussed in this section or are very difficult to solve. In such cases, we must resort to approximating the solutions using a graphing calculator.

EXAMPLE 4   ■   **The Gateway Arch**

When a cable is suspended from two points, it takes the shape of a *catenary* curve. This same curve was used in the design of the Gateway Arch in St. Louis, Missouri. The centers of mass of cross sections of the Arch (which happen to be equilateral triangles) are traced out by the graph of the equation

$$y = 694 - 34(e^{0.01x} + e^{-0.01x})$$

where x denotes the horizontal position measured from the central axis of the arch and y denotes the corresponding height in feet above the ground (see Figure 24). Roughly speaking, the graph of this equation traces out the path of the trams that are used to transport people up and down the Arch.

The Gateway Arch just before its completion in 1965.

a.   How high above the ground is a person riding in a tramcar when it is 50 feet from the central axis?

b.   How far from the central axis is a person riding in a tramcar when the car is at a height of 400 feet?

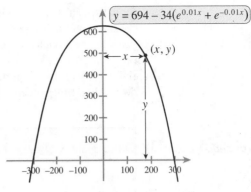

FIGURE 24

SOLUTION

a. We simply set $x = 50$ to obtain

$$y = 694 - 34(e^{0.01(50)} + e^{-0.01(50)}) \approx 617$$

Thus, the height above the ground is approximately 617 feet.

b. In this case, we wish to determine the value(s) of x for which $y = 400$. To do so, we must solve the equation

$$400 = 694 - 34(e^{0.01x} + e^{-0.01x})$$

Although this equation could be solved algebraically, the work is quite tedious, and so we opt for a graphical approximation. This can be done either by finding the point of intersection of $y = 400$ and $y = 694 - 34(e^{0.01x} + e^{-0.01x})$, or by writing the equation in the form $f(x) = 0$ and finding the x-intercepts of the graph of f. Choosing the former, we plot the graphs of $y = 694 - 34(e^{0.01x} + e^{-0.01x})$ and $y = 400$, as shown in Figure 25. Using the calculate-intersect feature, we obtain $x \approx 214$, as shown in Figure 26. Note that, because of symmetry, there is a second solution at $x = -214$. In either case, the tramcar is approximately 214 feet from the central axis when it is 400 feet above the ground.

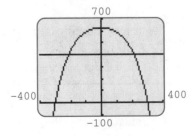

FIGURE 25

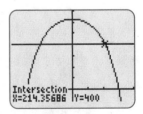

FIGURE 26

EXPONENTIAL GROWTH AND DECAY

As we have already seen, exponential functions are used to model diverse phenomena in a variety of disciplines. Indeed, exponential functions appear in virtually every natural and social science from physics to pharmacology, chemistry to criminology, and medicine to meteorology. This wide-ranging applicability stems from the fact that exponential functions increase (or decrease) at a constant percentage rate, a property shared by many real-world quantities. Thus, for example, both a population growing at the constant rate of 3% per year and a radioactive sample decaying by 50% every 1000 years can be modeled by exponential functions. For this reason, if the percentage growth rate of a quantity is constant, then the quantity is said to be either **growing exponentially** or **decaying exponentially,** according to whether the quantity is increasing or decreasing. More precisely, we have the following definitions.

Exponential Growth and Decay □

> If a quantity $Q = Q(t)$ changes at a constant percentage rate, then
> $$Q(t) = Q_0 e^{kt}$$
> where $Q_0 = Q(0)$ is the initial quantity, and k is called the **growth constant.**
> If $k > 0$, $Q(t)$ is increasing, and we say that Q is **increasing exponentially.**
> If $k < 0$, $Q(t)$ is decreasing, and we say that Q is **decaying exponentially.**

A few notes are in order. First, for several reasons, including the relative ease with which natural logarithms are computed, we have chosen to write the exponential growth and decay formulas with the base e. However, we could have chosen (as some authors do) to write the exponential growth formula as $Q(t) = Ca^t$. The equivalence of these seemingly different formulas is explored in Exercise 51. Second, our ''change at a constant percentage rate'' description of exponential growth and decay is by no means the only possible characterization. Indeed, when modeling real-world phenomena with exponential functions, it is essential that we recognize alternative characterizations of exponential growth and decay such as the following.

Characterizations of □
Exponential Growth and Decay

> 1. If a quantity is increasing at a constant percentage rate, then the quantity is growing exponentially.
> 2. If a quantity is decreasing at a constant percentage rate, then the quantity is decaying exponentially.
> 3. If the time it takes for a quantity to double (the doubling time) is constant, then the quantity is growing exponentially.
> 4. If the time it takes for a quantity to decrease to one-half its initial amount (the half-life) is constant, then the quantity is decaying exponentially.

EXAMPLE 5   ■   **Population Growth**

The population of Florida was 12.9 million in 1990 and 14.2 million in 1995. Assuming that Florida's population grows at a constant percentage rate, estimate the population of Florida in 2000 and 2025.

SOLUTION   Assuming exponential growth with an initial population of 12.9 million, we conclude that the population of Florida is modeled by a function of the form

$$P(t) = 12.9 e^{kt}$$

where t is the number of years after 1990 and $P(t)$ is the population in millions at time t. Since the population was 14.2 million in 1995, we can find k as follows:

Population after 5 years $= 14.2$

$$P(5) = 14.2$$

$$12.9e^{k \cdot 5} = 14.2$$

$$e^{5k} = \frac{14.2}{12.9}$$

$$\ln e^{5k} = \ln\left(\frac{14.2}{12.9}\right) \qquad \text{Taking the natural logarithm of both sides}$$

$$5k = \ln\left(\frac{14.2}{12.9}\right) \qquad \text{Using the logarithmic identity } \ln e^x = x$$

$$k = \frac{\ln\left(\dfrac{14.2}{12.9}\right)}{5} \approx 0.019203$$

Thus, we have $P(t) = 12.9e^{0.019203t}$. Since the year 2000 corresponds to $t = 10$, we predict a population of

$$P(10) = 12.9e^{0.019203(10)} \approx 15.6 \text{ million}$$

RULE OF THUMB In general, when using a calculator to solve an applied problem, you should retain all intermediate results in the calculator's memory. This is particularly advisable when working with exponential functions because small roundoff errors can have profound consequences (see Exercise 52). In the text, we often round intermediate results so that individual steps in the solution can be shown.

Since a radioactive substance decays at a constant percentage rate, it follows that the mass of a radioactive substance decays exponentially according to $Q(t) = Q_0 e^{kt}$, where $k < 0$. It is customary to report the rate of decay of a substance in terms of its **half-life**: the time required for one-half of the substance to decay. The half-lives of a few common isotopes are given in Table 5.

TABLE 5 Half-life

Element	Atomic weight	Half-life
Uranium	235	7×10^8 years
Carbon	14	5760 years
Lead	210	22.3 years
Barium	133	10.5 years
Iron	59	44.5 days
Gold	198	64.6 hours
Arsenic	76	26.5 hours
Copper	64	12.7 hours
Nitrogen	13	9.97 minutes

EXAMPLE 6 ■ Carbon-14 Dating

The Dead Sea Scrolls, consisting mostly of thousands of fragments such as this, have been carbon dated to around the time of Christ.

Carbon-14 (C-14) decays exponentially with a half-life of approximately 5760 years. If an artifact originally had 12 grams of C-14, and now has only 4 grams remaining, how old is the artifact?

SOLUTION   Let $Q(t)$ represent the quantity of C-14 remaining at time t. Since there were initially 12 grams of C-14, and the quantity is decaying exponentially, we may write $Q(t) = 12e^{kt}$. We next find the constant k by using the half-life of 5760 years. Since there were originally 12 grams, there would be only 6 grams remaining after 5760 years.

Quantity of C-14
remaining after 5760 years = 6

$$Q(5760) = 6$$

$$12e^{5760k} = 6 \qquad \text{Setting } t = 5760 \text{ in } Q(t)$$

$$e^{5760k} = \frac{1}{2} \qquad \text{Dividing by 12}$$

$$\ln(e^{5760k}) = \ln\left(\frac{1}{2}\right) \qquad \begin{array}{l}\text{Taking the natural} \\ \text{logarithm of both sides}\end{array}$$

$$5760k = \ln\left(\frac{1}{2}\right) \qquad \begin{array}{l}\text{Using the logarithmic} \\ \text{identity } \ln e^a = a\end{array}$$

$$k = \frac{\ln\left(\frac{1}{2}\right)}{5760} \approx -0.00012034 \qquad \text{Solving for } k$$

Thus, $Q(t) = 12e^{-0.00012034t}$. Because there are currently 4 grams remaining, we have

Quantity of C-14
remaining after t years = 4

$$Q(t) = 4$$

$$12e^{-0.00012034t} = 4$$

$$e^{-0.00012034t} = \frac{1}{3} \qquad \text{Dividing both sides by 12}$$

$$\ln(e^{-0.00012034t}) = \ln\left(\frac{1}{3}\right) \qquad \begin{array}{l}\text{Taking the natural} \\ \text{logarithm of both sides}\end{array}$$

$$-0.00012034t = \ln\left(\frac{1}{3}\right) \qquad \begin{array}{l}\text{Using the logarithm} \\ \text{identity } \ln e^a = a\end{array}$$

$$t = \frac{\ln\left(\frac{1}{3}\right)}{-0.00012034} \approx 9129 \text{ years} \qquad \text{Solving for } t$$

CARBON-14 DATING AND THE SHROUD OF TURIN

T he Shroud of Turin, a linen relic bearing the image of a man, first appeared in 14th-century France. Although the Roman Catholic Church took no official position on its authenticity, it was widely believed to be the burial shroud of Jesus. The origins of the shroud were questioned from the very beginning, but there was no way to disprove its authenticity. In fact, the Catholic Church allowed a team of investigators to conduct a scientific investigation on the shroud in 1978, and no conclusive evidence of fraud was produced, although one member of the team concluded that the image on the shroud was, in fact, a very thin watercolor.

In order to protect the possibly sacred shroud, the Church had not allowed **carbon-14 dating** to be performed. This test, if conducted using the technology available in 1978, would have caused the destruction of an unacceptably large portion of the shroud. By 1988, however, technological developments made it possible to conduct carbon-14 dating on the shroud with only very minimal destruction. The results demonstrated that the shroud dated from the 14th century, and hence could not have been the burial shroud of Jesus.

Carbon-14 is a radioactive isotope of carbon that is present in all living organisms. After death, the carbon-14 gradually decays to an isotope of nitrogen. As a consequence, the quantity of carbon-14 slowly decreases. By measuring the remaining carbon-14, scientists can determine how much time has passed since the organism died, and thus they can date pieces of wood, cloth, hides, bones, and other organic matter.

Since carbon-14 decays exponentially, we know that the amount remaining t years after the death of the organism is given by $Q(t) = Q_0 e^{kt}$, where Q_0 is the amount of carbon-14 prior to death. It can be shown (see Example 6) that since the half-life of carbon-14 is 5760 years, $k = -0.00012034$. Thus, $Q(t) = Q_0 e^{-0.00012034t}$.

EXAMPLE 7 ▒ Pollution in Lake Michigan

Under suitable conditions, and assuming all incoming pollution is stopped, the concentration of pollution in a lake decays exponentially over time. If it takes 5 years for the pollution concentration in Lake Michigan to decrease to 85% of its current level, how many years would it take for the pollution concentration to decrease to 50% of its current level?

SOLUTION Let $Q(t) = Q_0 e^{kt}$ denote the concentration of pollution in Lake Michigan at time t. We must first determine the value of k. Since the concentration in 5 years will be 85% of the initial concentration Q_0, we obtain the following equation.

$$\text{Concentration after 5 years} = 85\% \text{ of } Q_0$$
$$Q(5) = 0.85 Q_0$$
$$Q_0 e^{k(5)} = 0.85 Q_0$$

Now we solve for k.

$$Q_0 e^{k(5)} = 0.85 Q_0$$
$$e^{5k} = 0.85 \qquad \text{Dividing both sides by } Q_0$$
$$\ln e^{5k} = \ln 0.85 \qquad \text{Taking the natural logarithm of both sides}$$
$$5k = \ln 0.85$$
$$k = \frac{\ln 0.85}{5} \approx -0.032504$$

Thus, we have $Q(t) = Q_0 e^{-0.032504t}$. To find how long it takes for the concentration to reach 50% of its present level, we proceed as follows:

$$\text{Concentration after } t \text{ years} = 50\% \text{ of } Q_0$$
$$Q_0 e^{-0.032504t} = 0.50\, Q_0$$
$$e^{-0.032504t} = 0.5 \qquad \text{Dividing both sides by } Q_0$$
$$\ln e^{-0.032504t} = \ln 0.5 \qquad \text{Taking the natural logarithm of both sides}$$
$$-0.032504t = \ln 0.5$$
$$t = \frac{\ln 0.5}{-0.032504} \approx 21.3 \text{ years}$$

UNDERSTANDING AND MASTERY CHECKLISTS

CONCEPTS TO UNDERSTAND

☐ Exponential equations

☐ Exponential growth and decay

☐ Half-life

SKILLS TO MASTER

☐ Solve an exponential equation algebraically.

☐ Approximate the solutions of an exponential equation graphically.

☐ Construct an exponential function that models exponential growth or decay.

☐ Compute half-life and doubling time.

EXERCISES 5.4

EXERCISES 1–18 *Solve the exponential equation. When approximating, give answers to five decimal places of accuracy. You may wish to check your solution graphically.*

1. $2^x = 16$

2. $3^{-x} = \frac{1}{9}$

3. $25^{3x} = 125^{x+1}$

4. $3^{x^2+1} = 27$

5. $7^x = 10$

6. $5^x = 10$

7. $3^{-x} = 11$

8. $3^{x-1} = 2^x$

9. $\left(\frac{3}{5}\right)^x = 6^{2-x}$

10. $0.4^{2+x} = 1.5^{3x-1}$

11. $e^x = 2^{x+1}$

12. $5e^{-2.3x} = 4$

13. $e^{2x-1} = \pi^{2x}$

14. $e^{x/2} = 3^{x-4}$

15. $\frac{1}{2^{x+2}} = \frac{1}{8}$

16. $\left(\frac{2}{3}\right)^{-x+1} = \frac{81}{16}$

17. $1.06^t = 1000$

18. $1000e^{0.05t} = 2000$

EXERCISES 19–24 *Approximate the solution(s) of the given equation to the nearest hundredth.*

19. $e^{2x} - 5e^x = -6$

20. $4^x - 3 \cdot 2^x + 2 = 0$

21. $e^{-x} = x$

22. $4 - x^2 = e^{x-2}$

23. $\frac{1}{25}x^4 + 4 = x^2 + e^{x+1}$

24. $\frac{1}{8}x^3 + \frac{7}{8}x^2 = x + 2 - e^x$

APPLICATIONS

25. *Compound Interest* An account is established with a principal of $200 paying 4% interest compounded semiannually. How long will it take for the balance to reach $300?

26. *Tripling Money* How long does it take for money to triple in an account paying 5% interest compounded annually?

27. *Doubling Money* An account paying interest compounded continuously at a rate r doubles in size every 12 years. Find r.

28. *Competing Accounts* Suppose that Bob deposits $1000 in an account paying 6% interest compounded quarterly at the same time that Janet deposits $1200 in an account paying 6% interest compounded annually. How long will it take for Bob's balance to exceed Janet's?

29. *Competing Accounts* Suppose that $200 was deposited on January 1, 1999, into an account that earned 5% interest compounded annually. Suppose further that $200 was deposited on January 1, 2000, into a different account that earned 6% interest compounded annually. In what month of what year will the balance in the account earning 6% interest overtake the balance in the account earning 5%?

30. *Competing Accounts* On January 1, 2000, $1000 is deposited in an account paying 15% simple interest, and $100 is deposited in an account paying 5% interest compounded annually. In what month of what year will the balance in the simple interest account be overtaken by that in the compound interest account?

31. *Compact Disk Inflation* Suppose that the average price of a compact disk was $12.00 on January 1, 1992, and that the price increases at the rate of 3% per year. Give the month and the year in which the average price of a compact disk will be $100.

32. *Annual Inflation* Suppose that inflation is 3.5% annually. Then goods or services that cost P dollars today will, on the average, cost $P(1.035)^t$ dollars in t years. If tuition at a certain prestigious university is $25,000 per year in 2000, in what year will tuition reach $40,000?

33. *Consumer Price Index* The Consumer Price Index (CPI) was 99.6 in 1983 and 140.3 in 1992. In other words, goods and services that cost $99.60 in 1983 cost $140.30 in 1992. Assuming exponential growth, what will the CPI be in 2001? When will it reach 230?

34. *Dollar Devaluation* One dollar in 1992 was worth approximately 21¢ in 1960 dollars. Assuming this is an exponential decrease, when will a dollar be worth 1¢ in 1960 dollars?

35. *Population Growth* At 8:00 A.M. July 4, 2001, a colony of 1000 of the smallest free-living entity, *Mycoplasma laidlawii* (average weight ≈ 10^{-16} grams), is established. If the population were to double every hour, at what time would the collective mass of the colony exceed that of the largest organism on earth, the 190-ton blue whale? (*Hints:* If P represents the population of the colony and t the number of hours since the establishment of the colony, then $P = 1000 \cdot 2^t$. A ton is 2000 pounds, and 1 pound is 454 grams.)

36. *Population Decline* The population of a certain endangered species of owl is declining exponentially. There are currently 500 living specimens, whereas just 10 years ago there were 10,000. If the population continues to decline exponentially, how long will it be until there is only a single owl left?

37. *Population Decline* Gary, Indiana, tops the list of cities in the United States with the largest percentage loss in population between 1980 and 1990 (*Data source:* U.S. Bureau of the Census). The population of Gary was 152,968 in 1980 and 116,646 in 1990. Assuming the population is decreasing exponentially, what will the population be in the year 2000?

38. *Population Growth* The fastest growing city in the United States between the years 1980 and 1990 was Moreno Valley, California (*Data source:* U.S. Bureau of the Census). The population of Moreno Valley was 28,309 in 1980 and 118,779 in 1990. Assuming exponential growth, what will the population be in the year 2000?

39. *Carbon Dating* Physicists found that a certain Egyptian mummy contained only 25% of its original carbon-14. Given that the half-life of carbon-14 is 5760 years, how old is the mummy?

40. *Carbon Dating* The well-preserved corpse (see photo) of a Bronze Age man was found frozen in ice in the Austrian Alps in September 1991. If only 53.3% of the carbon-14 is found, and the half-life of carbon-14 is 5760 years, how old is the body?

41. *Pollution Turnover* The rate of pollution turnover in Lake Erie is quite rapid. Suppose that if all pollution inflow into Lake Erie were stopped, the concentration of pollution would decrease to 80% of its present level in 6 months. Assuming the pollution concentration decays exponentially, estimate how many years it would take for the concentration to decrease to 10% of its present level.

42. *Styrofoam Decay* Suppose that Styrofoam cups are biodegraded at an exponential rate and that after 10 years 99.5% of the cup remains. Compute the half-life of a Styrofoam cup.

◼ CONCEPTS AND CRITICAL THINKING

EXERCISES 43–46 *Answer true or false.*

43. For a given real number y, the equation $2^x = y$ will either have a unique solution or no solution at all.

44. The half-life of a substance is equal to one-half the age of the substance.

45. A quantity will grow exponentially if it grows at a constant rate.

46. A quantity will grow exponentially if it grows at a constant percentage rate.

EXERCISES 47–50 *Give an example of each.*

47. An exponential equation having no solution

48. An exponential equation that is more easily solved by using common logarithms than natural logarithms

49. A real-world context in which exponential growth occurs

50. A real-world context in which exponential decay occurs

51. In this text, we have chosen to represent exponential growth and decay models with functions of the form $Q = Q_0 e^{kt}$, but it is not uncommon to see them represented with functions of the form

$Q = Ca^t$. Show that, in fact, these two representations are equivalent. [*Hint:* $e^{kt} = (e^k)^t$]

52. In Example 5, round the value of k to two decimal places and compare the population you find with this new value to that found in the example. What does your result suggest about rounding the k-value in exponential growth and decay models?

QUESTIONS FOR DISCUSSION OR ESSAY

53. When a substance is decaying exponentially, the percentage of substance lost for any time period remains constant. It follows that for a radioactive substance, such as radium, a 1-gram sample will take exactly the same amount of time to decay as two 0.5-gram samples or ten 0.1-gram samples. Does ice melt exponentially? Support your answer. What does the rate of melting of ice depend on?

54. In Exercise 33, we made an assumption about the exponential growth of the Consumer Price Index and observed that the costs for goods and services increase dramatically over time. What additional information would be necessary to test this assumption? What would be the inevitable effect of exponential CPI growth on our monetary system? The yearly rise in the CPI is

known more commonly as inflation. Do you think inflation is one of the certainties in life, along with death and taxes? Why or why not? Have there ever been periods of deflation?

55. Suppose that the function $f(t) = 60,000e^{0.2t}$ gives the population of a city at time t, and the function $h(t) = 3t + 85,000$ measures the number of "people units" of housing available in the city at time t. Thus, the population is growing exponentially, whereas the available housing is growing at a constant rate (that is, linearly). Plot the two functions and approximate the time at which their graphs intersect. What does this point represent? What do these functions imply about the housing situation in the city over time? Is this realistic? Explain.

PROJECTS FOR ENRICHMENT

56. *A Medical Mystery* A patient was admitted to the hospital early in the morning complaining of chest pains. An initial dose of the heart medicine digoxin was administered at 9 A.M., although, due to an oversight, the amount was not recorded. Subsequent doses in a quantity sufficient to immediately raise the digoxin level in the bloodstream an additional 0.5 ng/ml were administered at 9 P.M. that evening and again at 9 A.M. the next morning. Unfortunately, the patient's condition did not improve, and at 9:05 A.M. he had a fatal heart attack. A test run at that time showed a level of 1.5 ng/ml in his bloodstream. In order to avoid a large malpractice settlement, the hospital must prove that the level of the drug was always within the therapeutic range, which is between 0.8 and 1.6 ng/ml. It is known from many clinical observations that a given dose of digoxin decays exponentially over time and is reduced to half its original level in the body in 36 hours.

a. Determine the level of digoxin in the bloodstream after the initial dose.

b. Illustrate graphically whether the level was always within the therapeutic range.

57. *Comparing Interest Rates and Compounding Periods* Suppose you are trying to decide whether to invest some money in an account that pays 5% compounded quarterly or 4.9% compounded monthly. Naturally, one way would be to determine which would yield the highest balance at the end of a given

period of time. Another way would be to convert both rates to their equivalent rates compounded annually. In other words, we would like to find the annually compounded interest rates that would yield the same balance as 5% compounded quarterly and 4.9% compounded monthly. For this task, it would be helpful to have a formula that "converts" an interest rate r compounded n times a year to its equivalent rate s compounded annually. To that end, suppose a principal P is deposited at an interest rate r compounded n times a year. Suppose that at the same time, a principal P is deposited at an interest rate s compounded annually. For the rates r and s to be equivalent, they must yield equal balances at the end of any time t. Thus, we must have

(1) $$P\left(1 + \frac{r}{n}\right)^{nt} = P(1 + s)^t$$

a. Solve equation (1) for s to show that

(2) $$s = \left(1 + \frac{r}{n}\right)^n - 1$$

The quantity s in equation (2) is called the **effective annual interest rate.**

b. Compute the effective annual interest rate that corresponds to 5% compounded quarterly and 4.9% compounded monthly. Which of the two interest-earning schemes would you choose? Why?

c. Solve equation (2) for r and use the resulting equation to find the interest rate r that, when compounded daily, would be equivalent to an effective annual rate of 6%.

d. Use equation (2) and your graphing calculator to estimate the number of compounding periods that would be needed for an interest rate of 5.85% to have an effective rate of 6%.

A continuously compounded interest rate can also be converted to an effective annual interest rate. If a principal P is deposited at an interest rate r compounded continuously, the effective annual interest rate is the value s that satisfies the equation

$$(3) \qquad Pe^{rt} = P(1 + s)^t$$

e. Solve equation (3) for s to show that

$$(4) \qquad s = e^r - 1$$

f. Find the effective annual rate corresponding to 4.8% compounded continuously.

g. Solve equation (4) for r and use the resulting formula to find the continuously compounded interest rate r that would be equivalent to an effective annual rate of 6%.

h. What does your solution to part g say about the number of compounding periods that would be necessary for an interest rate of 5.8% to have an effective rate of 6%?

SECTION 5.5   MODELING WITH EXPONENTIAL AND LOGARITHMIC FUNCTIONS

- ☐ Is hunting necessary to limit the exponential growth of a deer population?
- ☐ How many Internet hosts will there be in 2005?
- ☐ How much faster do pedestrians walk in Mexico City than in New York City?
- ☐ When will the number of transistors on a microprocessor exceed the number of neurons in the human brain?
- ☐ How much slower are the reflexes of a 60-year-old than those of a 19-year-old?

CONSTRUCTING MATHEMATICAL MODELS

Through the text, we have seen examples of how mathematics can be used to model real-world phenomena. However, in most of these instances, we were concerned primarily with using a given model to solve a real-world problem. With few exceptions, we allowed ourselves the luxury of assuming that a given mathematical model would provide accurate results. We now wish to consider in more detail the work that goes on "behind the scenes" in the construction and validation of a mathematical model.

The construction of a mathematical model is often preceded by detailed observations of some naturally occurring event, as was the case with the bacteria population example in Section 5.1. The actual construction may then be as simple as drawing a graph based on those observations or as complex as finding a formula that provides results similar to the observations. The goal is to provide a model that will accurately describe trends, predict future outcomes, and perhaps even provide insight into how future outcomes can be controlled or at least influenced. Unfortunately, because of the compromise that must often be made between accuracy and simplicity, even the best mathematical models have limitations. Thus, it is important to test and possibly modify models as the need arises. We discuss these issues as we study the development and testing of mathematical models.

EXAMPLE 1   ■   **Modeling a Deep Population Graphically**

Consider the following scenario: In 1994, a certain state's Department of Wildlife recommended a ban on deer hunting because of a dangerously low deer population. Now, 5 years later, the department wishes to reevaluate its recommendation. Table 6

TABLE 6

Year	0	1	2	3	4	5
Population	10,000	11,500	13,200	15,100	17,400	20,100

Increasing deer populations and human encroachment on the natural habitat force deer into urban and suburban settings.

gives estimates of the deer population since 1994 ($t = 0$). Construct a graphical model to illustrate any trends in the data and to predict what might happen in future years. Use a graph to find a rough approximation for the population in 2000.

SOLUTION The graph in Figure 27 shows that the population growth is occurring in a "smooth" way. Unless a dramatic change occurs, it is likely that the deer population will continue to grow—at least for a few years—according to the pattern that has been established. By fitting a smooth curve to the data, as shown in Figure 28, we obtain a rough estimate of 24,000 deer in 2000 ($t = 6$).

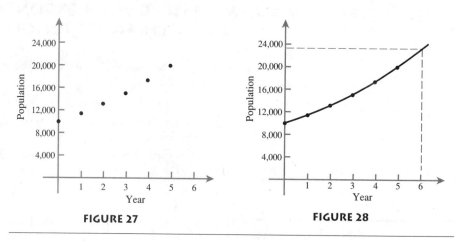

FIGURE 27   **FIGURE 28**

As useful as graphs are for displaying data, describing trends, and suggesting future behaviors, they do have limitations when it comes to making precise predictions. The

TABLE 7

Year	Population	Percentage increase over previous year
0	10,000	
1	11,500	$\dfrac{11,500 - 10,000}{10,000} = 0.15 = 15\%$
2	13,200	$\dfrac{13,200 - 11,500}{11,500} \approx 0.148 = 14.8\%$
3	15,100	$\dfrac{15,100 - 13,200}{13,200} \approx 0.144 = 14.4\%$
4	17,400	$\dfrac{17,400 - 15,100}{15,100} \approx 0.152 = 15.2\%$
5	20,100	$\dfrac{20,100 - 17,400}{17,400} \approx 0.155 = 15.5\%$

curve-fitting approach used in Example 1 involves some uncertainty: What does the graph look like beyond the known data points? Is it linear, parabolic, exponential, or something altogether different? It would be useful to have an explicit mathematical formula that captures the relationship between the quantities involved. Consider, for example, the data given in Table 6 and graphed in Figure 28. It appears that the growth is more rapid than linear. To verify this, we compute in Table 7 the percentage increase from one year to the next. Based on these computations, it appears that the percentage rate of growth is roughly constant (at 15% per year). This suggests exponential growth, and so we expect a function of the form $P(t) = ae^{bt}$.

EXAMPLE 2 ■ **Constructing an Exponential Model**

Use the first two data values in Table 6 to construct a population model of the form $P(t) = ae^{bt}$, and then estimate the population after 6 years and again after 10. Here $P(t)$ denotes the population after t years.

SOLUTION   We first use the fact that the population was 10,000 when $t = 0$.

$$\text{Population at time } 0 = 10{,}000$$
$$P(0) = 10{,}000$$
$$ae^{b(0)} = 10{,}000$$
$$a = 10{,}000$$

Thus, $P(t) = 10{,}000e^{bt}$. Next we use the population value 11,500 for $t = 1$.

$$\text{Population at year } 1 = 11{,}500$$
$$P(1) = 11{,}500$$
$$10{,}000e^{b(1)} = 11{,}500$$
$$e^b = \frac{11{,}500}{10{,}000} = 1.15 \qquad \text{Dividing both sides by } 10{,}000$$
$$b = \ln 1.15 \approx 0.1398 \qquad \text{Taking the natural logarithm of both sides}$$

Thus, $P(t) = 10{,}000e^{0.1398t}$. At 6 years, we have $P(6) = 10{,}000e^{0.1398(6)} \approx 23{,}100$. At 10 years, $P(10) \approx 40{,}500$.

Notice that the model in Example 2 was constructed using only two of the six available data points. Slightly different models would result if we were to use different combinations of data points. For example, if we were to use the points (0, 10,000) and (3, 15,100), we would obtain the exponential model $P(t) = 10{,}000e^{0.1374t}$. This raises an important question: Is there a way to construct an exponential model that takes all available data points into consideration? Fortunately, the answer is yes. The **exponential regression** feature of your graphing calculator is capable of finding a model of the form $P(t) = ae^{bt}$ that fits all available data points as closely as possible.

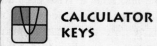

CALCULATOR KEYS

Exponential Regression

Exponential regression can be used to find an exponential function of the form $y = ae^{bx}$ (on some calculators, $y = ab^x$) that best fits a collection of data points (x, y). There are several steps to the process.

1. Clear any existing statistical data.
2. Enter the points (x, y) as statistical data.
3. Select exponential regression from the list of available regression options.

After step 3 has been completed, the calculator will compute values for a and b. Note that if your calculator finds a and b for a model of the form $y = ab^x$, you can obtain a model of the form $y = ae^{cx}$ by letting $c = \ln b$.

EXAMPLE 3 ■ Using Exponential Regression to Fit a Model

Use exponential regression to find a model of the form $P(t) = ae^{bt}$ that best fits the data in Table 6.

SOLUTION Following the steps just outlined, we clear any existing statistical data in the calculator, enter the population data from Table 6 as statistical data (using time as the x-coordinate and population as the y-coordinate), and then select the exponential regression option. The computed values for a and b are $a \approx 9992$ and $b \approx 0.1391$. Thus, the best fit model is $P(t) = 9992e^{0.1391t}$.

TESTING MATHEMATICAL MODELS

Before a model can be used with confidence, it must be tested for accuracy. When actual data are available, we can test a mathematical model for accuracy by comparing its predictions to the actual data. With models such as the ones in Examples 2 and 3, this can be done either by computing tables of actual and predicted data values or by graphing the model along with the actual data points.

EXAMPLE 4 ■ Testing Models

Compare the population data in Table 6 with the populations predicted by the following models.

a. $P(t) = 10,000e^{0.1398t}$ **b.** $P(t) = 9992e^{0.1391t}$

SOLUTION

a. Table 8 shows how the actual data values compare with the ones predicted by $P(t) = 10,000e^{0.1398t}$ for $t = 0$ through $t = 5$. Figure 29 shows the actual population data points plotted along with the graph of $P(t) = 10,000e^{0.1398t}$. The data and the graph show very close agreement. In fact, because of the scale that was chosen, the graph appears to pass through all the data points.

TABLE 8

t	Actual values from Table 6	$P(t) = 10{,}000e^{0.1398t}$
0	10,000	10,000
1	11,500	11,500
2	13,200	13,226
3	15,100	15,210
4	17,400	17,493
5	20,100	20,117

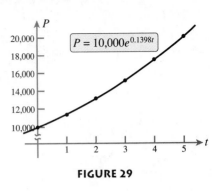

FIGURE 29

b. Table 9 shows how the actual data values compare with the ones predicted by $P(t) = 9992e^{0.1391t}$ for $t = 0$ through $t = 5$. In Figure 30, we used a graphing calculator to plot both $P(t) = 9992e^{0.1391t}$ and the actual population data. The data and graph show very close agreement.

TABLE 9

t	Actual values from Table 6	$P(t) = 9992e^{0.1391t}$
0	10,000	9,992
1	11,500	11,483
2	13,200	13,197
3	15,100	15,166
4	17,400	17,430
5	20,100	20,031

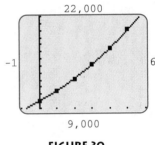

FIGURE 30

If a more precise measure of accuracy is desired, one can compute the *absolute error* or *deviation* between the actual data and the predicted values. For example, using the model $P(t) = 10{,}000e^{0.1398t}$ for the population data in Table 6, the absolute error at $t = 2$ is

$$|(\text{Actual population at } t = 2) - (\text{Predicted population at } t = 2)|$$
$$= |13{,}200 - 13{,}226| = 26$$

If we compute the absolute error for all known data values, then we can use either the sum of the errors or the largest of the errors as a single measure of the accuracy of the model.

EXAMPLE 5   ■   **Computing Absolute Error**

Determine the absolute errors between the data values given in Table 6 and those predicted by the following two models. Find the maximum absolute error and also the sum of the absolute errors. Which of the two models is more accurate?

a. $P(t) = 10{,}000e^{0.1398t}$ **b.** $P(t) = 9992e^{0.1391t}$

SOLUTION

a. Using Table 8, we obtain the following absolute errors for the model $P(t) = 10,000e^{0.1398t}$.

t	Actual values from Table 6	$P(t) = 10,000e^{0.1398t}$	Absolute error
0	10,000	10,000	$\lvert 10,000 - 10,000 \rvert = 0$
1	11,500	11,500	$\lvert 11,500 - 11,500 \rvert = 0$
2	13,200	13,226	$\lvert 13,200 - 13,226 \rvert = 26$
3	15,100	15,210	$\lvert 15,100 - 15,210 \rvert = 110$
4	17,400	17,493	$\lvert 17,400 - 17,493 \rvert = 93$
5	20,100	20,117	$\lvert 20,100 - 20,117 \rvert = 17$

Thus, the maximum absolute error is 110 when $t = 3$. The sum of the errors is 246.

b. Using Table 9, we obtain the following absolute errors for $P(t) = 9992e^{0.1391t}$.

t	Actual values from Table 6	$P(t) = 9992e^{0.1391t}$	Absolute error
0	10,000	9,992	8
1	11,500	11,483	17
2	13,200	13,197	3
3	15,100	15,166	66
4	17,400	17,430	30
5	20,100	20,031	69

Thus, the maximum absolute error is 69 when $t = 5$. The sum of the errors is 193. Since the maximum absolute error and the sum of the absolute errors is smaller for the model $P(t) = 9992e^{0.1391t}$, we conclude that it is the more accurate of the two.

The absolute errors found in the preceding example are quite small, especially when the size of the population is taken into consideration. However, the acceptable error must also be weighed against the simplicity of a model. A model that is extremely accurate but difficult to use may not serve the purpose for which it was designed. In the case of our deer example, we were fortunate in finding a simple exponential model that was also extremely accurate. This is not completely unexpected. Numerous statistical studies have shown that uninhibited population growth tends to be exponential in nature. In fact, the British economist and sociologist Thomas Malthus (1766–1834) had proposed exponential population models as early as 1798. The Malthusian model, as it has come to be called, has the form $P(t) = P_0 e^{kt}$, where t denotes time, P_0 represents the **initial population,** and k is the **growth constant.**

Many mathematical models are not as easily verified as our Malthusian deer population model, however, especially when observed data is not readily available for testing purposes. For example, it would be difficult—if not impossible—to rigorously verify mathematical models that are intended to predict the effect of ozone depletion. In such cases, one must be careful about depending too heavily on the model's predictions. Even if a model has been thoroughly verified, one should be wary about placing too much confidence in its predictions.

PREDICTING WITH MATHEMATICAL MODELS

Once we have verified that a model is accurate and have judged it to be sufficiently simple, the next step is to use the model to make inferences about future behavior. After all, the purpose for constructing and verifying a model is not just to showcase its accuracy and simplicity, but to use it to gain insight into how the future will unfold.

EXAMPLE 6   ▨   **Predicting with Exponential Models**

Suppose that on the basis of historical data, the Department of Wildlife (see Example 1) estimates that the natural resources in the state can support at most 240,000 deer. Proponents of deer hunting argue that the ban on deer hunting should be lifted when the population reaches half of that limiting value in order to prevent the deaths by starvation and auto accidents that would likely occur as the population nears the limiting value. Use the population model $P(t) = 9992e^{0.1391t}$ to predict when the deer population will reach 120,000.

SOLUTION   We want to determine the value for t at which the population predicted by $P = 9992e^{0.1391t}$ will be 120,000. Thus, we set $P(t)$ equal to 120,000 and solve for t.

$$P(t) = 120{,}000$$

$$9992e^{0.1391t} = 120{,}000$$

$$e^{0.1391t} = \frac{120{,}000}{9992} \qquad \text{Dividing both sides by 9992}$$

$$\ln e^{0.1391t} = \ln\left(\frac{120{,}000}{9992}\right) \qquad \text{Taking the natural logarithm of both sides}$$

$$0.1391t = \ln\left(\frac{120{,}000}{9992}\right) \qquad \text{Using the identity } \ln(e^x) = x$$

$$t = \frac{\ln\left(\dfrac{120{,}000}{9992}\right)}{0.1391} \approx 17.9$$

Thus, our model predicts that it will take approximately 17.9 years for the deer population to reach 120,000.

Notice that although the Department of Wildlife has estimated that the natural resources can support a maximum of 240,000 deer, our model actually suggests that the population growth has no bound. For example, the model predicts that in 25 years the deer population will be $9992e^{(0.1391)25}$, or approximately 324,000 and still rising. Does this mean that the estimated limiting population of 240,000 is wrong? Or is our model wrong? Perhaps our estimate of the limiting population is too low, but clearly there is a limiting population: If the deer population grew without bound, then eventually the entire state (and the airspace above it) would be nothing but a gigantic mound of deer. Thus, the problem with the model is the assumption that the deer population will continue to grow as it has over the first 5-year period. Due to limited natural resources, however, a slower population growth would be a certain though gradual outcome. It is not that our model is "wrong"; it worked quite well for short time periods. But in order to extend its range of applicability, we must modify our Malthusian exponential model to reflect this new assumption of limited growth.

MODIFYING MATHEMATICAL MODELS

Although there has been human settlement on what is now Istanbul for thousands of years, the city has not reached its limiting population.

The inability of a model to make long-term predictions is a problem that is common to exponential models and many other mathematical models. Thus, we must be wary of predictions that are far from the known data. We must also be on the lookout for ways in which a model can be modified to make it more accurate for a larger domain. One such modification for population models is possible using the logistic model developed by the Dutch mathematical biologist Pierre-François Verhulst (1804–1849). It has the general form

$$P(t) = \frac{L}{1 + Ce^{-kt}}$$

The logistic model improves on our earlier Malthusian model in that it imposes a limit on the size of a population. Such a limit is a natural consequence of competition for food, living space, and other natural resources.

EXAMPLE 7 ■ A Logistic Model for Population Growth

Using techniques from calculus, we can construct the following logistic model of a deer population with a limiting value of 240,000.

$$P(t) = \frac{240,000}{1 + 23e^{-0.1398t}}$$

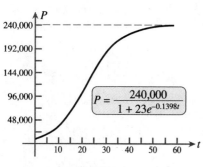

FIGURE 31

The graph of P versus t is shown in Figure 31. Plot the graph with a graphing calculator and predict when the population is growing the fastest and when the population will reach 90% of the limiting population of 240,000.

SOLUTION One view of the graph of $P(t)$ is shown in Figure 32. The population is growing the fastest where the graph is rising most rapidly. In other words, the growth rate is largest where the graph is the steepest. Using the trace feature, we determine that this occurs when t (the x-coordinate) is approximately 23, as shown in Figure 33. To find when the population will reach 90% of the limiting population, or 216,000 deer, we locate the point on the graph where the y-coordinate is approximately 216,000. After zooming in several times, we see in Figure 34 that $x \approx 38$.

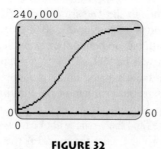

| **FIGURE 32** | **FIGURE 33** | **FIGURE 34** |

Although the logistic model for population growth appears to be more realistic than our earlier exponential model, it too does not account for all of the many factors that may affect population growth. For example, the logistic model suggests continued (though slower) growth, and so it would not accurately model the deer population if hunting were again allowed. We will ask you to supply some other overlooked factors in Exercise 32.

LOGARITHMIC MODELS

Logarithmic models, though not as common as exponential models, are useful for such tasks as measuring sound and shock intensity and modeling human response systems. In the following example, we consider a human memory model.

EXAMPLE 8 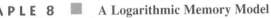 **A Logarithmic Memory Model**

In a study on human memory, subjects were given an initial spelling exam on obscure words in the English language and then retested monthly. The average scores for the initial exam and the next 3 months, out of a possible score of 20, are given in the following table.

Dominic O'Brien is listed in the Guinness Book of World Records *for memorizing 35 decks of playing cards (1820 cards) with only 2 errors.*

Month	0	1	2	3
Score	17	14	12	11

Construct a model of the form $y = a \log(t + 1) + b$ that approximates the average score y as a function of time t measured in months. Test the model for accuracy and use it to predict the average score after 6 months.

SOLUTION We wish to select constants a and b so that the resulting model $y = a \log(t + 1) + b$ most nearly fits the data. Close inspection of the data, or even plotting a graph of the data, probably will not give us much insight into an appropriate choice of constants: We just don't have enough familiarity with functions of the form $f(t) = a \log(t + 1) + b$ to predict the effect on the graph of adjusting the constants a and b. We can simplify things a great deal, however, by making the substitution $x = \log(t + 1)$ to obtain the linear equation $y = ax + b$. To find a and b, we first compute $x = \log(t + 1)$ for $t = 0$ to $t = 3$, as shown in Table 10, and then apply the linear regression feature of a graphing calculator to the points (x, y). We obtain $a \approx -10.12$ and $b \approx 16.99$. The resulting best-fit line, $y = -10.12x + 16.99$, is graphed in Figure 35 along with the data points (x, y) from Table 10. It appears that the line is a very good fit.

Returning now to the logarithmic model, we have

$$y = -10.12x + 16.99$$
$$= -10.12 \log(t + 1) + 16.99 \qquad \text{Using } x = \log(t + 1)$$

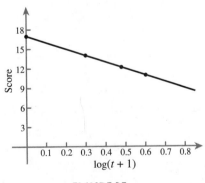

FIGURE 35

TABLE 10

Month (t)	0	1	2	3
$x = \log(t + 1)$	0	0.30103	0.47712	0.60206
Score (y)	17	14	12	11

The following table compares the actual test scores for the first three months with those predicted by the model $y = -10.12 \log(t + 1) + 16.99$.

Month	0	1	2	3
Actual score	17	14	12	11
Predicted	16.99	13.94	12.16	10.90
Absolute error	0.01	0.06	0.16	0.10

With a maximum absolute error of 0.16, this model is quite accurate. For $t = 6$, the model predicts an average score of $y = -10.12 \log(6 + 1) + 16.99 \approx 8.44$.

UNDERSTANDING AND MASTERY CHECKLISTS

CONCEPTS TO UNDERSTAND

- ☐ Graphical model
- ☐ Exponential model
- ☐ Exponential regression
- ☐ Absolute error
- ☐ Logistic model
- ☐ Logarithmic model

SKILLS TO MASTER

- ☐ Find a graphical model for a set of data and use it to make predictions.
- ☐ Find an exponential model for a set of data and use it to make predictions.
- ☐ Use exponential regression to fit an exponential model to a set of data.
- ☐ Compute the absolute error between actual and predicted values.
- ☐ Use a logistic model to make predictions.
- ☐ Find a logarithmic model for a set of data and use it to make predictions.

EXERCISES 5.5

APPLICATIONS

1. *Rabbit Population*   A population of rabbits has been observed over a 5-month period and the following data have been collected.

Month	0	1	2	3	4	5
Rabbits	20	24	30	36	45	54

a. Plot the data in the table on a coordinate system with an appropriate scale.

b. Visually fit a smooth curve to the data points you plotted in part a.

c. Use the curve in part b to estimate the population after 6 months.

2. *City Population*   The population of a city has been observed over a 5-year period and the following data have been collected.

Year	0	1	2	3	4	5
Population (in thousands)	125	127.5	130.1	132.7	135.4	138.2

a. Plot the data on a coordinate system with an appropriate scale.

b. Visually fit a smooth curve to the data points you plotted in part a.

c. Use the curve in part b to estimate the population after 6 years.

3. *Rabbit Population* The population of rabbits in Exercise 1 has been determined to be growing exponentially. Using the facts that there are 20 rabbits initially and 24 rabbits after 1 month, find an exponential function of the form $P(t) = ae^{bt}$ that estimates the number of rabbits after t months. Compare the values given by your function for the first 5 months with the values given in the table in Exercise 1. What is the largest absolute error? What is sum of the absolute errors?

4. *City Population* The population of the city in Exercise 2 has been determined to be growing exponentially. Using the fact that there are 125,000 people initially and 127,500 after 1 year, find an exponential function of the form $P(t) = ae^{bt}$ that estimates the population of the city after t years. Compare the values given by your function for the first 5 years with the values given in the table in Exercise 2. What is the largest absolute error? What is the sum of the absolute errors?

5. *Rabbit Population* Repeat Exercise 3, this time using the fact that there are 20 rabbits initially and 54 rabbits after 5 months. How does your model differ from that of Exercise 3? Is it more or less accurate?

6. *City Population* Repeat Exercise 4, this time using the fact that there are 125,000 people initially and 138,200 people after 5 years. How does your model differ from that of Exercise 4? Is it more or less accurate?

7. *Rabbit Population* Use the exponential regression feature of your graphing calculator to find a model of the form $P(t) = ae^{bt}$ that best fits the rabbit population data given in Exercise 1. How does the accuracy of this model compare to that of Exercise 3?

8. *City Population* Use the exponential regression feature of your graphing calculator to find a model of the form $P(t) = ae^{bt}$ that best fits the population data given in Exercise 2. How does the accuracy of this model compare to that of Exercise 4?

9. *Alaska Population* Population data for Alaska for the census years 1960, 1970, 1980, and 1990 are given in the table.

Year	1960	1970	1980	1990
Population	226,000	303,000	402,000	550,000

Data source: U.S. Bureau of the Census.

Construct and test an exponential model of the form $P(t) = ae^{bt}$ that estimates the population t years after 1960. According to the model, what will the population be in the year 2010? When will the population first exceed 1,500,000?

10. *New Hampshire Population* Population data for New Hampshire for the census years 1960, 1970, 1980, and 1990 are given in the table.

Year	1960	1970	1980	1990
Population	607,000	728,000	921,000	1,109,000

Data source: U.S. Bureau of the Census.

Construct and test an exponential model of the form $P(t) = ae^{bt}$ that estimates the population t years after 1960. According to the

model, what will the population be in the year 2010? When will the population first exceed 2,000,000?

11. *Sales Decline* In an effort to cut spending, a company decides to stop all advertising and promotions for one of its products. Over the next 4 months, the following monthly sales of this product are observed.

Month (t)	0	1	2	3	4
Sales (in thousands) (S)	80	72	66	61	58

Construct and test a model of the form $S(t) = ae^{bt}$ that estimates sales for month t. What does your model predict for sales for month 6? When will sales reach 20,000?

12. *Sales Increase* Due to unacceptable sales figures, a company decides to start a new promotion for one of its products. The following increases in monthly sales are observed over a 4-month period.

Month (t)	0	1	2	3	4
Sales (in thousands) (S)	58	61	65	68	73

Construct and test a model of the form $S(t) = ae^{bt}$ that estimates sales for month t. What does your model predict for sales for month 6? During what month will sales reach 100,000?

13. *Computer Viruses* In a study on computer viruses conducted in 1991, the following data were obtained concerning the percentage of American companies that encountered one or more computer viruses during 1990 and 1991.

Quarter	0 (Oct–Dec 1990)	1 (Jan–Mar 1991)	2 (Apr–Jun 1991)	3 (Jul–Sep 1991)
Percent	8	19	26	40

Data source: Dataquest/National Computer Security Association, 11/91.

Construct and test a model of the form $p(t) = ae^{bt}$ that estimates the percentage after t quarters. When does the model predict that 90% of American companies will have encountered a computer virus? Why is this model not useful for very long? What modification would you suggest?

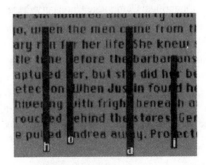

The Falling Letters virus causes characters to fall to the bottom of the screen.

14. **Internet Growth** The following table shows the approximate number of Internet hosts since 1991.

Year	1991	1993	1995	1997	1999
Internet hosts (thousands)	376	1313	5846	21,819	43,230

Data source: Network Wizards, http://www.nw.com.

Construct and test a model of the form $N(t) = ae^{bt}$ that estimates the number of hosts t years after 1991. How many hosts are predicted for the year 2005?

15. **Transistor Growth** The following table shows the approximate number of transistors in various Intel microprocessors introduced since 1971.

Year	1971	1974	1978	1982	1985	1989	1993	1997
Processor	4004	8080	8086	80286	80386	80486	Pentium	Pentium II
Transistors (in thousands)	2	6	29	134	275	1200	3100	7500

Data source: Intel Technology Briefing.

Construct and test a model of the form $N(t) = ae^{bt}$ that estimates the number of transistors t years after 1971. How many transistors are predicted for the year 2001? The human brain is estimated to have between 10 and 100 billion neurons. According to the model, when will the number of transistors exceed 100 billion?

16. **Spanish Flu Pandemic** The Spanish flu, first reported in the United States in late 1917 and carried to Europe by American troops sent to fight in World War I, is estimated to have caused the deaths of 25 to 40 million people worldwide. For the 12-week period from October 6, 1918, to November 29, 1918, the spread of the Spanish flu among U.S. Army personnel can be approximated using the logistic model

$$N(t) = \frac{21,250}{1 + 395e^{-0.992t}}$$

where $N(t)$ is the number of deaths reported in week t ($t = 0$ corresponds to the week of October 6, 1918) [*Data source:* Alfred W. Crosby, *America's Forgotten Pandemic: The Influenza of 1918* (Cambridge: Cambridge University Press, 1989)].

a. During what week did the number of deaths first exceed 10,000?

b. According to this model, what was the upper bound on the number of deaths?

17. **Spreading a Rumor** A rumor begins to spread around a college campus. The number of people N who have heard the rumor after t hours can be approximated using the logistic model

$$N = \frac{2000}{1 + 499e^{-0.3t}}$$

a. How many students started the rumor?

b. How many students have heard the rumor after 10 hours?

c. How long before 500 students have heard the rumor?

d. Plot the graph of N and estimate the limiting number of students who will hear the rumor.

e. During which hour did the greatest number of students hear the rumor?

18. **Contagious Virus** A group of tourists returns home after a 2-week tour; all have unknowingly been infected by a contagious flu virus. Suppose that, if left untreated, the spread of the flu virus in their city can be approximated using the logistic model

$$Q = \frac{150,000}{1 + 2999e^{-0.05t}}$$

where Q is the number of people with the flu virus after t days.

a. How large was the group of tourists?

b. How many people will be infected with the virus after 14 days?

c. How long before 10,000 people are infected?

d. Plot the graph of Q and estimate the limiting number of people who will be infected.

e. During which day did the greatest number of people become infected?

19. **Automobile Speed Record** The 1-mile automobile speed records for the years 1906–1983 can be approximated with the logistic model

$$S = \frac{700}{1 + 6.8e^{-0.057t}}$$

where S is the 1-mile speed record in miles per hour for year t ($t = 0$ corresponds to 1906).

a. In what year did the speed record first exceed 400 mph?

b. According to this model, what is the upper bound on the speed record?

Richard Noble's Thrust SSC breaks the sound barrier, traveling at more than 760 miles per hour.

20. **Memory Loss** A high school French class is given a test covering newly learned vocabulary words and then retested monthly

on the same material. The average scores for the first test and those of the next 3 months are given in the table.

Month	0	1	2	3
Avg Score	75.0	70.5	67.8	66.0

Using t to denote the month and y to denote the score, construct a logarithmic model of the form $y = m \log(t + 1) + b$ that best fits the data in the table. Test the model against the data in the table and use it to predict the average score after 7 months.

21. *Walking Speed* Various studies have found a correlation between the size of a city and the average walking speed of pedestrians. One such study obtained the following data.

Population	5,500	14,000	71,000	138,000	342,000
Velocity (ft/sec)	3.3	3.7	4.3	4.4	4.8

Data source: Marc and Helen Bornstein, "The Pace of Life," *Nature* 259 (19 February 1976): 557–559.

Using P to denote population and v to denote velocity, construct a logarithmic model of the form $v = m \log P + b$ that best fits the data in the table. Test the model against the data in the table and use the model to predict the walking speed in Little Rock, Arkansas (population 176,000), New York (population 7,300,000), and Mexico City (population 20,000,000).

22. *Reflex Speed* Numerous studies have shown a relationship between age and reflex speed. One such study obtained the following data.

Median age	19	27	45	60
Reflex time (sec)	0.18	0.20	0.23	0.25

Data source: Jean Hodgkins, "Reaction Time and Speed of Movement in Males and Females of Various Ages," *Res. Quart. Amer. Assoc. Hlth. Phys. Educ. Recr.* 34, no. 3 (1963): 335–343.

Using a to denote age and t to denote reflex time, construct a logarithmic model of the form $t = m \log a + b$ that best fits the data in the table. Test the model against the data in the table and use it to predict your reflex time.

■ CONCEPTS AND CRITICAL THINKING

EXERCISES 23–26 *Answer true or false.*

23. If a mathematical model fits known data exactly, then we can have 100% confidence in any predictions made that are based on the model.

24. The Malthusian exponential model tends to be accurate over very long time periods.

25. Exponential regression is used to fit a model of the form $P(t) = ae^{bt}$ to a collection of data.

26. The logistic model of population growth imposes an upper limit on the population.

EXERCISES 27–30 *Give an example of each.*

27. A reason why populations tend not to continue to grow exponentially over long time periods

28. A set of three data points such that the graph of the function obtained by exponential regression passes through all three data points

29. A graphical feature of a logistic model that distinguishes it from an exponential model

30. A logistic function with a limiting value of one million

■ QUESTIONS FOR DISCUSSION OR ESSAY

31. Discuss the pros and cons of graphical models such as the one we constructed in Example 1.

32. What are some factors that have not been considered in the logistic model for the deer population discussed in this section? What modifications would have to be made in the model to account for these factors? What are the dangers in using mathematical models such as the ones considered here to justify or condemn deer hunting?

33. Describe in your own words the various steps involved in the construction, testing, and use of an exponential model.

34. Exponential models for population growth have been validated using both theoretical and experimental observations. The theoretical explanation is based on the fact that, left unchecked, populations tend to increase at a constant percentage rate. Experimental observations have confirmed that exponential population models often work quite well for limited periods of time, but that eventually the growth rate levels off. Describe several factors that prevent unlimited growth of human populations.

■ PROJECTS FOR ENRICHMENT

35. *Comparing Models* In this section, we considered methods for constructing and testing exponential models of the form $y = ae^{bx}$. Similar methods can be used to construct and test models of many different forms. Two specific examples are **linear models** of the form $y = ax + b$ and **power models** of the form $y = ax^b$. In some cases, where it is not known which model form is needed, it may be necessary to construct and test several different models for the same set of data. Then it is possible to select the model that works the best.

a. As of 1990, the three largest cities in the United States were New York, Los Angeles, and Chicago. Census data for each of these cities for the years 1880–1930 are given in the following table.

Year	New York	Los Angeles	Chicago
1880	1,912,000	11,000	503,000
1890	2,507,000	50,000	1,100,000
1900	3,437,000	102,000	1,699,000
1910	4,767,000	319,000	2,185,000
1920	5,620,000	577,000	2,702,000
1930	6,930,000	1,238,000	3,376,000

For each city, use your graphing calculator to construct and test models of the form $y = ax + b$ and $y = ae^{bx}$ to see which fits the data best. Use the models that fit the best to predict the population of the city in 1940, and compare your prediction with the actual values (New York: 7,455,000; Los Angeles: 1,504,000; Chicago: 3,397,000). What is your conclusion as to the accuracy of the models?

b. Early in the 17th century, the German astronomer Johannes Kepler discovered three laws that govern the motion of the planets around the sun. The first states that the shape of each planet's orbit around the sun is an ellipse with the sun at one focus. The second states that a line from a planet to the sun sweeps out equal areas in equal times. Thus, referring to Figure 36, if it takes the same amount of time for a planet to travel from A to B as it does from A' to B', then the areas of ABF and $A'B'F$ are equal. Kepler's third law relates a planet's

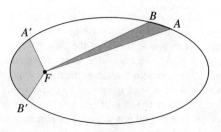

FIGURE 36

average distance from the sun to its orbital period. The data in the following table give the average distance and the orbital period for each of the nine planets.

Planet	Average distance (miles)	Period (days)
Mercury	36,000,000	88
Venus	67,000,000	225
Earth	93,000,000	365
Mars	142,000,000	687
Jupiter	484,000,000	4329
Saturn	887,000,000	10,753
Uranus	1,784,000,000	30,660
Neptune	2,795,000,000	60,150
Pluto	3,671,000,000	90,670

Construct and test models of the form $y = ae^{bx}$ and $y = ax^b$, where y is the period and x is the distance. Which model is more accurate? State Kepler's third law.

36. *Modeling the Spread of Lyme Disease* Lyme disease is a bacterial infection that is spread by the bite of certain species of ticks. The disease was first discovered in the United States in

TABLE 11 Lyme disease

Year	Cases*
1982	491
1983	1,086
1984	2,604
1985	5,352
1986	6,739
1987	9,131
1988	14,013
1989	22,816
1990	30,759
1991	40,229
1992	50,137
1993	58,394
1994	71,437
1995	83,137
1996	99,169

*Cumulative total since 1982
Data source: U.S. Centers for Disease Control.

1975, after a mysterious outbreak of arthritis near Lyme, Connecticut. Since then, cases have been reported in all states except Alaska and Montana, with a cumulative total of over 99,000 cases between 1983 and 1996 (see Table 11).

In this project, we investigate various exponential models for the spread of Lyme disease. We begin with a standard exponential growth model.

a. Use the data for the years 1982 and 1983 from Table 11 to construct a model of the form $N(t) = Ce^{kt}$, where N is the number of cases and t is the year (with $t = 0$ corresponding to 1982). How well does the model predict the number of cases in 1990 and 1995? What do you suspect about a prediction for 2000?

b. Use exponential regression on the data for the years 1982–1985 from Table 11 to construct a model of the form $N(t) = Ce^{kt}$, where N is the number of cases and t is the year (with $t = 0$ corresponding to 1982). How well does the model predict the number of cases in 1990 and 1995? What do you suspect about a prediction for 2000?

c. Based on your findings in parts a and b, justify the need for a model $N(t)$ that has a limiting value for the size of N.

Next we consider a logistic exponential model of the form

$$(1) \qquad N(t) = \frac{L}{1 + Ce^{-kt}}$$

where L, C, and k are positive constants, and $N(t)$ denotes the number of people infected with a disease at time t. The constant L is often referred to as the limiting value of the logistic model since $N(t)$ approaches (but never exceeds) L as t increases without bound. A typical graph of a model of this form is given in Figure 37.

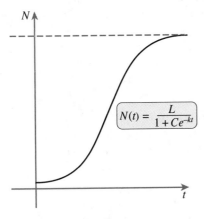

FIGURE 37

A detailed analysis of the data in Table 11 suggests that the number of cases had begun to level off by 1994. After some initial trial and error, we settle on a limiting value of 120,000 for the number of Lyme cases. We use this value to construct a

logistic model of the form given in equation (1). We first rewrite the logistic equation in a form for which we can apply an exponential regression.

d. Show that the logistic equation

$$N = \frac{120,000}{1 + Ce^{-kt}}$$

can be written in the form

$$Ce^{-kt} = \frac{120,000}{N} - 1$$

Note that this equation is of the form $ae^{bx} = y$ and is suitable for an exponential regression.

e. Use the data given in Table 11 to complete the following table for t ranging from 0 to 14. Note that $t = 0$ corresponds to 1982. For example, for $t = 0$ we have $N = 491$ so that

$$\frac{120,000}{N} - 1 = 245.4$$

You may find a spreadsheet to be extremely useful for this task.

t	N	$\dfrac{120{,}000}{N} - 1$
0	491	243.4
1	1,086	109.5
2	2,604	45.1
⋮	⋮	
14	99,169	0.2

f. Perform an exponential regression on the data in the columns labeled t and $120{,}000/N - 1$ in the preceding table, with those columns representing x and $y = ae^{bx}$, respectively. What are the computed values for k and C?

g. Write out the complete logistic model for the data given in Table 11. Use the model to predict the number of cases in the years 2000 and 2005.

h. Some graphing calculators have logistic regression capabilities. If yours does, use it to find a logistic model for the data given in Table 11. How does this model compare to the one you found in part g? In particular, how does the limiting value compare?

CHAPTER 5 REVIEW

EXERCISES 1–2 *Sketch the graph of f and use it to sketch the graph of g.*

1. $f(x) = 2^x$
   a. $g(x) = 2^{-x}$
   b. $g(x) = 2^x + 2$
   c. $g(x) = 2^{x+1}$

2. $g(x) = (\frac{1}{3})^x$
   a. $g(x) = -(\frac{1}{3})^x$
   b. $g(x) = (\frac{1}{3})^x - 4$
   c. $g(x) = (\frac{1}{3})^{x-2}$

EXERCISES 3–6 *Solve the given equation without using logarithms.*

3. $3^x = 27$

4. $\frac{1}{4^x} = 16$

5. $5^{1-2x} = 25$

6. $3^{x^2+4x} = \frac{1}{27}$

EXERCISES 7–12 *Evaluate each logarithmic expression without using a calculator.*

7. $\log_2 \frac{1}{4}$

8. $\log_3 81$

9. $\log 0.01$

10. $\ln e^3$

11. $\log_5 \sqrt{5}$

12. $\log_\pi 1$

EXERCISES 13–16 *Write an equivalent exponential equation for the given logarithmic equation.*

13. $\log_2 \frac{1}{16} = -4$

14. $\ln(3 - x) = 1$

15. $\log(2x + 1) = 4$

16. $\log_3 18 = x$

EXERCISES 17–20 *Write an equivalent logarithmic equation for the given exponential equation.*

17. $e^5 = x$

18. $3^{x-5} = 7$

19. $2^{x+3} = 5$

20. $10^{1/x} = \frac{1}{2}$

EXERCISES 21–24 *Sketch the graph of the given function and find its domain.*

21. $f(x) = \log_5 x$

22. $f(x) = 3 + \log x$

23. $f(x) = \ln(x - 2)$

24. $f(x) = -\ln(5x)$

EXERCISES 25–28 *Use logarithmic identities to expand each expression, rewriting it as a sum, difference, or multiple of logarithms.*

25. $\log_2(xy^2)$

26. $\log_3\left(\frac{x^2}{z}\right)$

27. $\log \sqrt[3]{x^2 y}$

28. $\ln\left(\frac{x^3 y^2}{\sqrt{z}}\right)$

EXERCISES 29–32 *Use logarithmic identities to rewrite each expression as a single logarithm.*

29. $2 \log x + 3 \log y$

30. $4 \log_2 p - 3 \log_2 q$

31. $\frac{1}{3} \ln x + \frac{2}{3} \ln y - \ln z$

32. $3 \log 4 - \log 6 + 2 \log 2$

EXERCISES 33–34 *Estimate the given logarithmic expression. Give answers accurate to 5 decimal places.*

33. $\log_4 10$

34. $\log_{2/5} 130$

EXERCISES 35–42 *Solve the logarithmic equation.*

35. $\log_2 x = -1$

36. $\log_8 x = \frac{1}{3}$

37. $\log_5(2x + 4) = \log_5 10$

38. $\ln(3x + 5) = \ln x$

39. $\log(5y - 2) = 1$

40. $2 \log_4 x - \log_4(x + 1) = 1 - \log_4 3$

41. $\ln(x - 1) + \ln(x + 2) = \ln 4$

42. $\log \sqrt{x + 1000} = 2$

EXERCISES 43–50 *Solve the exponential equation. When approximating, give answers to five decimal places of accuracy. You may wish to check your solution graphically.*

43. $5^x = 10$

44. $e^x = 2$

45. $1.05^t = 10$

46. $3^{-2x} = 9$

47. $4e^{0.1x} = 20$

48. $350 - 280e^{-0.05t} = 180$

49. $3^{x+2} = 2^{2x}$

50. $2^{2x} - 5 \cdot 2^x = 0$

EXERCISES 51–54 *Estimate the solution to the given equation to the nearest hundredth.*

51. $e^x = 8 - x^2$

52. $\ln x = x^3 - 3x^2$

53. $\ln(3x) + 0.02x^3 + 4x = x^2$

54. $-\frac{x^3}{9} + e^{x+2} = 2x^2 + 5x$

EXERCISES 55–66 *Parts a and b are connected: Part a involves an elementary concept, whereas part b involves related material from this chapter. First answer part a, and then use this result to answer part b.*

55. a. Simplify $\frac{2^x}{3^x}$.
    b. Solve $\frac{2^x}{3^x} = 2$.

56. a. Simplify $\sqrt[3]{x^6 y^9}$.
    b. Write $\ln(\sqrt[3]{x^6 y^9})$ as a sum of logarithms.

57. a. Solve $x^2 + x - 2 = 0$.
    b. Solve $3^{x^2+x} = 9$.

58. a. Simplify $\dfrac{x^2 - y^2}{x - y}$.

  b. Write $\log(x^2 - y^2) - \log(x - y)$ as a single logarithm.

59. a. Graph $y = e^{-x}$ and $y = x$ on the same set of coordinate axes.

  b. How many solutions does $e^{-x} - x = 0$ have?

60. a. Simplify $\dfrac{x^2 - 5x + 6}{x - 3}$.

  b. Solve $\log(x^2 - 5x + 6) - \log(x - 3) = \log 2$.

61. a. Solve $2x - 3 \le 0$.

  b. Find the domain of $f(x) = \ln(2x - 3)$.

62. a. Solve $u^2 - 11u + 10 = 0$.

  b. Solve $10^{2x} - 11(10^x) + 10 = 0$.

63. a. Solve $u + \frac{1}{u} = 2$.

  b. Solve $e^x - e^{-x} = 2$.

64. a. Simplify $f(f(f^{-1}(x)))$.

  b. Simplify $e^{(e^{\ln x})}$.

65. a. Graph $y = 4x - x^2$.

  b. Find the domain of $g(x) = \ln(4x - x^2)$.

66. a. Compute $16(\frac{3}{2})^4$.

  b. How long will it take $16 to grow to $81 with 50% interest compounded annually?

67. Compound Interest Suppose that $50 is deposited into an account paying 4% interest compounded quarterly. Find the balance in the account after 3 years.

68. Compound Interest If $42 is deposited into an account paying 3% compounded continuously, then what is the balance after 4 years?

69. Doubling Money How long does it take money that is deposited into an account paying 3% interest compounded monthly to double?

70. Sound Intensity The decibel rating of a sound of intensity I is given by $D = 10 \log(I/I_0)$, where I_0 is the intensity of a sound that is just perceptible. How many times more intense is a sound of 100 decibels than one of 50 decibels?

71. Population Growth A certain bacteria colony is growing exponentially. Initially there are 10 bacteria. After 1 hour, there are 25. How long will it take until the population of the colony surpasses the 1 million mark?

72. Radioactive Decay Suppose that a sample of a radioactive substance having a half-life of 12,000 years was placed in a time capsule 2000 years ago. What percentage of the original sample remains today?

73. Infectious Disease A community of astronauts living on a base on Mars is exposed to an infectious disease. The number of people who have contracted the disease t days from the initial exposure is given by

$$Q = \frac{200{,}000}{1 + 1999e^{-0.08t}}$$

  a. How many astronauts have the disease after 5 days?

  b. How many days will it take before 1000 astronauts have become infected?

  c. Plot the graph of Q and estimate the limiting number of astronauts that will become infected.

  d. During what day did the greatest number of astronauts become infected: day 5, day 96, or day 120?

74. Memory Model It is hypothesized that if a subject were introduced to 100 guests at a cocktail party, the number of people whose names the subject would remember after t months would be given by $y = m \log(t + 1) + b$. A certain subject remembered 50 people immediately after the party and 36 people 2 years after the party. How many guests will the subject remember 10 years after the party? How many guests did the subject remember 1 year after the party?

75. Car Depreciation A car purchased new for $14,000 in 1995 depreciates in value each year, as shown in the following table.

Year	1995	1996	1997	1998	1999
Value	$14,000	$12,000	$10,400	$8900	$7700

Construct and test an exponential model of the form $V(t) = ae^{bt}$ that estimates the value of the car t years after 1995. When will the car be worth $2000?

76. U.S. Population Growth U.S. census data for the years 1790–1840 are given in the table. Construct and test an exponential model of the form $P(t) = ae^{bt}$ that estimates the population t years after 1790. Use the model to predict the population in 1850. How does the predicted value compare to the actual population of 23.2 million in 1850?

Year	1790	1800	1810	1820	1830	1840
Population (in millions)	3.9	5.3	7.2	9.6	12.9	17.1

Data source: U.S. Bureau of the Census.

CHAPTER 5 TEST

PROBLEMS 1–8   *Answer true or false.*

1. $\log_a x \cdot \log_a y = \log_a x + \log_a y$ for all $x > 0$ and $y > 0$.

2. $\log_a(x^y) = y \log_a x$ for all $x > 0$ and $y > 0$.

3. If $f(x) = a^x$, then $f^{-1}(x) = \dfrac{1}{a^x}$.

4. The logarithm of a number cannot be negative.

5. The logarithm of a negative number is not defined.

6. Natural logarithms are simply logarithms with base e.

7. It is possible that a sample of a substance undergoing exponential decay will take longer to decay from 100 grams to 50 grams than from 50 grams to 25 grams.

8. Both $f(x) = a^x$ and $g(x) = \log_a x$ are 1–1 functions.

PROBLEMS 9–12   *Give an example of each of the following.*

9. An exponential function that is always decreasing

10. A number not in the domain of $\ln(x - 5)$ but in the domain of $\ln x$

11. A real number whose common logarithm is a negative integer

12. A reason why exponential growth of populations of bacteria colonies cannot continue indefinitely

PROBLEMS 13–16   *Solve the given equation. When approximating, give answers to five decimal places of accuracy.*

13. $2^{x^2+x} = 4$

14. $\log(x + 1) = 12$

15. $3^{x-1} = 2^x$

16. $5e^{0.3x} = 10$

PROBLEMS 17–18   *Evaluate the given logarithm.*

17. $\log_2 32$

18. $\log_3 \frac{1}{27}$

PROBLEMS 19–21   *Graph the given function.*

19. $f(x) = e^{x-2}$

20. $g(x) = \left(\frac{2}{3}\right)^x$

21. $f(x) = \ln(x + 3)$

22. Express $\log\left(\dfrac{x^4 y^2}{z^{-2}}\right)$ in terms of logarithms of x, y, and z.

23. Write $3 \ln x - 4 \ln y + 5 \ln z^2$ as a single logarithm.

24. Suppose that $300 is deposited into an account paying 6% interest compounded annually. How long does it take the balance to grow to $500?

25. Suppose that a radioactive sample is decaying exponentially and that 10 grams of the sample were present on January 1, 1997, and 7 grams on January 1, 2002. In what year will the quantity of the sample first drop below 1 gram?

26. The approximate number of new AIDS cases reported in each of the years 1983–1986 is given in the following table.

Year	1983	1984	1985	1986
New AIDS cases	2100	4400	8200	13,100

Data source: U.S. Centers for Disease Control.

Construct and test a model of the form $N(t) = ae^{bt}$ that estimates the number of new cases reported t years after 1983. Use the model to predict the number of new cases reported in 1987. How does the prediction compare to 21,100, the approximate number of new cases reported in 1987?

6 RELATIONS AND CONIC SECTIONS

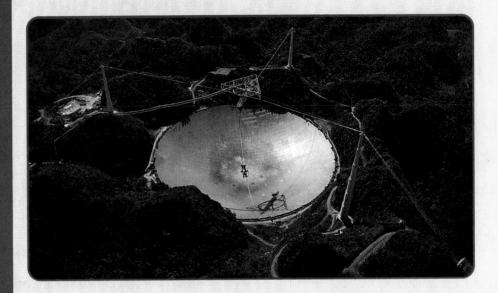

The Arecibo radio telescope located near Arecibo, Puerto Rico, is the largest single telescope of any kind. Electromagnetic transmissions from celestial objects are reflected from the 1000-foot-diameter dish to the antenna suspended above it. Radio telescopes, although employed for such exotic purposes as identifying black holes and quasars, determining the chemical composition of distant galaxies, and even searching for extraterrestrial life, are based on the same principle as the automobile headlight: the reflective properties of conic sections.

SECTION 6.1   RELATIONS AND THEIR GRAPHS

- ☐ How can symmetry be used in data compression?
- ☐ What does the message "ti bib dod" mean to Inspector Magill?
- ☐ How can you construct a face with two left sides?
- ☐ In what sense can one relation be the "mirror image" of another?
- ☐ What types of symmetry are possessed by the human body?

Throughout the text, we have considered equations of many different varieties and in many different contexts. We have studied equations of lines, such as $2x + y = 3$; equations of circles, such as $x^2 + y^2 = 9$; and equations that define functions, such as $y = e^{3x}$. We have also encountered equations arising from real-world contexts; examples include $A = \pi r^2$, $B = P(1 + r)^t$, and $F = \frac{9}{5}C + 32$. All of these equations are examples of **relations.** In this text, we deal with a variety of relations, nearly all of which are defined by equations relating two or more variables.

An equation involving only the variables x and y, such as $2x + y = 3$, $x^2 + y^2 = 9$, or $y = e^{3x}$, defines a **relation in x and y.** The graph of such a relation is simply the graph of the corresponding equation. In the case where the equation defines a function, we have many tools and techniques at our disposal for studying the graph, not the least of which is the graphing calculator. You may recall, however, that not all equations in x and y define functions. For example, the graph of $x^2 + y^2 = 9$ is a circle, which fails the vertical line test. It follows that the graph of $x^2 + y^2 = 9$ could not be the graph of a function $y = f(x)$.

In this section, we generalize some of the graphing techniques developed in Chapter 3 for functions, and apply them to more general relations.

TRANSLATIONS

Recall from Section 3.3 that the graph of $y = f(x - h) + k$ or, equivalently, $y - k = f(x - h)$, is a translation of the graph of $y = f(x)$ h units horizontally and k units vertically. In other words, replacing x with $x - h$ and y with $y - k$ in the equation $y = f(x)$ has the effect of translating the graph h units horizontally and k units vertically. This same result holds for relations in general. Consider, for example, the graphs of $x^2 + y^2 = 9$ and $(x - 1)^2 + (y - 2)^2 = 9$ shown in Figures 1 and 2. Both are circles of radius 3, but the first is centered at the origin, and the second at $(1, 2)$. Thus, the second graph is a translation of the first, 1 unit to the right and 2 units up. By comparing the two equations, we see that replacing x with $x - 1$ and y with $y - 2$ in the equation $x^2 + y^2 = 9$ has the effect of translating the graph 1 unit to the right and 2 units up.

In general, substituting $x - h$ for x results in a translation of the graph h units to the right. For example, if we substitute $x - 2$ for x, the graph will be translated 2 units to the right; if we substitute $x + 3$ for x, the graph will be translated -3 units to the right (3 units to the left), and so on. Similarly, substituting $y - k$ for y translates the graph k units upward, whereas substitution of $y + k$ for y translates the graph k units downward.

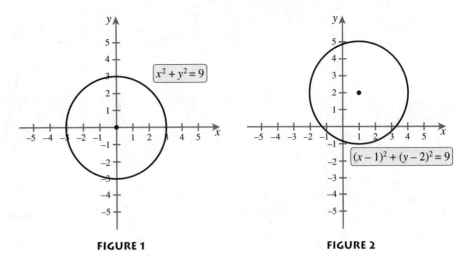

FIGURE 1   **FIGURE 2**

Translations ☐

Substitution	Resulting translation
$x - h$ in place of x	h units to the right
$x + h$ in place of x	h units to the left
$y - k$ in place of y	k units up
$y + k$ in place of y	k units down

E X A M P L E 1   ▨   **Translating a Parabola**

The graph of $y = x^2$ is given in Figure 3. Use it to graph $y + 4 = (x - 7)^2$.

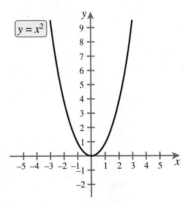

FIGURE 3

SOLUTION   The relation $y + 4 = (x - 7)^2$ is the result of substituting $x - 7$ for x and $y + 4$ for y. Thus, its graph can be obtained by translating the graph of $y = x^2$, as shown in Figure 4, 7 units to the right and 4 units down.

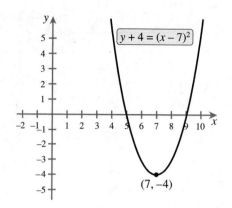

FIGURE 4

EXAMPLE 2 ■ **Graphing the Translation of a Relation**

Use the graph of $\dfrac{x^2}{4} + \dfrac{y^2}{9} = 1$ shown in Figure 5 to graph

$$\frac{(x+2)^2}{4} + \frac{(y-1)^2}{9} = 1$$

SOLUTION The relation $\dfrac{(x+2)^2}{4} + \dfrac{(y-1)^2}{9} = 1$ is obtained from $\dfrac{x^2}{4} + \dfrac{y^2}{9} = 1$ by substituting $x + 2$ for x and $y - 1$ for y. Thus, the desired graph can be found by translating the graph of $\dfrac{x^2}{4} + \dfrac{y^2}{9} = 1$, as shown in Figure 6, 2 units to the left and 1 unit up.

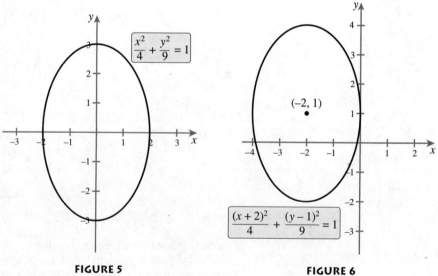

FIGURE 5   **FIGURE 6**

REFLECTIONS

From our work in Section 3.3, we know that the graph of $y = f(-x)$ can be obtained by reflecting the graph of $y = f(x)$ about the y-axis. The same is true for graphs of relations. More precisely, if we substitute $-x$ for x in a relation, the graph of the new relation will be a reflection of the graph of the original relation about the y-axis. Similarly, substitution of $-y$ for y will cause a reflection about the x-axis. Reflection about the origin is defined by reflecting about both the x- and y-axes, and it is achieved by substituting both $-x$ for x and $-y$ for y in a relation. These observations can be confirmed by considering the changes that occur in the coordinates of a point (x, y) upon reflection, as suggested by Figure 7.

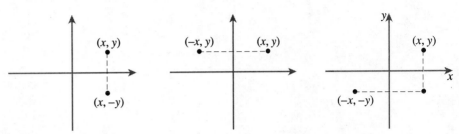

Reflection about the x-axis   **Reflection about the y-axis**   **Reflection about the origin**

FIGURE 7

Reflections □

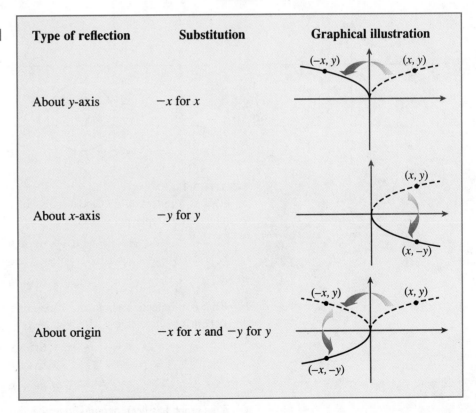

Type of reflection	Substitution	Graphical illustration
About y-axis	$-x$ for x	
About x-axis	$-y$ for y	
About origin	$-x$ for x and $-y$ for y	

EXAMPLE 3   ■   Using Reflections to Find Relations

The graph of $x^2 + y^2 - 4x - 6y + 9 = 0$ is given in Figure 8. Parts a, b, and c each show a reflection of this graph. Use this information to find the equation corresponding to each of the reflected graphs.

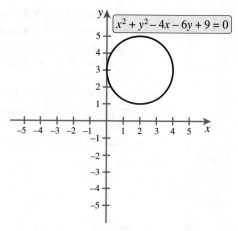

FIGURE 8

a.

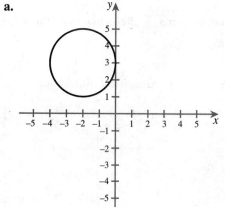

b.

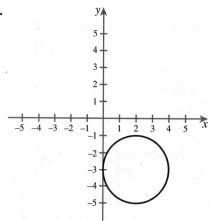

c.

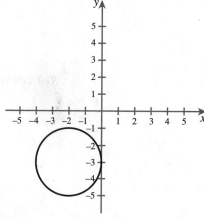

SOLUTION

a. This is a reflection about the y-axis of the original graph. Thus, its equation is obtained by replacing x with $-x$.

$$x^2 + y^2 - 4x - 6y + 9 = 0 \qquad \text{The original relation}$$
$$(-x)^2 + y^2 - 4(-x) - 6y + 9 = 0 \qquad \text{Replacing } x \text{ with } -x$$
$$x^2 + y^2 + 4x - 6y + 9 = 0 \qquad \text{Simplifying}$$

b. This is a reflection about the x-axis of the original graph. Thus, its equation is obtained by replacing y with $-y$.

$$x^2 + y^2 - 4x - 6y + 9 = 0 \qquad \text{The original relation}$$
$$x^2 + (-y)^2 - 4x - 6(-y) + 9 = 0 \qquad \text{Replacing } y \text{ with } -y$$
$$x^2 + y^2 - 4x + 6y + 9 = 0 \qquad \text{Simplifying}$$

c. This graph has been obtained by reflecting the original graph about the origin. We find its equation by replacing x with $-x$ and y with $-y$.

$$x^2 + y^2 - 4x - 6y + 9 = 0 \qquad \text{The original relation}$$
$$(-x)^2 + (-y)^2 - 4(-x) - 6(-y) + 9 = 0 \qquad \text{Replacing } x \text{ with } -x \text{ and } y \text{ with } -y$$
$$x^2 + y^2 + 4x + 6y + 9 = 0 \qquad \text{Simplifying}$$

Certain graphs can be obtained by recognizing them as reflections of known graphs, as illustrated in the following examples.

EXAMPLE 4 ▦ **Graphing a Reflection**

The graph of $y = x^2 + 3$ is given in Figure 9. Use it to graph $y = -x^2 - 3$.

SOLUTION   We can rewrite $y = -x^2 - 3$ as $-y = x^2 + 3$. This relation can be obtained from $y = x^2 + 3$ by substituting $-y$ for y. Thus, its graph is a reflection about the x-axis of the graph of $y = x^2 + 3$; that is, we simply flip the graph of $y = x^2 + 3$ upside down, as shown in Figure 10.

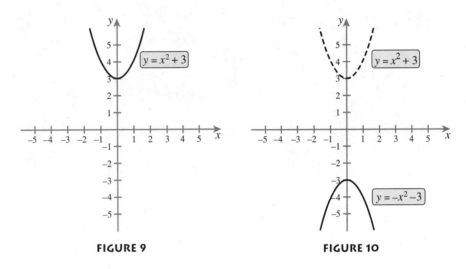

FIGURE 9   **FIGURE 10**

EXAMPLE 5 ▦ **Graphing a Combination of Reflections and Translations**

The graph of $x = y^2$ is given in Figure 11. Use it to sketch a graph of each of the following.

a. $x = -y^2$
b. $x - 3 = -(y + 2)^2$

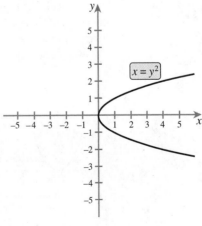

FIGURE 11

SOLUTION

a. The equation $x = -y^2$ is equivalent to $-x = y^2$ and so it can be obtained from the equation $x = y^2$ by replacing x with $-x$. Thus, the graph of $x = -y^2$ can be obtained from the graph of $x = y^2$ by reflecting about the y-axis, as shown in Figure 12.

b. The equation $x - 3 = -(y + 2)^2$ can be obtained from $x = -y^2$, the equation in part a, by substituting $x - 3$ in place of x and $y + 2$ in place of y. Thus, we translate the graph from part a, as shown in Figure 13, 3 units to the right and 2 units down.

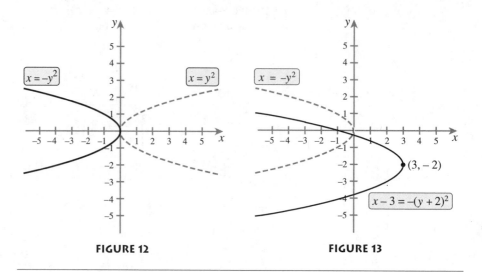

FIGURE 12   FIGURE 13

SYMMETRY

As we have seen, the reflection about the y-axis of the graph of a relation can be obtained by substituting $-x$ for x. Now suppose that we wish to reflect the graph of $y = x^2$ about the y-axis. If we substitute $-x$ for x, we obtain $y = (-x)^2 = x^2$. Thus, the reflection about the y-axis of the graph of $y = x^2$ is itself! Whenever the graph of a relation remains the same after undergoing some transformation (such as a reflection), we say that the graph is **symmetric** with respect to that transformation. Thus, we say that the graph of $y = x^2$ is symmetric with respect to reflection about the y-axis, or simply **symmetric with respect to the y-axis.**

The symmetry of the graph of $y = x^2$ with respect to the y-axis can be seen graphically as well. As Figure 14 shows, reflecting the graph of $y = x^2$ about the y-axis leaves the graph unchanged. This is because the right-hand portion of the graph is just a mirror image of the left-hand portion.

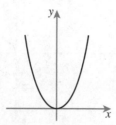

The graph of $y = x^2$

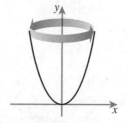

Reflection about the y-axis

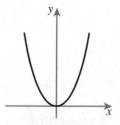

The original graph

FIGURE 14

More generally, the graph of a relation will be symmetric with respect to the y-axis if reflecting about the y-axis leaves the graph unchanged. Algebraically, this means that the relation resulting from substituting $-x$ for x is equivalent to the original relation. In other words, $(-x, y)$ is on the graph of the relation whenever (x, y) is on the graph. Geometrically, this means that the left- and right-hand sides of the graph are mirror images of one another. Figure 15 shows several graphs that are symmetric with respect to the y-axis.

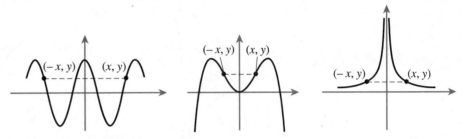

FIGURE 15   *Graphs symmetric with respect to the y-axis*

EXAMPLE 6   ▨   **Testing for Symmetry with Respect to the *y*-Axis**

Without graphing, indicate whether or not the relation $x^2 + 2y^2 = 4y - x^4$ is symmetric with respect to the y-axis.

SOLUTION   We replace x by $-x$ to test for symmetry with respect to the y-axis.

$$x^2 + 2y^2 = 4y - x^4 \qquad \text{The original relation}$$
$$(-x)^2 + 2y^2 = 4y - (-x)^4 \qquad \text{Substituting } -x \text{ for } x$$
$$x^2 + 2y^2 = 4y - x^4 \qquad \text{Simplifying}$$

Since the resulting relation is equivalent to the original, the graph of $x^2 + 2y^2 = 4y - x^4$ is symmetric with respect to the y-axis.

In a similar fashion, the graph of a relation is said to be **symmetric with respect to the *x*-axis** if the graph is unchanged after reflecting about the x-axis. In other words, a graph is symmetric with respect to the x-axis if the top half and the bottom half are mirror images of one another. Algebraically, this means that substituting $-y$ for y in the relation yields an equivalent relation. Figure 16 shows the graphs of several relations that are symmetric with respect to the x-axis.

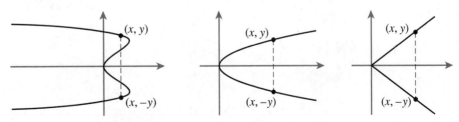

FIGURE 16   *Graphs symmetric with respect to the x-axis*

If the graph of a relation stays the same after reflection about the origin, then it is said to be **symmetric with respect to the origin.** Thus, the graph of a relation will be symmetric with respect to the origin if substitution of $-x$ for x *and* $-y$ for y yields an equivalent equation. Figure 17 shows the graphs of some relations that are symmetric with respect to the origin.

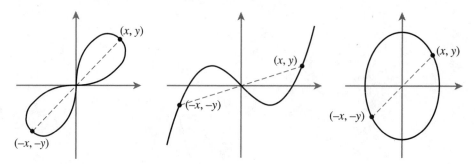

FIGURE 17 *Graphs symmetric with respect to the origin*

Symmetry can be a powerful tool for graphing relations. If we are aware of the symmetries that a graph possesses, then we can greatly reduce the number of points that must be plotted. For instance, if we know that a relation is symmetric with respect to the y-axis, then we need only plot enough points to obtain the right half of the graph; the left half of the graph will just be the reflection about the y-axis of the right half. The following examples illustrate this technique.

EXAMPLE 7 ▓ **Using Symmetry to Complete the Graph of a Relation**

A portion of the graph of $\dfrac{x^2}{16} + \dfrac{y^2}{9} = 1$ is shown in Figure 18. Use symmetry to complete the graph.

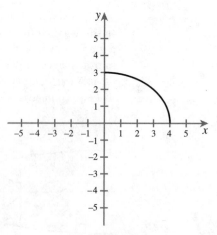

FIGURE 18

SOLUTION   We begin by noting that if either x is replaced by $-x$ or y is replaced by $-y$, we will obtain an equivalent relation. Thus, the graph must be symmetric with respect to the x-axis and the y-axis. Since the graph is symmetric with respect to the y-axis, the left-hand portion and the right-hand portion will be mirror images of one another about the y-axis. Reflecting the given portion of the graph about the y-axis, we obtain Figure 19. Finally, since the graph is symmetric with respect to the x-axis, the bottom portion of the graph will be a mirror image of the top portion. Reflecting about the x-axis gives us the final graph in Figure 20.

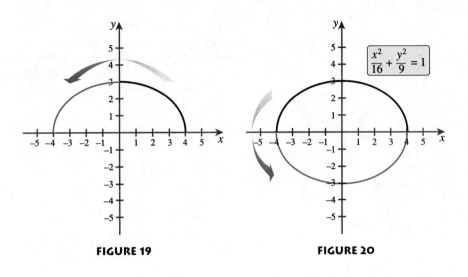

FIGURE 19   **FIGURE 20**

EXAMPLE 8 ■   Using Symmetry to Complete the Graph of a Relation

A portion of the graph of $y = x^3 - x$ is shown in Figure 21. Use symmetry to complete the graph.

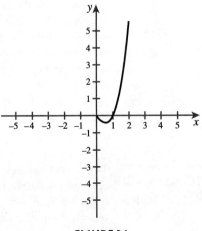

FIGURE 21

SOLUTION We test $y = x^3 - x$ for the three types of symmetry as follows.

y-axis (Replace x with $-x$)	*x*-axis (Replace y with $-y$)	Origin (Replace x with $-x$ and y with $-y$)
$y = (-x)^3 - (-x)$	$(-y) = x^3 - x$	$(-y) = (-x)^3 - (-x)$
$y = -x^3 + x$	$y = -x^3 + x$	$-y = -x^3 + x$
		$y = x^3 - x$

Since only the test for symmetry with respect to the origin yields the same equation, the relation is symmetric with respect to the origin, but not to the *y*-axis or *x*-axis. To obtain the graph, we reflect about the origin by first reflecting about the *y*-axis to obtain the red curve shown in Figure 22, and then reflecting the red curve about the *x*-axis to obtain the final graph shown in Figure 23.

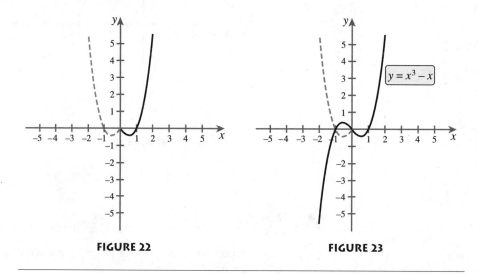

FIGURE 22   **FIGURE 23**

Just as with translations and reflections, the observations we have made about symmetry for graphs of relations build on observations we made in Section 3.3 about symmetry for graphs of functions. Recall that a function whose graph is symmetric with respect to the *y*-axis is said to be even, whereas one whose graph is symmetric with respect to the origin is said to be odd. One fundamental difference between symmetry for relations and symmetry for functions is that no function has a graph that is symmetric with respect to the *x*-axis. Graphs that are symmetric with respect to the *x*-axis do not pass the vertical line test and so cannot be graphs of functions.

RELATIONS AND GRAPHING CALCULATORS

Throughout the text, we have encountered instances in which a graphing calculator has been an invaluable tool for relations and their graphs. Unfortunately, a relation must be in the form $y = \square$, where $\square$ is an expression involving only the variable *x*, in order to be graphed directly using a graphing calculator. In other words, a relation must define *y* as a function of *x* in order for its graph to be obtained by entering and plotting a single equation. Other relations in x and y either cannot be graphed at all or can only be graphed in several pieces. The latter case arises when, in solving an equation for the variable y, we obtain two or more solutions—called **branches**—of the form $y = f(x)$. Each of these branches can then be graphed, and the union of the graphs gives the graph of the relation.

EXAMPLE 9 ■ Plotting a Relation with a Graphing Calculator

Use a graphing calculator to plot the graph of $4x^2 + 9y^2 = 36$, and determine the highest and lowest points on the graph.

SOLUTION We begin by solving for y.

$$4x^2 + 9y^2 = 36$$
$$9y^2 = 36 - 4x^2$$
$$y^2 = 4 - \frac{4}{9}x^2$$
$$y = \pm\sqrt{4 - \frac{4}{9}x^2}$$

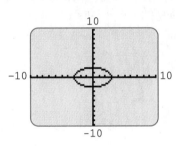

FIGURE 24

Thus, there are two branches, $y = \sqrt{4 - \frac{4}{9}x^2}$ and $y = -\sqrt{4 - \frac{4}{9}x^2}$. The combined graph of these branches forms the ellipse shown in Figure 24. Note that the top half is the graph of $y = \sqrt{4 - \frac{4}{9}x^2}$ and the bottom half is that of $y = -\sqrt{4 - \frac{4}{9}x^2}$. The highest and lowest points lie on the y-axis, at the points $(0, 2)$ and $(0, -2)$, respectively. We will study ellipses in detail in Section 6.3.

UNDERSTANDING AND MASTERY CHECKLISTS

CONCEPTS TO UNDERSTAND

- ☐ Relation
- ☐ Translation
- ☐ Reflection
- ☐ Symmetry

SKILLS TO MASTER

- ☐ Given the graph of a relation, sketch the graph of its translation.
- ☐ Given the graph of a relation, sketch the graph of its reflection.
- ☐ Given a relation and its graph, together with a translation of its graph, find an equation for the translated graph.
- ☐ Given a relation and its graph, together with a reflection of its graph, find an equation for the reflected graph.
- ☐ Determine the symmetries of a relation.
- ☐ Use symmetry as an aid in graphing a relation.

EXERCISES 6.1

EXERCISES 1–4 *A relation and its graph are given. Use translation to sketch the graph of the indicated relation.*

1. $y = x^3$

   a. $y = (x + 2)^3$

   b. $y + 1 = (x - 3)^3$

   c. $y = (x + 4)^3 - 2$

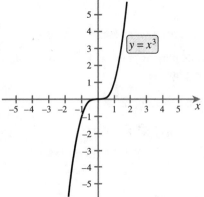

4. $\dfrac{y^2}{4} - x^2 = 1$

   a. $\dfrac{y^2}{4} - (x - 2)^2 = 1$

   b. $\dfrac{(y + 1)^2}{4} - (x + 3)^2 = 1$

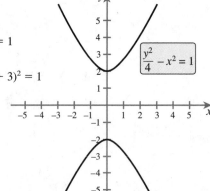

2. $x = y^2$

   a. $x = (y - 3)^2$

   b. $x - 4 = (y + 1)^2$

   c. $x = (y - 5)^2 + 2$

EXERCISES 5–8 *A relation and its graph are given, together with a translated graph. Find an equation for the translated graph.*

5. Original relation: $y = \sqrt{x}$

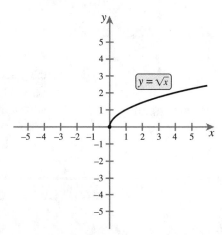

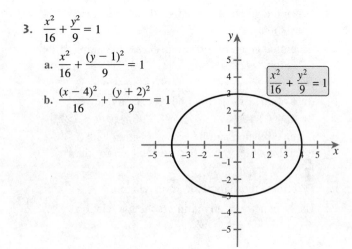

3. $\dfrac{x^2}{16} + \dfrac{y^2}{9} = 1$

   a. $\dfrac{x^2}{16} + \dfrac{(y - 1)^2}{9} = 1$

   b. $\dfrac{(x - 4)^2}{16} + \dfrac{(y + 2)^2}{9} = 1$

Translated graph:

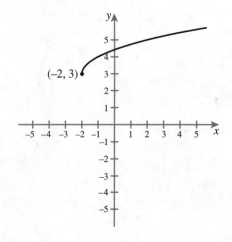

6. Original relation: $x = \sqrt[3]{y}$

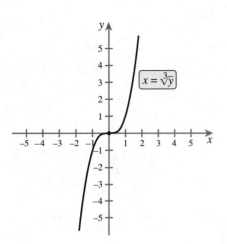

Translated graph:

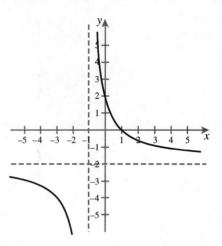

Translated graph:

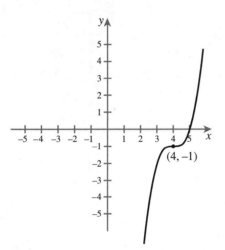

8. Original relation: $y^2 x = 1$

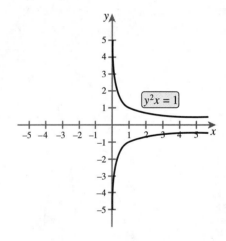

7. Original relation: $xy = 4$

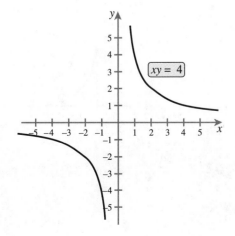

Translated graph:

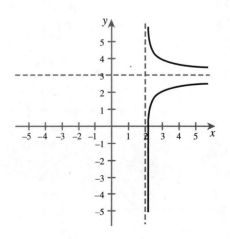

EXERCISES 9–12 *A relation and its graph are given. Use reflection to sketch the graph of the indicated relation.*

9. $x = \sqrt{y} - 2$
   a. $-x = \sqrt{y} - 2$
   b. $x = \sqrt{-y} - 2$

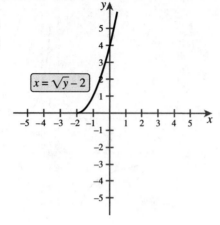

12. $\sqrt[3]{x} + \sqrt[3]{y} = \frac{1}{2}$
   a. $\sqrt[3]{x} - \sqrt[3]{y} = \frac{1}{2}$
   b. $\sqrt[3]{y} = \frac{1}{2} + \sqrt[3]{x}$

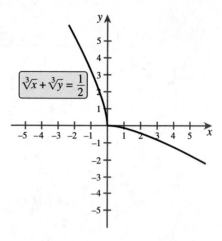

EXERCISES 13–14 *A relation and its graph are given, together with a reflected graph. Find an equation for the reflected graph.*

13. Original relation: $\sqrt{-x} + \sqrt{y} = 2$

10. $y = x^3 + 1$
   a. $-y = x^3 + 1$
   b. $y = -x^3 + 1$

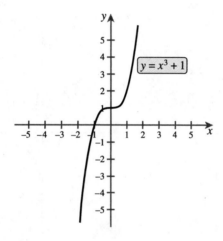

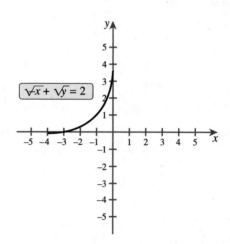

Reflected graph:

11. $x^3 - y^3 = 1$
   a. $y^3 - x^3 = 1$
   b. $x^3 + y^3 = 1$

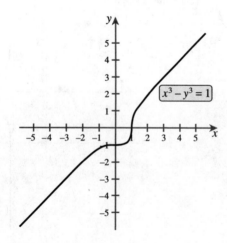

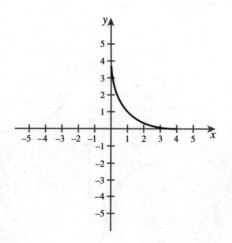

14. Original relation: $y + 1 = \dfrac{x + 5}{x^2 - 8x + 17}$

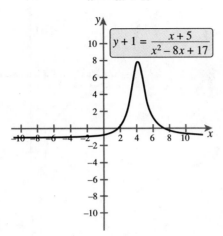

Reflected graph:

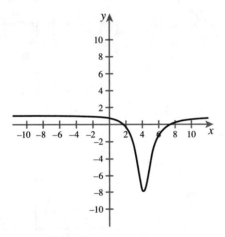

EXERCISES 15–16   *A relation and its graph are given. Use translation and/or reflection to sketch the graph of the indicated relation.*

15. $x = y^3$

   a. $x = -(y + 2)^3$

   b. $x + 3 = -(y - 4)^3$

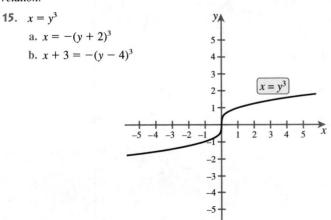

16. $y = x^2$

   a. $y = -(x - 1)^2$

   b. $y - 1 = -(x + 3)^2$

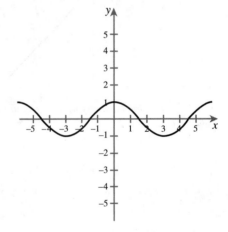

EXERCISES 17–20   *Check the given graph for symmetry with respect to the y-axis, the x-axis, and the origin.*

17.

18.

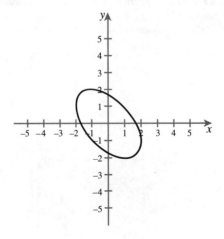

19.

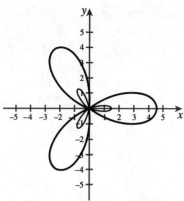

20.

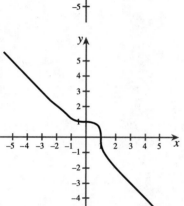

EXERCISES 21–28 *Test for symmetry with respect to the x-axis, the y-axis, and the origin.*

21. $x = y^2$

22. $x^2 + y^3 = 10$

23. $y = \frac{1}{x}$

24. $y^2 = x^4 + x^2 + 2$

25. $x^3 - y^3 = 1$

26. $y = |x|$

27. $y = e^x - e^{-x}$

28. $x = e^y + e^{-y}$

EXERCISES 29–36 *A portion of the graph of a relation is shown. Test for symmetry with respect to the x-axis, the y-axis, and the origin, and then use symmetry to complete the graph.*

29. $y = x^4$

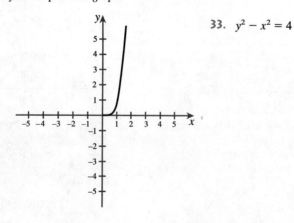

30. $xy = 5$

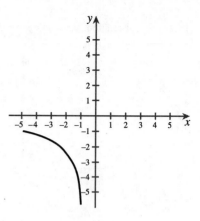

31. $10y = x^5$

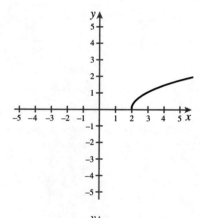

32. $y^2 = x - 2$

33. $y^2 - x^2 = 4$

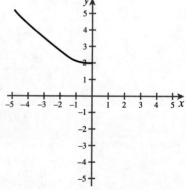

34. $x^6 + y^6 = 729$

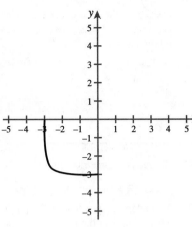

35. $x^2 y + 2y = 10$

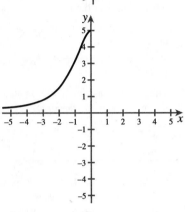

36. $|x| + |y| = 4$

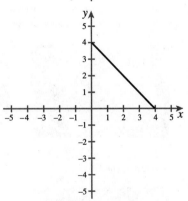

EXERCISES 37–40 *Use a graphing calculator to determine how the relations in parts a through c have been obtained from the given relation using translation and/or reflection.*

37. $y = x^2 + x - 2$

　　a. $y = x^2 + x - 4$

　　b. $y = x^2 - 5x + 4$

　　c. $y = -x^2 - x$

38. $y = x^3 - x$

　　a. $y = x^3 - 9x^2 + 26x - 24$

　　b. $y = x^3 - x + 4$

　　c. $y = -x^3 + x + 3$

39. $y = \dfrac{5}{x^2 + 1}$

　　a. $y = \dfrac{3x^2 + 8}{x^2 + 1}$

　　b. $y = \dfrac{-5}{x^2 + 2x + 2}$

　　c. $y = \dfrac{2x^2 + 8x + 15}{x^2 + 4x + 5}$

40. $y = \sqrt[3]{x^2 + 1}$

　　a. $y = \sqrt[3]{x^2 - 4x + 5}$

　　b. $y = \sqrt[3]{-x^2 - 1} - 2$

　　c. $y = \sqrt[3]{x^2 + 2x + 2} + 2$

EXERCISES 41–44 *Solve the relation for y and use a graphing calculator to plot the graph and to estimate the coordinates of the indicated point(s) to the nearest integer.*

41. $16x^2 + 25y^2 = 400$; highest and lowest points on the graph

42. $9y^2 - 16x^2 = 144$; point(s) closest to the origin

43. $y^2 + x^3 + x - 5 = 0$; x- and y-intercepts

44. $\dfrac{y^2}{x^4 - x^3 + 3} = 1$; x- and y-intercepts

■ **APPLICATIONS**

The graph shown in Figure 25 (the graph of the standard normal probability function) is extremely important in probability theory. The area under the curve between $x = a$ and $x = b$ (the shaded region in Figure 25) is the probability that the variable x lies in the interval (a, b). Although a detailed discussion of this graph is beyond the scope of this book, the graph is, as the figure suggests, symmetric with respect to the y-axis. Also, the area under the entire curve is 1. These facts may be useful in Exercises 45–48.

45. Find the area under the curve to the right of $x = 0$.

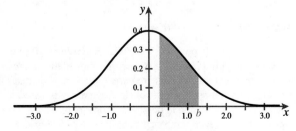

FIGURE 25

46. Find the area under the curve to the left of $x = 0$.

47. Given that the area under the curve between $x = 0$ and $x = 1$ is approximately 0.3413, compute the areas under the curve over the following intervals.

   a. $(1, \infty)$ **b.** $(-\infty, -1)$

   c. $(-1, 1)$ **d.** $(-\infty, 1)$

48. Given that the area under the curve to the left of $x = 2$ is approximately 0.9772, compute the area under the curve over the following intervals.

   a. $(2, \infty)$ **b.** $(-2, 2)$

   c. $(-\infty, -2)$ **d.** $(0, 2)$

▨ CONCEPTS AND CRITICAL THINKING

EXERCISES 49–54 *Answer true or false.*

49. If the graph of the relation obtained by replacing x with $-x$ is the same as the original graph, then we say that the graph is symmetric with respect to the y-axis.

50. Substitution of both $-x$ for x and $-y$ for y will result in a reflection about the x-axis.

51. Substituting $x + 3$ for x translates the graph 3 units to the right.

52. If a graph is symmetric with respect to both the x- and y-axes, then it will also be symmetric with respect to the origin.

53. If a graph is symmetric with respect to the origin, then it will also be symmetric with respect to both the x- and y-axes.

54. No line is symmetric with respect to both the x- and y-axes.

EXERCISES 55–58 *Give an example of each.*

55. Two relations such that the graph of one is the translation of the other, 5 units to the right

56. The graph of a relation that remains the same after translating 2 units to the right

57. The graph of a relation that is symmetric with respect to the line $y = x$

58. The graph of a relation that is also the graph of a function and is symmetric with respect to the origin

▨ QUESTIONS FOR DISCUSSION OR ESSAY

59. Suppose we wanted to electronically transmit a graph over a data line. Explain how the symmetry of a graph could be used to condense the transmission.

60. A face with two left sides can be constructed from the photo in Figure 26 by taking the mirror image of the left side of the face and attaching it on top of the right side. The result is shown in Figure 27. The same photo is used to create a face with two right sides in Figure 28. What do these two-faced photos suggest about symmetry in the human body? Give some additional evidence that supports your conclusion.

  FIGURE 26   **FIGURE 27**   **FIGURE 28**

61. Most forms of life possess symmetries. For example, the human body is roughly *bilaterally symmetric,* starfish are *rotationally symmetric,* and sand dollars are *radially symmetric.* Perhaps the most significant and the least obvious symmetry is *symmetry of scale.* An object is said to possess a symmetry of scale if a small piece of the object resembles the object as a whole. For example, if the branching patterns of limbs on the trunk, branches on a limb, leaves on a branch and veins on a leaf are similar, then we

would say that the tree possessed a certain symmetry of scale. The human circulatory, respiratory, and nervous systems all possess a symmetry of scale. Discuss the meaning of symmetry of scale in the context of each of these systems.

 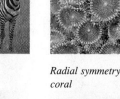

Bilateral symmetry in a Burchell's zebra   *Radial symmetry in organ-pipe coral*

Rotational symmetry in a starfish   *Symmetry of scale in a tree fern*

62. Inspector Magill has just been called to the scene of a homicide. The victim is lying in the middle of a carpeted floor, and the only objects near him are a pen and a hand-held mirror. A strange message is written on his forehead. It reads ''ti bib dod.'' Almost immediately Magill knows the first name of the murderer. Explain how the notion of reflection might have helped Magill decipher the message. Who is the murderer?

63. A graph is said to be **periodic** if it is symmetric with respect to a translation. That is, it remains unchanged when a certain translation takes place. For example, the given graph is periodic since it remains unchanged when translated 2 units to the left (or right). The term periodic is most often used when one of the variables in a relation is time; real-world phenomena are said to be periodic if they recur at regular time intervals. The time interval required for such a phenomenon to recur is called a **period**. For each of the following areas, give an example of a periodic phenomenon and its corresponding period.

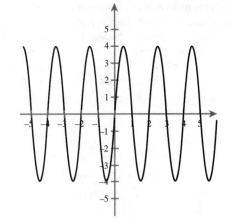

a. Biology

b. Astronomy

c. Business

PROJECTS FOR ENRICHMENT

64. *Inverses of Relations*   The **inverse of a relation** in x and y is obtained by switching the variables in the equation that defines the relation. For example, the inverse of the relation defined by $x^3 + 3y^2 = 12$ is given by $y^3 + 3x^2 = 12$.

a. Find the inverses of the following relations.

   i. $y = x^2 + 2x - 3$ ii. $\dfrac{3x - 4y}{x^2 + 3y^3} = 2$

   iii. $3x^2 + 5xy + 3y^2 = 10$

In Figure 29, we have plotted the points $A(-4, 2)$, $B(1, 3)$, and $C(4, 5)$. We then plotted the points $A'(2, -4)$, $B'(3, 1)$, and $C'(5, 4)$ obtained from A, B, and C, respectively, by exchanging the x- and y-coordinates. It is easily seen that A', B', and C' are obtained from A, B, and C by reflecting about the line $y = x$. Thus, whenever we exchange the x- and y-coordinates, we reflect the point (or points) about the line $y = x$. Since the inverse of a relation is obtained by exchanging the x- and y-coordinates, the graph of a relation and its inverse are reflections about the line $y = x$. Shown in Figure 30 is the graph of a relation and its inverse.

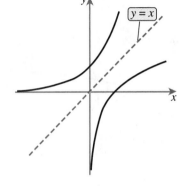

FIGURE 30

b. For each of the following graphs, sketch the graph of the inverse.

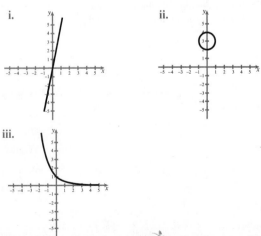

i.

ii.

iii.

Many relations cannot be graphed directly using a graphing calculator because they cannot easily be solved for y. If, however, the relation can be solved for x, a graphing calculator *can* produce a graph of its inverse. This graph can then be reflected about the line $y = x$ to obtain a graph of the original relation. Consider, for example, the relation $x - y^3 + 3y = 0$. Although it is difficult

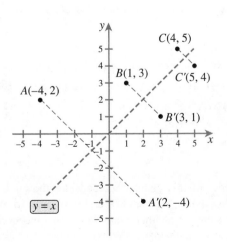

FIGURE 29

to solve this relation for *y*, solving for *x* gives us $x = y^3 - 3y$. Now the inverse of this relation is $y = x^3 - 3x$, which can be plotted using a graphing calculator. With a viewing window defined by `Xmin=-15`, `Xmax=15`, `Ymin=-10`, and `Ymax=10`, we obtain the plot shown in Figure 31. Finally, we reflect about the line $y = x$ to obtain the graph of $x = y^3 - 3y$, as shown in Figure 32.

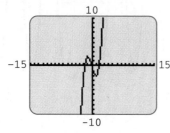

FIGURE 31

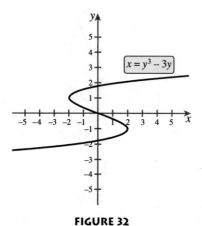

FIGURE 32

c. Use a graphing calculator and your understanding of inverse relations to graph the following relations.

i. $x = \sqrt{36 - y^2}$ ii. $x = \sqrt{y^2 - 36}$

iii. $x = \dfrac{4}{y^2 + 1}$ iv. $x - y^3 + 4y^2 = 0$

65. *Rotation by 90°* In this project, we explore rotations by 90° and their connections with reflections. Pictured in Figure 33 is an arbitrary point *P* with coordinates (x, y), and a point *P′* with coordinates (x', y'), the result of rotating the point *P* 90° counterclockwise.

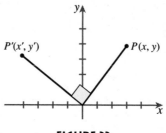

FIGURE 33

a. Use elementary geometry, trial and error, or any other technique to find a simple formula for x' and y' in terms of x and y.

b. Use the formula obtained in part a to find the coordinates of the point obtained by rotating (4, 6) 90° counterclockwise.

c. Find a formula for the coordinates of the points obtained by rotating (x, y) 270° counterclockwise by applying three consecutive 90° counterclockwise rotations. Then give a formula for a 90° *clockwise* rotation.

d. Use the formula obtained in part a to show that a 90° counterclockwise rotation followed by a reflection about the *y*-axis is equivalent to a reflection about the line $y = x$.

e. Show that reflecting about the *y*-axis followed by a 90° counterclockwise rotation is equivalent to reflecting about the line $y = x$ and then reflecting about the origin. Compare your results to those from part d.

S E C T I O N **6.2** **PARABOLAS**

☐ **How can parabolas help a Hollywood stunt driver or a cliff diver in Acapulco?**

☐ **What's the maximum height attained by human cannonball Emanual Zacchini?**

☐ **How might cones be connected with the disappearance of the dinosaur?**

☐ **Where on a satellite dish is the receiver placed?**

THE CONIC SECTIONS

Throughout much of the remainder of this chapter we consider the conic sections: circles, parabolas, ellipses, and hyperbolas. First studied and developed for purely theoretical reasons, the conic sections are yet another example of what has been referred to as the unreasonable effectiveness of mathematics. Who could have guessed that this collection of abstract curves, the intellectual offspring of ancient Greeks more philosophers than mathematicians, would provide the mathematical weaponry for communicating across continents, beaming moving images from Earth to space and back again, guiding spacecraft to distant planets, and painlessly obliterating kidney stones with little more than a well-placed whisper? From astronomy to economics, mass communication to geolocation, modern applications of this ancient subject abound. A few specific examples are cited in Table 1.

TABLE 1

Applied context	Type of conic
Path of a projectile	Parabola
Graph of a revenue function	Parabola
Shape of a satellite dish	Parabola
Shape of a shock wave lithotripsy reflector	Ellipse
Shape of a whispering room	Ellipse
LORAN navigation	Hyperbola
Orbit of a comet	Parabola, ellipse, or hyperbola
Shape of a telescope mirror	Parabola, ellipse, or hyperbola

There are several ways to study conic sections. A purely geometric approach was developed by the ancient Greeks around 350 B.C. They described the conic sections as curves that are formed by the intersection of a plane with a double cone, as shown in Figure 34.

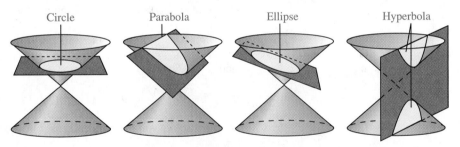

Circle   Parabola   Ellipse   Hyperbola

FIGURE 34

An algebraic approach to the study of conic sections was made possible after the development of analytic geometry in the 17th century. Using this approach, the conic sections are described as relations of the form

$$Ax^2 + Bxy + Cy^2 + Dx + Ey + F = 0$$

A third approach describes each conic as a *locus* or set of points satisfying certain geometric properties. This is the approach we will use to develop the standard equations of the conics. We begin with parabolas.

PARABOLAS

As suggested by Figure 34, a parabola is a curve that is formed when a cone is cut with a plane parallel to the side of the cone. It can be shown that any parabola formed in this manner can also be formed by applying the following definition.

Definition of a Parabola □

A **parabola** is the set of all points (x, y) that are equidistant from a fixed line (called the **directrix**) and a fixed point (called the **focus**) not on the line, as shown in Figure 35. The point midway between the focus and directrix is called the **vertex,** and the line passing through the focus and vertex is called the **axis of symmetry.**

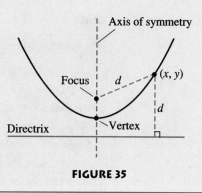

FIGURE 35

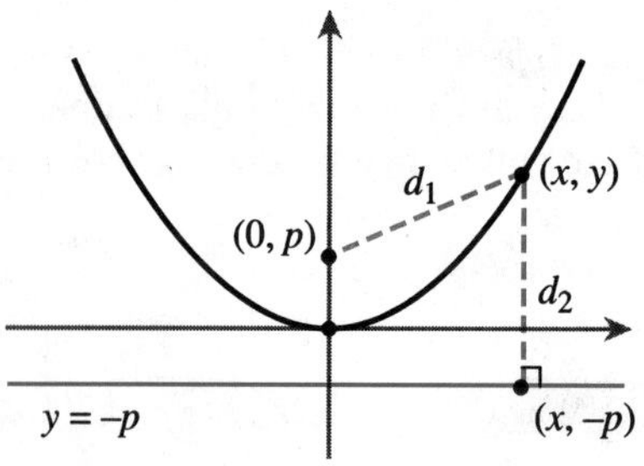

Let us initially consider a parabola with focus $(0, p)$ and directrix $y = -p$, as shown in Figure 36. Notice that in this case the vertex is at the origin and the axis of symmetry is vertical. Moreover, if $p > 0$ the parabola opens upward, whereas if $p < 0$ the parabola opens downward. Now according to the definition, for any point (x, y) on the parabola, the distance d_1 from $(0, p)$ to (x, y) must be the same as the distance d_2 from (x, y) to the point $(x, -p)$ on $y = -p$. Applying the distance formula, we have

$$d_1^2 = (x - 0)^2 + (y - p)^2 = x^2 + (y - p)^2$$

FIGURE 36

and

$$d_2^2 = (x - x)^2 + (y - (-p))^2 = (y + p)^2$$

We now set $d_1^2 = d_2^2$ and proceed as follows:

$$d_1^2 = d_2^2$$
$$x^2 + (y - p)^2 = (y + p)^2$$
$$x^2 + y^2 - 2yp + p^2 = y^2 + 2yp + p^2 \qquad \text{Multiplying out}$$
$$x^2 = 4yp \qquad \text{Simplifying}$$
$$y = \frac{1}{4p} x^2 \qquad \text{Solving for } y$$

Finally, we set $a = \frac{1}{4p}$ to obtain the equation $y = ax^2$. Thus, a parabola with vertex $(0, 0)$ and a *vertical* axis of symmetry has an equation of the form $y = ax^2$. A similar process could be used to show that a parabola with vertex $(0, 0)$ and a *horizontal* axis of symmetry has an equation of the form $x = ay^2$. In both cases, we have $a = \frac{1}{4p}$ or, equivalently, $p = \frac{1}{4a}$, and so the coefficient a gives us information about the location of the focus and directrix. We summarize these observations as follows.

Parabolas with Vertex □
at the Origin

The standard forms for the equation of a parabola with vertex $(0, 0)$ are as follows:

Equation	**Description**
$y = ax^2$	Vertical axis of symmetry Opens upward if $a > 0$, downward if $a < 0$ See Figure 37.
$x = ay^2$	Horizontal axis of symmetry Opens to the right if $a > 0$, to the left if $a < 0$ See Figure 38.

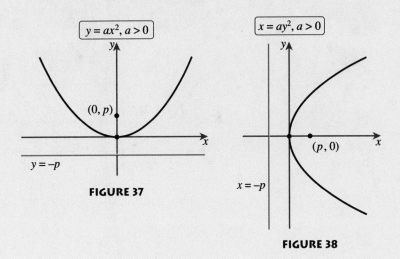

FIGURE 37

FIGURE 38

The focus is on the axis of symmetry in the *interior* of the parabola, $|p|$ units from the vertex, where

$$p = \frac{1}{4a}$$

The directrix is perpendicular to the axis of symmetry and is $|p|$ units from the vertex on the side of the parabola opposite the focus.

EXAMPLE 1 ▦   Finding the Focus and Directrix of a Parabola with Vertex $(0, 0)$

Sketch a graph of the parabola with equation $y = \frac{1}{8}x^2$, and find its focus and directrix.

SOLUTION We first note that the equation is of the form $y = ax^2$, and so the parabola has a vertical axis of symmetry and vertex $(0, 0)$. Moreover, since $a = \frac{1}{8} > 0$, the parabola opens upward. To add some detail, we locate two points on the graph by choosing 4 and -4 as convenient x-coordinates and computing $y = \frac{1}{8}(\pm 4)^2 = 2$. The final graph is shown in Figure 39. From the shape of the graph, we expect the focus to be above the vertex and the directrix to be below. To locate both precisely, we compute the value of p.

$$p = \frac{1}{4a}$$

$$= \frac{1}{4(\frac{1}{8})}$$

$$= 2$$

Thus, the focus is 2 units above the vertex, at the point $(0, 2)$, and the directrix is 2 units below the vertex, with equation $y = -2$ (see Figure 40).

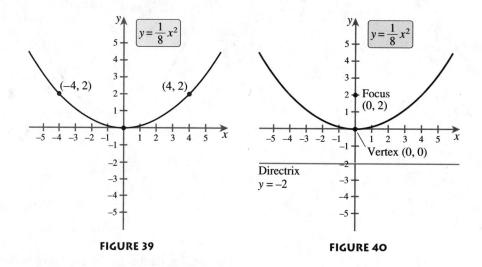

FIGURE 39   **FIGURE 40**

EXAMPLE 2 ■ **Finding the Focus and Directrix of a Parabola with Vertex (0, 0)**

Sketch a graph of the parabola with equation $5y^2 + 24x = 0$, and find its focus and directrix.

SOLUTION We begin by writing the equation in standard form.

$$5y^2 + 24x = 0$$

$$24x = -5y^2$$

$$x = -\frac{5}{24}y^2$$

This is the equation of a parabola with a horizontal axis of symmetry and vertex $(0, 0)$. Since $a = -\frac{5}{24} < 0$, the parabola opens to the left. To add detail to the graph, we find two points by taking $y = \pm 4$ and computing $x = -\frac{5}{24}(\pm 4)^2 = -\frac{10}{3}$. The graph is shown in Figure 41. Based on the graph, we know the focus will lie to the left of the vertex, and the directrix will lie to the right. Computing p, we obtain

$$p = \frac{1}{4a}$$

$$= \frac{1}{4\left(-\frac{5}{24}\right)}$$

$$= -\frac{6}{5}$$

Thus, the focus is $\frac{6}{5}$ units to the left of the vertex, at the point $\left(-\frac{6}{5}, 0\right)$. The directrix, lying $\frac{6}{5}$ units to the right of the vertex, has equation $x = \frac{6}{5}$. Refer to Figure 42.

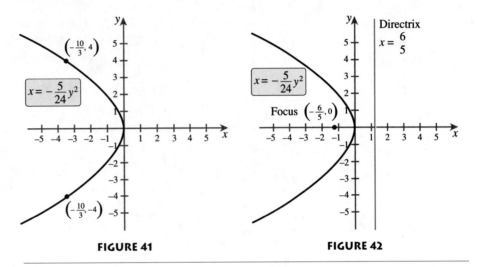

FIGURE 41   **FIGURE 42**

So far, we have considered only parabolas with vertex at the origin. Parabolas with vertical and horizontal axes and vertex *away* from the origin have standard equations that can be obtained using translations. Recall from Section 6.1 that substituting $x - h$ for x and $y - k$ for y has the effect of translating the graph h units horizontally and k units vertically. For example, the graph of the relation $x - 1 = (y - 3)^2$ can be obtained from the graph of $x = y^2$ by translating 1 unit to the right and 3 units up, as shown in Figure 43. From this, we see that the graph of $x - 1 = (y - 3)^2$ is a parabola with a

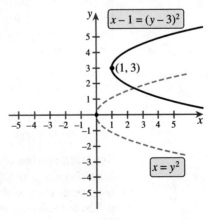

FIGURE 43

horizontal axis of symmetry and vertex (1, 3). More generally, we have the following standard equations.

Parabolas with Vertex (h, k) □

The standard forms for the equation of a parabola with vertex (h, k) are as follows:

Equation	Description
$y - k = a(x - h)^2$	Vertical axis of symmetry Opens upward if $a > 0$, downward if $a < 0$ See Figure 44.
$x - h = a(y - k)^2$	Horizontal axis of symmetry Opens to the right if $a > 0$, to the left if $a < 0$ See Figure 45.

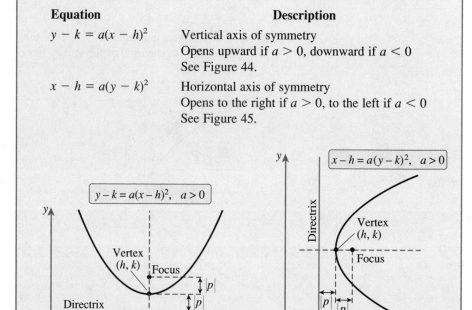

FIGURE 44   **FIGURE 45**

As with parabolas having vertex (0, 0), the focus and directrix can be found using

$$p = \frac{1}{4a}$$

Specifically, the focus is on the axis of symmetry in the interior of the parabola, $|p|$ units from the vertex. The directrix is perpendicular to the axis of symmetry and is $|p|$ units from the vertex on the side of the parabola opposite the focus.

EXAMPLE 3 ■ **Finding the Directrix and Focus of a Parabola**

Sketch a graph of the parabola with equation $y - 1 = -\frac{1}{4}(x + 2)^2$. Identify the vertex, focus, and directrix.

SOLUTION The equation is already in standard form, with $h = -2$, $k = 1$, and the x term being squared. Thus, we know that the vertex is $(-2, 1)$ and the axis of symmetry is vertical. Moreover, since $a = -\frac{1}{4} < 0$, the parabola opens downward. By plotting points as necessary, we obtain the graph shown in Figure 46. Since $p = \frac{1}{4a} = -1$, the focus is located 1 unit below the vertex at the point $(-2, 0)$. The directrix is located 1 unit above the vertex and has equation $y = 2$. Refer to Figure 47.

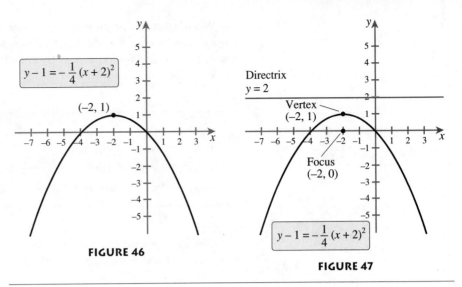

FIGURE 46

FIGURE 47

EXAMPLE 4   Graphing a Parabola

Sketch a graph of the parabola with equation $x = y^2 + 2y - 3$. Identify the vertex, focus, and directrix.

SOLUTION   We must first write the equation in standard form. We do this by completing the square on the y terms.

$$x = y^2 + 2y - 3$$
$$x + 3 = y^2 + 2y \qquad \text{Adding 3 to both sides to eliminate the constant on the right}$$
$$x + 3 + 1 = y^2 + 2y + 1 \qquad \text{Adding 1 to both sides to complete the square on the right}$$
$$x + 4 = (y + 1)^2 \qquad \text{Writing the right-hand side as a perfect square}$$

This is the equation of a parabola with horizontal axis of symmetry and vertex $(-4, -1)$. Moreover, since $a = 1 > 0$, the parabola opens to the right. The graph is shown in Figure 48. Since $p = \frac{1}{4a} = \frac{1}{4}$, the focus is located $\frac{1}{4}$ unit to the right of the vertex, at the point $(-3.75, -1)$. The directrix is located $\frac{1}{4}$ unit to the left of the vertex and has equation $x = -4.25$. See Figure 49.

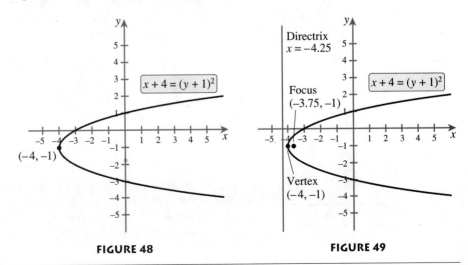

FIGURE 48

FIGURE 49

When equations of parabolas are given in the general form $y = ax^2 + bx + c$ or $x = ay^2 + by + c$, as was the case in Example 4, it is sometimes convenient to have a simple rule for finding the vertex. By completing the square on the general equation, we obtain the following rule.

Finding the Vertex
of a Parabola ☐

Parabola form	Axis of symmetry	Vertex
$y = ax^2 + bx + c$	Vertical	$\left(-\frac{b}{2a}, k\right)$, where k is found by setting $x = -\frac{b}{2a}$
$x = ay^2 + by + c$	Horizontal	$\left(h, -\frac{b}{2a}\right)$, where h is found by setting $y = -\frac{b}{2a}$

Consider, for example, the parabola with equation $x = y^2 + 2y - 3$ from Example 4. Since $a = 1$ and $b = 2$, the y-coordinate of the vertex is given by

$$y = -\frac{b}{2a} = -\frac{2}{2(1)} = -1$$

The x-coordinate is then found by substituting $y = -1$ into $x = y^2 + 2y - 3$, yielding

$$x = (-1)^2 + 2(-1) - 3 = 1 - 2 - 3 = -4$$

Thus, the vertex is the point $(-4, -1)$. We use this technique in the following example.

EXAMPLE 5 ■ Finding the Maximum Height of a Human Cannonball

The muzzle velocity of the renowned human cannonball Emanual Zacchini was estimated to be as much as 36 feet per second (*Source: The Guinness Book of World Records*). If we assume that this muzzle velocity was possible with the cannon pointed straight up, and if we assume that Zacchini was initially 5 feet above the ground, then his height in feet after t seconds would have been given by $s = -16t^2 + 36t + 5$. Find Zacchini's maximum height given these assumptions.

SOLUTION   In the coordinate system having height s marked along the vertical axis and time t on the horizontal, the graph of the equation $s = -16t^2 + 36t + 5$ is a parabola opening downward. Thus, the maximum value of the height s will occur at the vertex. We find the t-coordinate of the vertex—that is, the time at which the maximum height is attained—as follows:

$$t = -\frac{b}{2a} = -\frac{36}{-32} = \frac{9}{8} = 1.125 \text{ seconds}$$

The maximum height is then found by setting $t = \frac{9}{8}$ in the equation $s = -16t^2 + 36t + 5$.

$$s = -16\left(\frac{9}{8}\right)^2 + 36\left(\frac{9}{8}\right) + 5 = 25.25 \text{ feet}$$

Thus, the maximum height is 25.25 feet. Note that the coordinates of the vertex can easily be verified using a graphing calculator. The graph of $s = -16t^2 + 36t + 5$ is shown in Figure 50; the approximate coordinates of the vertex, as found using the calculate-maximum feature, are shown in Figure 51.

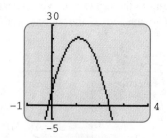

FIGURE 50

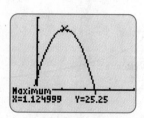

FIGURE 51

In all the examples to this point, we obtained information about the graphical features of a parabola from its equation. Now we consider how the equation of a parabola can be determined from its graphical features.

EXAMPLE 6   ■   Finding the Equation of a Parabola

Find the standard equation of the parabola that has vertex $(2, 3)$, and focus $(2, 0)$.

SOLUTION   Since the vertex and focus have the same x-coordinate, the parabola must have a vertical axis of symmetry. Combining this with the fact that the vertex is $(2, 3)$, we know that the equation must have the form

$$y - 3 = a(x - 2)^2$$

Since the focus is 3 units below the vertex, we also know that $p = -3$, and so $a = \frac{1}{4p} = -\frac{1}{12}$. Thus, the equation of the parabola is

$$y - 3 = -\frac{1}{12}(x - 2)^2$$

EXAMPLE 7   ■   Finding the Equation of a Parabola from Its Graph

Find a possible equation for the parabola shown in Figure 52. Verify with a graphing calculator.

SOLUTION   Figure 52 suggests a horizontal axis of symmetry and a vertex at $(-1, 2)$. These assumptions lead to an equation of the form

$$x + 1 = a(y - 2)^2$$

Since there appears to be an x-intercept at -3, we set $x = -3$ and $y = 0$ to obtain

$$-3 + 1 = a(0 - 2)^2$$
$$-2 = 4a$$
$$a = -\frac{1}{2}$$

The resulting equation is

$$x + 1 = -\frac{1}{2}(y - 2)^2$$

To plot the graph on a graphing calculator, we must solve for y.

$$x + 1 = -\frac{1}{2}(y - 2)^2$$
$$-2x - 2 = (y - 2)^2$$
$$\pm\sqrt{-2x - 2} = y - 2$$
$$2 \pm \sqrt{-2x - 2} = y$$

Graphing the branches $y = 2 + \sqrt{-2x - 2}$ and $y = 2 - \sqrt{-2x - 2}$, we obtain the parabola shown in Figure 53. It appears to be the same as that shown in Figure 52.

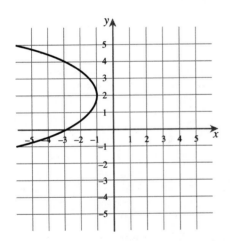

FIGURE 52

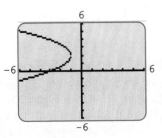

FIGURE 53

REFLECTIVE PROPERTY OF PARABOLAS

If sound or light is emitted from the focus of a conic section, certain important reflective properties can be observed. In the case of a parabola, the reflected sound or light will follow a path parallel to the axis of the parabola, as shown in Figure 54. This property of parabolas is utilized in the construction of searchlights. The reflector for the searchlight is formed by revolving a parabola around its axis, and the light source is placed at the focus of the parabola. The result is a light beam with little dissapation.

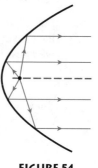

FIGURE 54

The Very Large Array of radio antennas in New Mexico

The reflective property of parabolas is also used in telescopes, satellite dishes, microwave antennas, and solar energy devices. When the parabolic reflector of these devices is positioned so that the incoming light rays, sound waves, or radio waves are (approximately) parallel to the axis, they will be reflected toward the focus. This results in an image or signal that is ''focused'' at a single point.

EXAMPLE 8 ■ **Determining Spotlight Dimensions**

Find the depth of a parabolic reflector in a spotlight with a 6-inch-wide beam given that the light source is located 1 inch away from the vertex of the reflector.

SOLUTION We begin by placing a cross section of the parabolic reflector on an xy-coordinate system in such a way that its vertex is at the origin and its focus (that is, the light source) is 1 unit to the right of the vertex, at the point $(1, 0)$. See Figure 55. Now this parabola opens to the right and has vertex $(0, 0)$, so its equation must be of the form $x = ay^2$. Moreover, since the focus is 1 unit to the right of the vertex, we must have $p = 1$. It follows that $a = \frac{1}{4p} = \frac{1}{4}$, so that the parabola has equation $x = \frac{1}{4}y^2$. Since the light beam is 6 inches wide, the point P in Figure 55 on the "edge" of the reflector must have y-coordinate 3. Setting $y = 3$ in the equation $x = \frac{1}{4}y^2$ gives $x = \frac{9}{4}$, which implies that the reflector is $\frac{9}{4} = 2.25$ inches deep.

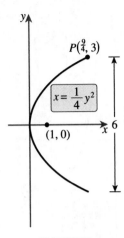

FIGURE 55

UNDERSTANDING AND MASTERY CHECKLISTS

CONCEPTS TO UNDERSTAND

☐ Conic section

☐ Parabola

☐ Vertex of a parabola

☐ Focus of a parabola

☐ Directrix of a parabola

☐ Axis of symmetry of a parabola

☐ Reflective property of a parabola

SKILLS TO MASTER

☐ Sketch the graph of a parabola.

☐ Determine the vertex, focus, and directrix of a parabola.

☐ Find the equation of a parabola given information about its graph.

EXERCISES 6.2

EXERCISES 1–12 *Find the vertex, focus, and directrix for the given parabola, and sketch its graph.*

1. $y = 2x^2$

2. $x = -4y^2$

3. $y^2 - 3x = 0$

4. $x^2 + 5y = 0$

5. $4y^2 + 7x = 0$

6. $2x^2 - 3y = 0$

7. $y - 3 = 3(x + 5)^2$

8. $x + 4 = -(y - 2)^2$

9. $x^2 - 4x - 4y + 8 = 0$

10. $y^2 + x - 6y + 10 = 0$

11. $x = 2y^2 + 6y + 2$

12. $y = -x^2 - x + 1$

EXERCISES 13–22 *Find an equation for the parabola with the given properties.*

13. Vertex at $(0, 0)$ and focus at $(2, 0)$

14. Vertex at $(0, 0)$ and focus at $(0, 4)$

15. Focus $(0, -4)$ and directrix $y = 4$

16. Vertex at $(0, 0)$ and directrix $x = 1$

17. Vertex at $(-2, 0)$, passing through $(2, 4)$, and a horizontal axis of symmetry

18. Vertex at $(-1, 3)$, y-intercept 5, and vertical axis of symmetry

19. Focus at $(4, -1)$ and directrix $y = 3$

20. Vertex at $(-2, -3)$ and focus $(-6, -3)$

21. The following graph

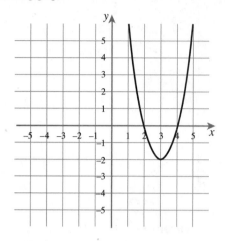

22. The following graph

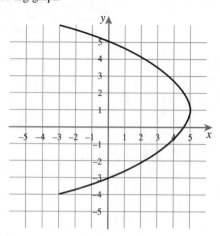

EXERCISES 23–30   *Solve the equation for y and plot the resulting equation on a graphing calculator. Identify the coordinates of the vertex.*

23. $x^2 - 4x - y = 0$ **24.** $-x^2 + 5x + y = 0$

25. $x^2 + 8x - 2y + 22 = 0$ **26.** $x^2 - 6x + 3y + 12 = 0$

27. $-y^2 + 4x + 8 = 0$ **28.** $y^2 + 5x - 5 = 0$

29. $y^2 - 2y + x + 2 = 0$ **30.** $4x + y^2 - 6y = 0$

▦ APPLICATIONS

31. *Path of a Ball*   A ball is thrown at an angle of 45° and with an initial velocity of 80 feet per second. Its path is parabolic and satisfies the equation

$$y = -\frac{x^2}{200} + x$$

where y is the height in feet when the ball is a horizontal distance of x feet from where it was thrown. Find the maximum height of the ball and the total horizontal distance it travels before hitting the ground.

32. *Stunt Car*   The stunt coordinator for a movie is planning a stunt with a car speeding off the top floor of a parking garage. It can be shown that the path of the falling car will be parabolic with an equation of the form

$$y = h - \frac{16}{v^2} x^2$$

where h is the height of the garage in feet, v is the initial speed of the car in feet per second, and y is the height of the car in feet when its horizontal distance away from the edge of the garage is x feet. If the height of the garage is 100 feet and the car must land 200 feet from the base of the garage, determine the required speed of the car.

33. *Cliff Diving*   The highest regularly performed dives are off La Quebrada in Acapulco, Mexico. (*Source: The Guinness Book of World Records*). The takeoff point is 87.5 feet above the water

and, because of the presence of rocks at the base, the divers must land 27 feet out from the takeoff point (see Figure 56).

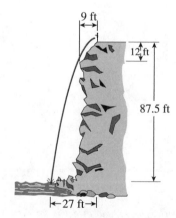

FIGURE 56

a. Assuming the path of a diver is a parabola with the vertex located at the takeoff point, choose a coordinate system with the origin at water level directly below the takeoff point and find an equation for the path.

b. Suppose that 12 feet below the takeoff point, a large rock juts out 9 feet from the vertical line passing through the takeoff point. By what horizontal distance will the diver clear the rock?

34. *Suspension Bridge* When a cable is suspended between two towers of a suspension bridge in such a way that it bears a uniform load, the shape of the cable will be parabolic.

    a. Find the equation of the shape of the cable if the towers are 200 feet apart and 40 feet high and the cable touches the road midway between the towers (see Figure 57).

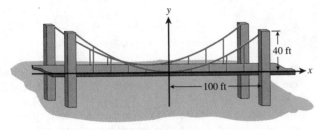

FIGURE 57

b. Use the equation from part a to determine the total length of cable needed for the vertical sections of cable spaced at 20 foot intervals between the towers.

35. *Radio Telescope* Suppose that the cross-sectional view of a radio telescope is parabolic and, with an appropriately positioned coordinate system, has an equation of the form $y = ax^2$ (see Figure 58). If the dish is 1000 feet across the top and 167 feet deep at the center, where must the receiver be located?

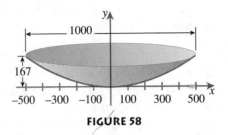

FIGURE 58

▦ CONCEPTS AND CRITICAL THINKING

EXERCISES 36–39 *Answer true or false.*

36. The focus of a parabola is a point on the graph of the parabola that is the same distance away from the vertex as is the directrix.

37. The graph of $x = ay^2 + by + c$ is a parabola with horizontal axis of symmetry.

38. The vertex of a parabola lies on its axis of symmetry.

39. The directrix of a parabola is a point lying on the parabola, halfway between the focus and the vertex.

EXERCISES 40–43 *Give an example of each.*

40. A parabola with a vertical axis of symmetry and vertex (2, 1)

41. A parabola with axis of symmetry $y = 4$

42. An application of the reflective property of parabolas

43. An equation of a parabola with its focus at the origin

▦ QUESTIONS FOR DISCUSSION OR ESSAY

44. We have seen three methods for finding the vertex of a parabola with equation $y = ax^2 + bx + c$: completing the square, using the formula $x = -b/(2a)$, and graphical estimation. Discuss the relative advantages and disadvantages of these three techniques.

45. Describe a procedure for plotting the graph of a parabola of the form $x = ay^2 + by + c$ with a graphing calculator. Is this procedure any easier than sketching the graph by hand? Why or why not?

46. In the examples and exercises of this section, we've seen many problems in which the equation of a parabola was to be determined from a few characteristics, such as (1) the focus, (2) the directrix, (3) the vertex, (4) the axis of symmetry, and (5) a point other than the vertex on the parabola. In Example 6, we deter-

mined the equation of a parabola having a given vertex and focus. Are any other pairs of characteristics from among the five listed sufficient to determine the equation of the parabola? What are they? Are any three of these five sufficient to determine the equation? Explain your answers.

47. The parabolic model of projectile motion results from the assumption that the gravitational field is uniform: everywhere pointing in the same direction with the same strength. Since Earth's gravitational field points toward the center of Earth and weakens with altitude, the path of a projectile thrown from Earth's surface will *not* be a parabola, as we have assumed throughout this section, but rather an ellipse. How do the authors get away with committing such a seemingly profound error?

▦ PROJECTS FOR ENRICHMENT

48. *Similarity Among Circles and Parabolas* Geometric objects that are the same shape but possibly different sizes are said to

be **similar.** For example, two triangles are said to be similar if corresponding sides are proportional. Clearly, all circles are the same shape. Thus, we say that all circles are similar to one another. Surprisingly, all parabolas are similar to one another as well.

I. Similarity of Circles

a. Using graph paper, graph the circle with equation $x^2 + y^2 = 1$.

b. Beginning with the equation $x^2 + y^2 = 1$, substitute $\frac{x}{2}$ for x, and $\frac{y}{2}$ for y, and simplify the resulting equation.

c. Now graph $x^2 + y^2 = 1$ and the equation obtained in part b on the same set of coordinate axes. What effect did the substitution have on the size and shape of the resulting graph?

d. How would the graphs of $x^2 + y^2 = 1$ and $\left(\frac{x}{a}\right)^2 + \left(\frac{y}{a}\right)^2 = 1$ compare in size and shape?

e. Substituting $\frac{x}{a}$ for x and $\frac{y}{a}$ for y is equivalent to changing the scale of the graph by a factor of a. We refer to such substitutions as **scale substitutions.** Show that the circle with equation $x^2 + y^2 = r^2$ is similar to the circle with equation $x^2 + y^2 = 1$ by showing that one can be obtained from the other by a scale substitution.

f. In part e, we showed that all circles centered at the origin are similar to the circle centered at the origin with radius 1, and hence to each other. Explain why we can conclude from this that all circles, whether centered at the origin or not, are similar to one another.

II. Similarity of Parabolas

a. Find the general equation of a parabola with a vertical axis of symmetry and vertex at the origin.

b. Show that all such parabolas are similar to the parabola with equation $y = x^2$.

c. Conclude that all parabolas (no matter what axis of symmetry or vertex they might have) are similar to one another.

d. Explain the following statement, "All parabolas are the same shape; it is as if we are viewing the same parabola from different distances."

49. *The Path of a Baseball* If a baseball is hit toward center field from a height of 3 feet, at an initial angle of 45°, and with an

initial velocity of 114 feet per second, will it clear a 12-foot-high center field fence 400 feet away? To answer this question, we must know something about the path of the baseball. The location of the ball at a given point in time can be expressed as an ordered pair (x, y), where x is the horizontal distance of the ball from home plate in feet and y is its height in feet. Since both x and y depend upon time, we must give equations for both. If we ignore air resistance, it can be shown that after t seconds, the position of the ball will be given by

(1) $$x = 57\sqrt{2}\,t$$

(2) $$y = -16t^2 + 57\sqrt{2}\,t + 3$$

For example, when $t = 0$, then $x = 0$ and $y = 3$, indicating that the ball leaves home plate at a height of 3 feet. When $t = 1$, $x = 57\sqrt{2} \approx 80.6$ feet and $y = -16 + 57\sqrt{2} + 3 \approx 67.6$ feet. Thus, after 1 second, the ball is 80.6 feet from home plate (as measured from a point on the ground beneath the ball) and is 67.6 feet high. (Note that equations (1) and (2) are called parametric equations.)

a. Complete the following table, plot the points (x, y), and connect the points with a smooth curve. What shape does the graph have?

t	0	1	2	3	4	5
x						
y						

We wish to find a relation in x and y whose graph gives the path of the ball.

b. Solve equation (1) for t and substitute into equation (2) to obtain an equation of the form $y = ax^2 + bx + c$.

c. Sketch the graph of the equation you found in part b. What shape is the path of the ball?

d. How far is the ball from home plate when it reaches its maximum height? What is the maximum height?

e. What is the height of the ball when it is 400 feet from home plate? Will the ball clear the 12-foot-high center field fence?

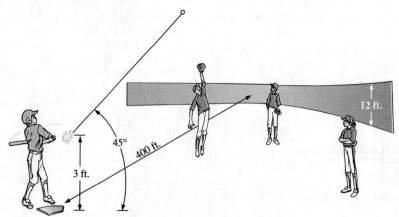

SECTION 6.3 ELLIPSES

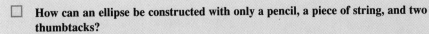

☐ **How can an ellipse be constructed with only a pencil, a piece of string, and two thumbtacks?**

☐ **It is virtually impossible to shine a flashlight in a room without seeing an ellipse. Where is it?**

☐ **How do whispering galleries work?**

☐ **If the center of Earth's orbit is *not* the center of the sun, then what is it?**

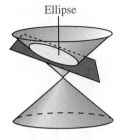

Ellipse

FIGURE 59

In Section 6.2, we described the conic sections as curves formed when a double cone is cut by a plane. If the plane is not parallel to the central axis or the side of the cone, we obtain an ellipse, as shown in Figure 59. Using geometric techniques, it can be shown that the following characterization of an ellipse is equivalent.

Definition of an Ellipse ☐

An **ellipse** is the set of all points (x, y) such that the sum of the distances from two fixed points (called the **foci**) is constant, as shown in Figure 60. The line through the foci intersects the ellipse at two points, called the **vertices.** The line segment connecting the two vertices is called the **major axis,** and the midpoint of the major axis is called the **center.** The line segment perpendicular to the major axis, passing through the center, and connecting two points on the ellipse is called the **minor axis.**

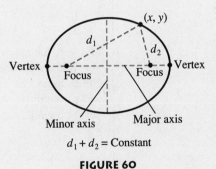

$d_1 + d_2 = \text{Constant}$

FIGURE 60

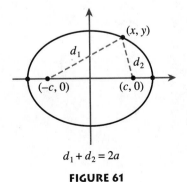

$d_1 + d_2 = 2a$

FIGURE 61

The standard equation of an ellipse centered at the origin is actually quite simple $\left(\dfrac{x^2}{a^2} + \dfrac{y^2}{b^2} = 1 \right)$, but deriving the result from the definition takes a little work. We begin by considering an ellipse with a horizontal major axis and foci at $(-c, 0)$ and $(c, 0)$, as shown in Figure 61. According to the definition, for any point (x, y) on the ellipse, the sum of the distances to the two foci must be constant. To simplify later computations, we denote this constant sum by $2a$. Using the distance formula for each of the two distances d_1 and d_2, we have

$$d_1 = \sqrt{(x + c)^2 + y^2} \quad \text{and} \quad d_2 = \sqrt{(x - c)^2 + y^2}$$

Setting $d_1 + d_2 = 2a$, we proceed as follows:

$$\sqrt{(x + c)^2 + y^2} + \sqrt{(x - c)^2 + y^2} = 2a$$

$$\sqrt{(x + c)^2 + y^2} = 2a - \sqrt{(x - c)^2 + y^2}$$

Now we square both sides and collect terms.

$$(x + c)^2 + y^2 = 4a^2 - 4a\sqrt{(x - c)^2 + y^2} + (x - c)^2 + y^2$$

$$x^2 + 2cx + c^2 + y^2 = 4a^2 - 4a\sqrt{(x - c)^2 + y^2} + x^2 - 2cx + c^2 + y^2$$

$$4a\sqrt{(x - c)^2 + y^2} = 4a^2 - 4cx$$

$$a\sqrt{(x - c)^2 + y^2} = a^2 - cx$$

Once again we square both sides and collect terms.

$$a^2[(x - c)^2 + y^2] = (a^2 - cx)^2$$

$$a^2(x^2 - 2cx + c^2 + y^2) = a^4 - 2a^2cx + c^2x^2$$

$$a^2x^2 - 2a^2cx + a^2c^2 + a^2y^2 = a^4 - 2a^2cx + c^2x^2$$

$$a^2x^2 - c^2x^2 + a^2y^2 = a^4 - a^2c^2$$

$$(a^2 - c^2)x^2 + a^2y^2 = a^2(a^2 - c^2)$$

Finally, since $a^2 - c^2 > 0$ (see Exercise 47), we set $b^2 = a^2 - c^2$ and substitute.

$$b^2x^2 + a^2y^2 = a^2b^2$$

$$\frac{x^2}{a^2} + \frac{y^2}{b^2} = 1$$

The standard equation of an ellipse with a vertical major axis can be derived in a similar manner.

Ellipses Centered at the Origin □

The standard form for the equation of an ellipse centered at $(0, 0)$ is

$$\frac{x^2}{a^2} + \frac{y^2}{b^2} = 1, \quad a \neq b$$

If $a > b$, the major axis is horizontal with length $2a$ (see Figure 62), whereas if $b > a$, the major axis is vertical with length $2b$ (see Figure 63). The vertices are the endpoints of the major axis. The foci lie on the major axis a distance of c units from the center, where $c = \sqrt{|a^2 - b^2|}$. Note that if $a = b$, the equation is that of a circle with radius a.

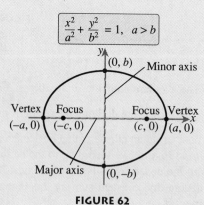

FIGURE 62

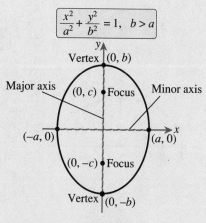

FIGURE 63

The graph of an ellipse centered at the origin can most easily be obtained by first locating the x- and y-intercepts and then connecting the intercepts with a smooth curve. To find the x-intercepts, we set $y = 0$ and solve for x. This yields $x = \pm a$. The y-intercepts are found by setting $x = 0$ and solving for y. We obtain $y = \pm b$. It will be readily apparent from the orientation of the graph whether the ellipse has a horizontal or vertical major axis.

EXAMPLE 1   ■   **An Ellipse Centered at the Origin**

Sketch the graph of the ellipse with equation

$$\frac{x^2}{25} + \frac{y^2}{16} = 1$$

Identify the vertices and foci.

SOLUTION   We first rewrite in standard form as

$$\frac{x^2}{5^2} + \frac{y^2}{4^2} = 1$$

Since $a = 5$, the x-intercepts are $(-5, 0)$ and $(5, 0)$. Similarly, since $b = 4$, the y-intercepts are $(0, -4)$ and $(0, 4)$. Connecting these points with a smooth curve leads to the graph shown in Figure 64. It is evident from the graph that the major axis is horizontal, and so the vertices are $(-5, 0)$ and $(5, 0)$. The foci are on the major axis a distance of c units from the origin, where

$$\begin{aligned} c &= \sqrt{|a^2 - b^2|} \\ &= \sqrt{|25 - 16|} \\ &= 3 \end{aligned}$$

Thus, the foci are $(-3, 0)$ and $(3, 0)$.

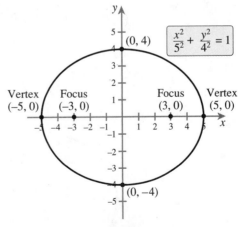

FIGURE 64

EXAMPLE 2   ■   **An Ellipse Centered at the Origin**

Sketch the graph of the ellipse with equation $9x^2 + 4y^2 - 36 = 0$. Identify the vertices and foci.

SOLUTION   We first write the equation in standard form.

$$9x^2 + 4y^2 - 36 = 0$$

$$9x^2 + 4y^2 = 36$$

$$\frac{9x^2}{36} + \frac{4y^2}{36} = 1$$

$$\frac{x^2}{4} + \frac{y^2}{9} = 1$$

$$\frac{x^2}{2^2} + \frac{y^2}{3^2} = 1$$

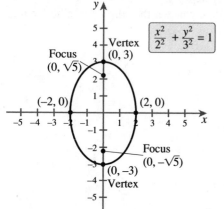

FIGURE 65

Thus, $a = 2$ and $b = 3$, giving us x-intercepts of ± 2 and y-intercepts of ± 3. The graph is shown in Figure 65. Since the major axis is vertical, the vertices have coordinates $(0, -3)$ and $(0, 3)$. To find the foci, we compute c as follows:

$$c = \sqrt{|a^2 - b^2|} = \sqrt{|4 - 9|} = \sqrt{5}$$

Thus, the foci are $\sqrt{5}$ units from the center on the major axis, at the points $(0, -\sqrt{5})$ and $(0, \sqrt{5})$.

Just as with parabolas with vertex away from the origin, ellipses centered away from the origin have standard equations that can be obtained using translations. For example, the graph of

$$\frac{(x - 1)^2}{25} + \frac{(y - 3)^2}{16} = 1$$

can be obtained from the graph of

$$\frac{x^2}{25} + \frac{y^2}{16} = 1$$

by translating 1 unit to the right and 3 units up, as shown in Figures 66 and 67. From

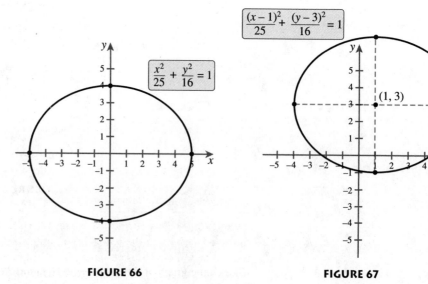

FIGURE 66   **FIGURE 67**

this, we see that the graph of $\dfrac{(x-1)^2}{25} + \dfrac{(y-3)^2}{16} = 1$ is an ellipse with a horizontal major axis and center $(1, 3)$. More generally, we have the following.

Ellipses Centered at (h, k) ☐

The standard form for the equation of an ellipse centered at (h, k) is

$$\frac{(x-h)^2}{a^2} + \frac{(y-k)^2}{b^2} = 1, \quad a \ne b$$

If $a > b$, the major axis is horizontal with length $2a$ (see Figure 68), whereas if $b > a$ the major axis is vertical with length $2b$ (see Figure 69). The vertices are the endpoints of the major axis. The foci lie on the major axis a distance of c units from the center, where $c = \sqrt{\,|a^2 - b^2|\,}$. Note that if $a = b$, the equation is that of a circle centered at (h, k) with radius a.

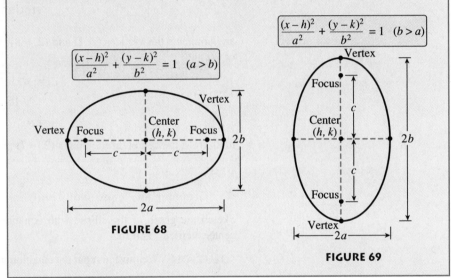

FIGURE 68

FIGURE 69

The graph of an ellipse centered at (h, k) can be obtained by viewing (h, k) as the origin of a new coordinate system and then applying the process used for an ellipse centered at the origin.

EXAMPLE 3 ■ Graphing an Ellipse and Identifying Key Features

Sketch the graph of the ellipse with equation

$$\frac{(x-2)^2}{9} + \frac{(y+3)^2}{25} = 1$$

Identify the center, vertices, and foci.

SOLUTION This ellipse has center $(2, -3)$. Since $a = \sqrt{9} = 3$, we plot points on the ellipse 3 units to the left and right of the center. Similarly, since $b = \sqrt{25} = 5$, we plot points on the ellipse 5 units above and below the center. Connecting these points with a smooth curve yields the graph in Figure 70. We see that the major axis is vertical

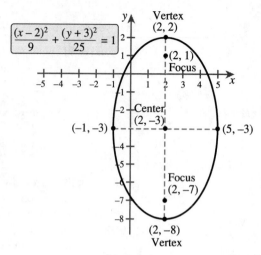

FIGURE 70

and connects the vertices $(2, 2)$ and $(2, -8)$. The foci are on the major axis a distance of

$$c = \sqrt{|9 - 25|}$$
$$= \sqrt{16}$$
$$= 4$$

units from the center, at the points $(2, -7)$ and $(2, 1)$.

EXAMPLE 4   ▓   **Graphing an Ellipse and Identifying Key Features**

Sketch the graph of the ellipse with equation $9x^2 + y^2 + 36x + 27 = 0$. Identify the center, vertices, and foci.

SOLUTION   We must first put the equation in standard form by completing the square.

$$9x^2 + y^2 + 36x + 27 = 0$$
$$(9x^2 + 36x +) + y^2 = -27$$
$$9(x^2 + 4x +) + y^2 = -27$$
$$9(x^2 + 4x + 4) + y^2 = -27 + 9(4) \qquad \text{Adding } 9(4) = 36 \text{ to both sides}$$
$$9(x + 2)^2 + y^2 = 9$$
$$\frac{(x + 2)^2}{1} + \frac{y^2}{9} = 1 \qquad \text{Dividing both sides by 9}$$

The center of the ellipse is $(-2, 0)$. Since $a = \sqrt{1} = 1$ and $b = \sqrt{9} = 3$, we plot points on the ellipse 1 unit to the left and right of the center and 3 units above and below the center. The graph is shown in Figure 71. The major axis is vertical, and the vertices have coordinates $(-2, 3)$ and $(-2, -3)$. The foci are on the major axis a distance of

$$c = \sqrt{|1 - 9|}$$
$$= \sqrt{8}$$
$$= 2\sqrt{2}$$

units from the center, at the points $(-2, -2\sqrt{2})$ and $(-2, 2\sqrt{2})$.

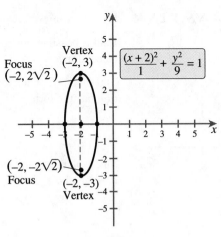

FIGURE 71

E X A M P L E 5   ■   Finding the Equation of an Ellipse

Find the equation of the ellipse with foci (± 4, 0) and major axis length of 12.

SOLUTION   Since the foci are on the x-axis at equal distances from the origin, the ellipse is centered at the origin and has a horizontal major axis (see Figure 72). Moreover, since the major axis has length 12, we have $2a = 12$ or $a = 6$. The coordinates of the foci, (± 4, 0), imply that $c = 4$. Since $b < a$, $c = \sqrt{a^2 - b^2}$ and so

$$4 = \sqrt{6^2 - b^2}$$
$$16 = 36 - b^2$$
$$b^2 = 20$$

Thus, the equation of the ellipse is

$$\frac{x^2}{36} + \frac{y^2}{20} = 1$$

and its graph is shown in Figure 73.

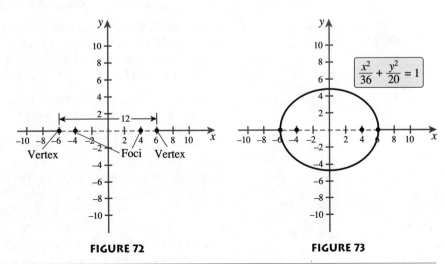

FIGURE 72   **FIGURE 73**

E X A M P L E 6   ▓   Finding the Equation of an Ellipse from Its Graph

Find a possible equation for the ellipse shown in Figure 74. Verify with a graphing calculator.

SOLUTION   The graph in Figure 74 appears to be centered at $(-3, -2)$, so the equation has the form

$$\frac{(x + 3)^2}{a^2} + \frac{(y + 2)^2}{b^2} = 1$$

Since the horizontal axis has length 6, and the vertical axis has length 4, it follows that $a = 3$ and $b = 2$. This gives us the equation

$$\frac{(x + 3)^2}{3^2} + \frac{(y + 2)^2}{2^2} = 1$$

To graph this equation on a graphing calculator, we must first solve for y.

$$\frac{(x + 3)^2}{9} + \frac{(y + 2)^2}{4} = 1$$

$$\frac{(y + 2)^2}{4} = 1 - \frac{(x + 3)^2}{9}$$

$$(y + 2)^2 = 4 - \frac{4}{9}(x + 3)^2 \qquad \text{Multiplying both sides by 4}$$

$$y + 2 = \pm\sqrt{4 - \frac{4}{9}(x + 3)^2} \qquad \text{Taking the square root of both sides}$$

$$y = -2 \pm \sqrt{4 - \frac{4}{9}(x + 3)^2}$$

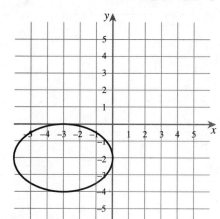

FIGURE 74

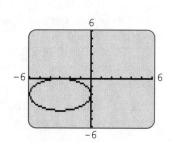

FIGURE 75

Graphing the branches $y = -2 + \sqrt{4 - \frac{4}{9}(x + 3)^2}$ and $y = -2 - \sqrt{4 - \frac{4}{9}(x + 3)^2}$, we obtain the ellipse shown in Figure 75. It appears to be the same as that shown in Figure 74.

E X A M P L E 7   ▓   An Application Involving an Ellipse

A train tunnel through a mountain is semielliptical, with height 18 feet at the center and width 24 feet at the base (see Figure 76). A train is hauling a load of flatcars that are piled high with freight. If the flatcars are 10 feet wide and the load reaches a height of 15 feet above the ground, will the flatcars clear the opening of the tunnel?

SOLUTION   To determine the clearance, we must know the height of the tunnel at the edge of the flatcar. We first construct a coordinate system with the x-axis on the ground and the origin at the center of the tracks. Next we find the equation of the ellipse that forms the tunnel opening. Using the dimensions of the tunnel opening, we know that the ellipse has a vertical major axis of length 2(18) and a horizontal minor axis of length 24. Thus $a = 12$ and $b = 18$. The equation is as follows:

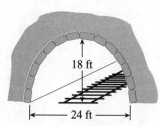

FIGURE 76

$$\frac{x^2}{12^2} + \frac{y^2}{18^2} = 1$$

The edge of a 10-foot wide flatcar corresponds to $x = 5$, and so the height of the tunnel at the edge of the flatcar is given by the y-value of the point on the ellipse when $x = 5$. Thus, we set $x = 5$ and solve for y.

$$\frac{5^2}{12^2} + \frac{y^2}{18^2} = 1$$

$$\frac{y^2}{324} = 1 - \frac{25}{144}$$

$$y^2 = 324\left(1 - \frac{25}{144}\right)$$

$$y = \pm\sqrt{324\left(1 - \frac{25}{144}\right)} \approx \pm 16.36$$

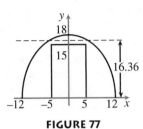

FIGURE 77

Thus, the height of the tunnel above the edge of the flatcar is approximately 16.36 feet, which gives 1.36 feet of clearance. See Figure 77.

COMETARY COLLISIONS

From ancient times, astronomers and amateur stargazers alike have been fascinated with comets. One reason for this must certainly be the fact that many comets have been visible to the naked eye. Indeed, comet sightings have been recorded as early as the 15th century B.C., long before the invention of the telescope. Ironically, early comet sightings were almost always associated with catastrophe: the assassination of Julius Caesar in 44 B.C., the fall of Jerusalem to the Romans in A.D. 70, and the Norman conquest of England in A.D. 1066, to name a few. In more modern times, as comets came to be understood as astronomic phenomena, such superstitious beliefs began to wane. However, one fear has remained even to this day—that a comet may one day collide with Earth.

How likely is such a collision? Most astronomers would likely argue that the chance of a collision with any specific comet is virtually zero. However, at least one astronomer, Brian Marsden, has suggested that there is "a small but not negligible chance" that the comet Swift-Tuttle will collide with Earth on August 14, 2126. Other astronomers, whether or not they agree with Marsden, would likely admit that if the known number of comets is taken into account, a collision at some time in the future is inevitable. As evidence of this likelihood, one need only look to the collision of the comet Shoemaker-Levy 9 with Jupiter in July 1994. Further evidence can perhaps be found closer to home. Some scientists have theorized that a cometary collision caused the extinction of the dinosaur. There are even some that suggest a more recent collision. In 1908, a powerful explosion leveled thousands of acres of trees in the Tunguska River basin in central Siberia. The accompanying

Devastation at Tunguska, Siberia

aftershock was detected by seismic recording equipment over 500 miles away. Eyewitnesses described a bright blue fireball, trailing smoke, that some experts have concluded was a small fragment from the comet Encke. Because no conclusive evidence of a crater was found, it has been suggested that the fragment exploded several miles above the surface of Earth.

The threat of a comet or asteroid collision has been taken seriously by astronomers and government officials, and it has led to the formation of NASA's Near-Earth Object (NEO) program. The purpose of this program is to detect, catalogue, and monitor potentially troublesome comets and asteroids. One essential ingredient of an early detection system is a thorough understanding of the path of a comet. Here is where conic sections come into play. Over 650 comets have been observed in our solar system, and each has been identified as having a path that is either elliptical, parabolic, or hyperbolic, with the sun as a focus. Comets with elliptical orbits reappear at regular (though perhaps quite lengthy) time intervals. Certainly the most famous comet with an elliptical orbit is Halley's comet, which has a period of approximately 76 years. It last appeared in 1986, and it will appear again in 2061. The Swift-Tuttle comet referenced earlier also has an elliptical orbit. It last passed Earth in 1992, and it will pass again in 2126. Comets with parabolic or hyperbolic paths are seen only once, and so it is likely that many have been missed and many have yet to be seen. The threat of a collision is greatest for those comets that have parabolic or hyperbolic paths since they are more difficult to detect and track.

REFLECTIVE PROPERTY OF ELLIPSES

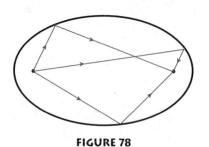

FIGURE 78

The reflective property of an ellipse involves both foci. When sound or light is emitted from one focus of an ellipse, it is reflected toward the other focus, as shown in Figure 78. An interesting application of this property can be found in "whispering rooms," such as the ones located at the Museum of Science and Industry in Chicago or the Capitol building in Washington, D.C. These rooms have walls (or ceilings) shaped like an ellipse. If you stand at one of the foci and whisper something to a friend standing at the other focus, you will be heard clearly by your friend but nobody else.

A more useful application of the reflective property of an ellipse can be found in a medical procedure called *shock wave lithotripsy* used in the treatment of kidney stones. If a shock wave transmitter and an elliptic reflector are placed in such a way that the transmitter and kidney stone are located at the two foci of the reflector, the shock waves will be directed at the kidney stone causing it to shatter, while leaving the rest of the body unharmed.

UNDERSTANDING AND MASTERY CHECKLISTS

CONCEPTS TO UNDERSTAND

- ☐ Ellipse
- ☐ Foci of an ellipse
- ☐ Vertices of an ellipse
- ☐ Center of an ellipse
- ☐ Major axis of an ellipse
- ☐ Minor axis of an ellipse
- ☐ Reflective property of an ellipse

SKILLS TO MASTER

- ☐ Sketch the graph of an ellipse.
- ☐ Identify the center, vertices, and foci of an ellipse.
- ☐ Find the equation of an ellipse given information about its graph.

EXERCISES 6.3

EXERCISES 1–14 *Find the center, vertices, and foci for the given ellipse, and sketch its graph.*

1. $\dfrac{x^2}{9} + \dfrac{y^2}{25} = 1$

2. $\dfrac{x^2}{16} + \dfrac{y^2}{9} = 1$

3. $4x^2 + 9y^2 = 36$

4. $16x^2 + 9y^2 = 144$

5. $9x^2 + 4y^2 = 1$

6. $16x^2 + 25y^2 = 1$

7. $5x^2 + 8y^2 = 40$

8. $3x^2 + 2y^2 = 6$

9. $\dfrac{(x + 3)^2}{4} + \dfrac{(y - 1)^2}{16} = 1$

10. $\dfrac{(x - 2)^2}{25} + \dfrac{(y + 2)^2}{9} = 1$

11. $9x^2 + 4y^2 - 18x - 24y + 9 = 0$

12. $25x^2 + 16y^2 - 200x - 32y + 16 = 0$

13. $x^2 + 36y^2 + 4x - 72y + 4 = 0$

14. $16x^2 + y^2 + 4y - 12 = 0$

EXERCISES 15–26 *Find an equation for the ellipse with the given properties.*

15. Center $(0, 0)$, horizontal major axis of length 8, and vertical minor axis of length 6

16. Vertices $(0, \pm 4)$ and minor axis of length 2

17. Vertices $(0, \pm\sqrt{6})$ and foci $(0, \pm\sqrt{2})$

18. Foci $(\pm\sqrt{2}, 0)$ and major axis of length 4

19. Vertices $(\pm 6, 0)$ and passes through the point $(-4, 2)$

20. Foci $(0, \pm 2)$ and passes through the point $(3, 2)$

21. Center $(2, -1)$, vertical major axis of length 6, horizontal minor axis of length 4

22. Vertices $(-2, 6)$ and $(-2, 0)$ and minor axis of length 1

23. Vertices $(-4, -2)$ and $(6, -2)$ and foci $(-2, -2)$ and $(4, -2)$

24. Foci (0, 4) and (6, 4) and major axis of length $\sqrt{10}$

25. The following graph

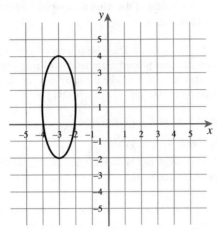

26. The following graph

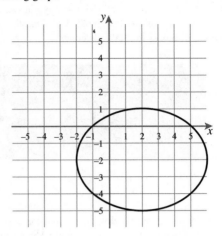

EXERCISES 27–30 *Solve for y and plot the resulting equation(s) with a graphing calculator. Identify the coordinates of the vertices.*

27. $x^2 + 4y^2 + 4x = 0$ **28.** $4x^2 + y^2 - 16x = 0$

29. $4x^2 + y^2 - 2y - 3 = 0$ **30.** $x^2 + 4y^2 + 16y = 0$

▪ APPLICATIONS

31. *Earth's Orbit*   Earth's orbit is elliptical with major axis length 186 million miles and minor axis length 185.8 million miles. Write an equation for the path of the orbit, assuming the center is located at the origin. It may surprise you to find out the sun is not at the center of the orbit. The sun is, in fact, on the major axis, a distance of 4.3 million miles from the center. Sketch a graph of the orbit of Earth and locate the sun on the major axis.

32. *Mars's Orbit*   The orbit of Mars is elliptical with major axis length 283.5 million miles and minor axis length 278.5 million miles. Write an equation for the path of the orbit, assuming the center is located at the origin. The sun is on the major axis, a distance of 26.5 million miles from the center. Sketch a graph of the orbit of Mars and locate the sun on the major axis.

33. *Railroad Bridge*   The arch of a railroad bridge over a two-lane highway has the shape of a semiellipse (see Figure 79). The distance from the base of the arch on one side of the road to the base on the other is 50 feet, and the height of the arch at the center is 20 feet.

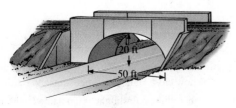

FIGURE 79

a. Find the equation of the ellipse that gives the shape of the arch. Assume that the *x*-axis is on the ground and runs perpendicular to the centerline of the highway, the origin is on the centerline, and the *y*-axis is vertical.

b. What is the height of the arch 5 feet from the base?

c. Could a tractor-trailer with a height of 14 feet and a width of 10 feet pass under the bridge without going over the centerline of the road?

34. *Whispering Room*   An elliptical whispering room is to be constructed with a maximum width of 30 feet and a maximum length of 50 feet. Locate the positions at which two people should stand if they wish to whisper to each other from opposite sides of the room without others hearing.

EXERCISES 35–36 *An Astronomical Unit (A.U.) refers to the mean distance from Earth to the sun. When dealing with elliptical orbits of celestial bodies with the sun located at one focus, the **perihelion** distance is the distance from the sun to the closest vertex, while the **aphelion** distance is the distance from the sun to the most distant vertex.*

35. *Comet Encke*   Some scientists have speculated that a piece of comet Encke fell to the Earth in 1908, causing massive destruction in a remote region of Siberia. (See mathematical note on page 497 for more details.) The orbit of comet Encke is an ellipse, with the sun located at one focus. The major axis has length 4.4 A.U., and the minor axis has length 2.2 A.U. Find the perihelion and aphelion distances for comet Encke.

36. *Halley's Comet*   The orbit of Halley's comet is an ellipse, with the sun located at one focus. The major axis has length 36.2 A.U., and the minor axis has length 9.1 A.U. Find the perihelion and aphelion distances for Halley's comet.

Halley's comet in space

37. *Anderson Elliptical Window*   The Anderson Window company manufactures a glass window in the shape of a semiellipse (see Figure 80), with a major axis of length 168.3 centimeters. The area of the glass is advertised as 3995 cm².

— 168.3 cm —

FIGURE 80

a. The area of an ellipse with equation $\dfrac{x^2}{a^2} + \dfrac{y^2}{b^2} = 1$ is given by $A = \pi ab$. Find the equation of the ellipse that forms the boundary of the Anderson window, assuming that the origin is at the center and the major axis lies on the x-axis.

b. Find the height of the window at a point 30 centimeters from the center.

38. *Comet Path Length*   There is no simple formula for the circumference, or more properly, arc length, of an ellipse. However, a crude approximation can be made by pretending that the ellipse is made up of line segments connecting points on the ellipse.

a. Find a point P on the ellipse $\dfrac{x^2}{16} + \dfrac{y^2}{9} = 1$ roughly halfway between $(4, 0)$ and $(0, 3)$.

b. Find the distance between P and $(4, 0)$, and the distance between P and $(0, 3)$.

c. Use the results from part b to estimate the arc length of the portion of the ellipse between $(4, 0)$ and $(0, 3)$. Is this result greater than or less than the actual arc length?

d. Use the result from part c to estimate the arc length of the entire ellipse. Is this result greater or less than the actual arc length?

e. How can this technique be modified to produce a more precise estimate of arc length?

f. Comet Hale-Bopp, which was visible to the naked eye during much of the spring of 1997, travels in an elliptical orbit through the solar system. Estimate the length of its orbit given that its major axis is 3.5×10^{10} miles and its minor axis is 1.0×10^{9} miles.

CONCEPTS AND CRITICAL THINKING

EXERCISES 39–42   *Answer true or false.*

39. An ellipse centered at the origin with horizontal major axis is symmetric with respect to the origin.

40. It is possible for the plot of an ellipse to be entirely contained within the viewing rectangle of a graphing calculator.

41. The reflective property of ellipses states that light emanating from one focus will be reflected through the other focus.

42. If the foci of an ellipse are very close together, then the ellipse will appear nearly circular.

EXERCISES 43–46   *Give an example of each.*

43. The equation of an ellipse centered at $(0, 3)$ with major axis of length 4

44. The equations of two ellipses such that one is a 90° rotation of the other

45. A real-world application of the reflective property of ellipses

46. The equation of an ellipse with foci 2 units apart

47. If an ellipse has a horizontal major axis of length $2a$ and foci at $(\pm c, 0)$, explain why $a^2 - c^2 > 0$.

QUESTIONS FOR DISCUSSION OR ESSAY

48. An elliptical "pool" table with only one pocket is being used for a trick-shooting demonstration. The pool shark carefully places a ball on the table but hits it in a seemingly arbitrary direction. Time after time, the ball bounces off the cushion directly into the pocket. Explain how this could happen.

49. Using only the information given in Exercise 36, is it possible to determine how close Halley's comet comes to Earth? Explain.

50. Light from an ordinary flashlight is contained within a cone. Explain how an ellipse is formed by shining a flashlight inside a room.

■ **PROJECTS FOR ENRICHMENT**

51. *Drawing Ellipses*   In this project we investigate how to draw ellipses of varying sizes. For materials, locate a large piece of cardboard, blank paper, two thumbtacks, several long pieces of string, and a pencil. Lay a piece of paper on top of the cardboard. Cut a piece of string a little longer than 4 inches and attach each end around a thumbtack so that 4 inches of string is between the tacks. Stick the tacks into the cardboard 3 inches apart so there is some slack in the string. Place the tip of the pencil as shown in Figure 81 and, keeping the string taut and the tip of the pencil on the paper, carefully move the pencil along the string. The pencil will trace out the path of an ellipse with foci located at the thumbtacks. The length of the major axis is equal to the total length of the string between tacks.

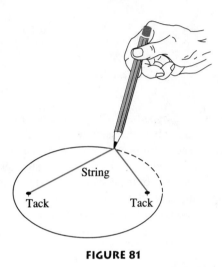

FIGURE 81

a. Use the fact that the string is 4 inches long and the tacks are 3 inches apart to find the equation of the ellipse you drew. Assume the *x*-axis passes through the foci and the origin is midway between them.

b. Draw an ellipse with foci a distance of 3 inches apart and major axis length of 5 inches.

c. Draw an ellipse with foci 3 inches apart and major axis length of 6 inches.

d. What general observation can you make about the shape of the ellipse if the distance between the foci is held constant but the length of the major axis (that is, the length of the string) is increased?

e. Compute the ratio of the distance between the foci to the length of the major axis for each of the three ellipses you have constructed. This ratio is called the *eccentricity* of the ellipse. What is the largest possible value of the eccentricity? Why? What is the connection between the eccentricity and the shape of the ellipse? What is the smallest the eccentricity could be? What would the shape of the ellipse be then?

SECTION 6.4   **HYPERBOLAS**

☐ **What is the path of a sonic boom?**

☐ **How can hyperbolas help explain shadows on a wall?**

☐ **How can hyperbolas help navigate a ship?**

☐ **Why are the shapes of conic sections often found in the mirrors of reflective telescopes?**

☐ **How can hyperbolas be used to help Inspector Magill locate a stolen car?**

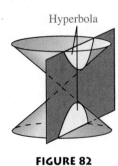

Hyperbola

FIGURE 82

In Section 6.2, we described the conic sections as curves formed when a double cone is cut by a plane. If the plane is parallel to the axis of the cone, we obtain a hyperbola, as shown in Figure 82. Using geometric techniques, it can be shown that the following characterization of a hyperbola is equivalent.

Definition of a Hyperbola   ☐

A **hyperbola** is the set of all points (x, y) such that the difference of the distances from two fixed points (called the **foci**) is constant, as shown in Figure 83. The line through the foci intersects the hyperbola at two points, called the **vertices**. The line segment connecting the two vertices is called the **transverse axis,** and the midpoint of the transverse axis is called the **center**.

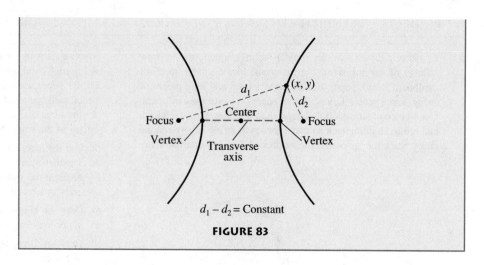

FIGURE 83

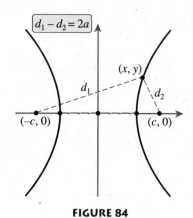

FIGURE 84

The standard form for the equation of a hyperbola with a horizontal transverse axis and centered at the origin can be derived in much the same way as that for the ellipse. Although the derivation is lengthy, the end result is quite simple. We begin by placing the foci at $(-c, 0)$ and $(c, 0)$, as shown in Figure 84. According to the definition, for any point (x, y) on the hyperbola, the difference of the distances from the two foci must be constant. To simplify later computations, we denote this constant difference by $2a$. Using the distance formula for each of the two distances d_1 and d_2, we have

$$d_1 = \sqrt{(x + c)^2 + y^2} \quad \text{and} \quad d_2 = \sqrt{(x - c)^2 + y^2}$$

Setting $d_1 - d_2 = 2a$, we proceed as follows:

$$\sqrt{(x + c)^2 + y^2} - \sqrt{(x - c)^2 + y^2} = 2a$$
$$\sqrt{(x + c)^2 + y^2} = 2a + \sqrt{(x - c)^2 + y^2}$$

Now we square both sides and collect terms.

$$(x + c)^2 + y^2 = 4a^2 + 4a\sqrt{(x - c)^2 + y^2} + (x - c)^2 + y^2$$
$$x^2 + 2cx + c^2 + y^2 = 4a^2 + 4a\sqrt{(x - c)^2 + y^2} + x^2 - 2cx + c^2 + y^2$$
$$4cx - 4a^2 = 4a\sqrt{(x - c)^2 + y^2}$$
$$cx - a^2 = a\sqrt{(x - c)^2 + y^2}$$

Once again, we square both sides and collect terms.

$$(cx - a^2)^2 = a^2[(x - c)^2 + y^2]$$
$$c^2x^2 - 2a^2cx + a^4 = a^2(x^2 - 2cx + c^2 + y^2)$$
$$c^2x^2 - 2a^2cx + a^4 = a^2x^2 - 2a^2cx + a^2c^2 + a^2y^2$$
$$c^2x^2 - a^2x^2 - a^2y^2 = a^2c^2 - a^4$$
$$(c^2 - a^2)x^2 - a^2y^2 = a^2(c^2 - a^2)$$

Finally, since $c^2 - a^2 > 0$ (see Exercise 55), we set $b^2 = c^2 - a^2$ and substitute.

$$b^2x^2 - a^2y^2 = a^2b^2$$
$$\frac{x^2}{a^2} - \frac{y^2}{b^2} = 1$$

The standard equation for a hyperbola with a vertical transverse axis can be derived in a similar manner.

Hyperbolas Centered at the Origin

The standard forms for the equation of a hyperbola centered at $(0, 0)$ are as follows:

Equation	Description
$\dfrac{x^2}{a^2} - \dfrac{y^2}{b^2} = 1$	Horizontal transverse axis
$\dfrac{y^2}{b^2} - \dfrac{x^2}{a^2} = 1$	Vertical transverse axis

The vertices are the endpoints of the transverse axis: $(\pm a, 0)$ in the case of a horizontal transverse axis (see Figure 85) and $(0, \pm b)$ for a vertical transverse axis (see Figure 86). The foci are a distance of $c = \sqrt{a^2 + b^2}$ units from the center on the same line as the vertices.

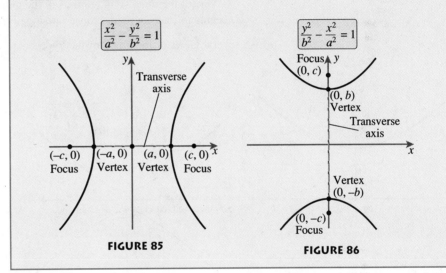

FIGURE 85   **FIGURE 86**

EXAMPLE 1 ■ **A Hyperbola Centered at the Origin**

Determine the orientation and find the vertices and foci of the hyperbola with equation

$$\frac{y^2}{36} - \frac{x^2}{64} = 1$$

SOLUTION This equation is in the standard form of a hyperbola with vertical transverse axis. Thus, since $b = \sqrt{36} = 6$, the vertices are $(0, -6)$ and $(0, 6)$. To find the foci, we compute c.

$$c = \sqrt{a^2 + b^2} = \sqrt{36 + 64} = 10$$

So the foci are 10 units away from the center, on the same line as the vertices, at the points $(0, -10)$ and $(0, 10)$. The graph is shown in Figure 87.

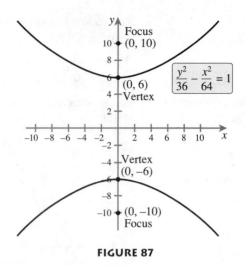

FIGURE 87

One of the interesting features of the graph of a hyperbola is that as we zoom out, it begins to look more and more like a pair of intersecting lines. Figures 88–90 show several views of the graph of the hyperbola $\dfrac{y^2}{36} - \dfrac{x^2}{64} = 1$ from Example 1.

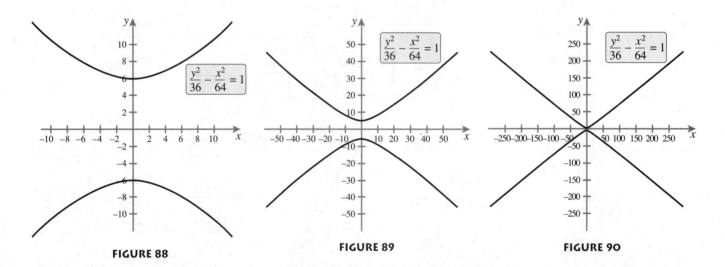

FIGURE 88   **FIGURE 89**   **FIGURE 90**

The lines that the graph of a hyperbola approaches are called its **asymptotes,** and they serve as useful aids in sketching the graph of the hyperbola. To determine the equations of the lines, let us look more closely at a hyperbola with a vertical axis. Solving the equation

$$\frac{y^2}{b^2} - \frac{x^2}{a^2} = 1$$

for y^2 we obtain

$$y^2 = \frac{b^2}{a^2} x^2 + b^2$$

Now, for very large values of x, the first term will dwarf the second. Thus, for large values of x we have

$$y^2 \approx \frac{b^2}{a^2} x^2 \qquad \text{or} \qquad y \approx \pm \frac{b}{a} x$$

Thus, the asymptotes for a hyperbola centered at the origin with a vertical axis are $y = \pm \frac{b}{a}x$. In similar fashion, we can see that a hyperbola with a horizontal axis has asymptotes $y = \pm \frac{b}{a}x$. In either case, the asymptotes can be located by constructing a rectangle with sides passing through the points $(-a, 0)$, $(a, 0)$, $(0, b)$, and $(0, -b)$. The extended diagonals of this rectangle are the asymptotes. See Figures 91 and 92 for details.

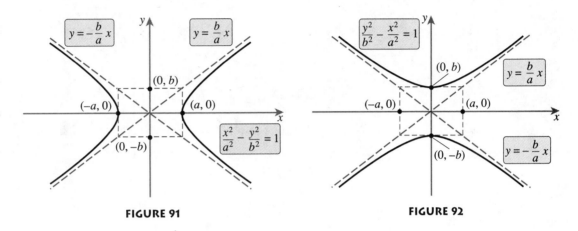

FIGURE 91   **FIGURE 92**

Asymptotes of a Hyperbola Centered at the Origin   ☐

The asymptotes for a hyperbola centered at the origin are

$$y = \frac{b}{a} x \qquad \text{and} \qquad y = -\frac{b}{a} x$$

EXAMPLE 2   ▣   A Hyperbola Centered at the Origin

Find the vertices, foci, and asymptotes for the hyperbola with equation

$$9x^2 - 16y^2 = 144$$

and sketch its graph.

SOLUTION   We first divide through by 144 to rewrite the equation in standard form.

$$9x^2 - 16y^2 = 144$$

$$\frac{9x^2}{144} - \frac{16y^2}{144} = 1$$

$$\frac{x^2}{16} - \frac{y^2}{9} = 1$$

From the standard form of the equation, we know that the hyperbola has a horizontal transverse axis with $a = \sqrt{16} = 4$ and $b = \sqrt{9} = 3$. Thus, the vertices are $(-4, 0)$ and $(4, 0)$ and the transverse axis is the line segment connecting these two points. The distance from the center to each focus is $c = \sqrt{16 + 9} = 5$, and so the foci are located at $(-5, 0)$ and $(5, 0)$. To find the asymptotes, we construct the rectangle whose sides pass through the vertices $(-4, 0)$ and $(4, 0)$ on the x-axis and the points $(0, -3)$ and $(0, 3)$ on the y-axis. The asymptotes are then formed by extending the diagonals of the rectangle. The equations of the asymptotes are $y = \frac{3}{4}x$ and $y = -\frac{3}{4}x$. The graph is shown in Figure 93.

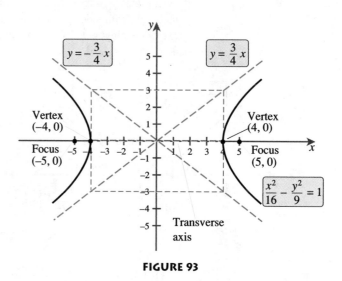

FIGURE 93

As with both parabolas and ellipses, the standard equations of hyperbolas centered away from the origin can be found by applying translations. These equations are summarized as follows.

Hyperbolas Centered at (h, k) ☐

The standard forms for the equation of a hyperbola centered at (h, k) are as follows:

Equation	Description
$\dfrac{(x - h)^2}{a^2} - \dfrac{(y - k)^2}{b^2} = 1$	Horizontal transverse axis
$\dfrac{(y - k)^2}{b^2} - \dfrac{(x - h)^2}{a^2} = 1$	Vertical transverse axis

The vertices are the endpoints of the transverse axis: a units from the center in the case of a horizontal transverse axis (see Figure 94) and b units from the center in the case of a vertical transverse axis (see Figure 95). The foci are a distance of $c = \sqrt{a^2 + b^2}$ units from the center on the same line as the vertices.

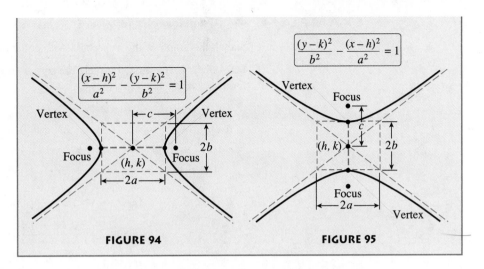

FIGURE 94   FIGURE 95

E X A M P L E 3 ▨ Graphing a Hyperbola and Identifying Key Features

Find the center, vertices, foci, and asymptotes for the hyperbola with equation

$$\frac{(x-2)^2}{4} - \frac{(y-1)^2}{16} = 1$$

and sketch its graph.

SOLUTION The center of the hyperbola is the point $(2, 1)$. The axis is horizontal with $a = 2$ and $b = 4$. Thus, the vertices are 2 units to the left and to the right of the center, at $(0, 1)$ and $(4, 1)$. The transverse axis is the line segment connecting the two vertices. The foci are located $c = \sqrt{4 + 16} = \sqrt{20} = 2\sqrt{5}$ units from the center on the same line as the vertices, at the points $(2 - 2\sqrt{5}, 1)$ and $(2 + 2\sqrt{5}, 1)$. To locate the asymptotes, we construct a rectangle with horizontal sides 4 units above and below the center and vertical sides 2 units to the left and right of the center. The asymptotes are the extended diagonals of this rectangle. They have slope $m = \pm\frac{b}{a} = \pm 2$, and they pass through the center $(2, 1)$. Their equations are thus $y - 1 = \pm 2(x - 2)$. The graph is shown in Figure 96.

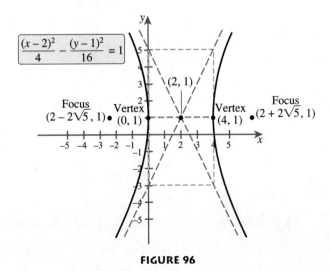

FIGURE 96

EXAMPLE 4 ■ Graphing a Hyperbola

Sketch the graph of the hyperbola with equation $16x^2 - y^2 + 64x - 2y + 67 = 0$.

SOLUTION We first complete the square in x and y to write the equation in standard form.

$$16x^2 - y^2 + 64x - 2y + 67 = 0$$

$$(16x^2 + 64x +) + (-y^2 - 2y +) = -67$$

$$16(x^2 + 4x +) - (y^2 + 2y +) = -67$$

$$16(x^2 + 4x + 4) - (y^2 + 2y + 1) = -67 + 64 - 1 \qquad \text{Adding } 16(4) \text{ and } -(1) \text{ to both sides}$$

$$16(x + 2)^2 - (y + 1)^2 = -4$$

$$-4(x + 2)^2 + \frac{(y + 1)^2}{4} = 1 \qquad \text{Dividing both sides by } -4$$

$$\frac{(y + 1)^2}{4} - 4(x + 2)^2 = 1$$

$$\frac{(y + 1)^2}{4} - \frac{(x + 2)^2}{1/4} = 1$$

$$\frac{(y + 1)^2}{2^2} - \frac{(x + 2)^2}{(1/2)^2} = 1$$

The center of the hyperbola is at $(-2, -1)$. The axis is vertical with $a = \frac{1}{2}$ and $b = 2$. The vertices are 2 units above and below the center, at $(-2, -3)$ and $(-2, 1)$. To sketch the hyperbola, we construct a rectangle with center $(-2, -1)$, horizontal sides 2 units above and below the center, and vertical sides $\frac{1}{2}$ unit to the left and to the right of the center. The asymptotes are formed by extending the diagonals, as shown in Figure 97.

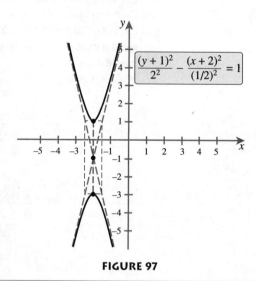

$$\frac{(y + 1)^2}{2^2} - \frac{(x + 2)^2}{(1/2)^2} = 1$$

FIGURE 97

EXAMPLE 5 ■ Finding the Equation of a Hyperbola

Find the equation of the hyperbola with vertices $(\pm 1, 0)$ and foci $(\pm\sqrt{5}, 0)$.

SOLUTION We begin by plotting the vertices and foci, as shown in Figure 98. It is clear from the locations of these points that the center is at the origin (halfway between the vertices) and the transverse axis is horizontal (since the vertices are on the x-axis). Thus, the standard form of the equation is

$$\frac{x^2}{a^2} - \frac{y^2}{b^2} = 1$$

The vertices of a hyperbola with this equation form are a units from the center, and so it must follow that $a = 1$. Furthermore, since the foci are located $\sqrt{a^2 + b^2}$ units from the center, we have

$$\sqrt{a^2 + b^2} = \sqrt{5}$$
$$\sqrt{1 + b^2} = \sqrt{5}$$
$$b^2 = 4$$

Thus, the hyperbola has equation

$$\frac{x^2}{1} - \frac{y^2}{4} = 1$$

The graph is shown in Figure 99.

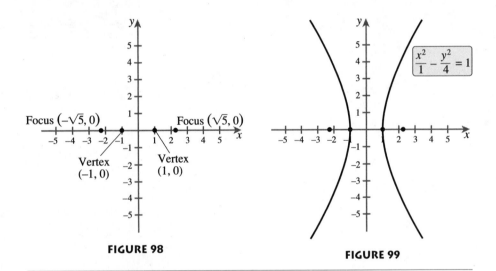

FIGURE 98

FIGURE 99

REFLECTIVE PROPERTY OF HYPERBOLAS

Just as with ellipses, the reflective property of hyperbolas involves its foci. If an incoming light ray or sound wave is aimed at one focus of a hyperbola, it is reflected toward the other focus, as shown in Figure 100. Hyperbolic reflectors are often used as

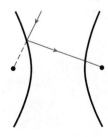

FIGURE 100

auxiliary mirrors in telescopes to direct the image to the eyepiece of the telescope, where it can be magnified. Two schematic diagrams of telescopes are given in Figures 101 and 102. Notice that in both the main reflector is parabolic. Also in both, the hyperbolic reflector is positioned so that its focus coincides with that of the parabolic mirror. In Figure 101, the eyepiece of the telescope is located at the other focus of the hyperbolic reflector. In Figure 102, an elliptic reflector is added, with one of its foci at the second focus of the hyperbolic reflector and its other focus at the eyepiece.

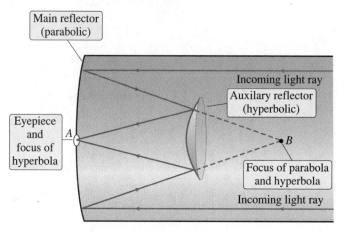

FIGURE 101

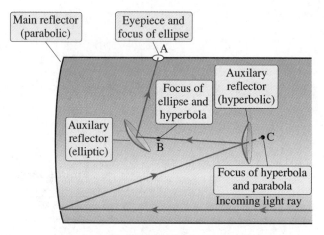

FIGURE 102

THE GENERAL EQUATION OF A CONIC SECTION

Table 2 gives some examples of conic sections from earlier in the text.

TABLE 2

Equation	Conic section
$x^2 + y^2 - 2x + 4y - 20 = 0$	Circle
$x = y^2 + 2y - 3$	Parabola
$9x^2 + y^2 + 36x + 27 = 0$	Ellipse
$16x^2 - y^2 + 64x - 2y + 67 = 0$	Hyperbola

In every case, the equation of the conic section can be written in the form $Ax^2 + Cy^2 + Dx + Ey + F = 0$. In fact, all conic sections we have seen thus far have equations of this form. More generally, we have the following.

The General Equation of a ☐
Conic Section

> The general equation of a circle, or any parabola, ellipse, or hyperbola with a vertical or horizontal axis is $Ax^2 + Cy^2 + Dx + Ey + F = 0$.

Note, however, that not all equations of the form $Ax^2 + Cy^2 + Dx + Ey + F = 0$ correspond to the familiar conic sections. Consider, for example, the equations

$$x^2 - y^2 = 0 \qquad \text{and} \qquad x^2 + y^2 + 1 = 0$$

The first equation can be rewritten as $x = \pm y$, and so it yields the equations of two lines. The second can be rewritten as $x^2 + y^2 = -1$, which has no solutions and hence no graph. These two equations are examples of **degenerate** conic sections. Throughout the remainder of the section, we consider only the general equations of **nondegenerate** conic sections—namely, circles, parabolas, ellipses, and hyperbolas. In these cases, it is possible to recognize the type of conic by inspecting the coefficients of x^2 and y^2. The following observations can be verified by rewriting the standard equations of the conics in general form.

Identifying Conic Sections ☐

> A nondegenerate conic section of the form $Ax^2 + Cy^2 + Dx + Ey + F = 0$ can be identified as follows:
>
> | Condition | Verbal description | Resulting conic |
> |---|---|---|
> | $A = C$ | The squared terms have equal coefficients. | Circle |
> | $AC = 0$ | There is only one squared term. | Parabola |
> | $AC > 0, A \neq C$ | The coefficients of the squared terms are unequal but have the same sign. | Ellipse |
> | $AC < 0$ | The coefficients of the squared terms have opposite sign. | Hyperbola |

EXAMPLE 6 ▓ Identifying Conic Sections

Identify each of the following nondegenerate conics as a circle, a parabola, an ellipse, or a hyperbola.

a. $4x^2 - 9y^2 - 8x - 36y - 68 = 0$
b. $y^2 + 8x + 6y + 25 = 0$
c. $9x^2 - 36x + 31 = -4y^2 - 8y$

SOLUTION
a. Since both squared terms are present and the coefficients have opposite sign, this is a hyperbola.
b. Since only one squared term is present, this is a parabola.

c. We first rewrite the equation in general form as

$$9x^2 + 4y^2 - 36x + 8y + 31 = 0$$

Since the coefficients of the squared terms have the same sign but are not equal, this is an ellipse.

If the equation of a conic section is given in general form, its graph can be obtained either by writing the equation in standard form and using the techniques described earlier, or by solving the equation for y and using a graphing calculator.

EXAMPLE 7 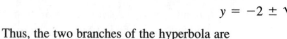   Graphing a Conic Section Using a Graphing Calculator

Identify the conic $-x^2 + y^2 + 3x + 4y - 5 = 0$ and obtain its plot with a graphing calculator. In the case of a parabola, ellipse, or hyperbola, approximate the coordinates of any vertices.

SOLUTION   Since the coefficients of x^2 and y^2 have opposite signs, this is a hyperbola. We solve for y by completing the square in y.

$$-x^2 + y^2 + 3x + 4y - 5 = 0$$
$$(y^2 + 4y + \quad) = x^2 - 3x + 5$$
$$(y^2 + 4y + 4) = x^2 - 3x + 5 + 4$$
$$(y + 2)^2 = x^2 - 3x + 9$$
$$y + 2 = \pm\sqrt{x^2 - 3x + 9}$$
$$y = -2 \pm \sqrt{x^2 - 3x + 9}$$

Thus, the two branches of the hyperbola are

$$y = -2 + \sqrt{x^2 - 3x + 9} \quad \text{and} \quad y = -2 - \sqrt{x^2 - 3x + 9}$$

After entering both equations and graphing, we obtain the graph shown in Figure 103. The coordinates of the vertices, found using either the trace or calculate-maximum/minimum feature, are approximately $(1.5, 0.6)$ and $(1.5, -4.6)$.

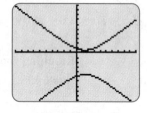

FIGURE 103

UNDERSTANDING AND MASTERY CHECKLISTS

CONCEPTS TO UNDERSTAND

- ☐ Hyperbola
- ☐ Foci of a hyperbola
- ☐ Vertices of a hyperbola
- ☐ Center of a hyperbola
- ☐ Transverse axis of a hyperbola
- ☐ Asymptotes of a hyperbola
- ☐ Reflective property of a hyperbola
- ☐ General equation of a conic section

SKILLS TO MASTER

- ☐ Sketch the graph of a hyperbola.
- ☐ Identify the center, vertices, and foci of a hyperbola.
- ☐ Find the asymptotes of a hyperbola.
- ☐ Find the equation of a hyperbola given information about its graph.
- ☐ Identify a nondegenerate conic section from its equation.
- ☐ Graph a conic section with a graphing calculator.
- ☐ Identify a conic section from its graph.

EXERCISES 6.4

EXERCISES 1–12 *Find the center, vertices, foci, and asymptotes for the given hyperbola, and sketch its graph.*

1. $\dfrac{x^2}{9} - \dfrac{y^2}{16} = 1$

2. $\dfrac{y^2}{25} - \dfrac{x^2}{16} = 1$

3. $4y^2 - 9x^2 = 36$

4. $25x^2 - 16y^2 = 400$

5. $12x^2 - 3y^2 = -24$

6. $5x^2 - 3y^2 = 15$

7. $\dfrac{(x-1)^2}{4} - \dfrac{(y-4)^2}{9} = 1$

8. $\dfrac{(y+3)^2}{16} - \dfrac{(x+2)^2}{25} = 1$

9. $4x^2 - 9y^2 - 16x + 54y - 101 = 0$

10. $25x^2 - 16y^2 - 50x + 160y + 25 = 0$

11. $4x^2 - 25y^2 - 32x + 164 = 0$

12. $x^2 - 4y^2 + 4x - 24y - 36 = 0$

EXERCISES 13–18 *Find an equation for the hyperbola with the given properties.*

13. Vertices $(0, \pm 4)$ and foci $(0, \pm 5)$

14. Vertices $(\pm 6, 0)$ and asymptotes $y = \pm\frac{4}{3}x$

15. Foci $(0, \pm 2\sqrt{5})$ and asymptotes $y = \pm\frac{1}{2}x$

16. Vertices $(\pm 3, 0)$ and passes through the point $(5, 4)$

17. Vertices $(-1, 3)$ and $(5, 3)$ and foci $(-3, 3)$ and $(7, 3)$

18. Vertices $(5, 2)$ and $(5, -6)$ and foci $(5, 3)$ and $(5, -7)$

EXERCISES 19–26 *Identify the conic section as a circle, parabola, ellipse, or hyperbola.*

19. $2x^2 - x - 3y + 4 = 0$

20. $3x^2 + 2y^2 = 7x - 7y - 5$

21. $4x^2 + 4y^2 = 2x - 6y + 10$

22. $4y^2 - 9x + 23y - 9 = 0$

23. $-5x^2 - 11x = y^2$

24. $x^2 - 2y^2 + 5 = 0$

25. $-2x^2 + 4y^2 - 15x + 8 = 0$

26. $-3x^2 + 2 = 3y^2 - 5y$

EXERCISES 27–36 *Solve the equation for y and plot the resulting equation(s) with a graphing calculator. Identify the conic section, and approximate the coordinates of any vertices.*

27. $-x^2 + 5x + y = 0$

28. $x^2 - 6x + 3y + 12 = 0$

29. $5x^2 + 8y^2 = 40$

30. $11y^2 - 3x^2 = 33$

31. $y^2 - 3x^2 + 4x + 9 = 0$

32. $y^2 - 4y - x = 0$

33. $2x^2 + y^2 - 5x - 8 = 0$

34. $2x^2 + y^2 + 3x - 6y + 4 = 0$

35. $3x^2 - y^2 + 8x - 8y - 7 = 0$

36. $y^2 + 8y - 2x + 22 = 0$

EXERCISES 37–44 *Plot both branches with a graphing calculator. Identify the resulting conic section and find the general form of its equation.*

37. $y = \pm\sqrt{x+3}$

38. $y = \pm\sqrt{8x - x^2}$

39. $y = 2 \pm\sqrt{16 - 6x - x^2}$

40. $y = -4 \pm\sqrt{2x - 5}$

41. $y = \pm\sqrt{\dfrac{x^2}{3} - 2}$

42. $y = \pm\sqrt{9 - 3x^2}$

43. $y = 1 \pm\sqrt{2 - 8x - 2x^2}$

44. $y = -3 \pm\sqrt{x^2 - 10x}$

■ APPLICATIONS

45. *Long-Range Navigation*   The long-range navigation (LORAN) system determines the locations of ships by timing radio signals. If signals are simultaneously transmitted to a ship from two stations, they will arrive at the ship at slightly different times. By assuming the ship lies on a hyperbola with foci at the transmitting stations, this time difference can be used to determine the position of the ship. Suppose transmitting stations *A* and *B* are located 200 miles apart on a coastline and a ship is located at a point *P*, as shown in Figure 104.

   a. If the ship receives the signal from station *A* 800 microseconds (1 microsecond $= 10^{-6}$ second) before the signal from station *B*, find the equation of the hyperbola shown in Figure 104. Assume the signals travel at a rate of 0.186 mile per microsecond. [*Hint:* If the equation is

$$\frac{y^2}{b^2} - \frac{x^2}{a^2} = 1$$

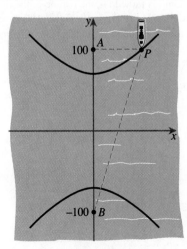

FIGURE 104

then the difference of the distances from any point (x, y) on the hyperbola to the foci is $2b$.]

b. If it is known that the ship is due east of station A, determine the location of the ship.

46. *Sonic Boom*   The British-French Concorde travels faster than sound. As a consequence, it creates a conical shock wave more commonly known as a "sonic boom." The region on the ground that is affected by the sonic boom has a boundary that is hyperbolic, as can be seen in Figure 105. When the Concorde is traveling at Mach 2 (twice the speed of sound) and at an altitude of 65,000 feet, the vertex of the hyperbolic boundary is approximately 24 miles from the point on the ground directly beneath the nose of the plane, as shown in Figure 105. If we were to set up a coordinate system with the origin on the ground directly beneath the nose of the plane, the x-axis lying on the ground parallel to the path of the plane, and the y-axis also lying on the ground, then the asymptotes for the hyperbolic boundary would

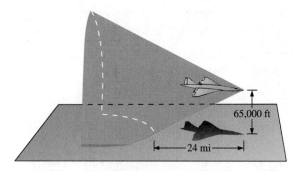

FIGURE 105

have equations $y = \pm\frac{1}{2}x$. Find the equation of the hyperbolic boundary. The region affected by the sonic boom extends as far as 55 miles behind the point on the ground below the plane. What is the width of the affected region when $x = -55$?

CONCEPTS AND CRITICAL THINKING

EXERCISES 47–50   *Answer true or false.*

47. The points $(\pm a, 0)$ and $(0, \pm b)$ lie on the graph of the hyperbola with equation

$$\frac{x^2}{a^2} - \frac{y^2}{b^2} = 1$$

48. Hyperbolas never intersect their asymptotes.

49. A hyperbola is the set of all points such that the sum of the distances from two fixed points (called the foci) is constant.

50. The foci of a hyperbola lie on the line determined by the vertices.

EXERCISES 51–54   *Give an example of each.*

51. A hyperbola for which the transverse axis is horizontal

52. A hyperbola with asymptotes $y = \pm 4x$

53. An application of the reflective property of hyperbolas

54. A characteristic of the graph of a hyperbola not shared by the graphs of the other conic sections

55. If a hyperbola has a horizontal transverse axis of length $2a$ and foci at $(\pm c, 0)$, explain why $c^2 - a^2 > 0$.

QUESTIONS FOR DISCUSSION OR ESSAY

56. A lamp with a shade is placed close to a wall in a dark room. When the lamp is turned on, the outline of the light on the wall will look similar to that shown in Figure 106. Consult Figure 82 to help you explain why the outline forms a hyperbola.

57. A rough sketch of a hyperbola can be drawn in the following way. Attach two strings of different lengths near the tip of a ballpoint pen, and attach the other ends of the strings to a piece of cardboard using thumbtacks, as shown in Figure 107. Keeping

FIGURE 106

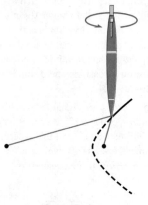

FIGURE 107

the string taut and the tip of the pen on the cardboard, slowly twirl the pen so the string begins to wind around the tip. As the lengths of string shorten, the pen will follow a hyperbolic path. Explain why this is so.

58. In Figure 102, a reflective telescope is shown that uses a parabolic main mirror and both elliptical and hyperbolic auxiliary

mirrors to route the incoming light to the eyepiece. Can a telescope be constructed with a parabolic main mirror and only one auxiliary mirror so that it still directs the incoming light to the eyepiece? Explain, using a sketch if necessary.

PROJECTS FOR ENRICHMENT

59. *Degenerate Conics* Recall that the geometric definitions of the conic sections are based on the intersection of a plane with a double cone. The degenerate conics—a line, a pair of intersecting lines, and a point—arise geometrically when the plane passes through the vertex of the cone. The equations of the degenerate conics are all special cases of the general equation $Ax^2 + Cy^2 + Dx + Ey + F = 0$. For example, if A and C are both zero, the resulting equation $Dx + Ey + F = 0$ is a line.

a. Explain why each of the following equations has the indicated description.

   i. $2x^2 + 3y^2 = 0$; a point

   ii. $4x^2 - 9y^2 = 0$; a pair of intersecting lines

   iii. $x^2 + 2y^2 + 1 = 0$; no graph

b. State some general conditions under which equations of the form $Ax^2 + Cy^2 + F = 0$ will be

   i. a point   ii. a pair of intersecting lines

c. Complete the square on the equation $Ax^2 + Cy^2 + Dx + Ey + F = 0$ to determine a test for characterizing the equation as

   i. a point   ii. a pair of intersecting lines

[*Hint:* Your test should involve the expression $\dfrac{D^2}{4A} + \dfrac{E^2}{4C} - F$, and it should take into account the signs of A and C.]

d. Use the test you developed in part c to determine which, if any, of the following equations are degenerate conics.

   i. $2x^2 - 3y^2 - 4x + 6y - 1 = 0$

   ii. $2x^2 + y^2 - 4x + 2y + 10 = 0$

   iii. $2x^2 + 3y^2 - 4x + 6y + 5 = 0$

60. *The Case of the Stolen Car* A victim of a carjacking calls Inspector Magill's office at 9:00 A.M. seeking help in recovering a stolen car. The following telephone conversation takes place.

Victim: This morning on my way to work, I was forced out of my car by two masked men. I overheard one say to the other that they would take the car to "the usual place" and store it there until noon. My car was equipped with an electronic homing device; unfortunately, it isn't working quite right and only emits a signal every 5 minutes. Is there any way to locate the source of the signal before the car is moved again?

Magill: Yes, but I will need time to round up three electronic receivers that can determine the precise time at which the

signal is received from the homing device. I will also need three mobile phones and the help of two assistants.

Victim: You've got it!!

At 10:15, the victim and a friend arrive at Magill's office with the three receivers. After synchronizing the timing devices in the receivers, Magill takes the two friends to positions labeled A and B in Figure 108, 10 miles apart, and he drives to a third position C, 10 miles from B. His instructions to the two friends are simply to record the exact time at which the first signal after 11:00 is received from the homing device and then call him on the mobile phones. By 11:30, the car has been recovered and the criminals have been apprehended. We will reconstruct the method used by Magill to locate the car.

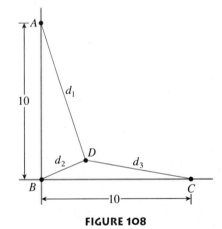

FIGURE 108

The times recorded by the receivers gave the following information.

■ It took 3.22×10^{-5} second longer to receive the signal at point A than at point B.

■ It took 2.15×10^{-5} second longer to receive the signal at point C than at point B.

a. If D denotes the position of the car and d_1, d_2, and d_3 denote the distances (in miles) to A, B, and C, respectively, determine the differences $d_1 - d_2$ and $d_3 - d_2$ (round to the nearest integer).

Assume that the signal travels at the speed of light, which is 186,000 miles per second. Recall that a hyperbola is the set of all points such that the difference of the distances from the foci is constant. Point D must thus be one of the points on the hyperbola for which the difference of the distances from A and B is $d_1 - d_2$, and it must also be a point on the hyperbola for which the difference of the distances from C and B is $d_3 - d_2$. In other words, if we treat A and B as the foci of one hyperbola and B and C as the foci of a second hyperbola, point D will be a point of intersection. We simply need to find the equations of the two hyperbolas to determine their points of intersection. We will demonstrate how to find the one with foci at B and C and leave the other for you. We first assign a coordinate system to Figure 108, choosing $\overrightarrow{BC}$ as the positive x-axis, $\overrightarrow{BA}$ as the positive y-axis, and the origin at point B. Since B and C are 10 miles apart, the center of the hyperbola is at $(5, 0)$. Thus, we are looking for an equation of the form

$$\frac{(x - 5)^2}{a^2} - \frac{y^2}{b^2} = 1$$

Since the foci are 5 miles from the center, $c = 5$. The distance $d_3 - d_2$ must be twice the distance from the center to the vertices, so $2a = d_3 - d_2$. Finally, $b = \sqrt{c^2 - a^2} = \sqrt{25 - a^2}$.

b. Use the results from part a to show that the equation of the hyperbola with vertices B and C is

$$\frac{(x - 5)^2}{4} - \frac{y^2}{21} = 1$$

c. Show that the equation of the second hyperbola is

$$\frac{(y - 5)^2}{9} - \frac{x^2}{16} = 1$$

d. Estimate the coordinates of the intersection points of these two hyperbolas and determine the location of the car.

CHAPTER 6 REVIEW

EXERCISES 1–4   *A relation and its graph are given. Use translation and/or reflection to sketch the graph of the indicated relation.*

1. $y = x^2$

   a. $y = (x - 2)^2$

   b. $y = x^2 + 3$

   c. $y - 4 = (x - 2)^2$

   d. $y = -x^2$

2. $x = y^3$

   a. $x = (y - 1)^3$

   b. $x = y^3 + 2$

   c. $-x = y^3$

   d. $-x = (y - 1)^3$

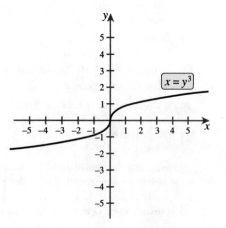

3. $xy^2 + x = 4$

   a. $x(y + 2)^2 + x = 4$

   b. $xy^2 + x = -4$

   c. $(x - 1)(y - 3)^2 + x - 1 = 4$

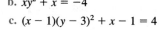

4. $x^2 + y^3 = 16$

   a. $(x - 2)^2 + y^3 = 16$

   b. $(x - 2)^2 + (y - 1)^3 = 16$

   c. $x^2 - y^3 = 16$

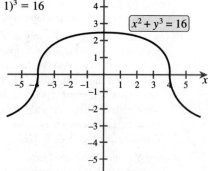

EXERCISES 5–6 *A relation and its graph are given, together with a translated graph. Find an equation for the translated graph.*

5. Original relation: $x^2 + y^4 = 16$

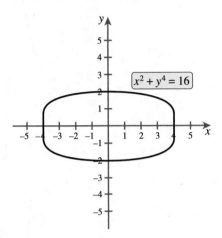

Translated graph:

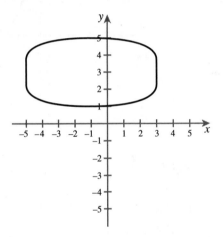

6. Original relation: $y = x^3$

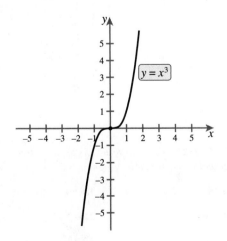

Translated graph:

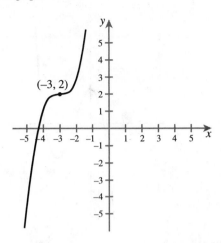

EXERCISES 7–8 *A relation and its graph are given, together with a reflected graph. Find an equation for the reflected graph.*

7. Original relation: $y = x^4 - 4x^2 + 2$

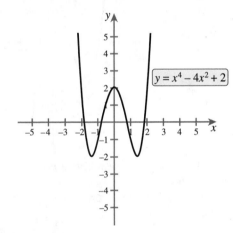

Reflected graph:

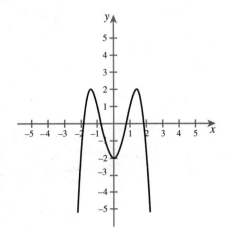

8. Original relation: $y = x^3 - 3x^2 + 2$

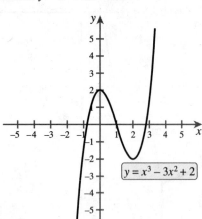

$$y = x^3 - 3x^2 + 2$$

Reflected graph:

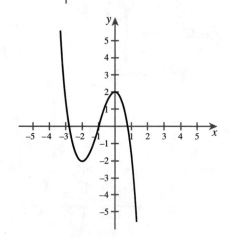

EXERCISES 9–12   *Test for symmetry with respect to the x-axis, the y-axis, and the origin.*

9. $y^2 - 2x = y^4$ **10.** $|x| + |y| = 1$

11. $x - y = 1$ **12.** $y|x| = x$

EXERCISES 13–16   *A portion of the graph of a relation is shown. Test for symmetry with respect to the x-axis, the y-axis, and the origin, and then use symmetry to complete the graph.*

13. $x = y^4 - 4y^2$

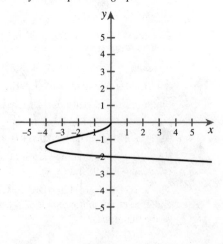

14. $y = 4x^2 - x^4$

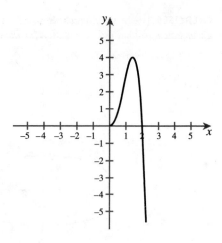

15. $xy + x^3y = 5$

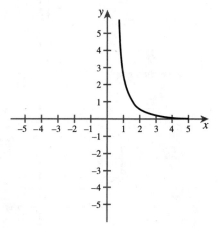

16. $9x^2 + y^4 = 81$

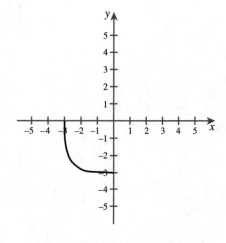

EXERCISES 17–22   *Find the vertex, focus, and directrix for the given parabola, and sketch its graph.*

17. $x = 2y^2$ **18.** $x^2 + 3y = 0$

19. $y + 4 = -2(x - 3)^2$ **20.** $2y - 4 = (x + 1)^2$

21. $x = 3y^2 + 6y + 7$ **22.** $y = 2x^2 + 6x + 1$

EXERCISES 23–26 *Find an equation for the parabola with the given properties.*

23. Vertex at the origin and focus at $(0, \frac{1}{4})$

24. Focus at $(-1, 0)$ and directrix $x = 1$

25. Vertex at $(-1, 2)$ and directrix $y = 6$

26. Horizontal axis, vertex at $(4, 3)$, and passing through $(2, 4)$

EXERCISES 27–32 *Find the center, vertices, and foci for the given ellipse, and sketch its graph.*

27. $\dfrac{x^2}{49} + \dfrac{y^2}{64} = 1$

28. $2x^2 + 3y^2 = 1$

29. $9(x - 3)^2 + 25(y - 2)^2 = 225$

30. $\dfrac{(x + 2)^2}{64} + \dfrac{(y - 3)^2}{100} = 1$

31. $x^2 + 4y^2 - 2x + 24y + 21 = 0$

32. $16x^2 + 9y^2 + 128x + 18y + 121 = 0$

EXERCISES 33–36 *Find an equation for the ellipse with the given properties.*

33. Horizontal major axis of length 6, vertical minor axis of length 4, and center at the origin.

34. Vertices at $(\pm 4, 0)$ and foci at $(\sqrt{5}, 0)$ and $(-\sqrt{5}, 0)$

35. Foci at $(0, \pm 4)$ and minor axis of length 6

36. Vertices at $(2, \pm 4)$ and minor axis of length 6

EXERCISES 37–42 *Find the center, vertices, foci, and asymptotes for the given hyperbola, and sketch its graph.*

37. $\dfrac{x^2}{25} - \dfrac{y^2}{9} = 1$

38. $8y^2 - 12x^2 = 1$

39. $4(y - 1)^2 - 9(x + 5)^2 = 36$

40. $\dfrac{(x + 1)^2}{4} - \dfrac{(y + 2)^2}{16} = 1$

41. $25x^2 - 16y^2 - 150x + 64y - 239 = 0$

42. $x^2 - 4y^2 - 6x + 32y = 51$

EXERCISES 43–44 *Find an equation for the hyperbola with the given properties.*

43. Vertices at $(\pm 5, 0)$ and foci at $(\pm \sqrt{34}, 0)$

44. Vertices at $(0, \pm \sqrt{5})$ and passing through $(4, 5)$

EXERCISES 45–50 *Identify the conic section as a circle, parabola, ellipse, or hyperbola.*

45. $4x^2 + y^2 - 1 = 0$

46. $4x^2 - 8x - y^2 + 8y - 16 = 0$

47. $y + 3x^2 + 2x = 0$

48. $-x^2 - y^2 + 2x + 4y = 0$

49. $x - 2y^2 - 2y + 10 = 0$

50. $x^2 - 8x + 4y^2 + 8y - 16 = 0$

EXERCISES 51–54 *Solve the equation for y and plot the resulting equation(s) on a graphing calculator. Identify the conic section, and approximate the coordinates of any vertices.*

51. $y^2 + 4x = 6$

52. $4y - 8 = -x^2 + 2x$

53. $2x^2 - 7x + y^2 = 6$

54. $y^2 + 6y = x^2 + 2$

EXERCISES 55–58 *Use a graphing calculator to plot the pair of equations on the same coordinate system. Identify the resulting conic section and find its general equation.*

55. $y = \pm\sqrt{16 - x^2}$

56. $y = \pm\sqrt{x^2 - 16}$

57. $y = 2 \pm\sqrt{x - 4}$

58. $y = 4 \pm\sqrt{16 - 4x^2}$

EXERCISES 59–64 *Parts a and b are connected: Part a involves an elementary concept, whereas part b involves related material from this chapter. First answer part a, and then use this result to answer part b.*

59. a. Classify $f(x) = \dfrac{x^2 + 2}{x^2 + 4}$ as even, odd, or neither.

   b. One portion of the graph of $y(x^2 + 4) = x^2 + 2$ is shown in Figure 109. Use symmetry to complete the graph.

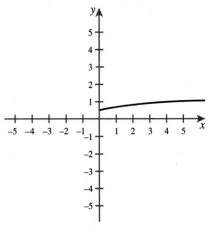

FIGURE 109

60. a. Expand $(x + 3)^2 + 2(y - 1)^2$.

   b. Find the center of the ellipse with equation $x^2 + 6x + 2y^2 - 4y + 11 = 1$.

61. a. Estimate the coordinates of the vertex of $y = -x^2 + 3x + 1$.

   b. Estimate the coordinates of the vertex of $x = -y^2 + 3y + 1$.

62. a. Find the distance between $(4, 5)$ and $(0, 2)$.

   b. Find a point in the first quadrant lying on the parabola having $(0, 2)$ as its focus and the x-axis as its directrix.

63. a. Find the midpoint and the slope of the line segment having endpoints $(-1, 0)$ and $(3, 4)$.

   b. The vertices of an ellipse are $(-1, 0)$ and $(3, 4)$, and one endpoint of the minor axis is $(2, 1)$. Find the other endpoint of the minor axis.

64. a. Eliminate the radical in the equation $y = \sqrt{1 - qx^2} + 2$ to rewrite it in the form $Ax^2 + Cy^2 + Dx + Ey + F = 0$.

   b. Find the values of q for which the graph of $f(x) = \sqrt{1 - qx^2} + 2$ is a portion of (i) a parabola, (ii) an ellipse, or (iii) a hyperbola.

65. *Pressure and Altitude*   The graph of the equation relating pressure P (measured in pounds per square inch) and altitude above sea level h (measured in feet) is given in Figure 110. Sketch the graph of the equation relating pressure and depth below sea level d.

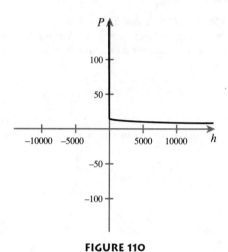

FIGURE 110

66. *Whispering Room*   A whispering room is in the shape of an ellipse that is twice as long as it is wide. If the foci are 1.86 feet from the vertices, find the distance from each focus to the center of the room.

67. *Home Run Path*   If air resistance is ignored, the path of a baseball hit with an initial angle of 45° and an initial velocity of 120 feet per second is given by

$$450y = -x^2 + 450x + 1350$$

where x is the horizontal distance from home plate (in feet) and y is the height (in feet).

   a. Find the maximum height of the ball.

   b. How far away from home plate does the ball land?

CHAPTER 6 TEST

PROBLEMS 1–6 *Answer true or false.*

1. If a graph is reflected first about the x-axis and then about the y-axis, the result is the same as when it is reflected first about the y-axis and then about the x-axis.

2. The graph of a relation can be symmetric with respect to both the x- and y-axes.

3. The distance from the focus to the vertex of a parabola is equal to the distance from the vertex to the directrix.

4. A point on a hyperbola is equidistant from the two foci of the hyperbola.

5. All conic sections with vertical or horizontal axes have equations of the form $Ax^2 + Cy^2 + Dx + Ey + F = 0$.

6. The foci of an ellipse always lie on the minor axis.

PROBLEMS 7–10 *Give an example of each.*

7. A relation whose graph is symmetric with respect to the x-axis

8. The equation of a parabola that opens to the left

9. The equation of a conic section with two vertices on the x-axis

10. An application of a reflective property of a conic section

11. The graph of $8x = y^4$ is given in Figure 111. Sketch the graphs of the relations in parts a and b that are formed by translating and/or reflecting the graph of $8x = y^4$.

    a. $8(x - 1) = (y + 3)^4$

    b. $8x = -y^4$

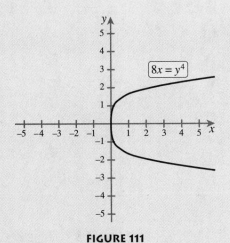

FIGURE 111

12. A portion of the graph of $x^3 - y^2 = 0$ is shown in Figure 112. Test the relation for symmetry and then use symmetry to complete the graph.

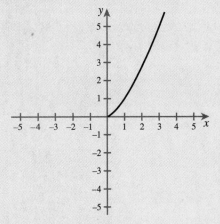

FIGURE 112

PROBLEMS 13–15 *Identify and sketch the graph of the given conic section. Find any of the following features that apply: center, vertices, foci, and asymptotes.*

13. $x = y^2 + 4y$

14. $4x^2 - 9y^2 = 36$

15. $4x^2 + y^2 - 16x + 6y + 9 = 0$

16. A comet travels in an elliptical orbit, with the sun at one focus. The major and minor axes are of lengths 12 A.U. and 10 A.U., respectively. Find the distance from the center of the comet's orbit to the sun.

7 SYSTEMS OF EQUATIONS AND INEQUALITIES

Hurricane Andrew struck southern Florida and Louisiana in August 1992 causing over $20 billion in damage. With enough destructive force to topple homes, uproot trees, and change the course of rivers, hurricanes provide meteorologists with a strong incentive to accurately model Earth's weather systems: the difference between 2 hours notice and 2 days notice of an impending hurricane can be the difference between life and death. Mathematical models of Earth's atmosphere are incredibly complex, often involving the solution of systems consisting of thousands of linear equations. In this chapter we will explore several methods of solving such systems.

SECTION **7.1 MATRICES**

☐ When is $AB \neq BA$?

☐ How can $AB = 0$ if neither A nor B is 0?

☐ How can you find the coordinates of a point after it undergoes a rotation?

☐ If 10% of nonsmokers take up smoking every year, while 20% of smokers quit smoking each year, then in a town of 2000 people, how many people will smoke after 32 years, and why doesn't the answer depend on the current number of smokers?

Based on evidence from ancient clay tablets, one might well imagine a Babylonian student of the 21st century B.C. struggling through a nonalgebraic solution to a word problem involving two or more unknowns. A student of the 21st century A.D., on the other hand, could, with little difficulty, express such problems algebraically as systems of equations, and then easily solve them with the aid of matrices and a calculator. Compared to most tools for doing mathematics, the calculator is in its infancy. Matrices, although ancient compared to the calculator, were formally defined only 150 years ago, and so they are also relatively new to mathematics. In this section, we investigate the fundamental properties of matrices, and in later sections we will see how matrices can be used to solve systems of equations.

FUNDAMENTALS

A matrix (plural: *matrices*) is a rectangular array used to store and manipulate data. For example, the linear system of equations

$$2x + 3y + 2z = 3$$
$$4x - 7y - z = 2$$
$$2x - 4y + 3z = -4$$

can be encoded as the 3×4 matrix

$$\begin{bmatrix} 2 & 3 & 2 & 3 \\ 4 & -7 & -1 & 2 \\ 2 & -4 & 3 & -4 \end{bmatrix}$$

The performance of a company whose domestic division has quarterly earnings (in millions of dollars) of 2.3, 2.7, 3.2, and 4.1, and whose international division has quarterly earnings of 1.8, 1.2, 2.0, and 1.7 can be expressed using the 2×4 matrix

$$\begin{bmatrix} 2.3 & 2.7 & 3.2 & 4.1 \\ 1.8 & 1.2 & 2.0 & 1.7 \end{bmatrix}$$

We give a formal definition of a matrix as follows.

Matrix Definitions □

- A **matrix of order $m \times n$** is a rectangular array consisting of entries in m (horizontal) **rows** and n (vertical) **columns,** as shown in Figure 1.

$$\underbrace{\begin{bmatrix} a_{11} & a_{12} & a_{13} \cdots a_{1n} \\ a_{21} & a_{22} & a_{23} \cdots a_{2n} \\ a_{31} & a_{32} & a_{33} \cdots a_{3n} \\ \vdots & \vdots & \vdots \quad\; \vdots \\ a_{m1} & a_{m2} & a_{m3} \cdots a_{mn} \end{bmatrix}}_{n \text{ columns}} \left.\vphantom{\begin{bmatrix} a \\ a \\ a \\ \vdots \\ a \end{bmatrix}}\right\} m \text{ rows}$$

FIGURE 1

- The $m \times n$ matrix in which every entry is zero is called the **$m \times n$ zero matrix.**
- Two $m \times n$ matrices are said to be equal if corresponding entries are equal.

Note that the entries of the matrix in Figure 1 are written using a double subscript. The first subscript indicates the row of the entry, and the second indicates the column of the entry. Rows are numbered from top to bottom, and columns are numbered from left to right. Thus, the entry in the ith row from the top and the jth column from the left is denoted by a_{ij}, as shown in Figure 2.

$$\begin{array}{c|cccccccc|} & \multicolumn{8}{c}{\text{Column}} \\ & 1 & 2 & 3 & & j & & n & \\ 1 & a_{11} & a_{12} & a_{13} & \cdots & a_{1j} & \cdots & a_{1n} & \\ 2 & a_{21} & a_{22} & a_{23} & \cdots & a_{2j} & \cdots & a_{2n} & \\ 3 & a_{31} & a_{32} & a_{33} & \cdots & a_{3j} & \cdots & a_{3n} & \\ \text{Row} & \vdots & \vdots & \vdots & & \vdots & & \vdots & \\ i & a_{i1} & a_{i2} & a_{i3} & & a_{ij} & & a_{in} & \\ & \vdots & \vdots & \vdots & & \vdots & & \vdots & \\ m & a_{m1} & a_{m2} & a_{m3} & \cdots & a_{mj} & \cdots & a_{mn} & \end{array}$$

Entry a_{ij} is in the ith row and the jth column.

FIGURE 2

The matrix for which the entry in row i and column j is a_{ij} is often denoted $[a_{ij}]$. This is an especially convenient way of representing matrices with unknown or arbitrary entries. Thus, for example, we can write

$$[a_{ij}] = \begin{bmatrix} a_{11} & a_{12} & a_{13} & \cdots & a_{1n} \\ a_{21} & a_{22} & a_{23} & \cdots & a_{2n} \\ a_{31} & a_{32} & a_{33} & \cdots & a_{3n} \\ \vdots & \vdots & \vdots & & \vdots \\ a_{m1} & a_{m2} & a_{m3} & \cdots & a_{mn} \end{bmatrix} \quad \text{and} \quad [b_{ij}] = \begin{bmatrix} b_{11} & b_{12} & b_{13} & \cdots & b_{1n} \\ b_{21} & b_{22} & b_{23} & \cdots & b_{2n} \\ b_{31} & b_{32} & b_{33} & \cdots & b_{3n} \\ \vdots & \vdots & \vdots & & \vdots \\ b_{m1} & b_{m2} & b_{m3} & \cdots & b_{mn} \end{bmatrix}$$

EXAMPLE 1 ▉ **Identifying Entries Using Subscript Notation**

Let

$$A = \begin{bmatrix} 5 & 6 & \frac{1}{2} \\ -2 & 3 & -7 \end{bmatrix}$$

a. What is the order of A?
b. If $A = [a_{ij}]$, identify a_{21} and a_{13}.

SOLUTION

a. Since A has 2 rows and 3 columns, it is of order 2×3.
b. The entry a_{21} is in the second row and the first column. Thus, $a_{21} = -2$. The entry a_{13} is in the first row and the third column, and so $a_{13} = \frac{1}{2}$.

If matrices only served as a convenient way of tabulating data, then they would be nothing more than glorified tables. It is the *algebraic* properties of matrices that set them apart from simple tables and make them especially useful as a tool for solving linear systems of equations. Under certain conditions, matrices can be added, subtracted, multiplied, and even divided.

ADDITION AND SUBTRACTION OF MATRICES

The definitions of addition and subtraction of matrices are very intuitive; we simply add or subtract corresponding entries. Thus, if

$$A = \begin{bmatrix} 2 & -4 \\ 3 & -5 \end{bmatrix} \quad \text{and} \quad B = \begin{bmatrix} -1 & 7 \\ 2 & 3 \end{bmatrix}$$

then $A + B$ and $A - B$ are computed as follows:

$$A + B = \begin{bmatrix} 2 & -4 \\ 3 & -5 \end{bmatrix} + \begin{bmatrix} -1 & 7 \\ 2 & 3 \end{bmatrix} \qquad A - B = \begin{bmatrix} 2 & -4 \\ 3 & -5 \end{bmatrix} - \begin{bmatrix} -1 & 7 \\ 2 & 3 \end{bmatrix}$$

$$= \begin{bmatrix} 2 + (-1) & -4 + 7 \\ 3 + 2 & -5 + 3 \end{bmatrix} \qquad = \begin{bmatrix} 2 - (-1) & -4 - 7 \\ 3 - 2 & -5 - 3 \end{bmatrix}$$

$$= \begin{bmatrix} 1 & 3 \\ 5 & -2 \end{bmatrix} \qquad = \begin{bmatrix} 3 & -11 \\ 1 & -8 \end{bmatrix}$$

More formally, we have the following definition.

Matrix Addition and Subtraction ☐

For $m \times n$ matrices $A = [a_{ij}]$ and $B = [b_{ij}]$,

$A + B$ is the $m \times n$ matrix $[c_{ij}]$, where $c_{ij} = a_{ij} + b_{ij}$
$A - B$ is the $m \times n$ matrix $[d_{ij}]$, where $d_{ij} = a_{ij} - b_{ij}$

We cannot define matrix addition or subtraction unless the matrices are the same order.

EXAMPLE 2 ▨ Adding and Subtracting Matrices

Perform the following matrix operations.

a. $\begin{bmatrix} 0 & 4 & 5 \\ 1 & 6 & 7 \end{bmatrix} + \begin{bmatrix} -2 & 2 & 4 \\ -2 & 5 & 4 \end{bmatrix}$

b. $\begin{bmatrix} 2 & 3 \\ 4 & 5 \end{bmatrix} - \begin{bmatrix} 1 & 2 \\ 3 & 2 \end{bmatrix}$

c. $\begin{bmatrix} 1 & 2 \\ 4 & 5 \\ 3 & 6 \end{bmatrix} + \begin{bmatrix} 0 & 2 \\ 5 & 2 \end{bmatrix}$

SOLUTION

a. $\begin{bmatrix} 0 & 4 & 5 \\ 1 & 6 & 7 \end{bmatrix} + \begin{bmatrix} -2 & 2 & 4 \\ -2 & 5 & 4 \end{bmatrix} = \begin{bmatrix} 0 + (-2) & 4 + 2 & 5 + 4 \\ 1 + (-2) & 6 + 5 & 7 + 4 \end{bmatrix}$

$$= \begin{bmatrix} -2 & 6 & 9 \\ -1 & 11 & 11 \end{bmatrix}$$

b. $\begin{bmatrix} 2 & 3 \\ 4 & 5 \end{bmatrix} - \begin{bmatrix} 1 & 2 \\ 3 & 2 \end{bmatrix} = \begin{bmatrix} 2 - 1 & 3 - 2 \\ 4 - 3 & 5 - 2 \end{bmatrix}$

$$= \begin{bmatrix} 1 & 1 \\ 1 & 3 \end{bmatrix}$$

c. Since the first matrix is of order 3×2 and the second is of order 2×2, the matrices are of different orders, and thus their sum is not defined.

Matrices are often useful in applications where large quantities of data are involved. In the following example, we see how matrices can be used to organize data, and also how matrix addition can be used to "combine" the information stored in matrices.

EXAMPLE 3 ▨ An Application of Matrix Addition

A telecommunications company has two divisions, one that deals primarily with computer networking and the second with satellite data transmission. The costs and revenues (in millions of dollars) for the two divisions for the four quarters of 1999 are as follows: The networking division had costs of 1.5, 0.5, 1.2, and 1.1 million for the first through fourth quarters, respectively, and revenues of 1.8, 0.7, 1.4, and 1.2 million, respectively. The satellite transmission division had costs of 2.8, 1.4, 1.5, and 1.7 million for the first through fourth quarters, respectively, and revenues of 3.3, 1.8, 2.0, and 2.1 million, respectively. Organize this information in matrix form and use matrix addition to find the company's total cost and total revenue for each of the four quarters of 1999.

SOLUTION We define two 4×2 matrices, one for each of the divisions. The 4 rows correspond to the 4 quarters of 1999 and the 2 columns correspond to cost and revenue (in millions of dollars). Thus, if N and S denote the matrices for the networking and satellite divisions, respectively, then

$$N = \begin{bmatrix} 1.5 & 1.8 \\ 0.5 & 0.7 \\ 1.2 & 1.4 \\ 1.1 & 1.2 \end{bmatrix} \quad \text{and} \quad S = \begin{bmatrix} 2.8 & 3.3 \\ 1.4 & 1.8 \\ 1.5 & 2.0 \\ 1.7 & 2.1 \end{bmatrix}$$

The company's total cost and revenue can be obtained by adding the matrices N and S.

$$N + S = \begin{bmatrix} 1.5 + 2.8 & 1.8 + 3.3 \\ 0.5 + 1.4 & 0.7 + 1.8 \\ 1.2 + 1.5 & 1.4 + 2.0 \\ 1.1 + 1.7 & 1.2 + 2.1 \end{bmatrix} = \begin{bmatrix} 4.3 & 5.1 \\ 1.9 & 2.5 \\ 2.7 & 3.4 \\ 2.8 & 3.3 \end{bmatrix}$$

Thus, the company's total cost and revenue for each of the quarters are as follows:

Quarter	Total cost	Total revenue
1	$4,300,000	$5,100,000
2	$1,900,000	$2,500,000
3	$2,700,000	$3,400,000
4	$2,800,000	$3,300,000

SCALAR MULTIPLICATION

In the context of matrices, it is customary to refer to ordinary real or complex numbers as **scalars.** To multiply a matrix by a scalar, we simply multiply each entry of the matrix by the scalar. For example,

$$2 \cdot \begin{bmatrix} 3 & 5 \\ 7 & 8 \end{bmatrix} = \begin{bmatrix} 2 \cdot 3 & 2 \cdot 5 \\ 2 \cdot 7 & 2 \cdot 8 \end{bmatrix}$$
$$= \begin{bmatrix} 6 & 10 \\ 14 & 16 \end{bmatrix}$$

More formally, we define multiplication by a scalar as follows.

Scalar Multiplication ☐ | If c is a real or complex number and $A = [a_{ij}]$, then $cA = [b_{ij}]$, where $b_{ij} = ca_{ij}$.

EXAMPLE 4 ■ Performing Scalar Multiplication

Compute $-\dfrac{1}{2} \cdot \begin{bmatrix} 2 & -4 \\ 8 & 5 \\ -6 & 7 \end{bmatrix}$.

SOLUTION   We simply multiply each entry of the matrix by $-\frac{1}{2}$:

$$-\frac{1}{2} \cdot \begin{bmatrix} 2 & -4 \\ 8 & 5 \\ -6 & 7 \end{bmatrix} = \begin{bmatrix} (-\frac{1}{2})2 & (-\frac{1}{2})(-4) \\ (-\frac{1}{2})8 & (-\frac{1}{2})5 \\ (-\frac{1}{2})(-6) & (-\frac{1}{2})7 \end{bmatrix} = \begin{bmatrix} -1 & 2 \\ -4 & -\frac{5}{2} \\ 3 & -\frac{7}{2} \end{bmatrix}$$

EXAMPLE 5 ▩   **Combining Scalar Multiplication with Subtraction**

Let

$$A = \begin{bmatrix} 1 & 4 \\ 5 & 3 \end{bmatrix} \quad \text{and} \quad B = \begin{bmatrix} 3 & 6 \\ 4 & 2 \end{bmatrix}$$

Compute $3A - 2B$.

SOLUTION

$$3A - 2B = 3 \cdot \begin{bmatrix} 1 & 4 \\ 5 & 3 \end{bmatrix} - 2 \cdot \begin{bmatrix} 3 & 6 \\ 4 & 2 \end{bmatrix}$$

$$= \begin{bmatrix} 3 & 12 \\ 15 & 9 \end{bmatrix} - \begin{bmatrix} 6 & 12 \\ 8 & 4 \end{bmatrix} \qquad \text{Performing the scalar multiplications}$$

$$= \begin{bmatrix} -3 & 0 \\ 7 & 5 \end{bmatrix} \qquad \text{Subtracting corresponding entries}$$

EXAMPLE 6 ▩   **An Application of Scalar Multiplication and Addition**

Suppose that financial forecasters for the telecommunications company from Example 3 predict that all quarterly revenues and costs will increase by 5% in 2000 for the networking division and by 8% for the satellite division. Estimate the total revenue and cost for each quarter of 2000.

SOLUTION   Increasing a quantity by 5% is equivalent to multiplying the quantity by 1.05. Thus, the quarterly networking costs and revenues for 2000 are given by

$$1.05N = 1.05 \cdot \begin{bmatrix} 1.5 & 1.8 \\ 0.5 & 0.7 \\ 1.2 & 1.4 \\ 1.1 & 1.2 \end{bmatrix}$$

$$= \begin{bmatrix} (1.05)1.5 & (1.05)1.8 \\ (1.05)0.5 & (1.05)0.7 \\ (1.05)1.2 & (1.05)1.4 \\ (1.05)1.1 & (1.05)1.2 \end{bmatrix}$$

$$= \begin{bmatrix} 1.575 & 1.890 \\ 0.525 & 0.735 \\ 1.260 & 1.470 \\ 1.155 & 1.260 \end{bmatrix}$$

Similarly, the satellite costs and revenues are given by

$$1.08S = 1.08 \cdot \begin{bmatrix} 2.8 & 3.3 \\ 1.4 & 1.8 \\ 1.5 & 2.0 \\ 1.7 & 2.1 \end{bmatrix}$$

$$= \begin{bmatrix} (1.08)2.8 & (1.08)3.3 \\ (1.08)1.4 & (1.08)1.8 \\ (1.08)1.5 & (1.08)2.0 \\ (1.08)1.7 & (1.08)2.1 \end{bmatrix}$$

$$= \begin{bmatrix} 3.024 & 3.564 \\ 1.512 & 1.944 \\ 1.620 & 2.160 \\ 1.836 & 2.268 \end{bmatrix}$$

The company's total cost and revenue can be obtained by adding $1.05N$ and $1.08S$.

$$1.05N + 1.08S = \begin{bmatrix} 1.575 & 1.890 \\ 0.525 & 0.735 \\ 1.260 & 1.470 \\ 1.155 & 1.260 \end{bmatrix} + \begin{bmatrix} 3.024 & 3.564 \\ 1.512 & 1.944 \\ 1.620 & 2.160 \\ 1.836 & 2.268 \end{bmatrix} = \begin{bmatrix} 4.599 & 5.454 \\ 2.037 & 2.679 \\ 2.880 & 3.630 \\ 2.991 & 3.528 \end{bmatrix}$$

Thus, the company's total cost and revenue for each quarter of 2000 would be as follows.

Quarter	Total cost	Total revenue
1	$4,599,000	$5,454,000
2	$2,037,000	$2,679,000
3	$2,880,000	$3,630,000
4	$2,991,000	$3,528,000

MATRIX MULTIPLICATION

As we have seen, to add or subtract two matrices, we simply add or subtract corresponding entries. The product of two matrices, however, is *not* found by multiplying corresponding entries. Instead, each entry of the product matrix is the "product" of a *row* of the first matrix and a *column* of the second matrix.

Multiplication of a Row and a Column ☐

If

$$R = [a_1 \quad a_2 \quad a_3 \quad \cdots \quad a_n] \quad \text{and} \quad C = \begin{bmatrix} b_1 \\ b_2 \\ \cdot \\ \cdot \\ b_n \end{bmatrix}$$

then $RC = a_1b_1 + a_2b_2 + \cdots + a_nb_n$. Note that the product of a row and a column is a scalar.

> **WARNING!**   A row and column must have the same number of entries in order to be multiplied.

EXAMPLE 7   ▓   **Multiplying a Row by a Column**

Let

$$A = \begin{bmatrix} 2 & -5 & 4 \\ -1 & 7 & 5 \end{bmatrix} \quad \text{and} \quad B = \begin{bmatrix} 1 & 2 & -3 & 5 \\ 3 & -2 & 1 & 5 \\ 5 & 4 & 0 & -7 \end{bmatrix}$$

Compute the product of the first row of A with the third column of B.

SOLUTION   We must compute

$$[2 \quad -5 \quad 4] \cdot \begin{bmatrix} -3 \\ 1 \\ 0 \end{bmatrix}$$

According to the definition, we multiply corresponding entries and then add. Thus, we have

$$[2 \quad -5 \quad 4] \cdot \begin{bmatrix} -3 \\ 1 \\ 0 \end{bmatrix} = 2(-3) + (-5)1 + 4 \cdot 0 = -11$$

Note that in the preceding example, matrix A has the same number of columns as matrix B has rows. This ensures that the rows of A have the same number of entries as the columns of B. In fact, the matrix product AB is defined only if the number of columns of A is the same as the number of rows of B. If so, then the entry in row i and column j of AB is simply the product of the ith row of A with the jth column of B.

Multiplication of Matrices   ☐

> If A has the same number of columns as B has rows, then $AB = [c_{ij}]$, where
>
> $$c_{ij} = [\text{entries from row } i \text{ of } A] \cdot \begin{bmatrix} \text{entries} \\ \text{from} \\ \text{column} \\ j \text{ of } B \end{bmatrix}$$
>
> It follows that if the order of A is $m \times n$ and the order of B is $n \times p$, then AB has order $m \times p$.

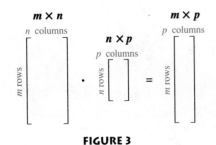

FIGURE 3

Figure 3 depicts the relationships between the orders of the factors and the resulting product matrix.

EXAMPLE 8   █   **Multiplication of Matrices**

Multiply

$$\begin{bmatrix} 1 & 2 & 3 \\ -2 & 0 & 5 \end{bmatrix} \cdot \begin{bmatrix} 2 & 1 \\ -3 & 4 \\ 2 & 1 \end{bmatrix}$$

SOLUTION   Since the first factor is a 2×3 matrix and the second is a 3×2 matrix, the product will be a 2×2 matrix. Let us denote the product by $[p_{ij}]$, so that we have

$$\begin{bmatrix} 1 & 2 & 3 \\ -2 & 0 & 5 \end{bmatrix} \cdot \begin{bmatrix} 2 & 1 \\ -3 & 4 \\ 2 & 1 \end{bmatrix} = \begin{bmatrix} p_{11} & p_{12} \\ p_{21} & p_{22} \end{bmatrix}$$

Table 1 illustrates the process of computing the product $[p_{ij}]$. Note that the product of row i from the first matrix and column j from the second gives p_{ij}.

Thus, the product is $\begin{bmatrix} 2 & 12 \\ 6 & 3 \end{bmatrix}$.

TABLE 1

Big picture	Computations	Update
$\begin{bmatrix} \boxed{1 \ \ 2 \ \ 3} \\ -2 \ \ 0 \ \ 5 \end{bmatrix} \cdot \begin{bmatrix} \boxed{\begin{matrix}2\\-3\\2\end{matrix}} \ \ \begin{matrix}1\\4\\1\end{matrix} \end{bmatrix} = \begin{bmatrix} p_{11} & p_{12} \\ p_{21} & p_{22} \end{bmatrix}$ Row 1 Column 1	$[1 \ \ 2 \ \ 3] \cdot \begin{bmatrix} 2 \\ -3 \\ 2 \end{bmatrix} = 1 \cdot 2 + 2(-3) + 3 \cdot 2$ $= 2 - 6 + 6$ $= 2$	$\begin{bmatrix} 2 & p_{12} \\ p_{21} & p_{22} \end{bmatrix}$
$\begin{bmatrix} \boxed{1 \ \ 2 \ \ 3} \\ -2 \ \ 0 \ \ 5 \end{bmatrix} \cdot \begin{bmatrix} 2 \ \ \boxed{\begin{matrix}1\\4\\1\end{matrix}} \\ -3 \\ 2 \end{bmatrix} = \begin{bmatrix} 2 & p_{12} \\ p_{21} & p_{22} \end{bmatrix}$ Row 1 Column 2	$[1 \ \ 2 \ \ 3] \cdot \begin{bmatrix} 1 \\ 4 \\ 1 \end{bmatrix} = 1 \cdot 1 + 2 \cdot 4 + 3 \cdot 1$ $= 1 + 8 + 3$ $= 12$	$\begin{bmatrix} 2 & 12 \\ p_{21} & p_{22} \end{bmatrix}$
$\begin{bmatrix} 1 \ \ 2 \ \ 3 \\ \boxed{-2 \ \ 0 \ \ 5} \end{bmatrix} \cdot \begin{bmatrix} \boxed{\begin{matrix}2\\-3\\2\end{matrix}} \ \ \begin{matrix}1\\4\\1\end{matrix} \end{bmatrix} = \begin{bmatrix} 2 & 12 \\ p_{21} & p_{22} \end{bmatrix}$ Row 2 Column 1	$[-2 \ \ 0 \ \ 5] \cdot \begin{bmatrix} 2 \\ -3 \\ 2 \end{bmatrix} = -2 \cdot 2 + 0(-3) + 5 \cdot 2$ $= -4 + 0 + 10$ $= 6$	$\begin{bmatrix} 2 & 12 \\ 6 & p_{22} \end{bmatrix}$
$\begin{bmatrix} 1 \ \ 2 \ \ 3 \\ \boxed{-2 \ \ 0 \ \ 5} \end{bmatrix} \cdot \begin{bmatrix} 2 \ \ \boxed{\begin{matrix}1\\4\\1\end{matrix}} \\ -3 \\ 2 \end{bmatrix} = \begin{bmatrix} 2 & 12 \\ 6 & p_{22} \end{bmatrix}$ Row 2 Column 2	$[-2 \ \ 0 \ \ 5] \cdot \begin{bmatrix} 1 \\ 4 \\ 1 \end{bmatrix} = -2 \cdot 1 + 0 \cdot 4 + 5 \cdot 1$ $= -2 + 0 + 5$ $= 3$	$\begin{bmatrix} 2 & 12 \\ 6 & 3 \end{bmatrix}$

RULE OF THUMB When multiplying matrices, first write their orders side by side in the order in which the product is to be computed. If the inner numbers are the same, the product is defined and the order of the product is given by the outer numbers. If the inner numbers are different, then the product is not defined.

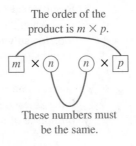

The order of the product is $m \times p$.

These numbers must be the same.

The order in which matrices are multiplied is crucial. For example, suppose that A is a 2×4 matrix and B is a 4×3 matrix. Then AB is a 2×3 matrix, where BA isn't even defined! Even if both AB and BA are defined, they are generally not equal.

E X A M P L E 9 ■ **Multiplying 2×2 Matrices in Both Orders**

Let

$$A = \begin{bmatrix} -1 & 2 \\ 3 & 4 \end{bmatrix} \quad \text{and} \quad B = \begin{bmatrix} 2 & -1 \\ 3 & 2 \end{bmatrix}$$

Show that $AB \neq BA$.

SOLUTION

$$AB = \begin{bmatrix} -1 & 2 \\ 3 & 4 \end{bmatrix} \cdot \begin{bmatrix} 2 & -1 \\ 3 & 2 \end{bmatrix}$$

$$= \begin{bmatrix} (-1)2 + 2 \cdot 3 & (-1)(-1) + 2 \cdot 2 \\ 3 \cdot 2 + 4 \cdot 3 & 3(-1) + 4 \cdot 2 \end{bmatrix}$$

$$= \begin{bmatrix} 4 & 5 \\ 18 & 5 \end{bmatrix}$$

$$BA = \begin{bmatrix} 2 & -1 \\ 3 & 2 \end{bmatrix} \cdot \begin{bmatrix} -1 & 2 \\ 3 & 4 \end{bmatrix}$$

$$= \begin{bmatrix} 2(-1) + (-1)3 & 2 \cdot 2 + (-1)4 \\ 3(-1) + 2 \cdot 3 & 3 \cdot 2 + 2 \cdot 4 \end{bmatrix}$$

$$= \begin{bmatrix} -5 & 0 \\ 3 & 14 \end{bmatrix}$$

Thus, $AB \neq BA$.

At this point, it is natural to question our motivation for defining matrix multiplication in such a seemingly unnatural and convoluted way. Why don't we simply multiply corresponding entries? For that matter, why bother to define multiplication of matrices at all? What's the point? Mathematical concepts and their accompanying definitions and notation don't appear out of thin air, nor are they handed down from on high. Instead, they are forged from the fires of utility: We construct mathematical tools to solve real-world problems or assist in mathematical investigations, but we keep them only when they continue to be useful. Matrix multiplication, born as a purely algebraic tool for solving systems of linear equations, has proven to be a potent weapon for attacking problems in virtually all branches of the mathematical, physical, and social sciences, including geometry, statistics, chemistry, biology, physics, astronomy, economics, and psychology. The common thread in each area of application is an underlying linear relationship between a collection of variables, which returns us to the birthplace of matrix multiplication—linear systems.

EXAMPLE 10 ■ Expressing a System of Equations in Matrix Form

Show that the system of equations

$$2x + 3y + 4z = 5$$
$$3x - 5y + z = 6$$
$$4x + 7y - z = 3$$

is equivalent to the matrix equation

$$\begin{bmatrix} 2 & 3 & 4 \\ 3 & -5 & 1 \\ 4 & 7 & -1 \end{bmatrix} \cdot \begin{bmatrix} x \\ y \\ z \end{bmatrix} = \begin{bmatrix} 5 \\ 6 \\ 3 \end{bmatrix}$$

SOLUTION We begin by multiplying the matrices on the left side of the equation.

$$\begin{bmatrix} 2 & 3 & 4 \\ 3 & -5 & 1 \\ 4 & 7 & -1 \end{bmatrix} \cdot \begin{bmatrix} x \\ y \\ z \end{bmatrix} = \begin{bmatrix} 5 \\ 6 \\ 3 \end{bmatrix}$$

$$\begin{bmatrix} 2x + 3y + 4z \\ 3x - 5y + z \\ 4x + 7y - z \end{bmatrix} = \begin{bmatrix} 5 \\ 6 \\ 3 \end{bmatrix}$$

Since two matrices are equal if and only if corresponding entries are equal, we have

$$2x + 3y + 4z = 5$$
$$3x - 5y + z = 6$$
$$4x + 7y - z = 3$$

CALCULATOR KEYS

Matrix Operations

The operations of matrix addition, subtraction, and multiplication, as well as scalar multiplication, can be performed by your graphing calculator. First, for each matrix involved in the computation, it is necessary to specify the order and then key in the entries. Once this is done, the matrix expression can be entered and evaluated. For details, consult your calculator's manual.

In the following example, we see an illustration of matrix multiplication's applicability to contexts seemingly unrelated to systems of equations. As you work through this example, note how the early investment in expressing a quantity with matrix multiplication pays off when recomputing the quantity with new data.

EXAMPLE 11 ■ An Application of Matrix Multiplication

Lachelle has $500 withheld out of each month's paycheck and placed in a 401k retirement plan. The plan offers two fund choices: a high-risk growth fund and a low-risk income fund. Lachelle has chosen to place $350 in the growth fund and $150 in the income fund. The manager of each fund determines how investments are distributed among three categories of stocks: blue chip, biotech, and international. The distributions are summarized in Table 2.

TABLE 2

	Blue chip stocks	Biotech stocks	International stocks
Growth fund	30%	50%	20%
Income fund	50%	35%	15%

For example, the growth fund manager distributes 30% of all investments to blue chip stocks. Express the distribution of Lachelle's $500 monthly withholding as a 2×3 matrix, and use matrix multiplication to find the earnings of each fund for a single month given that the rate of return is

a. 8% for blue chip stocks, 10% for biotech stocks, and 9% for international stocks
b. 9% for blue chip stocks, 13% for biotech stocks, and 7% for international stocks

SOLUTION We define a 2×3 matrix A representing Lachelle's investment portfolio, with each row corresponding to a fund (growth and income, in that order) and each column corresponding to a stock option (blue chip, biotech, and international, in that order). Thus, we have

$$A = \begin{bmatrix} 30\% \text{ of } 350 & 50\% \text{ of } 350 & 20\% \text{ of } 350 \\ 50\% \text{ of } 150 & 35\% \text{ of } 150 & 15\% \text{ of } 150 \end{bmatrix} = \begin{bmatrix} 105 & 175 & 70 \\ 75 & 52.5 & 22.5 \end{bmatrix}$$

a. Multiplying the amounts invested by the corresponding rates of return, we see that the total earnings of Lachelle's growth fund account is given by

$$8\% \cdot 105 + 10\% \cdot 175 + 9\% \cdot 70$$

Similarly, her income account generates earnings of

$$8\% \cdot 75 + 10\% \cdot 52.5 + 9\% \cdot 22.5$$

In matrix form, we could express her earnings for the two funds as

$$\begin{bmatrix} 0.08 \cdot 105 + 0.10 \cdot 175 + 0.09 \cdot 70 \\ 0.08 \cdot 75 + 0.10 \cdot 52.5 + 0.09 \cdot 22.5 \end{bmatrix}$$

or equivalently,

$$\begin{bmatrix} 105 & 175 & 70 \\ 75 & 52.5 & 22.5 \end{bmatrix} \begin{bmatrix} 0.08 \\ 0.10 \\ 0.09 \end{bmatrix}$$

Note that this is the product of the matrix A and the column matrix consisting of the rates of return. Evaluating the product with a graphing calculator, we obtain

$$\begin{bmatrix} 32.2 \\ 13.275 \end{bmatrix}$$

Thus, on Lachelle's investment of \$500, the growth fund earns \$32.20 and the income fund earns \$13.28.

b. In this case, the total earnings are given by

$$\begin{bmatrix} 105 & 175 & 70 \\ 75 & 52.5 & 22.5 \end{bmatrix} \begin{bmatrix} 0.09 \\ 0.13 \\ 0.07 \end{bmatrix} = \begin{bmatrix} 37.1 \\ 15.15 \end{bmatrix}$$

Thus, the growth fund earns \$37.10 and the income fund earns \$15.15.

We now summarize the fundamental algebraic properties of matrices. Proofs of some of these properties for 2×2 matrices will be developed in Exercises 73–76.

Properties of Matrix Operations ☐

For scalars x and y and matrices A, B, and C (of appropriate orders):

Algebraic description	Verbal description
1. $A + B = B + A$	Matrix addition is commutative.
2. $(A + B) + C = A + (B + C)$	Matrix addition is associative.
3. $x(yA) = (xy)A$	Scalar multiplication is associative.
4. $A(BC) = (AB)C$	Matrix multiplication is associative.
5. $x(AB) = (xA)B = A(xB)$	Scalar/matrix multiplication is associative.
6. $x(A + B) = xA + xB$	Scalar multiplication is distributive.
7. $A(B + C) = AB + AC$	Matrix multiplication is distributive (on the left).
8. $(A + B)C = AC + BC$	Matrix multiplication is distributive (on the right).

Note that, in general, $AB \neq BA$, so that matrix multiplication is *not* commutative.

UNDERSTANDING AND MASTERY CHECKLISTS

CONCEPTS TO UNDERSTAND

☐ Matrix of order $m \times n$
☐ Subscript notation for matrices
☐ Matrix addition and subtraction
☐ Scalar multiplication
☐ Matrix multiplication
☐ Matrix form for a system of equations
☐ Properties of matrix operations

SKILLS TO MASTER

☐ Determine the order of a matrix.
☐ Identify an entry of a matrix by its row and column.
☐ Add or subtract matrices.
☐ Find a scalar multiple of a matrix.
☐ Determine whether a matrix product can be computed.
☐ Compute the product of two matrices.
☐ Express a system of equations in matrix form.

EXERCISES 7.1

EXERCISES 1–4 *Identify the order of the given matrix.*

1. $\begin{bmatrix} 1 & 2 & 3 \\ 4 & 5 & 6 \end{bmatrix}$

2. $\begin{bmatrix} 1 & 0 \\ -1 & 0 \\ 0 & 1 \end{bmatrix}$

3. $\begin{bmatrix} a \\ b \\ c \\ d \end{bmatrix}$

4. $[i \quad j \quad k]$

EXERCISES 5–18 *Perform the indicated matrix operation, if possible.*

5. $2 \cdot \begin{bmatrix} 1 & 0 \\ 3 & 5 \end{bmatrix}$

6. $-3 \cdot \begin{bmatrix} 1 & 5 & 3 \\ 4 & 0 & 2 \end{bmatrix}$

7. $\begin{bmatrix} 2 & 3 \\ 6 & 4 \end{bmatrix} + \begin{bmatrix} -4 & 8 \\ 0 & 2 \end{bmatrix}$

8. $[0 \quad 1 \quad 2 \quad 3] + [4 \quad 3 \quad 1 \quad 0]$

9. $\begin{bmatrix} 1 & -1 \\ 0 & 3 \end{bmatrix} + \begin{bmatrix} 2 \\ 5 \end{bmatrix}$

10. $\begin{bmatrix} 2 & -3 & 5 \\ -1 & 7 & 0 \end{bmatrix} - \begin{bmatrix} 3 & -5 & 2 \\ 1 & 0 & 4 \end{bmatrix}$

11. $\begin{bmatrix} -3 \\ 1 \\ 6 \end{bmatrix} + 2 \cdot \begin{bmatrix} 4 \\ -1 \\ 0 \end{bmatrix}$

12. $\begin{bmatrix} 0 & 3 \\ 6 & 7 \end{bmatrix} + \begin{bmatrix} -2 & 4 \\ 0 & 1 \end{bmatrix} - \begin{bmatrix} 2 & -5 \\ 3 & -5 \end{bmatrix}$

13. $\begin{bmatrix} a & b \\ c & d \end{bmatrix} + \begin{bmatrix} 3a & 2b \\ c & 5d \end{bmatrix}$

14. $\begin{bmatrix} 1 & 0 & -1 \\ 3 & 2 & 1 \end{bmatrix} - \begin{bmatrix} 4 & 6 \\ 6 & 4 \end{bmatrix}$

15. $2 \cdot \begin{bmatrix} 1 & 2 & 0 \\ 2 & 6 & 9 \\ 3 & -1 & -4 \end{bmatrix} + \begin{bmatrix} -1 & 3 & 4 \\ 3 & -3 & 0 \\ 1 & 4 & 5 \end{bmatrix}$

16. $3 \cdot \begin{bmatrix} 2 & 3 & 0 \\ 1 & -1 & 1 \\ -1 & -2 & 2 \end{bmatrix} - 2 \cdot \begin{bmatrix} 0 & -6 & -1 \\ 2 & -4 & 3 \\ -1 & 2 & -2 \end{bmatrix}$

17. $\begin{bmatrix} 2 & a & 1 \\ 0 & 1 & 2 \end{bmatrix} - a \cdot \begin{bmatrix} 1 & 1 & 1 \\ 1 & 1 & 0 \end{bmatrix}$

18. $\begin{bmatrix} 1 & x \\ x^2 & 0 \\ 2 & 1 \end{bmatrix} + \begin{bmatrix} x & 2 \\ -x^2 & 3 \\ x^2 & -2 \end{bmatrix}$

EXERCISES 19–22 *Simplify the given quantity for*

$$A = \begin{bmatrix} 1 & 2 \\ 3 & 5 \end{bmatrix} \quad and \quad B = \begin{bmatrix} 2 & 1 \\ 4 & 9 \end{bmatrix}$$

19. $A + B$

20. $A - B$

21. $2A - 5B$

22. $3A + 2B$

EXERCISES 23–26 *Solve the given equation for the unknown matrix X. (Hint: Use matrix operations to solve for X, much as you would for an ordinary linear equation.)*

23. $2X = \begin{bmatrix} 4 & 6 \\ 8 & 2 \end{bmatrix}$

24. $3X + \begin{bmatrix} 1 & 0 \\ 0 & 1 \end{bmatrix} = \begin{bmatrix} 10 & 12 \\ 3 & 16 \end{bmatrix}$

25. $2X + \begin{bmatrix} 1 & 2 & 3 \\ 6 & 5 & 4 \end{bmatrix} = \begin{bmatrix} 0 & 0 & 0 \\ 0 & 0 & 0 \end{bmatrix}$

26. $-2\left(X + \begin{bmatrix} 2 & 6 \\ 3 & 3 \end{bmatrix}\right) = -4X + \begin{bmatrix} 0 & 0 \\ 2 & 4 \end{bmatrix}$

EXERCISES 27–32 *Let*

$$A = \begin{bmatrix} 1 & 2 & 4 \\ 2 & 3 & 0 \end{bmatrix}, \quad B = \begin{bmatrix} 1 & 2 \\ 3 & 4 \end{bmatrix}, \quad and \quad C = \begin{bmatrix} 3 & 7 & 9 \\ 4 & 5 & 1 \\ 2 & 6 & 4 \end{bmatrix}$$

Indicate whether the given product is defined. If so, give the order of the product matrix. You need not compute the product.

27. AB

28. AC

29. BA

30. BC

31. CA

32. CB

EXERCISES 33–54 *Compute the product, if possible.*

33. $\begin{bmatrix} 3 & 5 \\ 4 & 2 \end{bmatrix}\begin{bmatrix} 3 \\ -3 \end{bmatrix}$

34. $\begin{bmatrix} a & b \\ c & d \end{bmatrix}\begin{bmatrix} x \\ y \end{bmatrix}$

35. $\begin{bmatrix} 1 & 0 \\ 0 & 1 \end{bmatrix}\begin{bmatrix} 1 \\ 2 \\ 3 \end{bmatrix}$

36. $\begin{bmatrix} 2 & 3 \\ 4 & 6 \end{bmatrix}\begin{bmatrix} 1 & 2 \\ 0 & 1 \end{bmatrix}$

37. $\begin{bmatrix} 1 & 2 \\ 0 & 1 \end{bmatrix}\begin{bmatrix} 2 & 3 \\ 4 & 6 \end{bmatrix}$

38. $\begin{bmatrix} 1 & -2 \\ -5 & 7 \end{bmatrix}\begin{bmatrix} a & b \\ c & d \end{bmatrix}$

39. $\begin{bmatrix} 2 & 1 \\ 3 & 2 \end{bmatrix}\begin{bmatrix} 2 & -1 \\ -3 & 2 \end{bmatrix}$

40. $\begin{bmatrix} 1 & -1 & 0 \\ 2 & 0 & 3 \end{bmatrix}\begin{bmatrix} 4 & -1 \\ 1 & 0 \end{bmatrix}$

41. $\begin{bmatrix} 2 & 0 \\ 1 & 4 \\ 2 & 1 \end{bmatrix}\begin{bmatrix} 3 & 5 \\ 1 & 7 \end{bmatrix}$

42. $\begin{bmatrix} 0 & 2 & 0 \\ 0 & 0 & 3 \\ 0 & 0 & 0 \end{bmatrix}\begin{bmatrix} 0 & 2 & 0 \\ 0 & 0 & 3 \\ 0 & 0 & 0 \end{bmatrix}$

43. $\begin{bmatrix} 1.8 & 3.5 & 4.6 \\ 4.8 & 1.7 & 3.2 \\ 1.7 & 2.5 & 0.7 \end{bmatrix}\begin{bmatrix} 2.6 \\ 3.4 \\ 4.9 \end{bmatrix}$

44. $\begin{bmatrix} 2.75 & 3.01 \\ 4.07 & 0.07 \end{bmatrix}\begin{bmatrix} 1.7 & 3.4 \\ 6.8 & 5.9 \end{bmatrix}$

45. $\begin{bmatrix} 1 & 0 \\ x & 1 \end{bmatrix}\begin{bmatrix} a & b \\ c & d \end{bmatrix}$

46. $\begin{bmatrix} x & 0 \\ 0 & 1 \end{bmatrix}\begin{bmatrix} a & b \\ c & d \end{bmatrix}$

47. $\begin{bmatrix} a \\ b \end{bmatrix}\begin{bmatrix} c \\ d \end{bmatrix}$

48. $[a \quad b \quad c]\begin{bmatrix} i \\ j \\ k \end{bmatrix}$

49. $\begin{bmatrix} 0 & \frac{1}{b} \\ \frac{1}{a} & 0 \end{bmatrix}\begin{bmatrix} 0 & a \\ b & 0 \end{bmatrix}$

50. $[3 \quad 6][\frac{1}{3} \quad \frac{1}{6}]$

51. $\begin{bmatrix} 1 & 0 & 0 \\ 0 & 1 & 0 \\ 0 & 0 & 1 \end{bmatrix} \cdot \begin{bmatrix} p & q \\ r & s \\ t & u \end{bmatrix}$

52. $\begin{bmatrix} a & 0 & 0 & 0 \\ 0 & b & 0 & 0 \\ 0 & 0 & c & 0 \\ 0 & 0 & 0 & d \end{bmatrix} \cdot \begin{bmatrix} p & 0 & 0 & 0 \\ 0 & q & 0 & 0 \\ 0 & 0 & r & 0 \\ 0 & 0 & 0 & s \end{bmatrix}$

53. $\begin{bmatrix} 0 & 0 & 1 \\ 0 & 1 & 0 \\ 1 & 0 & 0 \end{bmatrix}\begin{bmatrix} a_{11} & a_{12} & a_{13} \\ a_{21} & a_{22} & a_{23} \\ a_{31} & a_{32} & a_{33} \end{bmatrix}$

54. $\begin{bmatrix} a_{11} & a_{12} & a_{13} \\ a_{21} & a_{22} & a_{23} \\ a_{31} & a_{32} & a_{33} \end{bmatrix}\begin{bmatrix} 0 & 0 & 1 \\ 0 & 1 & 0 \\ 1 & 0 & 0 \end{bmatrix}$

EXERCISES 55–60 *Express the given system of equations as a matrix equation.*

55. $2x - 3y = 5$
$-x + y = 6$

56. $a - b + c = 0$
$2a + 3b = 0$

57. $r + s = 1$
$2t - r = 3$
$r + s + t = 4$

58. $p + q + r = 0$
$p + 3r = -1$
$q - r - 2 = 0$

59. $x_1 + x_2 + x_3 + x_4 = 4$
$x_2 + x_3 + x_4 = 3$
$x_3 + x_4 = 2$

60. $a_1 - a_4 = 1$
$a_2 - a_1 = 0$
$a_3 - a_2 = 0$
$a_4 - a_3 = -1$

EXERCISES 61–64 *If A is an $n \times m$ matrix, then the **transpose** of A, denoted A^T, is the matrix whose rows are formed from the columns of A. More precisely, if $A = [a_{ij}]$, then $A^T = [b_{ij}]$, where $b_{ij} = a_{ji}$. For example, if*

$$A = \begin{bmatrix} 2 & 4 \\ 6 & 8 \end{bmatrix}, \quad then \quad A^T = \begin{bmatrix} 2 & 6 \\ 4 & 8 \end{bmatrix}$$

Compute the transposes of the given matrices.

61. $\begin{bmatrix} 1 & 2 & 3 \\ 4 & 5 & 6 \end{bmatrix}$

62. $\begin{bmatrix} a & b & c \\ d & e & f \\ g & h & i \end{bmatrix}$

63. $\begin{bmatrix} 1 & 2 & 3 \\ 2 & 5 & 6 \\ 3 & 6 & 4 \end{bmatrix}$

64. $\begin{bmatrix} x & y \\ 2x & 3y \\ -x & 4y \end{bmatrix}$

EXERCISES 65–68 *Perform the given operation. Note that for a square matrix A, A^2 is defined as $A \cdot A$. Similarly, $A^3 = A \cdot A \cdot A$, and so on.*

65. $\begin{bmatrix} 1 & 2 \\ 0 & 1 \end{bmatrix}^2$

66. $\begin{bmatrix} 0 & 1 \\ 0 & 0 \end{bmatrix}^3$

67. $\begin{bmatrix} 1 & 0 \\ 0 & 0 \end{bmatrix}^4$

68. $\begin{bmatrix} a & b \\ c & d \end{bmatrix}^2$

EXERCISES 69–72 *Compute the indicated powers, given that*

$$A = \begin{bmatrix} 1 & 2 & 3 \\ 0 & 1 & 4 \\ 0 & 0 & 1 \end{bmatrix}$$

69. A^2

70. A^3

71. A^{32} (*Hint:* If you enter 3 in your calculator and then press $\boxed{x^2}$, you will obtain 3^2. If you press $\boxed{x^2}$ again, you will obtain $(3^2)^2 = 3^4$. If you press $\boxed{x^2}$ once again, you obtain 3^8, and so on.)

72. A^{33} (*Hint:* $A^{33} = A^{32} \cdot A$.)

EXERCISES 73–76   *Confirm the given property for 2 × 2 matrices. Do so by letting*

$$A = \begin{bmatrix} a_1 & a_2 \\ a_3 & a_4 \end{bmatrix}, \quad B = \begin{bmatrix} b_1 & b_2 \\ b_3 & b_4 \end{bmatrix}, \quad and \quad C = \begin{bmatrix} c_1 & c_2 \\ c_3 & c_4 \end{bmatrix}$$

and directly verifying the given identity.

73. $A + B = B + A$

74. $(xA)B = x(AB)$, where x is a scalar

75. $(A + B)C = AC + BC$

76. $x(A + B) = xA + xB$, where x is a scalar

■ APPLICATIONS

77. *Health-Care Claims*   All employees of a computer software company, whether hourly workers or salaried personnel, select one of two health-care plans: major medical or comprehensive. Claims totals for 1998 are shown in Table 3. Express the breakdown of claims filed by hourly workers as a 2 × 2 matrix, and do likewise for claims filed by salaried personnel. Find and interpret the sum of these two matrices.

TABLE 3

	Hourly		Salaried	
	Employees	**Dependents**	**Employees**	**Dependents**
Major medical	$230,000	$320,000	$125,000	$250,000
Comprehensive	$280,000	$300,000	$400,000	$500,000

78. *Birth Distributions*   A city has two major hospitals: St. Mary's and Parkview. St. Mary's recorded births of 200 boys and 210 girls for the first half of 1978 and 185 boys and 177 girls for the second half. Parkview Hospital recorded 320 girls and 307 boys for the first half and 300 girls and 295 boys for the second half. Express the births for each 6-month period as a 2 × 2 matrix, and compute and interpret the sum.

79. *Point Breakdown*   The team statistician of the Charlotte Hornets recorded the following statistics from the team's last game against the Orlando Magic. The Hornets made 5 three-point shots, 40 two-point field goals, and 18 free throws. The Magic made 4 three-point shots, 37 two-point field goals, and 26 free throws. Express the shot distribution for the game as a 2 × 3 matrix S. Find a 3 × 1 matrix T such that the final score is given by the matrix product ST. What is the final score?

80. *Congressional Districts*   Since the number of congressional districts is fixed at 435 and representation in the House of Representatives is proportional to population, it is occasionally necessary to redefine congressional districts. Population growth in a certain state has lagged behind that of the rest of the nation; thus, the state will be combining portions of two districts into one. The first district is 40% Republican and 30% Democrat, whereas the second is 25% Republican and 55% Democrat. There are 200,000 registered voters in the first district and 150,000 in the second. Express the political party percentage distribution of the two districts as a 2 × 2 matrix, and then find the total number of Democrats and Republicans in the new district using matrix multiplication, assuming that the new district will be formed from:

   a. 200,000 voters from the first district and 150,000 from the second

   b. 175,000 voters from each district

■ CONCEPTS AND CRITICAL THINKING

EXERCISES 81–89   *Answer true or false.*

81. The sum $A + B$ is defined for any pairs of matrices A and B.

82. The product AB is defined only for matrices A and B of the same order.

83. If A and B are of the same order, then $A + B$ is defined.

84. If A and B are of the same order, then AB is defined.

85. A 5 × 3 matrix has 5 rows and 3 columns.

86. Matrix addition is commutative; that is, for any pair of matrices A and B for which $A + B$ is defined, $A + B = B + A$.

87. Matrix multiplication is commutative; that is, for any pair of matrices A and B for which AB is defined, $AB = BA$.

88. Multiplication of matrices is accomplished by multiplying corresponding entries.

89. Addition of matrices is accomplished by adding corresponding entries.

EXERCISES 90–95   *Give an example of each.*

90. A 2 × 3 matrix

91. Two matrices A and B such that AB is defined but BA is not defined

92. A matrix A such that A^2 is not defined

93. Two matrices A and B such that $A + B$ is defined but neither AB nor BA is defined

94. Two matrices A and B such that AB is defined but $A + B$ is not defined

95. A 3 × 2 matrix such that all of its columns are identical but no two of its rows are the same

96. For real numbers a and b, the zero-product property holds: If $ab = 0$, then either $a = 0$ or $b = 0$. Show that the zero-product property does *not* hold for matrices by finding a 1×2 matrix A and a 2×1 matrix B such that $AB = 0$ but neither A nor B consists entirely of zero entries.

QUESTIONS FOR DISCUSSION OR ESSAY

97. How does a matrix differ from a simple table of numbers?

98. Most of the matrix products in this section can be determined using a graphing calculator. Why bother to learn how to multiply matrices by hand? For that matter, is there any point in learning how to multiply numbers by hand?

99. We are free to define the product of matrices in any way we want. So why have we chosen to define them in such a seemingly bizarre way? Why not simply define the product of the matrices $[a_{ij}]$ and $[b_{ij}]$ to be $[c_{ij}]$, where $c_{ij} = a_{ij}b_{ij}$?

100. It's reasonable to wonder how much we really gain by having the graphing calculator perform matrix operations such as matrix multiplication. To multiply a pair of 2×2 matrices by hand, a total of 12 arithmetic operations (addition or multiplication) must be performed. On the other hand, $4 + 4 = 8$ entries must be entered into the calculator. Develop a formula for the number of operations (additions and multiplications count as one each) required to multiply a pair of $n \times n$ matrices. (*Hint:* The product of two $n \times n$ matrices is an $n \times n$ matrix with n^2 entries. So compute the number of multiplications and additions required to determine each entry of the product matrix, and then multiply this number by n^2.) How many entries must be entered into a calculator when multiplying two $n \times n$ matrices? In the case of multiplication of a pair of 6×6 matrices, is the calculator worthwhile?

PROJECTS FOR ENRICHMENT

101. *Linear Transformations: An Application of Matrix Multiplication*   If the line segment $\overline{OP}$ in Figure 4 is rotated by an angle α about the origin, the point $P(x, y)$ is transformed to a new point, $P'(x', y')$. We can express x' and y' in terms of x and y using matrix multiplication.

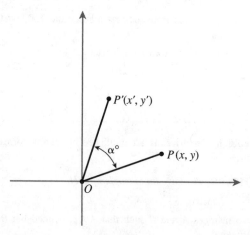

FIGURE 4

The point P in the Cartesian plane with coordinates (x, y) can also be represented as the coordinate matrix

$$\begin{bmatrix} x \\ y \end{bmatrix}$$

The coordinates of P' can be obtained by multiplying the coordinate matrix by a certain 2×2 matrix A. Thus, if the coordinates of P' are (x', y'), then we have

$$A\begin{bmatrix} x \\ y \end{bmatrix} = \begin{bmatrix} x' \\ y' \end{bmatrix}$$

To find the appropriate matrix A, we use the following rule.

- The first column of A represents the rotated coordinates of the point $(1, 0)$.
- The second column of A represents the rotated coordinates of the point $(0, 1)$.

For example, consider a rotation by 90° counterclockwise. The point $(1, 0)$ will be rotated to the point $(0, 1)$ and so the first column of A is

$$\begin{bmatrix} 0 \\ 1 \end{bmatrix}$$

The point $(0, 1)$ will be rotated to the point $(-1, 0)$ so the second column of A is

$$\begin{bmatrix} -1 \\ 0 \end{bmatrix}$$

Thus,

$$A = \begin{bmatrix} 0 & -1 \\ 1 & 0 \end{bmatrix}$$

a. Where does the point $(2, 3)$ end up if it is rotated counterclockwise by 90°?

b. Where does the point $(-1, 2)$ end up if it is rotated counterclockwise by 90°?

c. Define A to be the 90° counterclockwise rotation matrix just discussed. Compute A^4. Explain your answer.

d. Find the matrix B for a 45° counterclockwise rotation. [*Hint:* The point $(1, 0)$ will be rotated onto a point 1 unit from the origin on the line $y = x$; the point $(0, 1)$ will be rotated onto a point 1 unit from the origin on the line $y = -x$.]

e. Show directly that $B^2 = A$. Explain why this is logical.

f. What is the smallest whole number n so that $B^n = \begin{bmatrix} 1 & 0 \\ 0 & 1 \end{bmatrix}$? Explain your answer.

102. *Transition Matrices: An Application of Matrix Multiplication* Suppose that in a certain town, 10% of non-smokers take up smoking each year, whereas 20% of smokers quit each year. If, in a given year, there are x smokers and y nonsmokers, then the smoking distribution of the population can be represented as

$$\begin{bmatrix} x \\ y \end{bmatrix}$$

The next year, the smokers will consist of the 80% of smokers who did *not* quit plus the 10% of former nonsmokers who take up smoking. Thus, the number of smokers in the next year will be $0.8x + 0.1y$. Similarly, the number of nonsmokers will be $0.2x + 0.9y$. Thus, after 1 year, the distribution of the population is given by

$$\begin{bmatrix} 0.8x + 0.1y \\ 0.2x + 0.9y \end{bmatrix}$$

a. Find a 2×2 matrix A such that

$$A \begin{bmatrix} x \\ y \end{bmatrix} = \begin{bmatrix} 0.8x + 0.1y \\ 0.2x + 0.9y \end{bmatrix}$$

This is called a **transition matrix.**

b. Suppose that the town has a smoking distribution of

$$\begin{bmatrix} 600 \\ 1400 \end{bmatrix}$$

Compute the smoking distribution of the town after 4 years. (*Hint:* The smoking distribution after 1 year is given by

$$A \begin{bmatrix} 600 \\ 1400 \end{bmatrix}, \quad \text{after 2 years by } A \left(A \begin{bmatrix} 600 \\ 1400 \end{bmatrix} \right)$$

and so on.)

c. Find the smoking distribution for the town after 32 years, assuming its initial distribution is

$$\begin{bmatrix} 600 \\ 1400 \end{bmatrix}$$

Now find the smoking distribution after 64 years with the same initial distribution. What relationship do you find between your answers, and how can this be explained?

d. Compare the smoking distribution of two towns after 64 years, both with initial populations of 2000 people, but one having 100 smokers and the other having 1900 smokers. Explain the relationship between your results.

e. We are making numerous assumptions about the population when we employ transition matrices. For instance, we are assuming that there is no migration either into or out of the town. What other assumptions are we making, and how reasonable are they?

f. We have assumed that the smoking transition matrix for every year is the same. Of course, this is unlikely to be true. What sorts of trends would we be likely to find if we checked historical records? How would a smoking transition matrix for 1950 compare to one for the year 2000?

SECTION 7.2   **LINEAR SYSTEMS AND MATRICES**

☐ **How can diets be balanced by solving systems of equations?**

☐ **How can an athlete choose an appropriate combination of activities as part of a cross-training routine?**

☐ **Why are there never exactly two solutions to a linear system of equations?**

☐ **How can we *quickly* find the equation of the parabola passing through three given points?**

In Section 2.8, we investigated systems of two equations in two variables and introduced the notion of a *linear* system. Throughout much of the rest of this chapter, we develop a handful of systematic methods—algorithms—that can be applied to solve linear systems of arbitrary size. In each case, the first step to solving a linear system is to translate the system into matrices, the language of linear systems.

LINEAR SYSTEMS OF EQUATIONS

Recall from Section 2.8 that a system of equations is said to be linear if each equation is linear. Thus, a linear system of two equations in x and y is of the form

$$ax + by = c$$
$$dx + ey = f$$

For example, the system

$$3x - 4y = 12$$
$$2x + 9y = 1$$

is linear since both equations are linear, whereas the system

$$2x^2 - 3y = 5$$
$$x + 47y = 9$$

is nonlinear since one of the equations (the first in this case) is not linear.

We can consider linear systems consisting of any number of equations in any number of variables. For example, each of the following is a linear system of equations.

$$2x - 3y + 4z = 10 \qquad\qquad 2p - 3q + r - 5s = 3$$
$$x - y + z = 3 \qquad\qquad\quad p + q + 2r + 3s = -4$$
$$3x + 5y - 2z = 9$$

(3 linear equations in 3 variables)   (2 linear equations in 4 variables)

More precisely, we have the following definition.

Definition of a Linear System of Equations ☐

> A **linear system of m equations in the n variables $x_1, x_2, x_3, \ldots, x_n$** is a system that can be expressed in the form
>
> $$a_{11}x_1 + a_{12}x_2 + a_{13}x_3 + \cdots + a_{1n}x_n = b_1$$
> $$a_{21}x_1 + a_{22}x_2 + a_{23}x_3 + \cdots + a_{2n}x_n = b_2$$
> $$a_{31}x_1 + a_{32}x_2 + a_{33}x_3 + \cdots + a_{3n}x_n = b_3$$
> $$\vdots \qquad\quad \vdots \qquad\quad \vdots \qquad\qquad\quad \vdots \qquad \vdots$$
> $$a_{m1}x_1 + a_{m2}x_2 + a_{m3}x_3 + \cdots + a_{mn}x_n = b_m$$
>
> where the a_{ij}'s and b_i's are real numbers. The a_{ij}'s are called **coefficients.**

SOLUTION SETS OF LINEAR EQUATIONS

If a point (x, y) is a solution of the linear system

$$ax + by = c$$
$$dx + ey = f$$

then (x, y) must satisfy each equation. Moreover, because the graph of each equation is a line, a solution point (x, y) must lie on both lines and, hence, must be a point of intersection. Now a pair of lines may intersect in a single point, infinitely many points (if the lines are the same), or no points at all (if the lines are parallel). Thus, we see that the number of solutions to a system of two linear equations in two variables is either 0, 1, or infinity. Table 4 illustrates the range of possibilities for the solution set of a system of two equations in two variables.

TABLE 4 Solution possibilities

System	Graph	Points of intersection
$2x - y = 1$ $x - 2y = -4$	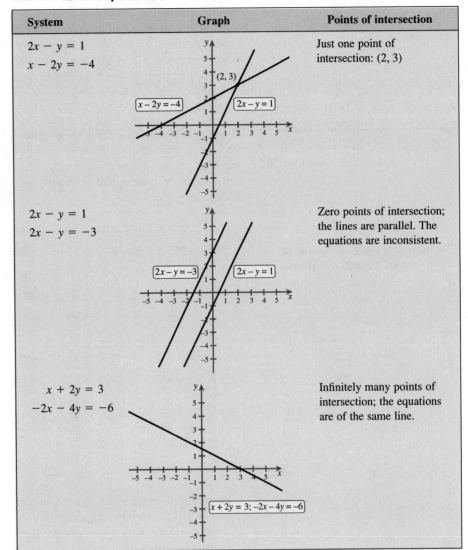	Just one point of intersection: (2, 3)
$2x - y = 1$ $2x - y = -3$		Zero points of intersection; the lines are parallel. The equations are inconsistent.
$x + 2y = 3$ $-2x - 4y = -6$		Infinitely many points of intersection; the equations are of the same line.

The results in Table 4 hold true for general linear systems regardless of the number of equations or variables.

Number of Solutions to a Linear   □   A linear system of equations has 0, 1, or infinitely many solutions.
System of Equations

EXAMPLE 1   ▪   A Linear System with Infinitely Many Solutions

Solve the system of equations

$$x - y = 6$$
$$2x - 2y = 12$$

SOLUTION Since the second equation can be obtained by multiplying both sides of the first equation by 2, there is, in essence, only one equation to be satisfied. Thus, the solution set consists of all pairs (x, y) such that $x - y = 6$. If we solve this equation for y, we obtain $y = x - 6$, so that the solution set to the system could also be described as the set of pairs $(x, x - 6)$. Thus, for example, when $x = 1$, we obtain the point $(1, -5)$ as a solution; when $x = -3$, we obtain $(-3, -9)$, and so on.

REPRESENTING SYSTEMS WITH MATRICES

The notation used in defining linear systems is suggestive of matrix notation. In fact, we have already seen (in Example 10 of Section 7.1) that matrices are a convenient way of expressing linear equations. As we will soon see, matrices are also an invaluable tool for *solving* linear systems of equations. The following definitions are central to our discussion.

Definition of Coefficient Matrix and Augmented Matrix □

The **coefficient matrix** of the linear system of equations

$$
\begin{aligned}
a_{11}x_1 + a_{12}x_2 + a_{13}x_3 + \cdots + a_{1n}x_n &= b_1 \\
a_{21}x_1 + a_{22}x_2 + a_{23}x_3 + \cdots + a_{2n}x_n &= b_2 \\
a_{31}x_1 + a_{32}x_2 + a_{33}x_3 + \cdots + a_{3n}x_n &= b_3 \\
&\vdots \\
a_{m1}x_1 + a_{m2}x_2 + a_{m3}x_3 + \cdots + a_{mn}x_n &= b_m
\end{aligned}
$$

is the matrix

$$
[a_{ij}] = \begin{bmatrix}
a_{11} & a_{12} & a_{13} & \cdots & a_{1n} \\
a_{21} & a_{22} & a_{23} & \cdots & a_{2n} \\
a_{31} & a_{32} & a_{33} & \cdots & a_{3n} \\
\vdots & \vdots & \vdots & & \vdots \\
a_{m1} & a_{m2} & a_{m3} & \cdots & a_{mn}
\end{bmatrix}
$$

The matrix formed by adjoining a column consisting of the b_i's to the coefficient matrix $[a_{ij}]$ is called the **augmented matrix** of the system. The augmented matrix is often written with a vertical bar separating the column of the b_i's from the coefficient matrix, as follows.

Augmented matrix

$$
\left[\begin{array}{ccccc|c}
a_{11} & a_{12} & a_{13} & \cdots & a_{1n} & b_1 \\
a_{21} & a_{22} & a_{23} & \cdots & a_{2n} & b_2 \\
a_{31} & a_{32} & a_{33} & \cdots & a_{3n} & b_3 \\
\vdots & \vdots & \vdots & & \vdots & \vdots \\
a_{m1} & a_{m2} & a_{m3} & \cdots & a_{mn} & b_m
\end{array}\right]
$$

E X A M P L E   2 ■ Finding a Coefficient and Augmented Matrix

Find the coefficient and augmented matrix for the following linear system.

$$
\begin{aligned}
2x - 4y + 2z &= 6 \\
3y + 5x &= 8
\end{aligned}
$$

SOLUTION   We begin by writing the variables in the same order in each equation of the system.

$$2x - 4y + 2z = 6$$
$$5x + 3y \qquad = 8$$

The coefficient matrix is then

$$\begin{bmatrix} 2 & -4 & 2 \\ 5 & 3 & 0 \end{bmatrix}$$

The augmented matrix is formed by "augmenting" the coefficient matrix with a column formed from the constants to the right of the equal sign. Thus, the augmented matrix is given by

$$\left[\begin{array}{ccc|c} 2 & -4 & 2 & 6 \\ 5 & 3 & 0 & 8 \end{array}\right]$$

Conversely, given an augmented matrix (and an ordering of the variables), we can reproduce the original system of equations, as the following examples illustrate.

EXAMPLE 3   ■   **Producing a System of Equations from an Augmented Matrix**

Find the linear system of equations in the variables x and y that corresponds to the following augmented matrix. (Assume that the entries correspond to the order x, y.)

$$\left[\begin{array}{cc|c} 2 & 5 & 1 \\ 1 & 3 & 0 \end{array}\right]$$

SOLUTION

$$2x + 5y = 1$$
$$x + 3y = 0$$

EXAMPLE 4   ■   **Finding and Solving a System of Equations**

Find and solve the linear system of equations corresponding to the following augmented matrix. (Assume that the variables are $x, y,$ and $z,$ respectively.)

$$\left[\begin{array}{ccc|c} 2 & 2 & -1 & -5 \\ 0 & 1 & 4 & 13 \\ 0 & 0 & 3 & 9 \end{array}\right]$$

SOLUTION   The augmented matrix corresponds to the following system of equations.

(1) $\qquad 2x + 2y - z = -5$
(2) $\qquad y + 4z = 13$
(3) $\qquad 3z = 9$

Solving equation (3) for z we have

$$3z = 9$$
$$z = 3$$

Substituting $z = 3$ into equation (2), we obtain

$$\begin{aligned} y + 4z &= 13 \\ y + 4(3) &= 13 \\ y + 12 &= 13 \\ y &= 1 \end{aligned}$$

Now we can find x by substituting the known values of y and z into equation (1).

$$\begin{aligned} 2x + 2y - z &= -5 \\ 2x + 2(1) - 3 &= -5 \\ 2x - 1 &= -5 \\ 2x &= -4 \\ x &= -2 \end{aligned}$$

Thus, the solution is given by $x = -2$, $y = 1$, and $z = 3$.

UPPER TRIANGULAR SYSTEMS AND BACK-SUBSTITUTION

In Example 4 we saw a system of equations that was especially easy to solve. The technique we employed is called **back-substitution.** In effect, we simply solve the last equation for the last variable, and then substitute this value into the second to the last equation to solve for the second to the last variable, continuing in this fashion until we have solved for all the variables. Of course, back-substitution only works for systems of a very special form. The system must have as many equations as unknowns. Moreover, the last equation must involve only the last variable, the second to the last equation must involve only the last two variables, and so on. In other words, the coefficient matrix must have zeros below the *main diagonal,* the diagonal running from the upper left corner of the matrix to the lower right corner. Whenever the coefficient matrix is of this form, we say that the matrix is *upper triangular.* We summarize these ideas as follows.

Back-Substitution and Upper Triangular Matrices □

- The **main diagonal** of a square matrix (a matrix with the same number of rows as columns) runs from the upper left corner to the lower right corner, as shown in Figure 5.
- Square matrices with zeros below the main diagonal are called **upper triangular.** See Figure 6.
- A system of equations for which the coefficient matrix is upper triangular is said to be an **upper triangular system,** and it can be solved using **back-substitution.**

Main diagonal

$$\begin{bmatrix} a_{11} & a_{12} & a_{13} & \cdots & a_{1n} \\ a_{21} & a_{22} & a_{23} & \cdots & a_{2n} \\ a_{31} & a_{32} & a_{33} & \cdots & a_{3n} \\ \vdots & \vdots & \vdots & \ddots & \vdots \\ a_{n1} & a_{n2} & a_{n3} & \cdots & a_{nn} \end{bmatrix}$$

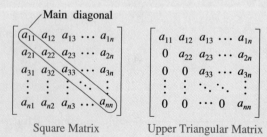

Square Matrix

FIGURE 5

$$\begin{bmatrix} a_{11} & a_{12} & a_{13} & \cdots & a_{1n} \\ 0 & a_{22} & a_{23} & \cdots & a_{2n} \\ 0 & 0 & a_{33} & \cdots & a_{3n} \\ \vdots & \vdots & \ddots & \ddots & \vdots \\ 0 & 0 & \cdots & 0 & a_{nn} \end{bmatrix}$$

Upper Triangular Matrix

FIGURE 6

EXAMPLE 5   ■   **Planning a Meal**

Table 5 gives nutritional information for a serving of lean ham, potatoes, and green beans. Suppose you wish to prepare a meal consisting of ham, potatoes, and green beans with 650 calories, 80 grams of carbohydrates, and 1200 I.U. of vitamin A. How many servings of each type of food must be used?

TABLE 5

	Ham	Potatoes	Green beans
Calories	245	145	35
Carbohydrates (in grams)	0	33	8
Vitamin A (in I.U.)	0	0	780

SOLUTION   Let x be the number of servings of ham, y the number of servings of potatoes, and z the number of servings of green beans. Then we have

$$\underset{\text{from ham}}{\text{Calories}} + \underset{\text{from potatoes}}{\text{Calories}} + \underset{\text{green beans}}{\text{Calories from}} = 650$$

$$245x \quad + \quad 145y \quad + \quad 35z \quad = 650$$

$$\underset{\text{from ham}}{\text{Carbohydrates}} + \underset{\text{from potatoes}}{\text{Carbohydrates}} + \underset{\text{from green beans}}{\text{Carbohydrates}} = 80$$

$$0x \quad + \quad 33y \quad + \quad 8z \quad = 80$$

$$\underset{\text{from ham}}{\text{Vitamin A}} + \underset{\text{from potatoes}}{\text{Vitamin A}} + \underset{\text{green beans}}{\text{Vitamin A from}} = 1200$$

$$0x \quad + \quad 0y \quad + \quad 780z \quad = 1200$$

Combining the three equations, we obtain the following system of equations.

(1) $\qquad\qquad\qquad 245x + 145y + 35z = 650$

(2) $\qquad\qquad\qquad\qquad\quad 33y + 8z = 80$

(3) $\qquad\qquad\qquad\qquad\qquad\quad 780z = 1200$

Note that this is an upper triangular system that can be solved by back-substitution. Solving equation (3) for z gives

$$780z = 1200$$

$$z \approx 1.5$$

Substituting $z = 1.5$ into equation (2) and solving for y, we have

$$33y + 8z = 80$$

$$33y + 8(1.5) = 80$$

$$33y = 68$$

$$y \approx 2.1$$

Substituting $z = 1.5$ and $y = 2.1$ into equation (1) and solving for x, we obtain

$$245x + 145y + 35z = 650$$
$$245x + 145(2.1) + 35(1.5) = 650$$
$$245x = 293$$
$$x \approx 1.2$$

Thus, the meal should be made with approximately 1.2 servings of ham, 2.1 servings of potatoes, and 1.5 servings of green beans.

UNDERSTANDING AND MASTERY CHECKLISTS

CONCEPTS TO UNDERSTAND

☐ Linear system of equations
☐ Solution possibilities for a linear system of equations
☐ Coefficient matrix of a linear system of equations
☐ Augmented matrix of a linear system of equations
☐ Back-substitution
☐ Square matrix
☐ Main diagonal
☐ Upper triangular matrix

SKILLS TO MASTER

☐ Determine whether a system of equations is linear.
☐ Determine whether a linear system has 0, 1, or infinitely many solutions.
☐ Find the augmented matrix corresponding to a linear system.
☐ Solve a system of equations using back-substitution.

EXERCISES 7.2

EXERCISES 1–6   *Indicate whether the given system of equations is linear.*

1. $2x - 3y - 2z = \dfrac{1}{10}$
$x + y + z = 2$

2. $x^2 - y = 0$
$x - y = 1$

3. $\sqrt{x} - 3y = 7$
$\phantom{\sqrt{x}}2x + 5y = 9$

4. $\dfrac{1}{x} + y = 0$
$\dfrac{3}{x} - z = 2$

5. $x - 2y = 3$
$x + 3yz = 1$

6. $\sqrt{34}y - \dfrac{1}{7}x = z$
$x + 2y - 3z = 0$

EXERCISES 7–12   *Express the given system of equations as an augmented matrix.*

7. $3x + 4y = 10$
$2x - 7y = 4$

8. $-2x + 3y = 2$
$x - 6y = 0$

9. $x = 3$
$y = 4$

10. $2x - 3y + 4z = 2$
$x + 6y + 3z = 5$
$5x - 7y + 2z = 1$

11. $x - y + 2z = 1$
$2y + 6z = 3$

12. $x + 2y - 3z = 4$
$y + 3z = 0$
$z = 6$

EXERCISES 13–18   *Determine whether the given system of equations has 0, 1, or infinitely many solutions.*

13. $2x - 3y = 4$
$4x - 6y = 7$

14. $3x + 2y = 4$
$2x + 3y = 1$

15. $x - y = 7$
$2x - 2y = 14$

16. $x + \dfrac{1}{2}y = 3$
$3x + \dfrac{3}{2}y = 7$

17. $5x - 6y = 1$
$x - y = 3$

18. $4x + 2y = 6$
$20x + 10y = 30$

EXERCISES 19–24   *Solve the given system using back-substitution.*

19. $2x + 3y = 7$
$2y = 2$

20. $3x - 6y = 4$
$\frac{1}{2}y = 3$

21. $2x + y - z = 12$
$y + z = 8$
$3z = 3$

22. $2a + 3b + 4c = 2$
$2b - c = 5$
$2c = -2$

23. $q - 2r + 3s + t = 3$
$4r + s - t = 18$
$2s + 2t = 4$
$3t = -6$

24. $a + b + c + d = 2$
$b + c + d = 2$
$c + d = 1$
$d = 1$

EXERCISES 25–26   *An upper triangular augmented matrix is given. Find the solution to the corresponding system using back-substitution. Assume that the variable names are x and y, in that order.*

25. $\begin{bmatrix} 2 & -3 & | & -5 \\ 0 & 4 & | & 12 \end{bmatrix}$

26. $\begin{bmatrix} -1 & -2 & | & -11 \\ 0 & \frac{1}{3} & | & 3 \end{bmatrix}$

EXERCISES 27–30   *An upper triangular augmented matrix is given. Find the solution to the corresponding system using back-substitution. Assume that the variable names are x, y, and z, in that order.*

27. $\begin{bmatrix} 1 & 3 & 6 & | & 0 \\ 0 & 1 & 2 & | & 2 \\ 0 & 0 & 1 & | & -4 \end{bmatrix}$

28. $\begin{bmatrix} 1 & 2 & -3 & | & 1 \\ 0 & 1 & 3 & | & 5 \\ 0 & 0 & 1 & | & 0 \end{bmatrix}$

29. $\begin{bmatrix} 3 & -2 & 4 & | & 3 \\ 0 & 1 & -3 & | & 1 \\ 0 & 0 & 2 & | & -2 \end{bmatrix}$

30. $\begin{bmatrix} 4 & -6 & 5 & | & 4 \\ 0 & 2 & -3 & | & 0 \\ 0 & 0 & 3 & | & 3 \end{bmatrix}$

APPLICATIONS

31. *Income Tax*   A family has income totaling \$58,000 from three different sources: salary, payments from municipal bonds, and payments from U.S. Treasury bonds. The municipal and U.S. Treasury bonds are free from state income tax, whereas salary is subject to a 6% state income tax. The U.S. Treasury bonds are free from federal tax, but both the salary and the income from the municipal bonds are subject to 20% federal tax. If the family's state tax totals \$2400 and their federal tax totals \$10,000, compute the income from each of the three sources.

32. *Sports Drink*   An athlete wishes to concoct a 165-calorie 16-ounce sports beverage by mixing together orange juice (which has 13.75 calories per ounce) and water. How many ounces of each ingredient should be used?

33. *Power Breakfast*   A breakfast consisting of milk, bananas, and whole wheat toast is to have 1200 units of vitamin A, 1.5 units of vitamin B_6, and 1.35 units of vitamin B_{12}. Table 6 shows the number of units of vitamins A, B_6, and B_{12} that are provided by a single serving of each of the three breakfast items. How many servings of each breakfast item should be consumed?

TABLE 6

	Milk	Bananas	Whole wheat toast
Vitamin A	500	226	0
Vitamin B_6	0.1	0.6	0.04
Vitamin B_{12}	0.9	0	0

34. *Cross Training*   A 170-pound athlete cross-trains during the off season by cycling (at 10 miles per hour), playing volleyball, and weightlifting. He determines that he needs to burn off an extra 6000 calories per week through exercise, and that he needs to ride 80 miles per week. Because of his full-time job, he will be unable to devote more than 20 hours per week to exercise. His trainer tells him that he will burn approximately 552 calories per hour while cycling at 10 miles per hour and 234 calories per hour while playing volleyball, whereas the calories burned through weightlifting are negligible. Determine the number of hours that the athlete should spend in each of the three activities.

CONCEPTS AND CRITICAL THINKING

EXERCISES 35–38   *Answer true or false.*

35. A system of equations is said to be linear if at least one of the equations is linear.

36. A system of equations is said to be nonlinear if at least one of the equations is nonlinear.

37. The augmented matrix $\begin{bmatrix} 1 & 0 & 0 & | & 3 \\ 2 & 1 & 0 & | & 2 \\ 1 & 1 & 0 & | & 3 \end{bmatrix}$ corresponds to the system

$$x = 3z$$
$$2x + y = 2z$$
$$x + y = 3z$$

38. Some linear systems have exactly two solutions.

EXERCISES 39–43   *Give an example of each.*

39. A nonlinear system of equations

40. A system of 3 equations in 3 unknowns such that the corresponding augmented matrix is upper triangular

41. An equation that, when paired with $2x + 3y = 5$, forms a system of equations with infinitely many solutions

42. An equation that, when paired with $2x + 3y = 5$, forms a system of equations with no solutions

43. An equation that, when paired with $2x + 3y = 5$, forms a system of equations with exactly one solution

44. Suppose that you were asked to write an exam for this class and you wished to produce a linear system of three equations with infinitely many solutions. If the first two equations are $2x + 3y - z = 5$ and $x + 2y + z = 3$, what might the third equation be? What if you wanted a system with no solution?

▓ QUESTIONS FOR DISCUSSION OR ESSAY

45. It turns out that the graph of a linear equation in three variables is a plane in three-dimensional space. Explain why three planes will meet in 0, 1, or infinitely many points. Give an example from the room you are in now that suggests three planes meeting in a single point; also describe a situation in which three planes meet in a line. Sketch both.

46. Give an example of a system of two equations in two unknowns that has exactly two solutions. Explain why this doesn't contradict the statement that the number of solutions to a linear system of equations is 0, 1, or infinity.

47. Suppose you are given a system of three linear equations and the equations are given in no particular order. In how many different ways can the system be represented using augmented matrices? Without checking all of these different matrix representations one by one, how can you tell just by looking at the equations whether one of the matrix representations will be upper triangular?

▓ PROJECTS FOR ENRICHMENT

48. *Linear Systems with More Than One Solution* In the text, it was stated that the solution set to a linear system of equations has 0, 1, or infinitely many solutions. In this project, we prove this statement by employing matrix notation.

a. Suppose we are given a linear system of m equations in the n variables $x_1, x_2, x_3, \ldots, x_n$:

$$a_{11}x_1 + a_{12}x_2 + a_{13}x_3 + \cdots + a_{1n}x_n = b_1$$
$$a_{21}x_1 + a_{22}x_2 + a_{23}x_3 + \cdots + a_{2n}x_n = b_2$$
$$a_{31}x_1 + a_{32}x_2 + a_{33}x_3 + \cdots + a_{3n}x_n = b_3$$
$$\vdots \qquad \vdots \qquad \vdots \qquad \qquad \vdots$$
$$a_{m1}x_1 + a_{m2}x_2 + a_{m3}x_3 + \cdots + a_{mn}x_n = b_m$$

i. Show that the system is equivalent to the matrix equation $AX = B$, where

$$A = \begin{bmatrix} a_{11} & a_{12} & a_{13} & \cdots & a_{1n} \\ a_{21} & a_{22} & a_{23} & \cdots & a_{2n} \\ a_{31} & a_{32} & a_{33} & \cdots & a_{3n} \\ \vdots & \vdots & \vdots & & \vdots \\ a_{m1} & a_{m2} & a_{m3} & \cdots & a_{mn} \end{bmatrix},$$

$$B = \begin{bmatrix} b_1 \\ b_2 \\ b_3 \\ \vdots \\ b_m \end{bmatrix}, \quad \text{and} \quad X = \begin{bmatrix} x_1 \\ x_2 \\ x_3 \\ \vdots \\ x_n \end{bmatrix}$$

ii. Use a property of matrices to show that if Y and Z are both solutions to $AX = B$, then $Y - Z$ is a solution to $AX = 0$.

iii. Now show that if Y and Z are solutions to $AX = B$, then so is $Y + t(Y - Z)$, where t is any real number.

b. For the given system of equations, two solutions are provided. Use the results from part a to give three more solutions.

i. $2x + 3y - z = 12$
$\quad x - y + z = 0$
$\quad 3x + 2y = 12$

Solution 1: $x = 4, y = 0, z = -4$
Solution 2: $x = 2, y = 3, z = 1$

Hint: In matrix form these solutions correspond to

$$Y = \begin{bmatrix} 4 \\ 0 \\ -4 \end{bmatrix} \quad \text{and} \quad Z = \begin{bmatrix} 2 \\ 3 \\ 1 \end{bmatrix}$$

ii. $p + q + 3r = 3$
$\quad 2q + 4r = 2$

Solution 1: $p = 2, q = 1, r = 0$
Solution 2: $p = 1, q = -1, r = 1$

c. **i.** Use the result of part a to show that if a linear system of equations has two solutions, then it has infinitely many solutions.

ii. Use the result of part c-i to explain why a linear system of equations has 0, 1, or infinitely many solutions.

49. *Interpolating with Polynomials*   It can be shown that given any collection of n points, there is a polynomial of degree $n - 1$ or lower that passes through the points. Suppose we wish to find the equation of the quadratic polynomial passing through three points, say $(1, 4)$, $(2, 6)$, and $(4, 22)$. Since the polynomial is quadratic, it is of the form $p(x) = ax^2 + bx + c$, and we have $p(1) = 4$, $p(2) = 6$, and $p(4) = 22$. From the fact that $p(1) = 4$, we have

$$p(1) = a \cdot 1^2 + b \cdot 1 + c = 4$$

which gives us the equation

$$a + b + c = 4$$

In similar fashion, from $p(2) = 6$ and $p(4) = 22$, we obtain the equations

$$4a + 2b + c = 6$$
$$16a + 4b + c = 22$$

thus giving us the linear system

$$a + b + c = 4$$
$$4a + 2b + c = 6$$
$$16a + 4b + c = 22$$

This system could be solved using substitution, elimination, or any of the other techniques that are described in this chapter. In this project, we discover an alternative way to find a polynomial passing through a given set of points that involves solving a system of equations by back-substitution.

a. We find the equation of the quadratic polynomial passing through the points $(1, 4)$, $(2, 6)$, and $(4, 22)$. We begin by assuming that the polynomial is of the form

$$p(x) = a + b(x - 1) + c(x - 1)(x - 2)$$

Note that the red and blue numbers in the polynomial are two of the x-coordinates of the given points.

   i. Set up a linear system involving the three variables a, b, and c.

   ii. Solve the system obtained in part i by back-substitution.

   iii. Now multiply out all products appearing in the expression for $p(x)$ in order to write $p(x)$ in standard form.

b. Use the technique developed in part a to find the quadratic polynomial passing through the points (x_1, y_1), (x_2, y_2), and (x_3, y_3).

c. Extend the approach developed in part a to find the cubic polynomial passing through

   i. $(-2, 25)$, $(0, 1)$, $(1, 5)$, and $(2, 19)$

   ii. $(0, 0)$, $(1, 0)$, $(4, 12)$, and $(2, -2)$

d. The U.S. Census Bureau has recorded the following data for the population of Arizona:

1960: 1,302,161
1970: 1,775,399
1980: 2,716,546
1990: 3,665,228

Let t represent the year (with $t = 0$ corresponding to 1960) and P the population of Arizona. Use the approach developed in part c to express P as a cubic polynomial in the variable t. According to this model, when will the population of Arizona reach 40 million? What are the limitations of this model?

SECTION 7.3 **GAUSSIAN AND GAUSS-JORDAN ELIMINATION**

> ☐ **What is the sum of the squares of the first 10,000 natural numbers: $1^2 + 2^2 + 3^2 + \cdots + 10,000^2$?**
>
> ☐ **A baseball is hit so it is at a height of 9 feet when it passes over the pitcher and clears the 10-foot center field wall by 4 feet. Is the ball rising or falling as it leaves the field?**
>
> ☐ **How can a system of equations be solved without ever writing down a single variable?**

GAUSSIAN ELIMINATION

In Section 2.8, we presented two methods for solving systems of equations: substitution and elimination. Substitution works well when the number of equations is small or if the equations have an especially simple form. Indeed, we saw in the previous section that if the coefficient matrix of a system of equations is upper triangular (all zeros below the main diagonal), then the system can be solved easily using back-substitution. In this section, we see how systematic elimination can be used to convert augmented matrices to a form that is closely related to upper triangular—namely, **row-echelon form.**

Row-Echelon Form   ☐

> A matrix is said to be in row-echelon form if the following conditions are satisfied.
>
> ■ The first nonzero entry of each row is a 1. It is called the **leading 1.**
> ■ Below each leading 1 are only 0s.
> ■ Each leading 1 is to the right of the leading 1 in the row above.
> ■ Any rows consisting entirely of 0s are at the bottom of the matrix.

EXAMPLE 1   ▦   Identifying Row-Echelon Form

Indicate whether the given matrix is in row-echelon form.

a. $\begin{bmatrix} 1 & 4 & 6 \\ 0 & 1 & 1 \\ 0 & 0 & 1 \end{bmatrix}$

b. $\begin{bmatrix} 1 & 2 & 3 \\ 0 & 1 & 5 \\ 0 & 6 & 2 \end{bmatrix}$

c. $\left[\begin{array}{cccc|c} 1 & 0 & 3 & 4 & 5 \\ 0 & 2 & 0 & 1 & 4 \end{array}\right]$

d. $\left[\begin{array}{ccc|c} 1 & 0 & 2 & 5 \\ 0 & 1 & 4 & 2 \\ 0 & 0 & 0 & 0 \end{array}\right]$

SOLUTION

a. This matrix is in row-echelon form.
b. This matrix is not in row-echelon form because there is a nonzero number (6) in the third row and second column, directly below the leading 1 of the second row. Also, the first nonzero entry of the third row is 6 and not 1.
c. This augmented matrix is not in row-echelon form because the first nonzero entry of the second row is not 1.
d. This augmented matrix is in row-echelon form. Note that there is one row consisting entirely of 0s, and it is the last row.

Systems of equations represented by augmented matrices in row-echelon form can often be solved quite easily using back-substitution. This is especially true for upper triangular systems, as we see in the following example.

EXAMPLE 2   ▦   Solving a System Corresponding to a Matrix in Row-Echelon Form

Solve the system of equations corresponding to the following augmented matrix in row-echelon form. (Assume the variables are x, y, and z, in that order.)

$$\left[\begin{array}{ccc|c} 1 & 3 & 2 & 11 \\ 0 & 1 & 4 & 6 \\ 0 & 0 & 1 & 1 \end{array}\right]$$

SOLUTION   Expressing the augmented matrix as a system of equations, we have

(1) $$x + 3y + 2z = 11$$
(2) $$y + 4z = 6$$
(3) $$z = 1$$

Using back-substitution, we substitute $z = 1$ into equation (2).

$$y + 4z = 6$$
$$y + 4(1) = 6$$
$$y = 2$$

Now that y and z have been determined, we can find x by substituting their values into equation (1).

$$x + 3y + 2z = 11$$
$$x + 3(2) + 2(1) = 11$$
$$x = 3$$

Thus, the solution is given by $x = 3$, $y = 2$, and $z = 1$.

In Example 2, we were fortunate to have an augmented matrix that was already in row-echelon form. If the augmented matrix is *not* in row-echelon form, we can convert it to row-echelon form using a systematic procedure known as **Gaussian elimination.** At each stage of Gaussian elimination, one of three **elementary row operations** is performed on the augmented matrix. These operations are interchanging two rows of the matrix, multiplying a row by a constant, and adding a multiple of one row to another. It can be shown that performing an elementary row operation on an augmented matrix results in an equivalent system of equations. The elementary row operations, their notation, and their effect on the underlying system are summarized as follows.

Elementary Row Operations □

Operation	Notation	Effect on system
Interchanging rows i and j	R_{ij}	Equivalent. (Two equations change places.)
Multiplying row i by c, where $c \neq 0$	cR_i	Equivalent. (An equation is multiplied by a nonzero constant.)
Adding c times row i to row j	$cR_i + R_j$	Equivalent. (A multiple of an equation is added to another.)

Note that the operation $cR_i + R_j$ changes the jth row, *but not the ith row*. In other words, the jth row is replaced by $cR_i + R_j$, but the ith row is left alone. Also, elementary row operations may be performed on any matrix, augmented or not.

EXAMPLE 3 ■ Performing Elementary Row Operations

For each matrix, perform the indicated elementary row operation.

a. $\begin{bmatrix} 1 & 3 \\ 2 & 5 \end{bmatrix}$, $-2R_1 + R_2$ **b.** $\begin{bmatrix} 3 & 6 & 5 & | & 3 \\ 4 & 2 & 1 & | & 7 \end{bmatrix}$, $-\frac{1}{3}R_1$

c. $\begin{bmatrix} 0 & 1 & 2 & | & 3 \\ 3 & 6 & 1 & | & 5 \\ 1 & 2 & 0 & | & 4 \end{bmatrix}$, R_{13}

SOLUTION

a. Multiplying the first row by -2 and adding to the second row gives us

$$\begin{bmatrix} 1 & 3 \\ 2 + (-2)1 & 5 + (-2)3 \end{bmatrix} = \begin{bmatrix} 1 & 3 \\ 0 & -1 \end{bmatrix}$$

b. Multiplying the first row by $-\frac{1}{3}$ gives us

$$\begin{bmatrix} (-\frac{1}{3})3 & (-\frac{1}{3})6 & (-\frac{1}{3})5 & | & (-\frac{1}{3})3 \\ 4 & 2 & 1 & | & 7 \end{bmatrix} = \begin{bmatrix} -1 & -2 & -\frac{5}{3} & | & -1 \\ 4 & 2 & 1 & | & 7 \end{bmatrix}$$

c. Exchanging rows 1 and 3 gives us

$$\begin{bmatrix} 1 & 2 & 0 & | & 4 \\ 3 & 6 & 1 & | & 5 \\ 0 & 1 & 2 & | & 3 \end{bmatrix}$$

Gaussian elimination proceeds column by column from left to right. In each column, a leading 1 is obtained in the diagonal position, and then the entries under the leading 1 are cleared out, one by one, using elementary row operations. The following example illustrates Gaussian elimination.

EXAMPLE 4 ■ Solving a System of Equations by Gaussian Elimination

Solve the following system of equations using Gaussian elimination.

$$2y - 3z = 12$$
$$2x + 7y - 11z = 27$$
$$x + 2y - z = 0$$

SOLUTION   We begin by writing the system as an augmented matrix.

$$\begin{bmatrix} 0 & 2 & -3 & | & 12 \\ 2 & 7 & -11 & | & 27 \\ 1 & 2 & -1 & | & 0 \end{bmatrix}$$

Next we perform Gaussian elimination to convert to row-echelon form, as shown in Table 7. For emphasis, we shade the rows at each stage that are modified.

TABLE 7

Beginning matrix	Operation	Explanation	Resulting matrix
$\begin{bmatrix} 0 & 2 & -3 & \vert & 12 \\ 2 & 7 & -11 & \vert & 27 \\ 1 & 2 & -1 & \vert & 0 \end{bmatrix}$	R_{13}	Interchange rows 1 and 3 to obtain a leading 1 in row 1, column 1.	$\begin{bmatrix} 1 & 2 & -1 & \vert & 0 \\ 2 & 7 & -11 & \vert & 27 \\ 0 & 2 & -3 & \vert & 12 \end{bmatrix}$
$\begin{bmatrix} 1 & 2 & -1 & \vert & 0 \\ 2 & 7 & -11 & \vert & 27 \\ 0 & 2 & -3 & \vert & 12 \end{bmatrix}$	$-2R_1 + R_2$	Add -2 times the first row to the second row to obtain a 0 in row 2, column 1.	$\begin{bmatrix} 1 & 2 & -1 & \vert & 0 \\ 0 & 3 & -9 & \vert & 27 \\ 0 & 2 & -3 & \vert & 12 \end{bmatrix}$

The first column is now completed; we begin on the second.

Beginning matrix	Operation	Explanation	Resulting matrix
$\begin{bmatrix} 1 & 2 & -1 & \vert & 0 \\ 0 & 3 & -9 & \vert & 27 \\ 0 & 2 & -3 & \vert & 12 \end{bmatrix}$	$\frac{1}{3}R_2$	Multiply row 2 by $\frac{1}{3}$ in order to obtain a leading 1 in row 2, column 2.	$\begin{bmatrix} 1 & 2 & -1 & \vert & 0 \\ 0 & 1 & -3 & \vert & 9 \\ 0 & 2 & -3 & \vert & 12 \end{bmatrix}$
$\begin{bmatrix} 1 & 2 & -1 & \vert & 0 \\ 0 & 1 & -3 & \vert & 9 \\ 0 & 2 & -3 & \vert & 12 \end{bmatrix}$	$-2R_2 + R_3$	Add -2 times the second row to the third row to obtain a 0 in row 3, column 2.	$\begin{bmatrix} 1 & 2 & -1 & \vert & 0 \\ 0 & 1 & -3 & \vert & 9 \\ 0 & 0 & 3 & \vert & -6 \end{bmatrix}$

The second column is now completed; on to the third.

Beginning matrix	Operation	Explanation	Resulting matrix
$\begin{bmatrix} 1 & 2 & -1 & \vert & 0 \\ 0 & 1 & -3 & \vert & 9 \\ 0 & 0 & 3 & \vert & -6 \end{bmatrix}$	$\frac{1}{3}R_3$	Multiply the third row by $\frac{1}{3}$ in order to obtain a leading 1 in row 3, column 3.	$\begin{bmatrix} 1 & 2 & -1 & \vert & 0 \\ 0 & 1 & -3 & \vert & 9 \\ 0 & 0 & 1 & \vert & -2 \end{bmatrix}$

The matrix is now in row-echelon form. Translating back into equation form, we have

(1) $$x + 2y - z = 0$$
(2) $$y - 3z = 9$$
(3) $$z = -2$$

We now use back-substitution with $z = -2$ to find y.

$$y - 3z = 9$$
$$y - 3(-2) = 9$$
$$y = 3$$

Finally, we use back-substitution with $z = -2$ and $y = 3$ to find x.

$$x + 2y - z = 0$$
$$x + 2(3) - (-2) = 0$$
$$x = -8$$

Thus, our solution is given by $x = -8$, $y = 3$, $z = -2$.

> **WARNING!** Although the process of Gaussian elimination always yields a matrix in row-echelon form, this matrix is not unique. The final row-echelon matrix depends on choices made during Gaussian elimination. For instance, in Example 4, if we were to swap rows 1 and 2 instead of rows 1 and 3, we would obtain
>
> $$\begin{bmatrix} 1 & \frac{7}{2} & -\frac{11}{2} & \bigm| & \frac{27}{2} \\ 0 & 1 & -\frac{3}{2} & \bigm| & 6 \\ 0 & 0 & 1 & \bigm| & -2 \end{bmatrix}$$

Recall from the previous section that a linear system of equations will have 0, 1, or infinitely many solutions. Thus far, we have used Gaussian elimination to solve systems of equations with exactly one solution, but it is also effective for expressing the solution sets of systems with infinitely many solutions and for exposing systems with no solution, as the following examples illustrate.

EXAMPLE 5 ■ **Solving a Linear System with Infinitely Many Solutions**

Describe the solution set of the system

$$x + 3y - 5z = 2$$
$$x - y + z = 4$$
$$x + y - 2z = 3$$

SOLUTION We first express the system as an augmented matrix and then use Gaussian elimination to convert the matrix to row-echelon form. In moving from one matrix to the next, we shade the row that is modified and indicate the elementary row operation that is used.

$$\begin{bmatrix} 1 & 3 & -5 & \bigm| & 2 \\ 1 & -1 & 1 & \bigm| & 4 \\ 1 & 1 & -2 & \bigm| & 3 \end{bmatrix} \xrightarrow{-R_1 + R_2} \begin{bmatrix} 1 & 3 & -5 & \bigm| & 2 \\ 0 & -4 & 6 & \bigm| & 2 \\ 1 & 1 & -2 & \bigm| & 3 \end{bmatrix}$$

$$\begin{bmatrix} 1 & 3 & -5 & \bigm| & 2 \\ 0 & -4 & 6 & \bigm| & 2 \\ 1 & 1 & -2 & \bigm| & 3 \end{bmatrix} \xrightarrow{-R_1 + R_3} \begin{bmatrix} 1 & 3 & -5 & \bigm| & 2 \\ 0 & -4 & 6 & \bigm| & 2 \\ 0 & -2 & 3 & \bigm| & 1 \end{bmatrix}$$

$$\begin{bmatrix} 1 & 3 & -5 & \bigm| & 2 \\ 0 & -4 & 6 & \bigm| & 2 \\ 0 & -2 & 3 & \bigm| & 1 \end{bmatrix} \xrightarrow{-\frac{1}{4}R_2} \begin{bmatrix} 1 & 3 & -5 & \bigm| & 2 \\ 0 & 1 & -\frac{3}{2} & \bigm| & -\frac{1}{2} \\ 0 & -2 & 3 & \bigm| & 1 \end{bmatrix}$$

$$\begin{bmatrix} 1 & 3 & -5 & \bigm| & 2 \\ 0 & 1 & -\frac{3}{2} & \bigm| & -\frac{1}{2} \\ 0 & -2 & 3 & \bigm| & 1 \end{bmatrix} \xrightarrow{2R_2 + R_3} \begin{bmatrix} 1 & 3 & -5 & \bigm| & 2 \\ 0 & 1 & -\frac{3}{2} & \bigm| & -\frac{1}{2} \\ 0 & 0 & 0 & \bigm| & 0 \end{bmatrix}$$

Converting to equations, we have

(1)
$$x + 3y - 5z = 2$$

(2)
$$y - \frac{3}{2}z = -\frac{1}{2}$$

(3)
$$0 = 0$$

Note that equation (3) gives us no further restrictions on x, y, or z. We use equations (1) and (2) to express x and y in terms of z. Solving equation (2) for y gives us

$$y - \frac{3}{2}z = -\frac{1}{2}$$

$$y = \frac{3}{2}z - \frac{1}{2}$$

Substituting $y = \frac{3}{2}z - \frac{1}{2}$ into equation (1) and solving for x, we obtain

$$x + 3\left(\frac{3}{2}z - \frac{1}{2}\right) - 5z = 2$$

$$x - \frac{1}{2}z - \frac{3}{2} = 2$$

$$x = \frac{1}{2}z + \frac{7}{2}$$

Thus, the solution set consists of triples of numbers (x, y, z), where $x = \frac{1}{2}z + \frac{7}{2}$, $y = \frac{3}{2}z - \frac{1}{2}$, and z is arbitrary.

EXAMPLE 6 ■ **A Linear System with No Solution**

Solve the linear system of equations

$$a - b + 3c - 2d = 1$$
$$-2a + 2b - 6c + 4d = 4$$

SOLUTION   Using Gaussian elimination, we have

$$\begin{bmatrix} 1 & -1 & 3 & -2 & | & 1 \\ -2 & 2 & -6 & 4 & | & 4 \end{bmatrix} \xrightarrow{2R_1 + R_2} \begin{bmatrix} 1 & -1 & 3 & -2 & | & 1 \\ 0 & 0 & 0 & 0 & | & 6 \end{bmatrix}$$

The second row of this matrix corresponds to the equation $0 = 6$, which is a contradiction. Thus, the system has no solution.

Once an augmented matrix has been converted to row-echelon form using Gaussian elimination, the corresponding system of equations can easily be characterized as consistent or inconsistent, as described by the following rule of thumb.

> **RULE OF THUMB**   Suppose that a system has been expressed as a matrix in row-echelon form. The system is inconsistent (has no solution) if there is a row with all zeros to the left of the partition and a nonzero number to the right. Otherwise, the system is consistent.

GAUSS-JORDAN ELIMINATION AND REDUCED ROW-ECHELON FORM

We have seen that a system of equations can be solved by first converting the augmented matrix to row-echelon form using Gaussian elimination and then converting back to equation form and solving using back-substitution. It is possible, however, to avoid reverting to equations altogether by performing additional matrix operations that mirror the process of back-substitution, thus producing an especially simple augmented matrix from which solutions are simply "read off." Such matrices are said to be in reduced row-echelon form.

Reduced Row-Echelon Form ☐

A matrix is said to be in **reduced row-echelon form** if the following two conditions are met.

■ The matrix is in row-echelon form.
■ There are only 0s directly above each leading 1.

E X A M P L E 7 ▨ Identifying Matrices in Reduced Row-Echelon Form

Determine whether each of the following matrices is in reduced row-echelon form.

$$\textbf{a.} \begin{bmatrix} 1 & 0 & 0 & | & 3 \\ 0 & 1 & 0 & | & 2 \\ 0 & 0 & 1 & | & 5 \end{bmatrix} \qquad \textbf{b.} \begin{bmatrix} 1 & 0 & 0 & | & 1 \\ 0 & 1 & 2 & | & 6 \\ 0 & 0 & 1 & | & 2 \end{bmatrix} \qquad \textbf{c.} \begin{bmatrix} 6 & 0 & 0 & | & 3 \\ 0 & 1 & 0 & | & 5 \\ 0 & 0 & 1 & | & 1 \end{bmatrix}$$

SOLUTION
a. This matrix is in reduced row-echelon form.
b. This matrix is in row-echelon form but not *reduced* row-echelon form. The culprit is the 2 in the second row and third column: It would need to be 0 for the matrix to be in reduced row-echelon form.
c. This matrix is not even in row-echelon form, let alone reduced row-echelon form. The culprit is the 6 in the first row and first column. If it were a 1, then this matrix would be in reduced row-echelon form.

It is easy to see why reduced row-echelon form is desirable. For example, suppose that a system of equations is represented by the following augmented matrix, which is in reduced row-echelon form (with columns corresponding to the variables x, y, and z, in that order).

$$\begin{bmatrix} 1 & 0 & 0 & | & 7 \\ 0 & 1 & 0 & | & -2 \\ 0 & 0 & 1 & | & 3 \end{bmatrix}$$

Converting this augmented matrix into equation form gives

$$x = 7$$
$$y = -2$$
$$z = 3$$

The system is solved, and no back-substitution is required!

Gauss-Jordan elimination is a systematic procedure for converting a matrix into reduced row-echelon form using elementary row operations. First, Gaussian elimination is used to produce a matrix in row-echelon form. Next, elementary row operations are used to eliminate the nonzero entries *above* each leading 1, in much the same way that nonzero entries below leading 1s are eliminated when converting the matrix to row-echelon form. It can be shown that the resulting matrix is unique: No matter what decisions are made during the course of Gauss-Jordan elimination, the final reduced row-echelon matrix will be the same.

EXAMPLE 8 ▨ **Solving a System of Equations Using Gauss-Jordan Elimination**

Solve the following system using Gauss-Jordan elimination.

$$2x - 4y - 10z = -12$$
$$2x - 2y - 2z = 2$$
$$-3x + 4y + 11z = 8$$

SOLUTION We first form the augmented matrix, and then begin applying elementary row operations to produce a reduced row-echelon matrix.

$$\begin{bmatrix} 2 & -4 & -10 & | & -12 \\ 2 & -2 & -2 & | & 2 \\ -3 & 4 & 11 & | & 8 \end{bmatrix} \xrightarrow{\frac{1}{2}R_1} \begin{bmatrix} 1 & -2 & -5 & | & -6 \\ 2 & -2 & -2 & | & 2 \\ -3 & 4 & 11 & | & 8 \end{bmatrix}$$

$$\begin{bmatrix} 1 & -2 & -5 & | & -6 \\ 2 & -2 & -2 & | & 2 \\ -3 & 4 & 11 & | & 8 \end{bmatrix} \xrightarrow{-2R_1 + R_2} \begin{bmatrix} 1 & -2 & -5 & | & -6 \\ 0 & 2 & 8 & | & 14 \\ -3 & 4 & 11 & | & 8 \end{bmatrix}$$

$$\begin{bmatrix} 1 & -2 & -5 & | & -6 \\ 0 & 2 & 8 & | & 14 \\ -3 & 4 & 11 & | & 8 \end{bmatrix} \xrightarrow{3R_1 + R_3} \begin{bmatrix} 1 & -2 & -5 & | & -6 \\ 0 & 2 & 8 & | & 14 \\ 0 & -2 & -4 & | & -10 \end{bmatrix}$$

$$\begin{bmatrix} 1 & -2 & -5 & | & -6 \\ 0 & 2 & 8 & | & 14 \\ 0 & -2 & -4 & | & -10 \end{bmatrix} \xrightarrow{\frac{1}{2}R_2} \begin{bmatrix} 1 & -2 & -5 & | & -6 \\ 0 & 1 & 4 & | & 7 \\ 0 & -2 & -4 & | & -10 \end{bmatrix}$$

$$\begin{bmatrix} 1 & -2 & -5 & | & -6 \\ 0 & 1 & 4 & | & 7 \\ 0 & -2 & -4 & | & -10 \end{bmatrix} \xrightarrow{2R_2 + R_3} \begin{bmatrix} 1 & -2 & -5 & | & -6 \\ 0 & 1 & 4 & | & 7 \\ 0 & 0 & 4 & | & 4 \end{bmatrix}$$

$$\begin{bmatrix} 1 & -2 & -5 & | & -6 \\ 0 & 1 & 4 & | & 7 \\ 0 & 0 & 4 & | & 4 \end{bmatrix} \xrightarrow{\frac{1}{4}R_3} \begin{bmatrix} 1 & -2 & -5 & | & -6 \\ 0 & 1 & 4 & | & 7 \\ 0 & 0 & 1 & | & 1 \end{bmatrix}$$

$$\begin{bmatrix} 1 & -2 & -5 & | & -6 \\ 0 & 1 & 4 & | & 7 \\ 0 & 0 & 1 & | & 1 \end{bmatrix} \xrightarrow{2R_2 + R_1} \begin{bmatrix} 1 & 0 & 3 & | & 8 \\ 0 & 1 & 4 & | & 7 \\ 0 & 0 & 1 & | & 1 \end{bmatrix}$$

$$\begin{bmatrix} 1 & 0 & 3 & | & 8 \\ 0 & 1 & 4 & | & 7 \\ 0 & 0 & 1 & | & 1 \end{bmatrix} \xrightarrow{-3R_3 + R_1} \begin{bmatrix} 1 & 0 & 0 & | & 5 \\ 0 & 1 & 4 & | & 7 \\ 0 & 0 & 1 & | & 1 \end{bmatrix}$$

$$\begin{bmatrix} 1 & 0 & 0 & | & 5 \\ 0 & 1 & 4 & | & 7 \\ 0 & 0 & 1 & | & 1 \end{bmatrix} \xrightarrow{-4R_3 + R_2} \begin{bmatrix} 1 & 0 & 0 & | & 5 \\ 0 & 1 & 0 & | & 3 \\ 0 & 0 & 1 & | & 1 \end{bmatrix}$$

It follows that $x = 5$, $y = 3$, and $z = 1$.

As noted earlier, one of the main benefits of Gauss-Jordan elimination is that it does not require back-substitution. However, if we were to judge solely on the basis of efficiency, Gaussian elimination with back-substitution would almost always come out ahead of Gauss-Jordan elimination. For most systems of equations, the total number of arithmetic operations required by Gauss-Jordan elimination will be much larger than the number required by the combination of Gaussian elimination with back-substitution. (See Exercise 75.)

UNDERSTANDING AND MASTERY CHECKLISTS

CONCEPTS TO UNDERSTAND

- ☐ Row-echelon form of a matrix
- ☐ Elementary row operation
- ☐ Gaussian elimination
- ☐ Reduced row-echelon or reduced row-echelon form
- ☐ Gauss-Jordan elimination

SKILLS TO MASTER

- ☐ Recognize when a matrix is in row-echelon or reduced row-echelon form.
- ☐ Apply an elementary row operation to a matrix.
- ☐ Use Gaussian elimination to find the row-echelon form of a matrix.
- ☐ Use Gauss-Jordan elimination to find the reduced row-echelon form of a matrix.
- ☐ Solve a linear system of equations by applying either Gaussian or Gauss-Jordan elimination.

EXERCISES 7.3

EXERCISES 1–6 *Determine whether the given matrix is in row-echelon form. If not, support your answer.*

1. $\begin{bmatrix} 1 & 3 \\ 0 & 1 \end{bmatrix}$

2. $\begin{bmatrix} 1 & 4 & 0 \\ 0 & 2 & 1 \end{bmatrix}$

3. $\begin{bmatrix} 1 & 2 & 3 \\ 0 & 1 & 2 \\ 1 & 0 & 3 \end{bmatrix}$

4. $\left[\begin{array}{ccc|c} 1 & 0 & 2 & 2 \\ 0 & 1 & 1 & 3 \\ 0 & 0 & 1 & 4 \end{array}\right]$

5. $\left[\begin{array}{cccc|c} 1 & 2 & 0 & -1 & 0 \\ 0 & 1 & -3 & 1 & 1 \\ 0 & 1 & 2 & 0 & 0 \end{array}\right]$

6. $\left[\begin{array}{ccc|c} 1 & 0 & 2 & 0 \\ 0 & 0 & 0 & 0 \\ 0 & 1 & 4 & 0 \end{array}\right]$

EXERCISES 7–10 *Perform the indicated row operation.*

7. $\begin{bmatrix} 2 & 4 \\ 3 & 0 \end{bmatrix}$; $\frac{1}{2}R_1$

8. $\left[\begin{array}{ccc|c} 1 & 0 & 3 & 1 \\ 0 & 0 & 2 & -1 \\ 0 & 1 & 6 & 3 \end{array}\right]$; R_{23}

9. $\left[\begin{array}{cc|c} 1 & 4 & 6 \\ 2 & -1 & 3 \end{array}\right]$; $-2R_1 + R_2$

10. $\left[\begin{array}{ccc|c} 1 & 3 & 7 & 1 \\ 0 & 1 & 2 & 3 \\ 0 & 0 & 1 & -1 \end{array}\right]$; $-3R_2 + R_1$

EXERCISES 11–22 *Use Gaussian elimination to write the augmented matrix in row-echelon form. (There may be more than one correct answer.)*

11. $\left[\begin{array}{cc|c} 4 & 8 & 12 \\ 3 & 9 & 21 \end{array}\right]$

12. $\left[\begin{array}{cc|c} \frac{2}{3} & -\frac{2}{3} & \frac{4}{3} \\ -4 & 2 & -18 \end{array}\right]$

13. $\left[\begin{array}{cc|c} 0 & 2 & -4 & 8 \\ 3 & 0 & 9 & 12 \end{array}\right]$

14. $\left[\begin{array}{ccc|c} 2 & 0 & 1 & 0 \\ 2 & 1 & 4 & 1 \\ 3 & 1 & 8 & 1 \end{array}\right]$

15. $\left[\begin{array}{ccc|c} 2 & 6 & 8 & -2 \\ 2 & 4 & 12 & 5 \\ 1 & 7 & -6 & -21 \end{array}\right]$

16. $\left[\begin{array}{ccc|c} 1 & 2 & 1 & 0 \\ -2 & 1 & 3 & 4 \\ 1 & 7 & 6 & 4 \end{array}\right]$

17. $\left[\begin{array}{ccc|c} 1 & -3 & -1 & 2 \\ -2 & 6 & 4 & 3 \\ 2 & -1 & 2 & 5 \end{array}\right]$

18. $\left[\begin{array}{ccc|c} 2 & 4 & 2 & 4 \\ 3 & 4 & 1 & 6 \end{array}\right]$

19. $\left[\begin{array}{ccc|c} 1 & 3 & 0 & 4 \\ \frac{1}{3} & -1 & -4 & \frac{16}{3} \end{array}\right]$

20. $\left[\begin{array}{cccc|c} 2 & 4 & 6 & 2 & 6 \\ -2 & -3 & -7 & 0 & 0 \\ 4 & 6 & 14 & -2 & -4 \end{array}\right]$

21. $\left[\begin{array}{cccc|c} 1 & 0 & 1 & 2 & 1 \\ -1 & 1 & -1 & -1 & -1 \\ 0 & 2 & 1 & 3 & 0 \\ 1 & 0 & 2 & 4 & 3 \end{array}\right]$

22. $\left[\begin{array}{cccc|c} 1 & 1 & 0 & 0 & 2 \\ 2 & 3 & 0 & 0 & 4 \\ 3 & 3 & 1 & 0 & 7 \\ 4 & 4 & 0 & 4 & 8 \end{array}\right]$

EXERCISES 23–32 *Solve the system of equations by first expressing it as an augmented matrix and then applying Gaussian elimination and back-substitution.*

23. $2x - 6y = -2$
    $3x + 5y = 11$

24. $2u + 4v = -6$
    $5u - 3v = 24$

25. $\begin{aligned} 2p + q &= \dfrac{5}{3} \\ 3p - 6q &= -\dfrac{5}{2} \end{aligned}$

26. $\begin{aligned} 4y + 6z &= -\dfrac{8}{5} \\ 3y - 5z &= \dfrac{13}{5} \end{aligned}$

27. $\begin{aligned} x + y + 2z &= 7 \\ x + 2y + 3z &= 10 \\ x - 4y + z &= 4 \end{aligned}$

28. $\begin{aligned} x - y + z &= -2 \\ x + y - z &= 0 \\ -x + y + z &= -4 \end{aligned}$

29. $\begin{aligned} 2p \quad\;\; + r &= 2 \\ 2p + q + 4r &= 9 \\ 3p + q + 8r &= 17 \end{aligned}$

30. $\begin{aligned} 2u - 4v + 2w &= 18 \\ 3u - 5v + 2w &= 23 \\ 4u + 9v + 6w &= -13 \end{aligned}$

31. $\begin{aligned} w + x + y + 2z &= 1 \\ -w + x - y - z &= -2 \\ 2x + y + 3z &= -1 \\ w + x + 2y + 5z &= 1 \end{aligned}$

32. $\begin{aligned} 2q + 4r + 6s - 2t &= 8 \\ r - s + t &= 2 \\ r + s - t &= 3 \\ -r + s + t &= -5 \end{aligned}$

EXERCISES 33–38 *Use Gauss-Jordan elimination to express the given matrix in reduced row-echelon form.*

33. $\left[\begin{array}{cc|c} 1 & 2 & 3 \\ 0 & 1 & 2 \end{array}\right]$

34. $\left[\begin{array}{cc|c} 1 & 3 & -2 \\ 0 & 1 & 5 \end{array}\right]$

35. $\left[\begin{array}{cc|c} 2 & 4 & -8 \\ 3 & 3 & -3 \end{array}\right]$

36. $\left[\begin{array}{cc|c} 2 & 3 & 0 \\ 2 & 6 & -2 \end{array}\right]$

37. $\left[\begin{array}{ccc|c} 1 & 2 & 2 & 2 \\ 0 & 1 & 3 & -2 \\ 0 & 0 & 1 & -1 \end{array}\right]$

38. $\left[\begin{array}{ccc|c} 1 & 3 & 1 & \frac{7}{2} \\ 0 & 1 & -\frac{1}{6} & 0 \\ 0 & 0 & 1 & 2 \end{array}\right]$

EXERCISES 39–46 *Solve the system of equations by expressing it in matrix form and using Gauss-Jordan elimination to obtain a matrix in reduced row-echelon form.*

39. $\begin{aligned} 2x - 3y &= 12 \\ x + 2y &= -1 \end{aligned}$

40. $\begin{aligned} 3x - y &= 1 \\ 6x + 3y &= 7 \end{aligned}$

41. $\begin{aligned} 4x + 6y &= 0 \\ x - y &= \dfrac{5}{6} \end{aligned}$

42. $\begin{aligned} 2x - 4y &= -\dfrac{13}{2} \\ 4x + 2y &= 7 \end{aligned}$

43. $\begin{aligned} x - 2y + z &= 9 \\ y - 3z &= -10 \\ z &= 3 \end{aligned}$

44. $\begin{aligned} x + 3y - 2z &= -9 \\ y + 2z &= 10 \\ z &= 5 \end{aligned}$

45. $\begin{aligned} x + 2y + 4z &= -10 \\ 2x + 3y + 6z &= -15 \\ x - y + z &= -7 \end{aligned}$

46. $\begin{aligned} x - 3y + z &= -12 \\ x + y + z &= 0 \\ 2x - y + z &= -8 \end{aligned}$

EXERCISES 47–54 *Express the system as an augmented matrix and solve using Gaussian elimination. There may be 0, 1, or infinitely many solutions.*

47. $\begin{aligned} 2x + y &= 5 \\ 4x + 2y &= 10 \end{aligned}$

48. $\begin{aligned} 2x - 3y &= 3 \\ -4x + 6y &= 5 \end{aligned}$

49. $\begin{aligned} -x + 3y &= -7 \\ 3x - 2y &= 7 \end{aligned}$

50. $\begin{aligned} 3x - y &= 1 \\ \dfrac{3}{2}x - \dfrac{1}{2}y &= \dfrac{1}{2} \end{aligned}$

51. $\begin{aligned} -3x - y + z &= -1 \\ x + 4y - z &= 3 \\ -5x + 2y + z &= 2 \end{aligned}$

52. $\begin{aligned} x + 2y + z &= 5 \\ x - y + 3z &= 6 \\ 2x + y + 4z &= 11 \end{aligned}$

53. $\begin{aligned} 2x + 4y + 2z &= 0 \\ 3x - y + z &= 1 \\ x - 5y - z &= 1 \end{aligned}$

54. $\begin{aligned} x - 2y + 3z &= 0 \\ 2x + y - z &= 1 \\ -x + z &= 2 \end{aligned}$

APPLICATIONS

55. *Fitting a Parabola*   Find the equation of the parabola that passes through the points (1, 5), (2, 7), and (3, 13).

56. *Unknown Constants*   It is given that the variables y and x are related by the formula

$$ y = a|x| + b|x - 1| + c|x - 2| $$

If $y = 5$ when $x = 0$, $y = 6$ when $x = 1$, and $y = 1$ when $x = 2$, find a, b, and c.

57. *Counting Coins*   Your Uncle Bob pulls 15 coins (nickels, dimes, and quarters) from behind your ear. The total value of the coins is $2.10, and there are two more quarters than nickels. How many of each coin does Uncle Bob have?

58. *River Beautification*   Kyung-Bai, Antoine, and Vanessa are cleaning a river bank of trash. The time required for Vanessa to clear a 100-yard stretch is the average of the time required for Kyung-Bai and Antoine. The sum of their times is 27 hours, and the time required for Antoine to clear a 100-yard stretch is 7

hours less than the sum of the corresponding times for Kyung-Bai and Vanessa.

a. Find the time required for each of them to clear a 100-yard stretch of the river bank.

b. How long would it take them to clear a 100-yard stretch if they work together?

59. *Barbell Weights*   A certain weight room includes a set of pre-weighted barbells, with the total weight printed on some of the barbells. A barbell with 2 large disks and 1 small disk on each side weighs 70 pounds; a barbell with 1 small disk on each side weighs 30 pounds; and a barbell with 3 large disks on each side weighs 75 pounds. Find the weight of the bar and each of the disks. Also, compute the weight of an (unmarked) barbell with 4 large disks and 2 small disks on each side.

60. *Unknown Power*   A weightlifter is bench-pressing a total of 305 pounds using a 45-pound bar on which there are only 45-,

25-, and 5-pound weights. He uses a total of 12 weights, and the number of 5-pound weights is equal to the sum of the number of 25- and 45-pound weights. How many weights of each type did the weightlifter use?

61. *Ancient Faiths* When Martin Luther broke from the Roman Catholic Church in 1517, Buddhism and Hinduism were a combined 5059 years old. When Joseph Smith founded the Mormon Church in 1827, Islam and Buddhism were a combined 3557 years old. Hinduism is 975 years older than Buddhism. Find the years in which Islam, Buddhism, and Hinduism were founded.

62. *Nut Mixture* How can 30 pounds of a mixture that is 40% pecans, 40% almonds, and 20% cashews be formed if only the following cans of mixed nuts are available?

	Percent pecans	Percent almonds	Percent cashews
1-pound can	50	30	20
2-pound can	30	60	10
5-pound can	40	30	30

63. *School Integration* In a certain community, the middle schools fall into three categories:

(1) Schools of about 500 students in which roughly half the students are White and half the students are Black

(2) Schools with about 450 students who are approximately half Black and half Latino

(3) Schools with about 400 students who are 80% White and 20% Black

A new high school with a capacity of 3750 students is being built. It is desired that the ethnic composition of the school be consistent with that of the community as a whole, which is 46.8% black, 35.2% white, and 18% Latino. How many middle schools of each type should feed into the new high school in order to achieve the desired ethnic composition?

64. *Ancestral Heritage* A woman is $\frac{19}{32}$ Cherokee and $\frac{13}{32}$ Irish. Her mother is half Cherokee. Her maternal grandfather has the same percentage of Cherokee blood as does her paternal grandfather. Her maternal grandmother has half as much Cherokee blood as does her paternal grandmother. Compute the fractions of Cherokee and Irish blood of her father and all four of her grandparents. [*Hint:* Let the four variables be the fraction of Cherokee blood of each of the woman's grandparents. If a person's father is a% Cherokee and the mother is b% Cherokee, then the person's percentage of Cherokee blood is given by $(a + b)/2$.]

CONCEPTS AND CRITICAL THINKING

EXERCISES 65–68 *Answer true or false.*

65. Any matrix in which each row begins with a 1 is in row-echelon form.

66. Interchanging two rows is an example of an elementary row operation.

67. Multiplying each entry in a column by 5 is an example of an elementary row operation.

68. If a matrix in row-echelon form has all zeros in the first row, then every entry of the matrix must be a zero.

EXERCISES 69–72 *Give an example of each.*

69. A matrix in row-echelon but not reduced row-echelon form

70. An elementary row operation

71. A matrix consisting only of 1s and 0s that is not in row-echelon form

72. An upper triangular matrix that is not in row-echelon form

73. The given numbers are from the final row of an augmented matrix that has been converted to reduced row-echelon form using Gauss-Jordan elimination. What conclusion can be drawn as to whether the corresponding system is consistent or inconsistent? If the system is consistent, how many solutions does it have?

a. $[0 \quad 0 \quad \cdots \quad 0 \quad 1 \mid 0]$

b. $[0 \quad 0 \quad \cdots \quad 0 \quad 0 \mid 0]$

c. $[0 \quad 0 \quad \cdots \quad 0 \quad 0 \mid 1]$

QUESTIONS FOR DISCUSSION OR ESSAY

74. What similarities do you see between the techniques of Gaussian elimination and synthetic division?

75. If we define an arithmetic step to be the addition, subtraction, or multiplication of two numbers, then what is the maximum number of arithmetic steps required to solve a system of three equations in three unknowns by Gaussian elimination with back-substitution? What about for Gauss-Jordan elimination? Support your answers.

76. One of the elementary row operations is multiplying a row by a nonzero constant. This operation corresponds to multiplying both sides of an equation by the constant. Why do we insist that the constant be nonzero? After all, if we multiply both sides of an equation by 0, we obtain the equation $0 = 0$, which is certainly true.

77. Gaussian elimination and Gauss-Jordan elimination are examples of algorithms, step-by-step methods for solving a problem

for which every step is prescribed. Name two other mathematical algorithms and describe two algorithms from everyday life. Does the technique of elimination as described in Section 2.8 qualify as an algorithm? Why or why not?

■ **PROJECTS FOR ENRICHMENT**

78. *The Sum of the First n Squares*   In this project, we develop a formula for the sum S of the squares of the first n natural numbers, $S = 1^2 + 2^2 + 3^2 + \cdots + n^2$. We take advantage of the well-known fact that, in general, if $T = 1^k + 2^k + \cdots + n^k$, where k and n are natural numbers, then T is a polynomial of degree $k + 1$ in the variable n with constant term 0. In particular, S can be expressed as a third-degree polynomial in n with constant term 0; that is, $S = an^3 + bn^2 + cn$.

a. Evaluate $S = 1^2 + 2^2 + \cdots + n^2$ for $n = 1, 2,$ and 3.

b. Evaluate the expression $an^3 + bn^2 + cn$ for $n = 1, 2,$ and 3.

c. Use the results of parts a and b to form a system of three equations in the three variables $a, b,$ and c.

d. Use Gaussian elimination to solve the system of part c.

e. Express S as a third-degree polynomial in n.

f. Compute the sum of the squares of the first 10,000 natural numbers.

79. *Line Drive Home Runs*   Suppose a batter hits a line drive that passes over the pitcher (60 feet, 6 inches from home plate) at a height of 9 feet above the field and then clears the 10-foot center field wall (330 feet from home plate) by 4 feet. A broadcaster announces that the ball "was still rising when it cleared the center field wall."

In this project, we investigate the possible locations at which the ball reaches its zenith, and, in particular, we determine if it is possible that the ball was still rising when it cleared the wall.

a. Assuming that there is no wind resistance, the ball travels in a parabolic path. Thus, if we let y represent the height of the ball and x the horizontal distance from home plate, then $y = ax^2 + bx + c$ for some constants $a, b,$ and c. Set up a system of equations that reflects the facts given in the introductory paragraph of this project.

b. Solve the system from part a. Describe the solution set in terms of the variable c. Note that c represents the height of the ball when it was struck by the batter.

c. The vertex of the parabola occurs when $x = -b/(2a)$. Find an expression for the location of the vertex in terms of the variable c.

d. Compute the location of the vertex for values of c ranging from 2 to 6, in increments of 1.

e. Find the value of c required so that the ball's peak occurs as it crosses the fence. Is this reasonable?

f. Could the broadcaster have been correct?

Mark McGwire belts his record 70th home run.

SECTION 7.4   **INVERSES OF SQUARE MATRICES**

☐ **How can linear systems of equations be solved using a graphing calculator?**

☐ **How can you graph the letter M on a graphing calculator?**

☐ **How can the physical laws governing the universe be discovered using matrices?**

☐ **We can divide by any real number except 0. For which matrices is division defined?**

We've seen that linear systems of equations can be expressed as matrix equations of the form $AX = B$, and we have developed techniques such as Gaussian elimination for solving such systems. But suppose that instead we naively attempted to solve for X by "dividing" both sides by A, in much the same way that we would divide both sides by a when solving an ordinary algebraic equation of the form $ax = b$. Immediately we are confronted with a difficulty: What does it mean to divide one matrix by another? In the case of a real number, dividing by a is equivalent to multiplying by a^{-1}. But what does A^{-1} mean when A is a matrix? Can we define A^{-1} for any nonzero matrix A, or are there some matrices A for which A^{-1} doesn't exist? In this section, we clarify the notion of the inverse of a matrix and apply it to solve certain equations of the form $AX = B$.

THE IDENTITY MATRIX AND INVERSE MATRICES

Recall from Section 7.2 that a matrix for which the number of rows is the same as the number of columns is called a **square matrix.** The **main diagonal** of a square matrix runs from the upper left corner to the lower right corner, as shown in Figure 7. Thus, it consists of the entries for which the row and column are the same. The $n \times n$ matrix with 1s on the main diagonal and 0s elsewhere (see Figure 8) is called the nth-order **identity matrix,** and is denoted by I_n, or simply I if the order is apparent.

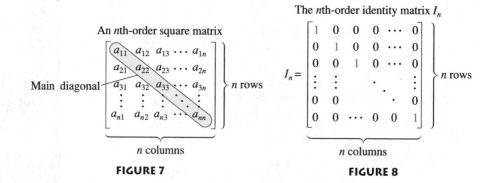

FIGURE 7

FIGURE 8

The following example suggests why I is called the *identity* matrix.

EXAMPLE 1 ■ **Multiplication of a Matrix by I**

Let

$$A = \begin{bmatrix} 1 & 2 & 3 \\ 4 & 5 & 6 \\ 7 & 8 & 9 \end{bmatrix}$$

Compute $A \cdot I$ and $I \cdot A$.

SOLUTION

$$A \cdot I = \begin{bmatrix} 1 & 2 & 3 \\ 4 & 5 & 6 \\ 7 & 8 & 9 \end{bmatrix} \cdot \begin{bmatrix} 1 & 0 & 0 \\ 0 & 1 & 0 \\ 0 & 0 & 1 \end{bmatrix}$$

$$= \begin{bmatrix} 1\cdot1 + 2\cdot0 + 3\cdot0 & 1\cdot0 + 2\cdot1 + 3\cdot0 & 1\cdot0 + 2\cdot0 + 3\cdot1 \\ 4\cdot1 + 5\cdot0 + 6\cdot0 & 4\cdot0 + 5\cdot1 + 6\cdot0 & 4\cdot0 + 5\cdot0 + 6\cdot1 \\ 7\cdot1 + 8\cdot0 + 9\cdot0 & 7\cdot0 + 8\cdot1 + 9\cdot0 & 7\cdot0 + 8\cdot0 + 9\cdot1 \end{bmatrix}$$

$$= \begin{bmatrix} 1 & 2 & 3 \\ 4 & 5 & 6 \\ 7 & 8 & 9 \end{bmatrix}$$

Thus, $A \cdot I = A$.

$$I \cdot A = \begin{bmatrix} 1 & 0 & 0 \\ 0 & 1 & 0 \\ 0 & 0 & 1 \end{bmatrix} \cdot \begin{bmatrix} 1 & 2 & 3 \\ 4 & 5 & 6 \\ 7 & 8 & 9 \end{bmatrix}$$

$$= \begin{bmatrix} 1 \cdot 1 + 0 \cdot 4 + 0 \cdot 7 & 1 \cdot 2 + 0 \cdot 5 + 0 \cdot 8 & 1 \cdot 3 + 0 \cdot 6 + 0 \cdot 9 \\ 0 \cdot 1 + 1 \cdot 4 + 0 \cdot 7 & 0 \cdot 2 + 1 \cdot 5 + 0 \cdot 8 & 0 \cdot 3 + 1 \cdot 6 + 0 \cdot 9 \\ 0 \cdot 1 + 0 \cdot 4 + 1 \cdot 7 & 0 \cdot 2 + 0 \cdot 5 + 1 \cdot 8 & 0 \cdot 3 + 0 \cdot 6 + 1 \cdot 9 \end{bmatrix}$$

$$= \begin{bmatrix} 1 & 2 & 3 \\ 4 & 5 & 6 \\ 7 & 8 & 9 \end{bmatrix}$$

Thus, $I \cdot A = A$.

In the previous example, we saw that the product of a third-order square matrix and I was the given third-order matrix. In fact, this is true for all square matrices. We leave the proof to Exercise 47.

The Identity Property of I_n □

> If A is an $n \times n$ matrix, then $A \cdot I_n = I_n \cdot A = A$.

The matrix identity I_n is similar to the multiplicative identity 1 in that multiplying a matrix by I_n leaves the matrix unchanged, just as multiplication by 1 leaves a number unchanged. Now, the multiplicative inverse of a number x is, by definition, a *number* that when multiplied by x gives the multiplicative identity 1; it is commonly denoted by x^{-1}. What then is the multiplicative inverse of an $n \times n$ matrix A? Of course, it should be a matrix that, when multiplied by A (on either side), gives the multiplicative identity I_n. We make the following definitions.

Definition of the Multiplicative Inverse of a Matrix □

> Let A be an $n \times n$ matrix. The **inverse** of A, if it exists, is an $n \times n$ matrix B satisfying $AB = I_n$ and $BA = I_n$. We say that A and B are inverses of one another and write $B = A^{-1}$. In this case, we say that the matrix A is **invertible**. If A^{-1} doesn't exist, then we say that A is **noninvertible**.

EXAMPLE 2 ■ **Confirming Inverses**

Show that A and B are inverses of one another, where

$$A = \begin{bmatrix} 2 & 0 \\ -4 & 1 \end{bmatrix} \quad \text{and} \quad B = \begin{bmatrix} \frac{1}{2} & 0 \\ 2 & 1 \end{bmatrix}$$

SOLUTION We must show that

$$AB = BA = \begin{bmatrix} 1 & 0 \\ 0 & 1 \end{bmatrix}$$

Thus, we proceed as follows:

$$AB = \begin{bmatrix} 2 & 0 \\ -4 & 1 \end{bmatrix} \cdot \begin{bmatrix} \frac{1}{2} & 0 \\ 2 & 1 \end{bmatrix}$$

$$= \begin{bmatrix} 2 \cdot \frac{1}{2} + 0 \cdot 2 & 2 \cdot 0 + 0 \cdot 1 \\ (-4)\frac{1}{2} + 1 \cdot 2 & (-4)0 + 1 \cdot 1 \end{bmatrix}$$

$$= \begin{bmatrix} 1 & 0 \\ 0 & 1 \end{bmatrix}$$

$$BA = \begin{bmatrix} \frac{1}{2} & 0 \\ 2 & 1 \end{bmatrix} \cdot \begin{bmatrix} 2 & 0 \\ -4 & 1 \end{bmatrix}$$

$$= \begin{bmatrix} \frac{1}{2} \cdot 2 + 0(-4) & \frac{1}{2} \cdot 0 + 0 \cdot 1 \\ 2 \cdot 2 + 1(-4) & 2 \cdot 0 + 1 \cdot 1 \end{bmatrix}$$

$$= \begin{bmatrix} 1 & 0 \\ 0 & 1 \end{bmatrix}$$

> **RULE OF THUMB**   Using more advanced techniques, it can be shown that
> if $AB = I$, then $BA = I$ also. Thus, it is not necessary to compute both AB and
> BA in order to show that A and B are inverses.

COMPUTING INVERSES

We will now see how the inverse of a 2×2 matrix can be computed by solving a pair
of systems of equations using Gauss-Jordan elimination. Let us consider the problem
of computing the inverse of the matrix A given by

$$A = \begin{bmatrix} 1 & 3 \\ 2 & 5 \end{bmatrix}$$

If we denote A^{-1} by

$$A^{-1} = \begin{bmatrix} a & b \\ c & d \end{bmatrix}$$

then we have

$$A \cdot A^{-1} = I$$

$$\begin{bmatrix} 1 & 3 \\ 2 & 5 \end{bmatrix} \cdot \begin{bmatrix} a & b \\ c & d \end{bmatrix} = \begin{bmatrix} 1 & 0 \\ 0 & 1 \end{bmatrix}$$

$$\begin{bmatrix} a + 3c & b + 3d \\ 2a + 5c & 2b + 5d \end{bmatrix} = \begin{bmatrix} 1 & 0 \\ 0 & 1 \end{bmatrix}$$

Equating corresponding entries gives two systems of equations—one involving a and
c, and the other involving b and d. These systems and the corresponding augmented
matrices are as follows:

	System	**Augmented matrix**
1	$\begin{aligned} a + 3c &= 1 \\ 2a + 5c &= 0 \end{aligned}$	$\left[\begin{array}{cc|c} 1 & 3 & 1 \\ 2 & 5 & 0 \end{array}\right]$
2	$\begin{aligned} b + 3d &= 0 \\ 2b + 5d &= 1 \end{aligned}$	$\left[\begin{array}{cc|c} 1 & 3 & 0 \\ 2 & 5 & 1 \end{array}\right]$

Each of these systems can be solved using Gauss-Jordan elimination.

SYSTEM 1:

$$\begin{bmatrix} 1 & 3 & | & 1 \\ 2 & 5 & | & 0 \end{bmatrix} \xrightarrow{-2R_1 + R_2} \begin{bmatrix} 1 & 3 & | & 1 \\ 0 & -1 & | & -2 \end{bmatrix} \xrightarrow{-1 \cdot R_2} \begin{bmatrix} 1 & 3 & | & 1 \\ 0 & 1 & | & 2 \end{bmatrix} \xrightarrow{-3R_2 + R_1} \begin{bmatrix} 1 & 0 & | & -5 \\ 0 & 1 & | & 2 \end{bmatrix}$$

Thus, $a = -5$ and $c = 2$.

SYSTEM 2:

$$\begin{bmatrix} 1 & 3 & | & 0 \\ 2 & 5 & | & 1 \end{bmatrix} \xrightarrow{-2R_1 + R_2} \begin{bmatrix} 1 & 3 & | & 0 \\ 0 & -1 & | & 1 \end{bmatrix} \xrightarrow{-1 \cdot R_2} \begin{bmatrix} 1 & 3 & | & 0 \\ 0 & 1 & | & -1 \end{bmatrix} \xrightarrow{-3R_2 + R_1} \begin{bmatrix} 1 & 0 & | & 3 \\ 0 & 1 & | & -1 \end{bmatrix}$$

Thus, $b = 3$ and $d = -1$.

Using the values of a, b, c, and d, we find that

$$A^{-1} = \begin{bmatrix} -5 & 3 \\ 2 & -1 \end{bmatrix}$$

Note that since systems 1 and 2 have the same coefficient matrix, the elementary row operations used at each stage of the Gauss-Jordan elimination are also the same. This suggests that we solve both systems simultaneously by applying Gauss-Jordan elimination to the following augmented matrix.

$$\begin{bmatrix} 1 & 3 & | & 1 & 0 \\ 2 & 5 & | & 0 & 1 \end{bmatrix}$$

When this matrix is in reduced row-echelon form, the first two columns will form the identity matrix I_2, the solution to system 1 will be given by the entries in the third column, and the solution to system 2 will be given by the entries in the fourth column. In other words, at the conclusion of the Gauss-Jordan elimination process, the matrix will be of the form

$$\begin{bmatrix} 1 & 0 & | & a & b \\ 0 & 1 & | & c & d \end{bmatrix}$$

with the columns to the right of the partition forming A^{-1}. We now compute A^{-1} with this procedure.

$$\begin{bmatrix} 1 & 3 & | & 1 & 0 \\ 2 & 5 & | & 0 & 1 \end{bmatrix} \xrightarrow{-2R_1 + R_2} \begin{bmatrix} 1 & 3 & | & 1 & 0 \\ 0 & -1 & | & -2 & 1 \end{bmatrix} \xrightarrow{-1 \cdot R_2} \begin{bmatrix} 1 & 3 & | & 1 & 0 \\ 0 & 1 & | & 2 & -1 \end{bmatrix} \xrightarrow{-3R_2 + R_1} \begin{bmatrix} 1 & 0 & | & -5 & 3 \\ 0 & 1 & | & 2 & -1 \end{bmatrix}$$

So

$$A^{-1} = \begin{bmatrix} -5 & 3 \\ 2 & -1 \end{bmatrix}$$

The preceding technique generalizes to square matrices of arbitrary order, as follows.

Computing Inverses with ☐
Gauss-Jordan Elimination

1. Form the augmented matrix whose first n columns constitute A and whose last n columns form I_n, symbolically $[A \mid I_n]$.

2. Use Gauss-Jordan elimination to write this matrix in reduced row-echelon form.

   A. If the first n columns of the reduced row-echelon matrix form the identity I_n, then the last n columns will form A^{-1}. Symbolically, the reduced row-echelon matrix is of the form $[I_n \mid A^{-1}]$.

   B. If the first n columns do not form the identity matrix I_n, then A is noninvertible.

E X A M P L E 3 ▨   **Computing Inverses of Matrices**

Compute the inverse of the given matrix and confirm using matrix multiplication.

a. $A = \begin{bmatrix} 1 & 0 & 1 \\ 1 & -1 & 0 \\ 0 & 2 & 1 \end{bmatrix}$ **b.** $B = \begin{bmatrix} 1 & 0 \\ 0 & 0 \end{bmatrix}$

SOLUTION

a. We form the augmented matrix $[A \mid I_3]$ and perform Gauss-Jordan elimination.

$$\left[\begin{array}{ccc|ccc} 1 & 0 & 1 & 1 & 0 & 0 \\ 1 & -1 & 0 & 0 & 1 & 0 \\ 0 & 2 & 1 & 0 & 0 & 1 \end{array}\right] \xrightarrow{-R_1 + R_2} \left[\begin{array}{ccc|ccc} 1 & 0 & 1 & 1 & 0 & 0 \\ 0 & -1 & -1 & -1 & 1 & 0 \\ 0 & 2 & 1 & 0 & 0 & 1 \end{array}\right]$$

$$\left[\begin{array}{ccc|ccc} 1 & 0 & 1 & 1 & 0 & 0 \\ 0 & -1 & -1 & -1 & 1 & 0 \\ 0 & 2 & 1 & 0 & 0 & 1 \end{array}\right] \xrightarrow{-1 \cdot R_2} \left[\begin{array}{ccc|ccc} 1 & 0 & 1 & 1 & 0 & 0 \\ 0 & 1 & 1 & 1 & -1 & 0 \\ 0 & 2 & 1 & 0 & 0 & 1 \end{array}\right]$$

$$\left[\begin{array}{ccc|ccc} 1 & 0 & 1 & 1 & 0 & 0 \\ 0 & 1 & 1 & 1 & -1 & 0 \\ 0 & 2 & 1 & 0 & 0 & 1 \end{array}\right] \xrightarrow{-2R_2 + R_3} \left[\begin{array}{ccc|ccc} 1 & 0 & 1 & 1 & 0 & 0 \\ 0 & 1 & 1 & 1 & -1 & 0 \\ 0 & 0 & -1 & -2 & 2 & 1 \end{array}\right]$$

$$\left[\begin{array}{ccc|ccc} 1 & 0 & 1 & 1 & 0 & 0 \\ 0 & 1 & 1 & 1 & -1 & 0 \\ 0 & 0 & -1 & -2 & 2 & 1 \end{array}\right] \xrightarrow{-1 \cdot R_3} \left[\begin{array}{ccc|ccc} 1 & 0 & 1 & 1 & 0 & 0 \\ 0 & 1 & 1 & 1 & -1 & 0 \\ 0 & 0 & 1 & 2 & -2 & -1 \end{array}\right]$$

$$\left[\begin{array}{ccc|ccc} 1 & 0 & 1 & 1 & 0 & 0 \\ 0 & 1 & 1 & 1 & -1 & 0 \\ 0 & 0 & 1 & 2 & -2 & -1 \end{array}\right] \xrightarrow{-R_3 + R_1} \left[\begin{array}{ccc|ccc} 1 & 0 & 0 & -1 & 2 & 1 \\ 0 & 1 & 1 & 1 & -1 & 0 \\ 0 & 0 & 1 & 2 & -2 & -1 \end{array}\right]$$

$$\left[\begin{array}{ccc|ccc} 1 & 0 & 0 & -1 & 2 & 1 \\ 0 & 1 & 1 & 1 & -1 & 0 \\ 0 & 0 & 1 & 2 & -2 & -1 \end{array}\right] \xrightarrow{-R_3 + R_2} \left[\begin{array}{ccc|ccc} 1 & 0 & 0 & -1 & 2 & 1 \\ 0 & 1 & 0 & -1 & 1 & 1 \\ 0 & 0 & 1 & 2 & -2 & -1 \end{array}\right]$$

Thus

$$A^{-1} = \begin{bmatrix} -1 & 2 & 1 \\ -1 & 1 & 1 \\ 2 & -2 & -1 \end{bmatrix}$$

We can check this result by multiplying A^{-1} by the original matrix A; if the product is I, then our result is correct.

$$A^{-1} \cdot A = \begin{bmatrix} -1 & 2 & 1 \\ -1 & 1 & 1 \\ 2 & -2 & -1 \end{bmatrix} \cdot \begin{bmatrix} 1 & 0 & 1 \\ 1 & -1 & 0 \\ 0 & 2 & 1 \end{bmatrix}$$

$$= \begin{bmatrix} (-1)1 + 2 \cdot 1 + 1 \cdot 0 & (-1)0 + 2(-1) + 1 \cdot 2 & (-1)1 + 2 \cdot 0 + 1 \cdot 1 \\ (-1)1 + 1 \cdot 1 + 1 \cdot 0 & (-1)0 + 1(-1) + 1 \cdot 2 & (-1)1 + 1 \cdot 0 + 1 \cdot 1 \\ 2 \cdot 1 + (-2)1 + (-1)0 & 2 \cdot 0 + (-2)(-1) + (-1)2 & 2 \cdot 1 + (-2)0 + (-1)1 \end{bmatrix}$$

$$= \begin{bmatrix} 1 & 0 & 0 \\ 0 & 1 & 0 \\ 0 & 0 & 1 \end{bmatrix}$$

b. We begin by forming the augmented matrix

$$\begin{bmatrix} 1 & 0 & | & 1 & 0 \\ 0 & 0 & | & 0 & 1 \end{bmatrix}$$

Since this matrix is already in reduced row-echelon form, and the first two columns do *not* constitute the identity matrix, B is noninvertible.

CALCULATOR KEYS

Inverse Matrices

If a matrix with numerical entries is invertible, its inverse can be computed with a graphing calculator. As with other matrix operations, you must first specify the order of the matrix and then key in its entries. Once this is done, the inverse can be computed using the $\boxed{x^{-1}}$ key.

SOLVING LINEAR SYSTEMS USING INVERSES

Although inverse matrices arise in many contexts, the most important application of the inverse for our purposes is as a tool for solving linear systems of equations. Consider the following system.

$$2x + 3y + z = 6$$
$$4x - y + z = -3$$
$$x + y + \frac{1}{2}z = 1$$

In matrix form we have

$$\begin{bmatrix} 2 & 3 & 1 \\ 4 & -1 & 1 \\ 1 & 1 & \frac{1}{2} \end{bmatrix} \cdot \begin{bmatrix} x \\ y \\ z \end{bmatrix} = \begin{bmatrix} 6 \\ -3 \\ 1 \end{bmatrix}$$

If we define

$$A = \begin{bmatrix} 2 & 3 & 1 \\ 4 & -1 & 1 \\ 1 & 1 & \frac{1}{2} \end{bmatrix}, \quad X = \begin{bmatrix} x \\ y \\ z \end{bmatrix}, \quad \text{and } B = \begin{bmatrix} 6 \\ -3 \\ 1 \end{bmatrix}$$

then we have the matrix equation $AX = B$. Now, if A is invertible, we can solve this equation for the matrix X as follows:

$AX = B$	The original matrix equation
$A^{-1} \cdot (AX) = A^{-1} \cdot B$	Multiplying both sides on the left by A^{-1}
$(A^{-1} \cdot A) \cdot X = A^{-1}B$	The associative law of matrix multiplication
$IX = A^{-1}B$	The definition of the multiplicative inverse
$X = A^{-1}B$	Using a property of the identity matrix. Note that the product on the right-hand side is defined since A^{-1} is a 3×3 matrix and B is a 3×1 matrix.

This technique is ideal for use with a graphing calculator because the solution involves only the operations of matrix inversion and multiplication, both of which are easily performed on a graphing calculator. After entering the matrices A and B into the calculator, and computing $A^{-1}B$, the calculator returns the answer shown in Figure 9. Thus, the solution to the original system is $x = 3.5$, $y = 4$, and $z = -13$.

The inverse matrix technique for solving linear systems is especially convenient if several systems must be solved for which the coefficient matrix is the same. In this case, it is only necessary to compute the inverse of the coefficient matrix once, and this inverse can then be multiplied by each of the matrices obtained from the right-hand sides of the systems. It should be noted, however, that the inverse matrix technique can be applied only to systems with an invertible coefficient matrix. If the coefficient matrix is not invertible, then some other technique, such as Gaussian or Gauss-Jordan elimination, must be employed to describe the solution set.

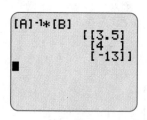

FIGURE 9

E X A M P L E   4   ■   **Finding the Equation of a Parabola Passing Through Three Given Points**

The graph of a quadratic function f includes the points $(-2, 25)$, $(1, 4)$, and $(3, 20)$. Find an expression for $f(x)$.

SOLUTION   Since f is a quadratic function, we know that $f(x) = ax^2 + bx + c$ for some choice of constants a, b, and c. Since $(-2, 25)$ is a point on the graph of f, we know that $f(-2) = 25$; thus, we have

$$f(-2) = a(-2)^2 + b(-2) + c = 25$$
$$4a - 2b + c = 25$$

Similarly, since $(1, 4)$ and $(3, 20)$ are points on the graph of f, we obtain the equations

$$f(1) = a + b + c = 4$$
$$f(3) = 9a + 3b + c = 20$$

Thus, we must solve the following system of equations.

$$4a - 2b + c = 25$$
$$a + b + c = 4$$
$$9a + 3b + c = 20$$

In matrix form, we have

$$\begin{bmatrix} 4 & -2 & 1 \\ 1 & 1 & 1 \\ 9 & 3 & 1 \end{bmatrix} \cdot \begin{bmatrix} a \\ b \\ c \end{bmatrix} = \begin{bmatrix} 25 \\ 4 \\ 20 \end{bmatrix}$$

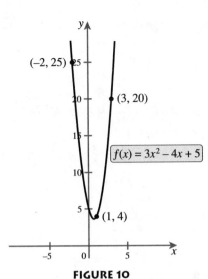

FIGURE 10

Solving for the unknown matrix we have

$$\begin{bmatrix} a \\ b \\ c \end{bmatrix} = \begin{bmatrix} 4 & -2 & 1 \\ 1 & 1 & 1 \\ 9 & 3 & 1 \end{bmatrix}^{-1} \cdot \begin{bmatrix} 25 \\ 4 \\ 20 \end{bmatrix}$$

Using a graphing calculator to evaluate the expression on the right-hand side, we obtain

$$\begin{bmatrix} a \\ b \\ c \end{bmatrix} = \begin{bmatrix} 3 \\ -4 \\ 5 \end{bmatrix}$$

Thus, $a = 3$, $b = -4$, and $c = 5$. It follows that the quadratic function f is defined by

$$f(x) = 3x^2 - 4x + 5$$

The graph of f is shown in Figure 10.

UNDERSTANDING AND MASTERY CHECKLISTS

CONCEPTS TO UNDERSTAND

- ☐ Square matrix
- ☐ Main diagonal of a square matrix
- ☐ Identity matrix
- ☐ Inverse of a square matrix
- ☐ Matrix product form of a linear system of equations

SKILLS TO MASTER

- ☐ Use matrix multiplication to determine if two matrices are inverses of each other.
- ☐ Use Gauss-Jordan elimination to find the inverse of a matrix or to show that it is not invertible.
- ☐ Use a graphing calculator to find the inverse of a matrix.
- ☐ Solve a linear system of equations using the inverse of the coefficient matrix.

EXERCISES 7.4

EXERCISES 1–4 *Use matrix multiplication to determine if A and B are inverses of one another.*

1. $A = \begin{bmatrix} 1 & -1 \\ 2 & -3 \end{bmatrix}$, $B = \begin{bmatrix} 3 & -1 \\ 2 & -1 \end{bmatrix}$

2. $A = \begin{bmatrix} 1 & 2 \\ -2 & 3 \end{bmatrix}$, $B = \begin{bmatrix} \frac{3}{7} & 0 \\ \frac{2}{7} & 0 \end{bmatrix}$

3. $A = \begin{bmatrix} 1 & 0 & -1 \\ 0 & 2 & 1 \\ 1 & 1 & 0 \end{bmatrix}$, $B = \begin{bmatrix} -1 & -1 & 2 \\ 1 & 1 & -1 \\ -2 & -1 & 3 \end{bmatrix}$

4. $A = \begin{bmatrix} 1 & 2 & 0 \\ 0 & -2 & -1 \\ 1 & -1 & -1 \end{bmatrix}$, $B = \begin{bmatrix} -1 & -2 & 2 \\ 1 & 1 & -1 \\ -2 & -3 & 2 \end{bmatrix}$

EXERCISES 5–26 *Find the inverse of the given matrix, if it exists. Use a graphing calculator as needed.*

5. $\begin{bmatrix} 1 & 0 \\ 2 & 1 \end{bmatrix}$

6. $\begin{bmatrix} \frac{1}{2} & \frac{1}{3} \\ 2 & 0 \end{bmatrix}$

7. $\begin{bmatrix} 0.2 & 0.4 \\ 0.5 & 0.5 \end{bmatrix}$

8. $\begin{bmatrix} 0.125 & 0.578 \\ 3.24 & 1.67 \end{bmatrix}$

9. $\begin{bmatrix} 1 & 2 \\ 2 & 4 \end{bmatrix}$

10. $\begin{bmatrix} a & 0 \\ 0 & b \end{bmatrix}$ $(a \neq 0, b \neq 0)$

11. $\begin{bmatrix} 0 & a \\ b & 0 \end{bmatrix}$ $(a \neq 0, b \neq 0)$

12. $\begin{bmatrix} x & y \\ 4x & 4y \end{bmatrix}$

13. $\begin{bmatrix} x & x + 1 \\ x & x - 1 \end{bmatrix}$ $(x \neq 0)$

14. $\begin{bmatrix} 1 & x \\ x^2 & x^3 \end{bmatrix}$

15. $\begin{bmatrix} -\frac{1}{5} & 0 & 1 \\ \frac{2}{5} & -1 & -1 \\ -\frac{1}{5} & 1 & 0 \end{bmatrix}$

16. $\begin{bmatrix} 1 & 2 & 0 \\ 0 & 1 & 4 \\ 0 & 0 & 1 \end{bmatrix}$

17. $\begin{bmatrix} 1 & 4 & -2 \\ 0 & 2 & -1 \\ -3 & 0 & 2 \end{bmatrix}$

18. $\begin{bmatrix} 1 & 2 & 3 \\ 4 & 5 & 6 \\ 7 & 8 & 9 \end{bmatrix}$

19. $\begin{bmatrix} -\frac{1}{12} & \frac{7}{12} & -\frac{1}{3} \\ -\frac{1}{12} & -\frac{5}{12} & \frac{2}{3} \\ \frac{5}{12} & \frac{1}{12} & -\frac{1}{3} \end{bmatrix}$

20. $\begin{bmatrix} 1.2 & -2.3 & 4.7 \\ 0 & 1.9 & -8.1 \\ 3.1 & 0 & 0 \end{bmatrix}$

21. $\begin{bmatrix} 1 & x & 0 \\ 0 & 1 & x \\ 0 & 0 & 1 \end{bmatrix}$

22. $\begin{bmatrix} 0 & x & x \\ x & 0 & x \\ x & x & 0 \end{bmatrix}$

23. $\begin{bmatrix} 1 & 0 & 0 & 1 \\ 0 & 1 & 0 & 1 \\ 0 & 0 & 1 & 1 \\ 1 & 1 & 1 & 3 \end{bmatrix}$

24. $\begin{bmatrix} 1 & 2 & 3 & 4 \\ 2 & 3 & 4 & 1 \\ 3 & 4 & 1 & 2 \\ 4 & 1 & 2 & 3 \end{bmatrix}$

25. $\begin{bmatrix} a & 0 & 0 & 0 & 0 \\ -a & a & 0 & 0 & 0 \\ 0 & -a & a & 0 & 0 \\ 0 & 0 & -a & a & 0 \\ 0 & 0 & 0 & -a & a \end{bmatrix}$

26. $\begin{bmatrix} a & a & a & a & a \\ 0 & a & a & a & a \\ 0 & 0 & a & a & a \\ 0 & 0 & 0 & a & a \\ 0 & 0 & 0 & 0 & a \end{bmatrix}$

EXERCISES 27–32 *Three systems of equations are given. Solve the systems by first expressing them in matrix form as $AX = B$ and then evaluating $X = A^{-1}B$. Note that the coefficient matrix A is the same for each of parts a–c.*

27.
 a. $\begin{aligned} x - y &= -1 \\ x + 2y &= 12 \end{aligned}$
 b. $\begin{aligned} x - y &= -4 \\ x + 2y &= -1 \end{aligned}$
 c. $\begin{aligned} x - y &= 1 \\ x + 2y &= -\frac{1}{2} \end{aligned}$

28.
 a. $\begin{aligned} -2x + 6y &= 6 \\ 3x + 5y &= 5 \end{aligned}$
 b. $\begin{aligned} -2x + 6y &= 8.4 \\ 3x + 5y &= 8.4 \end{aligned}$
 c. $\begin{aligned} -2x + 6y &= -2.5 \\ 3x + 5y &= 7.25 \end{aligned}$

29.
 a. $\begin{aligned} 1.5x - 4y &= -12.5 \\ 2.3x + 6y &= 41.5 \end{aligned}$
 b. $\begin{aligned} 1.5x - 4y &= 2 \\ 2.3x + 6y &= 15.2 \end{aligned}$
 c. $\begin{aligned} 1.5x - 4y &= -15 \\ 2.3x + 6y &= 13.4 \end{aligned}$

30.
 a. $\begin{aligned} \frac{2}{5}x + \frac{3}{5}y &= 8 \\ x - y &= -5 \end{aligned}$
 b. $\begin{aligned} \frac{2}{5}x + \frac{3}{5}y &= 2 \\ x - y &= -5 \end{aligned}$
 c. $\begin{aligned} \frac{2}{5}x + \frac{3}{5}y &= 1.4 \\ x - y &= -1.5 \end{aligned}$

31.
 a. $\begin{aligned} x + 2y - 4z &= -8 \\ 3x - y + z &= 7 \\ 2x + 2y - z &= 6 \end{aligned}$
 b. $\begin{aligned} x + 2y - 4z &= -6 \\ 3x - y + z &= -5 \\ 2x + 2y - z &= -5 \end{aligned}$
 c. $\begin{aligned} x + 2y - 4z &= -5 \\ 3x - y + z &= 15 \\ 2x + 2y - z &= 15 \end{aligned}$

32.
 a. $\begin{aligned} x + y + z &= 2 \\ x + 2y - z &= 1 \\ y + z &= -1 \end{aligned}$
 b. $\begin{aligned} x + y + z &= 6 \\ x + 2y - z &= 2 \\ y + z &= 5 \end{aligned}$
 c. $\begin{aligned} x + y + z &= \frac{7}{20} \\ x + 2y - z &= \frac{4}{5} \\ y + z &= \frac{3}{20} \end{aligned}$

APPLICATIONS

33. *Running Regimen* To safely increase aerobic capacity and burn fat, a man has developed a routine that consists of walking at 20 minutes per mile, jogging at 11 minutes per mile, and running at 8 minutes per mile. Given that walking burns 5.8 calories per minute, jogging burns 9.1 calories per minute, and running burns 14.1 calories per minute, how many minutes should he spend at each pace if he wants to

 a. Burn 1200 calories, work out for 2 hours, and travel 11 miles?

 b. Burn 750 calories, work out for 1.5 hours, and travel 7 miles?

34. *Investment Options* An investment company offers three different funds, each of which invests in highly volatile growth stocks, conservative but stable blue chip stocks, and income-generating bonds, but in differing proportions, as shown in Table 8.

TABLE 8

Fund content	Growth fund	Income fund	Security fund
Growth stocks	60%	20%	30%
Bonds	20%	50%	20%
Blue chips	20%	30%	50%

Determine the amount that should be invested in each of the three funds for an individual investor who wants to invest

 a. $2000 in growth stocks, $1000 in bonds, and $1000 in blue chips

 b. $1500 in growth stocks, $1500 in bonds, and $1500 in blue chips

35. *Fitting a Cubic* Find the equation of the cubic function $f(x) = ax^3 + bx^2 + cx + d$ passing through the points $(1, 5)$, $(3, 43)$, $(5, 201)$, and $(7, 575)$.

36. *Unknown Interest Rates* Suppose that a bank has three different money market accounts (growth, domestic, and international) paying annually compounded interest at the rates i, j, and k, respectively. Alberto, Brenda, and Carlos each invested $1000 in money market accounts at this bank, as indicated by Table 9.

 a. Find an expression involving i, j, and k for each person's balance at the end of 5 years. [*Hint:* If, for example, a principal of P is deposited into an account paying interest at the rate l for 4 years and then is switched to an account paying m for 3 years, at the end of 7 years the balance will be $P(1 + l)^4 (1 + m)^3$.]

 b. At the end of the 5 years, Alberto had $1240.05, Brenda had $1442.11, and Carlos had $1312.50. Set up a system of three equations in the three variables i, j, and k.

TABLE 9

Investor	Number of years in growth fund	Number of years in domestic fund	Number of years in international fund
Alberto	4	1	0
Brenda	0	1	4
Carlos	2	2	1

c. Take natural logarithms of both sides of each equation in part b. Let $I = \ln(1 + i)$, $J = \ln(1 + j)$, and $K = \ln(1 + k)$. Use properties of logarithms to write the resulting system as a *linear* system of three equations in the variables I, J, and K.

d. Use the techniques of this section to solve the linear system from part c for the variables I, J, and K.

e. Use the results of part d and the definitions of I, J, and K given in part c to find i, j, and k.

■ CONCEPTS AND CRITICAL THINKING

EXERCISES 37–40   *Answer true or false.*

37. Only square matrices can have inverses.

38. All invertible matrices can be row-reduced to the identity matrix.

39. Every square matrix has an inverse.

40. If A is an invertible matrix, then the solution to the matrix equation $AX = B$ is given by $X = BA^{-1}$.

EXERCISES 41–44   *Give an example of each.*

41. A noninvertible 1×1 matrix

42. A noninvertible 2×2 matrix with 1s on the main diagonal

43. A system of equations that cannot be solved using the technique of this section

44. A scalar multiple of the 3×3 identity matrix and its inverse

45. Let A and B be invertible matrices. Define $C = AB$ and $D = B^{-1}A^{-1}$. Show that C and D are inverses of one another. (*Hint:* Use the definitions of C and D to show that $CD = I$ and $DC = I$.)

46. Show that a matrix with either a row or a column of all zeros is not invertible.

47. Show that if A is an $n \times n$ matrix, then $I_n \cdot A = A$.

48. Show that $(aI)^{-1} = \frac{1}{a}I$ for any scalar a.

49. Let

$$A = \begin{bmatrix} a & 0 & 0 & 0 \\ 0 & b & 0 & 0 \\ 0 & 0 & c & 0 \\ 0 & 0 & 0 & d \end{bmatrix}$$

Under what conditions is A invertible? If A is invertible, what is its inverse?

50. Let A be an invertible matrix such that $A^2 = A^4$. Show that $A = A^{-1}$.

51. The **trace** of a square matrix is defined to be the sum of the elements on the main diagonal. Compute the trace for each of the following matrices.

a. $\begin{bmatrix} 2 & -1 \\ 3 & -4 \end{bmatrix}$

b. $\begin{bmatrix} 1 & 2 & 4 \\ 6 & 9 & 8 \\ 2 & 4 & 8 \end{bmatrix}$

c. aI_n, where a is a scalar

52. To find the coordinates of the point (x, y) after a counterclockwise rotation of $45°$, we multiply the matrix

$$\begin{bmatrix} x \\ y \end{bmatrix}$$

by

$$\begin{bmatrix} \dfrac{\sqrt{2}}{2} & -\dfrac{\sqrt{2}}{2} \\ \dfrac{\sqrt{2}}{2} & \dfrac{\sqrt{2}}{2} \end{bmatrix}$$

By what would we multiply

$$\begin{bmatrix} x \\ y \end{bmatrix}$$

in order to find the coordinates of the point (x, y) after undergoing a *clockwise* rotation of $45°$?

53. What is I^{-1}?

54. Consider the following system of equations.

$$x^2 y^3 z^4 = 10$$
$$x^3 y^2 z^5 = 9$$
$$x^4 y^6 z^6 = 5$$

a. Is the system linear?

b. Take the natural logarithm of both sides of each equation. The resulting system is linear in the variables $\ln x$, $\ln y$, and $\ln z$. Solve the system for these variables.

c. Find x, y, and z.

■ QUESTIONS FOR DISCUSSION OR ESSAY

55. Up to this point, we have considered several techniques for solving linear systems of equations. Describe each technique, and discuss the circumstances under which each is preferable. Consider such issues as the size of the system, the number of nonzero coefficients, the use of a calculator, and whether other systems are to be solved that have the same coefficient matrix.

56. Explain why a nonsquare matrix could not have an inverse.

57. If A is noninvertible but we attempt to compute A^{-1} by using

Gauss-Jordan elimination as outlined in the text, how will we discover that A^{-1} does not exist?

58. What similarities do you find between the inverse of a matrix and the inverse of a function? A function fails to have an inverse whenever it is not 1–1. What would it mean for a matrix to be 1–1? If f and g have inverses, then it can be shown that $(f \circ g)^{-1} = g^{-1} \circ f^{-1}$. What would the corresponding rule for matrices be?

■ PROJECTS FOR ENRICHMENT

59. *Dimensions*   Physical laws express relationships between quantities such as mass, length, time, velocity, acceleration, force, energy, work, and many others. Each of these quantities is assigned a unit of measurement, and the physical laws relating them are valid only if the units are consistent. When we are not concerned with the particular system of measurement—SI (metric) or U.S. customary—we can instead consider *dimensions*. To each of the three basic quantities—mass, length, and time—we associate the dimensions M, L, and T, respectively. Other quantities have dimensions that are products involving the three basic dimensions. For example, since velocity is equal to distance (dimension L) divided by time (dimension T), the dimension of velocity is $\frac{L}{T}$ or LT^{-1}.

a. Find the dimension of the following quantities. Express your answers in the form $M^a L^b T^c$. Assume that all constants have no dimension.

   i. Acceleration ($=$ velocity $\div$ time)

   ii. Force ($=$ mass $\times$ acceleration)

   iii. Kinetic energy ($= \frac{1}{2} \times$ mass $\times$ velocity^2)

   iv. Work ($=$ force $\times$ distance)

Equations that express physical laws must be *dimensionally compatible* in the sense that all terms must have the same dimension. Consider the equation $s = -\frac{1}{2}gt^2 + v_0t + s_0$, which gives the height of an object at time t if it is thrown into the air with initial velocity v_0 and from an initial height s_0. Both s and s_0 have dimension L. Since g is the acceleration due to gravity and $-\frac{1}{2}$ is a dimensionless constant, the term $-\frac{1}{2}gt^2$ has dimension $(LT^{-2})T^2 = L$. Likewise, v_0t has dimension $(LT^{-1})T = L$. So each term has dimension L and the equation is dimensionally compatible.

b. Determine whether the following equations are dimensionally compatible. Assume m denotes mass; s, distance; t, time; v, velocity; a, acceleration; F, force; and W, work.

   i. $F = mv + v^2$

   ii. $v^2 = v_0^2 + 2as$

   iii. $E = \frac{1}{2}mv^2 + mgs$ (E denotes total energy, which has dimension ML^2T^{-2})

   iv. $W = msv + Fs$

Some physical laws can be derived from a basic form and dimensional considerations. Let us consider the velocity of a falling object. It is reasonable to expect that velocity depends on the acceleration due to gravity and the total time the object has been falling. If we assume (perhaps on the basis of observations) that velocity varies directly as some power of the acceleration due to gravity and some power of time, we obtain the possible model $v = kg^a t^b$, where v is velocity, g is the acceleration due to gravity, t is time, and k, a, and b are constants. Now the dimensions of v, g, and t are LT^{-1}, LT^{-2}, and T, respectively, and the constant k is dimensionless. So the equation $v = kg^a t^b$ is dimensionally compatible only if

$$LT^{-1} = (LT^{-2})^a T^b$$

which implies

$$LT^{-1} = L^a T^{-2a+b}$$

Equating exponents, we obtain the system of equations

$$a = 1$$
$$-2a + b = -1$$

Solving the system, we find $a = 1$ and $b = 1$. Thus $v = kgt$ is dimensionally compatible.

c. Use the same process to find models of the given form.

   i. $T = kr^a g^b$, where T is the period of a pendulum with length r, k is a constant, and g is the acceleration due to gravity.

   ii. $a = kv^i r^j$, where a is the acceleration of a particle traveling with velocity v in a circular path of radius r, and k is a constant.

   iii. $F = km^a v^b r^c$, where F is the centrifugal force of a particle with mass m traveling in a circular path with radius r and velocity v, and k is a constant.

   iv. $F = k\mu^a v^b r^c$, where F is the force opposing a ball of radius r as it is falling with velocity v through air having viscosity coefficient μ (dimension $ML^{-1}T^{-1}$), and k is a constant.

   v. $V = kG^a m^b R^c$, where V is the velocity required for an object to escape the gravitational field of a planet with mass m and radius R, G is the universal gravitational constant with dimension $M^{-1}L^3T^{-2}$, and k is a constant.

60. *Drawing Letters with a Graphing Calculator*   In this project, we learn how matrix techniques can be used to draw letters on the display of a graphing calculator. In particular, we will illustrate how the letter M can be drawn on a graphing calculator. To do so, we use parametric equations. Instead of defining y as a function of x, we define both x and y as functions of the parameter t. For example, suppose we define

$$x = t^2 \qquad \text{and} \qquad y = t^3$$

In other words, to each real number t there corresponds the pair of numbers (x, y) of the form (t^2, t^3). Thus, if $t = -2$, the corresponding pair of numbers consists of $x = (-2)^2 = 4$ and $y = (-2)^3 = -8$, or the pair $(4, -8)$. Some additional values of this correspondence are shown in Table 10. If we plot the points given in the second column of Table 10 and connect the points with a smooth curve, we obtain the graph shown in Figure 11.

TABLE 10

t	(t^2, t^3)
-2	$(4, -8)$
-1	$(1, -1)$
0	$(0, 0)$
1	$(1, 1)$
2	$(4, 8)$

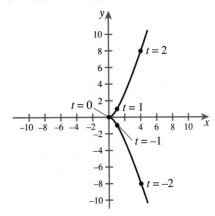

FIGURE 11

We now find functions $x(t)$ and $y(t)$ such that the graph of $(x(t), y(t))$ for t in an appropriate interval will resemble the letter M. Shown in Figure 12 is a depiction of the letter M superimposed on the coordinate axes.

Let us suppose that a bug begins crawling at time 0 from point P_1 to point P_2, arriving at point P_2 at time $t = 1$; continuing to point P_3, arriving at time $t = 2$; and so on, finally arriving at

TABLE 11

t	x	y
0	1	1
1	1	4
2	2	2
3	3	4
4	3	1

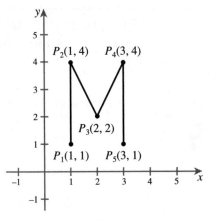

FIGURE 12

point P_5 at time $t = 4$. Table 11 gives the x- and y-coordinates of the bug at times t from 0 to 4.

a. With x as the vertical axis and t as the horizontal axis, plot the tabulated values of x versus t. Connect the points with line segments.

b. It can be shown that the figure obtained in part a is the graph of a function of the form

$$x(t) = a|t| + b|t - 1| + c|t - 2| + d|t - 3| + e|t - 4|$$

Find the constants a, b, c, d, and e by solving the system of equations arising from the fact that the graph must pass through the five indicated points.

c. Now, using the same process as that outlined in parts a and b, find constants f, g, h, i, and j so that

$$y(t) = f|t| + g|t - 1| + h|t - 2| + i|t - 3| + j|t - 4|$$

and the graph of $y(t)$ passes through the points (t, y) given in Table 10.

d. Using the parametric function capabilities of your graphing calculator, graph $(x(t), y(t))$ for t from 0 to 4.

e. Repeat the procedure outlined in steps a through d to draw the letter R.

SECTION 7.5   **DETERMINANTS AND CRAMER'S RULE**

☐ **How can you quickly determine whether a linear system having the same number of equations as unknowns has a unique solution?**

☐ **How can you quickly compute the value of a single variable in a system of equations without solving the entire system?**

☐ **How can matrices be used to determine whether three given points lie on a line?**

☐ **How can the area of a quadrilateral be computed from the coordinates of its vertices?**

DETERMINANTS OF 2 × 2 MATRICES

As mentioned in the previous section, not all square matrices have inverses. We now describe a test for determining *which* matrices have inverses. We begin by considering a matrix that we happen to know is noninvertible—namely,

$$A = \begin{bmatrix} 1 & 3 \\ 4 & 12 \end{bmatrix}$$

Let's see if we can gain insight into the "cause" of invertibility by naively attempting to compute A^{-1}. Our first step gives us

$$\begin{bmatrix} 1 & 3 & | & 1 & 0 \\ 4 & 12 & | & 0 & 1 \end{bmatrix} \xrightarrow{-4R_1 + R_2} \begin{bmatrix} 1 & 3 & | & 1 & 0 \\ 0 & 0 & | & -4 & 1 \end{bmatrix}$$

Already we are in trouble; there is no longer any possibility that the first two columns of this augmented matrix will form the identity matrix. This problem occurs because the second row of A is 4 times the first row of A, so the operation $-4R_1 + R_2$ "wipes out" the second row. In fact, this problem arises whenever one row is a multiple of another. We will show in Exercise 61 that one row of the matrix

$$\begin{bmatrix} a & b \\ c & d \end{bmatrix}$$

is a multiple of the other if and only if $ad - bc = 0$. Thus, the 2 × 2 matrix

$$\begin{bmatrix} a & b \\ c & d \end{bmatrix}$$

is *not* invertible whenever $ad - bc = 0$. Moreover, it can be shown that the matrix *is* invertible whenever $ad - bc \neq 0$. Since the value of $ad - bc$ *determines* the invertibility of A, this quantity is called the **determinant** of A and is denoted by $\det(A)$ or $|A|$. We summarize determinants of 2 × 2 matrices as follows.

Determinants of 2 × 2 Matrices ☐

1. The **determinant** of the matrix

$$A = \begin{bmatrix} a & b \\ c & d \end{bmatrix}$$

is the quantity $ad - bc$, which is denoted by $\det(A)$, $|A|$, or $\begin{vmatrix} a & b \\ c & d \end{vmatrix}$.

2. The determinant of a 2 × 2 matrix is the difference of the products of the diagonal entries, as suggested by Figure 13.

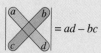

$$= ad - bc$$

FIGURE 13

3. A 2 × 2 matrix is invertible if and only if its determinant is nonzero.

In the following example, we see how the determinant can be used to determine whether a given matrix is invertible.

EXAMPLE 1 ▨ Using Determinants to Test for Invertibility

Use determinants to decide which of the following matrices are invertible.

a. $\begin{bmatrix} 4 & -2 \\ 3 & 5 \end{bmatrix}$ **b.** $\begin{bmatrix} 3 & 3 \\ 2 & 2 \end{bmatrix}$ **c.** $\begin{bmatrix} 1 & -1 \\ -1 & 0 \end{bmatrix}$

SOLUTION

a. The determinant is given by

$$\begin{vmatrix} 4 & -2 \\ 3 & 5 \end{vmatrix} = 4 \cdot 5 - (-2)3 = 26 \neq 0$$

Thus, the matrix is invertible.

b. In this case, we have

$$\begin{vmatrix} 3 & 3 \\ 2 & 2 \end{vmatrix} = 3 \cdot 2 - 3 \cdot 2 = 0$$

and so the matrix is noninvertible.

c. The determinant is given by

$$\begin{vmatrix} 1 & -1 \\ -1 & 0 \end{vmatrix} = 1 \cdot 0 - (-1)(-1) = -1 \neq 0$$

Thus, the matrix is invertible.

EVALUATING DETERMINANTS WITH MINORS

We have seen that the determinant of a 2×2 matrix indicates whether the matrix is invertible. More generally, we define determinants of higher-order square matrices so that the matrix is invertible if and only if its determinant is nonzero.

We define the determinant of a 3×3 matrix as a combination of 2×2 determinants; determinants of 4×4 matrices are expressed as a combination of 3×3 determinants; and so on. We describe the process of computing 3×3 determinants in some detail, but we leave the computation of higher-order determinants to the exercises.

The **minor of an entry** of a 3×3 matrix is defined as the determinant of the 2×2 matrix remaining after both the row and column of the entry are crossed out. For example, if

$$A = \begin{bmatrix} a_{11} & a_{12} & a_{13} \\ a_{21} & a_{22} & a_{23} \\ a_{31} & a_{32} & a_{33} \end{bmatrix}$$

then the minor of a_{11} is the determinant of the 2×2 matrix

$$\begin{bmatrix} a_{22} & a_{23} \\ a_{32} & a_{33} \end{bmatrix}$$

remaining after both the first row and the first column of A are crossed out. Table 12 shows how the minors of all the entries of the first row are computed.

EXAMPLE 2 ▨ Computing the Minor of an Entry

Find the minor of a_{23}, where

$$[a_{ij}] = \begin{bmatrix} -2 & 0 & 5 \\ 3 & 1 & -4 \\ 2 & -6 & -1 \end{bmatrix}$$

TABLE 12   Computing minors

The minor of a_{11}	Cross out row 1 and column 1 $\begin{vmatrix} a_{11} & a_{12} & a_{13} \\ a_{21} & a_{22} & a_{23} \\ a_{31} & a_{32} & a_{33} \end{vmatrix}$	. . . to obtain $\begin{vmatrix} a_{22} & a_{23} \\ a_{32} & a_{33} \end{vmatrix} = a_{22}a_{33} - a_{23}a_{32}$
The minor of a_{12}	Cross out row 1 and column 2 $\begin{vmatrix} a_{11} & a_{12} & a_{13} \\ a_{21} & a_{22} & a_{23} \\ a_{31} & a_{32} & a_{33} \end{vmatrix}$	. . . to obtain $\begin{vmatrix} a_{21} & a_{23} \\ a_{31} & a_{33} \end{vmatrix} = a_{21}a_{33} - a_{23}a_{31}$
The minor of a_{13}	Cross out row 1 and column 3 $\begin{vmatrix} a_{11} & a_{12} & a_{13} \\ a_{21} & a_{22} & a_{23} \\ a_{31} & a_{32} & a_{33} \end{vmatrix}$	. . . to obtain $\begin{vmatrix} a_{21} & a_{22} \\ a_{31} & a_{32} \end{vmatrix} = a_{21}a_{32} - a_{22}a_{31}$

SOLUTION   The minor of a_{23} is the determinant of the 2×2 matrix that results from crossing out row 2 and column 3.

$$\begin{bmatrix} -2 & 0 & 5 \\ 3 & 1 & -4 \\ 2 & -6 & -1 \end{bmatrix}$$

Thus, the minor of a_{23} is given by

$$\begin{vmatrix} -2 & 0 \\ 2 & -6 \end{vmatrix} = (-2)(-6) - 2 \cdot 0 = 12$$

The determinant of a 3×3 matrix can be obtained by multiplying the entries of the first row by their corresponding minors and then either adding or subtracting, as described in the following box. Determinants of higher-order matrices are defined in a similar fashion.

Definition of 3×3 Determinant   □

The determinant of

$$\begin{bmatrix} a_{11} & a_{12} & a_{13} \\ a_{21} & a_{22} & a_{23} \\ a_{31} & a_{32} & a_{33} \end{bmatrix}$$

is given by

$$\begin{vmatrix} a_{11} & a_{12} & a_{13} \\ a_{21} & a_{22} & a_{23} \\ a_{31} & a_{32} & a_{33} \end{vmatrix} = a_{11} \cdot (\text{Minor of } a_{11}) - a_{12} \cdot (\text{Minor of } a_{12}) \\ + a_{13} \cdot (\text{Minor of } a_{13})$$

$$= a_{11} \begin{vmatrix} a_{22} & a_{23} \\ a_{32} & a_{33} \end{vmatrix} - a_{12} \begin{vmatrix} a_{21} & a_{23} \\ a_{31} & a_{33} \end{vmatrix} + a_{13} \begin{vmatrix} a_{21} & a_{22} \\ a_{31} & a_{32} \end{vmatrix}$$

This technique of computing the determinant is called **expansion about the first row.**

E X A M P L E 3   ■   Computing a 3 × 3 Determinant by Expansion about the First Row

Find the determinant of the matrix

$$A = \begin{bmatrix} 1 & 4 & 3 \\ 0 & 2 & 1 \\ -2 & 6 & 0 \end{bmatrix}$$

SOLUTION   We multiply the entries of the first row by their corresponding minors and then add or subtract according to the definition.

$$\begin{vmatrix} 1 & 4 & 3 \\ 0 & 2 & 1 \\ -2 & 6 & 0 \end{vmatrix} = 1 \cdot \begin{vmatrix} 2 & 1 \\ 6 & 0 \end{vmatrix} - 4 \cdot \begin{vmatrix} 0 & 1 \\ -2 & 0 \end{vmatrix} + 3 \cdot \begin{vmatrix} 0 & 2 \\ -2 & 6 \end{vmatrix}$$

$$= 1(0 - 6) - 4[0 - (-2)] + 3[0 - (-4)]$$

$$= -6 - 8 + 12$$

$$= -2$$

Expansion about the first row is not the only technique for evaluating 3 × 3 determinants. In fact, expansion can be done about any row or column. To do so, we simply multiply the entries of the given row or column by their corresponding minors and then either add or subtract, depending on the position of the entry. Specifically, the sign of each term is given by the corresponding entry in the following **matrix of signs.**

$$\begin{bmatrix} + & - & + \\ - & + & - \\ + & - & + \end{bmatrix}$$

For example, to evaluate the determinant of the matrix

$$A = \begin{bmatrix} 1 & 4 & 3 \\ 0 & 2 & 1 \\ -2 & 6 & 0 \end{bmatrix}$$

from Example 3 by expanding about the second column, we would begin by forming the products of each of the entries of the second column with their corresponding minors, as shown.

$$4 \cdot \begin{vmatrix} 0 & 1 \\ -2 & 0 \end{vmatrix} \qquad 2 \cdot \begin{vmatrix} 1 & 3 \\ -2 & 0 \end{vmatrix} \qquad 6 \cdot \begin{vmatrix} 1 & 3 \\ 0 & 1 \end{vmatrix}$$

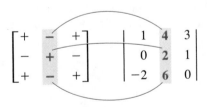

FIGURE 14

Next we affix the appropriate sign to each term using the matrix of signs, as suggested by Figure 14. Thus, we have

$$\begin{vmatrix} 1 & 4 & 3 \\ 0 & 2 & 1 \\ -2 & 6 & 0 \end{vmatrix} = -4 \cdot \begin{vmatrix} 0 & 1 \\ -2 & 0 \end{vmatrix} + 2 \cdot \begin{vmatrix} 1 & 3 \\ -2 & 0 \end{vmatrix} - 6 \cdot \begin{vmatrix} 1 & 3 \\ 0 & 1 \end{vmatrix}$$

$$= -4 \cdot 2 + 2 \cdot 6 - 6 \cdot 1$$

$$= -2$$

RULE OF THUMB   Since we may evaluate the determinant by expanding about any row or column, choose a row or column containing as many zeros as possible.

EXAMPLE 4   ■   Evaluating a 3 × 3 Determinant by Expansion

Compute

$$\begin{vmatrix} 2 & 1 & 3 \\ 1 & 0 & 0 \\ 4 & 5 & 6 \end{vmatrix}$$

SOLUTION   Although both the second and third columns contain a zero, the second row contains two zeros. We thus expand about the second row. Because the second row of the matrix of signs is $- + -$, we have

$$\begin{vmatrix} 2 & 1 & 3 \\ 1 & 0 & 0 \\ 4 & 5 & 6 \end{vmatrix} = (-1) \cdot \begin{vmatrix} 1 & 3 \\ 5 & 6 \end{vmatrix} + 0 \cdot \begin{vmatrix} 2 & 3 \\ 4 & 6 \end{vmatrix} - 0 \cdot \begin{vmatrix} 2 & 1 \\ 4 & 5 \end{vmatrix}$$

$$= (-1)(-9) + 0 - 0$$

$$= 9$$

Note that by choosing the second row, we had only one 2 × 2 determinant to evaluate.

CALCULATOR KEYS

Computing Determinants

If a matrix with numerical entries is invertible, its determinant can be computed with a graphing calculator. As with other matrix operations, you must first specify the order of the matrix and then key in its entries. The determinant can then be computed by selecting the determinant option from the menu of available matrix operations.

A SPECIAL RULE FOR COMPUTING DETERMINANTS OF 3 × 3 MATRICES

The technique of expanding a determinant about a row or a column works for square matrices of any order. We now describe a shortcut for computing 3 × 3 determinants; unfortunately, the process does not generalize to determinants of any other order. To compute the determinant

$$\begin{vmatrix} a_1 & b_1 & c_1 \\ a_2 & b_2 & c_2 \\ a_3 & b_3 & c_3 \end{vmatrix}$$

we begin by forming fourth and fifth columns consisting of copies of columns 1 and 2, respectively, as shown in Figure 15. Next, following the arrows going downward, we form three products to which we affix a + sign and, following the arrows going

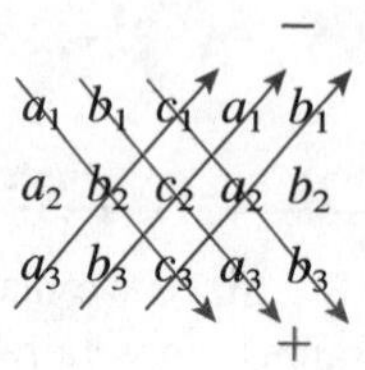

$$= a_1 b_2 c_3 + b_1 c_2 a_3 + c_1 a_2 b_3 - a_3 b_2 c_1 - b_3 c_2 a_1 - c_3 a_2 b_1$$

FIGURE 15

upward, we form three products to which we affix a $-$ sign. Finally, computing the sum of these products, we obtain the value of the determinant.

EXAMPLE 5 ▪ Computing a 3 × 3 Determinant Using a Shortcut

Evaluate

$$\begin{vmatrix} 1 & -2 & 3 \\ 2 & 1 & -1 \\ 1 & 0 & 4 \end{vmatrix}$$

SOLUTION We first form fourth and fifth columns consisting of copies of columns 1 and 2, respectively, and then apply the shortcut rule.

$$
\begin{aligned}
&= 1 \cdot 1 \cdot 4 + (-2)(-1)1 + 3 \cdot 2 \cdot 0 \\
&\quad - 1 \cdot 1 \cdot 3 - 0(-1)1 - 4 \cdot 2(-2) \\
&= 4 + 2 + 0 - 3 - 0 + 16 \\
&= 19
\end{aligned}
$$

CRAMER'S RULE FOR SOLVING SYSTEMS OF EQUATIONS

In the preceding section, we learned how certain systems of linear equations could be solved using inverse matrices. Of course, this method works only if the system is "square" (having the same number of equations as variables) and the coefficient matrix is invertible. It so happens that if precisely these same conditions are met, the solution to such a system of equations can be expressed using determinants according to **Cramer's Rule.**

Cramer's Rule ☐

> Consider a system of n equations in the n variables $x_1, x_2, \ldots, x_n$, expressible in matrix form as $AX = B$, where A is an invertible matrix. Let A_i be the matrix obtained by replacing the ith column of A with the $n \times 1$ matrix B. Then the solution to the system is given by
>
> $$x_1 = \frac{|A_1|}{|A|}, \quad x_2 = \frac{|A_2|}{|A|}, \ldots, x_n = \frac{|A_n|}{|A|}$$

EXAMPLE 6 ▪ Solving a System with Cramer's Rule

Solve the following system using Cramer's Rule.

$$
\begin{aligned}
x - y + 2z &= 3 \\
x + y - 3z &= -11 \\
2x + 3y + z &= 9
\end{aligned}
$$

SOLUTION This system can be expressed in matrix form as $AX = B$, with

$$A = \begin{bmatrix} 1 & -1 & 2 \\ 1 & 1 & -3 \\ 2 & 3 & 1 \end{bmatrix}, \quad X = \begin{bmatrix} x \\ y \\ z \end{bmatrix}, \quad \text{and } B = \begin{bmatrix} 3 \\ -11 \\ 9 \end{bmatrix}$$

Now, the matrices A_1, A_2, and A_3 are each defined by replacing the appropriate column of A with the matrix B. Thus,

$$A_1 = \begin{bmatrix} 3 & -1 & 2 \\ -11 & 1 & -3 \\ 9 & 3 & 1 \end{bmatrix}, \quad A_2 = \begin{bmatrix} 1 & 3 & 2 \\ 1 & -11 & -3 \\ 2 & 9 & 1 \end{bmatrix}, \quad \text{and } A_3 = \begin{bmatrix} 1 & -1 & 3 \\ 1 & 1 & -11 \\ 2 & 3 & 9 \end{bmatrix}$$

According to Cramer's Rule, we have

$$x = \frac{|A_1|}{|A|}, \quad y = \frac{|A_2|}{|A|}, \quad \text{and } z = \frac{|A_3|}{|A|}$$

Evaluation of the determinants, either by hand or with the aid of a graphing calculator, gives

$$|A| = 19, \quad |A_1| = -38, \quad |A_2| = 57, \quad \text{and } |A_3| = 76$$

Thus

$$x = \frac{-38}{19} = -2, \quad y = \frac{57}{19} = 3, \quad \text{and } z = \frac{76}{19} = 4$$

It should be noted that Cramer's rule is usually the least efficient of the methods we have seen for solving linear systems of equations. In Example 6, for example, it was necessary to compute four determinants, and this can be a tedious task even with the aid of a graphing calculator, and even with matrices as small as 3×3. The inverse matrix technique discussed in Section 7.4 would have been much more efficient, requiring us simply to enter the two matrices A and B into our calculator and then compute $A^{-1}B$. When solving linear systems without the aid of a graphing calculator, Gaussian elimination with back-substitution would likely be most efficient.

UNDERSTANDING AND MASTERY CHECKLISTS

CONCEPTS TO UNDERSTAND

- ☐ Determinant of a 2×2 matrix
- ☐ Minor of an entry of a matrix
- ☐ Determinant of a 3×3 matrix
- ☐ Matrix of signs
- ☐ Expansion about a row or column
- ☐ Cramer's Rule for solving linear systems of equations

SKILLS TO MASTER

- ☐ Compute a 2×2 determinant.
- ☐ Find the minor of a matrix entry.
- ☐ Compute a 3×3 determinant by expansion about a row or column.
- ☐ Compute a 3×3 determinant using the special rule.
- ☐ Compute a determinant using a graphing calculator.
- ☐ Use a determinant to determine whether a matrix is invertible.
- ☐ Solve a linear system of equations using Cramer's Rule.

EXERCISES 7.5

EXERCISES 1–22 *Evaluate the given determinant. (Try to select the most efficient technique.)*

1. $\begin{vmatrix} 2 & 3 \\ 4 & 5 \end{vmatrix}$

2. $\begin{vmatrix} 3 & 1 \\ -2 & 0 \end{vmatrix}$

3. $\begin{vmatrix} 1 & 6 \\ 2 & 12 \end{vmatrix}$

4. $\begin{vmatrix} 2 & 3 \\ -2 & 3 \end{vmatrix}$

5. $\begin{vmatrix} 3 & 7 \\ 4 & 1 \end{vmatrix}$

6. $\begin{vmatrix} -2 & 4 \\ -4 & 8 \end{vmatrix}$

7. $\begin{vmatrix} x & x-1 \\ x & x \end{vmatrix}$

8. $\begin{vmatrix} a & 2a \\ b & 2b \end{vmatrix}$

9. $\begin{vmatrix} 2 & 1 & 1 \\ 0 & 1 & 4 \\ 0 & 2 & 3 \end{vmatrix}$

10. $\begin{vmatrix} 1 & 0 & 4 \\ 3 & 5 & -1 \\ 2 & 0 & 6 \end{vmatrix}$

11. $\begin{vmatrix} -4 & 2 & 1 \\ -2 & 1 & 0 \\ 3 & -1 & 5 \end{vmatrix}$

12. $\begin{vmatrix} 3 & -1 & 6 \\ 2 & -4 & 1 \\ -1 & 7 & 0 \end{vmatrix}$

13. $\begin{vmatrix} 1.2 & 3.9 & 2.5 \\ 3.7 & 4.1 & 3.6 \\ 1.0 & 2.5 & 7.1 \end{vmatrix}$

14. $\begin{vmatrix} \frac{2}{3} & 4 & \frac{37}{55} \\ 1 & \frac{1}{2} & 0 \\ 0 & 0 & 3 \end{vmatrix}$

15. $\begin{vmatrix} a & b & c \\ 0 & 1 & d \\ 0 & 0 & 1 \end{vmatrix}$

16. $\begin{vmatrix} x-1 & 2 & 1 \\ 0 & x-2 & 2 \\ 0 & 2 & x-3 \end{vmatrix}$

17. $\begin{vmatrix} i & j & k \\ a_1 & a_2 & a_3 \\ b_1 & b_2 & b_3 \end{vmatrix}$

18. $\begin{vmatrix} a & a & a \\ 1 & 1 & 1 \\ x & y & z \end{vmatrix}$

19. $\begin{vmatrix} 1 & 2 & 3 & 4 \\ 2 & 2 & 5 & 7 \\ 8 & 7 & 0 & 2 \\ 8 & 3 & 9 & 0 \end{vmatrix}$

20. $\begin{vmatrix} 2 & 3 & 5 & 7 \\ 11 & 13 & 17 & 19 \\ 23 & 29 & 31 & 37 \\ 41 & 43 & 47 & 53 \end{vmatrix}$

21. $\begin{vmatrix} a & 1 & 1 & 1 & 1 \\ 0 & a & 1 & 1 & 1 \\ 0 & 0 & a & 1 & 1 \\ 0 & 0 & 0 & a & 1 \\ 0 & 0 & 0 & 0 & a \end{vmatrix}$

22. I_{100} (the 100×100 identity matrix)

EXERCISES 23–30 *Use determinants to test the given matrices for invertibility.*

23. $\begin{bmatrix} -3 & 5 \\ 1 & 0 \end{bmatrix}$

24. $\begin{bmatrix} 4 & -1 \\ -8 & 2 \end{bmatrix}$

25. $\begin{bmatrix} 2 & 5 \\ 4 & 10 \end{bmatrix}$

26. $\begin{bmatrix} 11 & 2 \\ 2 & 11 \end{bmatrix}$

27. $\begin{bmatrix} 2 & 0 & 5 \\ -1 & 3 & 0 \\ -1 & 9 & 5 \end{bmatrix}$

28. $\begin{bmatrix} 4 & -3 & 1 \\ -2 & 0 & 6 \\ 1 & 0 & 2 \end{bmatrix}$

29. $\begin{bmatrix} -12 & 1 & 14 \\ 1 & -16 & 2 \\ 15 & 0 & 0 \end{bmatrix}$

30. $\begin{bmatrix} 0 & 10 & -8 \\ 11 & -1 & 3 \\ 22 & 8 & -2 \end{bmatrix}$

EXERCISES 31–36 *Find the value(s) of x for which the given matrix is not invertible.*

31. $\begin{bmatrix} x & -1 \\ 1 & 2 \end{bmatrix}$

32. $\begin{bmatrix} 2 & x \\ 4 & -3 \end{bmatrix}$

33. $\begin{bmatrix} -3 & x \\ x & -3 \end{bmatrix}$

34. $\begin{bmatrix} x & 2 \\ 8 & x \end{bmatrix}$

35. $\begin{bmatrix} x & 0 & 1 \\ 1 & x & 0 \\ x & 1 & 1 \end{bmatrix}$

36. $\begin{bmatrix} 0 & x & 1 \\ x & 1 & 0 \\ 1 & 0 & x \end{bmatrix}$

EXERCISES 37–46 *Solve the system using Cramer's Rule.*

37. $x + 2y = 19$
    $3x - 7y = -8$

38. $3x - 8y = -2$
    $5x + 3y = 13$

39. $2x + z = 6$
    $x - y = 4$
    $-y + z = 7$

40. $x - y + 2z = 0$
    $4x + y = 11$
    $y - 3z = 5$

41. $x + 2y - z = -3$
    $3x - 8y - 5z = -3$
    $2x + 2y + z = 10$

42. $6x - 3y + 9z = -7.65$
    $x - 2y + 11z = -4.15$
    $3y - 4z = 9.5$

43. $x + 2y + 3z + 4w = 1$
    $y - z + w = 0$
    $2z + 3w = 0$
    $z - w = 0$

44. $p + q + r + s + t = 1$
    $q + r + s + t = 0$
    $r + s + t = 1$
    $s + t = 0$
    $s = 1$

45. $ax + by = c$
    $dx + ey = f$
    (Solve for x and y.
    Assume $ae - bd \neq 0$.)

46. $a_1x + a_2y + a_3z = 0$
    $b_1x + b_2y + b_3z = 0$
    $c_1x + c_2y + c_3z = 0$
    (Assume there is a unique solution.)

■ APPLICATIONS

EXERCISES 47–48 *Assume that production cost is a quadratic function of the number of units produced. Thus, if C denotes the cost and x the number of units produced, then $C = ax^2 + bx + c$ for certain values of a, b, and c. The value of c gives the fixed cost— the cost of production if no units are produced. Use the given data to set up a system of equations involving a, b, and c, and use Cramer's Rule to find the company's fixed costs.*

47.

x (units produced)	10	50	100
C (cost in dollars)	$8000	$5000	$10,000

48.

x (units produced)	150	200	300
C (cost in dollars)	$20,000	$25,000	$40,000

49. *Hospital Breakfast* A hospital patient is served a breakfast consisting of milk, oatmeal, and English muffins. The total mass, caloric value, and fat content were recorded as 427 grams, 311 calories, and 4.4 grams, respectively. It is later discovered that the patient is lactose intolerant, and so it becomes necessary to determine how many grams of milk were consumed. Use the information given in Table 13, together with Cramer's Rule, to determine the number of grams of milk consumed.

TABLE 13

	Oatmeal	Milk	English muffin
Calories per gram	0.54	0.42	2.37
Grams of fat per gram	0.0097	0.011	0.01

50. *Unknown Investment* A total of $10,000 is invested in three different funds that pay dividends and charge load fees according to Table 14.

TABLE 14

	First fund	Second fund	Third fund
Dividend	5%	10%	8%
Load fee	1%	1.5%	0.5%

The total paid in dividends is $680, and the total in load fees is $110. Use Cramer's Rule to determine the amount invested in the first fund.

■ CONCEPTS AND CRITICAL THINKING

EXERCISES 51–54 *Answer true or false.*

51. The determinant of a matrix is itself a matrix.

52. Only square matrices with nonzero determinants are invertible.

53. If a is a real number, then the matrix $\begin{bmatrix} a & -1 \\ 1 & a \end{bmatrix}$ is invertible no matter what the value of a.

54. Cramer's Rule is a method for solving linear systems of equations by evaluating determinants.

EXERCISES 55–58 *Give an example of each.*

55. A 3×3 matrix such that the easiest method of computing the determinant is probably expansion about the second column

56. An application of determinants

57. A 3×3 matrix with determinant 125 (*Hint:* $5^3 = 125$.)

58. A matrix for which the determinant isn't defined

59. Show that if A is a 3×3 matrix such that the matrix equation $AX = 0$ has a nonzero solution X, then $\det(A) = 0$. [*Hint:* Suppose $\det(A) \neq 0$. Then A is invertible. Apply A^{-1} to both sides of the equation $AX = 0$ to find a contradiction.]

60. Let

$$A = \begin{bmatrix} a & b \\ c & d \end{bmatrix}$$

Compute both $\det(A)$ and $\det(kA)$, where k is a scalar. If A is a 3×3 matrix, what is your guess as to how $\det(kA)$ compares to $\det(A)$?

61. Complete the following steps to show that one row of the matrix

$$\begin{bmatrix} a & b \\ c & d \end{bmatrix}$$

is a multiple of the other if and only if $ad - bc = 0$.

a. If $c = ka$ and $d = kb$ (that is, row 2 is a multiple of row 1), show that $ad - bc = 0$.

b. If $ad - bc = 0$, show that $a/c = b/d$ and so $a = kc$ and $b = kd$ for some k. Thus, row 1 is a multiple of row 2.

62. It is a well-known fact that $\det(AB) = \det(A)\det(B)$ for square matrices A and B. Use this fact to show that $\det(A^{-1}) = 1/\det(A)$.

■ **QUESTIONS FOR DISCUSSION OR ESSAY**

63. Since solving linear systems of equations with inverses or Cramer's Rule is relatively easy and can be performed by most graphing calculators, is there ever a reason to use Gauss-Jordan elimination? Explain. Are there any linear systems of equations that can be solved by Gauss-Jordan elimination but not by inverses or Cramer's Rule?

64. Suppose that ten people are to split a jackpot and that the shares correspond to the solution of a certain system of 10 equations in 10 unknowns. If you were given the task of computing each person's share, what method would you use? If you were one of the ten winners, what method would you use to compute *your* share? Explain your answers.

65. With Cramer's Rule, the solution values to a system of linear equations are expressed as quotients of determinants. Of course, Cramer's Rule will not provide us with a solution if a determinant appearing in the denominator is 0. If that should happen, would it make sense to attempt to solve the system using the inverse of the coefficient matrix?

66. What is the maximum total number of arithmetic operations (additions, subtractions, and multiplications) required to compute the determinant of a 3×3 matrix by expansion? What is the minimum? What about a 4×4 matrix?

■ **PROJECTS FOR ENRICHMENT**

67. *Performing Elementary Row Operations by Matrix Multiplication*   In this project, we see that for 3×3 matrices, each of the elementary row operations (switching rows, adding a multiple of one row to another, and multiplying a row by a constant) can be represented as multiplication by an appropriate matrix.

  a. Define

$$S_{12} = \begin{bmatrix} 0 & 1 & 0 \\ 1 & 0 & 0 \\ 0 & 0 & 1 \end{bmatrix}$$

  Show that if B is *any* 3×3 matrix, then $S_{12}B$ is the matrix obtained by switching the first and second rows of B.

  b. Find matrices S_{13} and S_{23} such that if B is an arbitrary 3×3 matrix, then $S_{13}B$ and $S_{23}B$ are the matrices obtained by switching the first and third rows of B, and the second and third rows of B, respectively.

  c. Find $M_1(x)$, $M_2(x)$, and $M_3(x)$ such that if B is an arbitrary 3×3 matrix, then $M_i(x)B$ is the matrix formed by multiplying row i of B by x.

  d. Define $A_{13}(x)$ to be the matrix

$$\begin{bmatrix} 1 & 0 & 0 \\ 0 & 1 & 0 \\ x & 0 & 1 \end{bmatrix}$$

  Show that if B is an arbitrary 3×3 matrix, then $A_{13}(x)B$ is the matrix obtained by adding x times row 1 of B to row 3 of B. Find expressions for $A_{12}(x)$ and $A_{23}(x)$.

  e. Compute the determinants of all of the matrices S_{ij}, $M_i(x)$, and $A_{ij}(x)$.

  f. It is a well-known fact that for square matrices C and D, $\det(CD) = \det(C)\det(D)$. Use this fact and the results of part

e to show that performing an elementary row operation on a matrix does not change its invertibility.

68. *Determinants and Plane Geometry*   In this project, we explore several geometric applications of determinants. The area of a triangle in the coordinate plane with vertices (x_1, y_1), (x_2, y_2), and (x_3, y_3) can be shown to be the absolute value of the quantity

$$\frac{1}{2} \begin{vmatrix} x_1 & y_1 & 1 \\ x_2 & y_2 & 1 \\ x_3 & y_3 & 1 \end{vmatrix}$$

  a. Find the area of the triangle having vertices with coordinates $(0, 0)$, $(1, 2)$, and $(3, 5)$.

  b. Use the determinant formula for the area of a triangle to develop a formula for the area of quadrilateral *PQRS* shown in Figure 16.

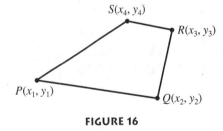

FIGURE 16

  c. Use the formula obtained in part b to find the area of the quadrilateral with vertices $(2, 3)$, $(4, 7)$, $(1, 8)$, and $(0, 5)$.

  d. If, for a given triple of points (x_1, y_1), (x_2, y_2), and (x_3, y_3), the area of the resulting triangle is 0, then what does this say about the three points?

  e. Suppose that (x, y) is a point on the line determined by (x_1, y_1) and (x_2, y_2). Use the fact that the area of the triangle determined by (x, y), (x_1, y_1), and (x_2, y_2) is 0 to find a determinant form of the equation of the line.

SECTION 7.6   **SYSTEMS OF INEQUALITIES**

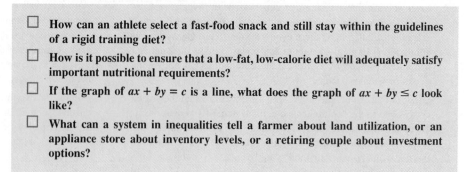

☐ **How can an athlete select a fast-food snack and still stay within the guidelines of a rigid training diet?**

☐ **How is it possible to ensure that a low-fat, low-calorie diet will adequately satisfy important nutritional requirements?**

☐ **If the graph of $ax + by = c$ is a line, what does the graph of $ax + by \leq c$ look like?**

☐ **What can a system in inequalities tell a farmer about land utilization, or an appliance store about inventory levels, or a retiring couple about investment options?**

In the first five sections of this chapter, we developed a handful of techniques for solving systems of linear equations. We now turn our attention to a closely related topic: systems of linear inequalities. Notationally, a system of linear inequalities appears nearly identical to a system of linear equations—the only difference is that the equals sign of a linear system of equations is replaced by one of the four standard inequality symbols. But the solution sets of systems of linear inequalities are considerably more complex than their equation counterparts. For example, the solution set to a linear system of equations in two variables is either empty, a single point, or a line in the plane. By contrast, the solution set to a linear system of inequalities could be virtually any planar region whose edges are linear, including the inside of a triangle, the outside of a hexagon, quadrants, half-planes, and so on. In fact, one of the primary applications of systems of linear inequalities is to mathematically define a region in the plane. In applied settings, this region might correspond to a subset of interest: men over 35 who weigh at least 160 pounds or students with IQs below 120 with a combined SAT score of 1000.

In this section, we solve systems of linear inequalities by exploiting the intimate connection between algebra and geometry. For the sake of simplicity, we confine our attention to systems involving just two variables.

LINEAR INEQUALITIES

A **linear inequality** in the variables x and y is an inequality that can be written in one of the forms

$$ax + by \leq c, \quad ax + by < c, \quad ax + by \geq c, \quad ax + by > c$$

where a, b, and c are real numbers. For example, $2x - 3y > 6$ is a linear inequality in x and y. A point (x, y) is a **solution** of an inequality in x and y if it satisfies the inequality. For example, $(4, -1)$ is a solution of $2x - 3y > 6$ since $2(4) - 3(-1) = 8 + 3 = 11$ and $11 > 6$. The **solution set** of an inequality is the set of all solutions.

EXAMPLE 1   ■   Checking Solutions of Linear Inequalities

Determine whether the following ordered pairs are solutions of the linear inequality $2x - 3y > 6$.

a. $(3, 0)$ **b.** $(-3, -5)$

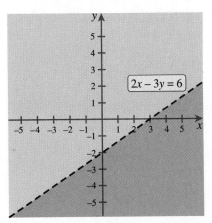

FIGURE 17

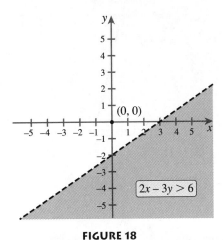

FIGURE 18

SOLUTION

a. $2(3) - 3(0) = 6$ and $6 \not> 6$. Thus, $(3, 0)$ is not a solution.

b. $2(-3) - 3(-5) = -6 - (-15) = 9$ and $9 > 6$. Thus, $(-3, -5)$ is a solution.

The **graph** of an inequality in the variables x and y is the set of all points (x, y) in the coordinate plane that are solutions of the inequality. We can obtain the graph of a linear inequality by first graphing the corresponding linear equation $ax + by = c$. This line divides the coordinate plane into two **half-planes,** exactly one of which contains solutions of the inequality.

Consider, for example, the inequality $2x - 3y > 6$. The graph of the corresponding linear equation $2x - 3y = 6$ is shown in Figure 17. Notice that the graph has been drawn with a dashed line. This is because the **strict inequality** $>$ in the expression $2x - 3y > 6$ indicates that none of the points on the line $2x - 3y = 6$ are actually part of the solution set. If the inequality had been $\geq$ or $\leq$, then the line itself would have been part of the solution set, and so a solid line would have been used. In any case, the line $2x - 3y = 6$ divides the coordinate plane into two half-planes (shaded green and blue in Figure 17), exactly one of which contains solutions of the inequality. To determine which half-plane is the correct one, we simply test a point on one side of the line to see if the inequality is satisfied. We select $(0, 0)$ as our test point and substitute $x = 0$, $y = 0$ into the original inequality.

$$2(0) - 3(0) \overset{?}{>} 6$$
$$0 \not> 6$$

Thus, $(0, 0)$ does not satisfy the inequality, and so we know that the solutions of the inequality are on the other side of the line. This is the region we have shaded in Figure 18.

The general procedure for graphing linear inequalities is summarized as follows.

Graphing a Linear Inequality ☐

> **1.** Replace the inequality symbol with an $=$ and sketch the graph of the resulting line. Use a dashed line for the strict inequalities $<$ or $>$ and a solid line for $\leq$ or $\geq$.
>
> **2.** Test a point on one side of the line. If the point satisfies the original inequality, then shade the half-plane containing that point. If the point does not satisfy the inequality, shade the half-plane on the other side of the line.

E X A M P L E 2 ▇ **Graphing a Linear Inequality**

Graph the linear inequality $4x + y \leq 0$.

SOLUTION First, we replace the inequality symbol with $=$ to obtain the equation $4x + y = 0$. Next, we rewrite the equation as $y = -4x$, which is the equation of a line passing through the origin with slope -4, as shown in Figure 19. Note that a solid line is used since the inequality symbol is $\leq$. To determine which side of the line to shade, we test a point on one side of the line. We choose $(0, -1)$ and substitute $x = 0, y = -1$ into the original inequality.

$$4(0) + (-1) \overset{?}{\leq} 0$$

$$-1 \leq 0$$

Since $(0, -1)$ satisfies the inequality, we shade the region containing $(0, -1)$, as shown in Figure 20.

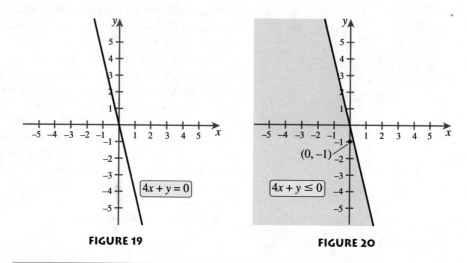

FIGURE 19   **FIGURE 20**

E X A M P L E   3   ■   Graphing a Linear Inequality

Graph the linear inequality $y < -3$.

SOLUTION   The graph of $y = -3$ is a horizontal line with y-intercept -3. The points that satisfy the inequality $y < -3$ are those having y-coordinate less than -3; that is, the points that satisfy the inequality are those that lie below the line, as shown in Figure 21.

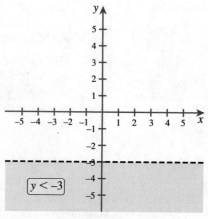

FIGURE 21

SYSTEMS OF LINEAR INEQUALITIES

A **system of linear inequalities** consists of a collection of two or more linear inequalities, such as

$$x + 2y \geq 2$$
$$-3x + 4y \leq 12$$

A **solution** of a system of inequalities in x and y is a point (x, y) that satisfies *all* of the inequalities in the system. For example, the point $(3, 2)$ is a solution of the system just given since $3 + 2(2) = 7 \geq 2$ and $-3(3) + 4(2) = -1 \leq 12$. The **graph** of a system of inequalities in x and y is the set of points (x, y) that satisfy all of the inequalities in the system. We can obtain the graph of a system of inequalities by first sketching the graph of each individual inequality on the same coordinate system and then determining the region common to the graphs of all the inequalities. In other words, we find the intersection of the graphs of all the inequalities of the system.

EXAMPLE 4   ■   Graphing a System of Linear Inequalities

Graph the system of inequalities

$$x + 2y \geq 2$$
$$-3x + 4y \leq 12$$

SOLUTION   We first graph $x + 2y \geq 2$. This is done by first plotting the graph of $x + 2y = 2$ as a solid line. Next, we test the point $(0, 0)$ to see whether it satisfies the inequality. Since $0 + 2(0) = 0 \not\geq 2$, we shade the side of the line that does not contain $(0, 0)$. The result is shown in Figure 22. Next, on the same coordinate system, we graph $-3x + 4y \leq 12$. To do this, we plot the graph of $-3x + 4y = 12$ as a solid line and test $(0, 0)$ to see whether it satisfies the inequality. Since $-3(0) + 4(0) = 0 \leq 12$, we shade the side of the line $-3x + 4y = 12$ that contains $(0, 0)$. The two shaded regions are shown in Figure 23. The graph of the system is the region that is common to the graphs of both inequalities, as shown in Figure 24.

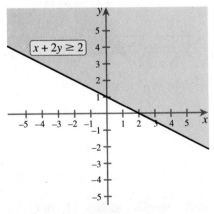

FIGURE 22

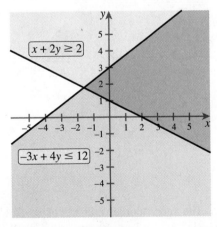

FIGURE 23

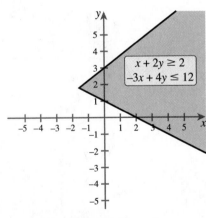

FIGURE 24

E X A M P L E 5 ■ Graphing a System of Linear Inequalities

Graph the system of inequalities

$$x + y \leq 4$$
$$-2x + y \leq 1$$
$$y \geq -1$$
$$x \leq 2$$

SOLUTION   Figures 25–28 illustrate the sequence of steps we took in graphing each of the inequalities on the same coordinate system. The graph of the system is the region that is common to the graphs of all four inequalities, as shown in Figure 29.

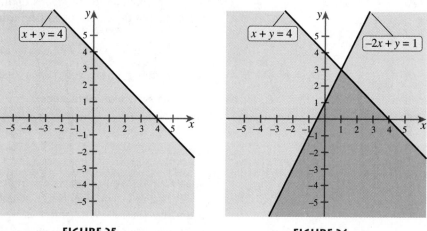

FIGURE 25   **FIGURE 26**

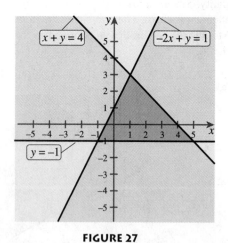

FIGURE 27

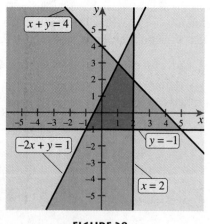

FIGURE 28

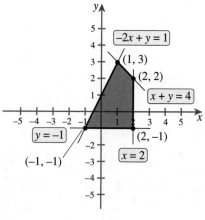

FIGURE 29

The four "corner points" of the region in Example 5—namely, (1, 3), (2, 2), (2, −1), and (−1, −1)—are called **vertices.** In general, the vertices of a system of linear inequalities are found by locating the intersection points of appropriate pairs of lines. Since the exact coordinates of the points of intersection may not be obvious from the graph, it is usually necessary to solve systems of linear equations. Thus, for example, the vertex $(1, 3)$ in Example 5 could be found by solving the equations $-2x + y = 1$ and $x + y = 4$ simultaneously, using any of the methods of Section 2.8 or of this chapter: We choose the elimination technique from Section 2.8.

$$-(-2x + y) = -1 \qquad \text{Multiplying the first equation by } -1 \text{ and}$$
$$+ \quad\;\; x + y = 4 \qquad\quad \text{adding to the second}$$
$$\overline{\qquad\qquad\quad}$$
$$3x = 3$$
$$x = 1$$
$$1 + y = 4 \qquad \text{Substituting } x = 1 \text{ into the second equation}$$
$$y = 3$$

EXAMPLE 6   ▨   **An Application of Systems of Linear Inequalities**

Andrea is training for a bike race and so wants to keep her weight down while maintaining a high intake of protein and carbohydrates. On the way home from a particularly hard workout, she decides to treat herself to a chocolate milk shake and an order of fries at her favorite fast-food restaurant. Each gram of chocolate milk shake and fries provides the following nutritional content.

	Milk shake	Fries
Calories	1.5	3
Grams of protein	0.04	0.04
Grams of carbohydrates	0.2	0.4

For this snack, she must not exceed 405 calories but would like to have at least 6 grams of protein and 40 grams of carbohydrates. Set up and graph a system of linear inequalities that describes how many grams of milk shake and fries she can have. Locate and interpret the vertices of the solution set.

SOLUTION   We assign the following variables to represent the number of grams of milk shake and fries.

$$x = \text{Number of grams of milk shake}$$
$$y = \text{Number of grams of fries}$$

To meet the described requirements, the following inequalities must be satisfied.

$$\text{Calories from milk shake} + \text{Calories from fries} \leq 405$$
$$1.5x + 3y \leq 405$$

$$\text{Protein from milk shake} + \text{Protein from fries} \geq 6$$
$$0.04x + 0.04y \geq 6$$

$$\text{Carbohydrates from milk shake} + \text{Carbohydrates from fries} \geq 40$$
$$0.2x + 0.4y \geq 40$$

Since x and y cannot be negative, we must also have $x \geq 0$ and $y \geq 0$. Thus, the system of inequalities is

$$1.5x + 3y \leq 405$$
$$0.04x + 0.04y \geq 6$$
$$0.2x + 0.4y \geq 40$$
$$x \geq 0$$
$$y \geq 0$$

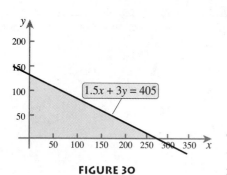

FIGURE 30

The last two inequalities tell us that we need to be concerned only with the first quadrant when we graph the first three inequalities. Refer to Figures 30–32 for these graphs. The resulting region is shaded in Figure 33.

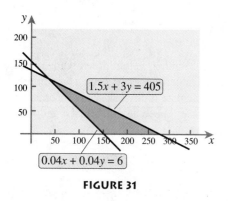

FIGURE 31

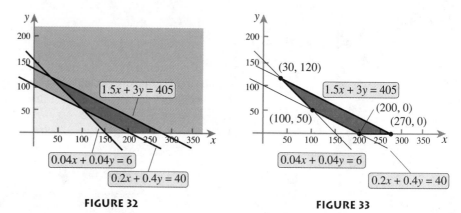

FIGURE 32

FIGURE 33

Any point (x, y) in the shaded region represents an acceptable combination of milk shake and fries for Andrea. The vertices $(200, 0)$ and $(270, 0)$ are the x-intercepts of the lines $0.2x + 0.4y = 40$ and $1.5x + 3y = 405$. The other two vertices are found by solving the following systems of equations. The details of these solutions are omitted.

$(100, 50)$: $0.04x + 0.04y = 6$ $(30, 120)$: $0.04x + 0.04y = 6$
 $0.2x + 0.4y = 40$ $1.5x + 3y = 405$

From the vertex $(100, 50)$, we conclude that one possible combination would be 100 grams of milk shake and 50 grams of fries. From the vertex $(30, 120)$, we conclude that Andrea can have 120 grams of fries if she drinks only 30 grams of milk shake. The values of the other two vertices, $(200, 0)$ and $(270, 0)$, suggest that if she skips the fries altogether, she can drink anywhere from 200 to 270 grams of milk shake.

NONLINEAR INEQUALITIES

The graph of a nonlinear inequality can be obtained in much the same way as the graph of a linear inequality. We first replace the inequality symbol with an $=$ and then sketch the graph of the resulting equation. For example, to graph $4 - x^2 \geq y$, we first graph $4 - x^2 = y$, as shown in Figure 34. As before, we use a dashed curve for the strict inequalities $<$ or $>$ and a solid curve for $\leq$ or $\geq$.

The graph of the equation normally divides the coordinate plane into two or more regions. In each region, either all of the points are solutions of the inequality or none of the points are solutions. Thus, we test a point in *each* of the regions to see which are solutions.

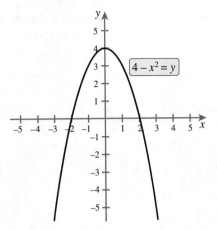

FIGURE 34

Test point	$4 - x^2 \overset{?}{\geq} y$	Conclusion
$(0, 5)$	$4 - (0)^2 \overset{?}{\geq} 5$ $4 \not\geq 5$	The region above the parabola is not in the solution set.
$(0, 0)$	$4 - (0)^2 \overset{?}{\geq} 0$ $4 \geq 0$	The region inside the parabola is in the solution set.

We conclude that the solution set consists of the region on and below the parabola $y = 4 - x^2$, as shown in Figure 35.

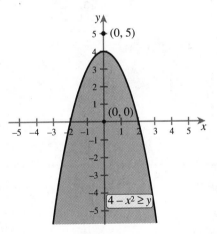

FIGURE 35

EXAMPLE 7   ■   **Graphing a Nonlinear Inequality**

Graph the inequality $\dfrac{x^2}{9} - \dfrac{y^2}{4} > 1$.

SOLUTION   We first plot the hyperbola

$$\frac{x^2}{9} - \frac{y^2}{4} = 1$$

with dashed curves, as shown in Figure 36. The hyperbola divides the plane into three regions, and so we pick a point in each to see which satisfies the inequality.

Test point	$\dfrac{x^2}{9} - \dfrac{y^2}{4} \overset{?}{>} 1$	Conclusion
$(-5, 0)$	$\dfrac{(-5)^2}{9} - \dfrac{(0)^2}{4} \overset{?}{>} 1$ $\dfrac{25}{9} > 1$	The region inside the left branch is in the solution set.
$(0, 0)$	$0 \not> 1$	The region between the branches is not in the solution set.
$(5, 0)$	$\dfrac{(5)^2}{9} - \dfrac{(0)^2}{4} \overset{?}{>} 1$ $\dfrac{25}{9} > 1$	The region inside the right branch is in the solution set.

We conclude that the solution set is inside the two branches of the hyperbola, as shaded in Figure 37.

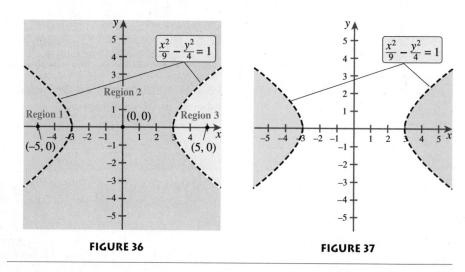

FIGURE 36   **FIGURE 37**

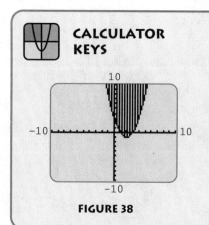

FIGURE 38

CALCULATOR KEYS

Graphing Inequalities

A graphing calculator can be used to plot the graph of an inequality of the form $y \le f(x)$ or $y \ge f(x)$ by indicating whether the region above or below the graph of $y = f(x)$ is to be shaded. For example, the graph of $y \ge x^2 - 4x + 3$ shown in Figure 38 was obtained by graphing $y = x^2 - 4x + 3$ with the "above-the-graph" shading option.

E X A M P L E 8   ■   **Graphing a System of Nonlinear Inequalities**

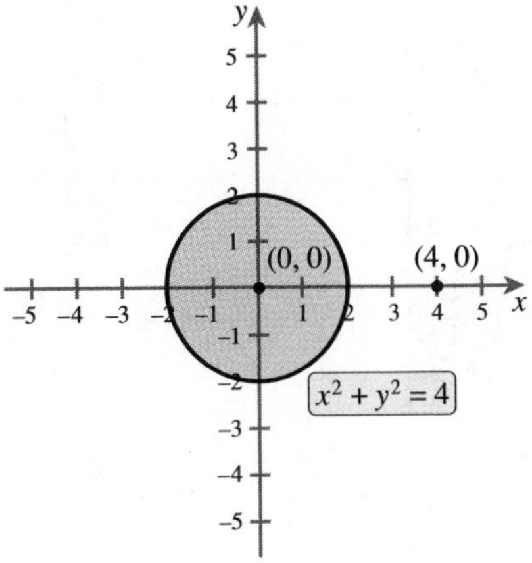

Graph the system of inequalities

$$x^2 + y^2 \leq 4$$
$$y \geq |x|$$

SOLUTION   For the inequality $x^2 + y^2 \leq 4$, we plot the circle $x^2 + y^2 = 4$ and test the points $(0, 0)$ and $(4, 0)$.

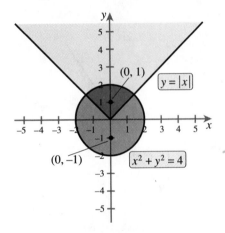

$(0, 0)$ $(4, 0)$

$x^2 + y^2 = 4$

FIGURE 39

Test point	$x^2 + y^2 \overset{?}{\leq} 4$	Conclusion
$(0, 0)$	$(0)^2 + (0)^2 \overset{?}{\leq} 4$ $0 \leq 4$	The region inside the circle is in the solution set.
$(4, 0)$	$(4)^2 + (0)^2 \overset{?}{\leq} 4$ $16 \nleq 4$	The region outside the circle is not in the solution set.

The resulting region is shaded in Figure 39. Similarly, for the inequality $y \geq |x|$, we plot $y = |x|$ (on the same coordinate system) and test the points $(0, -1)$ and $(0, 1)$.

$(0, 1)$ $y = |x|$

$(0, -1)$ $x^2 + y^2 = 4$

FIGURE 40

Test point	$y \overset{?}{\geq} |x|$	Conclusion
$(0, -1)$	$-1 \overset{?}{\geq} |0|$ $-1 \ngeq 0$	The region below $y = |x|$ is not in the solution set.
$(0, 1)$	$1 \overset{?}{\geq} |0|$ $1 \geq 0$	The region above $y = |x|$ is in the solution set.

The intersection of the two regions is shaded in dark blue in Figure 40.

UNDERSTANDING AND MASTERY CHECKLISTS

CONCEPTS TO UNDERSTAND

- ☐ Linear inequality in two variables
- ☐ Solution of a linear inequality in two variables
- ☐ Graph of a linear inequality in two variables
- ☐ System of linear inequalities
- ☐ Solution of a system of linear inequalities
- ☐ Graph of a system of linear inequalities
- ☐ All of the preceding concepts with ''nonlinear'' in place of ''linear''

SKILLS TO MASTER

- ☐ Determine whether a point is a solution of a system of inequalities.
- ☐ Graph an inequality in two variables.
- ☐ Graph a system of inequalities in two variables.

EXERCISES 7.6

EXERCISES 1–6 *Determine which of the ordered pairs are solutions of the inequality.*

1. $x - 2y \geq 5$; $(0, 0)$, $(4, 1)$, $(-1, -3)$
2. $y < 3x + 2$; $(0, 2)$, $(-1, 1)$, $(5, 5)$
3. $4x + 3y > 12$; $(1, 4)$, $(3, 0)$, $(-2, -1)$
4. $6x \geq 3y - 2$; $(-1, -5)$, $(1, 2)$, $(0, -3)$
5. $y + x^2 \leq 4$; $(0, 2)$, $(-1, 3)$, $(5, -1)$
6. $x^2 + y^2 > 25$; $(1, -2)$, $(-4, 3)$, $(5, 5)$

EXERCISES 7–12 *Match the inequality with its graph.*

7. $3x - 2y \geq 6$
8. $3x + 2y \leq 6$
9. $x < 3$
10. $y > 3$
11. $y - x^2 \geq 0$
12. $x + y^2 \leq 0$

iii.

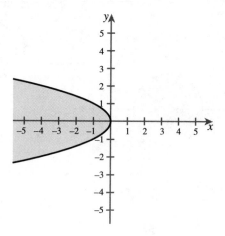

i.

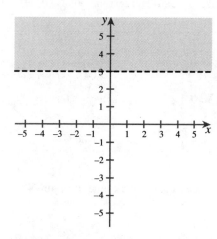

iv.

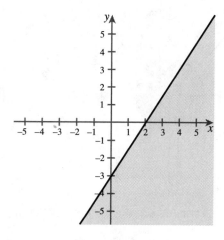

ii.

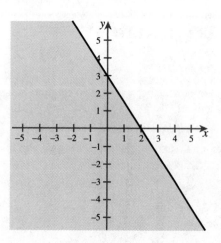

v.

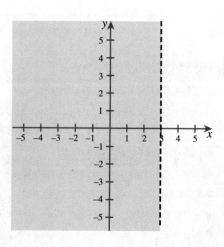

vi.

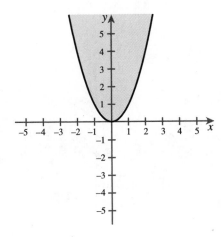

EXERCISES 13–28   *Graph the inequality.*

13. $x + y < 8$
14. $x - y \geq 2$
15. $6x - 2y \leq 12$
16. $5x + 4y > -20$
17. $x \geq -5$
18. $y < 4$
19. $x > -3y$
20. $2x - y \leq 0$
21. $y < x^2 + 1$
22. $x^2 - y \geq 4$
23. $y^2 \leq 1 - x^2$
24. $x^2 + y^2 > 9$
25. $y \geq \frac{1}{x}$
26. $x - y^2 < 0$
27. $y > e^x$
28. $y - \ln x \leq 0$

EXERCISES 29–32   *Match the graph display with the inequality that produced it.*

29. $y \leq -x^2 - 6x - 7$
30. $y \geq x^2 - 6x + 7$
31. $y \leq x^3 - 9x$
32. $y \geq -x^3 + 9x$

i.

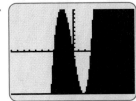

ii.

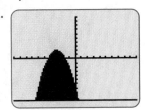

iii.

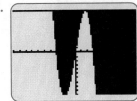

iv.

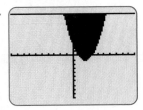

EXERCISES 33–54   *Graph the system of inequalities.*

33. $x + y \leq 5$
    $x - y \leq 1$
34. $2x + 3y > 6$
    $x - 3y < 3$
35. $y < 2x$
    $x + 2y < 5$
36. $x - 2y \geq -6$
    $x - 2y \leq 3$
37. $4x + y \leq 6$
    $-4x - y \leq 4$
38. $5x - 2y < 5$
    $2y > 5x - 20$
39. $x + y \leq 2$
    $-x + y \geq 2$
    $x \geq -2$
40. $x - 2y \leq 4$
    $y \geq -2x$
    $y \leq 4$
41. $-3x + 2y < 6$
    $-x + 3y > 2$
    $2x + y < 3$
42. $x - 2y > -4$
    $-3x + 4y > -18$
    $x - 3y > 6$
43. $2x + 5y \leq 10$
    $x + y \leq 3$
    $x \geq 0$
    $y \geq 0$
44. $2x + 3y \leq 15$
    $3x + y \leq 12$
    $x \geq 0$
    $y \geq 0$
45. $2x + y \geq 8$
    $2x - y \leq 8$
    $x \geq 3$
    $y \leq 4$
46. $y \geq 3x - 5$
    $y \leq 2x$
    $x \leq 3$
    $y \geq 1$
47. $-2x + y \geq 1$
    $y \leq 4 - x^2$
48. $x - y^2 > 0$
    $x + y < 2$
49. $xy < 1$
    $y < x$
50. $x^2 + y^2 \leq 9$
    $x - y \leq 0$
51. $x^2 + y^2 \leq 16$
    $x^2 + y^2 \geq 4$
52. $y \leq \sqrt{x}$
    $y \geq \frac{1}{4}(x^2 - 2x)$
53. $y > x^3 - x$
    $y < 6x$
    $x \geq 0$
54. $y \leq e^x$
    $y \geq 1$
    $x \geq 0$

EXERCISES 55–64   *Find an inequality or system of inequalities with the indicated region as the solution set.*

55.

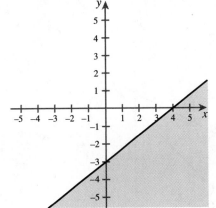

56.

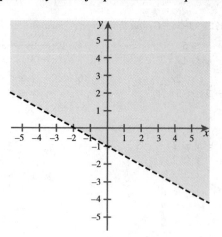

57.

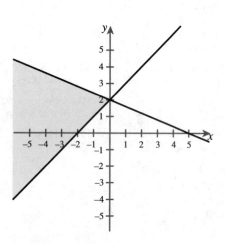

58.

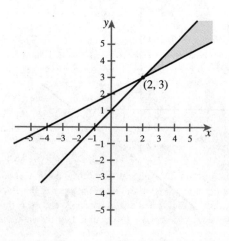

59.

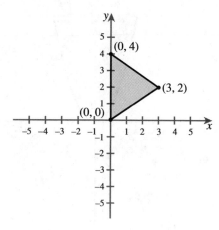

60.

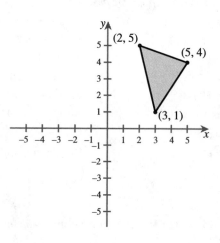

61.

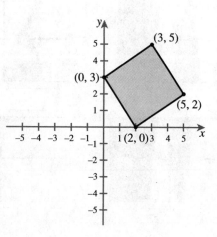

62.

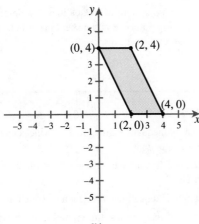

63.

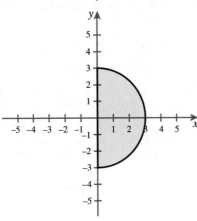

64.

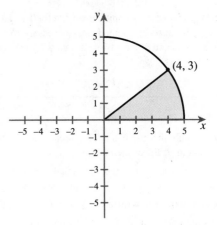

■ **APPLICATIONS**

---

65. *Retirement Planning*   Juan and Rosa are nearing retirement and want to invest up to $24,000 of their savings in two different mutual funds, Redbook Growth and Redbook Small Cap. Their financial adviser suggests they put at least $6000 in each fund, but because Redbook Small Cap has more risk, the amount they invest in it should be no more than half the amount invested in Redbook Growth. Set up and graph a system of inequalities that describes the various amounts they can invest in each fund. Locate the vertices of the solution set.

66. *Low-Fat Diet*   Aleshia is on a low-fat, low-calorie diet, but she prefers 2% milk over skim milk and won't give up her mozzarella cheese stick snacks. Each serving of 2% milk and mozzarella cheese contains the following.

	2% Milk	Mozzarella cheese
Calories	120	80
Grams of saturated fat	3	3
Grams of calcium	300	200

To stay within the guidelines of her diet, she must not exceed 480 calories and 15 grams of saturated fats from milk and cheese. Her daily goal for calcium from milk and cheese is at least 600 grams. Set up and graph a system of inequalities that describes how many servings of milk and cheese she can have. Locate the vertices of the solution set.

67. *Land Allocation*   Roy has 1000 acres of land available to raise corn and soybeans although he may leave some unplanted. Each acre of corn costs $100 and requires 2 hours of labor. Each acre of soybeans costs $80 and requires 1 hour of labor. Roy does not wish his costs for the two crops to exceed $88,000, and he has at most 1600 hours of labor available for the two crops. Also, to feed his own cattle, he must plant at least 200 acres of corn. Set up and graph a system of inequalities that describes how many acres of corn and soybeans Roy should plant. Locate the vertices of the solution set.

68. *Inventory Control*   A TV and appliance store carries a large inventory of TVs and refrigerators. Each TV requires 10 cubic feet of storage space and costs the store $500. Each refrigerator requires 50 cubic feet of storage space and costs the store $1000. The store has at most 1500 cubic feet of warehouse space and

$48,000 of inventory capital available for TVs and refrigerators. Finally, because of demand, it is necessary to stock at least 8 refrigerators and at least twice as many TVs as refrigerators. Set up and graph a system of inequalities that describes how many TVs and refrigerators should be kept in stock. Locate the vertices of the solution set.

CONCEPTS AND CRITICAL THINKING

EXERCISES 69–72 *Answer true or false.*

69. It is possible that the inside of a circle could be the solution set to a system of linear inequalities.

70. The solution set to a system of inequalities is the set of all points satisfying at least one of the inequalities in the system.

71. The solution set to a linear inequality is a half-plane.

72. The solution set of $mx + b < y$ is the half-plane lying entirely above the line $y = mx + b$.

EXERCISES 73–76 *Give an example of each.*

73. A system of inequalities, the graph of which consists of all points in the first quadrant

74. A system of inequalities, the graph of which is the inside of a semicircle

75. A system of inequalities, the graph of which is a right triangle

76. A system of inequalities, the graph of which is a square in the fourth quadrant

QUESTIONS FOR DISCUSSION OR ESSAY

77. When graphing systems of linear inequalities, it is helpful to be aware of the number of regions that are possible when one or more lines are drawn on the coordinate plane. For example, if one line is drawn on the coordinate plane, it divides the plane into two regions. If two lines are drawn, they divide the plane into either three or four regions, depending on whether the lines intersect. Discuss the connection between the number of regions formed by two lines and the solution sets of a system of two linear inequalities. What observations and connections can be made concerning the number of regions formed by three lines and the solutions of systems of three linear inequalities?

78. Describe the circumstances under which the solution set of a system of two linear inequalities is either the empty set or the entire coordinate plane. Give examples of such systems.

79. Explain in detail why a linear inequality involving $<$ or $>$ is graphed with a dashed line, whereas one involving $\leq$ or $\geq$ is graphed with a solid line.

PROJECTS FOR ENRICHMENT

80. *Systems of Inequalities and Flight Paths* Inspector Magill has been called in to settle a dispute between an airline company and a group of environmentalists. The airline company had been given a court order that prohibits planes from flying over a region that is the sole breeding ground for an endangered bird species. The environmentalists claim that certain flights are still flying over this region, and so Magill must determine whether this is indeed the case. His first task is to describe the region algebraically. By setting up a coordinate system with the origin at one of the vertices of the region, Magill has determined that the other two vertices are at (20, 80) and (120, 30), as shown in Figure 41.

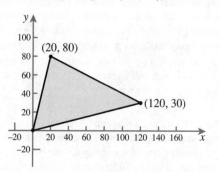

FIGURE 41

a. Find a system of three linear inequalities whose graph is the same as that shown in Figure 41.

Three daily flights are being contested by the environmentalists. The first is a direct flight from city A located at $(-22, 35)$ to city B located at $(110, -58)$.

b. Find the equation of the line connecting cities A and B, and sketch its graph on the same coordinate system as the triangular region. Does the flight violate the court order?

The second flight is a direct flight between city A and city C located at (140, 197). A plot of the line connecting these two points is shown in Figure 42. Since it is difficult to determine graphically if the line passes through the region, Magill decides to check it algebraically.

c. Find the equation of the line connecting cities A and C. Write the equation in the form $y = mx + b$.

d. Substitute the expression $mx + b$ found in part c in place of y in each of the three inequalities found in part a. After simplifying, the following three inequalities should be obtained.

$$x \geq -76$$
$$x \geq 19$$
$$x \leq 22$$

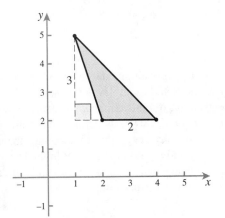

FIGURE 42

e. For the flight to pass through the region, there must be a value for x that satisfies each of the three inequalities given in part d. Does the flight pass through the region?

f. The third flight is between city B and city C. Use the technique described in parts c through e to determine if the third flight violates the court order.

81. *Systems of Inequalities and Area*   In this project, we consider the areas of closed regions determined by systems of inequalities. Consider, for example, the following linear system, whose solution is the triangular region shown in Figure 43.

$$x + y \leq 6$$
$$3x + y \geq 8$$
$$y \geq 2$$

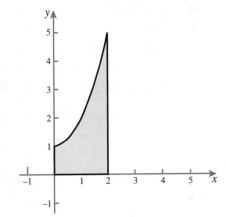

FIGURE 43

The area of the triangle in Figure 43 can be computed using the formula.

$$A = \frac{1}{2} bh = \frac{1}{2} (2)(3) = 3$$

a. Find the area of the region bounded by the given system of inequalities.

  i. $x + y \leq 7$
  $x + 2y \geq 8$
  $x \geq 2$

  ii. $x + 2y \leq 11$
  $2x + 3y \leq 1$
  $4x + y \geq 9$

  iii. $x - 2y \geq -6$
  $x - 2y \leq 0$
  $2x - y \leq 3$
  $2x - y \geq 0$

  iv. $x^2 + y^2 \leq 4$
  $x \geq 0$
  $y \geq 0$

The region in part iv is unusual in that one of the boundary curves is not linear and yet it is still possible to find the area using standard geometric formulas. In most cases, if one or more of the inequalities is nonlinear, it is not possible to find the area using standard formulas. Consider the following nonlinear system, which is graphed in Figure 44.

$$y \leq x^2 + 1$$
$$x \leq 2$$
$$x \geq 0$$
$$y \geq 0$$

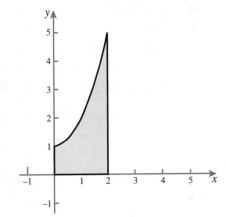

FIGURE 44

Because the top boundary curve is nonlinear, there is no standard formula for computing the area of the region. In fact, without the aid of calculus, we cannot compute the area exactly. Thus, we must settle for an approximation. In Figure 45, we have "inscribed" four rectangles of equal width in the region. The height of each rectangle is given by the y-coordinate of a point on the curve $y = x^2 + 1$. Thus, the first rectangle has height 1, the second has height $\frac{5}{4}$, and so on. Since each rectangle has width $\frac{1}{2}$, the area of each rectangle can be computed as follows:

$$A_1 = \frac{1}{2} \cdot 1 = \frac{1}{2} \qquad A_2 = \frac{1}{2} \cdot \frac{5}{4} = \frac{5}{8}$$

$$A_3 = \frac{1}{2} \cdot 2 = 1 \qquad A_4 = \frac{1}{2} \cdot \frac{13}{4} = \frac{13}{8}$$

The sum of the areas is

$$A = \frac{1}{2} + \frac{5}{8} + 1 + \frac{13}{8} = \frac{30}{8} = \frac{15}{4}$$

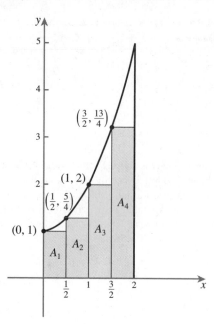

FIGURE 45

which is an approximation for the area of the shaded region in Figure 44. More accurate approximations could be found by inscribing more rectangles with smaller width.

b. Approximate the area of the region bounded by the given system of inequalities. Use the specified number of inscribed rectangles.

i. $y \leq x^2 + 1$
$x \leq 2$
$x \geq 0$
$y \geq 0$
8 rectangles

ii. $y \leq \dfrac{1}{x}$
$x \geq 1$
$x \leq 4$
$y \geq 0$
6 rectangles

iii. $y \leq x^2 - 2x$
$y \geq 0$
8 rectangles (2 with zero height)

SECTION 7.7 LINEAR PROGRAMMING

☐ **How can a business maximize its revenue without hiring additional labor?**

☐ **What combinations of stocks and bonds will yield the highest income for a fixed investment?**

☐ **What postwar mathematical discovery helped the U.S. Air Force allocate supplies more efficiently?**

☐ **How can a political party attract the largest number of TV viewers?**

A common problem in business, industry, agriculture, and many other areas is that of optimizing the use of limited resources. Some examples are a factory that wishes to maximize its revenues within the constraints of limited manpower and/or raw materials, a hospital food service that wishes to minimize its costs within the constraints of necessary nutritional requirements, or a university that wishes to maximize the number of students it can house in a new dormitory within the constraints of limited available land and funding. Linear programming is a mathematical method that can be used to solve such problems. We illustrate its basic ideas with an example.

A CASE STUDY

Willow Woods Furniture Factory makes tables and bookcases. In an effort to fine-tune production and increase profits, data have been collected on the labor requirements and profit margins on each piece of furniture. These data are summarized in Table 15. For example, each table requires 6 hours of woodworking, 3 hours of finishing, and yields a profit of $160. Without hiring additional labor, there are 48 hours available each day

TABLE 15

	Per table	**Per bookcase**	**Available**
Woodworking time (hours)	6	4	48
Finishing time (hours)	3	4	30
Profit (dollars)	160	200	—

for woodworking and 30 hours each day for finishing. Because of incoming orders, the factory must make at least 2 tables each day. Naturally, the owners of the factory would like to know how many tables and how many bookcases should be made each day to maximize profit. We now begin a step-by-step process that will lead us to the answer of this question.

STEP 1:   *Understanding the problem*

Among the various combinations of tables and bookcases, some bring in more profit than others. For example, if the factory sells 5 tables and 3 bookcases, the profit is $160(5) + 200(3) = \$1400$, whereas if it sells 4 tables and 6 bookcases, the profit is $160(4) + 200(6) = \$1840$. However, some combinations of tables and bookcases may exceed the constraints imposed by the available labor. The 5–3 combination would require $6(5) + 4(3) = 42$ hours of woodworking and $3(5) + 4(3) = 27$ hours of finishing, while the 4–6 combination would require $6(4) + 4(6) = 48$ hours of woodworking and $3(4) + 4(6) = 36$ hours of finishing. Thus, although the 4–6 combination brings in more profit, it is not allowable since it requires too many hours of finishing. We are looking for the combination that will provide the highest profit within the constraints of available labor.

STEP 2:   *Assigning variables and identifying the expression that is to be optimized*

The unknowns are the number of tables and bookcases to be made. Thus, we let

$$x = \text{Number of tables made each day}$$
$$y = \text{Number of bookcases made each day}$$

We wish to maximize the profit from the sale of x tables and y bookcases. Because each table sells for a profit of \$160, and each bookcase sells for a profit of \$200, the total profit P is given by

$$P = 160x + 200y$$

This equation defines the **objective function,** the quantity to be maximized.

STEP 3:   *Writing out inequalities that correspond to constraints*

Production at the furniture factory is constrained by the available labor for woodworking and finishing, and also by the perceived demand for tables. These constraints can be expressed algebraically using linear inequalities, as follows.

Woodworking constraint: Each table requires 6 hours, and each bookcase requires 4 hours. The total hours may not exceed 48.

Hours of woodworking for tables + Hours of woodworking for bookcases ≤ 48
$$6x \qquad\qquad + \qquad\qquad 4y \qquad\qquad \leq 48$$

Finishing constraint: Each table requires 3 hours, and each bookcase requires 4 hours. The total hours may not exceed 30.

$$\text{Hours of finishing for tables} + \text{Hours of finishing for bookcases} \leq 30$$
$$3x \qquad\qquad + \qquad\qquad 4y \qquad\qquad \leq 30$$

Demand constraint: At least 2 tables must be made each day.

$$x \geq 2$$

Implied constraint: The number of tables and bookcases may not be negative.

$$x \geq 0$$
$$y \geq 0$$

STEP 4: *Graphing the resulting system of inequalities*

The system of inequalities obtained in step 3 is

$$6x + 4y \leq 48$$
$$3x + 4y \leq 30$$
$$x \geq 2$$
$$y \geq 0$$

Notice that we have not included the constraint $x \geq 0$ since it is accounted for in the constraint $x \geq 2$. Moreover, the constraint $y \geq 0$ indicates that our region must be above the x-axis. Thus, we can restrict our attention to the first quadrant. The region that is common to the graph of all four inequalities is shaded dark blue in Figure 46.

The darkly shaded region in Figure 46 is called the **feasible set,** and it constitutes the set of points that satisfy all of the constraints. In the context of our problem, the feasible set consists of all the table–bookcase combinations that are within the labor and demand constraints. For example, the point $(5, 3)$ is in the feasible set. It corresponds to 5 tables and 3 bookcases. As we observed in step 1, this combination is within the constraints imposed by the available labor. But many other combinations are also within the constraints. What we need is a way to determine which of the table–bookcase combinations in the feasible set is the one that will yield the greatest profit. For this task, we note that if there is an optimum solution, *it must occur at one of the vertices of the feasible set.* This fact will be discussed more thoroughly later in the section. For now, we simply apply the result by locating and testing each of the vertices to see which yields the maximum profit.

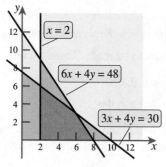

FIGURE 46

STEP 5: *Locating the vertices of the feasible set*

The four vertices are shown in Figure 47. The two on the x-axis are simply the x-intercepts of the lines $x = 2$ and $6x + 4y = 48$, namely $(2, 0)$ and $(8, 0)$, respectively. A third vertex is the intersection of the lines $x = 2$ and $3x + 4y = 30$. Setting $x = 2$ in the equation $3x + 4y = 30$ and solving for y gives $y = 6$. Thus, the vertex is $(2, 6)$. The fourth vertex is the intersection of the lines $6x + 4y = 48$ and $3x + 4y = 30$. Solving these simultaneously yields the solution $(6, 3)$.

STEP 6: *Evaluating the objective function and identifying the optimum solution*

We compute the profit at each vertex using the objective function $P = 160x + 200y$.

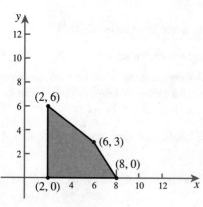

FIGURE 47

At $(2, 0)$: $P = 160(2) + 200(0) = 320$

At $(8, 0)$: $P = 160(8) + 200(0) = 1280$

At $(6, 3)$: $P = 160(6) + 200(3) = 1560$ Maximum profit

At $(2, 6)$: $P = 160(2) + 200(6) = 1520$

So a maximum (daily) profit of $1560 can be achieved by making 6 tables and 3 bookcases each day.

THE GENERAL LINEAR PROGRAMMING PROBLEM

The six steps we followed to determine the maximum profit for the Willow Woods Furniture Factory can be used as guidelines for solving a variety of optimization problems. We summarize the steps as follows.

Solving an Optimization Problem ☐

1. Analyze the problem carefully. If necessary, try some specific examples to help you understand the problem.
2. Assign variables to the unknowns and write out the objective function, the quantity that is to be maximized or minimized.
3. Write out the inequalities that correspond to the constraints in the problem.
4. Graph the system of inequalities to obtain the feasible set.
5. Locate the vertices of the feasible set.
6. Evaluate the objective function at each vertex, and identify the optimum solution.

EXAMPLE 1   ▓   A Nutrition Problem

Al is on a diet. The daily fruit portion of his diet must have as few calories as possible and yet provide at least 750 units of vitamin A, 0.72 unit of vitamin B_6, and 60 units of vitamin C. His fruit bowl contains a few small bananas and oranges. Use the nutritional information given in Table 16 to determine which combination of bananas and oranges will yield the fewest calories and yet meet his nutritional needs.

TABLE 16

	Banana	Orange
Calories	100	60
Units of vitamin A	250	250
Units of vitamin B_6	0.6	0.06
Units of vitamin C	12	60

SOLUTION   The unknowns are the number of bananas and oranges that Al should eat. Thus, we let

$$x = \text{Number of bananas}$$
$$y = \text{Number of oranges}$$

We wish to minimize the number of calories. Since each banana has 100 calories and each orange has 60 calories, the function to be minimized is

$$K = 100x + 60y$$

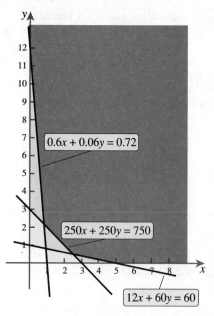

FIGURE 48

The nutritional requirements lead to the following inequalities.

Vitamin A: Units from bananas + Units from oranges ≥ 750

$$250x + 250y \geq 750$$

Vitamin B_6: Units from bananas + Units from oranges ≥ 0.72

$$0.6x + 0.06y \geq 0.72$$

Vitamin C: Units from bananas + Units from oranges ≥ 60

$$12x + 60y \geq 60$$

Since x and y must be nonnegative, we also have $x \geq 0$ and $y \geq 0$. Thus, we have the following constraints.

Constraints: $250x + 250y \geq 750$

$$0.6x + 0.06y \geq 0.72$$

$$12x + 60y \geq 60$$

$$x \geq 0$$

$$y \geq 0$$

The graph of this system of inequalities is shown in Figure 48. The portion shown in dark blue is the feasible set, and the vertices shown in Figure 49 are found by locating intercepts and the intersection points of appropriate pairs of lines.

To determine the minimum number of calories, we evaluate the objective function $K = 100x + 60y$ at each vertex.

At $(0, 12)$: $K = 100(0) + 60(12) = 720$

At $(1, 2)$: $K = 100(1) + 60(2) = 220$ Minimum

At $\left(\frac{5}{2}, \frac{1}{2}\right)$: $K = 100\left(\frac{5}{2}\right) + 60\left(\frac{1}{2}\right) = 280$

At $(5, 0)$: $K = 100(5) + 60(0) = 500$

Thus, to keep the fruit calories as low as possible and still obtain the required nutrients, Al should eat one banana and two oranges during the course of the day.

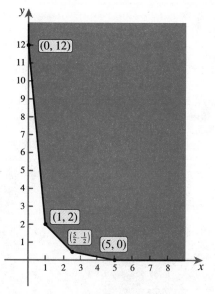

FIGURE 49

The two optimization problems we have considered so far have had several features in common. Each involved two variables, and each required us to maximize or minimize a linear objective function subject to two or more linear constraints. We refer to such problems as **linear programming problems** in two variables. In general, they are of the following form.

Linear Programming Problem ☐
in Two Variables

> Maximize or minimize an objective function $P = Ax + By + C$ subject to constraints of the form $ax + by \leq c$ or $ax + by \geq c$.

As we have seen, the solution of a linear programming problem can be found by testing the objective function at each of the vertices of the feasible set. This fact is known as the Fundamental Theorem of Linear Programming.

The Fundamental Theorem of   □   An optimal solution to a linear programming problem, if it exists, will occur at
Linear Programming   a vertex of the feasible set.

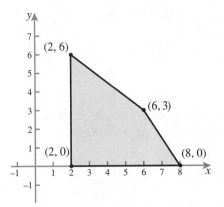

FIGURE 50

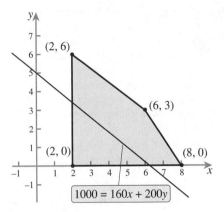

FIGURE 51

To see why the optimal solution occurs at a vertex, let us return briefly to the furniture factory problem. Recall that we wished to maximize the objective function $P = 160x + 200y$ on the feasible set shown in Figure 50. If we approach the problem naively, without any knowledge of the Fundamental Theorem of Linear Programming, we might choose to consider certain specified profit values and the corresponding ordered pairs (x, y) that would yield that profit. For example, to obtain a profit of $1000, we must have $1000 = 160x + 200y$. This is the equation of the line graphed in Figure 51.

Since the line $1000 = 160x + 200y$ passes through the feasible set, we can see that there are points (x, y) that satisfy the constraints of the problem and also yield a profit of $1000. To determine if there are points that yield a higher profit and still satisfy the constraints, we next consider profits of $1200 and $1500. The points that satisfy the constraints and yield profits of $1200 and $1500, respectively, are the points of intersection of the lines $1200 = 160x + 200y$ and $1500 = 160x + 200y$ with the feasible set, as shown in Figure 52.

Notice that the lines are all parallel and they progress further away from the origin as the profit values increase. This suggests that we look for the line that is parallel to the other three, intersects the feasible set, and is as far from the origin as possible. This line has equation $1560 = 160x + 200y$ and intersects the feasible set at the vertex $(6, 3)$, as shown in Figure 53. Thus, we see that the greatest profit, $1560, is attained at the vertex $(6, 3)$.

In general, specific values of the objective function lead to families of parallel lines. If the objective function has a maximum value, it will correspond to the line that intersects the feasible set and is furthest from, or nearest to, the origin, depending on the particular objective function $P = Ax + By + C$. This line must necessarily pass through a vertex. Likewise, if there is a minimum, it corresponds to the line that is closest to, or furthest from, the origin, again depending on the particular function $P = Ax + By + C$, and it must also pass through a vertex.

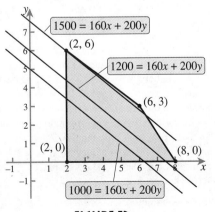

FIGURE 52

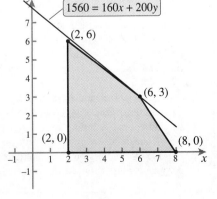

FIGURE 53

EXAMPLE 2   ■   Solving a Linear Programming Problem

Find the maximum and minimum values of $P = 3x + 6y$ subject to the following constraints.

$$x + y \leq 8$$
$$2x - y \leq 7$$
$$x + 2y \geq 6$$
$$x \geq 2$$
$$y \leq 5$$

Sketch several of the lines that correspond to specific values of the objective function, and show that the lines corresponding to the maximum and minimum values pass through vertices of the feasible set.

ADVANCED TECHNIQUES OF LINEAR PROGRAMMING

In this text, we investigate extremely simple linear programming problems involving several constraints and only two variables. We are using the geometric method: We graph the constraint inequalities and then evaluate the objective function at each of the vertices to determine where the maximum or minimum value is obtained. But what would we do if there were thousands of variables and tens of thousands of constraints? We are certainly not going to be able to graph the region in a thousand-dimensional space; indeed it would be difficult to solve even a three-dimensional problem in this fashion. Clearly we need an **algorithm,** or step-by-step method, that doesn't require geometric visualization.

Motivated by problems arising in the context of military planning and support for the U.S. Department of the Air Force, George Dantzig in 1947 developed the **simplex method,** a systematic technique for solving linear programming problems. The simplex method has found wide application in communications, airline scheduling, and inventory control, and has hence become an integral part of management, industrial engineering, and operations research training programs throughout the world.

In the two-dimensional problems that we consider in this section, the feasible set is a region in the plane bounded by line segments, and optimal solutions exist at vertices. In the sort of multidimensional problems in which the simplex method is applied, the feasible set is a "region" in a multidimensional space bounded by portions of **hyperplanes.** If we visualize the feasible set as a

Narendra K. Karmarkar

multifaceted gemstone, then the hyperplanes that form the boundary of the feasible set are the facets of the gemstone. Just as in the two-dimensional case, optimal solutions occur at the vertices (the corners of the gemstone). The simplex method begins with a vertex in the feasible set and then at the next step selects an adjacent vertex for which the objective function is greater (in the case of a maximization problem). In this manner, step by step, the simplex method propels us ever closer to an optimal solution.

In spite of the widespread success that the simplex method has enjoyed for half a century, problems have arisen, particularly in the communications industries, for which the simplex method has proven to be inadequate. For this reason, there is an ongoing search for more efficient methods for solving high-dimensional linear programming problems. Perhaps the most significant breakthrough in this area was made in the 1980s by Narendra K. Karmarkar of AT&T Laboratories (pictured here). Karmarkar's technique involves cutting a swath through the interior of the region rather than caroming from vertex to vertex on the surface. At each stage, another interior point, closer to the optimal solution than the last, is reached. Although the improvement in computation time is often modest for low-dimensional linear programming problems, the improvement can be quite dramatic when the dimension is large. The demand for more efficient techniques for solving linear programming problems will likely lead to the development of ever faster algorithms.

SOLUTION The feasible set is shown in Figure 54. Testing the objective function at each of the vertices yields the following results.

At $(2, 5)$: $P = 3(2) + 6(5) = 36$

At $(3, 5)$: $P = 3(3) + 6(5) = 39$ Maximum

At $(5, 3)$: $P = 3(5) + 6(3) = 33$

At $(4, 1)$: $P = 3(4) + 6(1) = 18$ Minimum

At $(2, 2)$: $P = 3(2) + 6(2) = 18$ Minimum

Thus, the maximum value of P is 39, which occurs at $(3, 5)$, and the minimum value of P is 18, which occurs at both $(4, 1)$ and $(2, 2)$. Figure 55 shows the lines corresponding to the objective function values 18, 24, 32, and 39. Note that the lines corresponding to the maximum $P = 39$ and minimum $P = 18$ pass through the vertices just found.

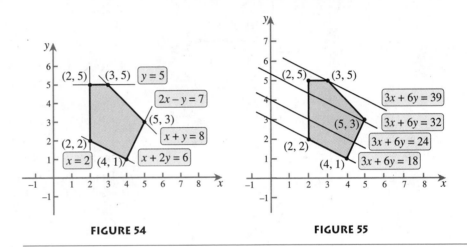

FIGURE 54   **FIGURE 55**

As can be seen in Example 2, the objective function can attain its maximum or minimum value at more than one vertex of a feasible set. Note also that the objective function need not have both a maximum and a minimum on a given feasible set. An example of this situation can be seen in Example 1, where the objective function has a minimum but not a maximum. See Exercise 38 for further discussion of this issue.

UNDERSTANDING AND MASTERY CHECKLISTS

CONCEPTS TO UNDERSTAND

☐ Objective function

☐ Constraint

☐ Feasible set

☐ Vertices of the feasible set

☐ Linear programming problem in two variables

☐ The Fundamental Theorem of Linear Programming

SKILLS TO MASTER

☐ Solve a linear programming problem in two variables.

☐ Formulate a linear programming problem from a real-world scenario.

EXERCISES 7.7

EXERCISES 1–6 *Find the maximum and minimum values of the objective function over the given feasible set.*

1. Objective function: $P = 8x + 10y$
   Feasible set:

   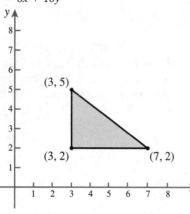

2. Objective function: $P = 10x + 4y$
   Feasible set:

   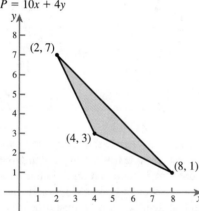

3. Objective function: $P = 4x + 5y$
   Feasible set:

   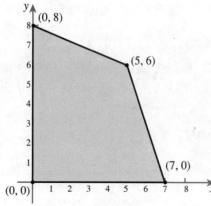

4. Objective function: $P = 6x - 2y$
   Feasible set:

   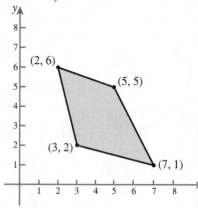

5. Objective function: $P = 4x - 4y$
   Feasible set: See Exercise 3.

6. Objective function: $P = 2x + 8y$
   Feasible set: See Exercise 4.

EXERCISES 7–16 *Solve the linear programming problem.*

7. Maximize $P = 10x + 15y$
   subject to $3x + 4y \le 24$
   $\qquad\quad x + 2y \le 10$
   $\qquad\qquad\quad x \ge 0$
   $\qquad\qquad\quad y \ge 0$

8. Maximize $P = 14x + 30y$
   subject to $3x + 5y \le 30$
   $\qquad\quad 2x + 4y \le 22$
   $\qquad\qquad\quad x \ge 0$
   $\qquad\qquad\quad y \ge 0$

9. Minimize $P = 15x + 40y$
   subject to $2x + 5y \ge 20$
   $\qquad\quad 3x + 2y \ge 19$
   $\qquad\qquad\quad x \ge 0$
   $\qquad\qquad\quad y \ge 0$

10. Minimize $P = 30x + 10y$
    subject to $2x + y \ge 8$
    $\qquad\quad 4x + y \ge 12$
    $\qquad\qquad\quad x \ge 0$
    $\qquad\qquad\quad y \ge 0$

11. Maximize $P = 10x + 12y$
    subject to $x + 2y \le 17$
    $\qquad\quad 2x + 3y \le 30$
    $\qquad\qquad\quad x \ge 3$
    $\qquad\qquad\quad y \ge 2$

12. Maximize $P = 12x + 8y$
    subject to $2x + y \le 18$
    $\qquad\quad x + 2y \le 21$
    $\qquad\qquad\quad x \ge 1$
    $\qquad\qquad\quad y \ge 4$

13. Minimize $P = 20x + 24y$
    subject to $2x + 3y \le 12$
    $\qquad\quad x + y \ge 5$
    $\qquad\quad x + 4y \ge 8$
    $\qquad\qquad\quad x \ge 0$
    $\qquad\qquad\quad y \ge 0$

14. Minimize $P = 10x + 4y$
    subject to $x + 3y \ge 16$
    $\qquad\quad x + y \le 10$
    $\qquad\quad 2x + y \ge 12$
    $\qquad\qquad\quad x \ge 0$
    $\qquad\qquad\quad y \ge 0$

15. Maximize $P = 2x + 10y$
    subject to $x + y \ge 9$
    $\qquad\quad 3x - 2y \le 24$
    $\qquad\quad x + 4y \le 36$
    $\qquad\qquad\quad x \ge 4$
    $\qquad\qquad\quad y \ge 3$

16. Minimize $P = 10x + 8y$
    subject to $x + 2y \ge 12$
    $\qquad\quad x - y \le 4$
    $\qquad\quad 3x + y \ge 16$
    $\qquad\qquad\quad x \le 8$
    $\qquad\qquad\quad y \le 10$

▮ APPLICATIONS

17. *Clothing Profits* A clothing company makes two styles of tailored suits. The labor requirements and profit margins for each style are given in Table 17. If 56 hours are available per day for cutting and 72 hours per day for sewing, how many suits of each style should be made to maximize profit?

TABLE 17

	Style A	Style B
Cutting (hours)	2	4
Sewing (hours)	3	2
Profit (dollars)	35	40

18. *Livestock Allocation* A farmer raises cattle and pigs. The investment requirements, labor requirements, and profit margins for each cow and pig are given in Table 18. If the farmer has 680 hours of labor and $52,000 of investment capital available, how many cows and pigs should be raised to maximize profit?

TABLE 18

	Cow	Pig
Investment (dollars)	480	1400
Labor (hours)	10	9
Profit (dollars)	120	400

19. *Refinery Production* An oil company owns two refineries. The daily production limits and operating costs for each refinery are given in Table 19. An order is received for 1000 barrels of high-grade oil, 1000 barrels of medium-grade oil, and 1800 barrels of low-grade oil. How many days should each refinery be operated so that the order can be filled at the least cost?

TABLE 19

	Refinery 1	Refinery 2
High-grade oil (barrels)	100	200
Medium-grade oil (barrels)	200	100
Low-grade oil (barrels)	300	200
Operating cost (dollars)	10,000	9000

20. *Furniture Profit* A furniture company makes chairs and sofas. The labor requirements and profit margins for each chair and sofa are given in Table 20. The total labor available each day for carpentry, finishing, and upholstery is 96 hours, 18 hours, and 72 hours, respectively. How many chairs and sofas should be made each day to maximize profit?

TABLE 20

	Chair	Sofa
Carpentry time (hours)	6	3
Finishing time (hours)	1	1
Upholstery time (hours)	2	6
Profit (dollars)	80	70

21. *Vehicle Allocation* Steve can borrow either his mother's car or his father's truck to commute to school, as long as he replaces the gas he uses. After averaging his costs for a period of time, he has determined that the car costs him 4¢ per mile, whereas the truck averages 6¢ per mile. Depending on the route he takes, he travels at least 300 miles each month but no more than 350. For insurance purposes, he must drive his father's truck at least twice as far as his mother's car. However, his father does not want him to use the truck for more than 250 miles each month. How many miles should Steve drive each vehicle in order to minimize his monthly cost? What is the least Steve will have to spend each month?

22. *Dormitory Construction* A university is making plans to build a new dormitory. Early estimates show that each single room will cost $20,000 and will require approximately 1200 cubic feet of space. Each double room will cost $25,000 and will require an average of 1800 cubic feet of space. The university can spend at most $2,200,000. Available space and zoning restrictions limit the total room space to 150,000 cubic feet. Finally, there must be at least as many double rooms as single rooms. How many of each type of room should the dormitory have in order to maximize the number of students it can hold?

23. *Campaign Strategy* A political party is planning a half-hour television show for its incumbent candidates for state governor and U.S. Senate. Based on a preshow survey, it is believed that 40,000 viewers will watch the show for each minute the senator is on and 60,000 viewers will watch for each minute the governor is on. The senator is a party "elder statesman" and so demands to be on the air at least one and a half times as long as the governor. The governor will not participate if her time is less than 10 minutes. And, of course, the sum of their speaking times may not exceed 30 minutes. Determine the time that should be allotted to each of the two candidates to maximize the number of viewers.

24. *Power Lunch* Sam's lunches consist primarily of peanut butter sandwiches and hamburgers from his favorite fast-food restaurant. Each sandwich has 1.5 grams of saturated fats, 40 grams of carbohydrates, 9 grams of protein, and 2 grams of iron. Each hamburger has 4 grams of saturated fats, 30 grams of carbohydrates, 12 grams of protein, and 4 grams of iron. Determine how many sandwiches and hamburgers Sam must eat to minimize the amount of saturated fats but still obtain at least 110 grams of carbohydrates, 30 grams of protein, and 7 grams of iron.

25. *Hospital Meal* A hospital food service is planning a meal that includes Swiss steak and peas. Each ounce of Swiss steak costs 9¢ and has 150 units of vitamin A, 0.06 unit of vitamin B_6, and 3 units of vitamin C. Each ounce of peas costs 4¢ and has 120 units of vitamin A, 0.02 unit of vitamin B_6, and 9 units of vitamin C. From the meat and vegetable portion of the meal, each patient must receive 1260 units of vitamin A, 0.35 unit of vitamin B_6, and 45 units of vitamin C. How many ounces of Swiss steak and peas should be served to minimize the cost?

26. *Troop Maneuvers* The leader of a UN peacekeeping unit wishes to lead his troops, by foot, to a hot point many miles away in as few days as possible. To save time, he decides that he will have his troops jog at a rate of 5 miles per hour for a portion of each day, but no more than 2 hours total. The rest of the 8-hour day, the troops will walk at a rate of 4 miles per hour. Since the food supply is limited, it is determined that each soldier can consume no more than 3200 calories during the daily hike. (Caloric consumption is given in the following table.) How many hours of each day should the troops jog and how many should they walk in order to travel as far as possible within these constraints?

	Walking	**Jogging**
Speed (mph)	4	5
Calories per hour	360	680

27. *Staffing Conferences* A high-tech company with two offices— one in Bozeman, MT, with 15 employees and another in New Haven, CT, with 20—wishes to send each of its employees to one of two conventions: the COMDEX computer trade show to be held in Orange County Convention Center in Orlando, FL, and a management training seminar to take place at the Fawcett Center on the campus of The Ohio State University in Columbus, Ohio. Management has concluded that at least 10 workers should attend COMDEX and at least 15 should travel to Columbus. Furthermore, the company president, who confiscates all frequent flyer mileage earned by his employees, insists that a minimum of 32,000 frequent flyer miles be accumulated to facilitate his upcoming around-the-world tour. One frequent flyer mile is given for each mile of the round-trip flights to Orlando only (see Table 21). How many employees from each office should be sent in order to meet these constraints while minimizing the total airfare (see Table 22)?

TABLE 21   Round-trip distances for frequent flier miles

	New Haven	**Bozeman**
Orlando	2000	4000

TABLE 22   Round-trip air fare

From: To:	**New Haven**	**Bozeman**
Orlando	$500	$750
Columbus	$300	$450

CONCEPTS AND CRITICAL THINKING

EXERCISES 28–31   *Answer true or false.*

28. Every point in the feasible set of a linear programming problem satisfies all of the constraints.

29. If the objective function of a linear programming problem has a maximum, then the maximum occurs at one of the vertices of the feasible set.

30. If the objective function of a linear programming problem has a minimum, then the minimum occurs at one of the vertices of the feasible set.

31. According to the Fundamental Theorem of Linear Programming, the minimum value of the function $R = x^2 + y^2$ subject to the constraints $-1 \le x \le 1$ and $-1 \le y \le 1$ will be obtained at one of the four points $(1, 1)$, $(-1, 1)$, $(-1, -1)$, or $(1, -1)$.

EXERCISES 32–35   *Give an example of each.*

32. A linear programming problem for which the feasible set is a half-plane

33. A linear programming problem for which the optimal solution is the origin

34. A linear programming problem for which the feasible set is an infinitely long horizontal strip with a width of 2 units

35. A linear programming problem for which the feasible set has just one vertex

QUESTIONS FOR DISCUSSION OR ESSAY

36. A friend relates the following story: "... so we're taking this test on Chapter 7, and there are only a couple minutes left in the hour, when 'Ivan the Terrible' announces that there is a typo in the linear programming problem I've just spent the last 10 minutes finishing. Since there isn't enough time to redo the problem, I erase everything and write a note explaining the steps I

would follow if given enough time. Can you believe I only got a couple of points for the problem?'' The friend shows you the test and the location of the typo. You immediately realize that, although your friend has the steps memorized, he doesn't really understand the linear programming process. Only a few simple calculations would have been necessary to correct the problem. Was the typo in the objective function or one of the constraints? Explain.

37. In examples such as the furniture factory problem, it is tempting to assume that the maximum value of the objective function will occur at the "obvious" vertex—that is, the one where the available resources have been completely utilized. Review the furniture factory problem, and explain why the solution $(6, 3)$ is the "obvious" one. Then show that it is possible, by changing only one of the coefficients in the objective function, to make the maximum occur at the vertex $(2, 6)$. At this vertex, which labor resource has not been fully utilized? In practice, do you think it's possible for a company to maximize profit without using all available resources? Explain.

38. In Example 1, we minimized the objective function $K = 100x + 60y$ over the feasible set shown in Figure 56.

TABLE 23

Vertex	$K = 100x + 60y$
$(0, 12)$	720
$(1, 2)$	220
$\left(\frac{5}{2}, \frac{1}{2}\right)$	280
$(5, 0)$	500

The values of the objective function at the vertices are given in Table 23.

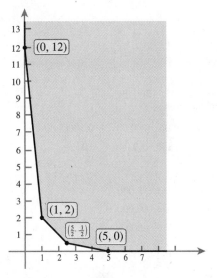

FIGURE 56

Thus, we found the minimum to be 220 at the vertex $(1, 2)$. Why can we not assume that the maximum is 720 at the vertex $(0, 12)$? What can be said about the maximum of the objective function over this feasible set? Does this contradict the Fundamental Theorem of Linear Programming? Explain.

PROJECTS FOR ENRICHMENT

39. *Linear Programming with Three Variables*   Linear programming problems can also involve more than two variables. In general, a linear programming problem in n variables would involve maximizing or minimizing an objective function $P = A_1x_1 + A_2x_2 + \cdots + A_nx_n$ subject to constraints of the form $a_1x_1 + a_2x_2 + \cdots + a_nx_n \leq b$ or $a_1x_1 + a_2x_2 + \cdots + a_nx_n \geq b$. Unfortunately, such problems usually cannot be solved geometrically unless it is possible to express $n - 2$ of the variables in terms of the remaining two. In the following example, we start with three variables but are able to replace one of the variables with an expression involving the other two.

Example: Suppose that $18,000 is to be invested in some combination of government bonds paying 6%, municipal bonds paying 8%, and stocks paying 10%. After analyzing the risks, it is decided that at most $15,000 is to be invested in government and municipal bonds, at least $10,000 in government bonds, and at most $6000 in stocks. What combination of investments will maximize the yearly income?

Solution: We initially assign the following variables.

$x =$ Amount invested in government bonds
$y =$ Amount invested in municipal bonds
$z =$ Amount invested in stocks

The yearly income from investing these amounts at the given rate is

$$I = 0.06x + 0.08y + 0.10z$$

We wish to maximize this objective function subject to the following constrains.

$x + y \leq 15,000$	At most $15,000 in government and municipal bonds
$x \geq 10,000$	At least $10,000 in government bonds
$z \leq 6000$	At most $6000 in stocks
$x \geq 0$	
$y \geq 0$	
$z \geq 0$	

To solve this problem geometrically, we must reduce it to only two variables. Since the total amount invested is $18,000, we know that

$$x + y + z = 18,000$$

so that

$$z = 18,000 - x - y$$

Substituting this expression into the objective function and the constraints gives

$$I = 0.06x + 0.08y + 0.10(18,000 - x - y)$$
$$= 1800 - 0.04x - 0.02y$$

and

$$x + y \le 15,000$$
$$x \ge 10,000$$

$18,000 - x - y \le 6000$ \hfill Or equivalently, $x + y \ge 12,000$

$$x \ge 0$$
$$y \ge 0$$

$18,000 - x - y \ge 0$ \hfill Or equivalently, $x + y \le 18,000$

The feasible set corresponding to these constraints is shown in Figure 57.

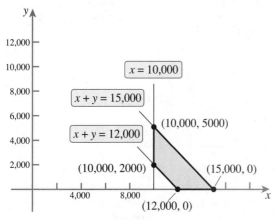

FIGURE 57

We now test the objective function at each of the vertices.

At (12,000, 0):
$$I = 1800 - 0.04(12,000) - 0.02(0) = 1320$$

At (15,000, 0):
$$I = 1800 - 0.04(15,000) - 0.02(0) = 1200$$

At (10,000, 2000):
$$I = 1800 - 0.04(10,000) - 0.02(2000) = 1360 \qquad \text{Maximum}$$
At (10,000, 5000):
$$I = 1800 - 0.04(10,000) - 0.02(5000) = 1300$$

So the maximum income is $1360 when $x = 10,000$ and $y = 2000$. For these values of x and y, we obtain

$$z = 18,000 - 10,000 - 2000 = 6000$$

Thus, $10,000 should be invested in government bonds, $2000 in municipal bonds, and $6000 in stocks to achieve a maximum income of $1360.

Use this method to solve the following linear programming problems involving three variables.

a. Michael has $20,000 to invest in some combination of mutual funds, stocks, and bonds. The projected annual rates of return are 14% for mutual funds, 16% for stocks, and 10% for bonds. He does not wish to invest more than $12,000 in mutual funds and stocks, but would like to invest at least $4000 in stocks. Moreover, he does not wish to invest more than $10,000 in bonds. What combination of investments will maximize Michael's income?

b. Eliza wishes to have $2000 from her monthly paycheck deposited automatically in some combination of three different accounts at her credit union. She would like to maximize the amount of earnings from each new deposit after one month. The savings account pays $\frac{1}{3}$% per month, the checking account pays $\frac{1}{2}$% per month on the average balance but has a fee of $5 per month, and the money market pays $\frac{2}{3}$% per month. Due to bills that must be paid during the course of the month, she must put at least $1000 in the checking account (assume that the average balance is half of what she deposits). She would like to deposit at least $800 in savings and the money market account although the money market account has restrictions on withdrawals, and so she does not wish to deposit more than $500 there. How should the deposit be distributed among the three accounts to maximize earnings?

c. A farmer wishes to plant exactly 1000 acres in some combination of corn, soybeans, and wheat. Each acre of corn costs $100 and requires 2 hours of labor. Each acre of soybeans costs $80 and requires 1 hour of labor. Each acre of wheat costs $70 and requires 1.5 hours of labor. The total cost for the three crops may not exceed $85,000 and the labor may not exceed 1600 hours. the profits from each acre of corn, soybeans, and wheat are $170, $110, and $90, respectively. How many acres of each should be planted to maximize profit?

CHAPTER 7 REVIEW

EXERCISES 1–4   *Perform the indicated matrix operation.*

1. $3 \cdot \begin{bmatrix} 2 & -1 \\ -5 & 3 \end{bmatrix}$

2. $-2 \cdot \begin{bmatrix} -3 & 0 \\ 4 & -1 \\ 0 & -2 \end{bmatrix}$

3. $\begin{bmatrix} 3 & 0 & -2 \\ 5 & -1 & 3 \end{bmatrix} + \begin{bmatrix} 0 & -1 & 4 \\ -2 & 6 & 2 \end{bmatrix}$

4. $\begin{bmatrix} 1 & 0 \\ 3 & -8 \end{bmatrix} - \begin{bmatrix} 0 & 5 \\ -1 & -2 \end{bmatrix}$

EXERCISES 5–12 *Let*

$$A = \begin{bmatrix} 1 & 0 \\ -2 & 3 \end{bmatrix}, \quad B = \begin{bmatrix} 2 & -1 \\ 4 & 3 \end{bmatrix},$$

$$C = \begin{bmatrix} 4 & 0 \\ -3 & 1 \\ 0 & 2 \end{bmatrix}, \quad and \ D = \begin{bmatrix} 3 & 1 & 0 \\ -2 & 0 & 4 \\ 1 & -2 & 1 \end{bmatrix}$$

Determine whether the given expression is defined. If so, evaluate it. If not, explain why not.

5. $A + 2B$

6. $2C - D$

7. CD

8. BA

9. DC

10. $C(A - B)$

11. C^2

12. A^2

EXERCISES 13–22 *Compute the given product, if possible.*

13. $\begin{bmatrix} -2 & 3 \\ 1 & 4 \end{bmatrix}\begin{bmatrix} 3 \\ -1 \end{bmatrix}$

14. $\begin{bmatrix} 3 & 0 \\ 2 & -4 \end{bmatrix}\begin{bmatrix} 2 & -1 \\ -4 & 6 \end{bmatrix}$

15. $\begin{bmatrix} 2 & 5 \\ 3 & 7 \end{bmatrix}[4 \quad 0]$

16. $\begin{bmatrix} 5 & -3 \\ 0 & 4 \end{bmatrix}\begin{bmatrix} 2 & 1 & -1 \\ -1 & 3 & 0 \end{bmatrix}$

17. $[3 \quad 4]\begin{bmatrix} 2 \\ -1 \end{bmatrix}$

18. $\begin{bmatrix} 2 & 3 & 5 \\ 7 & 11 & 13 \end{bmatrix}\begin{bmatrix} 2 & 4 \\ 8 & 16 \end{bmatrix}$

19. $\begin{bmatrix} 0 & 4 \\ -1 & 2 \\ 3 & 0 \end{bmatrix}\begin{bmatrix} 2 & 3 \\ -3 & 1 \end{bmatrix}$

20. $\begin{bmatrix} 1 & 1 & 0 \\ 0 & 1 & 1 \end{bmatrix}\begin{bmatrix} 1 & 2 & 3 \\ 0 & 1 & 2 \\ 0 & 0 & 1 \end{bmatrix}$

21. $\begin{bmatrix} 1 & -1 & 2 \\ 3 & 0 & -1 \\ 4 & 2 & 0 \end{bmatrix}\begin{bmatrix} b \\ c \\ a \end{bmatrix}$

22. $\begin{bmatrix} 1 & 2 & 4 \\ 2 & 4 & 1 \\ 4 & 1 & 2 \end{bmatrix}\begin{bmatrix} -1 & 0 & 2 \\ 0 & 2 & -1 \\ 2 & -1 & 0 \end{bmatrix}$

EXERCISES 23–24 *Write the given system as an augmented matrix.*

23. $5x + y = 3$
$\quad x - 3y = 4$

24. $x + 2y - 2z = 3$
$\quad 2x + y = -1$
$\quad 3y + 2z = 6$

EXERCISES 25–26 *Write out a system of equations in x, y, and z, in that order, that corresponds to the given augmented matrix, and then solve the system using back-substitution.*

25. $\begin{bmatrix} 1 & -4 & 1 & | & 3 \\ 0 & 1 & -2 & | & 5 \\ 0 & 0 & 1 & | & -2 \end{bmatrix}$

26. $\begin{bmatrix} 1 & -2 & -1 & | & 4 \\ 0 & 1 & \frac{1}{2} & | & 2 \\ 0 & 0 & 1 & | & 3 \end{bmatrix}$

EXERCISES 27–30 *Use Gaussian elimination to write the augmented matrix in row-echelon form. There may be more than one correct answer.*

27. $\begin{bmatrix} -1 & 3 & | & -2 \\ 4 & -9 & | & 1 \end{bmatrix}$

28. $\begin{bmatrix} 2 & 4 & -2 & | & 6 \\ 3 & 5 & -1 & | & 4 \\ -2 & -4 & 9 & | & 6 \end{bmatrix}$

29. $\begin{bmatrix} 2 & 4 & -3 & | & 1 \\ 3 & 0 & \frac{9}{2} & | & -\frac{3}{2} \end{bmatrix}$

30. $\begin{bmatrix} 3 & 10 & 4 & | & 0 \\ 1 & 3 & 1 & | & 1 \\ -1 & -6 & -3 & | & -2 \end{bmatrix}$

EXERCISES 31–38 *Express the system as an augmented matrix and solve using Gaussian elimination. There may be 0, 1, or infinitely many solutions.*

31. $-x + 2y = 7$
$\quad 2x - 3y = -9$

32. $3x + 6y = 1$
$\quad 2x + 2y = 1$

33. $-2x + 3y = 6$
$\quad 4x - 6y = -12$

34. $2x + 4y - 2z = 9$
$\quad\quad\quad y - z = 1$
$\quad -x - 3y + z = -6$

35. $x - y + z = -4$
$\quad 3x - 2y - z = -4$
$\quad x - 2z = 1$

36. $x + 2y + z = 3$
$\quad 2y + 3z = 2$
$\quad -x + 2z = 1$

37. $x - y = 4$
$\quad 2x - z = 5$
$\quad y - z = 1$

38. $x + y = 0$
$\quad -2x - y + z = -9$
$\quad 3x + y - z = 8$

EXERCISES 39–40 *Use Gauss-Jordan elimination to write the augmented matrix in reduced row-echelon form.*

39. $\begin{bmatrix} 1 & 2 & | & 4 \\ 2 & 3 & | & 5 \end{bmatrix}$

40. $\begin{bmatrix} 1 & -1 & 2 & | & 8 \\ 0 & 1 & -4 & | & -9 \\ 0 & 0 & 1 & | & 3 \end{bmatrix}$

EXERCISES 41–44 *Solve the system of equations by first writing it in matrix form and then using Gauss-Jordan elimination.*

41. $3x + 5y = -5$
$\quad -x - 2y = 3$

42. $x - 4y = -0.5$
$\quad -2x + 9y = 1.25$

43. $x - y + 3z = 9$
$\quad y - 2z = -3$
$\quad z = 4$

44. $x - y + z = 0$
$\quad -4x + 5y - 6z = 7$
$\quad 2x - y + z = 2$

EXERCISES 45–52 *Find the inverse of the given matrix, if it exists. Use a graphing calculator as needed.*

45. $\begin{bmatrix} 2 & 1 \\ -3 & -1 \end{bmatrix}$

46. $\begin{bmatrix} 0 & \frac{1}{2} \\ 8 & -\frac{1}{4} \end{bmatrix}$

47. $\begin{bmatrix} a & b \\ b & a \end{bmatrix}$

48. $\begin{bmatrix} 1 & x \\ x & x^2 \end{bmatrix}$

49. $\begin{bmatrix} 1 & -2 & 3 \\ 0 & 1 & -2 \\ 0 & 0 & 1 \end{bmatrix}$

50. $\begin{bmatrix} 0 & 4 & 8 \\ 4 & 0 & 4 \\ 8 & 4 & 0 \end{bmatrix}$

51. $\begin{bmatrix} 0 & 0 & a \\ 0 & 0 & 0 \\ a & 0 & 0 \end{bmatrix}$

52. $\begin{bmatrix} 1 & -1 & 0 & 0 \\ -1 & 1 & -1 & 0 \\ 0 & -1 & 1 & -1 \\ 0 & 0 & -1 & 1 \end{bmatrix}$

EXERCISES 53–56 *Solve the system of equations by first expressing it in matrix form as $AX = B$ and then evaluating $X = A^{-1}B$.*

53. **a.** $x + y = 3$
$\quad 2x + y = -4$

b. $x + y = 2$
$\quad 2x + y = 0$

54. a. $3x - 2y = 5$
$4x - y = -10$

b. $3x - 2y = -2$
$4x - y = 3$

55. a. $\dfrac{1}{2}x + \dfrac{3}{2}y = 3$

$-\dfrac{1}{4}x + \dfrac{5}{4}y = -4$

b. $\dfrac{1}{2}x + \dfrac{3}{2}y = 0$

$-\dfrac{1}{4}x + \dfrac{5}{4}y = -3$

56. a. $x + y - z = 3$
$-4x - 3y + 6z = -3$
$-x - 2y = 9$

b. $x + y - z = 1$
$-4x - 3y + 6z = 0$
$-x - 2y = 2$

EXERCISES 57–64 *Evaluate the given determinant.*

57. $\begin{vmatrix} 4 & 6 \\ 3 & 2 \end{vmatrix}$

58. $\begin{vmatrix} -2 & 1 \\ 3 & 5 \end{vmatrix}$

59. $\begin{vmatrix} \sqrt{5} & 2 \\ 2 & \sqrt{5} \end{vmatrix}$

60. $\begin{vmatrix} 3 & -1 & 0 \\ -1 & 3 & 0 \\ -2 & -2 & 3 \end{vmatrix}$

61. $\begin{vmatrix} x & 0 & 1 \\ 0 & x & 0 \\ 1 & 0 & x \end{vmatrix}$

62. $\begin{vmatrix} 1 & 2 & 3 \\ 1 & 1 & 1 \\ 3 & 2 & 1 \end{vmatrix}$

63. $\begin{vmatrix} 1 & -1 & 0 & 0 \\ 3 & 4 & 0 & 0 \\ 0 & 0 & 2 & -3 \\ 0 & 0 & 1 & 1 \end{vmatrix}$

64. $\begin{vmatrix} a & b & 0 & 0 & 0 \\ b & a & 0 & 0 & 0 \\ 0 & 0 & 1 & 0 & 0 \\ 0 & 0 & 0 & 1 & 0 \\ 0 & 0 & 0 & 0 & 1 \end{vmatrix}$

EXERCISES 65–68 *Find the value(s) of x for which the given matrix is not invertible.*

65. $\begin{bmatrix} 2 & 2 \\ 5 & x \end{bmatrix}$

66. $\begin{bmatrix} 2 & x \\ x & 8 \end{bmatrix}$

67. $\begin{bmatrix} 1 & 2x & -4 \\ 0 & x & -2 \\ 0 & 3 & x-5 \end{bmatrix}$

68. $\begin{bmatrix} 1 & 0 & 0 \\ 1 & x & x+1 \\ 5 & -4 & x \end{bmatrix}$

EXERCISES 69–72 *Solve the system of equations using Cramer's Rule.*

69. $3x + y = 0$
$5x + 3y = -1$

70. $2x - 3y = 16$
$5x + 2y = 2$

71. $x + y = 0$
$x + 2y - z = -6$
$-4x - 3y = -1$

72. $x + y + z - 2w = -6$
$-2x - y + 3w = 5$
$2x - z - 3w = -5$
$x - w = -2$

EXERCISES 73–80 *Graph the inequality.*

73. $x + y \geq 4$

74. $2x - y < -3$

75. $x < 2y$

76. $y \leq 3x - 4$

77. $x \geq -2$

78. $y \geq x^2 - 1$

79. $x^2 + y^2 < 4$

80. $x - y^2 \leq 0$

EXERCISES 81–90 *Graph the system of inequalities.*

81. $x - y \leq 6$
$x + y \leq 4$

82. $3x + y > 3$
$x < 3y$

83. $4x - 7y \geq 14$
$3y \leq x + 6$

84. $2x + y \leq 4$
$y \geq 2x - 1$
$y \geq -2$

85. $-x + 4y > 8$
$-x + y > 0$
$-x + 2y > 4$

86. $-4x + 3y \leq 24$
$-3x + 5y \leq 30$
$x \geq 2$
$y \geq 1$

87. $y \geq x^2 - 2$
$y \leq x$

88. $y + x^2 \leq 4$
$y \geq x^2$

89. $x^2 + y^2 \leq 25$
$x \geq 3$

90. $y \geq 2^x$
$y \leq 4$
$x \geq 0$

EXERCISES 91–92 *Find the maximum and minimum value(s) of the objective function over the given feasible set.*

91. $P = 5x + 8y$

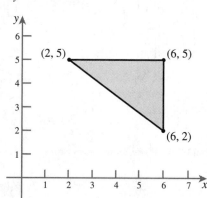

92. $P = 4x + 20y$

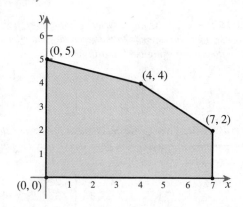

EXERCISES 93–96 *Solve the linear programming problem.*

93. Maximize $P = 12x + 10y$
    subject to $x + y \leq 6$
    $2x + y \leq 10$
    $x \geq 0$
    $y \geq 0$

94. Minimize $P = 6x + 12y$
    subject to $5x + 4y \geq 35$
    $5x + 3y \geq 30$
    $x \geq 0$
    $y \geq 0$

95. Minimize $P = 8x + 10y$
    subject to $2x + y \geq 10$
    $x + y \geq 7$
    $2x + 3y \geq 16$
    $x \geq 2$
    $y \geq 0$

96. Maximize $P = 10x + 15y$
    subject to $3x + 2y \leq 21$
    $x + y \leq 8$
    $y \leq 2x$
    $x \geq 0$
    $y \geq 2$

97. *Estate Management* An estate manager invested $100,000 in three funds: a money market fund paying 4% interest per year and charging an annual maintenance fee of 0.2%, a mutual fund (carrying substantial risk) averaging 12% per year with an annual load of 0.5%, and a savings account with no fees paying 3% interest per year. In one year, the investments earned a total of $6550 and $255 was charged in fees. How much was invested in each account?

98. *Balanced Breakfast* An 800-calorie breakfast is to consist of link sausages, pancakes, and orange juice and should contain 25 grams of protein and 90 grams of carbohydrates. Use the nutritional information given in Table 24 to determine how many servings of each breakfast item should be consumed.

TABLE 24

	Pancake	Orange	Sausage
Calories (per serving)	164	55	124
Carbohydrates (grams per serving)	23.7	13.2	0
Protein (grams per serving)	5.3	0.9	4.7

99. *Fitting a Cubic* Find the *y*-intercept of the graph of the cubic function $f(x)$ passing through the points $(-1, 5), (1, 5), (2, 10)$, and $(3, 31)$.

100. *Unknown Coins* A coin purse contains 10 coins (quarters, dimes, and nickels) with a total value of $1.60. There are as many quarters as there are nickels and dimes together. How many of each kind of coin are there?

101. *Land Allocation* A farmer intends to grow corn and hay. Between available cash and a bank line of credit, he has $50,000 available to invest in the two crops. Moreover, he and his son have a total of 1200 hours of labor available to devote to these two crops. Use the information given in Table 25 to determine how many acres of each crop should be planted to maximize revenue. How many acres of each should be planted to maximize profit?

TABLE 25

	Corn	Hay
Investment (dollars per acre)	125	100
Labor (hours per acre)	2	6
Revenue (dollars per acre)	245	210

102. *Food Service Planning* A university food service is planning a meal that includes broiled chicken breast and a baked potato. The chicken and potato portion of the meal must provide at least 31 units of protein, 80 units of vitamin B_6, and 16 units of iron. Use the information given in Table 26 to determine how many ounces of chicken and baked potato should be served to minimize the cost.

TABLE 26

	Chicken breast	Baked potato
Protein (units per ounce)	9	1
Vitamin B_6 (units per ounce)	18	6
Iron (units per ounce)	3	2
Cost (cents per ounce)	20	3

103. *Investment Options* Damon has $10,000 to invest in selected mutual funds and municipal bonds. For tax reasons, his accountant has advised that he invest at least twice as much in municipal bonds as in mutual funds. However, his financial adviser suggests that he invest at least $2500 in mutual funds. If the mutual funds average a 12% return and the municipal bonds average a 10% return, how much should he invest in each to maximize his earnings?

104. *Studying for Finals* Gina's final exams in college algebra and sociology are scheduled for the same day. She has 12 hours available to study for the two exams. Past experience suggests that each hour of studying for college algebra requires $\frac{1}{2}$ cup of

coffee, whereas each hour of studying for sociology requires $\frac{1}{4}$ cup of coffee. Based on previous exam scores, each hour of studying for college algebra results in 15 points, and each hour of studying for sociology results in 10 points. If she must score at least 60 points on the sociology exam, and she only has 4 cups of coffee left, how many hours should she study for each exam to maximize the total score?

CHAPTER 7 TEST

PROBLEMS 1–8 *Answer true or false.*

1. Matrices must be of the same order if they are to be added.

2. Matrices must be of the same order if they are to be multiplied.

3. All matrices have inverses.

4. If $\det(A) = 0$, then the system $AX = B$ can be solved by Cramer's Rule.

5. All upper triangular square matrices with 1s on the main diagonal are in row-echelon form.

6. Some linear systems of equations have exactly two solutions.

7. The solution set of a linear inequality is a half-plane.

8. According to the Fundamental Theorem of Linear Programming, the objective function will attain its maximum value at the vertex of the feasible region that is furthest from the origin.

PROBLEMS 9–13 *Give an example of each.*

9. A matrix of order 2×3

10. A noninvertible 2×2 matrix

11. An augmented matrix in row-echelon form but not reduced row-echelon form

12. A nonlinear system of equations

13. An elementary row operation

14. Let
$$A = \begin{bmatrix} 2 & -1 & 0 \\ 3 & 4 & -2 \\ 0 & -3 & 1 \end{bmatrix} \quad \text{and} \quad B = \begin{bmatrix} 0 & -2 & 5 \\ 6 & 3 & 1 \\ 4 & 0 & -1 \end{bmatrix}$$

Compute each of the following.

a. AB

b. $2A + B$

c. $A - B$

15. Express the following system as an equation involving matrices, and then solve it using an inverse matrix.
$$\begin{aligned} 3x - 5y + z &= 2 \\ 2x - 8y &= -2 \\ 3x - z &= 0 \end{aligned}$$

16. Consider the augmented matrix
$$\left[\begin{array}{ccc|c} 2 & -2 & 2 & 0 \\ 0 & 1 & -3 & -2 \\ 0 & 0 & 4 & 8 \end{array} \right]$$

a. Give an equivalent system of equations and solve by back-substitution.

b. Write the matrix in reduced row-echelon form.

17. Solve the following system of equations by Gaussian elimination.
$$\begin{aligned} x - y - z &= 4 \\ 3x + y + z &= 4 \\ x + 2y - z &= 13 \end{aligned}$$

18. Use Cramer's Rule to solve the following system for b.
$$\begin{aligned} 3a + b - c &= 1 \\ 2a &= -4 \\ a + c &= 2 \end{aligned}$$

19. In 1988, municipalities generated a total of 179.6 million tons of solid waste. Of this total, 58% was paper and yard waste; we'll classify the other 42% as miscellaneous waste. The total amount of paper waste was only 4.4 million tons less than the amount of miscellaneous waste. Find the amount of paper, yard, and miscellaneous waste. (*Data source:* Environmental Protection Agency.)

20. Find the maximum value of $P = 3x - 5y$ subject to the constraints $x \le 0$, $y \ge 0$, $y \le x + 6$, and $y \le \frac{1}{4}x + 3$. Graph the feasible set and clearly indicate all relevant vertices.

21. Find a system of inequalities such that the solution set is the interior of the triangle with vertices $(0, 0)$, $(2, 3)$, and $(-1, 6)$.

8 INTEGER FUNCTIONS AND PROBABILITY

Through personal experience we develop a sense of the frequency of many everyday events and hence of the probability of their occurrence. But personal experience doesn't provide us with intuition regarding events that rarely occur, such as becoming the victim of a terrorist when traveling abroad, winning a lottery, or being struck by lightning. Surprisingly, our chances of being struck by lightning are greater than winning certain state lotteries and much greater than becoming terrorist victims. In Section 4 of this chapter we will investigate such probabilities in detail.

SECTION 8.1 **SEQUENCES**

☐ What do flowers, pinecones, and sunflowers have in common with the family tree of a male honeybee?

☐ How is it possible to predict your salary 15 years into the future?

☐ If one checker is placed on the corner square of a checkerboard and the number of checkers in each successive square is doubled, how high will the pile of checkers in the 64th square reach?

☐ Why is 10 not necessarily the next number in the sequence 1, 4, 7, . . . ?

On the surface, a sequence is little more than a glorified list—of numbers, letters, or virtually anything. But the simplicity of this definition belies the significance of this fundamental notion and the role it plays in our lives. Sounds and images, documents and data are converted to binary sequences—strings of 1s and 0s—then chopped into pieces, pushed through conduits of copper, glass, and air, and reassembled at distant points on the vast network that is the World Wide Web. A digital cellular telephone reconstructs voices from sequences riding radio waves through space. Plastic and magnetic media, from CDs and DVDs to floppy disks and magnetic tape, archive our music, our cinema, and much of the collective knowledge of the human race in sequential form. Even our genetic code, the detailed blueprint for life itself replicated in nearly all of the 50 trillion cells that make up the human body, is, in essence, nothing more than a simple sequence.

In this section, we focus on the notation and terminology associated with sequences and on an intuitive development of central concepts. Though our tour will be little more than a high-altitude survey of the vast sequential landscape, we will take a closer pass at two areas of particular interest—arithmetic and geometric sequences.

NUMBER SEQUENCES

In mathematics, the term *sequence* refers to an ordered list, often consisting of real numbers. The following lists are all examples of sequences of real numbers.

Sequence a: $-2, 0, 5, 8$

Sequence b: $1, 4, 9, 16, 25, \ldots$

Sequence c: $\dfrac{1}{2}, \dfrac{1}{4}, \dfrac{1}{8}, \dfrac{1}{16}, \ldots$

Sequence d: $-1, 0, 1, 0, -1, 0, 1, \ldots$

The numbers that make up a sequence are called **terms.** For example, the terms of a are -2, 0, 5, and 8. Sequences like a that have a final term are called **finite,** whereas never-ending sequences like b, c, and d are said to be **infinite.** Sequences are ordered in the sense that they have a first term, a second term, a third term, and so on. This allows us to refer to specific terms of a sequence simply by attaching subscripts to the name of the sequence. For example, the fourth term of sequence b can be denoted by b_4, and so $b_4 = 16$. More generally, the nth term of sequence b can be denoted by b_n, and we observe that, since each term is the square of its position, $b_n = n^2$. We often define sequences in this way, by providing an expression for the nth term of the sequence.

THE INFORMATION REVOLUTION

We are currently witnessing nothing less than the most revolutionary change in the way information is stored and transmitted since Gutenberg invented the printing press. Compact disc players play back music free of hiss and distortion even after thousands of plays. Single CD-ROM discs contain entire encyclopedias. Visual images are transmitted across oceans via the World Wide Web. Entire libraries are accessible from personal computers, and interactive games are played with thousands of players living thousands of miles apart.

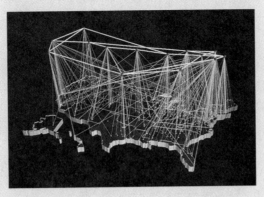

A visualization of data traffic over the National Science Foundation computer network.

Although this information revolution depends on technological developments such as computer chips, fiber-optic cable, and compact discs, it would have been impossible without mathematical innovations in the ways information can be encoded and decoded. By **digitizing** data—that is, converting information to a sequence of numbers, usually 1s and 0s—we are able to convert information into a language that computers can understand and manipulate, and we are able to bring to bear powerful mathematical techniques.

The accuracy of data transmission and storage is limited by the presence of **noise**—random errors with any number of possible causes: tiny flaws in the structure of a cable or a disc, electromagnetic spikes from sunspots, or even human error. Techniques for combating noise are the province of the branch of mathematics known as **error-correcting codes,** the basic principle of which is to introduce a redundancy in the way a message is encoded so that even with a small amount of noise, the correct message can still be determined. A simple (and too inefficient to be useful) example of an error-correcting code would be to simply repeat a digit 5 times. Thus, to transmit the digit 1, we would send "1 1 1 1 1"; to transmit 0, we would send "0 0 0 0 0." If the sequence received was "1 1 1 1 0," we would be fairly certain that the intended message was 1 and not 0.

An important key to sustaining progress in the information revolution is to continually increase the efficiency with which we store and send data. To this end, mathematicians have developed powerful **data compression** techniques that, in essence, allow us to store long sequences as shorter ones. As a simple example of a data compression technique, suppose we are dealing with sequences of 1s and 0s consisting of long runs in which a digit is repeated. Rather than send the original message, we could compress it by indicating the length of each run. Thus, the sequence consisting of 20 1s followed by 15 0s might be encoded as 20 15 (or in base 2: 10100 1111).

The exponential increase in access to information does not come without cost: The same automatic teller machine that allows us to perform financial transactions can be used by thieves to steal from our accounts; the cellular telephone system that enables us to call home from remote locations enables others to overhear our conversations; and the same medical records that can be downloaded by an emergency room physician also can be used by an unscrupulous insurance company to deny us coverage. Controlling who has access to what information poses one of the greatest challenges of the information revolution. **Cryptography** is a branch of mathematics that addresses how to transmit messages and have them remain secret, except to the intended receiver. Surprisingly, many encryption schemes involve factoring enormous numbers, and, for this reason, number theory, long viewed as pure mathematics with little direct applicability to the real world, is now an essential component of the information revolution.

EXAMPLE 1　■　Finding Terms of a Sequence

List the first four terms of the sequence with the given nth term.

a.　$a_n = 2n + 3$　　　　**b.**　$z_n = \dfrac{n}{n + 1}$　　　　**c.**　$r_n = \dfrac{(-1)^n}{2^n - 1}$

SOLUTION　In each case, we substitute $n = 1$, $n = 2$, $n = 3$, and $n = 4$.

a.　$a_1 = 2(1) + 3 = 5$
$a_2 = 2(2) + 3 = 7$
$a_3 = 2(3) + 3 = 9$
$a_4 = 2(4) + 3 = 11$

b.

$$z_1 = \frac{1}{1+1} = \frac{1}{2}$$

$$z_2 = \frac{2}{2+1} = \frac{2}{3}$$

$$z_3 = \frac{3}{3+1} = \frac{3}{4}$$

$$z_4 = \frac{4}{4+1} = \frac{4}{5}$$

c.

$$r_1 = \frac{(-1)^1}{2^1 - 1} = \frac{-1}{1} = -1$$

$$r_2 = \frac{(-1)^2}{2^2 - 1} = \frac{1}{3}$$

$$r_3 = \frac{(-1)^3}{2^3 - 1} = \frac{-1}{7}$$

$$r_4 = \frac{(-1)^4}{2^4 - 1} = \frac{1}{15}$$

It is no coincidence that the procedure for computing terms of a sequence is very much like that for evaluating a function. Indeed, a sequence can be viewed as a function whose domain is a set of integers. To emphasize this fact, we could define the sequence in part a of Example 1 by $a(n) = 2n + 3$ for $n = 1, 2, 3, \ldots$. However, it is customary to use the subscript notation a_n instead of the function notation $a(n)$. The formal definition of a sequence follows.

Definition of Sequence ☐

> A **sequence** a is a function whose domain is a set of integers. The function values $a(n)$ are called the **terms** of the sequence, and they are denoted a_n. Unless indicated otherwise, a_n is assumed to be defined for $n = 1, 2, 3, \ldots$.

EXAMPLE 2 ■ Finding the *n*th Term of a Sequence

Find an expression for the *n*th term of a sequence whose first few terms are given.

a. $3, 6, 9, 12, 15, \ldots$ **b.** $1, \frac{1}{2}, \frac{1}{4}, \frac{1}{8}, \frac{1}{16}, \ldots$

SOLUTION

a. Each term of the sequence is a multiple of 3. Specifically, $a_1 = 3(1)$, $a_2 = 3(2)$, $a_3 = 3(3)$, and so on. Thus, we define $a_n = 3n$.

b. Here each denominator is a power of 2. Thus, we first rewrite the sequence as

$$\frac{1}{2^0}, \frac{1}{2^1}, \frac{1}{2^2}, \frac{1}{2^3}, \frac{1}{2^4}, \ldots$$

Since the power of 2 in the denominator is one less than the position of the term,

$$a_n = \frac{1}{2^{n-1}}.$$

The sequences given in the previous example are representative of two important types of sequences. The sequence $3, 6, 9, 12, 15, \ldots$ has the property that each term after the first is 3 more than the one preceding it. It is an example of an *arithmetic sequence.* The sequence $1, \frac{1}{2}, \frac{1}{4}, \frac{1}{8}, \frac{1}{16}, \ldots$ has the property that each term after the first is half of the one preceding it. It is an example of a *geometric sequence.* We now investigate arithmetic and geometric sequences in more detail.

ARITHMETIC SEQUENCES

The distinctive feature of an arithmetic sequence is that each term after the first is obtained from the preceding term by adding a fixed number. In other words, each pair of consecutive terms has a common difference.

Definition of an ☐
Arithmetic Sequence

> A sequence $a_1, a_2, a_3, \ldots, a_n, \ldots$ is **arithmetic** if there is a number d, called the **common difference,** such that
>
> $$a_2 - a_1 = d, a_3 - a_2 = d, \ldots, a_n - a_{n-1} = d, \ldots$$
>
> Equivalently,
>
> $$a_2 = a_1 + d, a_3 = a_2 + d, \ldots, a_n = a_{n-1} + d, \ldots$$

E X A M P L E 3 ▧ **Examples of Arithmetic Sequences**

Compute the common difference d for each of the following arithmetic sequences.

a. $-10, -3, 4, 11, 18, \ldots$ **b.** $2, \frac{7}{4}, \frac{3}{2}, \frac{5}{4}, 1, \ldots$

SOLUTION Since we are given that the sequences are arithmetic, in each case we can compute the common difference using any pair of consecutive terms. We use the first two terms.

a. $d = -3 - (-10) = 7.$
b. $d = \frac{7}{4} - 2 = -\frac{1}{4}$

An arithmetic sequence is completely determined by the first term a_1 and the common difference d. In other words, if these two values are known, then all other terms can be computed. After all, the second term is obtained by adding d to the first term; the third term is obtained by adding d to the second term, and so forth. Algebraically, we have

$$a_1$$
$$a_2 = a_1 + d$$
$$a_3 = a_2 + d = (a_1 + d) + d = a_1 + 2d$$
$$a_4 = a_3 + d = (a_1 + 2d) + d = a_1 + 3d$$
$$\vdots$$

Thus, each term can be found by adding a multiple of d to a_1. Moreover, the multiple of d is one less than the subscript of the term. This leads us to the following formula for the nth term of the sequence.

The nth Term of an ☐
Arithmetic Sequence

> The nth term of an arithmetic sequence with first term a_1 and common difference d is given by
>
> $$a_n = a_1 + (n - 1)d$$

EXAMPLE 4   ■   **Finding the *n*th Term of an Arithmetic Sequence**

Find an expression for the *n*th term of the arithmetic sequence $8, -1, -10, \ldots$

SOLUTION   The common difference is $d = -1 - 8 = -9$. Using $a_1 = 8$, we apply the *n*th-term formula to find a_n.

$$a_n = a_1 + (n - 1)d$$
$$= 8 + (n - 1)(-9)$$
$$= -9n + 17$$

EXAMPLE 5   ■   **Finding the *n*th Term of an Arithmetic Sequence**

The first Saturday of 2001 falls on January 6. Find an expression for the day of the year on which the *n*th Saturday falls, and use this result to find the date of the tenth Saturday of 2001.

SOLUTION   Let a_n represent the day of the year on which the *n*th Saturday falls. From the given information, we know that $a_1 = 6$. Since Saturdays occur 7 days apart, it follows that $d = 7$. Applying the *n*th-term formula for an arithmetic sequence gives us

$$a_n = 6 + 7(n - 1)$$

To find the date of the tenth Saturday, we let $n = 10$, which gives us

$$a_{10} = 6 + 7(10 - 1) = 69$$

Now there are 31 days in January and 28 days in February (in a non-leap year like 2001), so the 69th day of the year falls in March on day

$$69 - (31 + 28) = 10$$

Thus, the tenth Saturday of 2001 falls on March 10.

EXAMPLE 6   ■   **Finding a Specified Term of an Arithmetic Sequence**

The fifth and eleventh terms of an arithmetic sequence are 28 and 64, respectively. Find the twentieth term.

SOLUTION   By substituting $n = 5$ and $a_5 = 28$ into the *n*th-term formula, we obtain

$$a_n = a_1 + (n - 1)d$$
$$a_5 = a_1 + (5 - 1)d$$
(1) $$28 = a_1 + 4d$$

Similarly, substituting $n = 11$ and $a_{11} = 64$ gives

(2) $$64 = a_1 + 10d$$

Subtracting equation (1) from equation (2) yields

$$64 = a_1 + 10d$$
$$-\ \ 28 = a_1 + 4d$$
$$\overline{\qquad\qquad\qquad}$$
$$36 = \qquad 6d$$
$$d = 6$$

Substituting $d = 6$ into equation (1) gives

$$28 = a_1 + 4(6)$$
$$a_1 = 4$$

Now that a_1 and d are known, we apply the nth-term formula once more with $n = 20$ to see that

$$a_{20} = 4 + (20 - 1)(6) = 118$$

GEOMETRIC SEQUENCES

We have defined an arithmetic sequence to be a sequence whose consecutive terms have a common difference. A *geometric sequence* is a sequence whose consecutive terms have a common ratio.

Definition of a Geometric Sequence   ☐

> A sequence $a_1, a_2, a_3, \ldots, a_n, \ldots$ is **geometric** if there is a number $r \neq 0$, called the **common ratio**, such that
>
> $$\frac{a_2}{a_1} = r, \frac{a_3}{a_2} = r, \ldots, \frac{a_n}{a_{n-1}} = r, \ldots$$
>
> Equivalently,
>
> $$a_2 = a_1 r, a_3 = a_2 r, \ldots, a_n = a_{n-1} r, \ldots$$

EXAMPLE 7 ▨   **Examples of Geometric Sequences**

Find the common ratio r for each of the following geometric sequences.

a. $6, 18, 54, 162, 486, \ldots$ **b.** $-\frac{4}{3}, \frac{8}{9}, -\frac{16}{27}, \frac{32}{81}, -\frac{64}{243}, \ldots$

SOLUTION   Since we are given that the sequences are geometric, in each case we can compute the common ratio using any pair of consecutive terms. We use the first two terms.

a. $r = \frac{18}{6} = 3$

b. $r = \dfrac{\frac{8}{9}}{-\frac{4}{3}} = \frac{8}{9} \cdot \left(-\frac{3}{4}\right) = -\frac{2}{3}$

A formula for the nth term of a geometric sequence can be found using a method similar to that used for arithmetic sequences. We begin with the first term a_1 and use the fact that each successive term is found by multiplying the preceding one by the common ratio r.

$$a_1$$
$$a_2 = a_1 r$$
$$a_3 = a_2 r = (a_1 r)r = a_1 r^2$$
$$a_4 = a_3 r = (a_1 r^2)r = a_1 r^3$$
$$\vdots$$

Thus, each term can be found by multiplying a_1 by an appropriate power of r. In fact, we see that the power is one less than the subscript of the term. Thus, we have the following formula.

The nth Term of a
Geometric Sequence □

> The nth term of a geometric sequence with first term a_1 and common ratio r is
>
> $$a_n = a_1 r^{n-1}$$

EXAMPLE 8 ▨ **Using the *n*th-Term Formula for Geometric Sequences**

Find the 25th term of the geometric sequence having common ratio $r = \frac{1}{4}$ and first term $a_1 = 3$.

SOLUTION Applying the nth-term formula for geometric sequences, we obtain

$$a_n = a_1 r^{n-1}$$
$$a_{25} = 3\left(\frac{1}{4}\right)^{25-1}$$
$$= \frac{3}{4^{24}}$$

EXAMPLE 9 ▨ **A Sequence of Checker Stacks**

A checker is placed on a corner square of a checkerboard. In each successive square, the number of checkers is doubled, as shown in Figure 1. How many checkers will be on the 64th square? If a checker is approximately 5 millimeters thick, how high will the stack of checkers reach? (*Note:* Figure 2 suggests the approximate height of the stack in the 32nd square.)

SOLUTION The number of checkers on each successive square is given by a term of the sequence 1, 2, 4, 8, 16, ... This is a geometric sequence with first term 1 and

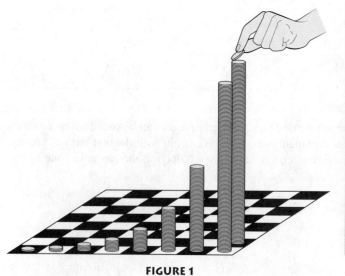

FIGURE 1

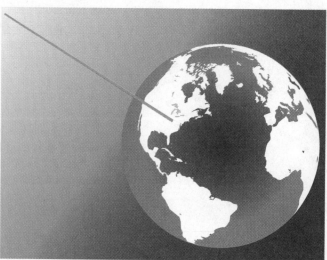

FIGURE 2

common ratio $r = 2$. Thus, by applying the nth-term formula for geometric sequences, we see that $a_n = (1)2^{n-1}$. The 64th term of this sequence is

$$a_{64} = 2^{63} \approx 9.22 \times 10^{18}$$

So there are approximately 9,220,000,000,000,000,000 checkers on the 64th square. If each checker is 5 millimeters thick, the pile of checkers would reach

$$(9.22 \times 10^{18})(5 \text{ millimeters})\left(\frac{1 \text{ kilometer}}{10^6 \text{ millimeters}}\right) = 4.61 \times 10^{13} \text{ kilometers}$$

which is approximately the distance to Proxima Centauri, the closest star (other than the sun) to Earth.

RECURSIVELY DEFINED SEQUENCES

Arithmetic and geometric sequences have the property that each term after the first can be computed from the preceding term. In the case of arithmetic sequences, we simply add the common difference d, so that $a_n = a_{n-1} + d$. For geometric sequences, we multiply by the common ratio r, obtaining $a_n = a_{n-1}r$. Such formulas for the nth term of a sequence are said to be **recursive**. Generally speaking, a **recursively defined sequence** is one for which the first few terms are given and every successive term can be computed using a formula involving one or more of the preceding terms.

EXAMPLE 10   Finding Terms of a Recursively Defined Sequence

Find the first four terms of the sequence defined recursively by $a_1 = 3$ and $a_n = 2a_{n-1} + 5$ for $n > 1$.

SOLUTION

$$a_1 = 3$$
$$a_2 = 2a_1 + 5 = 2(3) + 5 = 11$$
$$a_3 = 2a_2 + 5 = 2(11) + 5 = 27$$
$$a_4 = 2a_3 + 5 = 2(27) + 5 = 59$$

EXAMPLE 11   Finding Terms of a Recursively Defined Sequence

Find the first six terms of the sequence defined recursively by $a_1 = 1$, $a_2 = 1$, and $a_n = a_{n-1} + a_{n-2}$ for $n > 2$.

SOLUTION

$$a_1 = 1$$
$$a_2 = 1$$
$$a_3 = a_2 + a_1 = 1 + 1 = 2$$
$$a_4 = a_3 + a_2 = 2 + 1 = 3$$
$$a_5 = a_4 + a_3 = 3 + 2 = 5$$
$$a_6 = a_5 + a_4 = 5 + 3 = 8$$

The recursively defined sequence 1, 1, 2, 3, 5, 8, 13, 21, 34, 55, 89, ... of the previous example is known as the **Fibonacci sequence.** It is named after the Italian mathematician Leonardo of Pisa, whose nickname was Fibonacci. The sequence, which first appeared in Fibonacci's book *Liber Abaci* ("The Book of the Abacus") in 1202, can be described as the sequence for which the first two terms are 1 and each successive term is found by adding together the previous two terms. Despite its simplicity, the Fibonacci sequence has some fascinating mathematical properties. We will investigate some of these properties in Section 8.6.

The Fibonacci numbers occur with remarkable frequency in nature. Many species of flowers have petal arrangements consisting of 3, 5, 8, 13, 21, 34, 55, or 89 petals. The bracts of a pinecone spiral in sets of 8 and 13 rows. A sunflower has 34 petals, and its florets spiral out from the center in sets of 55 rows and 89 rows. The scales of a pineapple spiral in sets of 8, 13, and 21 rows.

The Fibonacci sequence is even present in the family tree of a male honeybee. As first observed by Johann Dzierzon in 1845, the male honeybee hatches from an unfertilized egg, and so has a mother but no father. Female honeybees, on the other hand, hatch from fertilized eggs, and so have a mother and a father. Thus, if we trace the family tree of the male honeybee, we see that each male "node" of the tree has only one parent branch (a mother), whereas each female "node" has two branches (a mother and a father). The number of nodes at each level of the family tree corresponds to the Fibonacci numbers, as shown in Figure 3.

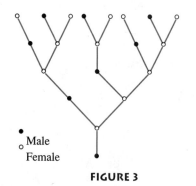

● Male
○ Female

FIGURE 3

8
5
3
2
1
1

UNDERSTANDING AND MASTERY CHECKLISTS

CONCEPTS TO UNDERSTAND

☐ Sequence

☐ Term of a sequence

☐ Sequence notation

☐ Arithmetic sequence

☐ Common difference of an arithmetic sequence

☐ nth-term formula for an arithmetic sequence

☐ Geometric sequence

☐ Common ratio of a geometric sequence

☐ nth-term formula for a geometric sequence

☐ Recursively defined sequence

☐ Fibonacci sequence

SKILLS TO MASTER

☐ Find the first few terms of a sequence given a formula for the nth term.

☐ Find an expression for the nth term of a sequence given the first few terms.

☐ Determine whether a sequence is arithmetic and, if so, find its common difference.

☐ Determine whether a sequence is geometric and, if so, find its common ratio.

☐ Find a specified term of an arithmetic or geometric sequence using partial information about the sequence.

☐ Find the first few terms of a recursively defined sequence.

EXERCISES 8.1

EXERCISES 1–16 *List the first four terms of the sequence with the given nth term.*

1. $a_n = 3n + 1$

2. $u_n = 4 - n$

3. $z_n = (-4)^n$

4. $c_n = -\dfrac{1}{2^n}$

5. $r_n = \dfrac{1}{n^2}$

6. $b_n = \dfrac{n-1}{n+1}$

7. $c_n = 1 + (-1)^n$

8. $x_n = (-1)^n n^2$

9. $v_n = \dfrac{(-1)^{n-1}}{n}$

10. $r_n = \dfrac{(-1)^{n+1}}{3^n}$

11. $a_n = \sqrt{n(n+1)}$

12. $y_n = \sqrt{n+1} - \sqrt{n}$

13. $b_n = \left(1 + \dfrac{1}{n}\right)^n$

14. $r_n = \dfrac{n^2}{2^n}$

15. $s_n = 1 + 2 + \cdots + n$

16. $a_n = 1 \cdot 2 \cdot \cdots \cdot n$

EXERCISES 17–24 *Find an expression for the nth term of a sequence whose first few terms are given.*

17. $2, 4, 6, 8, \ldots$

18. $3, 9, 27, 81, \ldots$

19. $-\dfrac{1}{2}, \dfrac{1}{4}, -\dfrac{1}{8}, \dfrac{1}{16}, \ldots$

20. $3, 7, 11, 15, \ldots$

21. $\dfrac{1}{2}, \dfrac{2}{3}, \dfrac{3}{4}, \dfrac{4}{5}, \ldots$

22. $2 \cdot 3, 3 \cdot 4, 4 \cdot 5, 5 \cdot 6, \ldots$

23. $1 - \dfrac{1}{1}, 1 - \dfrac{1}{2}, 1 - \dfrac{1}{3}, 1 - \dfrac{1}{4}, \ldots$

24. $1 + \dfrac{1}{2}, 1 - \dfrac{1}{4}, 1 + \dfrac{1}{8}, 1 - \dfrac{1}{16}, \ldots$

EXERCISES 25–34 *Indicate whether the sequence appears to be arithmetic, geometric, or neither. If you claim it is arithmetic, find the common difference. If you think it is geometric, find the common ratio.*

25. $5, 8, 11, 14, \ldots$

26. $1, 3, 6, 10, \ldots$

27. $3, 12, 60, 360, \ldots$

28. $3, -6, 12, -24, \ldots$

29. $5, 1, -3, -7, \ldots$

30. $3, 2, \dfrac{4}{3}, \dfrac{8}{9}, \ldots$

31. $\dfrac{1}{2}, \dfrac{3}{4}, \dfrac{7}{8}, \dfrac{15}{16}, \ldots$

32. $\dfrac{4}{3}, \dfrac{8}{3}, 4, \dfrac{16}{3}, \ldots$

33. $0.2, 0.02, 0.002, 0.0002, \ldots$

34. $1, 4, 9, 16, \ldots$

EXERCISES 35–40 *Find an expression for the nth term of the given arithmetic sequence.*

35. $7, 13, 19, 25, \ldots$

36. $8, 4, 0, -4, \ldots$

37. $100, 85, 70, 55, \ldots$

38. $2, 14, 26, 38, \ldots$

39. $-2, -\dfrac{3}{4}, \dfrac{1}{2}, \dfrac{7}{4}, \ldots$

40. $\dfrac{1}{2}, -\dfrac{1}{6}, -\dfrac{5}{6}, -\dfrac{3}{2}, \ldots$

EXERCISES 41–46 *Use the given information about the arithmetic sequence to find the indicated term.*

41. $a_1 = 3, d = 8; a_{10} = $ _____

42. $a_1 = 22, d = -4; a_{18} = $ _____

43. $a_1 = 10, a_5 = -10; a_{15} = $ _____

44. $a_1 = 9, a_4 = 30; a_{21} = $ _____

45. $a_8 = 38, a_{17} = 92; a_{31} = $ _____

46. $a_{12} = -16, a_{20} = -40; a_{26} = $ _____

EXERCISES 47–52 *Find an expression for the nth term of the given geometric sequence.*

47. $1, 3, 9, 27, \ldots$

48. $5, -10, 20, -40, \ldots$

49. $32, -16, 8, -4, \ldots$

50. $\dfrac{4}{3}, \dfrac{2}{3}, \dfrac{1}{3}, \dfrac{1}{6}, \ldots$

51. $4, 5, \dfrac{25}{4}, \dfrac{125}{16}, \ldots$

52. $-4, \dfrac{8}{3}, -\dfrac{16}{9}, \dfrac{32}{27}, \ldots$

EXERCISES 53–56 *Use the given information about the geometric sequence to find the indicated term.*

53. $a_1 = 2, r = 3; a_8 = $ _____

54. $a_1 = 6, r = -\dfrac{1}{2}; a_{10} = $ _____

55. $a_1 = 9, a_3 = 1; a_7 = $ _____

56. $a_1 = \dfrac{1}{8}, a_3 = 2; a_6 = $ _____

EXERCISES 57–64 *Find the first five terms of the given recursively defined sequence.*

57. $a_1 = 2$ and $a_n = 1 - 2a_{n-1}$ for $n \geq 2$

58. $a_1 = 1$ and $a_n = 3a_{n-1} + 1$ for $n \geq 2$

59. $a_1 = 1$ and $a_n = na_{n-1}$ for $n \geq 2$

60. $a_1 = 1$ and $a_n = n + a_{n-1}$ for $n \geq 2$

61. $a_1 = 2$ and $a_n = \sqrt{a_{n-1}}$ for $n \geq 2$

62. $a_1 = 2$ and $a_n = \sqrt{(a_{n-1})^2 + 1}$ for $n \geq 2$

63. $a_1 = 1, a_2 = 2,$ and $a_n = a_{n-1}a_{n-2}$ for $n \geq 3$

64. $a_1 = 1, a_2 = 2,$ and $a_n = \dfrac{a_{n-1}}{a_{n-2}}$ for $n \geq 3$

▪ APPLICATIONS

65. *Projecting Income* Juanita is considering a new job that offers a starting base salary of $28,500, a guaranteed $1200 raise each year, and an annual bonus equal to 3% of the base salary. Determine her potential income (salary plus bonus) for each of the first three years and show that these values form an arithmetic sequence. What would her income be in the seventh year?

66. *Projecting Income* Vincent is considering a new job that offers a starting salary of $27,000 and a guaranteed $1500 raise each year. The benefit package (health, retirement, and so on) is 11% of the annual salary. Determine the total value (salary plus benefits) for each of the first three years and show that these totals form an arithmetic sequence. What would the total be in the eighth year?

67. *Membership Increase* An organization claims that its membership has increased 20% each year since it was formed with 10 charter members. How many members are there at the beginning of the 12th year?

68. *Salary Increase* If you start a job with an annual salary of $30,000 and a guaranteed 5% raise each year, what will your salary be at the beginning of your 15th year?

69. *Accumulated Value* Suppose $1000 is deposited in an account in which interest is compounded annually. The balances at the beginning of each of the first four years are $1000.00, $1055.00, $1113.03, and $1174.25. Assuming this pattern continues, write out a formula for the nth term of this sequence. What will the balance be at the beginning of the tenth year? What is the annual interest rate?

70. *Amortization Schedule* The first four entries on the amortization schedule for a $120,000 mortgage indicate monthly interest charges of $800.00, $799.46, $798.92, and $798.38. Write out a formula for the nth term of this sequence. How much interest will be charged in the 24th month?

71. *Breeding Rabbits* A pair of baby rabbits (one male and one female) is placed inside an enclosed area for the purposes of breeding. During the first month, the pair does not produce any new rabbits, but each month thereafter it produces a pair (one male and one female) of rabbits. Assuming each new pair follows the same reproductive pattern, complete the first 6 rows of the following table and find the number of rabbits at the end of 12 months. (*Hint:* Because of the 1-month maturation period, the number of pairs of rabbits born in a given month equals the number of pairs of rabbits at the beginning of the previous month.)

Month	Pairs at the beginning of the month	Pairs born during the month	Pairs at the end of the month
1	1	0	1
2	1	1	2
3	2	1	3
4	3		

72. *Consumer Price Index* The Consumer Price Index, or CPI, measures the prices of consumer goods and services and is a measure of the pace of U.S. inflation. If the CPI is 150 in one year and 161 in the next, then it takes roughly $161 in the second year to purchase goods and services that could have been purchased for $150 in the first year. It follows that if i_n is the inflation rate for year n and C_n is the CPI for year n, then $C_n = C_{n-1}(1 + i_n)$.

a. Use this fact to complete the following table.

Year	1990	1991	1992	1993	1994	1995	1996
Inflation Rate	5.4%	4.2%	3.0%	3.0%	2.6%	2.8%	3.0%
CPI	130.7						

b. The last year in which there was double-digit inflation was 1981, with an inflation rate of 10.1% and a CPI of 90.9. Find the CPI for 1980.

CONCEPTS AND CRITICAL THINKING

EXERCISES 73–76 *Answer true or false.*

73. The sets $\{1, 2, 3\}$ and $\{3, 2, 1\}$ are identical.

74. The sequences 1, 2, 3 and 3, 2, 1 are identical.

75. The difference between consecutive terms of an arithmetic sequence is a constant.

76. The difference between consecutive terms of a geometric sequence is a constant.

EXERCISES 77–80 *Give an example of each.*

77. A recursively defined sequence

78. An arithmetic sequence with common difference 2

79. A geometric sequence with common ratio 3

80. A sequence that is both geometric and arithmetic

QUESTIONS FOR DISCUSSION OR ESSAY

81. Suppose the instructions for a set of algebra problems read "Find an expression for the nth term of *the* sequence whose first few terms are given." Explain why these are not valid instructions. To help you see the problem, compute the first four terms of the following sequences, and answer the question "Why is 10 not necessarily the next number in the sequence 1, 4, 7, . . . ?"

$$a_n = 3n - 2$$
$$a_n = n^3 - 6n^2 + 14n - 8$$

82. The nth term of an arithmetic or geometric sequence can be defined either *explicitly* or *recursively*. For example, the nth term of the arithmetic sequence 4, 10, 16, 22, . . . can be defined

explicitly by $a_n = 6n - 2$ for $n = 1, 2, 3, \ldots$ or recursively by $a_1 = 4$ and $a_n = a_{n-1} + 6$ for $n = 2, 3, \ldots$. Find both explicit and recursive definitions for the geometric sequence 5, 15, 45, ... Discuss the advantages and disadvantages of each definition.

83. In many of the application problems involving geometric sequences, we assumed that certain percentage rates of increase or decrease (for example, membership increase, salary increase, in-

terest rate, population growth or decline) were constant from one year to the next, and this allowed us to make projections well into the future. Which, if any, of these assumptions are valid? Justify your answer. If a rate cannot be assumed to be constant, are the projections of any use? Explain. Give some examples from the "real world" that involve rates of increase or decrease that could be assumed to be constant.

▪ PROJECTS FOR ENRICHMENT

84. *Limiting Behavior of Sequences*   Although a formal definition of the limit of a sequence is beyond the scope of this text, we will say that a sequence $a_1, a_2, a_3, \ldots$ has a limit L if the terms a_n approach L (and only L) as n increases without bound. In this case, we write $a_n \to L$ as $n \to \infty$. As an example, consider the sequence $a_n = \frac{1}{n}$ whose first few terms are $1, \frac{1}{2}, \frac{1}{3}, \frac{1}{4}, \ldots$. Clearly the terms of this sequence are getting closer and closer to zero. Thus, we conclude that $a_n \to 0$ as $n \to \infty$. It is sometimes possible to predict the limit of a sequence in this way, by inspecting the first "few" terms of the sequence.

a. Write out a sufficient number of terms for the given sequence and predict the limit.

i. $a_n = \dfrac{1}{n^2}$

ii. $a_n = \left(\dfrac{1}{2}\right)^n$

iii. $a_n = \dfrac{n}{n + 1}$

iv. $a_n = \dfrac{2^n}{2^n - 1}$

Note that the method just described can also be very unreliable. A sequence may converge to a limit so slowly that it is difficult to determine the limit.

b. Let

$$a_n = \frac{\ln n}{\sqrt{n}}$$

Find a_n for $n = 10$, $n = 20$, $n = 50$, $n = 100$, and $n = 1000$. What do you think the limit is?

Not all sequences have a limit. For example, the terms of the arithmetic sequence 5, 8, 11, 14, ... increase without bound and so do not approach any number. The terms of the sequence 1, $-1, 1, -1, 1, \ldots$ simply alternate between 1 and -1 and so do not approach a *single* number. For a sequence to have a limit, its terms must approach a unique finite number.

c. Determine whether the given sequence has a limit. If so, state the limit. If not, explain why not.

i. $a_n = \frac{1}{2}n$

ii. $a_n = \frac{1}{n} + 5$

iii. $a_n = (-1)^n + 1$

iv. $a_n = \dfrac{(-1)^n}{n}$

v. $a_n = \left(\frac{3}{2}\right)^n$

vi. $a_n = e^{-n}$

Sometimes a graphing calculator can be a useful tool for studying limits of sequences. Consider the sequence

$$a_n = \frac{n}{\sqrt{2n^2 + 1}}$$

If we define the function

$$f(x) = \frac{x}{\sqrt{2x^2 + 1}}$$

then $f(n) = a_n$ for all n. To investigate the limit of a_n, we plot the graph of f and use the trace feature to see what value $f(x)$ approaches as x gets very large (see Figure 4). The cursor shows that $f(16) \approx 0.706$ and so $a_{16} \approx 0.706$. Since the graph does not appear to be rising further, we conclude that the limit is approximately 0.706. Using calculus, we could show that the exact limit is $\sqrt{2}/2$.

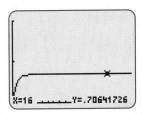

FIGURE 4

d. Use a graphing calculator to estimate the limit of the given sequence.

i. $a_n = \dfrac{2n}{3n + 1}$

ii. $a_n = \dfrac{4n^2 - 2n}{7n^2 + 3}$

iii. $a_n = \left(1 + \frac{1}{n}\right)^n$

e. Experiment with different viewing windows on a graphing calculator to investigate the limit of the sequence

$$a_n = \frac{0.001n^2 + n}{n + 1000}$$

What can you conclude? What potential hazards can you see in using a graphing calculator to estimate limits?

85. *Population Growth*   Suppose a certain city has population P_0 and a constant rate of growth of k people per thousand per year. In other words, for each group of 1000 people in the population, there will be an increase of k people after one year. We see in this project that constant growth rates lead to geometric sequences.

a. In 1990, Mesa, Arizona, had a population of approximately 288,000 and a growth rate of approximately 66 people per thousand per year. Estimate the population in 1991 and 1992.

b. Argue more generally that if there are P_n people at the beginning of year n, then there will be

$$P_n + k\left(\frac{P_n}{1000}\right)$$

people at the end of year $n + 1$.

c. Rewrite the expression in part b to show that if P_n denotes the number of people at the beginning of year n, then $P_{n+1} = (1 + 0.001k)P_n$.

d. Use $P_0 = 288,000$ (the population in 1990) and $k = 66$ to estimate the population of Mesa, Arizona, in each of the years 1991–1995. Explain why this sequence is a geometric sequence. Estimate the population of Mesa in the year 2010.

e. In 1990, Gary, Indiana, had a population of approximately 117,000 and a growth rate of -26 people per thousand. Estimate the population in each of the years 1991–1995. Do these values form a geometric sequence? Why or why not? What do you notice about the population of Gary? How should we interpret a negative growth rate?

SECTION 8.2   SERIES

☐ How would a 10-year-old child prodigy add up the first 100 positive integers?

☐ Would you buy a $9600 car on the following terms: no money down and 48 monthly payments of $200 plus 2% per month interest on the remaining balance?

☐ If you had a choice between a job with a starting salary of $32,000, a yearly raise of $3000, and a 4% year-end bonus, and one with a starting salary of $32,000 and a 6% yearly raise, which would you choose?

☐ How is it that a Greek philosopher's "proof" of the impossibility of motion took centuries to refute?

☐ If you were to win the grand prize in a sweepstakes, should you take the $1,000,000 lump-sum payment or the annuity option of $100,000 each year for the next 20 years?

SERIES AND SUMMATION NOTATION

The story is told that when the German mathematician Karl Friederich Gauss was only 10 years old, he amazed his teacher by adding together the first 100 positive integers in just seconds. Historians speculate that Gauss accomplished this feat by observing that the numbers in the sum $1 + 2 + 3 + \cdots + 98 + 99 + 100$ can be grouped in pairs that add up to 101, as follows:

$$
\underbrace{1 + 2 + 3 + \cdots + 49 + \overbrace{50 + 51}^{101} + 52 + \cdots + 98 + 99 + 100}
$$

Because there are 50 such pairs, the total is simply $50(101) = 5050$. The sum $1 + 2 + 3 + \cdots + 100$ is an example of a **series**. In general, a series is a sum of a sequence of numbers, as indicated by the following definition.

Definition of Series ☐

> Given an infinite sequence $a_1, a_2, a_3, \ldots, a_n, \ldots$, the sum
>
> $$a_1 + a_2 + a_3 + \cdots + a_n$$
>
> is called a **finite series,** whereas the sum
>
> $$a_1 + a_2 + a_3 + \cdots + a_n + \cdots$$
>
> is called an **infinite series.** Each a_j is called a **term** of the series.

EXAMPLE 1 ■ **Evaluating a Series**

Evaluate the series $a_1 + a_2 + \cdots + a_6$, where $a_n = \frac{1}{n}$.

SOLUTION

$$
\begin{aligned}
a_1 + a_2 + \cdots + a_6 &= \frac{1}{1} + \frac{1}{2} + \frac{1}{3} + \frac{1}{4} + \frac{1}{5} + \frac{1}{6} \\
&= \frac{60 + 30 + 20 + 15 + 12 + 10}{60} \\
&= \frac{147}{60} \\
&= \frac{49}{20}
\end{aligned}
$$

Since series involving many terms can be awkward to write, we introduce a convenient shorthand notation called **summation notation.**

Summation Notation ☐

> $$\sum_{k=l}^{n} a_k = a_l + a_{l+1} + \cdots + a_n$$
>
> $$\sum_{k=l}^{\infty} a_k = a_l + a_{l+1} + a_{l+2} + \cdots$$
>
> The symbol Σ is the capital Greek letter **sigma.** The variable k is called the **index variable.** The expression $k = l$ below the sigma defines the starting value of the index variable, otherwise known as the **lower limit.** The ending value of the index variable—the **upper limit**—is given above the sigma.

EXAMPLE 2 ■ **Evaluating Series**

Evaluate the given series.

a. $\displaystyle\sum_{k=1}^{5} 2^k$ 　　　　　　 **b.** $\displaystyle\sum_{i=1}^{6} i^2$ 　　　　　　 **c.** $\displaystyle\sum_{n=2}^{4} \left(-\frac{1}{10}\right)^n$

SOLUTION

a. With a lower limit of 1 and an upper limit of 5, we obtain

$$\sum_{k=1}^{5} 2^k = 2^1 + 2^2 + 2^3 + 2^4 + 2^5$$
$$= 2 + 4 + 8 + 16 + 32$$
$$= 62$$

b. Note that the choice of index variable is immaterial. We start with $i = 1$ and end with $i = 6$.

$$\sum_{i=1}^{6} i^2 = 1^2 + 2^2 + 3^2 + 4^2 + 5^2 + 6^2$$
$$= 1 + 4 + 9 + 16 + 25 + 36$$
$$= 91$$

c. Here we have a lower limit of 2 and an upper limit of 4. Thus,

$$\sum_{n=2}^{4} \left(-\frac{1}{10}\right)^n = \left(-\frac{1}{10}\right)^2 + \left(-\frac{1}{10}\right)^3 + \left(-\frac{1}{10}\right)^4$$
$$= \frac{1}{100} - \frac{1}{1000} + \frac{1}{10,000}$$
$$= \frac{100 - 10 + 1}{10,000}$$
$$= \frac{91}{10,000}$$

EXAMPLE 3 ▇ **Writing a Series with Summation Notation**

Express the given series using summation notation.

a. $1 + 4 + 9 + 16 + 25 + 36 + 49 + 64 + 81$

b. $\frac{1}{3} + \frac{1}{9} + \frac{1}{27} + \frac{1}{81} + \frac{1}{243} + \frac{1}{729}$

SOLUTION

a. The terms of this series are the squares of the numbers 1 through 9. Thus, the general kth term has the form $a_k = k^2$. The first term corresponds to $k = 1$ and the last term corresponds to $k = 9$. Thus, the lower and upper limits are 1 and 9, respectively. In summation notation we have

$$\sum_{k=1}^{9} k^2$$

b. The denominators of the terms of this series are successive powers of 3, starting with 3^1 and ending with $3^6 = 729$. Thus, the general kth term has the form $a_k = 1/3^k$. The first term corresponds to $k = 1$, and so the lower limit is 1. The last term corresponds to $k = 6$, and so the upper limit is 6. In summation notation we have

$$\sum_{k=1}^{6} \frac{1}{3^k}$$

So far, our emphasis has been on using and understanding summation notation rather than on techniques for evaluating series. We now turn our attention to methods for evaluating a few types of finite series. Infinite series will be considered in Exercise 79.

ARITHMETIC SERIES

Recall from Section 8.1 that an arithmetic sequence is one whose successive terms have a common difference. The sum of the terms of an arithmetic sequence is called an **arithmetic series.** We would like to develop a formula for evaluating *finite* arithmetic series. To that end, suppose $a_1, a_2, a_3, \ldots, a_n$ is an arithmetic sequence with common difference d, and let the sum S be given by

$$S = a_1 + a_2 + \cdots + a_{n-1} + a_n$$

Since $a_k = a_1 + (k - 1)d$, it follows that

(1) $\qquad S = a_1 + [a_1 + d] + \cdots + [a_1 + (n - 2)d] + [a_1 + (n - 1)d]$

By reversing the order of the terms of equation (1), we also have

(2) $\qquad S = [a_1 + (n - 1)d] + [a_1 + (n - 2)d] + \cdots + [a_1 + d] + a_1$

Adding equations (1) and (2) leads to the following:

$$\begin{array}{rl}
S = a_1 & + [a_1 + d] \qquad + \cdots + [a_1 + (n - 2)d] \; + [a_1 + (n - 1)d] \\
+ \; S = [a_1 + (n - 1)d] & + [a_1 + (n - 2)d] + \cdots + [a_1 + d] \qquad\quad + a_1 \\
\hline
2S = \underbrace{[2a_1 + (n - 1)d]}_{} & + [2a_1 + (n - 1)d] + \cdots + [2a_1 + (n - 1)d] + [2a_1 + (n - 1)d]
\end{array}$$

$$\underbrace{}_{n \text{ terms}}$$

$$2S = n[2a_1 + (n - 1)d]$$
$$2S = n[a_1 + a_1 + (n - 1)d]$$
$$2S = n(a_1 + a_n) \qquad\qquad \text{Substituting } a_n \text{ for } a_1 + (n - 1)d$$
$$S = \frac{n}{2}(a_1 + a_n) \qquad\qquad \text{Dividing both sides by 2}$$

Thus, we have the following formula.

Formula for Finite Arithmetic Series □

> If $a_1, a_2, \ldots, a_n$ is an arithmetic sequence, then
>
> $$a_1 + a_2 + \cdots + a_n = \frac{n}{2}(a_1 + a_n)$$
>
> In words, the sum of a finite arithmetic sequence is equal to the number of terms times the average of the first and last terms.

The formula for finite arithmetic series suggests another approach to the problem solved by the 10-year-old Gauss.

EXAMPLE 4   ■   Using the Formula for Finite Arithmetic Series

Find the sum of the first 100 positive integers.

SOLUTION The series $1 + 2 + 3 + \cdots + 100$ is a finite arithmetic series. Applying the formula for finite arithmetic series with $n = 100$, $a_1 = 1$, and $a_{100} = 100$, we obtain

$$1 + 2 + 3 + \cdots + 100 = \frac{100}{2}(1 + 100) = 5050$$

EXAMPLE 5 ▮ **Evaluating an Arithmetic Series**

Compute $\sum_{j=1}^{40} (3j + 5)$.

SOLUTION Substituting values of j and evaluating gives us

$$\sum_{j=1}^{40} (3j + 5) = 8 + 11 + 14 + \cdots + 125$$

We note that since consecutive terms differ by three, the series is arithmetic. Since the first term of the series is 8 and the last (40th) term is 125, we have

$$\sum_{j=1}^{40} (3j + 5) = \frac{40}{2}(8 + 125)$$
$$= 2660$$

EXAMPLE 6 ▮ **An Unusual Financing Arrangement**

A1 Used Cars is offering special financing. With no money down and approved credit, you can drive away in a $9600 car and pay the balance in 48 monthly payments of only $200, plus 2% interest per month on the remaining balance. How much interest will you pay over 4 years?

SOLUTION The 48 interest payments can be viewed as the terms of an arithmetic sequence, as shown in the following table.

Payment number	Balance before $200 payment	Interest charged	Total (payment plus interest)	Balance after $200 payment
1	$9600	$(0.02)9600 = \$192$	$200 + $192 = $392	$9400
2	$9400	$(0.02)9400 = \$188$	$200 + $188 = $388	$9200
3	$9200	$(0.02)9200 = \$184$	$200 + $184 = $384	$9000
⋮	⋮	⋮		
48	$200	$(0.02)200 = \$4$	$200 + $4 = $304	$0

Applying the formula for finite arithmetic series to the interest charges, we obtain

$$192 + 188 + 184 + \cdots + 4 = \frac{48}{2}(192 + 4) = 4704$$

Thus, the total interest paid over 4 years would be $4704. An unusual feature of this financing scheme is that the monthly payments decrease over the life of the loan since the amount of interest decreases each month.

GEOMETRIC SERIES

In Section 8.1, we defined a geometric sequence as one whose successive terms have a common ratio. The sum of the terms of a geometric sequence is called a **geometric series.** Just as we did with arithmetic series, we would like to develop a formula for evaluating geometric series. Thus, let $a_1, a_2, \ldots, a_n$ be a geometric sequence with common ratio r, and let the sum S be given by

$$S = a_1 + a_2 + \cdots + a_{n-1} + a_n$$

Since $a_k = a_1 r^{k-1}$, we have

(3) $$S = a_1 + a_1 r + a_1 r^2 + \cdots + a_1 r^{n-2} + a_1 r^{n-1}$$

If we multiply both sides of equation (3) by r, we obtain

(4) $$rS = a_1 r + a_1 r^2 + a_1 r^3 + \cdots + a_1 r^{n-1} + a_1 r^n$$

Subtracting equation (4) from equation (3) gives us

$$S = a_1 + a_1 r + a_1 r^2 + \cdots + a_1 r^{n-1}$$
$$- \quad rS = a_1 r + a_1 r^2 + \cdots + a_1 r^{n-1} + a_1 r^n$$

$$S - rS = a_1 - a_1 r^n$$
$$S(1 - r) = a_1(1 - r^n)$$
$$S = \frac{a_1(1 - r^n)}{1 - r} \qquad \text{Assuming } r \neq 1$$

Thus, we have the following formula.

Formula for Finite Geometric Series ☐

If $a_1, a_2, a_3, \ldots, a_n$ is a geometric sequence with common ratio r (so that $a_k = a_1 r^{k-1}$), and $r \neq 1$, then

$$a_1 + a_2 + \cdots + a_n = \frac{a_1(1 - r^n)}{1 - r}$$

EXAMPLE 7   ▇   **Finding the Sum of a Geometric Sequence**

Find the sum of the first 12 terms of the geometric sequence $5, 2, 0.8, 0.32, \ldots$.

SOLUTION   The formula for finite geometric series involves only the first term, the common ratio, and the number of terms. The first term is $a_1 = 5$. The common ratio is

$$r = \frac{a_2}{a_1} = \frac{2}{5} = 0.4$$

The number of terms is $n = 12$. Thus,

$$a_1 + a_2 + \cdots + a_{12} = \frac{a_1(1 - r^n)}{1 - r}$$
$$= \frac{5[1 - (0.4)^{12}]}{1 - 0.4}$$
$$\approx 8.33319$$

RULE OF THUMB   As suggested by Example 7, it is helpful to think of a_1 and n in the expression

$$\frac{a_1(1 - r^n)}{1 - r}$$

as simply the first term and the number of terms, respectively.

EXAMPLE 8 ■ **Evaluating a Geometric Series**

Evaluate $\displaystyle\sum_{n=3}^{10} 4(-\tfrac{1}{3})^n$.

SOLUTION   The first term of the series is $4(-\tfrac{1}{3})^3 = -\tfrac{4}{27}$. Since the starting index is 3 and the ending index is 10, there are 8 terms in the sum. Finally, the common ratio is $-\tfrac{1}{3}$. Thus,

$$\sum_{n=3}^{10} 4\left(-\frac{1}{3}\right)^n = \frac{(\text{First term})\left[1 - (-\tfrac{1}{3})^{(\text{Number of terms})}\right]}{1 - (-\tfrac{1}{3})}$$

$$= \frac{-\tfrac{4}{27}\left[1 - (-\tfrac{1}{3})^8\right]}{1 + \tfrac{1}{3}}$$

$$\approx -0.111094$$

EXAMPLE 9 ■ **Comparing Salary Contracts**

Suppose you are starting a new job that offers a choice between two contracts. The first has a starting base salary of \$32,000, yearly increases of \$3000 in the base salary, and a year-end bonus equal to 4% of the base salary for that year. The second has a starting salary of \$32,000 and a 6% raise each year. Which job would pay the higher total salary over a 26-year period?

SOLUTION   The following table shows the total income for both contract options for the first four years.

Year	Option 1: Bonus and fixed raise			Option 2: Percentage raise
	Base salary	Bonus of 4%	Total	Salary (including 6% raise)
1	32,000	(0.04)32,000 = 1280	33,280	32,000
2	35,000	(0.04)35,000 = 1400	36,400	32,000 + (0.06)32,000 = 32,000(1.06) = 33,920
3	38,000	(0.04)38,000 = 1520	39,520	33,920(1.06) = 35,955.20
4	41,000	(0.04)41,000 = 1640	42,640	35,955.20(1.06) = 38,112.51

The salary totals for Option 1 form an arithmetic sequence with common difference $d = 36{,}400 - 33{,}280 = 3120$. Using the formula for the nth term of an arithmetic sequence, the salary for year 26 will be $a_{26} = 33{,}280 + (26 - 1)3120 = 111{,}280$. Now we apply the formula for finite arithmetic series to determine the sum of the salaries for 26 years.

$$33{,}280 + 36{,}400 + \cdots + 111{,}280 = \frac{26}{2}(33{,}280 + 111{,}280)$$

$$= 1{,}879{,}280$$

The salaries for Option 2 form a geometric sequence with common ratio $r = 1.06$. Thus, using the formula for finite geometric series, we obtain the following salary total.

$$32{,}000 + 33{,}920 + \cdots + (\text{Salary for year 26}) = \frac{32{,}000(1 - 1.06^{26})}{1 - 1.06}$$

$$\approx 1{,}893{,}004.25$$

So even though Option 1 starts out with larger salaries, Option 2 results in a slightly higher total.

UNDERSTANDING AND MASTERY CHECKLISTS

CONCEPTS TO UNDERSTAND

- ☐ Series
- ☐ Term of a series
- ☐ Summation notation: index, lower limit, upper limit
- ☐ Arithmetic series
- ☐ Formula for a finite arithmetic series
- ☐ Geometric series
- ☐ Formula for a finite geometric series

SKILLS TO MASTER

- ☐ Evaluate a finite series given in summation form.
- ☐ Given the first few terms of a series, express the series in summation notation.
- ☐ Evaluate a finite arithmetic series.
- ☐ Evaluate a finite geometric series.

EXERCISES 8.2

EXERCISES 1–4 *Find the sum of the first five terms of the sequence.*

1. $a_n = 4n + 3$

2. $a_n = n^2$

3. $a_n = \dfrac{n}{n + 1}$

4. $a_n = \dfrac{4}{10^n}$

EXERCISES 5–14 *Find the indicated sum.*

5. $\displaystyle\sum_{k=1}^{6} (3k - 2)$

6. $\displaystyle\sum_{i=1}^{4} (1 - 5i)$

7. $\displaystyle\sum_{n=1}^{5} -3$

8. $\displaystyle\sum_{k=1}^{4} 8$

9. $\displaystyle\sum_{j=1}^{4} 2j^2$

10. $\displaystyle\sum_{k=2}^{5} k^3$

11. $\displaystyle\sum_{k=3}^{6} \frac{1}{k}$

12. $\displaystyle\sum_{n=0}^{4} \frac{1}{n + 1}$

13. $\displaystyle\sum_{i=0}^{4} 5\left(\tfrac{1}{2}\right)^i$

14. $\displaystyle\sum_{k=1}^{5} 3(-2)^k$

EXERCISES 15–24 *Express the series using summation notation.*

15. $\dfrac{1}{1^3} + \dfrac{1}{2^3} + \dfrac{1}{3^3} + \dfrac{1}{4^3} + \dfrac{1}{5^3} + \dfrac{1}{6^3}$

16. $\dfrac{1}{4^1} + \dfrac{1}{4^2} + \dfrac{1}{4^3} + \dfrac{1}{4^4} + \dfrac{1}{4^5}$

17. $[5(1) + 2] + [5(2) + 2] + [5(3) + 2] + \cdots + [5(12) + 2]$

18. $[7(1)^2 - 1] + [7(2)^2 - 1] + [7(3)^2 - 1] + \cdots + [7(15)^2 - 1]$

19. $3 + 9 + 27 + \cdots + 2187$

20. $2 - 4 + 8 - 16 + \cdots + 128$

21. $\frac{1}{1} - \frac{1}{4} + \frac{1}{9} - \frac{1}{16} + \cdots - \frac{1}{256}$

22. $\frac{1}{2} + \frac{2}{3} + \frac{3}{4} + \cdots + \frac{20}{21}$

23. $\frac{1}{2} + \frac{3}{4} + \frac{7}{8} + \cdots + \frac{63}{64}$

24. $\dfrac{1}{1 \cdot 2} + \dfrac{1}{2 \cdot 3} + \dfrac{1}{3 \cdot 4} + \cdots + \dfrac{1}{15 \cdot 16}$

EXERCISES 25–36 *Evaluate the given arithmetic series.*

25. $6 + 14 + 22 + \cdots + 158$ (20 terms)

26. $10 + 16 + 22 + \cdots + 154$ (25 terms)

27. $2 + 2.5 + 3 + \cdots + 12.5$

28. $47 + 44 + 41 + \cdots + 2$

29. $a_1 + a_2 + \cdots + a_{15}$, where $a_1 = 12$ and $a_2 = 8$

30. $a_1 + a_2 + \cdots + a_{18}$, where $a_1 = -3$ and $a_2 = 4$

31. $\displaystyle\sum_{n=1}^{50} (3n - 2)$

32. $\displaystyle\sum_{k=1}^{60} (8k + 2)$

33. $\displaystyle\sum_{k=1}^{30} (200 - 6k)$

34. $\displaystyle\sum_{n=1}^{80} (10 - 0.01n)$

35. $\displaystyle\sum_{i=10}^{25} \frac{5i + 3}{2}$

36. $\displaystyle\sum_{k=5}^{35} \left(\frac{1}{2}k + 3\right)$

43. $\displaystyle\sum_{k=1}^{8} 3(4^k)$

44. $\displaystyle\sum_{n=1}^{11} 10(0.6)^n$

45. $\displaystyle\sum_{n=0}^{9} 2\left(-\frac{1}{6}\right)^n$

46. $\displaystyle\sum_{k=0}^{8} \frac{1}{2}(-2)^k$

EXERCISES 37–48 *Evaluate the given geometric series.*

37. The sum of the first 10 terms of the sequence $5, 15, 45, \ldots$

38. The sum of the first 12 terms of the sequence $20, 4, 0.8, \ldots$

39. $a_1 + a_2 + \cdots + a_{15}$, where $a_1 = 3$ and $a_2 = 1$

40. $a_1 + a_2 + \cdots + a_{10}$, where $a_1 = -2$ and $a_2 = 4$

41. $4 - 0.4 + 0.04 - 0.004 + \cdots - 0.0000004$

42. $\frac{1}{25} + \frac{1}{5} + 1 + \cdots + 5^7$

47. $\displaystyle\sum_{k=5}^{16} -3(0.2)^k$

48. $\displaystyle\sum_{i=3}^{12} 9\left(\frac{2}{3}\right)^i$

EXERCISES 49–52 *Find the sum of the indicated integers.*

49. The first 1000 positive integers

50. The first 200 positive even integers

51. The first 100 positive odd integers

52. The first 100 positive multiples of 3

▪ APPLICATIONS

---

53. *Pipe Pile* A pile of sewer pipes has 12 layers. The first layer has 30 pipes, the second has 29, the third has 28, and so on, as shown in Figure 5. How many pipes are in the pile?

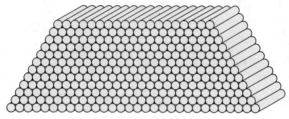

FIGURE 5

54. *Bricklaying* A bricklayer estimates that in an 8-hour shift she can complete 15 rows of bricks in a portion of a wall near the peak of a roof, as shown in Figure 6. The first row has 24 bricks, the second has 23 bricks, and so on. How many bricks will she need that day?

FIGURE 6

55. *Auditorium Seats* An auditorium has 25 rows of seats. The first row has 12 seats, the second row has 14 seats, the third row has 16 seats, and so on, as shown in Figure 7. How many seats are in the auditorium?

56. *Building Height* An object is dropped from the top of the First Interstate World Center in Los Angeles. During the first second, it falls 16 feet; during the next second, it falls 48 feet; during the third second, it falls 80 feet; and so on. The object hits the ground after approximately 8 seconds. How high is the World Center?

57. *Advertising Expenditures* The annual advertising expenditures by U.S. companies for the years 1980–1990 can be approximated with the model $a_n = 7.7n + 55$, where n denotes the year (with $n = 0$ corresponding to 1980) and a_n is the expenditure for that year (in billions of dollars) (*Data source: Advertising Age*). Find the total of all advertising expenditures for 1980 through 1990.

58. *Government Expenditures* The annual expenditures by the federal government for the years 1980–1992 can be approximated with the model $a_n = 69.3n + 580$, where n is the year (with $n = 0$ corresponding to 1980) and a_n is the outlay for that year (in billions of dollars) (*Data source:* U.S. Office of Management and Budget). Find the total of all expenditures for 1980–1992.

59. *Accumulated Value* Jamal deposits $100 at the beginning of each month in an account that pays 5% interest compounded

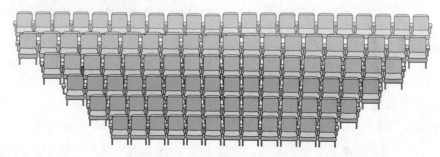

FIGURE 7

monthly. Evaluate the following geometric series to find the balance in his account at the end of 4 years.

$$B = 100\left(1 + \frac{0.05}{12}\right)^1 + 100\left(1 + \frac{0.05}{12}\right)^2$$
$$+ 100\left(1 + \frac{0.05}{12}\right)^3 + \cdots + 100\left(1 + \frac{0.05}{12}\right)^{48}$$

60. *Credit Card Debt*   Todd has run up a $2000 debt on a credit card that charges interest on the unpaid balance at a rate of 1.5% per month. He resolves to stop using the card and, 1 month later, begins making monthly payments of $50. His balance after the 60th payment can be found using the following formula. What is his balance?

$$B = 2000(1.015)^{60} - [50 + 50(1.015)^1$$
$$+ 50(1.015)^2 + \cdots + 50(1.015)^{59}]$$

61. *Automobile Expenses*   The fixed costs (insurance, license, depreciation, and so on) for owning and operating a car for the years 1980–1990 can be approximated with the model $a_n = 1960(1.06)^n$, where n denotes the year (with $n = 0$ corresponding to 1980) and a_n is the cost in dollars (*Data source: Motor Vehicle Facts and Figures,* Motor Vehicle Manufacturers Association of the United States, Detroit). Find the total cost for owning and operating a car from 1980 to 1990.

62. *Mail Delivery*   The number of pieces of mail delivered by the U.S. Postal Service from 1970 to 1990 can be approximated with the model $a_n = 82(1.036)^n$, where n denotes the year (with $n = 0$

corresponding to 1970) and a_n is the number of pieces of mail delivered (in billions) (*Data source:* U.S. Postal Service). Find the total number of pieces of mail delivered from 1970 through 1990.

63. *Loan Repayment*   Molly has just borrowed $2400 from her Uncle Dave and agrees to pay it back in 24 monthly payments of $100, plus 0.5% interest per month on the remaining balance. How much interest will Molly have to pay?

64. *Rumor Mill*   Suppose a disgruntled accountant in a large corporation begins his workday by telling four coworkers that the president of the corporation has been embezzling. After 1 hour, each of these four has spread the rumor to four of their friends. After the second hour, each of those has told four more, and so on. How many people have heard the rumor after 8 hours? If this pattern continues through the second day, how many will have heard the rumor after 16 hours?

65. *Contract Value*   Suppose a college basketball star signs a 15-year contract to play professional basketball. If the first year's salary is $1.6 million and the salary for each of the following 14 years is 10% more than the preceding year's, how much is the total contract worth?

66. *Job Options*   If you plan to stay with a job for 30 years, would you choose a job with a starting salary of $28,000, a yearly bonus of 3%, and a guaranteed annual raise of $4000, or one with a starting salary of $36,000 and a guaranteed raise of 4% per year? Explain.

CONCEPTS AND CRITICAL THINKING

EXERCISES 67–70   *Answer true or false.*

67. $\sum_{j=1}^{5} j$ is equivalent to $5j$.

68. $\sum_{j=1}^{5} k$ is equivalent to $5k$.

69. If successive terms in a series have a common ratio, then we say that the series is geometric.

70. A finite arithmetic series can be evaluated by multiplying the number of terms in the series by the average of the first and last terms.

EXERCISES 71–75   *Give an example of each.*

71. An arithmetic series

72. A geometric series

73. An arithmetic series with 5 terms that evaluates to 60

74. A series with m terms that evaluates to $3m$

75. A geometric series that evaluates to 0 and consists of ten terms, none of which is zero

QUESTIONS FOR DISCUSSION OR ESSAY

76. Expand and compare the terms of the following series.

a. $\sum_{k=1}^{5} \frac{1}{k}$

b. $\sum_{k=3}^{7} \frac{1}{k-2}$

Find a third way of writing the series, this time using $k = 0$ as the starting index. What can you conclude about the way in which a given series can be represented using summation notation? Explain.

77. In Example 9, we saw that a percentage raise eventually beat a fixed raise with a bonus. Do you think this is always the case?

How much effect does the length of time have on the situation? What are some factors that might lead you to choose the fixed-raise option even if the percentage option ends a little better? How would these factors be affected by the length of time you plan on keeping the job?

78. Financing terms such as those described in Example 6 are uncommon because federal law requires a lender to provide the loan's A.P.R. (annual percentage rate) as well as the total interest charge. This allows a borrower to make intelligent choices between loans with different terms. Consult a local newspaper or inquire with a local lending institution to see if you find the A.P.R. and interest charge information to be readily offered. Discuss some other factors that you would consider when choosing between different loan offers.

PROJECTS FOR ENRICHMENT

79. *Zeno's Paradox and Infinite Geometric Series* Around 460 B.C., the Greek philosopher Zeno of Elea posed a paradox that has intrigued philosophers and mathematicians for centuries. He stated that motion is paradoxical in that it is seemingly not possible to travel a finite distance in a finite amount of time. In order to do so, one would first have to travel half of the distance, then half of the remaining distance, then half of the remaining distance, and so on, as suggested in Figure 8.

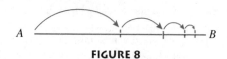

A B

FIGURE 8

Thus, one would have to cover an infinite number of distances, which is clearly not possible in a finite amount of time. Or is it? Suppose the distance between A and B is 1 mile and you start walking from A to B at a rate of 1 mile per hour. At that rate, you would cover the first half mile in $\frac{1}{2}$ hour, the next fourth mile in $\frac{1}{4}$ hour, the next eighth mile in $\frac{1}{8}$ hour, and so on. Thus, your total time in traveling from A to B would be given by the infinite sum

$$\frac{1}{2} + \frac{1}{4} + \frac{1}{8} + \cdots$$

Let us define S_n to be the sum of the first n terms of this infinite series. Then

$$S_n = \frac{1}{2} + \frac{1}{4} + \cdots + \left(\frac{1}{2}\right)^n = \sum_{k=1}^{n} \left(\frac{1}{2}\right)^k$$

a. Find $S_1, S_2, S_3,$ and S_4 and explain what each represents.

b. Use the formula for finite geometric series to show that

$$S_n = 1 - \frac{1}{2^n}$$

c. Compute $S_5, S_{10}, S_{20},$ and S_{40}. What do these values approach? Since each value is the sum of more and more terms of the infinite sum $\frac{1}{2} + \frac{1}{4} + \frac{1}{8} + \cdots$, what would be a reasonable guess for the value of the infinite sum? According to this value, how much time will it take to travel from A to B? Does this agree with the amount of time suggested by the standard formula $d = rt$?

Using calculus, a tool developed centuries after Zeno, it is possible to show that if $|r| < 1$, then the sum S of the infinite geometric series $a + ar + ar^2 + ar^3 + \cdots$ is given by

$$S = \frac{a}{1 - r}$$

d. Use this formula to verify the value you found in part *c*.

e. Find the sum of the following infinite geometric series.

   i. $\frac{1}{4} + \frac{1}{16} + \frac{1}{64} + \cdots$

  ii. $\frac{2}{3} + \frac{4}{9} + \frac{8}{27} + \cdots$

 iii. $\frac{1}{10} - \frac{1}{100} + \frac{1}{1000} - \frac{1}{10,000} + \cdots$

80. *Geometric Series and Annuities* An **annuity** is a sequence of regular payments or receipts, all subject to some underlying interest rate. Some examples of annuities include mortgage payments, installment purchases, premium payments on insurance policies, payroll deductions for pension plans, and lease payments. We limit our discussion to annuities for which payment is made at the end of the payment period. As a simple example, suppose you deposit $100 at the *end* of each year for the next 10 years at an interest rate of 5% compounded annually. The **future value** of a given payment is simply its value at the end of the last year, and this can be found using the standard formula for compound interest. Thus, the first payment, which earns 5% interest for 9 years, has a future value of $100(1 + 0.05)^9$. The second has a future value of $100(1 + 0.05)^8$, the third has a future value of $100(1 + 0.05)^7$, and so on up to the tenth, which has a future value of 100. The future value of the annuity is the sum of the future values of all the payments. If we denote this sum by S and write the terms of the sum in reverse order, we have

$$S = 100 + 100(1 + 0.05)^1$$
$$+ 100(1 + 0.05)^2 + \cdots + 100(1 + 0.05)^9$$

a. Use the formula for finite geometric series to find the future value of the annuity just described.

More generally, the future value of an annuity with periodic payments R made at the end of each of n periods with interest rate i is given by

$$S = R + R(1 + i)^1$$
$$+ R(1 + i)^2 + \cdots + R(1 + i)^{n-1}$$

b. Use the formula for finite geometric series to show that

$$S = R\left[\frac{(1 + i)^n - 1}{i}\right]$$

c. Find the future value for the following annuities.

   i. Eliza deposits $3000 at the end of each year for 20 years into a pension fund that earns 6% compounded annually.

   ii. Charlie deposits $500 at the end of each quarter for 5 years into a bank account that earns interest at a rate of 4% compounded quarterly. Note that the periodic interest rate is $i = \dfrac{0.04}{4} = 0.01$ and the number of periods is $n = 4(5) = 20$.

We next consider the notion of **present value**. As an example, suppose a parent wishes to invest a certain amount of money now, at 8% compounded annually, in order to provide a child with a yearly allowance of $1000 at the end of each year for the next 10 years. We can determine this amount by considering the present value of a single payment—that is, the amount that must be invested now to make a given payment. Using the formula for compound interest, the amount A_{10} that must be invested now to make the payment at the end of the tenth year is found by solving $1000 = A_{10}(1 + 0.08)^{10}$. Thus, $A_{10} = 1000/(1 + 0.08)^{10}$. Similarly, $A_9 = 1000/(1 + 0.08)^9$, $A_8 = 1000/(1 + 0.08)^8$, and so on down to $A_1 = 1000/(1 + 0.08)$. The present value of the annuity—that is, the entire amount that must be invested—is simply the sum of the present values of each payment. Thus,

$$A = A_1 + A_2 + \cdots + A_{10}$$

$$= 1000\left(\frac{1}{1 + 0.08}\right)^1 + 1000\left(\frac{1}{1 + 0.08}\right)^2$$

$$+ \cdots + 1000\left(\frac{1}{1 + 0.08}\right)^{10}$$

This is also a geometric series.

d. Use the formula for finite geometric series to find the present value of the annuity just described.

More generally, if an annuity has n periodic payments of R dollars and a periodic interest rate i, then its present value is given by

$$A = R\left(\frac{1}{1 + i}\right)^1 + R\left(\frac{1}{1 + i}\right)^2 + \cdots + R\left(\frac{1}{1 + i}\right)^n$$

e. Use the formula for finite geometric series to show that

$$A = R\left[\frac{1 - (1 + i)^{-n}}{i}\right]$$

f. Use the concept of present value to solve the following.

   i. An investment contract promises to pay $5000 at the end of each of the next 10 years. If you wish to earn 9% compounded annually, how much should you pay for the contract now?

   ii. Suppose you have just won the grand prize in a sweepstakes and must choose between receiving $1,000,000 now or $100,000 at the end of each of the next 20 years. Assuming an annual yield of 10% on the annuity option, which option should you choose? Explain. What additional factors might influence your decision?

SECTION **8.3**   **PERMUTATIONS AND COMBINATIONS**

☐ **Why should a combination lock really be called a permutation lock?**

☐ **Why is it not a good idea for a TV program manager to select a program schedule by simply trying all possibilities?**

☐ **How many credit cards does it take to make a single dollar bill?**

☐ **How many winning combinations are possible in a state lottery?**

As young children, we mastered the simplest and often most effective counting technique: Establish a one-to-one correspondence between the objects in a set and a subset of the natural numbers by pointing to the first object, announcing "one," pointing to the second object and saying "two," and so forth. As effective as this technique is for small sets, it fails miserably for larger sets. If you were capable of counting at the rate of one per second, then determining the number of possible five-card poker hands (working around the clock) would take a solid month; counting the number of possible 9-digit social security numbers would take 30 years; and if you're planning on enumerating the ways 15 people can form a line, don't make any other plans—you're going to be booked solid for the next 40,000 years.

More sophisticated counting techniques are the province of combinatorics, a branch of mathematics essential to the computation of probability and, hence, having widespread application throughout the physical and natural sciences. Though we only scratch the surface of this deep and broad subject, the three counting tools developed in this section—the Fundamental Counting Principle, permutations, and combinations—are sufficient to perform, in a matter of minutes, all the tasks referred to in the preceding paragraph.

THE FUNDAMENTAL COUNTING PRINCIPLE

Suppose you are on a committee that is planning a banquet. The committee has selected a caterer that offers a choice of two meats (chicken breast or roast beef), two potatoes (baked or mashed), and three vegetables (spinach, peas, or green beans). Your task is to prepare a list of all the meals consisting of one meat, one potato, and one vegetable. In the process of creating this list, you would likely discover the need for a systematic method for finding all the possibilities without any duplication. One such method involves the use of a **tree diagram.**

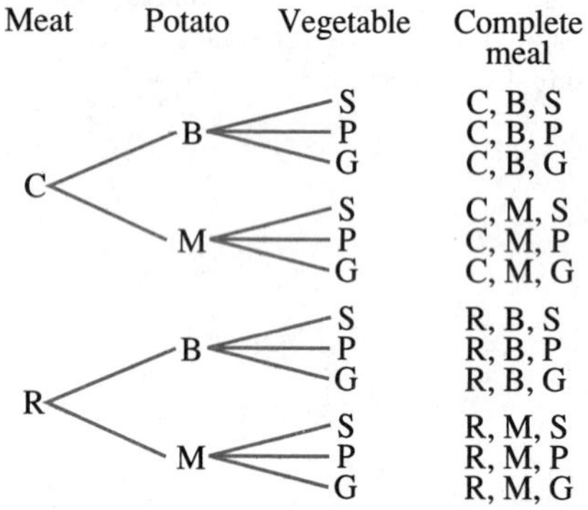

We first adopt the convention that each food item is denoted by the first letter of its name (that is, C for chicken breast, R for roast beef, and so on). We next list the two choices for meat in a column labeled "Meat," as shown in Figure 9. Since each meat choice has two choices for potato, we draw two lines, called **decision branches,** from each meat to the two possible choices for potato in the column labeled "Potato." Similarly, since each potato has three possible choices for vegetable, we draw three branches from each potato to the three vegetables in the column labeled "Vegetable." All the complete meals can be found by following the different paths from meat to vegetable. There are a total of 12 possible meals.

If it is only necessary to determine the *number* of possible meals, without actually listing them, there is a much easier way. Since each of the two meats can be paired with one of the two potatoes, there are $2 \cdot 2 = 4$ distinct meat and potato pairs. For each of these 4 meat and potato pairs, there are 3 possibilities for a vegetable, for a total of $4 \cdot 3 = 12$ meals. Thus, given a meal selection task involving 2 choices for meat, 2 choices for potato, and 3 choices of vegetable, there are $2 \cdot 2 \cdot 3 = 12$ possible meals. This observation generalizes into the **Fundamental Counting Principle.**

FIGURE 9

Fundamental Counting Principle □   If a task consists of k parts, the first of which can be completed in n_1 ways, the second of which can be completed in n_2 ways, and so on up to the kth, which can be completed in n_k ways, then the total number of ways in which the task can be completed is given by the product

$$n_1 n_2 \cdots n_k$$

EXAMPLE 1 ■ Counting License Plates

The state of Michigan has license plates of the form LLL DDD, where L denotes a letter and D denotes a digit. How many different license plates of this form are possible?

SOLUTION   It is convenient to first draw six blank slots to represent the letters and digits that must be selected.

_____   _____   _____   _____   _____   _____

Each of the first three slots must be filled with a letter, and so each has 26 possibilities. Each of the last three slots must be filled with one of the digits 0–9, and so each has 10 possibilities. We represent these choices as follows.

$$\underline{26} \quad \underline{26} \quad \underline{26} \quad \underline{10} \quad \underline{10} \quad \underline{10}$$

Applying the Fundamental Counting Principle, we see that there are a total of

$$26 \cdot 26 \cdot 26 \cdot 10 \cdot 10 \cdot 10 = 17{,}576{,}000$$

possible license plates.

EXAMPLE 2   ■   **Arranging TV Shows**

A TV program manager must choose among 6 shows for 4 available time slots. He decides to try each possible arrangement for 1 week to see which draws the highest ratings. How many weeks will this take?

SOLUTION   We first draw 4 blank lines to represent the 4 time slots that must be filled. Initially, there are 6 choices for filling the first time slot, and so we place a 6 in the first blank. If we assume that the same show may not be used twice, there are only 5 choices remaining for the second time slot. Continuing in this fashion, there are 4 choices for the third slot, and 3 for the fourth.

$$\underline{6} \quad \underline{5} \quad \underline{4} \quad \underline{3}$$

By viewing the selection process as a four-part task, each part of which can be completed in 6, 5, 4, and 3 ways, respectively, the Fundamental Counting Principle tells us that the total number of possibilities is

$$6 \cdot 5 \cdot 4 \cdot 3 = 360$$

Thus, it would take 360 weeks, or almost 7 years, to try all the possible arrangements.

PERMUTATIONS

In Example 2, the *ordering* of the TV shows was an essential part of the problem. That is, if the same four shows were arranged in a different order, we would consider it to be a completely different arrangement. We say that each of the ordered arrangements is a *permutation* of 6 shows taken 4 at a time. In general, we define a permutation as follows.

Definition of Permutation   □

> A **permutation** of n elements taken r at a time is an ordered arrangement, without repetitions, of r of the n elements. The number of permutations of n elements taken r at a time is denoted by $_nP_r$.

To compute $_nP_r$, we consider an ordered arrangement of r out of n elements without repetitions. Following the lead of Examples 1 and 2, we draw r blank slots to represent the r elements that must be chosen.

$$\underline{\hspace{1.5cm}}\ \underline{\hspace{1.5cm}}\ \underline{\hspace{1.5cm}}\ \cdots\ \underline{\hspace{1.5cm}}$$

$$r \text{ blank slots}$$

Since there are initially n elements to choose from, the first slot can be filled in n ways. Since repetitions are not allowed, the second slot can be filled in $n - 1$ ways. Continuing in this fashion, the third slot can be filled in $n - 2$ ways, and so on up to the rth slot, which can be filled in $n - (r - 1)$ or $n - r + 1$ ways.

$$\underbrace{\underline{\quad n \quad}\ \underline{\ n-1\ }\ \underline{\ n-2\ }\ \cdots\ \underline{\ n-r+1\ }}$$

$$r \text{ slots}$$

By applying the Fundamental Counting Principle, we obtain the following formula.

Permutation Formula ☐

> For integers n and r with $1 \le r \le n$,
>
> $$_nP_r = n(n - 1)(n - 2) \cdots (n - r + 1)$$
>
> For convenience, we define $_nP_0 = 1$.

> **RULE OF THUMB**   You do not need to memorize the permutation formula just given. Instead, just remember that the product starts with n and has r factors, each of which is one less than the preceding factor.

E X A M P L E 3 ▓   **Evaluating Permutations**

Evaluate the following permutations.

a. $_7P_3$ **b.** $_5P_4$

SOLUTION

a. To compute $_7P_3$, we start with 7 and write out 3 factors. Thus,

$$_7P_3 = \underbrace{7 \cdot 6 \cdot 5}_{3 \text{ factors}} = 210$$

b. We start with 5 and write out 4 factors to obtain

$$_5P_4 = \underbrace{5 \cdot 4 \cdot 3 \cdot 2}_{4 \text{ factors}} = 120$$

 CALCULATOR KEYS   *Permutations*

Most graphing calculators are capable of computing permutations. To compute $_nP_r$, first key in the number n, then select nPr from the menu of mathematical functions, and, finally, enter the number r.

EXAMPLE 4 ▨ Using Permutations to Count Batting Orders

The manager of a coed softball team must assign a batting order by selecting 10 players from a team consisting of 14 players: 8 men and 6 women. Find the number of batting orders that are possible if

a. there are no restrictions on the number of men and women
b. league rules require that 3 women must bat first

SOLUTION

a. If there are no restrictions, then the manager may simply select and arrange 10 out of the 14 players. Thus, there are $_{14}P_{10}$ different batting orders. Using a calculator, we obtain $_{14}P_{10} = 3{,}632{,}428{,}800$.

b. The number of ways to arrange 3 out of 6 women in the first 3 positions is $_6P_3 = 120$. Since no restrictions have been specified for the remaining 7 positions, the manager may select them from the 11 people that remain. This may be done in $_{11}P_7 = 1{,}663{,}200$ ways. By the Fundamental Counting Principle, the total number of ways to arrange the batting order is

$$_6P_3 \cdot {}_{11}P_7 = 120 \cdot 1{,}663{,}200 = 199{,}584{,}000$$

Notice that if we let $r = n$ in the permutation formula, we obtain

$$_nP_n = n(n - 1)(n - 2) \cdots (1)$$

which is the product of the first n positive integers. This product occurs so frequently that it is convenient to define a special notation for it: **factorial notation.**

Factorial Notation □

> For any natural number n, $n! = n \cdot (n - 1) \cdot \cdots \cdot 3 \cdot 2 \cdot 1$. For convenience, we define $0! = 1$. Note that $n!$ is read as "n factorial."

EXAMPLE 5 ▨ Evaluating Factorials

Evaluate the given expression.

a. $5!$

b. $\dfrac{20!}{18!}$

SOLUTION

a. $5! = 5 \cdot 4 \cdot 3 \cdot 2 \cdot 1 = 120$

b. $\dfrac{20!}{18!} = \dfrac{20 \cdot 19 \cdot \cancel{18} \cdot \cancel{17} \cdot \cancel{16} \cdot \cdots \cdot \cancel{2} \cdot \cancel{1}}{\cancel{18} \cdot \cancel{17} \cdot \cdots \cdot \cancel{2} \cdot \cancel{1}} = 20 \cdot 19 = 380$

Since $_nP_r = n(n - 1)(n - 2) \cdots (n - r + 1)$, we see that $_nP_r$ contains the first r factors of $n!$. Thus, it is perhaps no surprise that the permutation formula can be written using factorials. We simply multiply $_nP_r$ by a quotient whose numerator and denominator contain the factors of $n!$ that are missing from $_nP_r$.

$$_nP_r = \frac{n(n - 1)(n - 2) \cdots (n - r + 1)}{1} \cdot \frac{(n - r)!}{(n - r)!}$$

$$= \frac{n(n - 1)(n - 2) \cdots (n - r + 1)(n - r)(n - r - 1) \cdots (3)(2)(1)}{(n - r)!}$$

$$= \frac{n!}{(n - r)!}$$

This verifies the following factorial formula for $_nP_r$.

Factorial Formula for $_nP_r$ □

For integers n and r with $0 \le r \le n$,

$$_nP_r = \frac{n!}{(n - r)!}$$

WARNING! The factorial symbol cannot be distributed through parentheses. Thus, the expression $(n - r)!$ in the factorial formula for $_nP_r$ *cannot* be rewritten as $n! - r!$.

E X A M P L E 6 ■ **Applying the Factorial Formula for Permutations**

Use the factorial formula for $_nP_r$ to evaluate $_7P_3$.

SOLUTION

$$_7P_3 = \frac{7!}{(7 - 3)!} = \frac{7!}{4!} = \frac{7 \cdot 6 \cdot 5 \cdot \cancel{4} \cdot \cancel{3} \cdot \cancel{2} \cdot \cancel{1}}{\cancel{4} \cdot \cancel{3} \cdot \cancel{2} \cdot \cancel{1}} = 7 \cdot 6 \cdot 5 = 210$$

COMBINATIONS

We have seen that a permutation takes into account the order in which elements are selected. In many cases, however, the order of selection is not important. Suppose, for example, you are taking a course that requires you to read and report on three out of four unrelated journal articles. In this case, you would probably not be concerned about the order in which the articles are read. Thus, the list of possible reading combinations would not include all the different orderings of the same three articles. If we denote the four articles by the letters A, B, C, and D, the possible reading combinations are

$$\{A, B, C\}, \{A, B, D\}, \{A, C, D\}, \text{ and } \{B, C, D\}$$

Notice that these four combinations are simply the three-element subsets of $\{A, B, C, D\}$. More generally, a combination is defined in the following way.

Definition of Combination □

A **combination** of n elements taken r at a time is an r-element subset of a set of n elements. The number of combinations of n elements taken r at a time is denoted by $_nC_r$ or $\binom{n}{r}$.

Although the definitions of permutation and combination use similar wording, it is important to remember that permutations and combinations are not the same. The difference can be summed up in one word—**order.** A permutation takes order into account, whereas a combination does not. However, $_nC_r$, the *number* of combinations of n elements taken r at a time, is closely related to $_nP_r$, the *number* of permutations of

n elements taken r at a time. We first note that each combination of r elements can be rearranged in $r!$ distinct ways. Thus, by the Fundamental Counting Principle, there are $_nC_r \cdot r!$ ordered arrangements of r elements chosen from n. But $_nP_r$ also represents the number of ordered arrangements of r elements chosen from n. Thus,

$$_nC_r \cdot r! = {}_nP_r$$

and so

$$_nC_r = \frac{_nP_r}{r!}$$

By applying the two earlier formulas for $_nP_r$, we obtain the following formulas for $_nC_r$.

Combination Formulas □

For integers n and r with $0 \le r \le n$,

$$_nC_r = \frac{_nP_r}{r!} = \frac{n(n-1)(n-2)\cdots(n-r+1)}{r!}$$

or

$$_nC_r = \frac{n!}{r!(n-r)!}$$

EXAMPLE 7 ▨ **Evaluating Combinations**

Evaluate the following combinations.

a. $_5C_3$

b. $\begin{pmatrix} 8 \\ 6 \end{pmatrix}$

SOLUTION

a. $_5C_3 = \frac{_5P_3}{3!} = \frac{5 \cdot 4 \cdot 3}{3 \cdot 2 \cdot 1} = 10$

b. Recall that $\begin{pmatrix} 8 \\ 6 \end{pmatrix}$ is an alternate notation for $_8C_6$. Thus,

$$\begin{pmatrix} 8 \\ 6 \end{pmatrix} = \frac{8!}{6!(8-6)!} = \frac{8 \cdot 7 \cdot \not{6} \cdot \not{5} \cdot \not{4} \cdot \not{3} \cdot \not{2} \cdot \not{1}}{(\not{6} \cdot \not{5} \cdot \not{4} \cdot \not{3} \cdot \not{2} \cdot \not{1}) \cdot (2 \cdot 1)} = \frac{8 \cdot 7}{2 \cdot 1} = 28$$

EXAMPLE 8 ▨ **Using Combinations to Count Senate Subcommittees**

The U.S. Senate, consisting of 100 senators, wishes to form a 5-member subcommittee to study the selection process for Senate subcommittees.

a. How many different subcommittees are possible?

b. How many different subcommittees are possible if the Senate consists of 56 Republicans and 44 Democrats and each committee must have 3 Republicans and 2 Democrats?

SOLUTION

a. The order in which members are selected is unimportant. Thus, we wish to find the number of combinations of 100 senators taken 5 at a time.

$$_{100}C_5 = \frac{_{100}P_5}{5!} = \frac{100 \cdot 99 \cdot 98 \cdot 97 \cdot 96}{5 \cdot 4 \cdot 3 \cdot 2 \cdot 1} = 75{,}287{,}520$$

So there are a total of 75,287,520 possible subcommittees.

b. The number of ways of selecting 3 Republicans out of 56 without regard to order is $_{56}C_3$. The number of ways of selecting 2 Democrats out of 44 without regard to order is $_{44}C_2$. By the Fundamental Counting Principle, the total number of subcommittees is

$$_{56}C_3 \cdot _{44}C_2 = \frac{56 \cdot 55 \cdot 54}{3 \cdot 2 \cdot 1} \cdot \frac{44 \cdot 43}{2 \cdot 1} = 26{,}223{,}120$$

CALCULATOR KEYS

Combinations

Combinations can be computed with a calculator in much the same way as permutations. To compute $_nC_r$, first key in the number n, then select nCr from the menu of mathematical functions, and finally, enter the number r.

EXAMPLE 9   ▨   **Counting Card Hands**

A standard 52-card deck of playing cards consists of two black suits, clubs and spades, and two red suits, hearts and diamonds. Each suit has 13 cards of different face values—A (ace), 2, 3, 4, 5, 6, 7, 8, 9, 10, J (jack), Q (queen), and K (king). In certain varieties of poker, each player is dealt a hand of 5 cards.

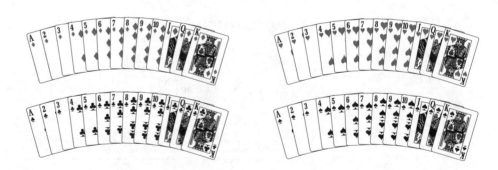

a. Find the number of 5-card poker hands that are possible.
b. A *full house* consists of a pair (two cards of the same face value from different suits) and three of a kind (three cards of the same face value from different suits), such as K-K-6-6-6. How many different full houses are possible?

SOLUTION

a. Since the order in which the cards are dealt is not important, the number of ways of being dealt 5 cards out of 52 is $_{52}C_5$. Using a calculator, we obtain $_{52}C_5 = 2{,}598{,}960$.

b. A pair can be formed by being dealt 2 out of 4 cards of the same face value, and there are $_4C_2$ ways in which this can be done. Since there are 13 different face values in the deck, we have

$$\text{The number of possible pairs} = 13 \cdot {_4C_2}$$

The three of a kind must consist of 3 out of 4 cards of the same face value, but a different face value from that of the pair. Since there are 12 other face values remaining, we see that

$$\text{The number of possible three of a kinds} = 12 \cdot {_4C_3}$$

Thus, applying the Fundamental Counting Principle, we see that the number of possible full houses is

$$(13 \cdot {_4C_2}) \cdot (12 \cdot {_4C_3}) = 3744$$

UNDERSTANDING AND MASTERY CHECKLISTS

CONCEPTS TO UNDERSTAND

- ☐ Tree diagram
- ☐ Fundamental Counting Principle
- ☐ Permutation definition and notation
- ☐ Factorial notation
- ☐ Combination definition and notation
- ☐ Factorial formulas for permutations and combinations

SKILLS TO MASTER

- ☐ Compute factorials.
- ☐ Compute permutations.
- ☐ Compute combinations.
- ☐ Solve a counting problem using a tree diagram.
- ☐ Solve a counting problem using the Fundamental Counting Principle.
- ☐ Solve a counting problem involving permutations.
- ☐ Solve a counting problem involving combinations.

EXERCISES 8.3

EXERCISES 1–16   *Evaluate the given expression.*

1. $6!$

2. $(9 - 5)!$

3. $\dfrac{10!}{7!}$

4. $\dfrac{8!}{2! \cdot 6!}$

5. $_6P_2$

6. $_9P_1$

7. $_5P_5$

8. $_8P_4$

9. $_6C_4$

10. $\dbinom{7}{2}$

11. $\dbinom{8}{1}$

12. $_4C_4$

13. $\dbinom{5}{0}$

14. $\dbinom{9}{8}$

15. $_8C_6$

16. $_{10}C_3$

▨ APPLICATIONS

EXERCISES 17–20   *Draw a tree diagram to illustrate the number of choices that are possible.*

17. The math and science requirement at a certain university includes a math course (College Algebra or Calculus I), a physics course (Elementary Physics or University Physics I), and a biology course (General Biology or Zoology). How many different course combinations will satisfy the math and science requirement?

18. The option packages for a popular pickup truck include a choice of cab style (regular or extended), bed size (short or long), and drive train (two-wheel or four-wheel). How many different option packages are possible?

19. A computer store offers three sizes of monitors (14″, 15″, or 17″), two styles of computer case (desktop or tower), and two types of keyboards (programmable or with a built-in calculator). How many different computer configurations are possible?

20. A restaurant offers a single-topping pizza special with a choice of appetizer (bread sticks or garlic bread), type of crust (thin, hand-tossed, or pan), and meat topping (pepperoni or sausage). How many different meals are possible?

EXERCISES 21–24   *Use the Fundamental Counting Principle to determine the number of possibilities.*

21. How many 3-digit numbers can be formed using the digits {1, 2, 3, 4, 5}
    a. if repetitions are allowed?
    b. if repetitions are not allowed?

22. How many 4-letter ''words'' can be formed using the letters {a, e, i, o, u, y}
    a. if repetitions are allowed?
    b. if repetitions are not allowed?

23. The state of California has license plates of the form NLLLDDD, where N denotes one of the digits 1–4, each L denotes a letter, and each D denotes one of the digits 0–9. How many license plates of this form are possible?

24. The state of Indiana has license plates of the form NNLDDDD, where NN is a number from 1 to 99 (depending on the county in which the plate was issued), L is one of the letters A–Z (corresponding to the license branch that issued the plate), and each D is one of the digits 0–9. How many license plates of this form are possible?

EXERCISES 25–28   *Express your answer using permutation notation and evaluate.*

25. In a survey of TV viewer preferences, respondents are asked to rank their favorite 3 out of 8 sitcoms. How many different survey responses are possible?

26. A club consisting of 12 members wishes to elect a president, vice president, secretary, and treasurer. In how many ways can this be done if no member may hold more than one office?

27. The manager of a Little League baseball team intends to select the starting 9 players by assigning a batting order. In how many ways can this be done if there are 13 players to choose from and
    a. there are no restrictions on the order of selection?
    b. the two children of the team's sponsor must bat first and second?

28. Four boys and six girls are attending a birthday party. In how many ways can they be lined up for ice cream
    a. if there are no restrictions on the order?
    b. if the girls are served before the boys?

EXERCISES 29–32   *Express your answer using combination notation and evaluate.*

29. A local election ballot allows each voter to select 3 out of 8 school board candidates. In how many ways can this be done?

30. A club consisting of 12 members wishes to appoint a 4-person committee. In how many ways can this be done?

31. The manager of a high school volleyball team must select 6 first-string players from a team of 12 girls. In how many ways can this be done if
    a. there are no restrictions?
    b. exactly 2 of the 6 first-string players must be chosen from the 5 seniors on the team?

32. A 5-card poker hand is dealt from a standard 52-card deck. How many hands are possible if
    a. at least three cards have the same face value?
    b. exactly three cards have the same face value?

EXERCISES 33–44   *Express your answer using permutation or combination notation, if possible, and evaluate.*

33. A student has 4 remaining time slots in her schedule. Ten different classes are available, each of which has a section open at those times. In how many ways can the time slots be filled?

34. An employment agency is asked to fill secretarial openings in 6 different companies. A total of 20 people have applied. In how many ways can the positions be filled?

35. A certain state lottery is played by selecting 5 different numbers from 1 through 40. In how many ways can this be done, assuming the order of selection is not important?

36. A certain state lottery is played by selecting 6 different numbers from 1 through 50. In how many ways can this be done, assuming the order of selection is not important?

37. The dial on a combination lock is marked with the numbers 1 through 40. The lock may be opened by selecting the correct sequence of 3 numbers. How many different combinations are possible
    a. if repetitions of numbers are not allowed?
    b. if repetitions of numbers are allowed?

38. The combination lock on a briefcase consists of three dials that are each marked with the digits 0–9. The lock may be opened by selecting the correct digit on each of the three dials. How many different combinations are possible
    a. if repetitions of digits are not allowed?
    b. if repetitions of digits are allowed?

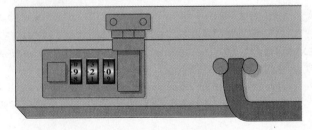

39. On a history exam, a student is asked to select 4 out of 6 essay questions. In how many ways can this be done? What if a student is also asked to select 10 out of 15 short-answer questions?

40. On a mathematics placement exam, a student is asked to select 3 out of 5 application problems. In how many ways can this be done? What if a student is also asked to select 6 out of 10 computational problems?

41. In a classroom of 30 students, how long would it take for every student to shake hands with each of the other students, assuming it takes 2 seconds for each shake and only two students can shake hands at a time? What about a class of 60 students?

42. If each of the 8 teams in a softball league must play every other team twice, what is the total number of games that must be played?

43. A 5-card poker hand is dealt from a standard 52-card deck. How many hands are possible if all 5 cards are of the same suit? (Such a hand is called a *flush*.)

44. A 5-card poker hand is dealt from a standard 52-card deck. How many hands are possible if 4 of the 5 cards have the same face value? (Such a hand is called *four of a kind*.)

45. *Photomosaic* The accompanying image is a photomosaic formed by placing Mastercards in a 23-by-49 array. Photomosaics are constructed with the aid of a powerful computer program written by Robert Silvers, MIT media lab graduate and founder of Photomosaics. The process begins with digitizing the larger image (Washington's portrait) and a collection of images (actual credit cards in this case) that will serve as tiles. Next, the program systematically selects and arranges the tiles according to color, tone, shapes within the tiling images, and countless other graphical features. For this image, Silvers had images of 5000 cards at his disposal.

a. Find an expression for the number of possible mosaics given that the cards used to form the mosaic are selected from a collection of 5000 Mastercards. Assume no duplicates are allowed. Note that your calculator will not be able to simplify this expression. It is approximately 8.86×10^{4108}, a number with enough digits to fill this page. It is reproduced here in a much smaller font.

8864288768500121732156523157192190546717011732794079263353574061150337286664005674192364
7483733556617462791986997278542652442114244423399552763937565019226469913332834127512736
7839859221860048919839762572769315859304692252489450665503536285307459622242254225422652
2833045605341753656666405670549055118613279478160061936243431454580129295532954408592028
9183921951364931769843307174921669496327667708792898352492493870148266087274708670602206
6325852338951830942816247048855057231516104397439120484780828299165826695828022437245818
9303666621917351982640809826912772574158990165635587499543519543465123918803197540390991
5350255671100288842220594118275655701389929442887048148380982499493907851708599850535806
2340557773557887007183762513775504832515832342265818033103649778523661239124075028751754
6226082190715419505243507195278517402194444285446495207894645908381376305820231873448536
6921132939992610199537463307429858815579653030322220820364738254708621227312196986470644
3779207020383258548112516092246103269294729799909671990111942532610748397355658634721641
7344345861009128721427847781654607099363042444702453916908432819389040735519190662840090
7712334610021278220716822318443001888854132446131250808053685887418582519347645586075242
5857800747675735834813729317663376150133017811923581918433223692926105713269427047103930
7365644563298239229558319934790454930332355303667505914486148457497744808441034459713406
1508871032078537649332463408535544070315120371419230362112196986447338271386806698335594
0403340666873762528032153706292317011895129780728799091209424780754567304825134792386537
2879147889987925839884934676438963683621330654092308747725737537402201661659954708652641
6618796110166675938940313552064949451555875763199055966821968716286105371870837893368791
2693425302500166109060646070356940059073342215111338584116319097615167986545608937837619
6786359009772050812535969892744868313854576577758890822590894676527325698076170342441717
3403300304168299593598036694909419100417321570895099378332287569544912652487257354851305
7499669760664396456088091485179477175095192773016147142355967061799134283410443841924226
7844601672223146049918355471068596413137759234053324634123645175339388318894893220547319
0974813123070886129621346312736314704652252074825799389160699499192801845117027755522516
6043878202546344540368699451815955717524752699759617307816217231249439687172194591827260
7300339964565915968324278001437487062785838394061459996448342118069765276709175266353213
2164788680679492783402636341979123296015793618635921597359931280046205758261325957506191
8674053302241147796184571232240123885879928178919284629413131844000257040216042573584684
3658627584374372473384548911658957233135189342534641484958165949257195271093586872200541
6030895277028062422000965984352273726413015503237914487343950725535159897069783619498658
5865119478358432999677937150508662464352165010418531358914331150297343267084725328916689
1649771620986605346774697207284769269491721780651786614298818438837740582808371037208339
8732323105091276594609786405080595392655112880748124294554777141567860184754578301525069
4540375272920370062036919345183934602999524774978040563269025969646234506899285943992018
2529654466884060500657475989368778875208487964457639736756684056369815826908154065789518
9876195198387369931558098867979081471142690800726465633819220528199933332924580165812308
0048774798358143243009720199210975039760849122616791077186955551704088727636705010868911
5091049771246777782388007119310929633208939617572471634611044867334786099074186864415602
0492441590596427321459153708313254999425678994281347173363000007744455878447821457780643
3441560555237726935819598420972353035650704762756325529894861765696104565711989464853571
5331388317786650956874966607136015905368238763481778792016534946164719041140385179751520
3462580631656964538867247538170087403333615616000000000000000000000000000000000000000000
0000000000000000000000000000000000000000000000000000000000000000000000000000000000000000
0000000000000000000000000000000000000000000000000000000000000000000000000000000000000000
00000000000000000000000000000000000000000000000000000000000000000000000

b. Find an expression for the number of possible ways of forming the mosaic if duplicates are allowed. (This number, 5.49×10^{4168}, will also be too large for your calculator to approximate.)

■ **CONCEPTS AND CRITICAL THINKING**

EXERCISES 46–49 *Answer true or false.*

46. In general, combinations arise when order is important, whereas permutations occur when the order is immaterial.

47. $0! = 0$

48. For all nonnegative integers n and r with $n \geq r$, ${}_nP_r \geq {}_nC_r$.

49. $31! - 13! = 18!$

EXERCISES 50–53   *Give an example of each.*

50. Two numbers n and r such that $_nP_r = {_n}C_r$

51. A distinction between combinations and permutations

52. An alternative way of expressing $_nC_r$

53. A number n such that $n! = n$

QUESTIONS FOR DISCUSSION OR ESSAY

54. Explain why a combination lock should really be called a per-mutation lock. What are some other instances in which the word *combination* should perhaps be replaced with the word *permutation*? Explain your examples. To get you started, think about a TV cooking show referring to "the right combination of in-gredients" or a coach referring to "a winning combination of players."

55. In Exercise 35, you were asked to determine the number of ways of selecting 5 numbers out of 40 in a state lottery. The answer to that exercise represents the number of possible winning com-binations. What does this say about your chances of winning a lottery? How many different lottery combinations would you have to play in order to give yourself a 50/50 chance of winning? Explain.

56. Suppose ten sprinters are competing in the 100-meter dash at a track meet. If the race is a qualifying round, the top three finishers advance to the next round. If the race is the final round, the top three finishers are awarded first, second, and third place. For each of these two scenarios, determine whether a permutation or com-bination would be used to count the number of ways of selecting 3 out of 10 sprinters. Explain in detail why you chose the method you did.

57. A basketball coach, notorious for changing his starting lineup, states that he will try every combination until he finds one that works. Assuming the team has 12 players, and that a season consists of 30 games, how many seasons will it take to try every combination? In practice, would a coach really try every possible starting lineup?

PROJECTS FOR ENRICHMENT

58. *Combinatorial Identities*   The combination formula

$$\binom{n}{r} = \frac{n!}{r!(n-r)!}$$

leads to many useful and sometimes surprising identities. We consider a few in this project, most of which can be verified by direct simplification. One example is the identity

$$\binom{n}{n} = 1$$

This can be verified by observing that

$$\binom{n}{n} = \frac{n!}{n!(n-n)!} = \frac{n!}{n!0!} = 1$$

The interpretation of this identity is that there is only one way to select n elements (without regard to order) out of a set of n elements.

a. Simplify the following combinations and give a verbal inter-pretation of the result.

   i. $\binom{n}{1}$ **ii.** $\binom{n}{0}$

   iii. $\binom{n}{n-1}$

b. Simplify the following combinations.

   i. $\binom{n}{2}$ **ii.** $\binom{n}{n-2}$

You should have found that the two combinations in part b sim-plify to the same expression. In general,

$$\binom{n}{r} = \binom{n}{n-r}$$

for any $r \le n$. In other words, a set of n elements has the same number of r-element subsets as it has $(n-r)$-element subsets.

c. Simplify $\binom{n}{n-r}$ to show that

$$\binom{n}{r} = \binom{n}{n-r}$$

   Interpret this identity.

d. The following identities are often useful when working with complicated calculations involving combinations. Verify each.

   i. $\binom{n}{m}\binom{m}{r} = \binom{n}{r}\binom{n-r}{m-r}$

   (*Hint:* Simplify both sides of the equation.)

   ii. $\binom{n}{r-1} + \binom{n}{r} = \binom{n+1}{r}$

   (*Hint:* Find a common denominator for the expression to the left of the equal sign.)

   iii. $\binom{n-1}{r} = \frac{r+1}{n}\binom{n}{r+1}$

59. *Counting Poker Hands*   In Example 9, we computed the number of possible 5-card poker hands that could be dealt from a standard 52-card deck, and we also computed the number of possible full houses. Table 1 lists all of the 5-card poker hands, together with a partial list of the number of possible such hands.

TABLE 1 5-Card Poker Hands

Hand	Number possible
Royal flush	4
Other straight flush	36
Four of a kind	
Full house	3,744
Flush	
Straight	10,200
Three of a kind	54,912
Two pairs	
One pair	
Nothing	1,302,540
Total	2,598,960

a. Look up and write out a description of each type of hand.

b. Show how combinations can be used to arrive at the number of possible hands for each type.

c. Although there are many varieties of 5-card poker, the general idea is that 5 cards are dealt to all players and there are one or more rounds of betting. After all betting is finished, the player with the hand that appears highest on the list wins the money. Why does it make sense that the hand appearing highest on the list in Table 1 should win?

d. Do you think it's necessary for a poker player to know how many hands of each type are possible? Why or why not?

60. *Seating Arrangements*   Three couples—Andy and Becky, Calvin and Dawn, and Eddie and Fran—are all good friends and decide to go to a concert together. They have reserved six adjoining seats next to an aisle. Determine the number of seating arrangements that are possible given the following conditions.

a. There are no restrictions.

b. The men must sit in the three seats closest to the aisle.

c. Each couple must sit together. (*Hint:* First find the number of pairs of seats where each couple can sit, then the number of arrangements for each of the three couples.)

d. The women must sit together. (*Hint:* First find the number of groups of 3 seats where the women can sit, then find the number of arrangements of the three women in those seats, and, finally, find the number of arrangements of the men in the remaining three seats.)

e. Andy and Becky had a fight and refuse to sit next to each other. (*Hint:* First find the number of ways in which Andy and Becky can be separated, and then find the number of ways of arranging the remaining four people.)

After the concert, the three couples decide to go out for a late dinner. Determine the number of ways in which they can be seated around a circular table, given the following conditions. Keep in mind that we are only concerned with their positions relative to each other, not with the particular seat each person occupies.

f. There are no restrictions.

g. Each couple must sit together.

h. The women must sit together.

i. Andy and Becky had a fight and refuse to sit next to each other.

SECTION **8.4 PROBABILITY**

☐ **How can it be that approximately 60% of the people testing positive for drug usage are actually not drug users, even though the test is 97% accurate?**

☐ **What is the probability that, in a class of 30 students, at least 2 will have the same birthday?**

☐ **How likely is it that a fatal accident will involve a driver between the ages of 18 and 21 who had been drinking?**

☐ **Which is more likely, getting struck by lightning or winning a state lottery?**

☐ **Did the world's smartest human give bad advice to game show contestants?**

Our language is replete with terms used to indicate the degree of certainty of uncertain events. We refer to outcomes as being "improbable," "unlikely," "a sure thing," "more likely than not," or a "long shot." **Probability** is the branch of mathematics in which degrees of certainty are quantified. Virtually every day of our lives we encounter uncertainties that can be expressed as probabilities. We are told that there is a 30% chance of thunderstorm activity, that there is a one in an million chance of winning a lottery, that one in four women will develop breast cancer, and that three out of four dentists recommend a certain sugarless gum. In addition to these everyday occurrences, probabilities arise in, and indeed are central to, such diverse areas as medicine, insurance, sociology, psychology, physics, and engineering.

As with the topics discussed earlier in the chapter, we provide here only a brief overview of the fundamental notions and broad applicability of mathematical probability. A more thorough treatment is left to later course work.

FUNDAMENTALS

In mathematical probability, a real number between 0 and 1 is associated with a possible outcome of an experiment: The greater the probability, the more likely it is that the outcome occurs. Events that have a probability of 1 are virtually certain to occur, whereas events with probability 0 virtually never occur. An event with a probability of 0.5 will occur about half the time. More formally, we have the following definitions.

Probability Terminology and Notation ☐

> A process of observation or measurement is referred to as an **experiment.** The **sample space** of an experiment is the set of all possible outcomes. The number of possible outcomes in a sample space S is denoted $n(S)$. An **event** is any subset of the sample space. The number of outcomes in an event E is denoted $n(E)$.

EXAMPLE 1 ▨ An Experiment Involving a Die

An experiment consists of rolling an ordinary six-sided die.

a. Find the sample space S, and determine $n(S)$.
b. Describe the event E that an even number is rolled, and determine $n(E)$.

SOLUTION

a. Since an outcome of the experiment consists of rolling one of the numbers 1–6, the sample space S can be written as $\{1, 2, 3, 4, 5, 6\}$; $n(S)$ is the number of outcomes in S and so $n(S) = 6$.

b. The event E that an even number is rolled is the set $\{2, 4, 6\}$. Since $n(E)$ is the number of outcomes in E, $n(E) = 3$.

EXAMPLE 2 ▨ A Coin-Tossing Experiment

An experiment consists of tossing a coin two times.

a. Find the sample space S, and determine $n(S)$.
b. Find $n(E)$, where E is the event that a tail is tossed both times.

SOLUTION

a. Let H denote the outcome that a head is tossed and let T denote the outcome that a tail is tossed. Thus, for example, the outcome consisting of tossing a head and then a tail can be written as HT. The sample space S is the set $\{HH, HT, TH, TT\}$, and $n(S) = 4$.

b. The event E is the set of outcomes for which a tail is tossed both times, or $\{TT\}$. Thus $n(E) = 1$.

EXAMPLE 3 ▓ Family Composition

A couple has two children.

a. Find the sample space S consisting of the possible outcomes with respect to the sexes of the children.

b. Describe the event E that the children are of different sexes.

SOLUTION

a. Let M represent a male birth and let F represent a female birth. Thus, for example, MF represents the outcome that the first child is male and the second child is female. The sample space S can then be expressed as the set $\{MM, MF, FM, FF\}$.

b. The two outcomes in which the children are of different sexes are MF and FM. Thus, E can be expressed as the set $\{MF, FM\}$.

When all outcomes are equally likely, we can find the probability of an event simply by dividing the number of outcomes that make up the event by the total number of outcomes in the sample space. For example, the probably of rolling a 2 on a fair die (one for which all outcomes are equally likely) is $\frac{1}{6}$, since one of six outcomes corresponds to rolling a 2. Similarly, the probability of obtaining a head with one toss of a fair coin is $\frac{1}{2}$. More generally, we have the following definition of the probability of an event.

Probability of an Event ☐

> If E is an event in the sample space S and if all of the outcomes of S are equally likely, then the **probability of E** is denoted by $P(E)$ and is defined by
>
> $$P(E) = \frac{n(E)}{n(S)}$$

EXAMPLE 4 ▓ Computing a Probability Involving a Die

If an ordinary six-sided die is rolled, what is the probability that an even number is rolled?

SOLUTION From Example 1, we have $S = \{1, 2, 3, 4, 5, 6\}$ and $E = \{2, 4, 6\}$. Thus,

$$P(E) = \frac{n(E)}{n(S)} = \frac{3}{6} = \frac{1}{2}$$

EXAMPLE 5 ▓ Computing a Probability Involving Two Coins

If a coin is tossed two times, what is the probability that a tail is tossed both times?

SOLUTION As we saw in Example 2, $S = \{HH, HT, TH, TT\}$ and $E = \{TT\}$. Thus,

$$P(E) = \frac{n(E)}{n(S)} = \frac{1}{4}$$

EXAMPLE 6 ■ Computing a Probability Involving Two Dice

Two ordinary six-sided dice are rolled. What is the probability that the total of the dice is 7?

SOLUTION There are six possible outcomes for the first die and six possible outcomes for the second die. Thus, by the Fundamental Counting Principle, there are 36 possible outcomes in the sample space S. These outcomes are shown as ordered pairs in Table 2. Now for each possible outcome of the first die, there is exactly one outcome for the second die that will result in a total of 7. Thus, there are six ways of rolling a total of 7. These appear in the circled diagonal of Table 2.

TABLE 2 Possible Outcomes

		Second die				
	1	**2**	**3**	**4**	**5**	**6**
1	(1, 1)	(1, 2)	(1, 3)	(1, 4)	(1, 5)	(1, 6)
2	(2, 1)	(2, 2)	(2, 3)	(2, 4)	(2, 5)	(2, 6)
3	(3, 1)	(3, 2)	(3, 3)	(3, 4)	(3, 5)	(3, 6)
4	(4, 1)	(4, 2)	(4, 3)	(4, 4)	(4, 5)	(4, 6)
5	(5, 1)	(5, 2)	(5, 3)	(5, 4)	(5, 5)	(5, 6)
6	(6, 1)	(6, 2)	(6, 3)	(6, 4)	(6, 5)	(6, 6)

(First die labels the rows.)

So if E is the event that a 7 is rolled, then $n(E) = 6$. Since $n(S) = 36$, we have

$$P(E) = \frac{n(E)}{n(S)} = \frac{6}{36} = \frac{1}{6}$$

EXAMPLE 7 ■ Computing Probabilities Involving Three Coins

Three coins are tossed. Find the probabilities of the following events.

  a. All three coins come up the same.
  b. Exactly two heads are tossed.
  c. At least one tail is tossed.

SOLUTION We first find $n(S)$. There are two possible outcomes for each of the three coins—namely, heads or tails. Thus, by the Fundamental Counting Principal, there are $2 \cdot 2 \cdot 2 = 8$ possible outcomes in the sample space S. Thus, $n(S) = 8$.

  a. Let E be the event that all three coins come up the same. Since the event E consists of two outcomes, *HHH* and *TTT*, $n(E) = 2$. Thus,

$$P(E) = \frac{n(E)}{n(S)} = \frac{2}{8} = \frac{1}{4}$$

  b. Let F be the event that exactly two heads are tossed. Then F consists of the outcomes *HHT*, *HTH*, and *THH*. So $n(F) = 3$ and we have

$$P(F) = \frac{n(F)}{n(S)} = \frac{3}{8}$$

c. Let G be the event that at least one tail is tossed. Of the 8 outcomes in S, the only one that doesn't have at least one tail is HHH. Thus, $n(G) = 7$ and we have

$$P(G) = \frac{n(G)}{n(S)} = \frac{7}{8}$$

EXAMPLE 8   ▨   Playing the Lottery

A certain state lottery is designed so that a player selects 6 different numbers from 1 to 40 (inclusive). If those 6 numbers match the 6 that are randomly drawn by the lottery (without regard to order), the player wins. Find the probability that a player wins the lottery after having selected 10 different combinations of 6 numbers.

SOLUTION   The sample space S consists of all possible combinations of 6 numbers chosen from the numbers 1 to 40. Thus $n(S) = {}_{40}C_6 = 3,838,380$. Let E be the event that 1 of the player's 10 combinations is a winner. Then $n(E) = 10$. Thus,

$$P(E) = \frac{n(E)}{n(S)} = \frac{10}{3,838,380} \approx 0.0000026$$

EXAMPLE 9   ▨   A Five-Card Poker Hand

A **straight flush** is a poker hand consisting of five cards in sequence from the same suit, such as A-2-3-4-5 or 8-9-10-J-Q. Find the probability of being dealt a straight flush from a standard 52-card deck.

SOLUTION   The sample space S consists of all possible 5-card hands from a deck of 52 cards. Thus,

$$n(S) = {}_{52}C_5 = 2,598,960$$

Let E be the event that a straight flush is dealt. Then $n(E)$ is the number of straight flushes that are possible in a 52-card deck. If we assume that the ace can be either low or high, then the possible straights in a given suit are A-2-3-4-5, 2-3-4-5-6, . . . , 10-J-Q-K-A. Thus, there are 10 possible straights in each of the four suits, for a total of 40 possible straights. So $n(E) = 40$ and we have

$$P(E) = \frac{n(E)}{n(S)} = \frac{40}{2,598,960} \approx 0.000015$$

EXAMPLE 10   ▨   Reliability of Drug Testing

The president of a university with an enrollment of 30,000 is contemplating mandatory cocaine testing for all students. The test is 97% accurate in the sense that 97% of all students who are cocaine users will test positive and 97% of students who are not cocaine users will test negative. If 2% of the student body are, in fact, cocaine users, what is the probability that someone who tests positive is a cocaine user?

SOLUTION   Consider an experiment consisting of selecting a student from the group that tested positive. Then the sample space S consists of the set of all students testing positive, whether they are cocaine users or not. Next we let E be the event that a person tests positive and is indeed a cocaine user. To find $P(E)$, we must determine

$n(S)$ and $n(E)$. Since 2% of the 30,000 university students are cocaine users, there are $(0.02)30{,}000 = 600$ cocaine users and $30{,}000 - 600 = 29{,}400$ nonusers. Now the test will come out positive for approximately 97% of the users, but it will also come out positive for about 3% of the nonusers. Thus,

$$
\begin{aligned}
n(S) &= \text{Number of students who test positive} \\
&= \text{Number of } users \text{ who test positive} \\
&\quad + \text{Number of } nonusers \text{ who test positive} \\
&\approx 97\% \text{ of } 600 + 3\% \text{ of } 29{,}400 \\
&= (0.97)600 + (0.03)29{,}400 \\
&= 1464
\end{aligned}
$$

and

$$
\begin{aligned}
n(E) &= \text{Number of students who test positive and are cocaine users} \\
&\approx (0.97)600 \\
&= 582
\end{aligned}
$$

Thus, we have

$$
P(E) = \frac{n(E)}{n(S)} = \frac{582}{1464} \approx 0.4
$$

So the probability that someone who tests positive for cocaine use is actually a user is less than $\frac{1}{2}$! We will ask you to discuss this surprising result in Exercise 66.

INTERSECTIONS, UNIONS, AND COMPLEMENTS OF EVENTS

In many cases, we are interested in finding probabilities that involve two or more events from the same sample space or that indirectly involve a given event. For these tasks, we make the following definitions.

Definitions of Intersection, Union, and Complement   ☐

> Let E and F be events from the same sample space S. Then
>
> **1.** The **intersection** of E and F, denoted $E \cap F$, is the event that both E and F occur.
>
> **2.** The **union** of E and F, denoted $E \cup F$, is the event that either E or F (or both) occur.
>
> **3.** The **complement** of E, denoted E' and read "E complement," is the event that E does *not* occur.

EXAMPLE 11   ■   **Finding Intersections, Unions, and Complements**

Suppose that a number from 1 to 10 (inclusive) is selected at random. Let E be the event that the number chosen is a multiple of 3, and let F be the event that the number chosen is even. Find expressions for each of the following events, and then compute the probability of the event.

a. $E \cap F$ **b.** $E \cup F$
c. E' **d.** F'

SOLUTION

a. $E \cap F$ consists of all numbers from 1 to 10 that are both even *and* multiples of 3. Thus, $E \cap F = \{6\}$, and $n(E \cap F) = 1$. Since $n(S) = 10$, we have

$$P(E \cap F) = \frac{n(E \cap F)}{n(S)} = \frac{1}{10}$$

b. $E \cup F$ consists of all numbers from 1 to 10 that are even, multiples of 3, or both. Thus, $E \cup F = \{2, 3, 4, 6, 8, 9, 10\}$. Hence $n(E \cup F) = 7$, and $P(E \cup F) = \frac{7}{10}$.

c. E' is the set of numbers from 1 to 10 that are *not* multiples of 3. Thus, $E' = \{1, 2, 4, 5, 7, 8, 10\}$, and $n(E') = 7$. Hence $P(E') = \frac{7}{10}$.

d. F' is the set of numbers from 1 to 10 that are not in F—that is, those numbers that are not even. Hence F' consists of the odd numbers between 1 and 10, and so $F' = \{1, 3, 5, 7, 9,\}$. Thus, $n(F') = 5$, and $P(F') = \frac{5}{10} = \frac{1}{2}$.

If E and F are events from the same sample space, and E and F have no outcomes in common, then $E \cap F = \{\ \}$, the empty set, and we say that the events are **mutually exclusive**. Consider, for example, an experiment in which a die is rolled. If E is the event that an odd number is rolled and F is the event that an even number is rolled, then E and F are mutually exclusive since no outcome belongs to both events.

When two events E and F are mutually exclusive, the number of outcomes that belong to one of the two events is simply the sum of the number of outcomes in each event. In other words, for mutually exclusive events E and F

$$n(E \cup F) = n(E) + n(F)$$

Dividing by $n(S)$ gives us

$$\frac{n(E \cup F)}{n(S)} = \frac{n(E)}{n(S)} + \frac{n(F)}{n(S)}$$

$$P(E \cup F) = P(E) + P(F)$$

As an example, consider the event E that a 3 is obtained by a single throw of a die and the event F that a 5 is thrown. Since these events are mutually exclusive (we can't throw both a 3 and a 5 on the same toss), the probability of $E \cup F$ is simply the sum of the probabilities of E and F. Since $P(E) = \frac{1}{6}$ and $P(F) = \frac{1}{6}$, we have

$$P(E \cup F) = P(E) + P(F) = \frac{1}{6} + \frac{1}{6} = \frac{1}{3}$$

EXAMPLE 12 ▨ Winning the World Series

Suppose that a preseason poll of sports writers suggests that the probability of the Chicago Cubs winning the World Series is 0.06 and the probability of the New York Mets winning is 0.13. Find the probability that one of the two teams will win.

SOLUTION Let E be the event that the Cubs win and let F be the event that the Mets win. We would like to determine $P(E \cup F)$, the probability that E or F occurs. Since it isn't possible for both teams to win the World Series in a given year, the events E and F are mutually exclusive. Thus,

$$P(E \cup F) = P(E) + P(F) = 0.06 + 0.13 = 0.19$$

Now let us consider two events that are *not* mutually exclusive. Suppose that 2 coins are tossed. Then the sample space S can be represented as the set $\{HH, HT, TH, TT\}$. Let E be the event that the first coin comes up heads, and let F be the event that the second coin comes up heads. Thus, $E = \{HH, HT\}$ and $F = \{TH, HH\}$, so that $n(E) = n(F) = 2$. The event $E \cup F$ consists of those outcomes for which either the first coin or the second coin is a head. That is, $E \cup F = \{HH, HT, TH\}$, and thus $n(E \cup F) = 3$. But $n(E) + n(F) = 2 + 2 = 4$, so that $n(E \cup F) \neq n(E) + n(F)$. The reason for the discrepancy is that the outcome HH belongs to both E and F so that when we add $n(E)$ and $n(F)$, we have counted this outcome twice.

In general, the sum $n(E) + n(F)$ counts the outcomes belonging to $E \cap F$ twice. To correct for this double counting, we must subtract the number of elements in the set $E \cap F$. Thus, we have

$$n(E \cup F) = n(E) + n(F) - n(E \cap F)$$

Figure 10 illustrates this formula with a **Venn diagram**—a collection of overlapping circles used to illustrate relationships among sets. In this diagram, the outcomes of the event E are represented by the squares and triangles lying inside the circle on the left. The outcomes of F consist of the solid circles and triangles lying inside the circle on the right. As the diagram suggests, the event $E \cup F$ consists of those outcomes lying in both E and F—in this case, the triangles.

Not surprisingly, a similar formula holds for finding the probability of the union of two events that may not be mutually exclusive. It is stated as follows.

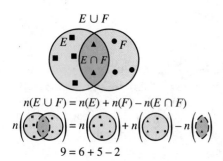

$E \cup F$

$n(E \cup F) = n(E) + n(F) - n(E \cap F)$

$9 = 6 + 5 - 2$

FIGURE 10

The Probability of the Union of Two Events   □

Let E and F be events in the same sample space. Then the probability of E or F occurring is given by $$P(E \cup F) = P(E) + P(F) - P(E \cap F)$$

Note that this formula applies even if E and F are mutually exclusive since, in that case, $E \cap F = \{\ \}$, and so $P(E \cap F) = 0$.

EXAMPLE 13   ■   Computing the Probability of the Union of Two Events

Suppose that a single card is selected at random from a standard 52-card deck. Find the probability that the card selected is either a face card (jack, queen, or king) or a heart.

SOLUTION   Let E be the event that the card selected is a face card, and let F be the event that the card selected is a heart. Then $E \cup F$ represents the event that the card selected is a face card or a heart. Thus, we are interested in $P(E \cup F)$. Since there are three face cards in each of the four suits, there are 12 face cards in all. Thus,

$$P(E) = \frac{n(E)}{n(S)} = \frac{12}{52} = \frac{3}{13}$$

On the other hand, $\frac{1}{4}$ of the cards in the deck are hearts, so that

$$P(F) = \frac{1}{4}$$

The event $E \cap F$ consists of those outcomes in which a face card of hearts is drawn. Since there are 3 hearts which are face cards, we have

$$P(E \cap F) = \frac{n(E \cap F)}{n(S)} = \frac{3}{52}$$

We can now compute $P(E \cup F)$ as follows:

$$\begin{aligned} P(E \cup F) &= P(E) + P(F) - P(E \cap F) \\ &= \frac{3}{13} + \frac{1}{4} - \frac{3}{52} \\ &= \frac{12 + 13 - 3}{52} \\ &= \frac{22}{52} = \frac{11}{26} \end{aligned}$$

EXAMPLE 14   ■   **Probabilities Involving Fatal Car Accidents**

According to the U.S. Federal Highway Administration, of the drivers involved in fatal motor vehicle accidents in 1989,

- ■ 14.1% were 18–21 years old
- ■ 31.9% had been drinking
- ■ 5.3% were 18–21 years old and had been drinking

If a driver involved in a fatal motor vehicle accident is selected at random, find the probability of each of the following events.

a. The driver was 18–21 and had been drinking.
b. The driver had not been drinking.
c. The driver had been drinking but was not 18–21.

SOLUTION   Let S be the sample space consisting of drivers involved in fatal accidents. Let E be the event that the selected driver was 18–21, and let F be the event that the driver had been drinking. According to the given data, 5.3% of the drivers in S are in $E \cap F$. Moreover, since 14.1% of the drivers are in E, there must be 14.1% − 5.3% = 8.8% of the drivers in E that aren't also in $E \cap F$. Similarly, there are 31.9% − 5.3% = 26.6% of the drivers in F that aren't also in $E \cap F$. It is convenient to summarize these percentages in a Venn diagram such as the one in Figure 11. Note that 59.3% of drivers involved in fatal accidents had not been drinking and were not 18–21 years old. Note also that the percentages themselves represent probabilities.

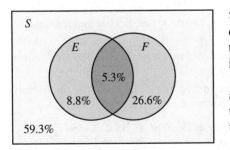

FIGURE 11

a. The event that the driver was 18–21 and had been drinking is $E \cap F$, and $P(E \cap F) = 5.3\% = 0.053$.
b. The event that the driver had not been drinking is F', and $P(F') = 59.3\% + 8.8\% = 0.593 + 0.088 = 0.681$.
c. The event that the driver had been drinking but was not 18–21 is $F \cap E'$. In the Venn diagram, this is the region inside F but outside E. Thus, $P(F \cap E') = 26.6\% = 0.266$.

UNDERSTANDING AND MASTERY CHECKLISTS

CONCEPTS TO UNDERSTAND

☐ Experiment
☐ Sample space
☐ Event
☐ Probability of an event
☐ Intersection of events
☐ Union of events
☐ Complement of an event
☐ Mutually exclusive events

SKILLS TO MASTER

☐ Find the sample space of an experiment.
☐ Compute the probability of an event.
☐ Describe the intersection of two events and find its probability.
☐ Describe the union of two events and find its probability.
☐ Describe the complement of an event and find its probability.

EXERCISES 8.4

EXERCISES 1–4   *Find the sample space S for the given experiment and determine n(S).*

1. A single ball is drawn from an urn containing red, blue, green, and white balls.

2. A button is pressed in an elevator in a 10-story building.

3. A coin is tossed three times.

4. A six-sided die is rolled and a coin is tossed.

EXERCISES 5–8   *A standard six-sided die is rolled. Find the probability of the given event.*

5. Rolling a 4

6. Rolling a number larger than 4

7. Rolling a number less than 7

8. Rolling an even number

EXERCISES 9–12   *A ball is selected at random from an urn containing 3 red balls, 4 blue balls, and 5 white balls. Find the probability of the given event.*

9. Selecting a red ball

10. Selecting a red or white ball

11. Selecting a ball that is not white

12. Selecting a yellow ball

EXERCISES 13–18   *A single card is drawn from a standard 52-card deck. Compute the probability of the given event.*

13. Drawing the ace of spades

14. Drawing a spade

15. Drawing an ace

16. Drawing an ace or a spade

17. Drawing a red face card

18. Drawing a 3 or a 4

EXERCISES 19–22   *Assume that the probability of a male birth is 0.5.*

19. A family has three children. Compute the probability that the children are of the same sex.

20. Compute the probability that in a family with three children, exactly one of the children is female.

21. Compute the probability that in a family with three children, exactly two of the children are male.

22. Compute the probability that in a family of three children, there is no girl with an older brother.

EXERCISES 23–26   *Balls are selected (without replacement) from an urn containing 5 red balls and 4 blue balls.*

23. If two balls are drawn, find the probability that both are red.

24. If two balls are drawn, find the probability that they are the same color.

25. If two balls are drawn, find the probability that they are different colors.

26. If three balls are drawn, find the probability that two of the balls are blue and the other is red.

EXERCISES 27–30   *A pair of standard six-sided dice are rolled. Find the probability of the given event.*

27. Rolling the same number on each die

28. Rolling "snake-eyes" (a 1 on each die)

29. Rolling a total of 10

30. Rolling a 6 on at least one of the dice

EXERCISES 31–34   *Compute the probability of being dealt the given 5-card poker hand from a standard 52-card deck.*

31. Four aces   32. Four of a kind

33. A flush (all cards of the same suit)

34. Three of a kind (three cards of the same face value and two cards of different face values)

EXERCISES 35–40 *A fair coin is tossed. Compute the probability of the given event.*

35. Tossing 4 heads in a row

36. Having a streak of 4 heads in a row if the coin is being tossed 5 times

37. Obtaining your first head on the third toss

38. Obtaining your second head on the fourth toss

39. Tossing exactly 3 heads out of the first 5

40. Alternating heads/tails or tails/heads in the first five tosses

EXERCISES 41–44 *An experiment is described and two events E and F are given. Find P(E) and P(F), describe the events E ∩ F, E ∪ F, E', and F', and find the probability of each.*

41. A single card is drawn from a standard 52-card deck; *E* is the event that a red card is drawn, and *F* is the event that a face card is drawn.

42. A six-sided die is rolled; *E* is the event that an odd number is rolled, and *F* is the event that a number less than 5 is rolled.

43. Two six-sided dice are rolled; *E* is the event that the sum is even, and *F* is the event that the sum is less than 8.

44. Two integers from 1 to 10 are chosen at random (repetitions are allowed); *E* is the event that the sum is odd, and *F* is the event that the sum is greater than 10.

■ **APPLICATIONS**

---

45. *NCAA Basketball Championship*   Suppose that at the beginning of the NCAA basketball tournament, it has been determined that the probability that Duke will win the championship is 0.2 and the probability that Michigan will win is 0.15. Find the probability that one of these two teams will win the championship.

46. *Math Grades*   Suppose you have determined that the probability of getting an A in this class is 0.25 and the probability of getting a B is 0.6. Find the probability that you will get an A or a B.

47. *Lottery Chances*   A state lottery is designed so that a player selects 5 different numbers from 1 to 40. What is the probability that a single choice of 5 numbers will win the lottery? What is the probability of winning if 20 combinations of numbers are chosen? How many combinations must be chosen so that the probability of winning is $\frac{1}{2}$?

48. *Lottery Chances*   A state lottery is designed so that a player selects 6 different numbers from 1 to 50. What is the probability that a single choice of 6 numbers will win? What is the probability of winning if 100 combinations of numbers are chosen? How many combinations must be chosen so that the probability of winning is $\frac{1}{2}$?

49. *Grade Averages*   The distribution of high school grade averages for college freshmen in the U.S. in 1990 is given in Figure 12 (*Data source: The American Freshman: National Norms,*

University of California, Los Angeles). If a college freshman were selected at random, find the probability that the student

a. had a C− to C+ average

b. did not have an A− to A+ average

50. *College Enrollment*   The distribution of college enrollment in the U.S. in 1990 is given in Figure 13 (*Data source: Digest of Education Statistics,* U.S. National Center for Education Statistics). If a college student were selected at random, find the probability that the student

a. was a minority U.S. citizen

b. was not a white U.S. citizen

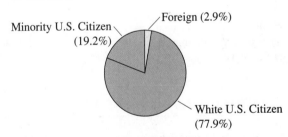

Minority U.S. Citizen (19.2%)   Foreign (2.9%)
White U.S. Citizen (77.9%)

FIGURE 13

51. *Polygraph Testing*   A large fast-food chain with 10,000 employees has been experiencing a rash of employee theft and so begins requiring employees to take a polygraph test (that is, a lie detector test). The test is 90% accurate in the sense that 90% of all employees who are stealing will fail the test and 90% of employees who are not stealing will pass the test. If 5% of the employees are in fact stealing, what is the probability that someone who fails the test is stealing?

52. *Smoking Distributions*   In 1989, 12.4% of the population of the United States was black, 28.8% of the entire population were smokers, and 3.8% of the population were black smokers. If an American is selected at random, find the probability of each of the following events.

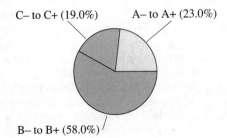

C− to C+ (19.0%)   A− to A+ (23.0%)
B− to B+ (58.0%)

FIGURE 12

a. The person selected is black or smokes.

b. The person selected is nonblack.

c. The person selected does not smoke.

d. The person selected is black but does not smoke.

e. The person selected is a nonblack smoker.

53. *Hispanic Californians* According to the United States census of 1990, approximately 12% of the population lives in California. Also, 25.8% of all Californians are Hispanic. Nationwide, 9% of the population is Hispanic. If an American is selected at random, find the probability of each of the following events.

a. The person is a Californian or Hispanic.

b. The person is a Californian Hispanic.

c. The person is a non-Californian.

d. The person is a Californian but is not Hispanic.

e. The person is Hispanic but is not a Californian.

54. *Extracurricular Activities* At a certain university, extracurricular activities consist entirely of athletics, Greek organizations (fraternities and sororities), and clubs. You are given the following information regarding the makeup of the student body.

- 10% participate in athletics.
- 40% belong to Greek organizations.
- 45% are club members.
- 25% are involved in both Greek organizations and clubs but not athletics.
- 2% are involved in athletics, Greek organizations, and clubs.
- 15% are non-Greek, nonathlete, club members.
- 4% are Greek athletes who do not belong to any clubs.

Assume that a student is selected at random from the student body of this university. Compute the probabilities of each of the following events.

a. The student is involved in an extracurricular activity.

b. The student is involved in exactly one extracurricular activity.

c. The student is involved in a club and is an athlete but is not a member of a Greek organization.

d. The student is involved in exactly two extracurricular activities.

CONCEPTS AND CRITICAL THINKING

EXERCISES 55–58 *Answer true or false.*

55. In general, for events E and F, $P(E \cup F) = P(E) + P(F)$.

56. There is no event that has a probability of 1.7.

57. If E and F are mutually exclusive events, then $P(E \cup F) = P(E) + P(F)$.

58. If a coin is loaded in such a way that heads come up about 7 times for every 10 times that tails come up, then the probability of heads is about $\frac{7}{10}$.

EXERCISES 59–62 *Give an example of each.*

59. A real-world event for which the probability is $\frac{1}{7}$

60. A number that isn't the probability of any event

61. Two events E and F such that $P(E \cup F) = P(E) + P(F)$

62. A real-world event for which the probability is 1

QUESTIONS FOR DISCUSSION OR ESSAY

63. Estimates for the probability of getting struck by lightning in a given year range as high as 0.000007. How do you think such estimates are obtained? Do you think such estimates are valid? Why or why not? In Example 8, we saw that the probability of winning a certain state lottery after having played 10 combinations of numbers was 0.0000026, which is lower than the estimate for getting struck by lightning. Can we conclude that one is more likely to get struck by lightning than to win a state lottery? Explain.

64. A weathercaster announces that there is a 70% chance of rain on Saturday and a 30% chance of rain on Sunday, and concludes that there is a 100% chance of rain for the weekend. Is his reasoning correct? If not, what would be a more reasonable estimate for the chance of rain over the weekend?

65. In the column "Ask Marilyn" in *Parade* magazine, Marilyn Vos Savant (who is listed in the *Guinness Book of World Records* as having the highest IQ) posed the following problem. Suppose that you are a contestant on a game show. You are asked to select one of three doors. Behind one of the doors is a new car and behind each of the other two doors are goats. After choosing your door, the host opens one of the remaining doors to reveal a goat. She then gives you the choice of staying with your original selection or switching to the other door that is still closed. Ms. Vos Savant claimed that you improve your probability of winning if you switch. She was promptly flooded with letters from irate readers (many of them educators) who argued that the probability of winning is 0.5 whether you choose to switch or not. They reasoned that there are two doors remaining—behind

one is a car, behind the other is a goat—and so the probability of correctly guessing the door with the car behind it is $\frac{1}{2} = 0.5$. Surprisingly, this is not the case. You would in fact improve your probability of winning to $\frac{2}{3}$ by switching to the other door. Explain why this seems counterintuitive. How would you design an experiment to test the strategy? How relevant is the fact that the host of the game show knows which door the car is behind, and so will always show you a door with a goat behind it? Suppose there were 10 doors, and after your selection the host showed you what was behind all but two, the one you chose and one other. Would you switch doors? What if there were 1000 doors, and again the host showed you what was behind all but two, the one you chose and one other? What light does this shed on the original problem?

66. In Example 10, we considered a drug test that claimed to be 97% accurate and discovered that the probability that a student who tested positive was actually a user was only 0.4. In other words, it was likely that 60% of the students who tested positive were not actually users. On the other hand, the test will fail to detect only 3% of the users. What do these observations suggest about the use of such tests or the interpretation of the results?

■ PROJECTS FOR ENRICHMENT

67. *Assessing Psychic Phenomena*   A group of 10,000 people, all of whom claim to be psychic, are assembled on New Year's Eve 2002. Each is asked to make four predictions regarding the new year:

   (1) The day in January of 2003 on which the highest temperature is achieved in Tempe, Arizona

   (2) The party (Democrat or Republican) that wins the presidential election of 2003

   (3) The state in which the percentage change in population is the greatest

   (4) Whether the economy will strengthen or weaken over the course of the year 2003

   a. Find the probability of correctly guessing the outcome to the above predictions, assuming all outcomes are equally likely.

Nostradamus, a French astrologer and physician of the 16th century, became famous for his prediction of the death of King Henry II.

   b. Find the probability of a given individual being correct on all four predictions, if he or she is guessing at random.

   c. Find the probability of an individual making at least one incorrect prediction.

   d. Find the probability of none of the 10,000 alleged psychics being correct on all four correct predictions.

   e. Find the probability that at least one of the 10,000 alleged psychics will make all correct predictions.

   f. Is it fair to say that any individual who makes four correct predictions is psychic?

   g. What is your opinion as to the credibility of psychic predictions? Can mathematics be used to prove or disprove the possibility of psychic phenomena? Explain.

68. *Coincidental Birthdays*   In this project, we investigate the probability that two people in a group have the same birthday. For convenience, we ignore leap years.

   a. In how many ways can two people have *different* birthdays? (*Hint:* Let us call the two people Alfonso and Belinda. If we assign Alfonso any of the 365 days of the year, then that leaves only 364 possible days for Belinda's birthday. Use the Fundamental Counting Principle.)

   b. Find the number of ways in which two people can have birthdays.

   c. Find the probability of two people having different birthdays.

   d. Find an expression for the number of ways in which n people can have different birthdays. Use the permutation symbol $_nP_r$.

   e. Find an expression for the probability that among a group of n people, no two have the same birthday.

   f. Use the result of part e to find the probability that among a group of 30 people, there are at least two people with a common birthday.

   g. By trial and error, determine the minimum number of people required to ensure that the probability of at least one common birthday is greater than 0.5.

SECTION 8.5 THE BINOMIAL THEOREM

- ☐ What is the coefficient of x^2 in the expansion of $(1 + x)^{100}$?
- ☐ How many subsets does a set with 50 elements have?
- ☐ How can $\sqrt{10}$ be approximated to ten decimal places using only the four basic arithmetic operations?
- ☐ How can you compute 11^7 *by hand* without ever multiplying?
- ☐ What is Pascal's triangle and how can it be used to expand $(a + b)^5$?

EXPANDING BINOMIALS

Polynomials with just two terms, such as $a + b$, are called *binomials*. In this section, we develop the **Binomial Theorem,** a general formula for the expansion of $(a + b)^n$. We are already familiar with some examples of the binomial formula. Several are listed here.

$$(a + b)^0 = 1$$
$$(a + b)^1 = 1a + 1b$$
$$(a + b)^2 = 1a^2 + 2ab + 1b^2$$
$$(a + b)^3 = 1a^3 + 3a^2b + 3ab^2 + 1b^3$$

Notice that in each case the expansion of $(a + b)^n$ is a polynomial in the variables a and b, and that each term in the polynomial is of degree n. For example, in the expansion of $(a + b)^3$, each term has degree 3; that is, the sum of the exponents of a and b is equal to 3. For convenience, we have written each expression so that the powers of a are decreasing and the powers of b are increasing. Thus, we would expect the expansion of $(a + b)^4$ to be of the form

$$(a + b)^4 = \boxed{}a^4 + \boxed{}a^3b + \boxed{}a^2b^2 + \boxed{}ab^3 + \boxed{}b^4$$

where the entries in each of the boxes are constants called **binomial coefficients.**

We can determine the binomial coefficients using basic counting principles. Suppose that we were to evaluate $(a + b)^4$ directly by multiplication using the formula

$$(a + b)^4 = (a + b)(a + b)(a + b)(a + b)$$

The terms of the product can be formed by repeatedly selecting either an a or a b from each of the four factors $(a + b)$. For example, the term a^4 is obtained by selecting a from each of the four factors. Since this can be done in only one way, the coefficient of a^4 is 1. The terms of the form a^3b arise by selecting b from one of the four factors, and a from the remaining three. This corresponds to choosing a subset of size 1 from a set with 4 elements. In Section 8.3, we saw that this can be done in

$$_4C_1 = \binom{4}{1} = \frac{4!}{1!3!} = 4$$

ways. We illustrate the four choices as follows.

Four ways of obtaining a^3b
$(a + \boxed{b})(\boxed{a} + b)(\boxed{a} + b)(\boxed{a} + b)$
$(\boxed{a} + b)(a + \boxed{b})(\boxed{a} + b)(\boxed{a} + b)$
$(\boxed{a} + b)(\boxed{a} + b)(a + \boxed{b})(\boxed{a} + b)$
$(\boxed{a} + b)(\boxed{a} + b)(\boxed{a} + b)(a + \boxed{b})$

Products of the form a^2b^2 arise by selecting a's from two of the four factors and b's from the other two factors. Thus, the coefficient of a^2b^2 is $\binom{4}{2}$. Similarly, the coefficients of ab^3 and b^4 are $\binom{4}{3}$ and $\binom{4}{4}$, respectively. Thus, we have

$$(a+b)^4 = \binom{4}{0}a^4 + \binom{4}{1}a^3b + \binom{4}{2}a^2b^2 + \binom{4}{3}ab^3 + \binom{4}{4}b^4$$

$$= \frac{4!}{0!4!}a^4 + \frac{4!}{1!3!}a^3b + \frac{4!}{2!2!}a^2b^2 + \frac{4!}{3!1!}ab^3 + \frac{4!}{4!0!}b^4$$

$$= a^4 + 4a^3b + 6a^2b^2 + 4ab^3 + b^4$$

More generally, the coefficient of $a^{n-k}b^k$ in the expansion of $(a+b)^n$ is $\binom{n}{k}$. Thus, we have the following formula.

Binomial Theorem ☐

For n a positive integer,

$$(a+b)^n = \binom{n}{0}a^n + \binom{n}{1}a^{n-1}b + \binom{n}{2}a^{n-2}b^2$$

$$+ \cdots + \binom{n}{n-1}ab^{n-1} + \binom{n}{n}b^n$$

Thus, the coefficient of $a^{n-k}b^k$ in the expansion of $(a+b)^n$ is $\binom{n}{k}$, where

$$\binom{n}{k} = {}_nC_k = \frac{n!}{k!(n-k)!}$$

EXAMPLE 1   ■   **Finding a Coefficient Using the Binomial Theorem**

Find the coefficient of x^4y^2 in the expansion of $(x+y)^6$.

SOLUTION   Applying the Binomial Theorem with $n=6$ and $k=2$, we find that the coefficient of x^4y^2 is given by

$$\binom{6}{2} = \frac{6!}{2!(6-2)!}$$

$$= \frac{6!}{2!4!}$$

$$= \frac{6\cdot5\cdot4\cdot3\cdot2\cdot1}{(2\cdot1)(4\cdot3\cdot2\cdot1)}$$

$$= \frac{6\cdot5}{2}$$

$$= 15$$

EXAMPLE 2   ■   **Finding a Given Coefficient with the Binomial Theorem**

Find the coefficient of x^2y^3 in the expansion of $(x - y)^5$.

SOLUTION   We begin by expressing $(x - y)^5$ as $[x + (-y)]^5$. Then, according to the Binomial Theorem, the coefficient of $x^2(-y)^3$ is

$$\binom{5}{3} = \frac{5!}{3!2!} = \frac{5 \cdot 4 \cdot 3 \cdot 2 \cdot 1}{(3 \cdot 2 \cdot 1)(2 \cdot 1)} = \frac{5 \cdot 4}{2} = 10$$

Thus, the term in question is $10x^2(-y)^3 = -10x^2y^3$, and so the coefficient of x^2y^3 is -10.

EXAMPLE 3   ■   **Expanding with the Binomial Theorem**

Expand $(x + 2)^5$.

SOLUTION   According to the Binomial Theorem,

$$(x + 2)^5 = \binom{5}{0}x^5 + \binom{5}{1}x^4 \cdot 2 + \binom{5}{2}x^3 \cdot 2^2 + \binom{5}{3}x^2 \cdot 2^3 + \binom{5}{4}x \cdot 2^4 + \binom{5}{5}2^5$$

$$= x^5 + \frac{5!}{1!4!} x^4 \cdot 2 + \frac{5!}{2!3!} x^3 \cdot 2^2 + \frac{5!}{3!2!} x^2 \cdot 2^3 + \frac{5!}{4!1!} x \cdot 2^4 + 2^5$$

$$= x^5 + 5x^4 \cdot 2 + 10x^3 \cdot 4 + 10x^2 \cdot 8 + 5x \cdot 16 + 32$$

$$= x^5 + 10x^4 + 40x^3 + 80x^2 + 80x + 32$$

TECHNIQUES FOR EVALUATING BINOMIAL COEFFICIENTS

Expanding powers of binomials using the Binomial Theorem can be quite tedious; fortunately, we can take advantage of several properties of binomial coefficients to reduce our work considerably. For example, if we evaluate the binomial coefficient $\binom{7}{3}$ using only factorials, we have

$$\binom{7}{3} = \frac{7!}{3!(7 - 3)!} = \frac{7!}{3!4!} = \frac{5040}{6 \cdot 24} = 35$$

which is a fairly tedious calculation. On the other hand, we can use the formula for combinations

$$\binom{n}{r} = {}_nC_r = \frac{{}_nP_r}{r!}$$

introduced in Section 8.3 to obtain

$$\binom{7}{3} = \frac{{}_7P_3}{3!}$$

$$= \frac{7 \cdot 6 \cdot 5}{3 \cdot 2 \cdot 1}$$

$$= 35$$

More generally, we have the following formula.

***Alternative Form for
Binomial Coefficients*** ☐

The binomial coefficient $\binom{n}{k}$ can be computed as follows:

$$\binom{n}{k} = \frac{n(n-1)(n-2)\cdots(n-k+1)}{k(k-1)(k-2)\cdots(2)(1)}$$

Note that there are a total of k factors in both the numerator and denominator.

E X A M P L E   4   ■   **Using the Alternative Form to Evaluate Binomial Coefficients**

Evaluate each of the following binomial coefficients using the alternative form.

a. $\binom{10}{3}$ 　　　　　　　　　**b.** $\binom{8}{4}$

SOLUTION

a. $\binom{10}{3} = \dfrac{10 \cdot 9 \cdot 8}{3 \cdot 2 \cdot 1}$

$= \dfrac{720}{6}$

$= 120$

b. $\binom{8}{4} = \dfrac{8 \cdot 7 \cdot 6 \cdot 5}{4 \cdot 3 \cdot 2 \cdot 1}$

$= 7 \cdot 2 \cdot 5$

$= 70$

The computation of binomial coefficients can be further simplified by noting that

$$\binom{n}{n-k} = \frac{n!}{(n-k)![n-(n-k)]!}$$

$$= \frac{n!}{(n-k)!k!}$$

$$= \binom{n}{k}$$

We have thus established the following property of binomial coefficients.

***Symmetry Property of
Binomial Coefficients*** ☐

For integers n and k with $0 \le k \le n$,

$$\binom{n}{k} = \binom{n}{n-k}$$

Thus, the coefficient of $a^k b^{n-k}$ in the expansion of $(a+b)^n$ is the same as the coefficient of $a^{n-k}b^k$.

It follows from the symmetry property that the first binomial coefficient is the same as the last; the second is the same as the second to the last, and so forth. As a consequence, the work required to compute the binomial coefficients of $(a + b)^n$ is essentially cut in half, as illustrated by the following example.

EXAMPLE 5 ■ Using the Symmetry Property to Simplify Binomial Expansions

Find the binomial expansion of $(a + b)^6$.

SOLUTION According to the Binomial Theorem, we have

$$(a + b)^6 = \binom{6}{0}a^6 + \binom{6}{1}a^5b + \binom{6}{2}a^4b^2 + \binom{6}{3}a^3b^3$$
$$+ \binom{6}{4}^2 a^2b^4 + \binom{6}{5}ab^5 + \binom{6}{6}b^6$$

Evaluating the "first half" of these coefficients, we have

$$\binom{6}{0} = 1, \quad \binom{6}{1} = \frac{6}{1} = 6, \quad \binom{6}{2} = \frac{6 \cdot 5}{2 \cdot 1} = 15, \quad \text{and} \quad \binom{6}{3} = \frac{6 \cdot 5 \cdot 4}{3 \cdot 2 \cdot 1} = 20$$

By the symmetry property, we have

$$\binom{6}{4} = \binom{6}{2} = 15, \quad \binom{6}{5} = \binom{6}{1} = 6, \quad \text{and} \quad \binom{6}{6} = \binom{6}{0} = 1$$

Thus,

$$(a + b)^6 = a^6 + 6a^5b + 15a^4b^2 + 20a^3b^3 + 15a^2b^4 + 6ab^5 + b^6$$

 **CALCULATOR KEYS**

Computing Binomial Coefficients

A graphing calculator can also be used to compute binomial coefficients, but note that, for most calculators, the combination notation nCr is used instead of $\binom{n}{r}$. To compute a binomial coefficient $\binom{n}{r}$, first key in the number n, then select nCr from the menu of mathematical functions, and finally enter the number r.

Another convenient method for evaluating binomial coefficients arises from the observation that the binomial coefficients form a triangular array of numbers known as *Pascal's triangle*. Consider the expansions of $(a + b)^n$ for $n = 1, 2, 3,$ and 4, shown in Figure 14 with the binomial coefficients boxed.

$$
\begin{aligned}
(a + b)^0 &= \boxed{1} \\
(a + b)^1 &= \boxed{1}a + \boxed{1}b \\
(a + b)^2 &= \boxed{1}a^2 + \boxed{2}ab + \boxed{1}b^2 \\
(a + b)^3 &= \boxed{1}a^3 + \boxed{3}a^2b + \boxed{3}ab^2 + \boxed{1}b^3 \\
(a + b)^4 &= \boxed{1}a^4 + \boxed{4}a^3b + \boxed{6}a^2b^2 + \boxed{4}ab^3 + \boxed{1}b^4
\end{aligned}
$$

FIGURE 14

Now if we write down *only* the binomial coefficients, we have the array known as Pascal's triangle, shown in Figure 15.

1
1   1
1   2   1
1   3   3   1
1   4   6   4   1

FIGURE 15

Note that any given binomial coefficient is the sum of the two binomial coefficients immediately above it in the preceding row. In other words,

$$\binom{n+1}{k} = \binom{n}{k-1} + \binom{n}{k}$$

This property, which was considered in Exercise 58 of Section 8.3, can be used to complete the fifth and sixth rows of Pascal's triangle, as shown in Figure 16. Note that the first and last entry of each row is 1 and that all other entries are obtained by adding together the entries immediately above the given entry, as suggested by the arrows.

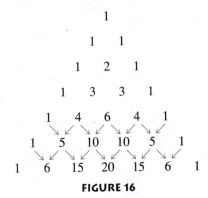

FIGURE 16

We refer to the top row of Pascal's triangle—the row consisting of a single 1—as row 0. The row consisting of two 1s is row 1, and so on. Since the nth row of Pascal's triangle consists of the binomial coefficients $\binom{n}{k}$, Pascal's triangle can be used to obtain binomial expansions, as illustrated in the following example.

EXAMPLE 6   ▧   **Expanding with Pascal's Triangle**

Expand using Pascal's triangle.

a. $(a + b)^6$ **b.** $(3x - 2y)^5$ **c.** $(1 + \sqrt{2})^4$

SOLUTION

a. The sixth row of Pascal's triangle in Figure 16 consists of the numbers 1, 6, 15, 20, 15, 6, and 1. Thus

$$(a + b)^6 = 1a^6 + 6a^5b + 15a^4b^2 + 20a^3b^3 + 15a^2b^4 + 6ab^5 + 1b^6$$

b. The fifth row of Pascal's triangle in Figure 16 consists of the numbers 1, 5, 10, 10, 5, and 1. Thus

$$(3x - 2y)^5 = 1(3x)^5 + 5(3x)^4(-2y) + 10(3x)^3(-2y)^2 + 10(3x)^2(-2y)^3 + 5(3x)(-2y)^4 + 1(-2y)^5$$
$$= 243x^5 - (5 \cdot 81 \cdot 2)x^4y + (10 \cdot 27 \cdot 4)x^3y^2 - (10 \cdot 9 \cdot 8)x^2y^3 + (5 \cdot 3 \cdot 16)xy^4 - 32y^5$$
$$= 243x^5 - 810x^4y + 1080x^3y^2 - 720x^2y^3 + 240xy^4 - 32y^5$$

c. From Figure 16, we see that the fourth row of Pascal's triangle consists of the numbers 1, 4, 6, 4, and 1. Thus, we have

$$(1 + \sqrt{2})^4 = 1 \cdot 1^4 + 4 \cdot 1^3 \cdot \sqrt{2} + 6 \cdot 1^2 \cdot (\sqrt{2})^2 + 4 \cdot 1 \cdot (\sqrt{2})^3 + 1 \cdot (\sqrt{2})^4$$
$$= 1 + 4\sqrt{2} + 6 \cdot 2 + 4 \cdot 2\sqrt{2} + 4$$
$$= 17 + 12\sqrt{2}$$

UNDERSTANDING AND MASTERY CHECKLISTS

CONCEPTS TO UNDERSTAND

☐ Binomial
☐ Binomial coefficient
☐ Binomial Theorem
☐ Symmetry property of binomial coefficients
☐ Pascal's triangle

SKILLS TO MASTER

☐ Compute a binomial coefficient using combinations.
☐ Find the coefficient of a specified term of a binomial expansion.
☐ Use the Binomial Theorem to expand a binomial.
☐ Use Pascal's triangle to find binomial coefficients.

EXERCISES 8.5

EXERCISES 1–12 *Find the coefficient of the indicated term in the expansion of the binomial.*

1. $(A + B)^4$; AB^3
2. $(C - D)^3$; CD^2
3. $(x + y)^5$; x^3y^2
4. $(x + y)^6$; x^2y^4
5. $(z + 1)^6$; z^5
6. $(x + 2)^3$; x^3
7. $(2x - 3y)^3$; xy^2
8. $(2a - b)^6$; a^3b^3
9. $(x + \frac{1}{x})^4$; the constant term
10. $(z - \frac{1}{z})^4$; z^2
11. $(1 + w^2)^5$; w^4
12. $(x + 2w^3)^4$; xw^9

EXERCISES 13–30 *Expand the given expression.*

13. $(x + y)^5$
14. $(a - b)^5$
15. $(2x - y)^4$
16. $(3a + b)^4$
17. $(3x + 5y)^3$
18. $(u - 2w)^4$
19. $(x + 1)^5$
20. $(2s + 1)^4$
21. $(10w - 1)^6$
22. $(z - 2)^6$
23. $(1 + \sqrt{2})^3$
24. $(3 + \sqrt{3})^4$
25. $(\sqrt{2} + \sqrt{3})^4$
26. $(\sqrt{2} - \sqrt{3})^4$
27. $(x^2 - 1)^3$
28. $(y^3 + 1)^3$
29. $(x + \frac{1}{x})^5$
30. $(y - \frac{1}{y})^4$

EXERCISES 31–36 *Simplify. (Recall that $i^2 = -1$.)*

31. $(1 + i)^3$
32. $(2 - 3i)^3$
33. $(2 + 2i)^4$
34. $(1 - 3i)^4$
35. $\left(\frac{\sqrt{2}}{2} + \frac{\sqrt{2}}{2}i\right)^2$
36. $\left(\frac{3}{5} + \frac{4}{5}i\right)^3$

EXERCISES 37–44 *Verify the given identity involving factorials or binomial coefficients.*

37. $n! \cdot n = (n + 1)! - n!$
38. $(n^2 - n)(n - 2)! = n!$
39. $\dfrac{(n + 1)!}{n!} = n + 1$
40. $\dfrac{(n - 1)!}{(n + 1)!} = \dfrac{1}{n^2 + n}$
41. $\dbinom{n - 1}{k} + \dbinom{n - 1}{k - 1} = \dbinom{n}{k}$
42. $\dbinom{n}{r} = \dfrac{n - r + 1}{r}\dbinom{n}{r - 1}$
43. $\dbinom{n}{r} = \dfrac{n}{n - r}\dbinom{n - 1}{r}$
44. $n\dbinom{n - 1}{r} = (r + 1)\dbinom{n}{r + 1}$

■ CONCEPTS AND CRITICAL THINKING

EXERCISES 45–48 *Answer true or false.*

45. The symbols $\dbinom{n}{r}$ and $_nP_r$ have the same meaning.
46. $\dbinom{1,000,000}{999,998} = \dbinom{1,000,000}{2}$
47. All binomial coefficients can be found within Pascal's triangle.
48. The symbols $\dbinom{n}{r}$ and $_nC_r$ have the same meaning.

EXERCISES 49–52 *Give an example of each.*

49. Another context in which binomial coefficients arise (other than in expanding powers of binomials)
50. Two numbers x and y such that $(x + y)^2 = x^2 + y^2$.
51. A method for evaluating binomials that does not involve computation of factorials
52. A binomial coefficient $\dbinom{n}{r}$ that evaluates to 17

53. An extremely common error is for students to simplify $(x + y)^n$ as $x^n + y^n$. Why do you think this error is so common? Are there any values of x and y such that $(x + y)^n = x^n + y^n$ for all n? Graph the set of points (x, y) such that $(x + y)^2 = x^2 + y^2$.

54. Explain how
$$(x + y)^4 - 4(x + y)^3 y + 6(x + y)^2 y^2 - 4(x + y) y^3 + y^4$$
can be simplified within 10 seconds.

55. Write 11^7 as $(10 + 1)^7$, and explain how the Binomial Theorem can be used to compute 11^7 without using a calculator. Will this technique work for other similar computations? If so, give some examples. If not, explain why not.

▨ QUESTIONS FOR DISCUSSION OR ESSAY

56. Under what circumstances is Pascal's triangle easier to use than the Binomial Theorem?

57. There was a time when nearly every high school student studying algebra would learn the Binomial Theorem. Recently, however, an increasing number of college students have never been ex-posed to the Binomial Theorem. In fact, many instructors using this text will not include this section in their course. How important do you feel this topic is? Why might it have been considered more important 40 years ago than it is today?

▨ PROJECTS FOR ENRICHMENT

58. *The Extended Binomial Theorem*   The Binomial Theorem tells us how to expand expressions of the form $(a + b)^n$, where n is a positive integer. In this project, we consider a generalization of the Binomial Theorem to the case where n is not an integer. We begin by extending our definition of binomial coefficients. For a fixed integer k with $0 \le k \le n$, the expression $\binom{n}{k}$ is a polynomial in the variable n. For example,
$$\binom{n}{2} = \frac{n(n - 1)}{2}$$

Clearly, this polynomial is defined whether n is an integer or not. Thus, for example, we could define
$$\binom{\frac{1}{2}}{2} = \frac{\frac{1}{2}(\frac{1}{2} - 1)}{2} = -\frac{1}{8}$$

Similarly, we define, for any real number n
$$\binom{n}{3} = \frac{n(n - 1)(n - 2)}{3 \cdot 2 \cdot 1},$$
$$\binom{n}{4} = \frac{n(n - 1)(n - 2)(n - 3)}{4 \cdot 3 \cdot 2 \cdot 1}, \cdots$$

a. Compute each of the following extended binomial co-efficients.

   i. $\binom{\frac{1}{3}}{2}$ **ii.** $\binom{\frac{3}{2}}{3}$

   iii. $\binom{\frac{6}{5}}{4}$

Using this extended notion of a binomial coefficient, it can be shown that if $\left| \frac{b}{a} \right| < 1$, then
$$(a + b)^n = a^n + \binom{n}{1}a^{n-1}b + \binom{n}{2}a^{n-2}b^2$$
$$+ \binom{n}{3}a^{n-3}b^3 + \binom{n}{4}a^{n-4}b^4 + \cdots$$

Notice here that if n is not a natural number, the expansion has an *infinite* number of terms of the form
$$\binom{n}{k}a^{n-k}b^k$$

As an example, consider the binomial expansion for $(a + b)^{1/2}$.
$$(a + b)^{1/2} = a^{1/2} + \binom{\frac{1}{2}}{1}a^{-1/2}b + \binom{\frac{1}{2}}{2}a^{-3/2}b^2$$
$$+ \binom{\frac{1}{2}}{3}a^{-5/2}b^3 + \cdots$$
$$= a^{1/2} + \frac{1}{2}a^{-1/2}b - \frac{1}{8}a^{-3/2}b^2 + \frac{1}{16}a^{-5/2}b^3 + \cdots$$

b. Find the first four terms of the expansion for the given bi-nomial expression.

   i. $(a + b)^{1/3}$ **ii.** $(a + b)^{3/2}$

   iii. $(a + b)^{6/5}$

One application of the extended Binomial Theorem is approxi-mating roots, such as $\sqrt{10}$. We first write 10 as the sum of a perfect square and a "small" number—namely, $9 + 1$. Now, if we set $a = 9$ and $b = 1$ in the binomial expansion of $(9 + 1)^{1/2}$, we obtain
$$(9 + 1)^{1/2} = (9)^{1/2} + \frac{1}{2}(9)^{-1/2}(1) - \frac{1}{8}(9)^{-3/2}(1)^2$$
$$+ \frac{1}{16}(9)^{-5/2}(1)^3 + \cdots$$
$$= 3 + \frac{1}{2} \cdot \frac{1}{3} - \frac{1}{8} \cdot \frac{1}{27} + \frac{1}{16} \cdot \frac{1}{243} + \cdots$$

Taking only the first four terms, we thus have
$$\sqrt{10} \approx 3 + \frac{1}{6} - \frac{1}{216} + \frac{1}{3888} \approx 3.162$$

Note that a and b were chosen so that a was a perfect square. If we were to approximate $(a + b)^{1/3}$ using this procedure, we would want a to be a perfect cube. In addition, it is necessary for a and b to be such that $\left|\frac{b}{a}\right| < 1$. Finally, as we will see shortly, the accuracy of the approximation is much better if b is small compared to a.

c. Use the first four terms of an appropriate binomial expansion to approximate the following roots.

    i. $\sqrt{26}$ **ii.** $\sqrt{102}$

    iii. $\sqrt[3]{220}$

Any square root between 9 and 16 can be approximated by writing it in the form $(9 + x)^{1/2}$ and then expanding, as we just did. For example, we can approximate $\sqrt{14}$ by expanding $(9 + 5)^{1/2}$. However, if we use only the first four terms of the expansion, the accuracy may not be very good. To see this, consider the expansion of $(9 + x)^{1/2}$, as follows:

$$(9 + x)^{1/2} = (9)^{1/2} + \frac{1}{2}(9)^{-1/2}(x) - \frac{1}{8}(9)^{-3/2}(x)^2$$
$$+ \frac{1}{16}(9)^{-5/2}(x)^3 + \cdots$$
$$= 3 + \frac{1}{6}x - \frac{1}{216}x^2 + \frac{1}{3888}x^3 + \cdots$$

The first four terms give us the polynomial

$$p(x) = 3 + \frac{1}{6}x - \frac{1}{216}x^2 + \frac{1}{3888}x^3$$

d. Use a graphing calculator to plot the graphs of

$$f(x) = \sqrt{9 + x}$$

and

$$p(x) = 3 + \frac{1}{6}x - \frac{1}{216}x^2 + \frac{1}{3888}x^3$$

on the same coordinate axes. Then use the trace feature to determine the values of x for which $p(x)$ approximates $f(x)$ to at least one decimal place.

59. *Finding the Number of Subsets of a Set*   A set with 3 elements, such as $S = \{a, b, c\}$, has 2^3 subsets. The subsets of S are $\{\ \}$ (the empty set), $\{a\}$, $\{b\}$, $\{c\}$, $\{a, b\}$, $\{a, c\}$, $\{b, c\}$, and $\{a, b, c\}$. In this project, we demonstrate that a set with n elements has 2^n subsets.

a. Recall from Section 8.3 that $\binom{n}{k}$ represents the number of subsets of size k of a set with n elements. Show that the *total* number of subsets of a set with n elements is given by

$$\binom{n}{0} + \binom{n}{1} + \binom{n}{2} + \cdots + \binom{n}{n-1} + \binom{n}{n}$$

b. Now use the Binomial Theorem to express the sum in part a as a binomial to the nth power.

c. Finally, simplify the expression obtained in part b to show that the number of subsets of a set with n elements is 2^n.

SECTION 8.6 MATHEMATICAL INDUCTION

☐ **What does a cascading line of dominoes have in common with a mathematical technique for proving statements involving the natural numbers?**

☐ **How can a statement such as $n^n \geq n!$ be shown to be true for *every* natural number n?**

☐ **How can it be shown that any natural number debt of \$8 or greater can be paid using only \$3 and \$5 bills?**

☐ **How can the incorrect use of a mathematical technique be used to prove that all horses are the same color?**

THE PRINCIPLE OF MATHEMATICAL INDUCTION

Many important mathematical formulas and theorems involving the natural numbers— that is, the numbers 1, 2, 3, . . .—can be proved using a technique known as *mathematical induction*. Before we give a formal statement of mathematical induction, let us consider the following nonmathematical examples of induction.

■ A group of first graders are standing in a line. It is known that any first grader who gets pushed will push the first grader in front of him. It is also known that a second-grade bully pushes the first grader at the back of the line. We can thus conclude that every first grader in the line gets pushed.

- Dominoes are lined up in such a way that if a domino falls, it will knock over the domino in front of it. The first domino is pushed over. It then follows that all of the dominoes will fall.

- Sufferers of a certain genetic disorder will pass the disorder along to all of their children. It is known that Stan Edbury has this disorder. We can thus conclude that all of Stan Edbury's descendents will have the disorder.

Notice how in each of these examples we conclude that every member of a list (a line of first graders, a row of dominoes, the descendents of Stan Edbury) satisfies a certain property (being pushed, falling down, suffering from a genetic disorder) provided that two conditions are satisfied.

1. The property is possessed by the first member of the list.

2. If a member of the list has the property, then the next member of the list will have the property as well.

Clearly, if either of the conditions is not satisfied, then our conclusion is invalid. For example, if the dominoes are spaced sufficiently far apart that a falling domino *doesn't* knock over the next domino, then even if the first domino is toppled, it doesn't follow that all of the dominoes will fall. Likewise, even if our genetic disorder is inherited by 100% of one's offspring, if Stan Edbury himself doesn't have the disorder, then we cannot conclude that all of his descendents will have the disorder.

Mathematical induction is the application of this line of reasoning to statements made about the natural numbers. If we can show that a statement is true for the first natural number (the number 1) and that the truth of the statement for a natural number implies the truth of the statement for the next natural number, then we can conclude that the statement is, in fact, true for all natural numbers. More formally, we have the following.

Principle of □
Mathematical Induction

> If a statement about the natural numbers is true for the number 1, and if the truth of the statement for the natural number k implies the truth of the statement for $k + 1$ (the next natural number), then the statement is true for all natural numbers.

Proofs by mathematical induction consist of two steps.

Step 1 Show that the statement is true for $n = 1$.

Step 2 Show that if the statement is true for a natural number k, then it is true for $k + 1$.

E X A M P L E   1   ■   **Establishing a Formula by Mathematical Induction**

Use mathematical induction to prove that

$$1 + 2 + 3 + \cdots + n = \frac{n(n + 1)}{2}$$

for all natural numbers n.

SOLUTION

STEP 1:   For $n = 1$, the statement reduces to

$$1 = \frac{1(1 + 1)}{2}$$

$$1 = 1$$

which is clearly true.

STEP 2:   Assume that the statement is true for the natural number k; that is, assume that

$$1 + 2 + 3 + \cdots + k = \frac{k(k + 1)}{2} \qquad \text{Replacing } n \text{ with } k \text{ in the original statement}$$

We must show that this implies the truth of the statement for $k + 1$; that is, we must show that

$$1 + 2 + \cdots + k + (k + 1) = \frac{(k + 1)[(k + 1) + 1]}{2} \qquad \begin{array}{l}\text{Replacing } n \text{ with } k + 1 \\ \text{in the original statement}\end{array}$$

We proceed as follows:

$$1 + 2 + \cdots + k + (k + 1) = (1 + 2 + \cdots + k) + (k + 1)$$

$$= \frac{k(k + 1)}{2} + (k + 1) \qquad \begin{array}{l}\text{Using the assumption that} \\ 1 + 2 + \cdots + k = \\ \dfrac{k(k + 1)}{2}\end{array}$$

$$= \frac{k(k + 1)}{2} + \frac{(k + 2)}{2}$$

$$= \frac{(k + 1)(k + 2)}{2}$$

$$= \frac{(k + 1)[(k + 1) + 1]}{2}$$

We can thus conclude from the Principle of Mathematical Induction that

$$1 + 2 + \cdots + n = \frac{n(n + 1)}{2}$$

for all natural numbers n.

EXAMPLE 2   ■   **Establishing Divisibility by Mathematical Induction**

Use mathematical induction to prove that $n^3 - n + 3$ is divisible by 3 for all natural numbers n.

SOLUTION

STEP 1:   For $n = 1$, our statement reads "$1^3 - 1 + 3$ is divisible by 3," which is true since $1^3 - 1 + 3 = 3$

STEP 2:   Assume that the statement is true for the natural number k. That is, assume $k^3 - k + 3$ is divisible by 3. We must show that $(k + 1)^3 - (k + 1) + 3$ is divisible by 3 also. We start by rewriting $(k + 1)^3 - (k + 1) + 3$.

$$(k + 1)^3 - (k + 1) + 3 = (k^3 + 3k^2 + 3k + 1) - k - 1 + 3$$
$$= (k^3 - k + 3) + (3k^2 + 3k)$$
$$= (k^3 - k + 3) + 3(k^2 + k)$$

Now $3(k^2 + k)$ is divisible by 3, and we are assuming that $k^3 - k + 3$ is divisible by 3. Thus, their sum, $(k^3 - k + 3) + 3(k^2 + k)$, is divisible by 3 also.

We can thus conclude from the Principle of Mathematical Induction that $n^3 - n + 3$ is divisible by 3 for all natural numbers n.

EXAMPLE 3   ■   **Establishing an Inequality by Mathematical Induction**

Use mathematical induction to prove that $n^n \geq n!$ for all natural numbers n.

SOLUTION
STEP 1:   For $n = 1$, we have $1^1 \geq 1!$ or $1 \geq 1$, which is true.

STEP 2:   Assume that the statement is true for the natural number k; that is, assume that $k^k \geq k!$. It must be shown that

$$(k + 1)^{k+1} \geq (k + 1)!$$

We proceed as follows:

$$
\begin{aligned}
(k + 1)^{k+1} &= (k + 1)^k(k + 1) \\
&> k^k(k + 1) && \text{Using the fact that } (k + 1)^k > k^k \\
&\geq k!(k + 1) && \text{Using the assumption that } k^k \geq k! \\
&= (k + 1)! && \text{Since } (k + 1)! = (k + 1) \cdot k!
\end{aligned}
$$

So $(k + 1)^{k+1} \geq (k + 1)!$.

Thus, we can conclude that $n^n \geq n!$ for all natural numbers n.

GENERALIZED PRINCIPLE OF MATHEMATICAL INDUCTION

Let us return to the scenario described earlier, in which dominoes are lined up so that falling is contagious: When one domino falls, the next one falls also. This time, however, instead of upsetting the *first* domino, we knock over the fifth. What happens? Well, we can't say anything about the first four dominoes, but since the fifth will knock over the sixth and the sixth will knock over the seventh, and so forth, we can conclude that all dominoes from the fifth on will fall. This line of reasoning suggests the following generalization of the Principle of Mathematical Induction.

Generalized Principle of ☐
Mathematical Induction

> If a statement about the natural numbers is true for n_0 and if the truth of the statement for any natural number $k \geq n_0$ implies the truth of the statement for $k + 1$, then the statement is true for all natural numbers greater than or equal to n_0.

The Generalized Principle of Mathematical Induction is invaluable when proving statements that do not hold true for small natural numbers n, but rather are true from some point on. As with the ordinary Principle of Mathematical Induction, we prove a statement using the Generalized Principle of Mathematical Induction by proceeding in two steps. In step 1, we show that the statement is true for a particular natural number n_0, which is usually stated in the problem. In step 2, we assume that the statement is true for some natural number k greater than n_0, and we then show that this implies the statement is also true for $k + 1$.

EXAMPLE 4 ■ Applying the Generalized Principle of Mathematical Induction

Show that $n! > 20n$ for $n \geq 5$.

SOLUTION

STEP 1: When $n = 5$, we have $n! = 5! = 120$, and $20n = 20 \cdot 5 = 100$. Since $120 > 100$, the statement is true for $n = 5$.

STEP 2: Now assume that the statement is true for some natural number $k > 5$; that is, that $k! > 20k$ for some $k > 5$. We must show that $(k + 1)! > 20(k + 1)$.

$$
\begin{aligned}
(k + 1)! &= (k + 1) \cdot k! \\
&> (k + 1) \cdot 20k \qquad \text{Using the assumption that } k! > 20k \\
&= k \cdot 20(k + 1) \\
&> 20(k + 1) \qquad \text{Since } k > 1
\end{aligned}
$$

Thus, $(k + 1)! > 20(k + 1)$.

We can thus conclude from the Generalized Principle of Mathematical Induction that $n! > 20n$ for all $n \geq 5$.

UNDERSTANDING AND MASTERY CHECKLISTS

CONCEPTS TO UNDERSTAND

☐ Principle of Mathematical Induction

☐ Generalized Principle of Mathematical Induction

SKILLS TO MASTER

☐ Use the Principle of Mathematical Induction to prove a statement about natural numbers.

☐ Use the Generalized Principle of Mathematical Induction to prove a statement about natural numbers.

EXERCISES 8.6

EXERCISES 1–24 *Use the Principle of Mathematical Induction to show that the statement is true for all natural numbers n.*

1. $1 + 3 + 5 + 7 + 9 + \cdots + (2n + 1) = (n + 1)^2$

2. $2 + 4 + 6 + 8 + \cdots + (2n) = n(n + 1)$

3. $1^2 + 2^2 + 3^2 + \cdots + n^2 = \dfrac{n(n + 1)(2n + 1)}{6}$

4. $1^3 + 2^3 + 3^3 + \cdots + n^3 = \dfrac{n^2(n + 1)^2}{4}$

5. $5 + 10 + 15 + \cdots + 5n = \dfrac{5n(n + 1)}{2}$

6. $6 + 10 + 14 + \cdots + (4n + 2) = 2n(n + 2)$

7. $1 + 2 + 2^2 + 2^3 + \cdots + 2^n = 2^{n+1} - 1$

8. $1 - 2 + 2^2 - 2^3 + \cdots - 2^{2n-1} = \dfrac{1 - 2^{2n}}{3}$

9. $1 - 2 + 2^2 - 2^3 + \cdots - 2^{2n-1} + 2^{2n} = \dfrac{1 + 2^{2n+1}}{3}$

10. $3 + 3^2 + 3^3 + \cdots + 3^n = \dfrac{3^{n+1} - 3}{2}$

11. $\dfrac{1}{1 \cdot 2} + \dfrac{1}{2 \cdot 3} + \dfrac{1}{3 \cdot 4} + \cdots + \dfrac{1}{n(n + 1)} = 1 - \dfrac{1}{n + 1}$

12. $\dfrac{1}{1 \cdot 3} + \dfrac{1}{3 \cdot 5} + \dfrac{1}{5 \cdot 7} + \cdots + \dfrac{1}{(2n - 1) \cdot (2n + 1)} = \dfrac{n}{2n + 1}$

13. $1 \cdot 2 + 3 \cdot 4 + 5 \cdot 6 + \cdots + (2n - 1) \cdot 2n = \dfrac{n(n + 1)(4n - 1)}{3}$

14. $1 \cdot 2 + 2 \cdot 3 + 3 \cdot 4 + \cdots + n \cdot (n + 1) = \dfrac{n(n + 1)(n + 2)}{3}$

15. $1 \cdot 2^1 + 2 \cdot 2^2 + 3 \cdot 2^3 + \cdots + n \cdot 2^n = 2^{n+1}(n - 1) + 2$

16. $\dfrac{1}{2} + \dfrac{3}{2^2} + \dfrac{5}{2^3} + \cdots + \dfrac{2n - 1}{2^n} = 3 - \dfrac{2n + 3}{2^n}$

17. $n^2 + n$ is even.

18. $n^2 - n + 1$ is odd.

19. $n^3 - n$ is divisible by 3.

20. $n^3 + 2n$ is divisible by 3.

21. $n \leq 2^{n-1}$

22. $1 + 2n \leq 3^n$

23. If $0 \leq a < 1$, then $a^n < 1$.

24. If $a > 1$, then $a^n > 1$.

EXERCISES 25–30 *Use the Generalized Principle of Mathematical Induction to show that the statement is true.*

25. $2^n < n!$ for $n \geq 4$

26. $4n < 2^n$ for $n \geq 5$

27. $n + 12 \leq n^2$ for $n \geq 4$

28. $n^2 + 18 \leq n^3$ for $n \geq 3$

29. $n^2 + 4 < (n + 1)^2$ for $n \geq 2$

30. $n^3 > (n + 1)^2$ for $n \geq 3$

EXERCISES 31–34 *These exercises deal with the Fibonacci sequence defined in Section 8.1 by $F_1 = 1$, $F_2 = 1$, and $F_n = F_{n-1} + F_{n-2}$ for $n > 2$. Use the Principle of Mathematical Induction (or the Generalized Principle) to prove the given property of the Fibonacci sequence.*

31. $F_1 + F_2 + \cdots + F_n = F_{n+2} - 1$

32. $F_2 + F_4 + \cdots + F_{2n} = F_{2n+1} - 1$

33. $F_1 + F_3 + \cdots + F_{2n-1} = F_{2n}$

34. $F_1^2 + F_2^2 + F_3^2 + \cdots + F_n^2 = F_n F_{n+1}$

CONCEPTS AND CRITICAL THINKING

EXERCISES 35–38 *Answer true or false.*

35. When proving statements by mathematical induction, we actually assume that what we are trying to prove is true.

36. If we prove that a statement is true for $n = 7$ and we prove that if it is true for $n = k$ then it is also true for $n = k + 1$, then we have proved that the statement is true for all positive integers.

37. If we prove that a statement is true for $n = 7$ and we prove that if it is true for $n = k$ then it is also true for $n = k + 1$, then we have proved that the statement is true for all positive integers greater than or equal to 7.

38. If it is the case that whenever a certain statement is true for $n = k$ it is also true for $n = k + 1$, then it follows that the statement is true for all positive integers n.

EXERCISES 39–42 *Give an example of each.*

39. A statement about the natural numbers that is true for $n = 1$ and 2 but is false for $n = 3$

40. A statement about the integers that is true for integers $n \geq 3$ but is false for $n = 1$ and 2

41. An alternative method (other than the induction proof given in Exercise 17) for establishing that $n^2 + n$ is even for all natural numbers n

42. An alternative method (other than the induction proof given in Exercise 1) for establishing that

$$1 + 3 + 5 + 7 + 9 + \cdots + (2n + 1) = (n + 1)^2$$

QUESTIONS FOR DISCUSSION OR ESSAY

43. What is wrong with the following proof that all odd numbers are even?

    Any odd number can be written in the form $2n + 1$. Thus, we must show that all numbers of the form $2n + 1$ are even. Suppose that the statement is true for n. Then $2n + 1$ is even and hence divisible by 2. There is, therefore, a natural number m such that $2n + 1 = 2 \cdot m$. Now we must show that $2 \cdot (n + 1) + 1 = 2n + 3$ is even as well. But $2n + 3 = (2n + 1) + 2 = 2 \cdot m + 2 = 2 \cdot (m + 1)$. So $2n + 3$ is divisible by 2 and is thus

    even. By induction, we have shown that all odd numbers are even.

44. In this exercise, we investigate a mysterious use of mathematical induction that seemingly proves that all horses are the same color! The "proof" of the statement *All horses are of the same color* is as follows:

    Suppose there is 1 horse. Obviously, it is of one color. Now suppose it is true that any group of n horses have the same color.

We will demonstrate that any group of $n + 1$ horses will also be of one color, as follows.

Consider a group of $n + 1$ horses. Remove 1 horse (whom we will call Silver) from the group so that n horses are left. By our earlier assumption, all n of these horses are the same color. We need only show that Silver is the same color as the rest of the horses.

To do this, return Silver to the group, and remove a different horse so that again n horses are left. By hypothesis, these n horses are all colored the same. Since Silver is one of the horses and the rest of the horses are part of the original group of n horses that had the same color, Silver is the same color as the other horses. Thus, by induction, all horses are the same color.

Since the theorem is obviously false, there must be an error somewhere in the proof. Explain the error in detail.

45. It has been said that mathematical induction is not a method for discovering mathematical statements but a technique for rigorously proving a statement that has already been discovered. Do you agree? Why or why not?

46. Inductive reasoning is a method of reasoning whereby specific examples lead to a general conclusion. An example would be the argument, "Trevor studied hard and got an A on the test; LaShonda studied hard and got an A on the test; therefore, everyone who studies hard will get an A on the test." A mathematical example would be the argument "$5^2 < 2^5$, $6^2 < 2^6$, and $7^2 < 2^7$; therefore, $n^2 < 2^n$ for all natural numbers n." What do you see as a major problem with inductive reasoning? Under what circumstances is inductive reasoning useful? What is the connection between inductive reasoning and mathematical induction?

47. It is obviously not legitimate to prove a theorem by assuming that which is to be proven. For example, consider the following "proof" that all mathematicians are males.

Suppose that all mathematicians are men. Let Pat be a mathematician. Since all mathematicians are men, Pat is a man. The same will be true of all other mathematicians. Thus, all mathematicians are men.

Of course, the reasoning is absurd, but isn't this precisely the same reasoning that is used when, in the course of a proof by mathematical induction, we assume that the statement is true for k? If not, how is it different?

■ PROJECTS FOR ENRICHMENT

48. *Proving the Binomial Theorem* In this project, we use the technique of mathematical induction to establish the Binomial Theorem:

$$(a + b)^n = \binom{n}{0}a^n + \binom{n}{1}a^{n-1}b + \binom{n}{2}a^{n-2}b^2$$
$$+ \cdots + \binom{n}{n-1}ab^{n-1} + \binom{n}{n}b^n$$

a. Show that the Binomial Theorem is true for $n = 1$.

b. Assume that the Binomial Theorem is true for $n = k$. By writing $(a + b)^{k+1}$ as $(a + b)^k(a + b)$, show that

$$(a + b)^{k+1} = \binom{k}{0}a^{k+1} + \binom{k}{1}a^k b + \binom{k}{2}a^{k-1}b^2$$
$$+ \cdots + \binom{k}{k-1}a^2 b^{k-1} + \binom{k}{k}ab^k$$
$$+ \binom{k}{0}a^k b + \binom{k}{1}a^{k-1}b^2 + \binom{k}{2}a^{k-2}b^3$$
$$+ \cdots + \binom{k}{k-1}ab^k + \binom{k}{k}b^{k+1}$$

c. Combine like terms in the expression obtained in part b. Then use this result to show that the coefficient of $a^{(k+1)-j}b^j$ in the expansion of $(a + b)^{k+1}$ is given by

$$\binom{k}{j} + \binom{k}{j-1}$$

d. Use the fact that

$$\binom{k}{j} + \binom{k}{j-1} = \binom{k+1}{j}$$

to complete the induction proof.

49. *Paying Debts with $3 and $5 Bills* In this project, we show that any $\$n$ debt, where n is a natural number greater than or equal to 8, can be paid using only $3 and $5 bills.

a. Prove by induction that any natural number debt of the form $3n + 5$, where $n \geq 1$, can be paid using only $3 and $5 bills.

b. Prove by induction that any natural number debt of the form $3n + 6$, where $n \geq 1$, can be paid using only $3 and $5 bills.

c. Prove by induction that any natural number debt of the form $3n + 7$, where $n \geq 1$, can be paid using only $3 and $5 bills.

d. Show that the results of parts a through c imply that any natural number debt greater than or equal to $8 can be paid using only $3 and $5 bills.

e. There is a slightly more general version of the induction principle that states that if a statement is true for a natural number k, and if it can be shown that the truth of the statement for all natural numbers between k and n implies the truth of the statement for $n + 1$, then the statement is true for all natural numbers $n \geq k$. Use this more general induction principle to prove that all $\$n$ debts can be paid using only $3 and $5 bills, provided that n is a natural number greater than or equal to 8.

CHAPTER 8 REVIEW

EXERCISES 1–8 *List the first five terms of the sequence with the given nth term. If the sequence is arithmetic, indicate its common difference. If it is geometric, indicate its common ratio.*

1. $u_n = 1 - 5n$

2. $a_n = n + \dfrac{(-1)^n}{n}$

3. $c_n = ne^{-n}$

4. $b_n = \dfrac{n + 3}{2}$

5. $x_n = \dfrac{1}{2} + (-1)^n$

6. $a_n = \dfrac{2^n}{3^n}$

7. $a_1 = 2$ and $a_n = 3a_{n-1}$ for $n \geq 2$

8. $u_1 = 256$ and $u_n = \sqrt{u_{n-1}}$ for $n \geq 2$

EXERCISES 9–16 *Find an expression for the nth term of a sequence whose first few terms are given. If the sequence is arithmetic, indicate its common difference. If it is geometric, indicate its common ratio.*

9. $2, 10, 50, 250, \ldots$

10. $e, 2e^2, 3e^3, 4e^4, \ldots$

11. $1, \frac{1}{2}, \frac{1}{3}, \frac{1}{4}, \ldots$

12. $-3, 2, 7, 12, \ldots$

13. $1, -\frac{3}{2}, -4, -\frac{13}{2}, \ldots$

14. $24, -6, \frac{3}{2}, -\frac{3}{8}, \ldots$

15. $1, -8, 27, -64, \ldots$

16. $-\frac{1}{2}, \frac{2}{3}, -\frac{3}{4}, \frac{4}{5}, \ldots$

EXERCISES 17–22 *Use the given information about the sequence to find the indicated term.*

17. $a_1 = 3$, common difference $d = 4$; $a_{12} = $ _____

18. $a_2 = 1$, common ratio $r = 2$; $a_{10} = $ _____

19. $a_1 = 1$, $a_2 = 4$, the sequence is geometric; $a_6 = $ _____

20. $a_1 = 1$, $a_2 = 4$, the sequence is arithmetic; $a_6 = $ _____

21. $a_1 = 1$, $a_{n+1} = (a_n + 1)^2$; $a_4 = $ _____

22. $a_1 = 2$, $a_2 = 3$, $a_{n+2} = a_n + a_{n+1}$; $a_7 = $ _____

EXERCISES 23–26 *Find the indicated sum.*

23. $\displaystyle\sum_{k=1}^{4} (2k + 5)$

24. $\displaystyle\sum_{n=1}^{5} 3(2^n)$

25. $\displaystyle\sum_{n=2}^{6} \dfrac{1}{n^2}$

26. $\displaystyle\sum_{k=0}^{4} \dfrac{k}{k + 1}$

EXERCISES 27–30 *Express the series using summation notation.*

27. $[5(1) - 2] + [5(2) - 2] + [5(3) - 2] + \cdots + [5(11) - 2]$

28. $1^3 + 2^3 + 3^3 + \cdots + 16^3$

29. $4\left(\frac{2}{3}\right)^3 + 4\left(\frac{2}{3}\right)^4 + 4\left(\frac{2}{3}\right)^5 + 4\left(\frac{2}{3}\right)^6 + \cdots + 4\left(\frac{2}{3}\right)^{13}$

30. $\frac{1}{2} - \frac{1}{3} + \frac{1}{4} - \frac{1}{5} + \cdots - \frac{1}{15}$

EXERCISES 31–38 *Evaluate the given arithmetic or geometric series.*

31. The sum of the first 10 terms of the arithmetic sequence 2, 8, 14, 20, . . .

32. The sum of the first 11 terms of the geometric sequence 2, 8, 32, 128, . . .

33. $10 + \dfrac{10}{5} + \dfrac{10}{5^2} + \cdots + \dfrac{10}{5^6}$

34. $10 + 2 - 6 - 14 - \cdots - 62$

35. $\displaystyle\sum_{n=1}^{9} 5(0.3)^n$

36. $\displaystyle\sum_{k=1}^{12} (8 + 0.6k)$

37. $\displaystyle\sum_{k=3}^{10} (1000 - 9k)$

38. $\displaystyle\sum_{n=8}^{20} \frac{1}{3}(-3)^n$

EXERCISES 39–46 *Evaluate the given expression.*

39. $_9C_2$

40. $\dfrac{6!}{(6 - 2)!}$

41. $_8P_5$

42. $\dbinom{5}{4}$

43. $\dfrac{5!}{3!(5 - 3)!}$

44. $_{10}P_6$

45. $\dbinom{10}{3}$

46. $_7C_3$

EXERCISES 47–50 *A single card is drawn from a standard 52-card deck. Compute the probability of the given event.*

47. Drawing a diamond

48. Drawing a face card

49. Drawing a diamond or face card

50. Drawing a diamond face card

EXERCISES 51–54 *A pair of standard six-sided dice are rolled. Compute the probability of the given event.*

51. Rolling an even number on both dice

52. Rolling the number 6 on exactly one die

53. Rolling a total of 5

54. Rolling a total that is either odd or less than 5

EXERCISES 55–56 *Find the probability that in a family of 3 children, the indicated outcome will occur. Assume that the probability of a male birth is 0.5.*

55. All of the children are boys.

56. At least 2 of the children are boys.

EXERCISE 57–60 *Use Figure 17 to determine the probability that a randomly selected member of the armed services will satisfy the given criteria.*

57. Is in the Army

58. Is in the Marines

59. Is in the Marines or Navy

60. Is not in the Air Force

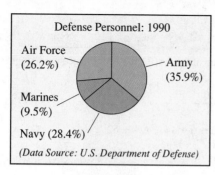

Defense Personnel: 1990

Air Force (26.2%)

Army (35.9%)

Marines (9.5%)

Navy (28.4%)

(Data Source: U.S. Department of Defense)

FIGURE 17

EXERCISES 61–64 *Find the coefficient of the indicated term in the expansion of the binomial.*

61. $(x + y)^8$; x^5y^3

62. $(a - b)^5$; a^3b^2

63. $(2a - 5b)^4$; ab^3

64. $(3 + v^2)^7$; v^6

EXERCISES 65–70 *Expand the given expression.*

65. $(x - y)^4$

66. $(2a + 3)^5$

67. $(r + \frac{1}{s})^3$

68. $(x^2 - 3y)^4$

69. $(\sqrt{z} + 2)^5$

70. $(4 + \sqrt{2})^3$

EXERCISES 71–76 *Use the Principle of Mathematical Induction to show that the statement is true for all natural numbers n.*

71. $1 + 4 + 7 + \cdots + (3n - 2) = \dfrac{n(3n - 1)}{2}$

72. $3 + 9 + 15 + \cdots + (6n - 3) = 3n^2$

73. $1^3 + 3^3 + 5^3 + \cdots + (2n - 1)^3 = n^2(2n^2 - 1)$

74. $\dfrac{1}{1 \cdot 4} + \dfrac{1}{4 \cdot 7} + \dfrac{1}{7 \cdot 10} + \cdots + \dfrac{1}{(3n - 2)(3n + 1)} = \dfrac{n}{3n + 1}$

75. $2^{2n} - 1$ is divisible by 3.

76. $5^n - 1$ is divisible by 4.

EXERCISES 77–80 *Express your answer first using permutation or combination notation, if possible, and then evaluate.*

77. A club consisting of 20 members wishes to elect a president, vice president, secretary, and treasurer. In how many ways can this be done if no member holds more than one office?

78. A club consisting of 20 members wishes to form a four-member committee. In how many ways can this be done?

79. Jermaine's CD wish list has 6 rock groups and 4 rap groups. Unfortunately, he has only enough money to buy 5 of the CDs on the list. Determine the number of different purchases he can make if

a. there are no other restrictions

b. he would like to buy 3 of the rock groups and 2 of the rap groups

80. Alia's CD player can hold 6 CDs at a time. Her CD collection consists of 10 classical and 8 jazz CDs. Determine the number of ways in which she can program her CD player to play 6 CDs if

a. there are no other restrictions

b. she would like to listen to 4 classical and 2 jazz CDs

81. *Exam Variations* A true-false portion of an exam consists of 10 questions. In how many different ways can this portion of the exam be completed if no questions are left blank?

82. *Exam Variations* A multiple-choice portion of an exam consists of 10 questions, each with 4 possible answers. In how many different ways can this portion of the exam be completed?

83. *Population Projection* The population of a certain city is increasing at a rate of 5% per year. Assuming a population of 200,000 in the year 2000 and the growth rate remains the same, find the population for the next 3 years, and show that these values form a geometric sequence. What will the population be in 2012?

84. *Car Price Inflation* The price of a certain model of car has increased 3% per year. If the price was initially $15,000, find the price for the next 3 years, and show that these values form a geometric sequence. What will the price of the car be in the 10th year?

85. *Income Projection* Suppose you have been offered a job with a starting salary of $26,000 and guaranteed annual raises of 6%. Use a geometric series to determine the total income for the first 10 years.

86. *Sales Projection* At a year-end stockholders' meeting, a company reports that sales for 2001 totaled $1.5 million. Based on a market analysis, the company predicts annual increases in sales of 8% through the year 2010. Use a geometric series to determine the total sales for the 10-year period from 2001 to 2010.

87. *Exam Possibilities* A mathematics instructor decides to write a final exam by choosing 20 of the 80 problems that appeared on exams during the semester. Use combination notation to write an expression for the number of final exams that are possible if

a. there are no additional restrictions

b. there were four exams during the semester, each one had 20 problems, and the final exam must have 5 problems from each exam

88. *Basketball Recruiting* A college basketball coach wishes to recruit 6 high school players to replace the current seniors on the team when they graduate. She has narrowed her list to a group of 15 prospective players, consisting of 4 guards, 6 forwards, and 5 centers. Determine the number of ways she can rank the top 6 players if

a. there are no restrictions on positions

b. she would like to recruit 2 guards, 3 forwards, and 1 center

89. *Game Show Probability* On a certain game show, you are given the 5 digits of the price of a car, but you must arrange them in the correct order to win the car. Find the probability that you will win the car if you order the digits at random. Assume no digits are repeated.

90. *Lottery Probability* In a certain state lottery, a player selects 4 numbers from 1 through 30. Find the probability that a single choice of 4 numbers will win the lottery, assuming the order of the numbers is not important.

CHAPTER 8 TEST

PROBLEMS 1–8 *Answer true or false.*

1. No sequence is both arithmetic and geometric.

2. Some sequences are neither arithmetic nor geometric.

3. A permutation takes order into account, whereas a combination does not.

4. If $n \geq r$, then the value of $_nC_r$ is a natural number.

5. If E is an event in a sample space S, then $0 < P(E) < 1$.

6. If E and F are events in the same sample space S, then $P(E \cup F) = P(E) + P(F) - P(E \cap F)$.

7. Each term in the expansion of $(x + y)^n$ has degree $n + 1$.

8. Mathematical induction can be used to prove that a statement is true for all real numbers.

PROBLEMS 9–14 *Give an example of each.*

9. A formula for an arithmetic sequence with common difference -4

10. A formula for a geometric sequence with common ratio $\frac{1}{3}$

11. A series that is neither arithmetic nor geometric

12. An experiment involving coins that has a sample space S with $n(S) = 8$

13. An event that has probability $\frac{1}{3}$

14. A row from Pascal's triangle with at least 5 numbers

15. Write out the first five terms of the sequence $a_n = 3n^2 + 1$.

16. Write an expression for the nth term of a sequence whose first few terms are $\frac{2}{1}, \frac{3}{2}, \frac{4}{3}, \frac{5}{4}, \ldots$

17. Evaluate $\displaystyle\sum_{k=1}^{7} (5k - 2)$.

18. Write the geometric series $\frac{5}{2} + \frac{5}{4} + \frac{5}{8} + \cdots + \frac{5}{64}$ using summation notation and find the sum.

19. How many different ways are there to answer the first 8 problems of this exam? What is the probability that you will get them all correct if you randomly assign answers?

20. A Little League baseball team consists of 7 second graders and 8 first graders. The coach must select 9 children to start the next game. Assuming the coach is not concerned with batting order or playing position, determine the number of ways this can be done if

   a. there are no restrictions

   b. there must be 5 second graders and 4 first graders on the field at all times

   How would your solutions in part a change if the coach were arranging the 9 children in a batting order?

21. A pair of standard six-sided dice are rolled. Find the probability that the sum is a multiple of 3.

22. A single card is drawn from a standard 52-card deck. Find the probability that the card is a black card or a face card.

23. Find the coefficient of z^4 in the expansion of $(z + 5)^9$.

24. Expand the expression $(2x - 3y)^4$.

25. Use the Principle of Mathematical Induction to show that $1 + 3 + 5 + \cdots + (2n - 1) = n^2$ for all natural numbers n.

ANSWERS

CHAPTER 1

Section 1.1 ■ *page 11*

1. Natural, whole, integer, rational, real, complex
3. Irrational, real, complex   5. Complex
7. Whole, integer, rational, real, complex
9. Rational, real, complex
11. Natural, whole, integer, rational, real, complex
13. Integer, rational, real, complex   15. Complex
17. Rational, real, complex   19. $>$ 21. $<$
23. $\geq$ 25. $-23, -22.9, 22.9, 23$
27. $3 + \frac{1}{10} + \frac{4}{100}, \pi, \frac{22}{7}, 3.15$
29. The number of miles from Earth to the sun, 1,000,000,000, the GNP of the United States in dollars, the number of living organisms on Earth
31. $(-2, 4]$![number line from −3 to 5 with open circle at −2 and bracket at 4]
33. $(-7 -\frac{1}{2})$![number line from −8 to 0]
35. $(-\infty, -3)$![number line from −7 to −1]
37. $[\sqrt{2}, \infty)$![number line from 0 to 8]
39. $\{x \mid 3 < x < 5\}$ 41. $\{x \mid 1 \leq x \leq 10^{100}\}$
43. $\{x \mid 1 \leq x \leq 2\}$ 45. $\{x \mid x > 100\}$
47. Answers may vary.   49. Answers may vary.
51. Answers may vary.   53. Answers may vary.
55. Answers may vary.   57. Answers may vary.
59. True   61. True   63. True

Section 1.2 ■ *page 21*

1.

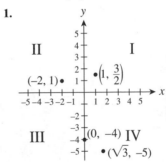

3.

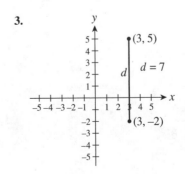

5.

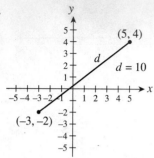

$d = 10$

7.

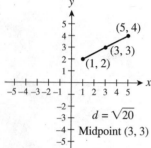

$d = \sqrt{20}$
Midpoint (3, 3)

9.

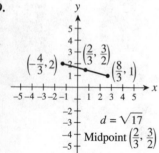

$d = \sqrt{17}$
Midpoint $\left(\frac{2}{3}, \frac{3}{2}\right)$

11.

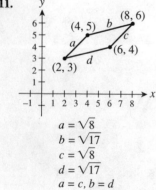

$a = \sqrt{8}$
$b = \sqrt{17}$
$c = \sqrt{8}$
$d = \sqrt{17}$
$a = c, b = d$

13.

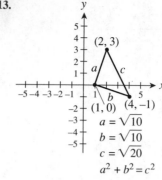

$a = \sqrt{10}$
$b = \sqrt{10}$
$c = \sqrt{20}$
$a^2 + b^2 = c^2$

15.

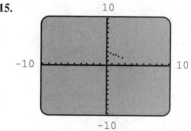

$y \approx 1.8$ (Answers may vary.)

17.

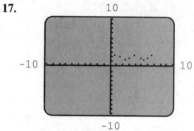

$y \approx 2.5$ (Answers may vary.)

19. (8, 0)

21. (2, 4)

23. Counterclockwise; $\sqrt{89} + \sqrt{104} + \sqrt{61} \approx 27.44$

25.

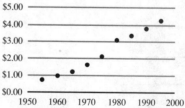

Approximately $4.50 in 1997 (answers may vary), which is $0.65 above the actual value of $5.15

27. $\sqrt{117} + 2 \approx 12.82$ blocks traveling on Lloyd Avenue from point A to the point (3, 4) and then right to point B; 17 blocks traveling from point A to the point (5, −2) and then to point B

29. In the middle of the fourth row of tiles, between the 4th and 5th tiles

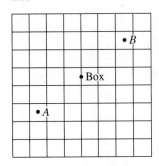

31. False   **33.** False

Section 1.3 ■ page 33

1. $a \cdot b = b \cdot a$

3. The product of a number and the sum of the two other numbers is the same as the sum of the products of the first number with each of the other two numbers.

5. $4 + 2i$ **7.** $4 - 2i$ **9.** $6 + 2i$ **11.** $-1 + i$

13. $5 + i$ **15.** $-\frac{1}{2}i$ **17.** $\frac{1}{2} - \frac{1}{2}i$ **19.** $2 - i$

21. $-4 + i$ **23.** $-\frac{3}{4} + 2i$ **25.** -7 **27.** -6.5

29. 16   **31.** 6   **33.** 0   **35.** 2   **37.** -1

39. $x - 1$ **41.** $-(y + 2)$ **43.** $|-3 - 7| = 10$

45. $\left|-\frac{1}{3} - \left(-\frac{2}{5}\right)\right| = \frac{1}{15}$ **47.** $|3x - 2|$

49. $\left|\left(x + \frac{2}{3}\right) - \left(-\frac{2}{3}\right)\right| = \left|x + \frac{4}{3}\right|$ **51.** \$9.60

53. a. \$5.67   **b.** \$5.67
   c. $0.1(3.50 + 2.80) = 0.1(3.50) + 0.1(2.80)$

55. False   **57.** True   **59.** False

Section 1.4 ■ page 47

1. 2^6 **3.** x^7 **5.** p^5q^2 **7.** -1 **9.** 16   **11.** $\frac{1}{9}$

13. $-\frac{1}{2}$ **15.** 3   **17.** 0.00001   **19.** $\frac{1}{1.44} \approx 0.6944$

21. -25 **23.** $\frac{49}{9}$ **25.** $\frac{1}{4096}$ **27.** 4096   **29.** 1

31. a^9 **33.** $2x^5$ **35.** $\frac{1}{r^2}$ **37.** y **39.** $\frac{a}{b^6}$

41. x^4y^6 **43.** p^6q^4 **45.** $\frac{ac^5}{b^4}$ **47.** $\frac{x + y}{x - y}$

49. $\frac{3x^6}{y^3}$ **51.** $\frac{z^3}{y^2}$ **53.** $\frac{4y^6z^{10}}{x^6}$ **55.** 0.00304

57. 300, 100   **59.** 7.5   **61.** 4×10^6

63. 9.43×10^{-2} **65.** 5.2×10^{13}

67. 2.031×10^4 **69.** 1×10^8 **71.** 1.3176×10^{25}

73. 7×10^{-6} **75.** \$148   **77.** \$58,558.82

79. The second option, since the first pays \$9.3 million for the whole month whereas the second pays \$10.7 million on the last day alone

81. a. Ben Franklin's deposit   **b.** Answers may vary.
   c. Answers may vary.

83. Approximately 3.19×10^{37} **85.** \$48.65

87. False   **89.** True

Section 1.5 ■ page 56

1. Trinomial; degree 2   **3.** Monomial; degree 4

5. Not a polynomial   **7.** Trinomial; degree 5

9. Not a polynomial   **11.** Monomial; degree 3

13. 33   **15.** $-\frac{15}{4}$ **17.** $\frac{7}{3} - \sqrt{2}$ **19.** 7

21. 0   **23.** -1 **25.** -64 **27.** $4x^2 + 3x - 8$

29. $-2y^3 - 2y^2 + 5y + 4$ **31.** $\frac{7}{2}x^2 + 4x$ **33.** $6a^2 - a - 2$

35. $-6y^2 + \frac{13}{2}y + 2$ **37.** $4t^2 - s^2$ **39.** $x^2 - 3$

41. $4x^4 - 1$ **43.** $t^2 + 8t + 16$ **45.** $9a^2 + 12ab + 4b^2$

47. $2x^3 - 11x^2 - 2x + 2$ **49.** $x^2 - y^2 + 4y - 4$

51. $n^3 + 6n^2 + 11n + 6$ **53.** $6a^2 + a - 7$

55. $a^3 - 3a^2b + 3ab^2 - b^3$ **57.** $3 - 2i$ **59.** $-6 + 17i$

61. 18   **63.** $-i$ **65.** i **67.** $z^2 + 4$ **69.** $x^2 + 5 + 12i$

71. $\frac{3(2x - 4)}{6} + 2$ simplifies to x.

73. $(x - y)(x + y) + y^2$ simplifies to x^2.   **75.** 455 cubic inches

77. 324 board feet; approximately 10,050 board feet

79. 64 feet; 96 feet; $t = 5$ seconds

81. Approximately 282.75 feet.

83. a. 6   **b.** 792 days (2 years and 62 days)

85. True   **87.** False

Section 1.6 ■ page 66

1. $4n(n^4 - 3n^2 + 6)$ **3.** $2ab(6a + 3b - 4)$

5. $(w + 2)(w + 5)$ **7.** $(a + 2)(9b + 5)$

9. $(4p + 3q)(2p - 3q)$ **11.** $(x - 6)(x + 6)$

13. $4(y - 2z)(y + 2z)$ **15.** $2(x - 2)(x + 2)(x^2 + 4)$

17. $(s - 3)(s^2 + 3s + 9)$ **19.** $(a + 2b)(a^2 - 2ab + 4b^2)$

21. $2(x + 4)(x^2 - 4x + 16)$ **23.** $(x - 4)(x + 3)$

25. $(y - 7)(y - 2)$ **27.** $(3m - 2)^2$ **29.** $(y + 3)(2y - 1)$

31. $2(t - 2)(t + 2)$ **33.** $(3n - 2)(2n - 5)$

35. $(z - 1)^2(z + 1)^2$ **37.** $(t - 2)(t + 2)(t^2 - 2t + 4)(t^2 + 2t + 4)$

39. $(p^2 - 5q)(p^4 + 5p^2q + 25q^2)$ **41.** $3(3x - 1)(9x^2 + 3x + 1)$

43. $(3x - y)(z + 2w)$ **45.** $(2x - 3y)^2$

47. $6(xy + 2)(x^2y^2 - 2xy + 4)$ **49.** 2 inches   **51.** False

53. False

Section 1.7 ■ page 76

1. $\frac{3a}{5c}$ **3.** $t^2(2t^2 - 5)$ **5.** $x - 2$ **7.** $\frac{r + 5}{2(r - 3)}$

9. y **11.** $x + y$ **13.** $\frac{9}{16x}$ **15.** $\frac{2a}{(a + 1)(a - 1)}$

17. $\frac{1 - 5z}{(z + 1)(z - 2)}$ **19.** $\frac{x - 2}{x + 1}$ **21.** $\frac{2}{x(x + 2)^2}$

23. $\frac{2(b - a)(a + 2b)}{a - 2b}$ **25.** $\frac{4t}{(t + 1)(t - 1)^2}$ **27.** $-\frac{1}{(r + s)^2}$

29. -1 **31.** $\frac{6}{t - 6}$ **33.** $\frac{b}{a}$ **35.** $-\frac{1}{3(h + 3)}$

37. $-\frac{1}{xc}$ **39.** $\frac{(x - 10)^2 - 5x}{x - 20} + 5$ simplifies to x if $x \neq 20$.

41. $\frac{19}{3}$ **43.** $\frac{abc + a + c}{bc + 1}$

45. a. $\frac{8}{3}$ ohms

47. a. 1.46×10^{-8} newton

  b. 1.59×10^{-8} newton

  c. The obstetrician

49. a. $\dfrac{45m - 495w - 495}{m}$ **b.** 5.4%

51. False   **53.** True

Section 1.8 ■ page 89

1. 3   **3.** $\frac{1}{2}$ **5.** $3\sqrt{2}$ **7.** $10\sqrt[4]{10}$

9. $243\sqrt[3]{9}$ **11.** $ab^2\sqrt{b}$ **13.** $2x^2yz\sqrt{7y}$

15. $5a^{15}b^{10}c^5\sqrt{2c}$ **17.** $x^3(y+z)^2\sqrt{x}$ **19.** $3s^2t^2\sqrt[3]{2t}$

21. $x + y$ **23.** $\sqrt[3]{x-y}$

25. $\dfrac{a^2}{b}$ **27.** $3\sqrt[3]{2}$ **29.** $7\sqrt{3}$ **31.** $\dfrac{3+\sqrt{5}}{2}$

33. $\sqrt{a} + \sqrt{b}$ **35.** $\dfrac{1}{3\sqrt{2}}$ **37.** $\dfrac{1}{\sqrt{x}+\sqrt{a}}$ **39.** $3\sqrt{5}$

41. $2xy\sqrt[3]{5y^2z^2}$ **43.** $\sqrt{5}$ **45.** $\dfrac{x\sqrt{y}}{y}$ **47.** $x - 9$

49. $3\sqrt{6}$ **51.** $\dfrac{5\sqrt{2}}{2}$ **53.** $\sqrt{a}\left(\dfrac{a+1}{a}\right)$ **55.** $\sqrt[6]{x}$

57. $x^{1/3}$ **59.** $17^{1/2}$ **61.** $a^{3/4}b^{3/2}$ **63.** $x^{-1/2}$

65. $5i$ **67.** $-\dfrac{\sqrt{2}}{2}i$ **69.** $2\sqrt{2}$ **71.** $\sqrt[3]{b^2}$

73. $(a-b)\sqrt{a-b}$ **75.** $x\sqrt[10]{x^3}$ **77.** $\frac{1}{4}$ **79.** 4

81. x **83.** $x^{1/3}$ **85.** $3ab^3$ **87.** $z^{1/2}$

89. $\dfrac{x^{3/2}}{y^{2/3}}$ **91.** Approximately 1461 feet per second   **93.** $\sqrt{2}x$

95. a. $v_0 = 16t$

  b. Approximately 16.6 feet per second; approximately 4.3 feet (51.5 inches)

  c. Yes

97. 0.0085 second   **99.** True   **101.** False   **103.** False

Chapter 1 Review ■ page 92

1. Rational, real, complex   **3.** Complex

5. $1.4, \sqrt{2}, 1.5, \sqrt{3}, 2$

7. $[-5, 1)$

9. $\left(-\infty, \frac{1}{2}\right]$

11.

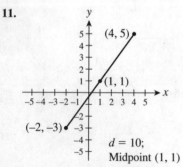

$d = 10$;
Midpoint $(1, 1)$

13.

1.8 (Answers may vary.)

15. False   **17.** False   **19.** $5 - i$ **21.** $\frac{3}{10} + \frac{1}{10}i$

23. 81   **25.** $\dfrac{a^2}{b^4}$ **27.** 3.14×10^7 **29.** 1.1×10^{10}

31. -10 **33.** $-t^3 - t^2 + 5t$ **35.** $2y^2 - 5y - 12$

37. $2x^2 - 2x - 2$ **39.** $(3u - 2)(3u + 2)$

41. $(t - 4)(t + 8)$ **43.** $(3x + 1)^2$

45. $(2s - 1)(2s + 1)(4s^2 - 2s + 1)(4s^2 + 2s + 1)$

47. $\dfrac{1}{x - 3}$ **49.** $\dfrac{t^2 + t + 1}{t(t + 1)}$ **51.** $\dfrac{2(y - 1)}{y(y + 2)}$

53. $-\dfrac{x}{(x + 1)(x + 3)}$ **55.** $\dfrac{1}{2t}$ **57.** $a^2b\sqrt[3]{b}$ **59.** $3\sqrt{3}$

61. $\dfrac{x + \sqrt{2}}{x^2 - 2}$ **63.** $2s^3t\sqrt{3}$ **65.** $11\sqrt{7}$ **67.** $t^{1/2}$

69. $\dfrac{1}{x^{1/3}}$ **71.** $\sqrt[3]{x^2}$ **73.** $x^{1/12}$ **75.** $\dfrac{b^2}{a}$

77. a. $\frac{7}{12}$ **b.** $\dfrac{x + y}{xy}$

79. a. $4x^2 - 12xy + 9y^2$ **b.** $2x - 3y$

81. a. $(0, 7)$ **b.** $(2, 11)$

83. a. $2x^3 - 2x$ **b.** $2x(x - 1)$

85. $1480

87. 84,375,000 cubic feet; approximately 1055

89. $N_2 = -\dfrac{1000(10i^2 + 15i - 2)}{(1 + i)^2}$; when $i = 0.10$, $N_2 \approx 330.58$;

when $i = 0.15$, $N_2 \approx -359.17$; profitable after 2 years when $i = 0.10$.

Chapter 1 Test ■ page 95

1. True   **2.** False   **3.** True   **4.** True

5. True   **6.** True   **7.** False   **8.** True

9. $\frac{2}{3}$, for example   **10.** $\sqrt{2}$, for example   **11.** i, for example

12. $x^3y^2 + 2xy - y$, for example   **13.** $x^2 + 2x + 1$, for example

14. $x + 3$, for example   **15.** $-8x^3$ **16.** $\dfrac{v^{1/5}}{u^{2/3}}$

17. $\dfrac{b^{10}}{a^{10}}$ **18.** $3x^2y^3\sqrt{3x}$ **19.** $2ab^2\sqrt[4]{15}$ **20.** $\sqrt[3]{2}$

21. $3z^4 - 7z^3 + z^2$ **22.** $6x^2 + 5x - 6$ **23.** $\dfrac{x - 1}{x(x - 2)}$

24. $\dfrac{y + 3}{y(y + 1)}$ **25.** $(2x - 3)(2x + 3)$ **26.** $(t + 5)(t - 3)$

27. $(u + 2v)(u^2 - 2uv + 4v^2)$ **28.** $(a - 2)(a + b)$

29. Distributive property   **30.** $|2x + 7|$

31. Approximately $14,588.79

32. 24; -12; the ball hits the ground at some time between 2 and 3 seconds.

CHAPTER 2

Section 2.1 ■ *page 109*

1. $8^2 + 6^2 = 100$
$0^2 + (-10)^2 = 100$
$(5\sqrt{2})^2 + (5\sqrt{2})^2 = 100$

3. $\dfrac{-2}{(-2) + 1} = 2$

$\dfrac{\left(-\frac{4}{3}\right)}{\left(-\frac{4}{3}\right) + 1} = 4$

$\dfrac{\sqrt{2}}{\sqrt{2} + 1} = 2 - \sqrt{2}$

5.

x	$y = 1000(x^3 + 1)$
-0.1	999
0.0	1000
0.1	1001
0.2	1008
0.3	1027

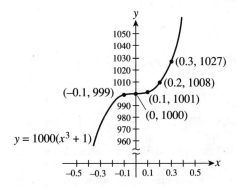

7.

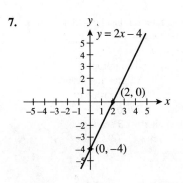

Linear

9.

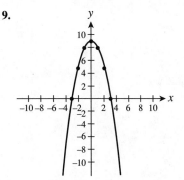

11.

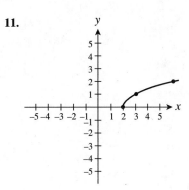

13. Center: $(0, 0)$; radius: 5

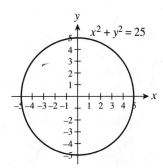

15. Center: $(0, 0)$; radius: $2\sqrt{2}$

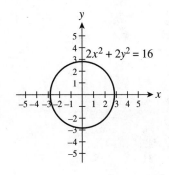

17. Center: $(-4, 1)$; radius: 2

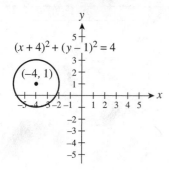

$(x+4)^2 + (y-1)^2 = 4$
$(-4, 1)$

19. Center: $(2, -3)$; radius: 2

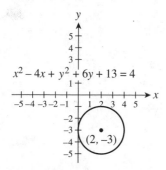

$x^2 - 4x + y^2 + 6y + 13 = 4$
$(2, -3)$

21. Center: $(4, -1)$; radius: $4\sqrt{2}$

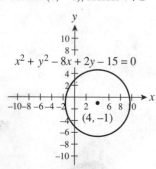

$x^2 + y^2 - 8x + 2y - 15 = 0$
$(4, -1)$

23. Center: $(-\frac{3}{2}, -1)$; radius: 2

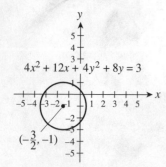

$4x^2 + 12x + 4y^2 + 8y = 3$
$(-\frac{3}{2}, -1)$

25. $(x+1)^2 + (y-3)^2 = 16$ **27.** $(x-4)^2 + (y+3)^2 = 25$
29. $(x+1)^2 + (y-5)^2 = 18$ **31.** ii   **33.** iii   **35.** vi

37. a. ii   **b.** i   **c.** iv   **d.** iii
39. a. i   **b.** iii   **c.** iv   **d.** ii
41. a. $x = -2$ **b.** $(-\infty, -2)$ **c.** $(-2, \infty)$
43. a. $x = -3, x = 2$ **b.** $(-3, 2)$ **c.** $(-\infty, -3) \cup (2, \infty)$
45. x-intercepts: $x \approx -10.41, x \approx 2.75, x \approx 10.05$
     y-intercept: $y = 12$
47. $(-4, -12)$ **49.** $(5.94, 28.51)$ **51.** $x \approx 10.24, x \approx 5.76$

53.

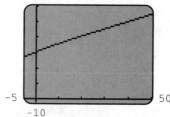

Approximately 1964

55.

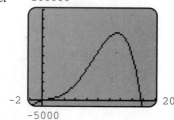

Approximately 1993

57. a. $x^2 + (y-35)^2 = 30^2$ **b.** Approximately 63.3 feet
59. False   **61.** True

Section 2.2 ■ *page 125*

 1. Yes   **3.** No   **5.** Yes   **7.** Yes   **9.** Yes   **11.** Yes
13. No   **15.** Yes   **17.** Yes   **19.** Yes
21. Not equivalent since $x = 0$ is a solution of the first equation but
     not the second
23. Not equivalent since $x = -3$ is a solution of the first equation
     but not the second
25. Equivalent
27. Not equivalent since the first equation has no real solutions, but
     the second equation has solution $x = 1$
29. Equivalent
31. a. Subtraction property   **b.** Division property
33. a. Multiplication property   **b.** Subtraction property
     c. Subtraction property   **d.** Division property
35. 1.5   **37.** $-3.1, -1.7, 19.7$ **39.** 0.2, 5.8
41. 1.8, 2.0, 2.2   **43.** Approximately 1.3 seconds
45. Approximately 8.3%   **47.** True   **49.** False

Section 2.3 ■ *page 136*

 1. $x = 2$ **3.** $x = 2$ **5.** $z = -21$ **7.** $x = \frac{4}{3}$
 9. $m = -\frac{15}{2}$ **11.** $x = 0.56$ **13.** $w = \frac{95}{13}$

15. $q = 3 \times 10^{-14}$ **17.** $x = -\frac{23}{36} \approx -0.6389$

19. No solution **21.** $x = 3$ **23.** $y = -2$

25. $(-\infty, 13)$

27. $(-1, \infty)$

29. $[-\frac{1}{4}, \infty)$

31. $(-\infty, \infty)$

33. $(\frac{4}{3}, \infty)$

35. $(-\infty, -8)$

37. $(-\infty, 3.1875)$

39. $(-\infty, -1)$

41. $t = \frac{d}{r}$ **43.** $r = \frac{c}{2\pi}$ **45.** $t = \frac{I}{pr}$ **47.** $l = \frac{P - 2w}{2}$

49. $m_1 = \frac{Fr^2}{Gm_2}$ **51.** $y = \frac{2}{3}x - 2; x = 3$

53. $y = x^2 - 2x - \frac{15}{4}; x \approx 3.18, x \approx -1.18$

55. $y = \frac{x^2 + 3x}{x^2 + 4}; x = 0, x = -3$ **57.** $10 = x - 6; x = 16$

59. $4x = 4 + x; x = \frac{4}{3}$ **61.** $2x - 1 < 5; x < 3$

63. 215 and 216

65. The length is 55 feet and the width is 35 feet. **67.** 97

69. The second offer has a combined salary and benefit package of $28,222.22.

71. a. \$471.70 **b.** Approximately 17.65%

73. 2 hours **75.** $3\frac{1}{3}$ miles; more than 183 miles per hour

77. Approximately 1.36 miles **79.** False **81.** True

Section 2.4 ■ *page 147*

1. $x = \pm 6$ **3.** $y = \pm \frac{5}{6}$ **5.** $u = 2, u = 3$ **7.** No solution

9. $x \approx -1.7, x \approx 2.3$ **11.** $|x - 5| = 7; x = -2, x = 12$

13. $|2x - (-\frac{2}{3})| = \frac{7}{3}; x = -\frac{3}{2}, x = \frac{5}{6}$

15. $|\frac{x}{2} - 2| = 5; x = -6, x = 14$

17. $(-2, 6)$

19. $[1, \frac{3}{2}]$

21. $[3, \frac{9}{2}]$

23. $[\frac{5}{2}, \infty)$

25. $(-5, 5)$ **27.** $(-3, 3)$ **29.** $(-\infty, -12) \cup (12, \infty)$

31. $[3, 7]$ **33.** $(-\infty, -\frac{3}{2}] \cup [-\frac{7}{6}, \infty)$ **35.** $(-\frac{7}{2}, \frac{9}{2})$

37. $(-\infty, \frac{19}{2}) \cup (\frac{13}{2}, \infty)$ **39.** No solution **41.** $[-1, \frac{7}{3}]$

43. $(-\frac{1}{6}, \frac{3}{2})$ **45.** $|x - 1| < 5; (-4, 6)$

47. $|2x + 3| > 6; (-\infty, -\frac{9}{2}) \cup (\frac{3}{2}, \infty)$

49.

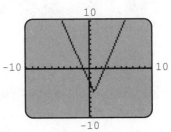

The x-intercepts; x-values where y is positive, x-values where y is negative

51. $x = 2.4, x = 6$ **53.** From 141.15 to 205.26 miles

55. $|W - 95| \leq \frac{1}{2}$ **57.** Between 13.4 and 13.6 inches

59. From 98.6 to 102.2 degrees Fahrenheit

61. False **63.** True

Section 2.5 ■ *page 157*

1. $x = -2, x = 1$ **3.** $y = -3, y = -\frac{1}{2}, y = \frac{2}{5}$

5. $t = -2, t = -1$ **7.** $x = -3, x = 1$

9. $x = -3, x = -2, x = 2$ **11.** $y = -5, y = 0$

13. $s = 0, s = 3$ **15.** $x = -3, x = \frac{1}{2}$ **17.** $t = \pm 2\sqrt{2}$

19. $y = -1 \pm \sqrt{3}$ **21.** $x = 3$ **23.** $z = \pm \sqrt[6]{5}$

25. $w = -\frac{3}{2}$ **27.** $x = -4, x = -2$ **29.** $w = -1, w = -\frac{1}{2}$

31. $x \approx 0.75$ **33.** $x \approx -1.67, x = 3$

35. $x \approx -11.99, x \approx 11.99$ **37.** $x \approx -11, x \approx 1.91, x \approx 2.09$

39. $p(-1) = 0, p(-\frac{1}{2}) = 0, p(2) = 0$

41. $p(-\sqrt{2}) = 0, p(\sqrt{2}) = 0, p(1) = 0$

43. $x = -2, x = 0, x = 3$ **45.** $t = -\frac{1}{2}, t = 0, t = \frac{1}{2}$

47. $c = -9$ **49.** $c = 2$ **51.** $c = 4$ or $c = -4$

53. $[-3, 1]$ **55.** $(-\infty, -2) \cup (\frac{1}{2}, \infty)$ **57.** $[-5, 1]$

59. $(-\infty, \frac{1}{2}) \cup (3, \infty)$ **61.** $(-2\sqrt{2}, 2\sqrt{2})$

63. Approximately $(-\infty, 1.86)$

65. Approximately $(-\infty, -1.22) \cup (0.72, \infty)$

67. $(-1, 0) \cup (3, \infty)$ **69.** $(-\infty, -1)$

71. Approximately $(-\infty, -3.12) \cup (0.36, 1.76)$

73. $x = -\frac{3}{2}, x = \frac{2}{7}$ **75.** $(-1, 0) \cup (0, \infty)$

77. $[-2, 0] \cup [2, \infty)$

79. Approximately $(-20.06, 15.24) \cup (26.82, \infty)$

81. $y = \pm \sqrt{4 - 3x}; x \approx 1.33$

83. $y = \pm \frac{1}{2} \sqrt[4]{9 - x^4}; x \approx 1.73$

85. $t = 10$ seconds **87.** Approximately 5.9%

89. Approximately 118.6 inches **91.** Approximately 19.8 miles

93. 2001 **95.** Between 2 and 3 seconds

97. Between 0 and 27 inches **99.** 1998

101. False **103.** False

Section 2.6 ■ *page 170*

1. $x = -3, x = 3$ **3.** $x = -7, x = 1$ **5.** $t = 1 \pm 2\sqrt{5}$

7. $x = -3, x = -2$ **9.** $x = -1, x = 3$ **11.** $x = \frac{-3 \pm \sqrt{5}}{2}$

13. $x = \frac{3 \pm \sqrt{17}}{2}$ **15.** $x = -2, x = 3$ **17.** $t = -\frac{1}{2}, t = \frac{2}{3}$

19. $x = -3$ **21.** $u = 3 \pm \sqrt{6}$ **23.** $x = \frac{-2 \pm \sqrt{2}}{2}$

25. $x = \dfrac{-1 \pm \sqrt{5}}{2}$ **27.** $x = -1, x = 1$

29. $w = 2 + i, w = 2 - i$

31. $y = \dfrac{1 \pm \sqrt{2}i}{3}$ **33.** $x = \pm\sqrt{\dfrac{3 - \sqrt{7}}{2}}, x = \pm\sqrt{\dfrac{3 + \sqrt{7}}{2}}$

35. $w = \dfrac{-21 \pm 5\sqrt{5}}{2}$ **37.** 2   **39.** 0   **41.** 2

43. 1   **45.** 2   **47.** $a = \dfrac{-x \pm \sqrt{x^2 - 4x}}{2}$

49. $t = \dfrac{-x \pm \sqrt{-7x^2}}{2}$ **51.** $y = -1 \pm \frac{1}{2}\sqrt{2x - 6}; x = 5$

53. $y = 3 \pm \sqrt{4x^2 + 8x + 3}; x \approx -2.58, x \approx 0.58, y \approx 4.73, y \approx 1.27$

55. $r = \dfrac{-1 + \sqrt{181}}{6} \approx 2.08$ inches

$h = \dfrac{1 + \sqrt{181}}{2} \approx 7.23$ inches

57. a. $t = \dfrac{-1 + \sqrt{2}}{4} \approx 0.1$ second

b. $t = \dfrac{2 - \sqrt{3}}{4} \approx 0.07$ second

59. $r = \frac{1}{1499} \approx 6.7 \times 10^{-4}$ feet (0.008 inch)

61. a. \$3152.50   **b.** $r = \dfrac{-3 + \sqrt{13}}{2} \approx 0.303 = 30.3\%$

63. True   **65.** False

Section 2.7 ■ *page 185*

1. $x = 2$ **3.** $y = 10$ **5.** $x = \frac{1}{2}$ **7.** $t = \frac{5}{6}$

9. $s = -6$ **11.** $x = -1$ **13.** $x = -\frac{1}{2}, x = 3$

15. $x = 1, x = 5$ **17.** No solution   **19.** $x = \frac{3}{2}$

21. $x = -5 \pm 2\sqrt{7}$ **23.** $x = -3$ **25.** $(-\infty, -4] \cup (-3, \infty)$

27. $(-\infty, -1) \cup (2, 3)$ **29.** $(-\infty, -3) \cup \left(-1, -\frac{1}{2}\right]$

31. $(-\infty, -3] \cup [2, \infty)$ **33.** $x = 9$ **35.** No solution

37. $x = \frac{7}{8}$ **39.** $u = 3$ **41.** $t = -\frac{1}{4}$ **43.** $r = 3$

45. $x = 4$ **47.** $x = 3$ **49.** $x = 1$ **51.** $y = \frac{5}{4}$

53. $x \approx 0.68$ **55.** $x \approx -2.08, x = -1$ **57.** $x \approx 4.44$

59. a. $s = \dfrac{dh}{l - h}$ **b.** $s = \frac{110}{23} \approx 4.78$ feet

61. a. $h = \frac{180}{41} \approx 4.39$ **b.** $z = \dfrac{hxy}{3xy - hx - hy}$

c. $z = 56$

63. a. $\dfrac{40}{140 - x}$ **b.** $\dfrac{40}{140 + x}$ **c.** 20 miles per hour

65. a. 6 hours   **b.** $5\frac{1}{3}$ hours   **c.** $13\frac{1}{3}$ hours

67. Either 128 yards or 55.3 yards to the left of point B

69. a. 19   **b.** 1.5 or 14   **71.** True   **73.** True

Section 2.8 ■ *page 199*

1. $x = 3, y = 4$ **3.** $x = -2, y = 1$ **5.** Inconsistent

7. $m = 3, n = 7$ **9.** $x = 2, y = -\frac{3}{5}$ **11.** Inconsistent

13. $a = \frac{12}{5}, b = -7$ **15.** $x = 1.3, y = 2.4$

17. $x = 3.4, y = 1.7$ **19.** $x = 1.05 \times 10^7, y = 1.2 \times 10^7$

21. $x = 3, y = 5$
$x = 5, y = 3$

23. $p = 3, q = -2$
$p = -\frac{9}{5}, q = \frac{14}{5}$

25. $a = 4, b = -6$

27. $x = 0, y = 3$
$x = -\frac{15}{13}, y = -\frac{36}{13}$

29. $x = 3, y = 2$
$x = -3, y = 2$
$x = 3, y = -2$
$x = -3, y = -2$

31. $(2, 3)$ **33.** $(-2, 0), (1, 3)$ **35.** $(1, 1), (1, -1)$

37. $(-1.25, 1.31), (0.45, 1.64), (1.80, 6.05)$

39. No points of intersection

41. All points on the line segment from $(2, 0)$ to $(4, 6)$

43. 8 nickels and 4 dimes

45. $\frac{50}{9}$ ounces of yogurt and $\frac{40}{9}$ ounces of dessert topping

47. Mark is 33 and Benjamin is 3.

49. 8000 \$20 tickets and 5000 \$30 tickets

51. Arnold's time would be approximately 25.8 hours, and Franco's time would be approximately 35.8 hours.

53. The older brother's time is 10 seconds, and the younger brother's time is 12.5 seconds.

55. Approximately 15.2 miles east and 13 miles north of point A

57. True   **59.** False

Chapter 2 Review ■ *page 204*

1. $(-1, 1)$ and $(-2, 0)$ are on the graph and $(2, -2)$ is not.

3.

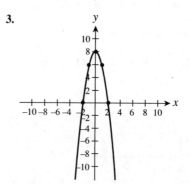

5. Center: $(0, 0)$; radius: 4

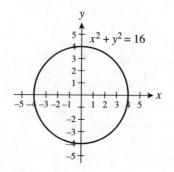

7. Center: $(2, -1)$; radius: 3

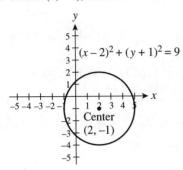

9. Center: $(-5, 4)$; radius: 3

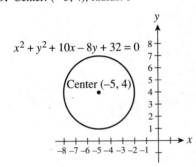

11. $(x - 2)^2 + (y - 5^2) = 9$ **13.** $(x - 2)^2 + (y - 4)^2 = 100$

15. iv   **17.** ii

19. x-intercepts: $x \approx -3.79$, $x \approx 1.03$, $x \approx 10.26$
y-intercept: $y = 5$

21. $(2, 3)$ **23.** Yes   **25.** No   **27.** Yes

29. No   **31.** $x \approx -1.14$ **33.** $x \approx -0.33$, $x \approx 1.54$

35. $x = \frac{5}{3}$ **37.** $x = \frac{7}{4}$

39. $(-4, \infty)$

41. $(-\infty, 10)$

43. $(-\infty, \infty)$

45. $x = \dfrac{c - by}{a}$ **47.** $x = 1, x = 4$ **49.** $\left(\frac{2}{3}, 2\right)$

51. $|3x - 4| = 8$; $x = -\frac{4}{3}$, $x = 4$

53. $|2x - 4| > 6$; $(-\infty, -1) \cup (5, \infty)$

55. $x = -3$, $x = 0$, $x = \frac{1}{2}$ **57.** $v = -2$, $v = 2$

59. $x = -2$, $x = \frac{3}{2}$ **61.** $(-\infty, -5) \cup (-1, \infty)$

63. $(-0.47, 1.40)$ **65.** $x = 3$, $x = -4$

67. $w = -2 \pm \sqrt{2}$ **69.** $y = -\frac{1}{2}$, $y = \frac{2}{3}$

71. 2   **73.** 1   **75.** $x = 2$ **77.** $x = -\frac{1}{3}$, $x = 2$

79. $(-\infty, 1) \cup (3, \infty)$ **81.** $x = 2$ **83.** $a = 2$

85. $y = 4 - \frac{2}{3}x$; $x = 6$ **87.** $y = \pm\sqrt[4]{x + 5}$; $x = -5$

89. $y = 2 \pm \sqrt{x - 1}$; $x = 5$ **91.** $x = 3$, $y = 2$

93. Inconsistent   **95.** $\left(\frac{9}{7}, \frac{8}{7}\right)$ **97.** $(-1, 6)$, $(3.5, 10.5)$

99. **a.** 3   **b.** 2

101. **a.** $-\frac{1}{9}$ **b.** $\frac{1}{3}$

103. **a.** 210   **b.** 5, 6, and 7

105. **a.** $x = -2$, $x = 1$ **b.** $[-2, 1]$

107. **a.** $x^3 + 9x^2 - 9x - 81$ **b.** $x = -9$, $x = -3$, $x = 3$

109. 117 and 119

111. Between 0 and 0.18 second, and after 1.69 seconds

113. Length 4 feet and width $\frac{3}{4}$ foot

115. Side lengths 24 and 45 inches

Chapter 2 Test ■ *page 208*

1. False   **2.** True   **3.** True   **4.** False

5. False   **6.** False   **7.** False   **8.** False

9. $x^2 - 6x + 9 = 0$, for example

10. $(x - 1)(x - 2)(x - 3) = 0$, for example

11. $x + y = 2$, for example
$\quad x + y = 1$

12. $|x + 1| < 0$, for example   **13.** $t = \frac{1}{2}$

14. $x = -\frac{11}{2}$, $x = \frac{3}{2}$ **15.** $(-\infty, 1) \cup (2, \infty)$

16. $u = -2$, $u = 0$, $u = 2$

17. $x \approx -1.71$, $x \approx 1.35$ **18.** $(-2, 0)$

19. $x = 1 \pm \sqrt{5}$

20. $[-1, 5]$ **21.** $x = -16$ **22.** $x = 4$

23. **a.** ii   **b.** i   **c.** iii

24. $x = 3$, $y = 5$ **25.** 100, 102, 104

26. 3 days for Habitat for Humanity and 6 days for the construction company

CHAPTER 3

Section 3.1 ■ *page 221*

1. 4   **3.** $-\frac{11}{5}$ **5.** $\frac{9}{5}$ **7.** -1 **9.** $\frac{22}{13} \approx 1.69$

11. 1,000,000   **13.** 3   **15.** -4 **17.** -3 **19.** $-\frac{5}{3}$

21. Slope of $l_1 = -2$
Slope of $l_2 = \frac{2}{3}$
Slope of $l_3 = 3$
Slope of $l_4 = 0$

23. Slope of $l_1 = \frac{3}{7}$
Slope of $l_2 = -\frac{7}{3}$
Slope of $l_3 = \frac{3}{7}$

25. $y = x + 2$ **27.** $y = -x + 10$ **29.** $x = 3$

31. $y = 5x - \frac{11}{6}$ **33.** $y = 1$ **35.** $y = \frac{1}{3}x + \frac{11}{3}$

37. $y = -2x + 4$ **39.** $x = 3$ **41.** $y = 0.7353x + 2.5176$

43. $y = \frac{3}{20}x + \frac{17}{40}$ **45.** $x = \pi$

47. $m = \frac{2}{3}, b = -\frac{8}{3}$

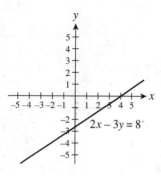

$2x - 3y = 8$

49. $m = \frac{1}{3}, b = -\frac{4}{3}$

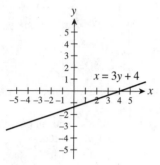

$x = 3y + 4$

51. $m = -2, b = -3$

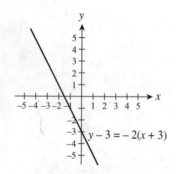

$y - 3 = -2(x + 3)$

53. $y = \frac{2}{3}x + 2$ **55.** $y = 4x + 5$ **57.** $y = -3x + 2$
59. $x = 3$ **61.** $3x - y = -4$ **63.** $2x - y = \frac{1}{3}$ or $6x - 3y = 1$
65. $\left(-\frac{13}{5}, -\frac{14}{5}\right)$ **67.** $(2, 5)$ **73.** $C = 5N + 1000$
75. a. $h = \frac{264}{67}t$ **b.** Approximately 3 inches
c. Approximately 50 inches
77. $128,000
79. a. 242.4 million   **b.** 271 million
81. False   **83.** True   **85.** False

Section 3.2 ■ *page 235*

 1. Function   **3.** Not a function   **5.** Function
 7. Function   **9.** y is not a function of x.
11. y is a function of x.   **13.** y is a function of x.
15. y is not a function of x.   **17.** y is a function of x.
19. a. 12   **b.** 2   **c.** 2
21. a. $\frac{14}{11}$ **b.** $\frac{17}{15}$ **c.** $\dfrac{-3t + 2}{-4t - 5}$
23. a. $3t^4$ **b.** $3t^2 - 6t + 3$ **c.** $\dfrac{75}{t^2}$
25. a. x^2 **b.** $x^4 + 2x^2 + 1$ **c.** $\dfrac{1}{a^2}$

27. a. $\frac{1}{8}$ **b.** $\dfrac{1}{t^2 - 1}$ **c.** $\dfrac{1}{x^2}$

29. $h(x) = |x|$ **31.** $f(x) = \dfrac{1}{x + 3}$ **33.** All real numbers

35. All real numbers except 4
37. All real numbers except -3 and 3   **39.** $[-1, \infty)$
41. All real numbers in $[-2, \infty)$ except 1 and 3
43. All real numbers except -3 and 3

45. $f(x) = 2x + 1$, for example   **47.** $f(x) = \dfrac{1}{x + 1}$, for example

49. $f(x) = \sqrt{7 - x}$, for example

51. $f(x) = \dfrac{1}{x(x - 3)}$, for example

53.

x	$f(x) = -2x + 5$
0	5
2	1
4	-3

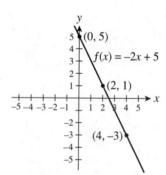

$(0, 5)$ $f(x) = -2x + 5$ $(2, 1)$ $(4, -3)$

55.

x	$f(x) = (x + 1)^2 - 3$
-3	1
-2	-2
-1	-3
0	-2
1	1

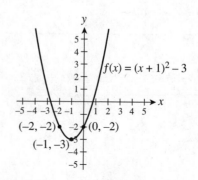

$f(x) = (x + 1)^2 - 3$
$(-2, -2)$ $(0, -2)$
$(-1, -3)$

57. a.

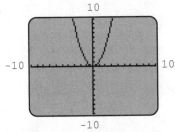

Domain: $(-\infty, \infty)$
Range: $[0, \infty)$

b.

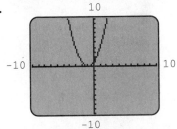

Domain: $(-\infty, \infty)$
Range: $[0, \infty)$

c.

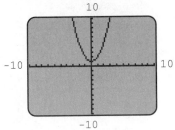

Domain: $(-\infty, \infty)$
Range: $[1, \infty)$

d.

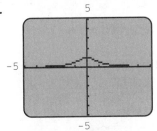

Domain: $(-\infty, \infty)$
Range: $(0, 1]$

59. a.

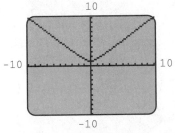

Domain: $(-\infty, \infty)$
Range: $[1, \infty)$

b.

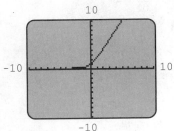

Domain: $(-\infty, \infty)$
Range: $(0, \infty)$

c.

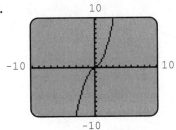

Domain: $(-\infty, \infty)$
Range: $(-\infty, \infty)$

d.

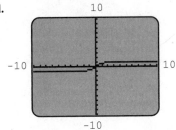

Domain: $(-\infty, \infty)$
Range: $(-1, 1)$

61. a. $x = -2 \pm \sqrt{y - 2}$ **b.** $y \geq 2$ **c.** $[2, \infty)$

63. $V = \frac{\pi}{6}d^3$ **65.** $A = 2r^2$ **67.** $A = \dfrac{a^2}{2(a - 2)}$

69. $V = x(16 - 2x)^2$; the domain is the set $(0, 8)$.

71. $P = \dfrac{2000}{(1 + r)^4}$

73. $V = (8.603 \times 10^{38})t^3$; 4.358×10^{43} cubic miles

75. $R(x) = \frac{2}{5}x - 220$

77. False   **79.** False

Section 3.3 ■ *page 255*

1. $x = -\frac{1}{3}$ **3.** $x = -2, x = 4$ **5.** $x = -\frac{5}{2}$

7. $x = -2, x = 2$ **9.** $a = 3$

11. $x \approx -2.56, x \approx 0.10, x \approx 3.96$

13. $x \approx -2.34, x \approx -0.61, x \approx 0.14$ **15.** $x \approx -1.24, x \approx 3.24$

17. a. $g(x) = |x - 2|$ **b.** $g(x) = |x| + 1$
   c. $g(x) = |x + 2| - 2$

19. a. $g(x) = -\frac{1}{4}x^4 + \frac{1}{3}x^3 + x^2$ **b.** $g(x) = \frac{1}{4}x^4 + \frac{1}{3}x^3 - x^2$

21. a.

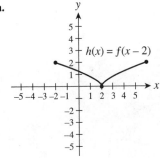

b.

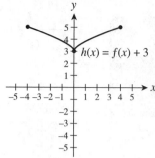

$h(x) = f(x) + 3$

c.

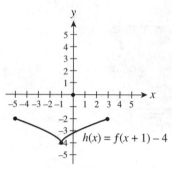

$h(x) = f(x + 1) - 4$

d.

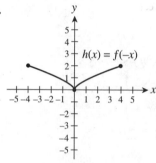

$h(x) = f(-x)$

e.

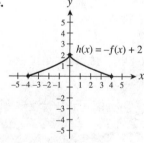

$h(x) = -f(x) + 2$

f.

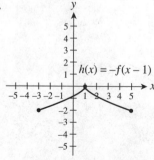

$h(x) = -f(x - 1)$

23. a.

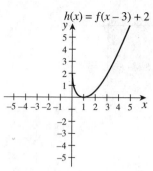

$h(x) = f(x - 3) + 2$

b.

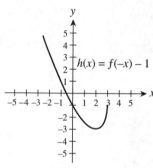

$h(x) = f(-x) - 1$

c.

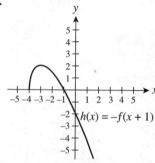

$h(x) = -f(x + 1)$

d.

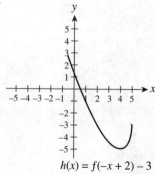

$h(x) = f(-x + 2) - 3$

25. Even   **27.** Odd   **29.** Neither

31. a.

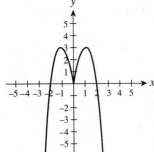

b.

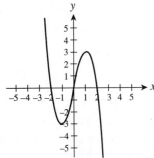

33. a.

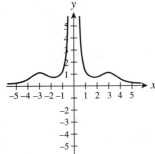

b.

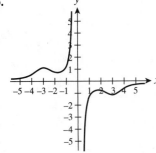

35. Increasing: $(2, \infty)$
Decreasing: $(-\infty, 2)$
Turning point: Local minimum at $(2, 1)$

37. Increasing: $(-\infty, -0.62), (1.62, \infty)$
Decreasing: $(-0.62, 1.62)$
Turning points: Local maximum at $(-0.62, 7.09)$, local minimum at $(1.62, -4.09)$

39. Increasing: $(-\infty, 2)$
Decreasing: $(2, \infty)$
Turning point: Local maximum at $(2, 5)$

41. Increasing: $(-1.30, 0.17), (1.13, \infty)$
Decreasing: $(-\infty, -1.30), (0.17, 1.13)$
Turning points: Local minimum at $(-1.3, -3.51)$, local maximum at $(0.17, 0.08)$, local minimum at $(1.13, -1.07)$

43. $Q(x) = (x + 5)^2 + 14(x + 5) + 100$

45. $g(t) = -16t^2 + 80t + 20$; the graph of g is the graph of f shifted 20 units upward.

47. Approximately 30 or 270 systems per month to break even; 150 systems per month to maximize profit

49. $\left(2, \frac{3}{2}\right)$ **51.** $50' \times 100'$

53. a.

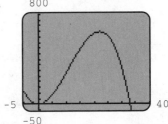

Approximately 645.33 feet

b. $g(t) = 33t$

c.

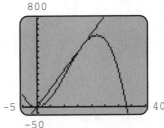

They meet at $t = 16.5$ seconds; the support car is furthest from the checkpoint at $t = 22$ seconds.

d. 33 feet per second

55. Turning points: $(0.05, 361.72), (0.43, 357.99), (1.02, 362.87), (1.54, 359.49), (1.97, 364.40)$; the concentration is lowest in late fall and highest in late spring.

57. Increasing: $[0, 1.54), (11.29, 15.72)$
Decreasing: $(1.54, 11.29), (15.72, 16]$
Reserves highest at $x \approx 1.54$, which corresponds to midway through 1974

59. True   **61.** True

Section 3.4 ■ *page 270*

1. a. $(f + g)(x) = x + 5$ **b.** $(f - g)(x) = x - 5$
c. $(fg)(x) = 5x$ **d.** $(f/g)(x) = \frac{x}{5}$

3. a. $(f + g)(x) = x^2 + 3x + 1$ **b.** $(f - g)(x) = x^2 - 3x - 1$
c. $(fg)(x) = 3x^3 + x^2$ **d.** $(f/g)(x) = \frac{x^2}{3x + 1}, x \neq -\frac{1}{3}$

5. a. $(f + g)(x) = 2x$ **b.** $(f - g)(x) = 0$
c. $(fg)(x) = x^2$ **d.** $(f/g)(x) = 1, x \neq 0$

7. a. $(f + g)(x) = 2x - 2 + \dfrac{2}{x + 5}, x \neq -5$

  b. $(f - g)(x) = 2x - 2 - \dfrac{2}{x + 5}, x \neq -5$

  c. $(fg)(x) = \dfrac{4x - 4}{x + 5}, x \neq -5$

  d. $(f/g)(x) = x^2 + 4x - 5, x \neq -5$

9. a. $(f + g)(x) = \sqrt{x - 2} + x - 4, x \geq 2$

  b. $(f - g)(x) = \sqrt{x - 2} - x + 4, x \geq 2$

  c. $(fg)(x) = (x - 4)\sqrt{x - 2}, x \geq 2$

  d. $(f/g)(x) = \dfrac{\sqrt{x - 2}}{x - 4}, x \geq 2, x \neq 4$

11.

x	$(f + g)(x)$	$(f - g)(x)$	$(fg)(x)$	$(f/g)(x)$
0	7	7	0	Undefined
1	4	−4	0	0
2	6	−10	−16	$-\frac{1}{4}$

Domain $f + g$: $\{0, 1, 2\}$
Domain $f - g$: $\{0, 1, 2\}$
Domain fg: $\{0, 1, 2\}$
Domain f/g: $\{1, 2\}$

13. $(f \circ g)(x) = 8x - 7$
  $(g \circ f)(x) = 8x + 7$

15. $(f \circ g)(x) = 4x^2 + 28x + 49$
  $(g \circ f)(x) = 2x^2 + 7$

17. $(f \circ g)(x) = 9x, x \neq 0$
  $(g \circ f)(x) = \frac{x}{9}, x \neq 0$

19. $(f \circ g)(x) = x^3 - 3x + 1$
  $(g \circ f)(x) = x^3 + 3x^2 - 2$

21. $(f \circ g)(x) = 7$
  $(g \circ f)(x) = 71$

23. $(f \circ g)(x) = x^2$
  $(g \circ f)(x) = x^2, x \geq 0$

25. $(f \circ g)(x) = 3\sqrt{x - 2} + 5, x \geq 2$
  $(g \circ f)(x) = \sqrt{3x + 3}, x \geq -1$

27. $(f \circ g)(x) = x, x \neq 0$
  $(g \circ f)(x) = x, x \neq -2$

29.

x	$(f \circ g)(x)$	$(g \circ f)(x)$
1	2	2
2	4	8
3	8	Undefined
4	Undefined	Undefined

Domain $f \circ g$: $\{1, 2, 3\}$
Domain $g \circ f$: $\{1, 2\}$

31.

x	$f(x)$	$g(x)$	$(f \circ g)(x)$	$(g \circ f)(x)$
0	2	1	0	0
1	0	2	1	1
2	1	0	2	2

33. $f^3(x) = x + 15$ **35.** $f^4(x) = 16x$

37. $f^{10}(0.98) \approx 1.03633 \times 10^{-9}$;
  $f^{10}(1.02) \approx 6.40583 \times 10^8$;
  $f^n(0.98)$ approaches 0;
  $f^n(1.02)$ approaches ∞

39. a.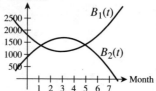

The first refinery gradually decreases production until about the third month, and then increases for the rest of the period. The second refinery does the opposite—gradually increasing production until about the third month and declining thereafter.

	Maximum	**Minimum**
Refinery 1	1800	1125
Refinery 2	1665	1325

b. $B(t) = -10t^2 + 60t + 2700$

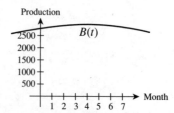

Total production is virtually constant.
Maximum: 2790
Minimum: 2700

41. a. $f(t) = 2t$

  b. $g(n) = 135 + \frac{1}{2}n$

  c. $(f \circ g)(t) = 270 + t$
    $(g \circ f)(t) = 135 + t$

  d. $(g \circ f)(t)$ models Nakajima's weight t minutes into the contest.

  e. Since meat is denser than bread, his weight might go up in "spurts."

43. False   **45.** False

Section 3.5 ■ *page 280*

1. Inverses **3.** Not inverses **5.** Inverses
7. Inverses **9.** Inverses **11.** 1–1 **13.** Not 1–1
15. 1–1 **17.** Not 1–1 **19.** Not 1–1
21. 1–1; $f^{-1}(x) = 3x + 3$ **23.** Not 1–1
25. 1–1; $f^{-1}(x) = \sqrt{x - 1}, x \ge 1$
27. 1–1; $f^{-1}(x) = \frac{1}{2}(x^2 - 5), x \ge 0$
29. Not 1–1

31. 1–1;

x	$f^{-1}(x)$
4	0
7	1
10	2
13	3
16	4

33. Not 1–1

35.

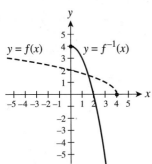

37. f^{-1} does not exist.

39.

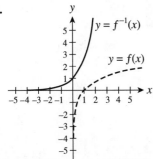

41. a. $M(t) = 98 + 0.12t$ **b.** 110°F
c. $M^{-1}(t) = 8.33t - 817$

d. $M^{-1}(t)$ gives the year (after 1935) in which the maximum temperature is $t°$F. For example, since $M^{-1}(120) \approx 183$, this model predicts that the maximum temperature will be 120°F in 2117.
43. a. 4 years
  b. $f^{-1}(t) = \frac{72}{t}$ approximates the interest rate at which an amount takes t years to double.
  c. 9%
45. a. Yes; the formula suggests a maximal weight of 195 pounds.
  b. Assume $h \ge 0$; $f^{-1}(w) = 5.30548\sqrt{w}$ gives the maximum height required for a person of a given weight to not be overweight.
  c. About $16\frac{1}{2}$ feet tall
47. False **49.** True

Section 3.6 ■ *page 292*

1. Quadratic **3.** None **5.** None
7. Linear **9.** Linear **11.** Piecewise

13.

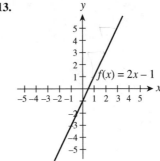

Slope 2, y-intercept -1

15.

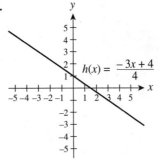

Slope $-\frac{3}{4}$, y-intercept 1

17. $h(x) = -\frac{1}{2}x + 5$ **19.** $g(x) = \frac{4}{3}x - 4$
21. $(0, -4)$ **23.** $\left(\frac{3}{2}, \frac{11}{2}\right)$ **25.** $(-8, -54)$ **27.** $(1, 0)$
29. $f(x) = 2(x - 1)^2 + 3$ **31.** $g(x) = -(x - 3)^2 - 1$

33. $f(-2) = -2$; $f(0) = -4$; $f(3) = 2$

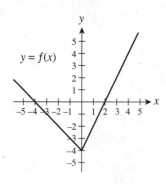

35. $h(0) = -1$; $h(2) = 2$; $h(4) = 3$

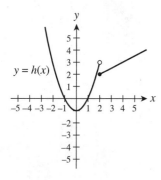

37. $f(-3) = -3$; $f(-1) = 1$; $f(4) = 6$

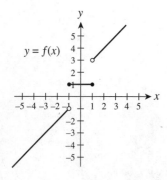

39.

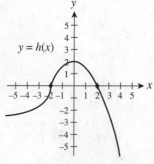

41.

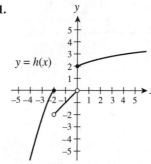

43.

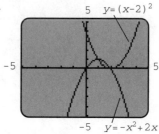

45.

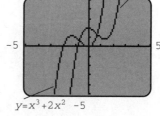

47.

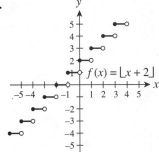

$f(x) = \lfloor x + 2 \rfloor$

49. a.

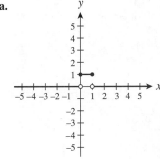

b.

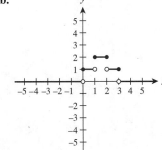

c.

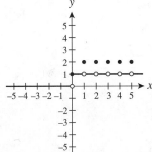

51. a. $0.8; 0.7; \frac{3}{5}$

b.

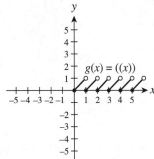

$g(x) = ((x))$

c. x

53. a. $C(x) = 80x + 5000$, where x is the number of units rented.
   b. $p = -5x + 550$
   c. $R(x) = -5x^2 + 550x$
   d. $P(x) = -5x^2 + 470x - 5000$
   e. 47 units at \$315 per month

55. a. 81.13 feet
   b. Yes

57. In 1958; approximately 1.16 billion acres

59.

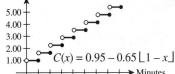

$C(x) = 0.95 - 0.65 \lfloor 1 - x \rfloor$

7 minutes

61. \$20,748.50 for filing separately; \$19,146.00 for filing jointly
63. TWA JULY TWO   **65.** False   **67.** True

Section 3.7 ■ *page 309*

1. 4   **3.** 48   **5.** 10   **7.** 12   **9.** 60

11. $\frac{243}{64}$ **13.** $f(x) = \dfrac{18}{x^2}$ **15.** $f(x) = x^2 - 4x$

17. $h(x) = x^2 - 3x + 5$ **19.** $y = 0.5x + 2.33$
21. $y = 1.6x - 4.8$ **23.** $y = 2.5x + 90$; $y = 152.5$ when $x = 25$
25. $y = -31.55x + 77.51$; $y = -17.14$ when $x = 3$
27. 2592 board feet   **29.** 800,000 joules
31. 5.75×10^{-25} newton   **33.** 3000 pounds
35. $y = 2.6x + 21.7$; 281.7 kilograms
37. $y = 9.57x + 177.1$; approximately 388 parts per trillion
39. $y = 2x^2 + 22x + 320$, where $x = 0$ corresponds to 1980;
   1,560,000 inmates
41. True   **43.** False

Chapter 3 Review ■ *page 313*

1. $\frac{1}{3}$ **3.** 4   **5.** $y = 3x + 2$ **7.** $x = 2$
9. $y = -2x + 4$ **11.** $y = \frac{9}{10}x + \frac{51}{20}$ **13.** $y = -2x + 2$
15. $y = -5x + 5$ **17.** $y = -3x + 22$ **19.** $y = -\frac{3}{2}x - 4$
21. Function   **23.** Not a function

25. Defines y as a function of x

27. Does not define y as a function of x

29. a. 21   **b.** 28

31. a. $\dfrac{x^2}{x^2 + 1}$ **b.** $\dfrac{x^2 + 2x + 1}{x^2 + 2x + 2}$

33. a. 4   **b.** -1 **c.** $-a + 4$

35. All real numbers   **37.** $\left[-\frac{1}{2}, \infty\right)$

39.

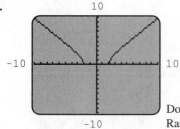

Domain: $(-\infty, -2] \cup [2, \infty)$
Range: $[0, \infty)$

41.

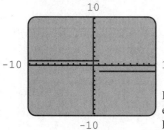

Domain: All real numbers
except 1
Range: $\{-1, 1\}$

43. a. $g(x) = -2x^3 + 6x - 1$ **b.** $g(x) = 2x^3 + 6x^2 - 3$

45. a. $h(x) = f(x + 3)$

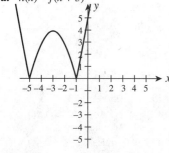

b. $h(x) = f(x - 2) + 1$

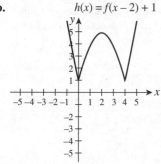

c.

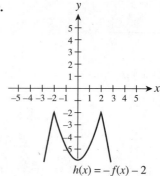

$h(x) = -f(x) - 2$

47. Odd   **49.** Neither   **51.** $x = \frac{9}{4}$ **53.** $x = \frac{8}{3}$

55. a. $x \approx -1.88$, $x \approx 0.35$, $x \approx 1.53$

b. Turning points: Local maximum at $(-1, 3)$, local minimum at $(1, -1)$

c. Increasing: $(-\infty, -1) \cup (1, \infty)$
Decreasing: $(-1, 1)$

57. a. $x \approx -1.73$, $x \approx 1.73$

b. Turning points: Local maximum at $(1, 1)$, local minimum at $(3, 3)$

c. Increasing: $(-\infty, 1) \cup (3, \infty)$
Decreasing: $(1, 2) \cup (2, 3)$

59. a. $(f + g)(x) = x^2 + 2x - 1$

b. $(fg)(x) = 2x^3 - x^2$

c. $(f/g)(x) = \dfrac{x^2}{2x - 1}$; $x \neq \frac{1}{2}$

d. $(f \circ g)(x) = 4x^2 - 4x + 1$

e. $(g \circ f)(x) = 2x^2 - 1$

61. a. $(f + g)(x) = \dfrac{1}{x - 4} + x^3$; $x \neq 4$

b. $(fg)(x) = \dfrac{x^3}{x - 4}$; $x \neq 4$

c. $(f/g)(x) = \dfrac{1}{x^4 - 4x^3}$; $x \neq 4$, $x \neq 0$

d. $(f \circ g)(x) = \dfrac{1}{x^3 - 4}$; $x \neq \sqrt[3]{4}$

e. $(g \circ f)(x) = \dfrac{1}{(x - 4)^3}$; $x \neq 4$

63. a. $(f + g)(x) = 2x^2 + \sqrt{2x + 4}$; $x \geq -2$

b. $(fg)(x) = 2x^2\sqrt{2x + 4}$; $x \geq -2$

c. $(f/g)(x) = \dfrac{2x^2}{\sqrt{2x + 4}}$; $x > -2$

d. $(f \circ g)(x) = 4x + 8$; $x \geq -2$

e. $(g \circ f)(x) = \sqrt{4x^2 + 4}$

65.

x	$(f + g)(x)$	$(fg)(x)$	$(f/g)(x)$	$(f \circ g)(x)$	$(g \circ f)(x)$
0	4	3	$\frac{1}{3}$	0	2
1	4	4	1	3	0
2	3	0	Undefined	1	1
3	1	0	0	2	3

Domain $f + g$: $\{0, 1, 2, 3\}$
Domain fg: $\{0, 1, 2, 3\}$
Domain f/g: $\{0, 1, 3\}$
Domain $f \circ g$: $\{0, 1, 2, 3\}$
Domain $g \circ f$: $\{0, 1, 2, 3\}$

67. Inverses   **69.** Not inverses   **71.** 1–1; $f^{-1}(x) = \dfrac{x + 2}{3}$

73. Not 1–1   **75.** 1–1; $f^{-1}(x) = x^2 + 4, x \geq 0$

77. 1–1;

x	$f^{-1}(x)$
2	1
5	2
10	3
17	4
26	5

79.

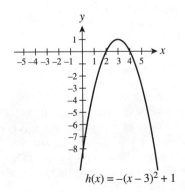

81. Quadratic

$h(x) = -(x - 3)^2 + 1$

x-intercepts $(2, 0)$ and $(4, 0)$; y-intercept $(0, -8)$

83. None

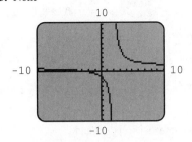

x-intercept $(-2, 0)$; y-intercept $(0, -1)$

85. Piecewise

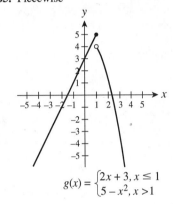

$g(x) = \begin{cases} 2x + 3, x \leq 1 \\ 5 - x^2, x > 1 \end{cases}$

x-intercepts $\left(-\frac{3}{2}, 0\right)$ and $(\sqrt{5}, 0)$; y-intercept $(0, 3)$

87.

$f(x) = (x + 1)^2 - 4$

$(-1, -4)$

Vertex $(-1, -4)$; x-intercepts $(-3, 0)$ and $(1, 0)$; y-intercept $(0, -3)$

89.

$g(x) = -x^2 + 8x - 17$

$(4, -1)$

Vertex $(4, -1)$; no x-intercepts; y-intercept $(0, -17)$

91. $f(x) = -\frac{1}{2}x + 3$ **93.** $h(x) = (x - 2)^2 - 1$ **95.** 15
97. 36   **99.** $f(x) = -x^3 + 3$ **101.** $y = 0.605x + 1.316$
103. $y = 2.37x + 42.6$; $y = 161.1$ when $x = 50$
105. a. $x = -3, x = -2$ **b.** $-3, -2$
107. a. $x = \frac{1}{y} - 2$ **b.** $f^{-1}(x) = \frac{1}{x} - 2$
109. a. $8x^3 - 12x^2 + 6x - 1$
   b. $f(x) = x^3$, $g(x) = 2x - 1$
111. a. $\dfrac{-x - 26}{(x + 5)(x - 2)}$
   b. Domain: All real numbers except -5 and 2
      Zero: $x = -26$
113. a. $k^4 x^2$
   b. z varies with the square of x.
115. $A = \frac{1}{4\pi}C^2$ **117.** $h = \sqrt{d^2 - 40,000}$, $d \geq 200$
119. a. 32 feet   **b.** Yes; approximately 117 feet
121. $y = 4.98x - 41.34$; 48.3%
123. $y = \frac{11}{3}x^2 + \frac{94}{3}x + 23$; 422 million

Chapter 3 Test ■ *page 319*

1. False   **2.** True   **3.** True   **4.** True
5. True   **6.** False   **7.** True   **8.** False
9. $f(x) = 2x$: $f^{-1}(x) = \frac{1}{2}x$, for example
10. $f(x) = x^2$; $(0, 0)$, for example
11. $f(x) = \begin{cases} -x, & x < -1 \\ 0, & -1 \leq x \leq 1, \text{ for example} \\ x & x > 1 \end{cases}$
12. $(0, 0), (1, 1), (2, 2)$, for example
13. $f(x) = x^3$, for example   **14.** $f(x) = x + 1$, for example
15. $x = -2$ **16.** $y = \frac{3}{4}x + \frac{19}{4}$
17. y defines a function of x; the set of all real numbers except 2
18. Inverses since $f(g(x)) = g(f(x)) = x$
19. a. 7   **b.** 7   **c.** 49
20. $[2, 3) \cup (3, \infty)$

21.

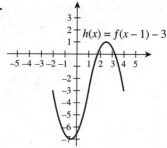

22. Zeros: $x = 0, x = 6$
Turning points: Local maximum at $(0, 0)$, local minimum at $(4, -32)$
Increasing: $(-\infty, 0) \cup (4, \infty)$
Decreasing: $(0, 4)$

23.

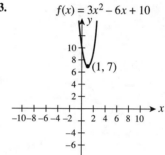

Vertex $(1, 7)$

24. $z = \frac{40}{9}$ **25.** -3 **26.** 1 minute 37.19 seconds

CHAPTER 4

Section 4.1 ■ *page 331*

1. $f(x) \to \infty$ as $x \to \infty$ or $-\infty$; graph: ii
3. $f(x) \to -\infty$ as $x \to \infty$ or $-\infty$; graph: i
5. $f(x) \to \infty$ as $x \to \infty$ or $-\infty$
7. $f(x) \to -\infty$ as $x \to \infty$ or $-\infty$
9. $h(x) \to \infty$ as $x \to -\infty$, $h(x) \to -\infty$ as $x \to \infty$
11. $g(x) \to -\infty$ as $x \to -\infty$, $g(x) \to \infty$ as $x \to \infty$
13. a. 1   **b.** 0
15. a. 5   **b.** 4
17. $f(x) = (x - 4)(x + 2)$
   Zeros: 4, -2
19. $f(x) = 3x(x - 1)(x + 1)$
   Zeros: 0, 1, -1
21. $f(x) = -x^2(x + 6)^2$
   Zeros: $x = 0, -6$
23. Zeros: $-1.05, 3.36, 3.69$
   Turning points: Local maximum at $(0.47, 14.13)$, local minimum at $(3.53, -0.13)$

25. Zero: -15.04
Turning points: Local maximum at $(-10.14, 58.04)$, local minimum at $(-0.53, 2.48)$
27. Zeros: $-1.70, 0, 1.78, 11.91$
Turning points: Local maximum at $(-1.00, -6.25)$, local minimum at $(1.00, 5.75)$, local maximum at $(9.00, -506.25)$
29. Zeros: $-4.16, -0.80, 11.96$
Turning points: Local maximum at $(-2.58, 4.07)$, local minimum at $(7.24, -43.29)$
31. Zeros: $1.64, 10.00$
Turning point: Local maximum at $(8.33, 668.79)$
33. Deaths peak in approximately 1994; deaths drop to 10,000 in approximately 1999; $f(x) \to -\infty$ as $x \to \infty$ and so f fails to be valid after $f(x)$ drops below 0; $f(1.5)$ could be interpreted as the number of deaths between July 1, 1984, and June 30, 1985.
35. Turning point: $(4.84, 14.85)$, corresponding to 1965; recycling percentage increased from 1965 to 1995
37. a. $f(t) \to -\infty$ as $t \to \infty$, whereas $g(t) \to \infty$ as $t \to \infty$

b. g; $f(t)$ eventually becomes negative, whereas $g(t)$ continues to increase

c. $t \approx 0.2$ (early 1990), $t \approx 2.6$ (mid-1992), $t \approx 4.6$ (mid-1994)

d. Approximately 5.3 in early 1991

39. True   **41.** False   **43.** True

Section 4.2 ■ *page 340*

1. $q(x) = 2x + 3, r(x) = 0$ **3.** $q(x) = 4x - 3, r(x) = 0$
5. $q(x) = x^2 - 3x - 3, r(x) = 18$ **7.** $q(x) = x - 1, r(x) = 2$
9. $q(x) = 3x, r(x) = 5x + 2$
11. $q(x) = x^2 - 1, r(x) = -x^2 + x + 1$
13. $q(x) = 2x - 3, r(x) = 0$ **15.** $q(x) = x^2 - 2, r(x) = 0$
17. $q(x) = 3x^3 + x^2 + 3x - 1, r(x) = 0$
19. $q(x) = x - 1, r(x) = -1$ **21.** $q(x) = x^2 - 2x, r(x) = 2$
23. $q(x) = -4x^3 - \frac{1}{3}x^2 - \frac{1}{9}x - \frac{1}{27}, r(x) = \frac{80}{81}$
25. $f(x) = (x + 5)(x + 2)(x - 1)$
Zeros: $-5, -2, 1$
27. $p(x) = (2x + 1)(x - 2)(x - 3)$
Zeros: $-\frac{1}{2}, 2, 3$
29. $g(x) = (x + 4)^2(x + 2)(x - 1)$
Zeros: $-4, -2, 1$
31. $h(x) = (3x + 2)(2x + 1)(x - 1)(x - 8)$
Zeros: $-\frac{2}{3}, -\frac{1}{2}, 1, 8$
33. a. 39   **b.** -6 **c.** 4
35. a. 12   **b.** 1608   **c.** 88.2663
37. a. 14   **b.** 33
39. False   **41.** False

Section 4.3 ■ *page 350*

1. $f(x) = (x + 3)(x + 2)(x - 2)$
Zeros: $-3, -2, 2$
3. $f(x) = (x + 1)(2x + 1)(x - 5)$
Zeros: $-1, -\frac{1}{2}, 5$
5. $f(x) = 3(x + 4)\left(x + \frac{1}{3}\right)(x - 1)^2$
Zeros: $-4, -\frac{1}{3}, 1$
7. $f(x) = (x + 2)(x - 1)(5x - 2)$
Zeros: $-2, 1, \frac{2}{5}$
9. $f(x) = (x - 2)(2x - 1)(3x + 1)$
Zeros: $2, \frac{1}{2}, -\frac{1}{3}$
11. $f(x) = (x + 1)(x - 2)(x^2 - 3)$
Zeros: $-1, 2, -\sqrt{3}, \sqrt{3}$
13. $f(x) = x^2 - x - 12$, for example
15. $f(x) = \frac{1}{2}(x - 5)(x - 1)(x + 2)$ **17.** $f(x) = \frac{1}{3}x^2(x - 2)^2$
19. $f(x) = (x + 1)^3\left(x - \frac{1}{2}\right)^2$, for example
21. $f(x) = -x^2 + 2x + 3$ **23.** $f(x) = \frac{1}{4}(x + 4)(x - 2)^2$
25. $f(x) = -(x + 2)(x + 1)(x - 1)(x - 2)$
27. $f(x) = (x - 3)(x + 2i)(x - 2i)$
Zeros: $3, -2i, 2i$
29. $f(x) = (x - 1 - i)(x - 1 + i)(x - 4)(x + 1)$
Zeros: $1 + i, 1 - i, 4, -1$
31. Zeros: $i, -i$
$f(x) = (x - i)(x + i)$
33. Zeros: $0, 1 + 2i, 1 - 2i$
$f(x) = x(x - 1 - 2i)(x - 1 + 2i)$
35. Zeros: $-1, 1, -2i, 2i$
$f(x) = (x + 1)(x - 1)(x + 2i)(x - 2i)$

37. Zeros: $0, 1, 2 - i, 2 + i$
$f(x) = x^2(x - 1)(x - 2 + i)(x - 2 - i)$
39. $f(x) = (x + 2)(x^2 + 4)$, for example
41. $f(x) = (x^2 + 1)(x^2 - 2x + 2)$, for example
43. $f(x) = x(x^2 - 2)(x^2 + 4x + 8)$, for example
45. Zeros: 0, 8, 10
Domain: (0, 8)
Original dimensions: $20'' \times 16''$
47. Profit is zero at 100 or 300 frames per month and is positive between these two values.
49. $s(t) = -16t^2 + 80t - 64$; 36 feet
51. False   **53.** False

Section 4.4 ■ *page 360*

1. Possible rational zeros: $\pm1, \pm3, \pm9, \pm27$
Actual rational zeros: $-3, 3$
3. Possible rational zeros: $\pm1, \pm2, \pm3, \pm4, \pm6, \pm12$
Actual rational zeros: $-2, 2, 3$
5. Possible rational zeros: $\pm1, \pm2, \pm\frac{1}{2}, \pm\frac{1}{4}$
Actual rational zeros: $-1, -\frac{1}{2}, \frac{1}{2}, 2$
7. Possible rational zeros: $\pm1, \pm3, \pm\frac{1}{2}, \pm\frac{3}{2}$
Actual rational zeros: $-\frac{3}{2}, -1, 1$
9. Possible rational zeros: $\pm1, \pm2, \pm3, \pm4, \pm6, \pm12, \pm\frac{1}{2}, \pm\frac{3}{2}$
Actual rational zeros: $-1, \frac{3}{2}, 4$
11. Number of positive zeros: 0
Number of negative zeros: 1
13. Number of positive zeros: 0
Number of negative zeros: 0
15. Possible number of positive zeros: 1 or 3
Actual number of positive zeros: 1
Number of negative zeros: 0
17. Number of positive zeros: 1
Possible number of negative zeros: 0 or 2
Actual number of negative zeros: 0
19. Possible number of positive zeros: 0 or 2
Actual number of positive zeros: 2
Number of negative zeros: 1
27. $-6, -1, 1$ **29.** 1, 12   **31.** $-3, \frac{1}{6}, \frac{1}{2}$
33. $-\frac{1}{8}, -\sqrt{2}, \sqrt{2}$ **35.** $-11, 11$
37. $x = -2, x = -1, x = 2$
39. $x = -3, x = 3$ **41.** $x = -1, x = -\frac{1}{2}, x = 1$
43. -0.32 **45.** $-1.49, 0.80$ **47.** 11.00
49. 1.5 inches $\times$ 1.5 inches $\times$ 4 inches
51. The square should have a side length of either 0.5 inch or approximately 1.7 inches.
53. 1968 and 1980
55. c. $v \approx 0.953$; $r \approx 4.9\%$
57. c. Either 1 or 3 positive zeros and 1 negative zero
d. $v \approx 1.189, v \approx -1.028$
$r \approx -15.9\%, r \approx 197.3\%$
59. True   **61.** False

Section 4.5 ■ *page 374*

1. Domain: all real numbers except 2
Zeros: none; graph: vii

3. Domain: all real numbers except 0
Zero: -1; graph: iii

5. Domain: all real numbers except -2
Zeros: none; graph: iv

7. Domain: all real numbers except 1 and -1
Zeros: ± 2; graph: ii

9. a.

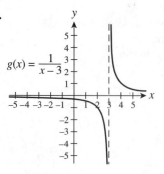

$g(x) = \dfrac{1}{x-3}$

b.

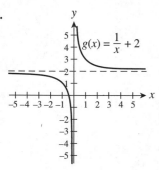

$g(x) = \dfrac{1}{x} + 2$

c.

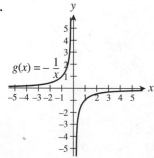

$g(x) = -\dfrac{1}{x}$

11. a.

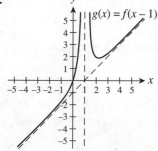

$g(x) = f(x-1)$

b.

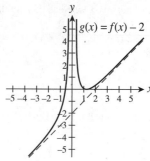

$g(x) = f(x) - 2$

c.

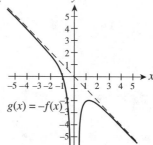

$g(x) = -f(x)$

13.

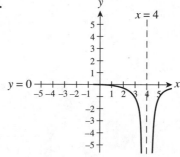

$x = 4$

$y = 0$

15.

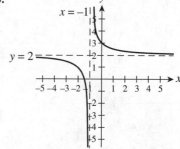

$x = -1$

$y = 2$

17. Vertical asymptote: $x = -3$
Horizontal asymptote: $y = 0$

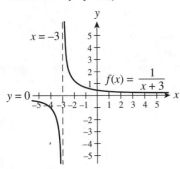

19. Vertical asymptote: $x = -2$
Horizontal asymptote: $y = 1$

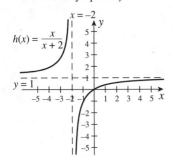

21. Vertical asymptote: $x = 2$
Horizontal asymptote: $y = 3$

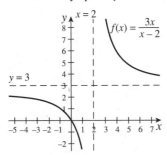

23. Vertical asymptotes: $x = 3$, $x = -3$
Horizontal asymptote: $y = 0$

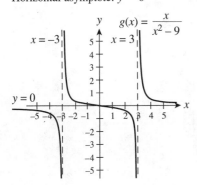

25. Vertical asymptote: none
Horizontal asymptote: $y = 5$

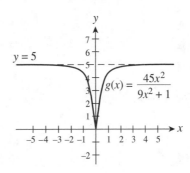

27. Vertical asymptote: $x = -1$
Horizontal asymptote: $y = 0$

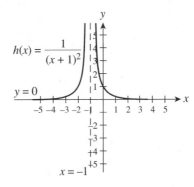

29. Vertical asymptote: $x = 2$
Horizontal asymptote: $y = 1$

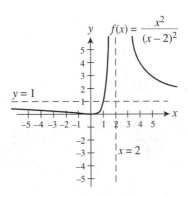

31. Vertical asymptote: $x = 0$
Inclined asymptote: $y = 2x$

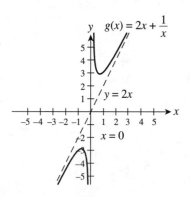

33. Vertical asymptote: $x = 1$
Inclined asymptote: $y = x + 1$

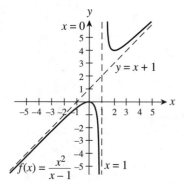

35. Vertical asymptote: $x = -2$
Inclined asymptote: $y = -3x$

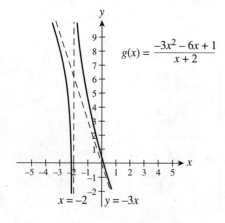

37. Horizontal asymptote: $y = 0$

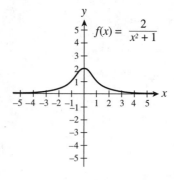

39. Vertical asymptote: $x = 15$
Horizontal asymptote: $y = 12$

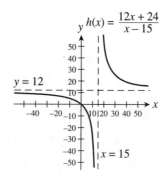

41. Vertical asymptote: $x = 3$
Inclined asymptote: $y = x - 1$

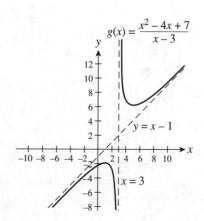

43. Vertical asymptotes: $x = 3$, $x = -3$
Horizontal asymptote: $y = 0$

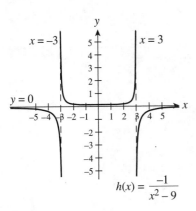

$$h(x) = \frac{-1}{x^2 - 9}$$

45. Vertical asymptote: $x = 3$

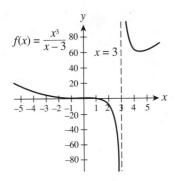

$$f(x) = \frac{x^3}{x - 3}$$

47. Vertical asymptote: $x = 100$
Horizontal asymptote: $y = -0.001$
A 75% reduction; 100% reduction is unattainable according to this model.

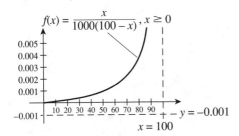

$$f(x) = \frac{x}{1000(100 - x)}, x \geq 0$$

49. Vertical asymptote: none
Horizontal asymptote: $y = 0$
Approximately 11.9 years; unemployment rate decreases as education increases; 0% unemployment is unattainable according to this model.

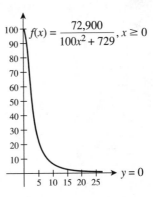

$$f(x) = \frac{72,900}{100x^2 + 729}, x \geq 0$$

51. Horizontal asymptote: $y = 100$
The average cost approaches \$100 as the number of bicycles increases.

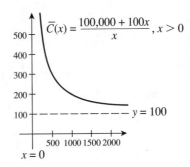

$$\overline{C}(x) = \frac{100,000 + 100x}{x}, x > 0$$

53. a. 58 balls should be ordered 10 times per year.
b. Inclined asymptote: $y = \frac{3}{2}x + 600$; as the number of balls ordered gets very large, the inventory cost increases by roughly \$1.50 for each additional ball ordered.

55. False   **57.** True

Chapter 4 Review ▒ *page 380*

1. $g(x) \to -\infty$ as $x \to \infty$ or $-\infty$
3. $f(x) \to -\infty$ as $x \to -\infty$, $f(x) \to \infty$ as $x \to \infty$
5. $h(x) \to -\infty$ as $x \to -\infty$ or ∞
7. Zero: 9.11
Turning points: Local maximum at $(0.00, -3.00)$, local minimum at $(6.00, -39.00)$
9. Zeros: -13.12, 3.12
Turning points: Local minimum at $(-9.61, -417.63)$, local maximum at $(0.13, -3.94)$, local minimum at $(1.98, -8.40)$
11. $q(x) = 2x - 1$, $r(x) = 0$
13. $q(x) = x^2 - 4x + 11$, $r(x) = -44x + 2$
15. $q(x) = 3x^2 - 5x + 2$, $r(x) = 0$
17. $q(x) = -2x^3 + 6x^2 - 14x + 43$, $r(x) = -132$
19. a. $\frac{255}{8}$ **b.** 10
21. a. -610 **b.** 2
23. $f(x) = (x + 3)(2x - 1)$
Zeros: $-3, \frac{1}{2}$
25. $h(x) = (x - 3)(x + 3)(x^2 - 2)$
Zeros: $3, -3, -\sqrt{2}, \sqrt{2}$

27. $h(x) = (x + 1)(2x + 1)(3x - 1)$
 Zeros: $-1, -\frac{1}{2}, \frac{1}{3}$
29. $f(x) = x^3 - 6x^2 - x + 30$, for example
31. $f(x) = 4(x - 1)^2(x + 1)^2$ **33.** $f(x) = x^3 - 3x^2 - x + 3$
35. $f(x) = x^3 + 3x^2 + 16x + 48$, for example
37. $f(x) = (x - 2 - i)(x - 2 + i)$
 Zeros: $2 - i, 2 + i$
39. $h(x) = (x - 3)(x + 3)(x - 4i)(x + 4i)$
 Zeros: $3, -3, -4i, 4i$
41. Possible rational zeros: $\pm 1, \pm 2, \pm 3, \pm 6$
 Actual rational zeros: $-2, 1, 3$
43. Possible rational zeros: $\pm 1, \pm 3, \pm\frac{1}{2}, \pm\frac{1}{4}, \pm\frac{3}{2}, \pm\frac{3}{4}$
 Actual rational zeros: $-1, \frac{1}{2}, 1, \frac{3}{2}$
45. Number of positive zeros: 0
 Number of negative zeros: 1
47. Number of positive zeros: 1
 Possible number of negative zeros: 2 or 0
 Actual number of negative zeros: 2
51. $x = -1, x = 2, x = 6$ **53.** $x = -\frac{2}{3}, x = -\frac{1}{4}, x = 4$
55. $-2.13, -0.20, 2.33$ **57.** $-21.00, 1.52$
59. Domain: all real numbers except 3
 Zeros: none; graph: iv
61. Domain: all real numbers except 3 and -3
 Zeros: -2; graph: i

63. Vertical asymptote: $x = -2$
 Horizontal asymptote: $y = 0$

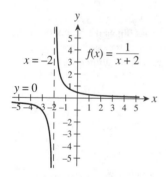

65. Vertical asymptote: $x = \frac{1}{3}$
 Horizontal asymptote: $y = \frac{2}{3}$

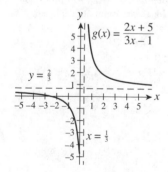

67. Vertical asymptote: $x = 0$
 Horizontal asymptote: $y = 0$

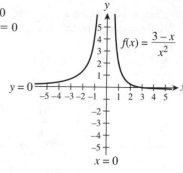

69. Horizontal asymptote: $y = 2$

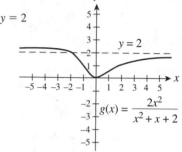

71. Vertical asymptote: $x = 1$
 Horizontal asymptote: $y = 0$

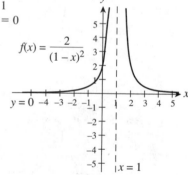

73. Vertical asymptote: $x = 0$
 Inclined asymptote: $y = -2x$

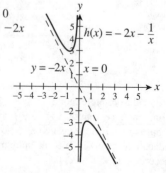

75. Vertical asymptote: $x = -3$
Inclined asymptote: $y = 3x - 1$

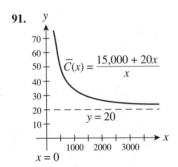

$$h(x) = \frac{3x^2 + 8x - 9}{x + 3}$$

$$y = 3x - 1$$

$x = -3$

77. a. $\dfrac{x^2 - 4x + 5}{x - 1}$ **b.** $y = x - 3$

79. a. $x = \dfrac{-3 \pm \sqrt{13}}{2}$ **b.** $x = \dfrac{-3 \pm \sqrt{13}}{2}, x = 0$

81. a. $x^2 + 4x + 4$ **b.** $(x + 5) + \dfrac{9}{x - 1}$

83. a. Vertical: $x = 3$
Horizontal: $y = -2$

b. $h(x) = \dfrac{1}{x - 3} - 2$

85. a. $(2 - 3i) + (2 + 3i) = 4$
$(2 - 3i)(2 + 3i) = 13$

b. $f(x) = x^2 - 4x + 13$

87. $f(x) = -\dfrac{3}{1000}x^2 + 18x - 15{,}000$
The largest profit of \$12,000 occurs when $x = 3000$.

89. Zeros: 2, 10.3
The months when the average temperature is zero

91.

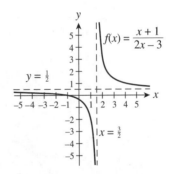

$$\overline{C}(x) = \frac{15{,}000 + 20x}{x}$$

$y = 20$

$x = 0$

The average cost approaches \$20.

Chapter 4 Test ■ *page 384*

1. True   **2.** True   **3.** False   **4.** False
5. True   **6.** True   **7.** False   **8.** True
9. $f(x) = -x$, for example
10. $f(x) = x^3 - 6x^2 + 5x + 12$, for example
11. $f(x) = x^3 - 2x^2 + 9x - 18$, for example
12. $f(x) = x^4 + 1$, for example
13. $f(x) = \dfrac{x}{x + 1}$, for example
14. $f(x) = x - 3 + \dfrac{1}{x - 2}$, for example
15. $f(x) = x^2(x - 4)(2x + 1)$
Zeros: $0, 4, -\frac{1}{2}$
16. $f(x) = (x + 2)(3x - 1)(2x - 3)$
Zeros: $-2, \frac{1}{3}, \frac{3}{2}$
17. $q(x) = 3x^3 - 6x^2 + 5, r(x) = 4$
18. $f(x) = x(5x - 2)(x + 3i)(x - 3i)$
Zeros: $0, \frac{2}{5}, -3i, 3i$
19. Possible rational zeros: $\pm 1, \pm 3, \pm 9, \pm \frac{1}{2}, \pm \frac{3}{2}, \pm \frac{9}{2}$
Actual rational zeros: $-3, 1, \frac{3}{2}$
20. Possible number of positive zeros: 3 or 1
Actual number of positive zeros: 1
Number of negative zeros: 1
21. Approximately 1.52

22. Vertical asymptote: $x = \frac{3}{2}$
Horizontal asymptote: $y = \frac{1}{2}$

$f(x) = \dfrac{x + 1}{2x - 3}$

$y = \frac{1}{2}$

$x = \frac{3}{2}$

23. a. Approximately 1988 and 1995
b. The rate peaked at approximately 9.5 per 100,000 in 1992.
c. The crime rate dropped for 1992 through 1997.

CHAPTER 5

Section 5.1 ▪ *page 395*

1.

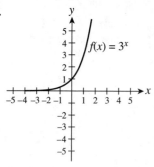

a.

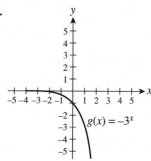

b.

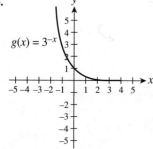

c.

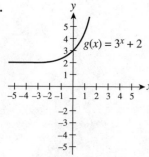

d.

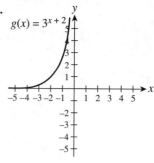

3. $a = \frac{1}{3}$ **5.** $a = 10$

7.

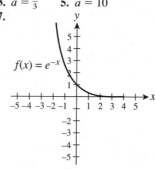

9.

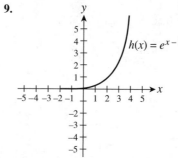

11. $t \approx 11.79$ **13.** $t \approx 49.70$ **15.** $x = 4$ **17.** $x = 3$
19. $x = -3$ **21.** $x = -5$ **23.** $x = -2, x = 1$
25. a. $1790.85   **b.** $1814.02   **c.** $1819.40
   d. $1822.03   **e.** $1822.12
27. 20 gerbils initially present;

t	$f(t)$
0	20
1	30
2	45
3	68
4	101

Approximately 228 gerbils after 6 months; approximately 735 billion gerbils after 5 years
29. $20,805.70; $21,103.48

31. Approximately 25.7 quadrillion Btu in 1940; approximately 76.4 quadrillion Btu in 1995; approximately 197% increase; approximately 102.8 quadrillion Btu in 2010

33. Approximately 4.64 hours

35.

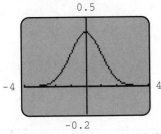

Maximum at approximately $(0, 0.40)$; $f(x)$ approaches 0 as x approaches $\pm\infty$

37. a. Approximately 130 students after 2 days; approximately 190 students after 5 days

b.

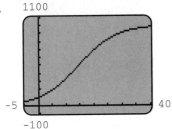

c. Approximately 24 days

d. $N(t)$ approaches 1000 as t approaches ∞.

39. True   **41.** True

Section 5.2 ▦ *page 408*

.23856

1. 2.72130   **3.** -0.17609 **5.** 0.53959   **7.** -0.70877
9. 4   **11.** Undefined   **13.** 2   **15.** -3
17. -2 **19.** -3 **21.** 100   **23.** 7   **25.** 2.24
27. 5.64   **29.** -6.75 **31.** $2^{-2} = \frac{1}{4}$ **33.** $10^3 = 1000$
35. $e^2 = x + 1$ **37.** $x^3 = 10$ **39.** $\log_3 9 = 2$
41. $\log_{1/2} 8 = -3$ **43.** $\ln 5 = x$ **45.** $\ln \frac{1}{2} = -0.013t$

47.

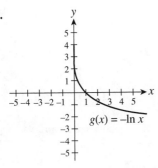

49.

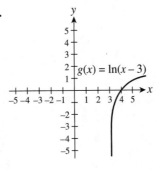

51.

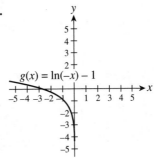

53. $f^{-1}(x) = 3^x$

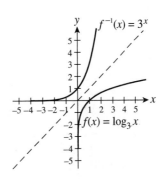

55. $f^{-1}(x) = \ln(x + 5)$

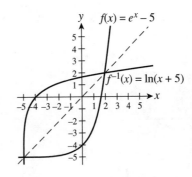

57. $f^{-1}(x) = \dfrac{10^x}{1000} = 10^{x-3}$

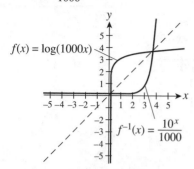

59.

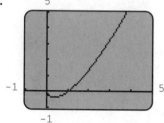

Domain: $(0, \infty)$

61.

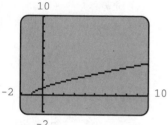

Domain: $(-1, \infty)$

63. Approximately 49.7 minutes   **65.** Approximately 13.4 hours
67. Approximately 20.4 decibels   **69.** 100 decibels
71. Approximately 3.55   **73.** Approximately 4.5
75. 25 times
77. a. 60 wpm in approximately 4.6 weeks; 80 wpm in
approximately 8 weeks
b. The time gets larger without bound; no

c.

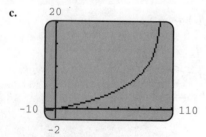

Domain: $[0, 100]$

79. True   **81.** False   **83.** True

Section 5.3 ■ *page 420*

1. $2 \ln x + \ln y$ **3.** $3 \log_5 x + 4 \log_5 y + 3 \log_5 z$
5. $1 - \ln 7 - 3 \ln x$ **7.** $\frac{1}{4} \log x + \frac{1}{2} \log y - \log z$
9. $\log \dfrac{x}{y^2}$ **11.** $\ln \dfrac{x^3 y^4}{z}$ **13.** $\ln \sqrt[3]{\dfrac{x - z}{(x + z)^2}}$ **15.** $\log \frac{64}{3}$
17. $\ln \dfrac{x + 3}{x + 2}$ **19.** $x = 27$ **21.** $x = 2$ **23.** $x = \sqrt[3]{25}$
25. $x = -\frac{15}{32}$ **27.** $x = -\frac{1}{3}$ **29.** $x = 2$ **31.** $x = \sqrt{6} - 1$
33. $x = 95$ **35.** $x = 1$ **37.** No solution   **39.** $x = \pm 999$
41. All $x > 0$ **43.** $x = e^e$ **45.** $\dfrac{\ln 16}{\ln 3} \approx 2.52372$
47. $\dfrac{\ln 0.341}{\ln 12} \approx -0.43296$ **49.** $\dfrac{\ln 10}{\ln \pi} \approx 2.01147$
51. 0.16, 3.15   **53.** 0.04, 1.50   **55.** $-11.42, -2.97, 0.59$
57. 10,000 times more intense   **59.** 30 decibels
61. Approximately 586.8 tons
63. Approximately 177.8 times more energy
65. False   **67.** True   **69.** True   **71.** True

Section 5.4 ■ *page 432*

1. $x = 4$ **3.** $x = 1$ **5.** $x = \dfrac{\log 10}{\log 7} \approx 1.18329$
7. $x = -\dfrac{\log 11}{\log 3} \approx -2.18266$
9. $x = \dfrac{\log 36}{\log \frac{18}{5}} \approx 2.79758$
11. $x = \dfrac{\ln 2}{1 - \ln 2} \approx 2.25889$
13. $x = \dfrac{1}{2(1 - \ln \pi)} \approx -3.45471$
15. $x = 1$ **17.** $t = \dfrac{3}{\log 1.06} \approx 118.54959$
19. $x \approx 0.69, x \approx 1.10$ **21.** $x \approx 0.57$
23. $x \approx -4.48, x \approx -2.11, x \approx 0.35$
25. Approximately 10.2 years   **27.** Approximately 5.78%
29. February 2005   **31.** August 2063
33. Approximately 197.6; the year 2005
35. At approximately 6.33 A.M. on July 7, 2001
37. Approximately 88,949   **39.** Approximately 11,520 years
41. Approximately 5.2 years   **43.** True   **45.** False

Section 5.5 ■ *page 444*

1. a.

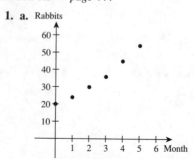

b. Rabbits

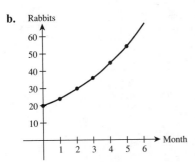

c. Rabbits

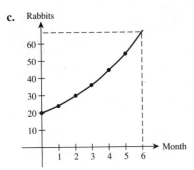

Approximately 66 rabbits (Answers may vary.)

3. $P(t) = 20e^{0.1823t}$

t	Actual data	$P(t) = 20e^{0.1823t}$	Absolute error
0	20	20	0
1	24	24	0
2	30	28.8	1.2
3	36	34.6	1.4
4	45	41.5	3.5
5	54	49.8	4.2

Largest absolute error: 4.2
Sum of the absolute errors: 10.3

5. $P(t) = 20e^{0.1987t}$

t	Actual data	$P(t) = 20e^{0.1987t}$	Absolute error
0	20	20	0
1	24	24.4	0.4
2	30	29.8	0.2
3	36	36.3	0.3
4	45	44.3	0.7
5	54	54	0

Largest absolute error: 0.7
Sum of the absolute errors: 1.6
This model is more accurate.

7. $P(t) = 19.9e^{0.201t}$

t	Actual data	$P(t) = 19.9e^{0.201t}$	Absolute error
0	20	19.9	0.1
1	24	24.3	0.3
2	30	29.7	0.3
3	36	36.4	0.4
4	45	44.5	0.5
5	54	54.4	0.4

Largest absolute error: 0.5
Sum of the absolute errors: 2.0
The largest error is smaller, but the sum of the errors is larger.
Both models are extremely accurate.

9. $P(t) = 225,300e^{0.0295t}$

t	Actual data	$P(t) = 225,300e^{0.0295t}$	Absolute error
0	226,000	225,300	700
10	303,000	302,600	400
20	402,000	406,400	4400
30	550,000	545,900	4100

Largest absolute error: 4400
Sum of the absolute errors: 9600
The population in 2010 will be approximately 985,000.
The population will first exceed 1,500,000 in 2024.

11. $S(t) = 78.7e^{-0.0809t}$

t	Actual data	$S(t) = 78.7e^{-0.0809t}$	Absolute error
0	80	78.7	1.3
1	72	72.6	0.6
2	66	66.9	0.9
3	61	61.7	0.7
4	58	56.9	1.1

Largest absolute error: 1.3
Sum of the absolute errors: 4.6
The sales for month 6 will be approximately $48,400.
The sales will reach $20,000 in month 17.

13. $p(t) = 9.2e^{0.5142t}$

t	Actual data	$p(t) = 9.2e^{0.5142t}$	Absolute error
0	8	9.2	1.2
1	19	15.4	3.6
2	26	25.7	0.3
3	40	43.0	3.0

Largest absolute error: 3.6
Sum of the absolute errors: 8.1
90% will have encountered the virus approximately halfway through the 4th quarter.

15. $N(t) = 2.7e^{0.3214t}$

t	**Actual data**	$N(t) = 2.7e^{0.3214t}$	**Absolute error**
0	2	2.7	0.7
3	6	7.1	1.1
7	29	25.6	3.4
11	134	92.6	41.4
14	275	243.0	32.0
18	1200	878.7	321.3
22	3100	3,178.1	78.1
26	7500	11,494.8	3994.8

Largest absolute error: 3994.8
Sum of the absolute errors: 4472.8
There will be approximately 41,575 thousand (41,575,000) transistors in 2001.
The number of transistors will exceed 100 billion in 2025.

17. a. 4 students   **b.** Approximately 77 students
 c. Approximately 17 hours

d.

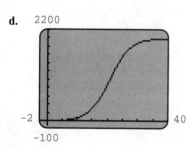

Approximately 2000 students

e. The 21st hour

19. a. 1944   **b.** 700 mph

21. $v = 0.81 \log P + 0.3$

P	**Actual**	$v = 0.81 \log P + 0.3$	**Absolute error**
5,500	3.3	3.33	0.03
14,000	3.7	3.66	0.04
71,000	4.3	4.23	0.07
138,000	4.4	4.46	0.06
342,000	4.8	4.78	0.02

Largest absolute error: 0.07

Sum of the absolute errors: 0.22
Little Rock: 4.5 feet/second
New York: 5.9 feet/second
Mexico City: 6.2 feet/second

23. False   **25.** True

Chapter 5 Review ▪ *page 450*

1.

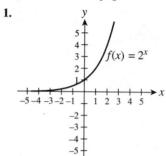

a.

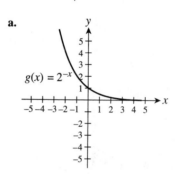

b.

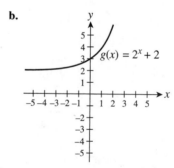

c.

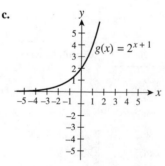

3. $x = 3$ **5.** $x = -\frac{1}{2}$ **7.** -2 **9.** -2 **11.** $\frac{1}{2}$
13. $2^{-4} = \frac{1}{16}$ **15.** $10^4 = 2x + 1$ **17.** $\ln x = 5$
19. $\log_2 5 = x + 3$

21.

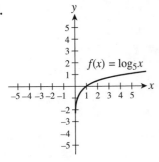

Domain: $(0, \infty)$

23.

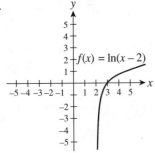

Domain: $(2, \infty)$

25. $\log_2 x + 2 \log_2 y$ **27.** $\frac{2}{3} \log x + \frac{1}{3} \log y$ **29.** $\log(x^2 y^3)$

31. $\ln\left(\dfrac{\sqrt[3]{xy^2}}{z}\right)$ **33.** $\dfrac{\ln 10}{\ln 4} \approx 1.66096$ **35.** $x = \frac{1}{2}$

37. $x = 3$ **39.** $y = \frac{12}{5}$ **41.** $x = 2$

43. $x = \dfrac{\ln 10}{\ln 5} \approx 1.43068$ **45.** $t = \dfrac{\ln 10}{\ln 1.05} \approx 47.1936$

47. $x = 10 \ln 5 \approx 16.0944$ **49.** $x = \dfrac{2 \ln 3}{2 \ln 2 - \ln 3} \approx 7.63768$

51. $x \approx -2.82$, $x \approx 1.66$ **53.** $x \approx 0.17$, $x \approx 5.05$, $x \approx 45.48$

55. a. $\left(\frac{2}{3}\right)^x$ **b.** $x = \dfrac{\ln 2}{\ln(2/3)} \approx -1.70951$

57. a. $x = -2, x = 1$ **b.** $x = -2, x = 1$

59. a.

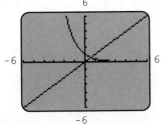

b. One solution

61. a. $x \leq \frac{3}{2}$ **b.** $x > \frac{3}{2}$
63. a. $u = 1$ **b.** $x = 0$

65. a.

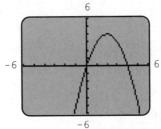

b. $(0, 4)$

67. $56.34   **69.** Approximately 23.1 years
71. Approximately 12.6 hours
73. a. Approximately 149 astronauts
   b. Approximately 28.8 days

c.

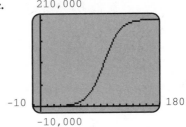

Approximately 200,000 astronauts

d. The 96th day

75. $V(t) = 13{,}978 e^{-0.1495t}$

t	Actual data	$V(t) = 13{,}978e^{-0.1495t}$	Absolute error
0	14,000	13,978	22
1	12,000	12,037	37
2	10,400	10,366	34
3	8,900	8,926	26
4	7,700	7,687	13

Largest absolute error: 37
Sum of the absolute errors: 132
The car will be worth $2000 in 2008 years.

Chapter 5 Test ▪ *page 452*

1. False   **2.** True   **3.** False   **4.** False
5. True   **6.** True   **7.** False   **8.** True
9. $f(t) = e^{-t}$, for example   **10.** 1, for example

11. $\frac{1}{10}$, for example   **12.** Limited food, for example
13. $x = -2, x = 1$ **14.** $x = 10^{12} - 1 = 999{,}999{,}999{,}999$
15. $x = \dfrac{\ln 3}{\ln 3 - \ln 2} \approx 2.70951$ **16.** $x = \dfrac{\ln 2}{0.3} \approx 2.31049$
17. 5   **18.** -3

19.

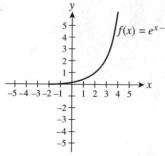
$f(x) = e^{x-2}$

20.

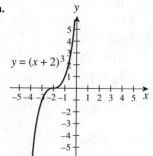
$g(x) = \left(\dfrac{2}{3}\right)^x$

21.

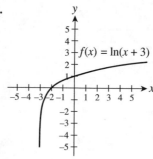
$f(x) = \ln(x + 3)$

22. $4 \log x + 2 \log y + 2 \log z$ **23.** $\ln\left(\dfrac{x^3 z^{10}}{y^4}\right)$
24. Approximately 8.8 years   **25.** 2029

26. $N(t) = 2243 e^{0.6115t}$

t	Actual data	$N(t) = 2243 e^{0.6115t}$	Absolute error
0	2,100	2,243	143
1	4,400	4,134	266
2	8,200	7,620	580
3	13,100	14,046	946

Largest absolute error: 946
Sum of the absolute errors: 1935
Approximately 25,900 new cases in 1987
The approximation overestimates the number of new cases by almost 5000.

CHAPTER 6

Section 6.1 ■ *page 466*

1. a.

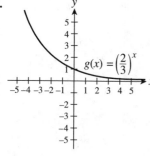
$y = (x + 2)^3$

b.

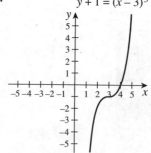
$y + 1 = (x - 3)^3$

c. $y = (x + 4)^3 - 2$

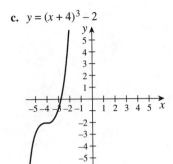

b.

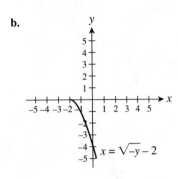

$x = \sqrt{-y} - 2$

3. a.

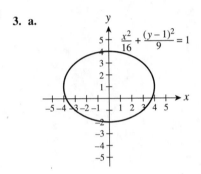

$\dfrac{x^2}{16} + \dfrac{(y-1)^2}{9} = 1$

11. a.

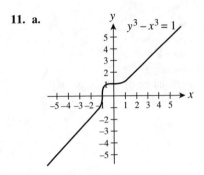

$y^3 - x^3 = 1$

b.

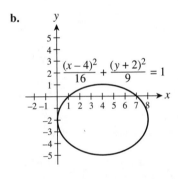

$\dfrac{(x-4)^2}{16} + \dfrac{(y+2)^2}{9} = 1$

b.

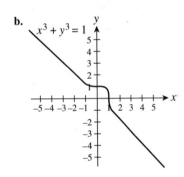

$x^3 + y^3 = 1$

5. $y - 3 = \sqrt{x + 2}$ **7.** $(x + 1)(y + 2) = 4$

13. $\sqrt{x} + \sqrt{y} = 2$

9. a.

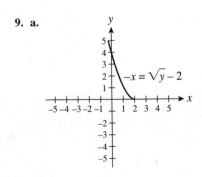

$-x = \sqrt{y} - 2$

15. a.

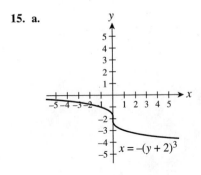

$x = -(y + 2)^3$

b.

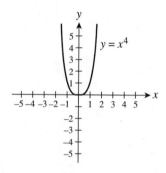

$x + 3 = -(y - 4)^3$

17. Symmetric with respect to the y-axis
19. Symmetric with respect to the x-axis
21. Symmetric with respect to the x-axis
23. Symmetric with respect to the origin
25. Not symmetric with respect to the x-axis, y-axis, or origin
27. Symmetric with respect to the origin

29. Symmetric with respect to the y-axis

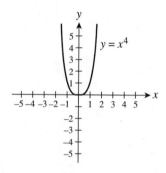

$y = x^4$

31. Symmetric with respect to the origin

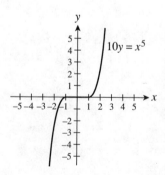

$10y = x^5$

33. Symmetric with respect to the x-axis, the y-axis, and the origin

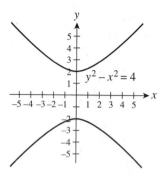

$y^2 - x^2 = 4$

35. Symmetric with respect to the y-axis

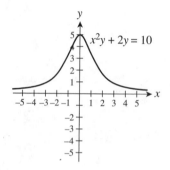

$x^2y + 2y = 10$

37. a. Translate 2 units down.
   b. Translate 3 units to the right.
   c. Reflect about the x-axis and translate 2 units down.
39. a. Translate 3 units up.
   b. Reflect about the x-axis and translate 1 unit to the left.
   c. Translate 2 units to the left and 2 units up.

41. $y = \pm\frac{4}{5}\sqrt{25 - x^2}$

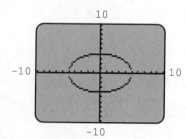

Highest point: $(0, 4)$
Lowest point: $(0, -4)$

43. $y = \pm\sqrt{5 - x - x^3}$

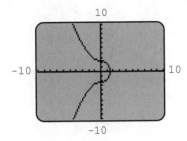

x-intercept: $x \approx 1.52$
y-intercepts: $y \approx -2.24$, $y \approx 2.24$

45. $\frac{1}{2}$

47. a. 0.1587   **b.** 0.1587
   c. 0.6826   **d.** 0.8413

49. True   **51.** False   **53.** False

Section 6.2 ▪ *page 485*

1. Vertex: $(0, 0)$; focus: $\left(0, \frac{1}{8}\right)$; directrix: $y = -\frac{1}{8}$

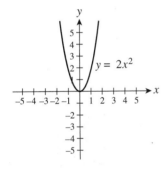

3. Vertex: $(0, 0)$; focus: $\left(\frac{3}{4}, 0\right)$; directrix: $x = -\frac{3}{4}$

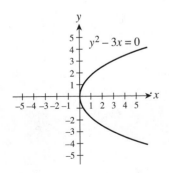

5. Vertex: $(0, 0)$; focus: $\left(-\frac{7}{16}, 0\right)$; directrix: $x = \frac{7}{16}$

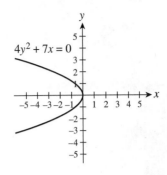

7. Vertex: $(-5, 3)$; focus: $\left(-5, \frac{37}{12}\right)$; directrix: $y = \frac{35}{12}$

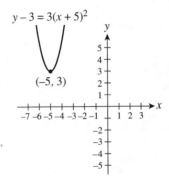

9. Vertex: $(2, 1)$; focus: $(2, 2)$; directrix: $y = 0$

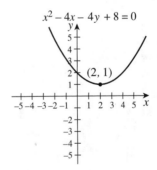

11. Vertex: $\left(-\frac{5}{2}, -\frac{3}{2}\right)$; focus: $\left(-\frac{19}{8}, -\frac{3}{2}\right)$; directrix: $x = -\frac{21}{8}$

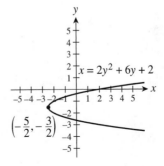

13. $x = \frac{1}{8}y^2$ **15.** $y = -\frac{1}{16}x^2$

17. $x + 2 = \frac{1}{4}y^2$

19. $y - 1 = -\frac{1}{8}(x - 4)^2$ **21.** $y + 2 = 2(x - 3)^2$

23. $y = x^2 - 4x$; vertex: $(2, -4)$

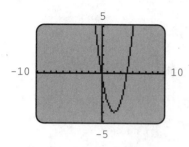

25. $y = \frac{1}{2}x^2 + 4x + 11$; vertex: $(-4, 3)$

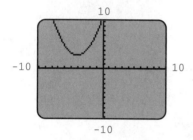

27. $y = \pm\sqrt{4x + 8}$; vertex: $(-2, 0)$

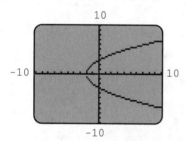

29. $y = 1 \pm \sqrt{-x - 1}$; vertex: $(-1, 1)$

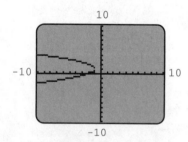

31. Maximum height: 50 feet
Horizontal distance traveled: 200 feet

33. a. $y - 87.5 = -0.12x^2$
    b. 1 foot

35. Approximately 374 feet above the vertex

37. True   **39.** False

Section 6.3 ■ *page 498*

1. Center: $(0, 0)$; vertices: $(0, -5)$, $(0, 5)$; foci: $(0, -4)$, $(0, 4)$

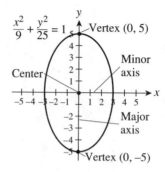

3. Center: $(0, 0)$; vertices: $(-3, 0)$, $(3, 0)$; foci: $(-\sqrt{5}, 0)$, $(\sqrt{5}, 0)$

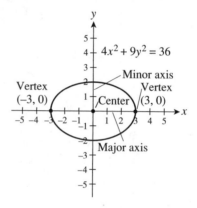

5. Center: $(0, 0)$; vertices: $(0, -\frac{1}{2})$, $(0, \frac{1}{2})$; foci: $\left(0, -\frac{\sqrt{5}}{6}\right)$, $\left(0, \frac{\sqrt{5}}{6}\right)$

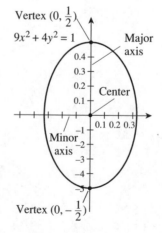

7. Center: $(0, 0)$; vertices: $(-2\sqrt{2}, 0)$, $(2\sqrt{2}, 0)$; foci: $(-\sqrt{3}, 0)$, $(\sqrt{3}, 0)$

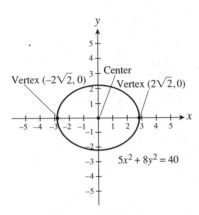

9. Center: $(-3, 1)$; vertices: $(-3, -3)$, $(-3, 5)$; foci: $(-3, 1 - 2\sqrt{3})$, $(-3, 1 + 2\sqrt{3})$

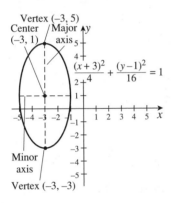

11. Center: $(1, 3)$; vertices: $(1, 0)$, $(1, 6)$; foci: $(1, 3 - \sqrt{5})$, $(1, 3 + \sqrt{5})$

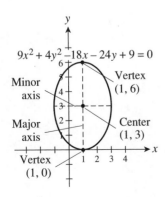

13. Center: $(-2, 1)$; vertices: $(-8, 1)$, $(4, 1)$; foci: $(-2 - \sqrt{35}, 1)$, $(-2 + \sqrt{35}, 1)$

$$x^2 + 36y^2 + 4x - 72y + 4 = 0$$

15. $\dfrac{x^2}{16} + \dfrac{y^2}{9} = 1$ **17.** $\dfrac{x^2}{4} + \dfrac{y^2}{6} = 1$

19. $\dfrac{x^2}{36} + \dfrac{y^2}{36/5} = 1$ **21.** $\dfrac{(x-2)^2}{4} + \dfrac{(y+1)^2}{9} = 1$

23. $\dfrac{(x-1)^2}{25} + \dfrac{(y+2)^2}{16} = 1$

25. $\dfrac{(x+3)^2}{1} + \dfrac{(y-1)^2}{9} = 1$

27. $y = \pm\sqrt{-\frac{1}{4}x^2 - x}$; vertices: $(-4, 0)$, $(0, 0)$

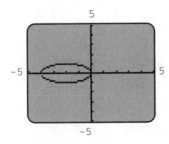

29. $y = 1 \pm \sqrt{4 - 4x^2}$; vertices: $(0, -1)$, $(0, 3)$

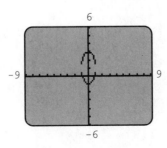

31. $\dfrac{x^2}{93^2} + \dfrac{y^2}{92.9^2} = 1$

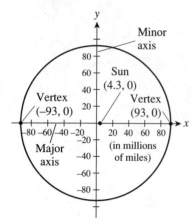

33. a. $\dfrac{x^2}{25^2} + \dfrac{y^2}{20^2} = 1$ **b.** 12 feet   **c.** Yes

35. Perihelion: approximately 0.3 A.U.
Aphelion: approximately 4.1 A.U.

37. a. $\dfrac{x^2}{85.15^2} + \dfrac{y^2}{15.1^2} = 1$

b. Approximately 14.1 centimeters

39. True   **41.** True

Section 6.4 ■ *page 513*

1. Center: $(0, 0)$; vertices: $(-3, 0)$, $(3, 0)$; foci: $(-5, 0)$, $(5, 0)$;
asymptotes: $y = \pm\frac{4}{3}x$

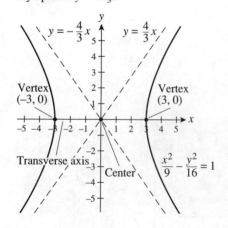

3. Center: $(0, 0)$; vertices: $(0, -3)$, $(0, 3)$; foci: $(0, -\sqrt{13})$,
$(0, \sqrt{13})$; asymptotes: $y = \pm\frac{3}{2}x$

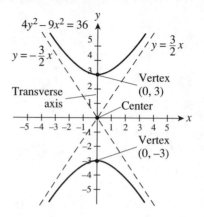

5. Center: $(0, 0)$; vertices: $(0, -2\sqrt{2})$, $(0, 2\sqrt{2})$; foci: $(0, -\sqrt{10})$,
$(0, \sqrt{10})$; asymptotes: $y = \pm 2x$

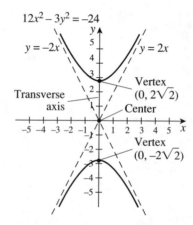

7. Center: $(1, 4)$; vertices: $(-1, 4)$, $(3, 4)$; foci: $(1 - \sqrt{13}, 4)$,
$(1 + \sqrt{13}, 4)$; asymptotes: $y = -\frac{3}{2}x + \frac{11}{2}$, $y = \frac{3}{2}x + \frac{5}{2}$

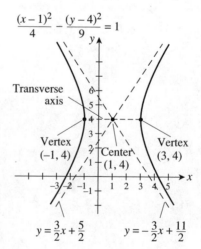

9. Center: (2, 3); vertices: (−1, 3), (5, 3); foci: $(2 - \sqrt{13}, 3)$, $(2 + \sqrt{13}, 3)$; asymptotes: $y = -\frac{2}{3}x + \frac{13}{3}$, $y = \frac{2}{3}x + \frac{5}{3}$

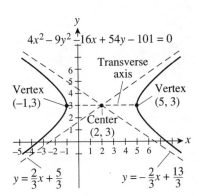

$4x^2 - 9y^2 - 16x + 54y - 101 = 0$

Transverse axis

Vertex (−1,3)

Vertex (5, 3)

Center (2, 3)

$y = \frac{2}{3}x + \frac{5}{3}$

$y = -\frac{2}{3}x + \frac{13}{3}$

11. Center: (4, 0); vertices: (4, −2), (4, 2); foci $(4, -\sqrt{29})$, $(4, \sqrt{29})$; asymptotes: $y = -\frac{2}{5}x + \frac{8}{5}$, $y = \frac{2}{5}x - \frac{8}{5}$

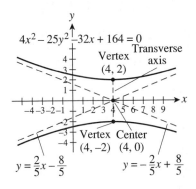

$4x^2 - 25y^2 - 32x + 164 = 0$

Vertex (4, 2)

Transverse axis

Vertex (4, −2)   Center (4, 0)

$y = \frac{2}{5}x - \frac{8}{5}$

$y = -\frac{2}{5}x + \frac{8}{5}$

13. $\dfrac{y^2}{16} - \dfrac{x^2}{9} = 1$ **15.** $\dfrac{y^2}{4} - \dfrac{x^2}{16} = 1$

17. $\dfrac{(x - 2)^2}{9} - \dfrac{(y - 3)^2}{16} = 1$ **19.** Parabola   **21.** Circle

23. Ellipse   **25.** Hyperbola

27. $y = x^2 - 5x$; parabola; vertex: (2.5, −6.25)

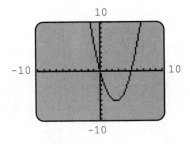

29. $y = \pm\sqrt{5 - \frac{5}{8}x^2}$; ellipse; vertices: approximately (±2.83, 0)

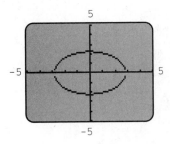

31. $y = \pm\sqrt{3x^2 - 4x - 9}$; hyperbola; vertices: approximately (−1.19, 0) and (2.52, 0)

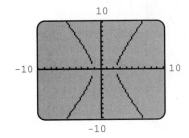

33. $y = \pm\sqrt{8 + 5x - 2x^2}$; ellipse; vertices: approximately (1.25, ±3.34)

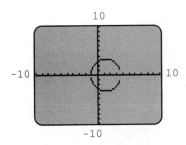

35. $y = -4 \pm \sqrt{3x^2 + 8x + 9}$; hyperbola; vertices: approximately (−1.33, −2.09), (−1.33, −5.91)

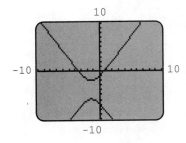

37. $y^2 - x - 3 = 0$; parabola

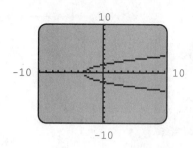

39. $x^2 + y^2 + 6x - 4y - 12 = 0$; circle

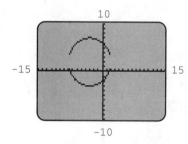

41. $x^2 - 3y^2 - 6 = 0$; hyperbola

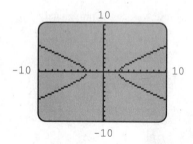

43. $2x^2 + y^2 + 8x - 2y - 1 = 0$; ellipse

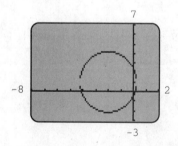

45. a. $\dfrac{y^2}{74.24^2} - \dfrac{x^2}{66.99^2} = 1$

    b. Approximately 60.5 miles east of transmitter A

47. False   **49.** False

Chapter 6 Review ■ *page 516*

1. a.

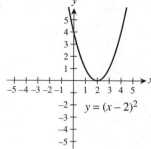

b.

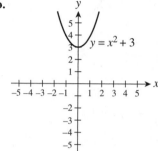

c.

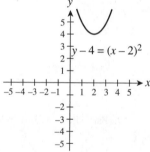

d.

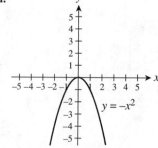

3. a.

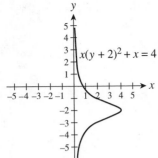

$x(y + 2)^2 + x = 4$

b.

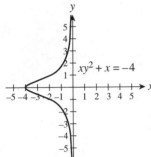

$xy^2 + x = -4$

c.

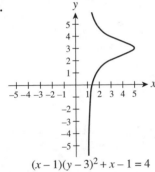

$(x - 1)(y - 3)^2 + x - 1 = 4$

5. $(x + 1)^2 + (y - 3)^4 = 16$
7. $-y = x^4 - 4x^2 + 2$
9. Symmetric with respect to the x-axis
11. No symmetry

13. Symmetric with respect to the x-axis

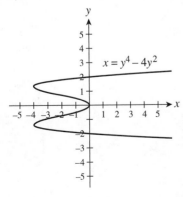

$x = y^4 - 4y^2$

15. Symmetric with respect to the origin

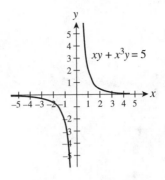

$xy + x^3y = 5$

17. Vertex: $(0, 0)$; focus: $\left(\frac{1}{8}, 0\right)$; directrix: $x = -\frac{1}{8}$

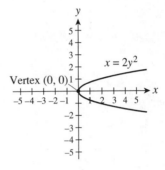

$x = 2y^2$

Vertex $(0, 0)$

19. Vertex: $(3, -4)$; focus: $\left(3, -\frac{33}{8}\right)$; directrix: $y = -\frac{31}{8}$

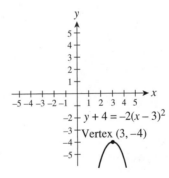

$y + 4 = -2(x - 3)^2$

Vertex $(3, -4)$

21. Vertex: $(4, -1)$; focus: $\left(\frac{49}{12}, -1\right)$; directrix: $x = \frac{47}{12}$

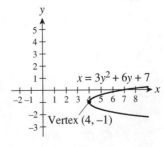

$x = 3y^2 + 6y + 7$

Vertex $(4, -1)$

23. $y = x^2$ **25.** $y - 2 = -\frac{1}{16}(x + 1)^2$

27. Center: $(0, 0)$; vertices: $(0, -8)$, $(0, 8)$; foci: $(0, \sqrt{15})$, $(0, -\sqrt{15})$

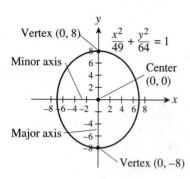

29. Center: $(3, 2)$; vertices: $(-2, 2)$, $(8, 2)$; foci: $(7, 2)$, $(-1, 2)$

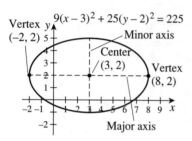

31. Center: $(1, -3)$; vertices: $(-3, -3)$, $(5, -3)$; foci: $(1 - 2\sqrt{3}, -3)$, $(1 + 2\sqrt{3}, -3)$

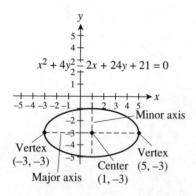

33. $\dfrac{x^2}{9} + \dfrac{y^2}{4} = 1$

35. $\dfrac{x^2}{9} + \dfrac{y^2}{25} = 1$

37. Center: $(0, 0)$, vertices: $(-5, 0)$, $(5, 0)$; foci: $(\sqrt{34}, 0)$, $(-\sqrt{34}, 0)$; asymptotes: $y = \pm\frac{3}{5}x$

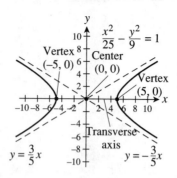

39. Center: $(-5, 1)$; vertices: $(-5, -2)$, $(-5, 4)$; foci: $(-5, 1 + \sqrt{13})$, $(-5, 1 - \sqrt{13})$; asymptotes: $y = -\frac{3}{2}x - \frac{13}{2}$, $y = \frac{3}{2}x + \frac{17}{2}$

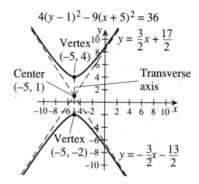

41. Center: $(3, 2)$; vertices: $(-1, 2)$, $(7, 2)$; foci: $(3 - \sqrt{41}, 2)$, $(3 + \sqrt{41}, 2)$; asymptotes: $y = -\frac{5}{4}x + \frac{23}{4}$, $y = \frac{5}{4}x - \frac{7}{4}$

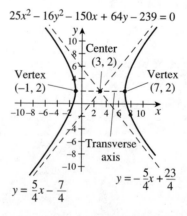

43. $\dfrac{x^2}{25} - \dfrac{y^2}{9} = 1$

45. Ellipse

47. Parabola

49. Parabola

51. $y = \pm\sqrt{6 - 4x}$; parabola; vertex: $(1.5, 0)$

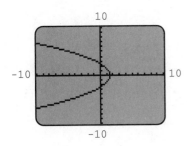

53. $y = \pm\sqrt{6 + 7x - 2x^2}$; ellipse; vertices: $(1.75, -3.48)$, $(1.75, 3.48)$

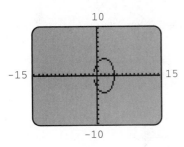

55. Circle; $x^2 + y^2 - 16 = 0$

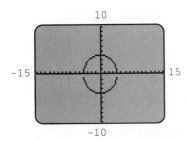

57. Parabola; $y^2 - x - 4y + 8 = 0$

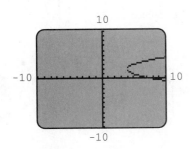

59. a. even

b.

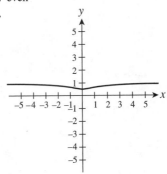

61. a. $(1.5, 3.25)$ **b.** $(3.25, 1.5)$

63. a. Midpoint: $(1, 2)$; slope: 1 **b.** $(0, 3)$

65.

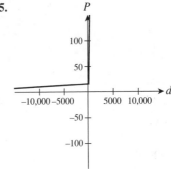

67. a. 115.5 feet   **b.** 453 feet

Chapter 6 Test ■ *page 521*

1. True   **2.** True   **3.** True   **4.** False

5. True   **6.** False

7. $x = y^2$, for example

8. $x = -y^2$, for example

9. $\dfrac{x^2}{16} + \dfrac{y^2}{4} = 1$, for example

10. A parabolic satellite dish, for example

11. a.

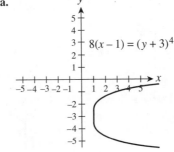

$8(x - 1) = (y + 3)^4$

b.

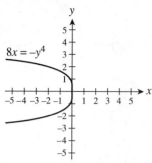

$8x = -y^4$

12. Symmetric with respect to the *x*-axis

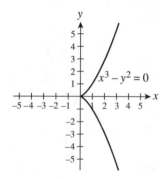

$x^3 - y^2 = 0$

13. Parabola; vertex: $(-4, -2)$; focus: $(-\frac{15}{4}, -2)$

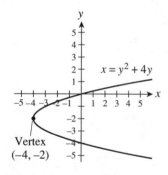

$x = y^2 + 4y$

Vertex $(-4, -2)$

14. Hyperbola; center: $(0, 0)$; vertices: $(-3, 0)$, $(3, 0)$; foci: $(-\sqrt{13}, 0)$, $(\sqrt{13}, 0)$; asymptotes: $y = \pm\frac{2}{3}x$

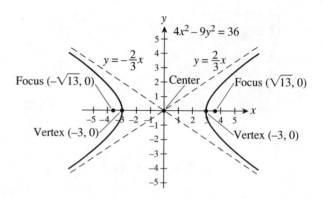

$4x^2 - 9y^2 = 36$

$y = -\frac{2}{3}x$ $y = \frac{2}{3}x$

Focus $(-\sqrt{13}, 0)$ Center   Focus $(\sqrt{13}, 0)$

Vertex $(-3, 0)$ Vertex $(-3, 0)$

15. Ellipse; center $(2, -3)$; vertices: $(2, -7)$, $(2, 1)$; foci: $(2, -3 - 2\sqrt{3})$, $(2, -3 + 2\sqrt{3})$

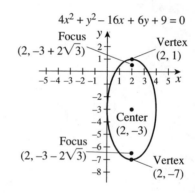

$4x^2 + y^2 - 16x + 6y + 9 = 0$

Focus $(2, -3 + 2\sqrt{3})$ Vertex $(2, 1)$

Center $(2, -3)$

Focus $(2, -3 - 2\sqrt{3})$ Vertex $(2, -7)$

16. Approximately 3.3 A.U.

CHAPTER 7

Section 7.1 ■ *page 537*

1. 2×3 **3.** 4×1 **5.** $\begin{bmatrix} 2 & 0 \\ 6 & 10 \end{bmatrix}$ **7.** $\begin{bmatrix} -2 & 11 \\ 6 & 6 \end{bmatrix}$

9. Not defined   **11.** $\begin{bmatrix} 5 \\ -1 \\ 6 \end{bmatrix}$ **13.** $\begin{bmatrix} 4a & 3b \\ 2c & 6d \end{bmatrix}$

15. $\begin{bmatrix} 1 & 7 & 4 \\ 7 & 9 & 18 \\ 7 & 2 & -3 \end{bmatrix}$ **17.** $\begin{bmatrix} 2 - a & 0 & 1 - a \\ -a & 1 - a & 2 \end{bmatrix}$

19. $\begin{bmatrix} 3 & 3 \\ 7 & 14 \end{bmatrix}$ **21.** $\begin{bmatrix} -8 & -1 \\ -14 & -35 \end{bmatrix}$ **23.** $X = \begin{bmatrix} 2 & 3 \\ 4 & 1 \end{bmatrix}$

25. $X = \begin{bmatrix} -\frac{1}{2} & -1 & -\frac{3}{2} \\ -3 & -\frac{5}{2} & -2 \end{bmatrix}$ **27.** Not defined

29. Defined; 2×3 **31.** Not defined   **33.** $\begin{bmatrix} -6 \\ 6 \end{bmatrix}$

35. Not defined   **37.** $\begin{bmatrix} 10 & 15 \\ 4 & 6 \end{bmatrix}$ **39.** $\begin{bmatrix} 1 & 0 \\ 0 & 1 \end{bmatrix}$

41. $\begin{bmatrix} 6 & 10 \\ 7 & 33 \\ 7 & 17 \end{bmatrix}$ **43.** $\begin{bmatrix} 39.12 \\ 33.94 \\ 16.35 \end{bmatrix}$ **45.** $\begin{bmatrix} a & b \\ ax + c & bx + d \end{bmatrix}$

47. Not defined   **49.** $\begin{bmatrix} 1 & 0 \\ 0 & 1 \end{bmatrix}$ **51.** $\begin{bmatrix} p & q \\ r & s \\ t & u \end{bmatrix}$

53. $\begin{bmatrix} a_{31} & a_{32} & a_{33} \\ a_{21} & a_{22} & a_{23} \\ a_{11} & a_{12} & a_{13} \end{bmatrix}$ **55.** $\begin{bmatrix} 2 & -3 \\ -1 & 1 \end{bmatrix}\begin{bmatrix} x \\ y \end{bmatrix} = \begin{bmatrix} 5 \\ 6 \end{bmatrix}$

57. $\begin{bmatrix} 1 & 1 & 0 \\ -1 & 0 & 2 \\ 1 & 1 & 1 \end{bmatrix}\begin{bmatrix} r \\ s \\ t \end{bmatrix} = \begin{bmatrix} 1 \\ 3 \\ 4 \end{bmatrix}$ **59.** $\begin{bmatrix} 1 & 1 & 1 & 1 \\ 0 & 1 & 1 & 1 \\ 0 & 0 & 1 & 1 \end{bmatrix}\begin{bmatrix} x_1 \\ x_2 \\ x_3 \\ x_4 \end{bmatrix} = \begin{bmatrix} 4 \\ 3 \\ 2 \end{bmatrix}$

61. $\begin{bmatrix} 1 & 4 \\ 2 & 5 \\ 3 & 6 \end{bmatrix}$ **63.** $\begin{bmatrix} 1 & 2 & 3 \\ 2 & 5 & 6 \\ 3 & 6 & 4 \end{bmatrix}$ **65.** $\begin{bmatrix} 1 & 4 \\ 0 & 1 \end{bmatrix}$ **67.** $\begin{bmatrix} 1 & 0 \\ 0 & 0 \end{bmatrix}$

69. $\begin{bmatrix} 1 & 4 & 14 \\ 0 & 1 & 8 \\ 0 & 0 & 1 \end{bmatrix}$ **71.** $\begin{bmatrix} 1 & 64 & 4064 \\ 0 & 1 & 128 \\ 0 & 0 & 1 \end{bmatrix}$

77. Hourly: $\begin{bmatrix} 230{,}000 & 320{,}000 \\ 280{,}000 & 300{,}000 \end{bmatrix}$

Salaried: $\begin{bmatrix} 125{,}000 & 250{,}000 \\ 400{,}000 & 500{,}000 \end{bmatrix}$

Sum: $\begin{bmatrix} 355{,}000 & 570{,}000 \\ 680{,}000 & 800{,}000 \end{bmatrix}$

Total major medical claims were \$355,000 for workers and \$570,000 for dependents, whereas total comprehensive claims were \$680,000 for workers and \$800,000 for dependents

79. $S = \begin{bmatrix} 5 & 40 & 18 \\ 4 & 37 & 26 \end{bmatrix}$; $T = \begin{bmatrix} 3 \\ 2 \\ 1 \end{bmatrix}$; $ST = \begin{bmatrix} 113 \\ 112 \end{bmatrix}$

So the final score is Hornets 113, Magic 112.
81. False   **83.** True   **85.** True   **87.** False
89. True

Section 7.2 ■ page 548

1. Linear   **3.** Not linear   **5.** Not linear

7. $\begin{bmatrix} 3 & 4 & 10 \\ 2 & -7 & 4 \end{bmatrix}$ **9.** $\begin{bmatrix} 1 & 0 & 3 \\ 0 & 1 & 4 \end{bmatrix}$

11. $\begin{bmatrix} 1 & -1 & 2 & 1 \\ 0 & 2 & 6 & 3 \end{bmatrix}$ **13.** 0 solutions

15. Infinitely many solutions   **17.** 1 solution
19. $x = 2, y = 1$ **21.** $x = 3, y = 7, z = 1$
23. $q = -1, r = 3, s = 4, t = -2$
25. $x = 2, y = 3$ **27.** $x = -6, y = 10, z = -4$
29. $x = 1, y = -2, z = -1$
31. Salary: \$40,000
Income from municipal bonds: \$10,000
Income from U.S. treasury bonds: \$8000
33. Approximately 2 bananas, 1.5 servings of milk, and 3.75 slices of wheat toast
35. False   **37.** False

Section 7.3 ■ page 560

1. Yes   **3.** No; $a_{31} \neq 0$ **5.** No; $a_{32} \neq 0, a_{33} \neq 1$

7. $\begin{bmatrix} 1 & 2 \\ 3 & 0 \end{bmatrix}$ **9.** $\begin{bmatrix} 1 & 4 & 6 \\ 0 & -9 & -9 \end{bmatrix}$ **11.** $\begin{bmatrix} 1 & 2 & 3 \\ 0 & 1 & 4 \end{bmatrix}$

13. $\begin{bmatrix} 1 & 0 & 3 & 4 \\ 0 & 1 & -2 & 4 \end{bmatrix}$ **15.** $\begin{bmatrix} 1 & 3 & 4 & -1 \\ 0 & 1 & -2 & -\frac{7}{2} \\ 0 & 0 & 1 & 3 \end{bmatrix}$

17. $\begin{bmatrix} 1 & -3 & -1 & 2 \\ 0 & 1 & \frac{4}{5} & \frac{1}{5} \\ 0 & 0 & 1 & \frac{7}{2} \end{bmatrix}$ **19.** $\begin{bmatrix} 1 & 3 & 0 & 4 \\ 0 & 1 & 2 & -2 \end{bmatrix}$

21. $\begin{bmatrix} 1 & 0 & 1 & 2 & 1 \\ 0 & 1 & 0 & 1 & 0 \\ 0 & 0 & 1 & 1 & 0 \\ 0 & 0 & 0 & 1 & 2 \end{bmatrix}$ **23.** $x = 2, y = 1$ **25.** $p = \frac{1}{2}, q = \frac{2}{3}$

27. $x = 1, y = 0, z = 3$ **29.** $p = 0, q = 1, r = 2$

31. $w = \frac{3}{2}, x = -\frac{1}{2}, y = 0, z = 0$ **33.** $\begin{bmatrix} 1 & 0 & -1 \\ 0 & 1 & 2 \end{bmatrix}$

35. $\begin{bmatrix} 1 & 0 & 2 \\ 0 & 1 & -3 \end{bmatrix}$ **37.** $\begin{bmatrix} 1 & 0 & 0 & 2 \\ 0 & 1 & 0 & 1 \\ 0 & 0 & 1 & -1 \end{bmatrix}$

39. $x = 3, y = -2$ **41.** $x = \frac{1}{2}, y = -\frac{1}{3}$
43. $x = 4, y = -1, z = 3$ **45.** $x = 0, y = 3, z = -4$

47. $\begin{bmatrix} 2 & 1 & 5 \\ 4 & 2 & 10 \end{bmatrix}$; $\left(\frac{5}{2} - \frac{1}{2}y, y\right)$

49. $\begin{bmatrix} -1 & 3 & -7 \\ 3 & -2 & 7 \end{bmatrix}$; $x = 1, y = -2$

51. $\begin{bmatrix} -3 & -1 & 1 & -1 \\ 1 & 4 & -1 & 3 \\ -5 & 2 & 1 & 2 \end{bmatrix}$; inconsistent

53. $\begin{bmatrix} 2 & 4 & 2 & 0 \\ 3 & -1 & 1 & 1 \\ 1 & -5 & -1 & 1 \end{bmatrix}$; $\left(\frac{2}{7} - \frac{3}{7}z, -\frac{1}{7} - \frac{2}{7}z, z\right)$

55. $y = 2x^2 - 4x + 7$ **57.** 3 nickels, 7 dimes, and 5 quarters
59. The bar weighs 15 pounds, the small disk weighs 7.5 pounds, and the large disk weighs 10 pounds. A barbell with 4 large disks and 2 small disks on each weighs 125 pounds.
61. Hinduism in 1500 B.C., Buddhism in 525 B.C., and Islam in A.D. 622
63. 4 schools of the first type, 3 schools of the second type, and one school of the third type
65. False   **67.** False

Section 7.4 ■ page 571

1. Yes   **3.** No   **5.** $\begin{bmatrix} 1 & 0 \\ -2 & 1 \end{bmatrix}$ **7.** $\begin{bmatrix} -5 & 4 \\ 5 & -2 \end{bmatrix}$

9. No inverse   **11.** $\begin{bmatrix} 0 & \frac{1}{b} \\ \frac{1}{a} & 0 \end{bmatrix}$ **13.** $\begin{bmatrix} \dfrac{1-x}{2x} & \dfrac{x+1}{2x} \\ \dfrac{1}{2} & -\dfrac{1}{2} \end{bmatrix}$

15. No inverse   **17.** $\begin{bmatrix} 1 & -2 & 0 \\ \frac{3}{4} & -1 & \frac{1}{4} \\ \frac{3}{2} & -3 & \frac{1}{2} \end{bmatrix}$ **19.** $\begin{bmatrix} 1 & 2 & 3 \\ 3 & 2 & 1 \\ 2 & 3 & 1 \end{bmatrix}$

13.

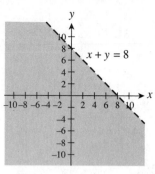

21. $\begin{bmatrix} 1 & -x & x^2 \\ 0 & 1 & -x \\ 0 & 0 & 1 \end{bmatrix}$ **23.** No inverse   **25.** $\begin{bmatrix} \frac{1}{a} & 0 & 0 & 0 & 0 \\ \frac{1}{a} & \frac{1}{a} & 0 & 0 & 0 \\ \frac{1}{a} & \frac{1}{a} & \frac{1}{a} & 0 & 0 \\ \frac{1}{a} & \frac{1}{a} & \frac{1}{a} & \frac{1}{a} & 0 \\ \frac{1}{a} & \frac{1}{a} & \frac{1}{a} & \frac{1}{a} & \frac{1}{a} \end{bmatrix}$

27. a. $x = \frac{10}{3}, y = \frac{13}{3}$
   b. $x = -3, y = 1$
   c. $x = \frac{1}{2}; y = -\frac{1}{2}$
29. a. $x = 5, y = 5$
   b. $x = 4, y = 1$
   c. $x = -2, y = 3$
31. a. $x = 2, y = 3, z = 4$
   b. $x = -2, y = 0, z = 1$
   c. $x = 5, y = 5, z = 5$
33. a. Approximately 35.1 minutes walking, 40.2 minutes jogging, and 44.7 minutes running
   b. Approximately 38.6 minutes walking, 39.7 minutes jogging, and 11.7 minutes running
35. $f(x) = 2x^3 - 3x^2 + 5x + 1$ **37.** True   **39.** False

15.

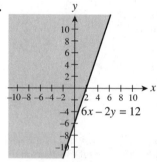

17.

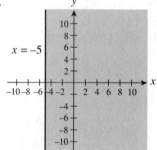

Section 7.5 ▪ *page 583*

1. -2 **3.** 0 **5.** -25 **7.** x
9. -10 **11.** -1 **13.** -51.406 **15.** a
17. $(a_2b_3 - a_3b_2)i - (a_1b_3 - a_3b_1)j + (a_1b_2 - a_2b_1)k$
19. -561 **21.** a^5 **23.** Invertible
25. Not invertible   **27.** Not invertible
29. Invertible   **31.** $-\frac{1}{2}$ **33.** $3, -3$
35. Invertible for all real x **37.** $x = 9, y = 5$
39. $x = 1, y = -3, z = 4$ **41.** $x = \frac{87}{23}, y = -\frac{25}{23}, z = \frac{106}{23}$
43. $x = 1, y = 0, z = 0, w = 0$
45. $x = \dfrac{ce - bf}{ae - bd}, y = \dfrac{af - cd}{ae - bd}$
47. Approximately $9722.22
49. Approximately 185.6 grams of milk
51. False   **53.** True

Section 7.6 ▪ *page 596*

1. $(0, 0)$: No; $(4, 1)$: No; $(-1, 3)$: Yes
3. $(1, 4)$: Yes; $(3, 0)$: No; $(-2, -1)$: No
5. $(0, 2)$: Yes; $(-1, 3)$: Yes; $(5, -1)$: No
7. iv   **9.** v   **11.** vi

19.

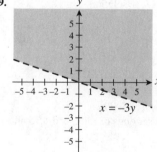

21.

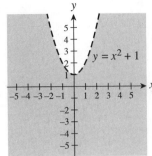

23.

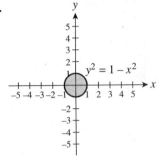

25.

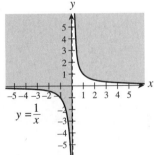

27.

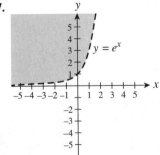

29. ii **31.** i

33.

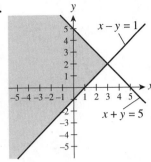

35.

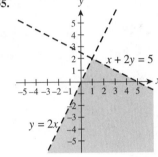

37.

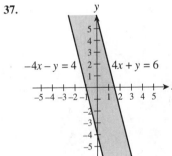

39.

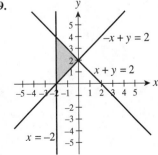

41.

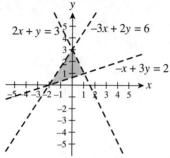

$2x + y = 3$ $-3x + 2y = 6$ $-x + 3y = 2$

43.

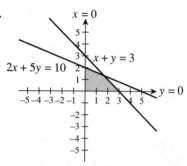

$x = 0$ $x + y = 3$ $2x + 5y = 10$ $y = 0$

45.

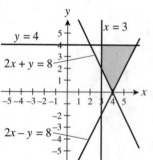

$y = 4$ $x = 3$ $2x + y = 8$ $2x - y = 8$

47.

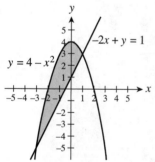

$-2x + y = 1$ $y = 4 - x^2$

49.

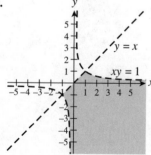

$y = x$ $xy = 1$

51.

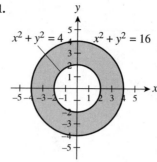

$x^2 + y^2 = 4$ $x^2 + y^2 = 16$

53.

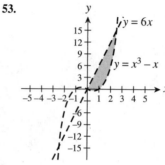

$y = 6x$ $y = x^3 - x$

55. $y \leq \frac{3}{4}x - 3$

57. $y \geq x + 2$
$y \leq -\frac{2}{5}x + 2$

59. $x \geq 0$
$y \leq -\frac{2}{3}x + 4$
$y \geq \frac{2}{3}x$

61. $y \geq \frac{2}{3}x - \frac{4}{3}$
$y \leq -\frac{3}{2}x + \frac{19}{2}$
$y \geq -\frac{3}{2}x + 3$
$y \leq \frac{2}{3}x + 3$

63. $x^2 + y^2 \leq 9$
$x \geq 0$

65. $x =$ Amount in Redbook Growth
$y =$ Amount in Redbook Small Cap
$x \geq 6000$
$y \geq 6000$
$y \leq \dfrac{x}{2}$
$x + y \leq 24,000$

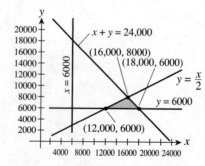

$x + y = 24,000$
$(16,000, 8000)$
$(18,000, 6000)$
$x = 6000$
$y = \dfrac{x}{2}$
$y = 6000$
$(12,000, 6000)$

67. x = Number of acres of corn
y = Number of acres of soybeans

$$x + y \leq 1000$$
$$100x + 80y \leq 88{,}000$$
$$2x + y \leq 1600$$
$$x \geq 200$$
$$y \geq 0$$

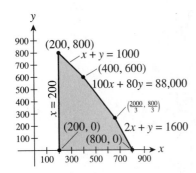

69. False   **71.** True

Section 7.7 ▪ *page 610*

1. Maximum of 76 at $(7, 2)$; minimum of 44 at $(3, 2)$
3. Maximum of 50 at $(5, 6)$; minimum of 0 at $(0, 0)$
5. Maximum of 28 at $(7, 0)$; minimum of -32 at $(0, 8)$
7. Maximum of 85 at $(4, 3)$ **9.** Minimum of 150 at $(10, 0)$
11. Maximum of 144 at $(12, 2)$ **13.** Minimum of 104 at $(4, 1)$
15. Maximum of 88 at $(4, 8)$ **17.** 22 of type A; 3 of type B
19. 4 days for Refinery 1; 3 days for Refinery 2
21. Steve should travel 100 miles by car and 200 miles by truck. This will cost $16.00.
23. 18 minutes for the senator; 12 minutes for the governor
25. 4 ounces of Swiss steak, $5\frac{1}{2}$ ounces of peas
27. 6 employees from the Bozeman office and 4 from New Haven should fly to Orlando; the remaining employees from each office should fly to Columbus.
29. True   **31.** False

Chapter 7 Review ▪ *page 614*

1. $\begin{bmatrix} 6 & -3 \\ -15 & 9 \end{bmatrix}$ **3.** $\begin{bmatrix} 3 & -1 & 2 \\ 3 & 5 & 5 \end{bmatrix}$ **5.** $\begin{bmatrix} 5 & -2 \\ 6 & 9 \end{bmatrix}$

7. Not defined; the number of columns of C is not equal to the number of rows of D.

9. $\begin{bmatrix} 9 & 1 \\ -8 & 8 \\ 10 & 0 \end{bmatrix}$

11. Not defined; the number of rows of C is not equal to the number of columns of C (C isn't square).

13. $\begin{bmatrix} -9 \\ -1 \end{bmatrix}$ **15.** Not defined   **17.** $[2]$ **19.** $\begin{bmatrix} -12 & 4 \\ -8 & -1 \\ 6 & 9 \end{bmatrix}$

21. $\begin{bmatrix} b - c + 2a \\ 3b - a \\ 4b + 2c \end{bmatrix}$ **23.** $\begin{bmatrix} 5 & 1 & | & 3 \\ 1 & -3 & | & 4 \end{bmatrix}$

25. $x - 4y + z = 3$
$y - 2z = 5$
$z = -2$
$x = 9, y = 1, z = -2$

27. $\begin{bmatrix} 1 & -3 & | & 2 \\ 0 & 1 & | & -\frac{7}{3} \end{bmatrix}$ **29.** $\begin{bmatrix} 1 & 2 & -\frac{3}{2} & | & \frac{1}{2} \\ 0 & 1 & -\frac{3}{2} & | & \frac{1}{2} \end{bmatrix}$

31. $x = 3, y = 5$ **33.** $\left(\frac{3}{2}y - 3, y\right)$
35. $x = -5, y = -4, z = -3$ **37.** $x = 0, y = -4, z = -5$

39. $\begin{bmatrix} 1 & 0 & | & -2 \\ 0 & 1 & | & 3 \end{bmatrix}$ **41.** $x = 5, y = -4$

43. $x = 2, y = 5, z = 4$ **45.** $\begin{bmatrix} -1 & -1 \\ 3 & 2 \end{bmatrix}$

47. $\frac{1}{a^2 - b^2} \cdot \begin{bmatrix} a & -b \\ -b & a \end{bmatrix}$ **49.** $\begin{bmatrix} 1 & 2 & 1 \\ 0 & 1 & 2 \\ 0 & 0 & 1 \end{bmatrix}$ **51.** No inverse

53. a. $x = -7, y = 10$ **b.** $x = -2, y = 4$
55. a. $x = \frac{39}{4}, y = -\frac{5}{4}$ **b.** $x = \frac{9}{2}; y = -\frac{3}{2}$
57. -10 **59.** 1   **61.** $x^3 - x$ **63.** 35
65. $x = 5$ **67.** $x = 2, x = 3$ **69.** $x = \frac{1}{4}, y = -\frac{3}{4}$
71. $x = 1, y = -1, z = 5$

73.

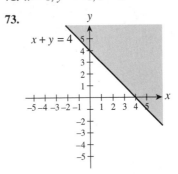

75.

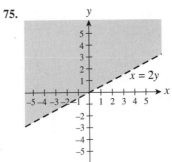

77.

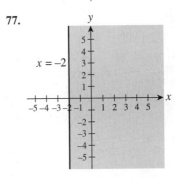

79.

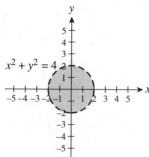

$x^2 + y^2 = 4$

81.

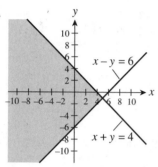

$x - y = 6$

$x + y = 4$

83.

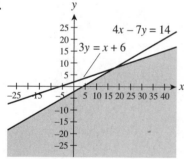

$4x - 7y = 14$

$3y = x + 6$

85.

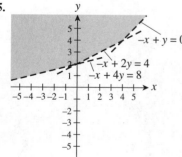

$-x + y = 0$

$-x + 2y = 4$

$-x + 4y = 8$

87.

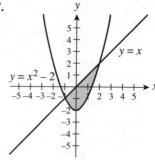

$y = x$

$y = x^2 - 2$

89.

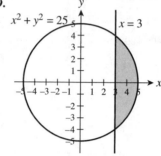

$x^2 + y^2 = 25$

$x = 3$

91. Minimum 46 at $(6, 2)$
Maximum 70 at $(6, 5)$

93. Maximum 68 at $(4, 2)$ **95.** Minimum 60 at $(5, 2)$

97. $40,000 money market, $35,000 mutual fund, $25,000 savings account

99. $\frac{13}{2}$

101. To maximize either revenue or profit, plant 327.3 acres of corn and 90.9 acres of hay.

103. $6666.67 in bonds, $3333.33 in mutual funds

Chapter 7 Test ■ *page 618*

1. True   **2.** False   **3.** False   **4.** False
5. True   **6.** False   **7.** True   **8.** False

9. $\begin{bmatrix} 1 & 2 & 3 \\ 4 & 5 & 6 \end{bmatrix}$, for example

10. $\begin{bmatrix} 0 & 0 \\ 0 & 0 \end{bmatrix}$, for example

11. $\left[\begin{array}{cc|c} 1 & 1 & 2 \\ 0 & 1 & 3 \end{array}\right]$, for example

12. $xy = 1$
$x - y = 2$, for example

13. Switching two rows, for example

14. a. $\begin{bmatrix} -6 & -7 & 9 \\ 16 & 6 & 21 \\ -14 & -9 & -4 \end{bmatrix}$

b. $\begin{bmatrix} 4 & -4 & 5 \\ 12 & 11 & -3 \\ 4 & -6 & 1 \end{bmatrix}$

c. $\begin{bmatrix} 2 & 1 & -5 \\ -3 & 1 & -3 \\ -4 & -3 & 2 \end{bmatrix}$

15. $\begin{bmatrix} 3 & -5 & 1 \\ 2 & -8 & 0 \\ 3 & 0 & -1 \end{bmatrix}\begin{bmatrix} x \\ y \\ z \end{bmatrix} = \begin{bmatrix} 2 \\ -2 \\ 0 \end{bmatrix}$; $x = \frac{13}{19}, y = \frac{8}{19}, z = \frac{39}{19}$

16. a. $2x - 2y + 2z = 0$
$\ y - 3z = -2$
$\ 4z = 8$
$\ x = 2, y = 4, z = 2$

b. $\begin{bmatrix} 1 & 0 & 0 & 2 \\ 0 & 1 & 0 & 4 \\ 0 & 0 & 1 & 2 \end{bmatrix}$

17. $x = 2, y = 3, z = -5$ **18.** $b = 11$

19. Approximately 71.0 million tons of waste paper, 33.2 million tons of yard waste, and 75.4 million tons of miscellaneous waste

20. Maximum 0 at $(0, 0)$

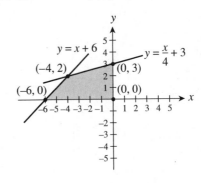

21. $y > \frac{3}{2}x$
$y < -x + 5$
$y > -6x$

CHAPTER 8

Section 8.1 ■ *page 629*

1. 4, 7, 10, 13   **3.** $-4, 16, -64, 256$ **5.** $1, \frac{1}{4}, \frac{1}{9}, \frac{1}{16}$

7. 0, 2, 0, 2   **9.** $1, -\frac{1}{2}, \frac{1}{3}, -\frac{1}{4}$ **11.** $\sqrt{2}, \sqrt{6}, 2\sqrt{3}, 2\sqrt{5}$

13. $2, \frac{9}{4}, \frac{64}{27}, \frac{625}{256}$ **15.** 1, 3, 6, 10   **17.** $a_n = 2n$

19. $a_n = \left(-\frac{1}{2}\right)^n$ **21.** $a_n = \frac{n}{n+1}$ **23.** $a_n = 1 - \frac{1}{n}$

25. Arithmetic; $d = 3$ **27.** Neither   **29.** Arithmetic; $d = -4$

31. Neither   **33.** Geometric; $r = \frac{1}{10}$ **35.** $a_n = 6n + 1$

37. $a_n = 115 - 15n$ **39.** $a_n = \frac{5n - 13}{4}$ **41.** $a_{10} = 75$

43. $a_{15} = -60$ **45.** $a_{31} = 176$ **47.** $a_n = 3^{n-1}$

49. $a_n = 32\left(-\frac{1}{2}\right)^{n-1}$ **51.** $a_n = 4\left(\frac{5}{4}\right)^{n-1}$ **53.** $a_8 = 4374$

55. $a_7 = \frac{1}{81}$ **57.** $2, -3, 7, -13, 27$ **59.** 1, 2, 6, 24, 120

61. $2, \sqrt{2}, \sqrt[4]{2}, \sqrt[8]{2}, \sqrt[16]{2}$ **63.** 1, 2, 2, 4, 8

65. $29,355; 30,591; 31,827; d = 1236; 36,771$

67. 74   **69.** $a_n = 1000(1.055)^{n-1}$; $1619.09; 5.5\%$

71.

Month	Pairs at the beginning of the month	Pairs born during the month	Pairs at the end of the month
1	1	0	1
2	1	1	2
3	2	1	3
4	3	2	5
5	5	3	8
6	8	5	13
⋮	⋮	⋮	⋮
12	144	89	233

73. True   **75.** True

Section 8.2 ■ *page 639*

1. 75   **3.** $\frac{71}{20}$ **5.** 51   **7.** -15

9. 60   **11.** $\frac{19}{20}$ **13.** $\frac{155}{16}$ **15.** $\sum\limits_{k=1}^{6} \frac{1}{k^3}$

17. $\sum\limits_{k=1}^{12} (5k + 2)$ **19.** $\sum\limits_{k=1}^{7} 3^k$ **21.** $\sum\limits_{k=1}^{16} \frac{(-1)^{k-1}}{k^2}$

23. $\sum\limits_{k=1}^{6} \frac{2^k - 1}{2^k}$ **25.** 1640   **27.** 159.5   **29.** -240

31. 3725   **33.** 3210   **35.** 724   **37.** 147,620

39. Approximately 4.5   **41.** Approximately 3.6364

43. 262,140   **45.** Approximately 1.7143

47. Approximately -0.0012 **49.** 500,500

51. 10,000   **53.** 294   **55.** 900

57. Approximately \$1028.5 billion

59. \$5323.58   **61.** \$29,344.42   **63.** \$150

65. Approximately \$50.8 million

67. False   **69.** True

Section 8.3 ■ *page 651*

1. 720   **3.** 720   **5.** 30   **7.** 120

9. 15   **11.** 8   **13.** 1   **15.** 28

17.

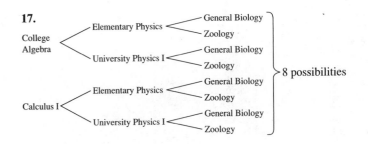

19.

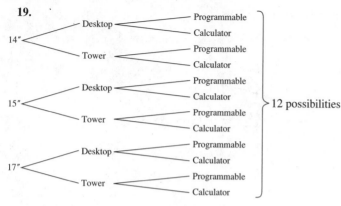

21. **a.** 125   **b.** 60
23. 70,304,000   25. $_8P_3 = 336$
27. **a.** $_{13}P_9 = 259,459,200$ **b.** $_2P_2 \cdot _{11}P_7 = 3,326,400$
29. 56
31. **a.** $_{12}C_6 = 924$ **b.** $_5C_2 \cdot _7C_4 = 350$
33. $_{10}P_4 = 5040$ 35. $_{40}C_5 = 658,008$
37. **a.** $_{40}P_3 = 59,280$ **b.** $40^3 = 64,000$
39. $_6C_4 = 15$; $_6C_4 \cdot _{15}C_{10} = 45,045$
41. $2 \cdot _{30}C_2 = 870$ seconds; $2 \cdot _{60}C_2 = 3540$ seconds
43. $4 \cdot _{13}C_5 = 5148$
45. **a.** $_{5000}P_{1127}$ **b.** 5000^{1127}
47. False   49. False

Section 8.4 ■ *page 664*

1. $S = \{\text{red, blue, green, white}\}$; $n(S) = 4$
3. $S = \{HHH, HHT, HTH, HTT, THH, THT, TTH, TTT\}$; $n(S) = 8$
5. $\frac{1}{6}$ 7. 1   9. $\frac{1}{4}$ 11. $\frac{7}{12}$ 13. $\frac{1}{52}$ 15. $\frac{1}{13}$
17. $\frac{3}{26}$ 19. $\frac{1}{4}$ 21. $\frac{3}{8}$ 23. $\frac{5}{18}$ 25. $\frac{5}{9}$ 27. $\frac{1}{6}$
29. $\frac{1}{12}$ 31. $\frac{1}{54,145} \approx 1.85 \times 10^{-5}$ 33. $\frac{33}{16,660} \approx 0.002$
35. $\frac{1}{16}$ 37. $\frac{1}{8}$ 39. $\frac{5}{16}$
41. $P(E) = \frac{1}{2}$; $P(F) = \frac{3}{13}$; $E \cap F$ is the event that a red face card is drawn; $P(E \cap F) = \frac{3}{26}$; $E \cup F$ is the event that a red card or a face card is drawn; $P(E \cup F) = \frac{8}{13}$; E' is the event that a black card is drawn; $P(E') = \frac{1}{2}$; F' is the event that a non–face card is drawn; $P(F') = \frac{10}{13}$
43. $P(E) = \frac{1}{2}$; $P(F) = \frac{7}{12}$; $E \cap F$ is the event that the sum is even and less than 8; $P(E \cap F) = \frac{1}{4}$; $E \cup F$ is the event that the sum is even or less than 8; $P(E \cup F) = \frac{5}{6}$; E' is the event that the sum is odd; $P(E') = \frac{1}{2}$; F' is the event that the sum is 8 or greater; $P(F') = \frac{5}{12}$
45. 0.35
47. $\frac{1}{658,008} \approx 1.52 \times 10^{-6}$; $\frac{5}{164,502} \approx 3.04 \times 10^{-5}$; 329,004
49. **a.** 0.19   **b.** 0.77
51. $\frac{9}{28} \approx 0.32$
53. **a.** 0.179   **b.** 0.031   **c.** 0.88   **d.** 0.089   **e.** 0.059
55. False   57. True

Section 8.5 ■ *page 674*

1. 4   3. 10   5. 6   7. 54   9. 6   11. 10
13. $x^5 + 5x^4y + 10x^3y^2 + 10x^2y^3 + 5xy^4 + y^5$
15. $16x^4 - 32x^3y + 24x^2y^2 - 8xy^3 + y^4$
17. $27x^3 + 135x^2y + 225xy^2 + 125y^3$
19. $x^5 + 5x^4 + 10x^3 + 10x^2 + 5x + 1$

21. $1,000,000w^6 - 600,000w^5 + 150,000w^4 - 20,000w^3 + 1500w^2 - 60w + 1$
23. $7 + 5\sqrt{2}$ 25. $49 + 20\sqrt{6}$ 27. $x^6 - 3x^4 + 3x^2 - 1$
29. $x^5 + 5x^3 + 10x + \frac{10}{x} + \frac{5}{x^3} + \frac{1}{x^5}$ 31. $-2 + 2i$
33. -64 35. i 45. False   47. True

Section 8.6 ■ *page 680*

35. False   37. True

Chapter 8 Review ■ *page 683*

1. $-4, -9, -14, -19, -24$; arithmetic; $d = -5$
3. $e^{-1}, 2e^{-2}, 3e^{-3}, 4e^{-4}, 5e^{-5}$ 5. $-\frac{1}{2}, \frac{3}{2}, -\frac{1}{2}, \frac{3}{2}, -\frac{1}{2}$
7. 2, 6, 18, 54, 162; geometric; $r = 3$
9. $a_n = 2(5)^{n-1}$; geometric; $r = 5$
11. $a_n = \frac{1}{n}$ 13. $a_n = \frac{7}{2} - \frac{5}{2}n$; arithmetic; $d = -\frac{5}{2}$
15. $a_n = (-1)^{n-1}n^3$ 17. 47   19. 1024
21. 676   23. 40   25. Approximately 0.4914
27. $\sum_{k=1}^{11} (5k - 2)$ 29. $\sum_{k=3}^{13} 4(\frac{2}{3})^k$ 31. 290
33. Approximately 12.5   35. Approximately 2.1428
37. 7532   39. 36   41. 6720   43. 10   45. 120
47. $\frac{1}{4}$ 49. $\frac{11}{26}$ 51. $\frac{1}{4}$ 53. $\frac{1}{9}$ 55. $\frac{1}{8}$
57. 0.359   59. 0.379   61. 56   63. -1000
65. $x^4 - 4x^3y + 6x^2y^2 - 4xy^3 + y^4$
67. $r^3 + \frac{3r^2}{s} + \frac{3r}{s^2} + \frac{1}{s^3}$
69. $z^2\sqrt{z} + 10z^2 + 40z\sqrt{z} + 80z + 80\sqrt{z} + 32$
77. $_{20}P_4 = 116,280$
79. **a.** $_{10}C_5 = 252$ **b.** $_6C_3 \cdot _4C_2 = 120$
81. 1024
83. The populations for 2001–2003 are 210,000, 220,500, and 231,525; the common ratio is 1.05; the population in 2012 will be approximately 359,171.
85. Approximately \$342,700.67
87. **a.** $_{80}C_{20}$ **b.** $(_{20}C_5)^4$
89. $\frac{1}{120}$

Chapter 8 Test ■ *page 685*

1. False   2. True   3. True   4. True
5. False   6. True   7. False   8. False
9. $a_n = -4n$, for example   10. $a_n = (\frac{1}{3})^n$, for example
11. $\sum_{k=1}^{5} \frac{1}{k}$, for example   12. Tossing three coins, for example
13. A red ball is drawn from a bag containing one red ball, one white ball, and one blue ball, for example.
14. 1, 4, 6, 4, 1, for example   15. 4, 13, 28, 49, 76
16. $a_n = \frac{n + 1}{n}$ 17. 126   18. $\sum_{k=1}^{6} 5(\frac{1}{2})^k \approx 4.9219$
19. 256; $\frac{1}{256}$
20. **a.** $_{15}C_9 = 5005$
    b. $_7C_5 \cdot _8C_4 = 1470$; the combinations would be replaced with permutations.
21. $\frac{1}{3}$ 22. $\frac{8}{13}$ 23. 393,750
24. $16x^4 - 96x^3y + 216x^2y^2 - 216xy^3 + 81y^4$

INDEX OF APPLICATIONS

INDEX

CREDITS

Chapter 1. 1 Left: Nashua River Watershed Association; **Right:** George Steinmetz. **6:** Courtesy of Steve Grohe/Thinking Machines, Corp. **9:** Photo Chamoix Initiative/Liaison Agency, Inc. **12:** NASA/SPL/Photo Researchers, Inc. **25:** Werner Krutein/Liaison Agency, Inc. **58:** Greg Vaughn/Tom Stack & Associates. **74:** NOVOSTI/Liaison Agency, Inc. **94:** Frank Siteman/Rainbow.

Chapter 2. 97: Antonio Ribiero/Liaison Agency, Inc. **111:** CORBIS/Oscar White. **115:** CORBIS/Yann Arthus-Bertrand. **143:** AP/WIDE WORLD PHOTOS. **152:** Harry J. Rockwell/Fundamental Photographs, NY. **159:** AP/WIDE WORLD PHOTOS. **161 Left:** © WOLFGANG KAEHLER/www.wkaehlerphoto.com; **Right:** © TSM/Viviane Moos, 1993. **168:** Professor Peter Goddard/SPL/Photo Researchers, Inc. **171:** USC Sports Information. **203:** Crown Copyright/Health & Safety Laboratory/SPL/Photo Researchers, Inc.

Chapter 3. 209: SYGMA/CORBIS. **223:** NASA/Phototake NYC. **234:** Peter Marlow/SYGMA/CORBIS. **239:** © 1996 Jonathan Bowen. Reprinted with permission. Reprinted from: http://www.cs.reading.ac.uk/archive/hypercubes/ **240 Left:** Reuters/CORBIS; **Right:** J.L. Altan/SYGMA/CORBIS. **269:** Monique Salaber/Liaison Agency, Inc. **282 Left:** NASA/National Geographical Society Image Collection; **Right:** GLOBE PHOTOS, Inc. **288:** © Bob Daemmrich/Stock Boston. **301:** CORBIS. **310 Left:** J.P. Varin/Jacana/Photo Researchers, Inc.; **Right:** NASA. **319:** AP/WIDE WORLD PHOTOS.

Chapter 4. 321: IBM Research/Peter Arnold, Inc. **328:** AP/WIDE WORLD PHOTOS. **329:** Kuroda/Lee/Superstock, Inc. **335 Figures 26-28:** Gregory Sams/SPL/Photo Researchers, Inc. **377:** Damian Strohmayer/Sports Illustrated.

Chapter 5. 385: National Snow & Ice Data Center/SPL/Photo Researchers, Inc. **397 Top:** James D.Wilson/Woodfin Camp & Associates; **Bottom:** Robert Frerck/Woodfin Camp & Associates. **408:** Flip Nicklin/Minden Pictures. **411:** Luiz C. Marigo/Peter Arnold, Inc. **417:** Photri. **420:** Alan Benainous/Liaison Agency, Inc. **426:** © Tom Ebenhoh/Image Quest. **430 Top:** Douglas Burrows/Liaison Agency, Inc.; **Bottom:** Patrick Mesner/Liaison Agency, Inc. **433:** Hintlerleitner/Liaison Agency, Inc. **436:** Marta Serra-Jovenich/Bruce Coleman. **442:** R&S Michaud/Woodfin Camp & Associates. **443:** Philip Crossley/Hurricane Entertainments Limited. **445:** Central Point Software Inc. **446:** Peter Brock/Liaison Agency, Inc.

Chapter 6. 453: Stephanie Maze/Woodfin Camp & Associates. **472 Left column:** Photri; **Right column clockwise left to right:** Tim Davis/Photo Researchers, Inc.; Gary Braasch/Woodfin Camp Associates; Norbert Wu; Norbert Wu. **484:** Doug Johnson/SPL/Photo Researchers, Inc. **486:** P. Delacroix/Liaison Agency, Inc. **497:** Sovfoto/Eastfoto. **500:** Photri.

Chapter 7. 523: Allan Tannenbaum/SYGMA/CORBIS. **563:** © Brian Spurlock/SportsChrome, USA. **608:** Courtesy of AT&T Archives.

Chapter 8. 619: Armen Kachaturian/Liaison Agency, Inc. **621:** NCSA, University of Illinois/SPL/Photo Researchers, Inc. **628 Left to right:** Rod Planck/Photo Researchers, Inc.; Davel Nagel/Liaison Agency, Inc.; Snowdon/Hoyer Focus/Woodfin Camp & Associates; Ned Haines/Photo Researchers, Inc. **653:** © Rob Silvers. **667:** CORBIS.